Utilisation and Management of Medicinal Plants

— Volume 3 —

The Editors

Dr. Vijay Kumar Gupta, Ph.D., FLS, London (born 1953-) Ex. Chief Scientist, CSIR- Indian Institute of Integrative Medicine, Jammu, India. He did his M.Sc. (1975) and Ph.D. (1979) in Zoology both from University of Jammu, Jammu-India. His research capabilities are substantiated by his excellent work on histopathology, ecology and reproductive biology of fishes, turtles, birds and mammals, which has already got recognition in India and abroad.

Dr. Gupta has to his credit more than 150 scientific publications and review articles which have appeared in internationally recognized Indian and foreign journals. Founder fellow, life member and office bearer of many national societies, academies and associations. He has successfully completed a number of research/consultancy projects funded by government, private and multinational agencies. His current areas of interest are histopathology, toxicology, pre-clinical safety pharmacology, reproductive efficacy studies of laboratory animals and biodiversity.

He is the Series Editor of the recently published multi-volume set of books, "**Comprehensive Bioactive Natural Products (Vols. 1-8)**", published by M/S Studium Press, LLC, USA. He is also Editor-in-Chief of the books, "**Utilisation and Management of Medicinal Plants (Vols. 1-3)**", "**Medicinal Plants: Phytochemistry, Pharmacology and Therapeutics (Vols. 1-4)**", "**Traditional and Folk Herbal Medicine (Vols. 1-3)**", "**Natural Products: Research Reviews (Vols. 1-4)**", "**Bioactive Phytochemicals: Perspectives for Modern Medicine (Vols. 1-3)**", "**Perspectives in Animal Ecology and Reproduction (Vols. 1-10)**" and "**Animal Diversity, Natural History & Conservation (Vols. 1-5)**". The Editor-in-chief of the American Biographical Institute, USA, has appointed him as *Consulting Editor* of *The Contemporary Who's Who*. Dr. Gupta also appointed as Nominee for the *Committee for the Purpose of Control and Supervision of Experiments on Animals* (CPCSEA, Govt. of India). The *Linnaean Society of London, U.K.* has awarded fellowship to him in November 2009 in recognition of his contribution towards the cultivation of knowledge in Science of Natural History. Recently, Modern Scientific Press, USA has nominated Dr. Gupta as the Editor of the *International Journal of Traditional and Natural Medicine*.

Dr. Anpurna Kaul (born 1956-) working as Scientist in Pharmacology Division of Indian Institute of Integrative Medicine (CSIR), Jammu, India. She has completed her Ph.D. in 1996 from University of Jammu, Jammu, India. Her field of research is immune-pharmacology.

Dr. Kaul has 35 research publications in national and international journals, 12 patents and also presented several research papers in various symposia/conferences. She has screened more than 4000 plant extracts/fractions and pure compounds on immune system by *in vivo* and *in vitro* methods. She has also actively participated in Indo-Malaysian research project as member and visited Malaysia in 2007 to attend *Women's Health and Asian Traditional Medicine* Conference.

Utilisation and Management of Medicinal Plants

— Volume 3 —

Editor-in-Chief
V.K. Gupta
Chief Scientist (Ex)
CSIR-Indian Institute of Integrative Medicine
Canal Road, Jammu – 180 001,
India

Editor
A. Kaul
Scientist
Division of Pharmacology
CSIR-Indian Institute of Integrative Medicine
Canal Road, Jammu – 180 001,
India

2016
Daya Publishing House®
A Division of
Astral International Pvt. Ltd.
New Delhi – 110 002

Cataloging in Publication Data--DK
Courtesy: D.K. Agencies (P) Ltd. <docinfo@dkagencies.com>

Utilisation and management of medicinal plants / editor-in-chief, V.K. Gupta; editor, A. Kaul.
volume3cm
Contributed articles.
Includes bibliographical references and indexes.

ISBN 978-93-5130-886-7 (International Edition)

1. Medicinal plants--Utilization. 2. Medicinal plants--Biotechnology. 3. Medicinal plants--Analysis. I. Gupta, V. K. (Vijay Kumar), 1953- editor. II. Kaul, A. (Anpurna), 1956- editor.

SB293.U85 2016 DDC 633.88 23

Published by : **Daya Publishing House®**
A Division of
Astral International Pvt. Ltd.
– ISO 9001:2008 Certified Company –
4760-61/23, Ansari Road, Darya Ganj
New Delhi-110 002
Ph. 011-43549197, 23278134
E-mail: info@astralint.com
Website: www.astralint.com

Laser Typesetting : **Classic Computer Services,** Delhi - 110 035

Editorial Board

Prof. Yukihiro Shoyama

Faculty of Pharmaceutical Science,
Nagasaki International University,
His Ten Bosh, Sasebo, Nagasaki, JAPAN
Tel: +81-956-20-5653
E-mail: shoyama@niu.ac.jp

Prof. Ian Fraser Pryme

Department of Biomedicine,
University of Bergen,
Jonas Lies vei 91, N-5009 Bergen, NORWAY
Tel: +47-55-586438
Fax: +47-55-586360
E-mail: ian.pryme@biomed.uib.no

Dr. George Qian Li

Herbal Medicines Research and Education Centre,
Faculty of Pharmacy, The University of Sydney,
NSW 2006, AUSTRALIA
Tel: 612 9351
Fax: 9351 8638
E-mail: george.li@sydney.edu.au

Dr. Julia Serkedjieva

Institute of Microbiology,
Bulgarian Academy of Sciences,
26, Academician Georgy Bonchev St.,
1113 Sofia, BULGARIA
Tel: +359 2 979 31 85
Fax: +359 2 870 01 09
E-mail: jserkedjieva@microbio.bas.bg

SEP
SECRETARÍA DE
EDUCACIÓN PÚBLICA

Instituto Politécnico Nacional
Escuela Nacional de Ciencias Biológicas

Foreword

The book series entitled "UTILISATION AND MANAGEMENT OF MEDICINAL PLANTS" volume-3, covers a wide range of aspects concerning the importance of pharmacological activities for the plants in their respective ecological niche. The authors evaluate species from different countries. The chapters include biological activities like radio-protective effect in mice exposed gamma radiation, anti-bacterial, anti-fungal, contraceptive, anti-inflammation, antipyretic, antioxidant, anti-proliferative, immunomodulation, antidiabetic, insecticidal, larvicidal, anti Trypanosomal , diuretic, hepatoprotective, wound healing and development of mosquito repellent, by illustrating some.

Four chapters have described optimization of condition for the extraction and chemical characterization of different secondary metabolites from herbs and two on physiological and growth conditions to obtain better yield and quality of two species. As well as, one chapter on inter-specific variation studies using histochemical and SDS-PAGE, other chapter on comparative study of five medicinally extracts against multi drug resistant human pathogenic bacteria and two chapters on plant culture and two chapters on Aloe barbadensis and other on ornamental medicinal treasure of orchids.

We feel that this important interdisciplinary topic should be represented in studies of medicinal plants and the biological activities. This volume-3 is based on a lecture on biology of natural substances, which is part important in the evaluation and management of medicinal plants. For this reason, we outline basic tests with relevant natural compounds.

The discussion about health-promoting either secondary metabolites or extracts from different herbs. It was describe natural species or compounds affecting health, the volume summarize the results of intensive research and discuss current attempts

Prolongación de Carpio y Plan de Ayala s/n, Col. Santo Tomás, Del. Miguel Hidalgo, México, D.F., C.P. 11340
Tel.: 5729 6300, exts.: 62491 y 46268, fax.: 46211
www.ipn.mx

to answer scientifically the empirical use of the medicinal plants from different countries where the use of herb are relevant in the human health. Finally, the inspiration in the study of pharmacological effects from natural products will be briefly addressed.

In view of the vast published literature on the subject, it is possible consider a great interest to continue to address every aspect concerning the biology of natural substances.

Thus we hope that this volume will be a useful book for students and researchers in science of Medicinal Plants and related areas.

Meléndez-Camargo M.E.

Silva-Torres R.

México, City

Preface

People around the world have used herbal medicines since the beginning of civilization. Yet regarding herbal medicines in the modern age has proved to be an enormous challenge. Consumers want to benefit from access to quality herbal products and at the same time be protected from the associated risks. The industrialized world views herbal medicines as drugs in some cases distinguishing them from pharmaceuticals. Herbal medicines have the potential to contribute positively to health of people.

It is generally estimated that there are approximately 300,000 species of higher plants. However, some report the number to be 250,000, others estimate the number to be as high as 750,000. The disparity in the numbers partly reflects a difference in philosophy among systematic botanists. It also reflects the more aggressive exploration of unusual environments, particularly diverse environments such as the tropical rainforests, where new species of higher plants are being encountered continually. Of the approximately 300,000 species of higher plants, about 1%, or roughly 3000, has been utilized for food. Of those 3000, about 150 have been commercially cultivated. In today's marketplaces throughout the world, unusual fruits and vegetables are beginning to appear because there is an increasing desire on the part of the world's populations for more "exotic" foodstuffs. However, the vast majority of caloric intake derives from about 20 species of plants. These plants represent the basis upon which the world's population is fed, representing a very narrow foundation supporting the world's human population.

On the other hand, approximately 10,000 of the world's plants have documented medicinal use. Looking specifically at the utilization of plant materials in western medicine (the US, Western Europe, etc.), it is found that roughly 150–200 of such agents are incorporated. This is still a very small percentage of all higher plants. Thus, there are potentially many more important discoveries in the plant kingdom to be exploited for pharmaceutical application.

The World Health Organization (WHO) estimates that 4 billion people, 80% of the world population, presently use herbal medicine for some aspect of primary health care. WHO notes that of 119 plant-derived pharmaceutical medicines, about 74% are used in modern medicine in some ways that correlate directly with their traditional use as medicines by native cultures. India having approximately eight percent of world's biodiversity has the potential of becoming a major global player in screening and developing herb-based formulations, medicines and products. About 17000 plant species are said to exist in India of which 7500 plant species find mention in the recorded folklore of India. Of these only 1700 have been documented in 'Ayurveda' and classical healing treatise.

This reflects the yawning gap that exists between the recorded and the unrecorded knowledge about plant species. There is still a good deal of knowledge that exists with local healers, herbalists and other practitioners of herbal knowledge which remains to be documented. An urgent need for inventory knowledge of the plant species and their uses is felt at the regional and national level.

Medicinal plants research has received renewed interests of scientists in India and abroad and there is a spurt of publications. New researches are coming up in medicinal plants and a number of active principles and biological activities information for wider use by scientific community globally. Continuing earlier volumes, the 3rd volume of the series ***"Utilisation and Management of Medicinal Plants"*** deals with recent progress on medicinal plants and exhibits a total of 34 chapters authored by scientists from across the world and representing different scientific disciplines, bringing different approaches and showing the multidisciplinary research in combating a number of diseases and conditions for which there is lesser satisfactory treatment in modern medicine. I am sure these findings will provide lead material for further investigations to researchers interested in the subject. We expect that the content of this book will be a source of inspiration for future developments in these fields and hope that readers will find it a useful and functional tool in medicinal plants research.

V.K. Gupta

Contents

Utilisation and Management of Medicinal Plants Vol. 3 (2015) *Pages* **1–26**
Editors: **V.K. Gupta and A. Kaul**
Published by: **DAYA PUBLISHING HOUSE, NEW DELHI**

1

Indian Medicinal Plant Jamun, *Syzygium cumini* Exerts its Radioprotective Effect by Reducing DNA Damage in Mice Exposed to γ-Gamma Radiation

Ganesh Chandra Jagetia[1*]

ABSTRACT

Radioprotective effect of 50 mg/kg body weight jamun (Syzygium cumini Linn. Skeels, family Myrtaceae) extract was studied in the cultured splenocytes of mice exposed to 0, 0.5, 1, 2, 3, or 4 Gy of γ-radiation. The spleens of irradiated mice were removed aseptically and the splenocytes were extracted from the individual spleens and cultured. The micronuclei were prepared 72 h after irradiation in binucleate splenocytes by blocking cytokinesis with cytochalasin-B. Irradiation of mice resulted in a dose dependent elevation in the micronucleated splenocytes. The irradiation of mice not only elevated splenocytes bearing one micronuclei but also cells bearing two and multiple (more than two) micronuclei indicating induction of complex DNA damage after irradiation. Treatment of mice with 50 mg/kg body weight of jamun leaf extract protected mice splenocytes against the radiation-induced micronuclei formation. Jamun extract also protected splenocytes against the formation of two and multiple micronuclei indicating repair or inhibition of complex DNA damage. The assessment of lipid peroxidation in mice brain homogenate has indicated a concentration dependent inhibition of lipid peroxidation by jamun extract. Studies in cell free system revealed that jamun extract inhibited the formation of OH, O_2^-, DPPH and $ABTS^+$ free radicals in a concentration dependent manner.

---

1 Department of Zoology, Mizoram University, Aizawl – 796 004, Mizoram, India.
* *Author*: E-mail: gc.jagetia@gmail.com

Our study demonstrates that jamun extract protected mice against the radiation-induced DNA damage and inhibition of radiation-induced free radical formation may be one of the mechanisms of radioprotection.

Keywords: Jamun, *Syzygium cumini*, Radiation, Radioprotection, Splenocytes, Micronuclei, Mice.

Introduction

Human exposure to ionizing radiations has increased due to their use for medical diagnosis and therapy, industrial applications, nuclear power generation, atomic weapons testing, deployment of nuclear arsenals, atomic detonations for peaceful purposes, nuclear accidents, stellar radiations and air and space travel (Cucinotta and Wilson, 1994; Jagetia, 2007). Irradiation has been known to cause genomic instability and subsequently mutagenesis and carcinogenesis (Wallace, 1994; Morgan *et al.*, 1996). However, potent mutagenic effect of ionizing radiation was first demonstrated more than 80 years ago by Mueller (1927). Mutagens can induce genomic changes by targeting DNA directly or/and indirectly, by binding to proteins involved in the maintenance of genome integrity. Ionizing radiations, including X-rays, γ- rays, α or β-particles, nuclear fission products and radionucleides induce various genetic damages and mutations, and are established carcinogens. Ionizing radiations induce a variety of DNA lesions by direct and indirect interactions including DNA base alterations, DNA-DNA and DNA-protein cross-links, single- and double strand breaks and immunosupressions, micronucleus formation, sister chromatid exchanges, chromosomal aberrations, and genomic instability (Natarajan *et al.*, 1986; Wallace, 1994; Littlefield *et al.*, 2000; Jagetia *et al.*, 2003; Jagetia and Venkatesha, $2006_{a,b}$). The induction of chromosomal aberrations depends on the stage of cell cycle during radiation exposure. Irradiation of cells in G_1 and early S phase cause chromosome aberrations, whereas chromatid aberrations are produced during G_2 or mid and late S phase (Goaz, 1987).

The cytogenetic damage induced by physical or chemical agents has been traditionally evaluated using chromosome aberration analysis after the challenge until 1973, when Heddle devised a test to detect micronuclei in human peripheral blood lymphocytes as marker for chromosomal damage. This was followed by micronucleus test in erythrocytes in mouse bone marrow (Schmid, 1975). The micronucleus test has become very popular amongst the investigators owing to its simplicity, speed and sensitivity. In addition, it has an advantage over chromosome aberration analysis as more number of cells could be scored in a relatively less time and it precluded the utilization of services of highly skilled personnel that is essential for chromosome aberration analysis (Krepinsky and Heddle, 1983). Owing to easy application, the micronucleus assay immediately found its use in screening of environmental toxicants, identification of individuals more sensitive to a clastogenic exposure (Krepinsky and Heddle, 1983), routine screening of radiation workers (Krepinsky and Heddle, 1983), prediction of radiocurability of tumors (Peter *et al.*, 1986), and assessment of genomic damage during prenatal development (Cole *et al.*, 1981). The use of cytochalsin-B by Fenech and Morley (1985) allowed screening of micronuclei in first division binucleate cells as presence of cytochalsin-B inhibited actin polymerization and formation of cytoplasmic bridge in the daughter cells (MacLean-Fletcher and Pollard, 1980). This test has been especially useful in cultured mammalian cells including human peripheral blood lymphocytes. Ionizing radiations induce DNA damage that in turn is converted into chromosome damage and subsequently to micronuclei after one cell division (Natarajan *et al.*, 1986; Wallace, 1994; Fenech, 2006). Therefore, assessment of micronuclei is an appropriate indicator to evaluate the radioprotective activities of agents *in vitro* as well as *in vivo.*

The drug/s which can differentially protect the sourrounding normal tissue and/or sensitize hypoxic tumor cells is/are desirable to improve the therapeutic index during radiotherapy of tumors. However, a radioprotector will also be useful during nuclear accidents, inadvertent release of radioactivity, space and air travel as well as nuclear terror attacks (Jagetia, 2007). Several chemical agents have been screened for their radioprotective potential since Patt and his coworkers (1949) first demonstrated that cysteine, an amino acid containing sulphydryl group protected rats against the radiation-induced lethality. However, majority of the sulphydryl containing compounds provided protection at very high but toxic dose, and their application for human use has been always limited (Sweeney, 1979). Therefore, it is necessary to screen new protectors in the hope to find a suitable agent that is cheap, non-toxic at protective dose in a wide dose range, orally administered, rapidly absorbed, provide higher degree of protection, possesses a reasonably good dose reduction factor and can act through multiple mechanisms. Since plants are biochemically similar than synthetic chemicals it is reasonable to screen plant or herbs for their radioprotective properties (Jagetia, 2007).

Jamun, *Syzygium cumini* Linn. Skeels (family Myrtaceae) is a, medium sized to large tree, and it has been attributed to posses several medicinal properties in the folklore system of medicine (Warrier *et al.*, 1996). The stem bark is astringent, sweet, refrigerant, carminative, diuretic, digestive, antihelminthic, febrifuge, constipating, stomachic and antibacterial. The fruits and seeds are used to treat diabetes, and splenopathy. Various studies have indicated that jamun is antioxidant, anti-inflammatory, antibacterial, antidiabetic and gastroprotective (Muruganandan *et al.*, 2002; Shafi *et al.*, 2002; Ramirez and Roa, 2003). The leaves of the plant have been reported to reduce the radiation-induced DNA damage in the cultured human peripheral blood lymphocytes and increase mouse survival (Jagetia and Baliga, 2002, 2003). Similarly, the seed extract of jamun has been found to protect against radiation-induced lethality in mice (Jagetia *et al.*, 2005). The aim of the present study was to evaluate the radioprotective activity of leaf extract of *Syzygium cumini* on the radiation-induced micronuclei formation in the cultured mouse splenocytes exposed to different doses of γ-radiation.

Materials and Methods

Chemicals

Phytohemagglutinin (PHA) was procured from Difco Laboratories (Detroit, MI), while cytochalasin-B, RPMI, fetal calf serum (FCS), L-glutamine, concanavalin-A, dimethylsulfoxide (DMSO), deoxyribose, EDTA, ascorbic acid, nitrobluetetrazolium, 2,2-diphenyl-1-picrylhydrazyl(DPPH), 2,2-azinobis(3-ethylbenzothiazoline-6-sulphonic acid) diammonium salt (ABTS), 2-thiobarbutaric acid (TBA), and malondialdehyde (MDA), were procured from Sigma Chemical Co. (St Louis, MO); trichloroacetic acid (TCA) was procured from Acros Organics (Geel, Belgium). The other chemicals were procured from Ranbaxy Fine Chemicals (New Delhi, India).

Preparation of Extract

The fresh, mature leaves of *Syzygium cumini* Linn. Skeels (or Eugenia cumini. Druce) families Myrtaceae were collected locally during the month of April-May of the year.The leaves were cleaned, shade dried, and powdered in a mixer and the extract was prepared as described earlier (Suffness, and Douros, 1979). Briefly, the leaf powder was extracted in petroleum ether and chloroform and finally in 1:1 dichloromethane and methanol at 50-60°C using a Soxhlet apparatus. The extract was cooled and concentrated by evaporating its liquid contents in vacuo and freeze dried. The extract was stored at -70°C until further use. Henceforth the extract of *Syzygium cumini* will be called as SCE.

Preparation of Drug and other Solutions

SCE extract was dissolved in 1 per cent carboxymethyl cellulose (CMC) in sterile physiological saline. The animals were given 0.01 ml/g body weight of CMC or SCE through an oral gavage (p.o) consecutively for five days before irradiation. Cytochalasin B was dissolved in DMSO at a concentration of 10 mg/ml, stored at -80°C and diluted with the RPMI immediately before use. Concanavalin-A was dissolved in PBS and sterilized by filtration through a 0.22 μm membrane.

Animal Handling

Animal care and handling were performed according to guidelines issued by the World Health Organization (Geneva, Switzerland), the Indian National Science Academy, New Delhi, India and the "Guide for the care and use of Laboratory Animals" (NIH publication #86-23, revised in 1985). Ten to 12 week old male Swiss albino mice weighing 30-36 g were selected from an inbred colony maintained under controlled conditions of temperature (23±2s C), humidity (50±5 per cent) and light (12 of light and dark cycle). Animals had free access to sterile food and water. Four to six animals were housed in a polypropylene cage containing sterile paddy husk (procured locally) as bedding throughout the experiment. The author's Institutional Animal Ethical Committee approved the study.

Radioprotective Effect

The radioprotective effect of 50 mg/kg body weight SCE was studied on the radiation-induced micronuclei formation in the cultured splenocytes of mouse exposed to different doses of gamma radiation, where the animals were divided into the following groups:

CMC + Irradiation

The animals of this group received 0.01ml/g body weight of CMC before exposure to different doses of gamma radiation through an oral gavage.

SCE + Irradiation

The animals of this group were administered with 50 mg/kg body weight of SCE p.o. once daily for 5 consecutive days before exposure to different doses of gamma radiation.

Irradiation

One hour after the last administration of CMC or SCE, animals from both groups were exposed to 0, 0.5, 1, 2, 3 and 4 Gy ^{60}Co γ -radiation from a Tele cobalt therapy source (Theratron, Atomic Energy Agency, Ontario, Canada) at a dose rate of 1.66 Gy/min.

Splenocyte Culture

The animals were humanely killed by cervical dislocation 30 min post irradiation. The animals were thoroughly wiped with sterillium disinfectant (Bode Chemie, Hamburg, Germany) and their abdominal cavity was opened with the help of sterile scissors and forceps under aseptic conditions. The spleens of the animals were removed aseptically and washed twice in the sterile phosphate buffered saline (PBS). The splenocytes were extracted as described earlier (Jagetia *et al.*, 2001) and cultured in RPMI-1640 medium containing 10 per cent fetal calf serum (FCS) and concanavalin-A as mitogen, at 37°C for 72 h. Duplicate cultures were set from each animal for the each drug/radiation dose.

Micronucleus Assay

Micronuclei were prepared according to the method of Fenech and Morley (1985) with minor modifications. Briefly, 40 h after initiation of splenocyte culture, 5 µg/ml of cytochalasin-B was added to each culture and the cells were allowed to grow for next 32 h. The splenocytes were harvested at 72 h after initiation of the cultures by centrifugation. The splenocytes were subjected to mild hypotonic treatment (0.7 per cent ammonium oxalate) so as to retain the cytoplasm, fixed in carnoy's fixative (3:1 methanol: acetic acid), centrifuged again and resuspended in a small volume of fixative. The cell suspension was placed onto precleaned coded wet slides to avoid observer's bias and the cells were spread by gentle blowing.

Cells were stained with acridine orange (25 mg/50 ml) (BDH, England, Gurr Cat. No. 34001 9704640E) in Sorensen's buffer (pH 6.8). The slides were washed twice in Sorensen's buffer, mounted and observed under a fluorescent microscope, equipped with 450-490 nm BP filter set with excitation at 453 nm (Photomocroscope III, Carl Zeiss, Oberkohn, Germany), using a 40X Neofluar objective. A minimum of thousand binucleate splenocytes (BNC) with well-preserved cytoplasm was scored from each culture and the frequency of micronucleated binucleate cells (MNBNC) was determined. The micronuclei identification was done as described earlier Fenech (2006). A total of 48 animals were used for this experiment.

Free Radical Scavenging

The effect of SCE was studied on the scavenging of following free radicals in cell free system *in vitro*:

Hydroxyl Radical

Scavenging of hydroxyl (•OH) free radical was measured by the method of Halliwell *et al.* (1987). Briefly, the reaction mixture contained deoxyribose (2.8mM), KH_2PO_4-NaOH buffer, pH 7.4(0.05M), $FeCl_3$ (0.1 mM), EDTA (0.1mM), H_2O_2 (1 mM), ascorbate (0.1mM) and SCE (50-500 μg) in a final volume of 2 ml. The reaction mixture was incubated for 30 minutes at ambient temperature, followed by the addition of 2 ml of TCA (2.8 per cent w/v) and TBA. The reaction mixture was kept in a boiling water bath for another 30 min, cooled and the absorbance was read at 532 nm in a UV-visible double beam spectrophotometer (UV-260; Shimadzu Corp., Tokyo, Japan). All the analyses were done in triplicates and average values were taken. Inhibition (I) of deoxyribose degradation in percent was calculated according to the equation:

$$I= (A_o - A_1/A_o) \times 100$$

where,

A_0 is the absorbance of the control reaction and A_1 is the absorbance of the test compound.

Superoxide Anion Radical

Scavenging of superoxide ($O_2^{\bullet -}$) anion free radical was determined as described earlier (Hyland *et al.*, 1983). Briefly, the reaction mixture consisted of various concentrations of SCE (50-400 μg), nitroblue tetrazolium and alkaline DMSO. The pure DMSO was used as a blank instead of alkaline DMSO. The absorbance was read at 560 nm using a UV-Visible double beam spectrophotometer. Decrease in absorbance of the reaction mixture indicates increased $O_2^{\bullet -}$ scavenging activity. The inhibition of $O_2^{\bullet -}$ generation was calculated by using the following formula:

$$\text{Per cent Inhibition}= [(A_0 - A_1/A_0) \times 100]$$

where,

A_0 is the absorbance of the control and A_1 is the absorbance of the test compound.

DPPH Scavenging Activity

The ability of SCE to scavenge the stable free radical DPPH was measured as a decrease in absorbance at 517 nm (Mensor *et al.*, 2001). The methanolic solution of DPPH (0.05 mM) was mixed with an equal volume of SCE (10-400 μg), to make up the final volume up to 1.0 ml. An equal amount of methanol was added to the control. After 20 min, absorbance was recorded at 517 nm in a UV-Visible double beam spectrophotometer. Lower absorbance of the reaction mixture indicates higher free radical scavenging activity.

Total Antioxidant Activity Assay

Total antioxidant potential was determined by the ABTS assay, as described earlier (Miller *et al.*, 1996). This technique measures the relative ability of antioxi dant substances to scavenge the $ABTS^{\bullet +}$ cation radical generated in the aqueous phase. The reaction mixture contained ABTS (0.00017M), SCE (10-100 μg) and buffer in a total volume of 3.5 ml. The absorbance was measured at 734 nm in a UV-Visible double beam spectrophotometer.

Preparation of Brain Homogenate

A separate experiment was undertaken to study the effect of SCE on $FeCl_3$-induced lipid peroxidation in mice brain homogenate where the animals were anesthetized with diethyl ether, decapitated, perfused transcardially with ice cold normal saline to prevent contamination of brain tissue with blood before. The brain was removed and weighed, homogenized (10 per cent w/v) in 0.15 M potassium chloride and centrifuged at 13,000 rpm for 10 min. The supernatant was immediately used for the estimation of *in vitro* lipid peroxidation (Sharma, 1976).

Inhibition of Lipid Peroxidation-by SCE

The reaction mixture contained 0.5 ml of brain homogenate, 100 µl of 150 mM potassium chloride and ethanol (10 µl) or various concentrations of SCE, and 100 µM $FeCl_3$ (to initiate lipid peroxidation) in a final volume of 1 ml incubation mixture. The reaction mixture without SCE was considered as control. The above mixture was incubated for 20 min at 37 °C, and the reaction was stopped by adding 2 ml ice cold 0.25 N hydrochloric acid containing 15 per cent trichloroacetic acid, 0.38 per cent thiobarbituric acid and 0.05 per cent BHT. Finally the mixture was heated at 80° C for 15 min, cooled, centrifuged at 1000 g for 10 min and the absorbance of the supernatant was measured at 532 nm. The amount of lipid peroxidation was determined using the molar extinction coefficient of $1.56{\bullet}10^5\ M^{-1}\ cm^{-1}$ and expressed as TBARS (Braughler *et al.*, 1986). Percent inhibition of TBARS formed was calculated by comparing control samples. All solutions were freshly prepared in distilled water immediately before use. Since most buffers trap hydroxyl radical or interfere with iron (Braughler *et al.*, 1986), the reactions were unbuffered and carried out in 0.15 M KCl. Results are means of triplicate experiments done.

Statistical Analyses

The statistical significance was determined by one-way ANOVA with application of Bonferroni's post-hoc test was used for multiple comparisons. The Student's't'-test was used for biochemical studies. Solo 4 (BMDP Statistical Software, Los Angeles, CA) was used for statistical analysis.

Results

The results are expressed as frequency of MNBNCs per 1000 ± SEM (standard error of the mean) and percent inhibition of TBARS and various free radicals in Figures 1.1–1.9 and Tables 1.1–1.2.

Radioprotective Effect

The dose of 50 mg/kg SCE orally once daily for five days before irradiation was selected on the basis of survival studies, where a maximum number of survivors were observed (data not shown). The base line MNBNC frequency in the CMC + sham-irradiated mice was 14.33±1.45 and administration of 50 mg/kg SCE orally once daily for five consecutive days before sham-irradiation did not alter the base line frequency of MNBNC (13.30±1.73) significantly (Table 1.1). The frequency of MNBNC increased in a dose-dependent manner in CMC + irradiation group and the dose-response relationship was linear (Figure 1.1). Pretreatment of animals with SCE before exposure to different doses of gamma radiation resulted in a significant decline in the frequency of MNBNC when compared with CMC+irradiation group and the greatest protection factor was obtained for 0.5 Gy (2.6), which declined with increase in radiation dose and the lowest protection factor of 1.6 was observed for 4 Gy (Table 1.2). The dose effect relationship was linear (Table 1.1).

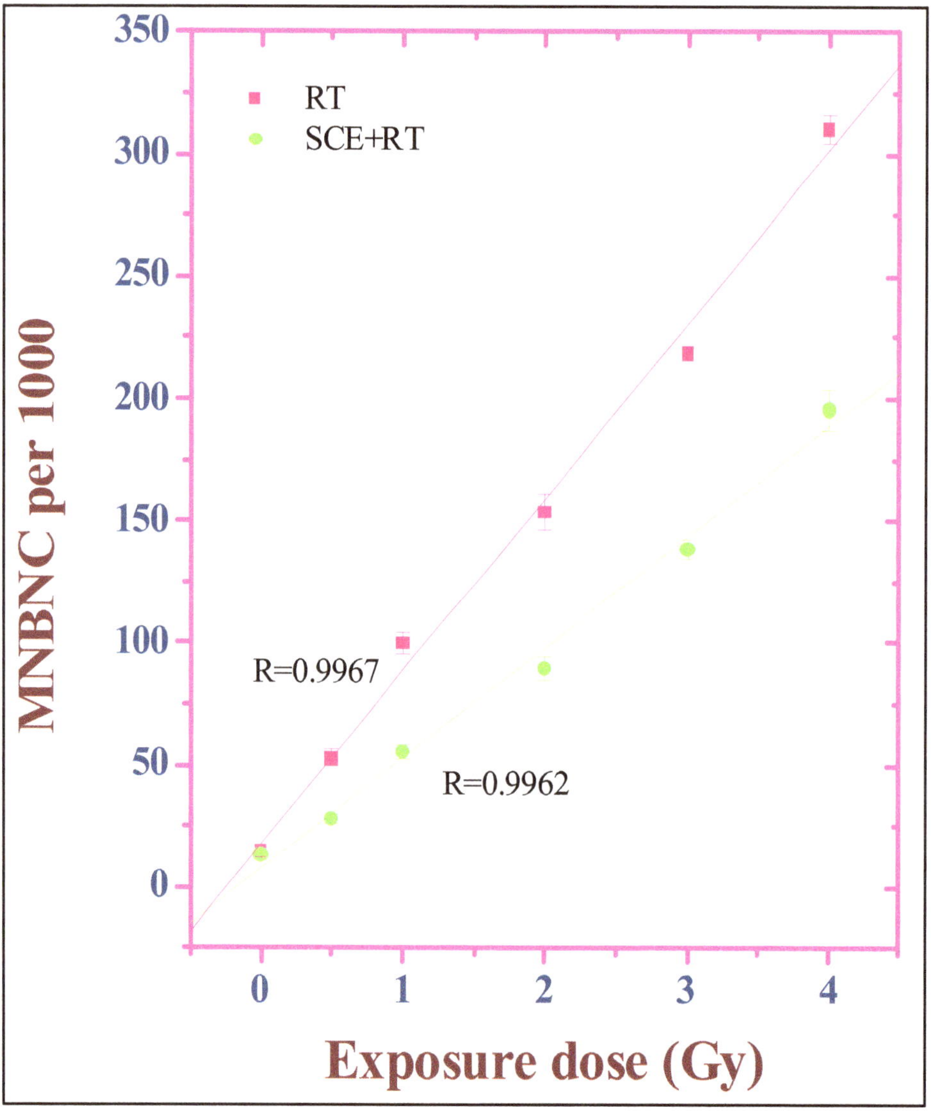

Figure 1.1: Effect of SCE on the Induction of MNBNC (total) in Cultured Splenocytes Exposed to different Doses of γ-irradiation. ● CMC + irradiation; ■ SCE + irradiation.

Table 1.1: Alteration in the Radiation-Induced Micronuclei Formation in the Cultured Splenocytes of Mouse Orally Administered with 50 mg/kg SCE Once Daily for Five Days Before Exposure to Various Doses of γ-radiation

Dose (Gy)	*Frequency of Binucleated Cells (BNC) Bearing Micronuclei (MN) (± SEM)*							
	One		*Two*		*Multiple (>3)*		*Total*	
	CMC+IR	*SCE+IR*	*CMC+IR*	*SCE+IR*	*CMC+IR*	*SCE+IR*	*CMC+IR*	*SCE+IR*
0	14.33±1.45	13.30±1.73	0.00	0.00	0.00	0.00	14.33±1.45	13.30±1.73
0.5	49.00±2.65^{a}	26.33±1.45	3.67±0.88	2.0±0.58	0.00	0.00	52.67±3.48	28.33±1.76
1.0	86.67±2.60^{b}	47.00±1.15	9.0±1.53	5.67±1.76	4.33±1.20	2.67±0.88	100.00±4.04^{cx}	55.33±2.85*
2.0	129.33±7.06^{cy*}	73.67±5.36§	15.33±1.20^{a}	11.0±1.53	8.67±0.67^{ax}	5.0±0.58	153.33±7.21^{cz*}	89.67±5.17^{p}
3.0	185.00±4.62$^{cz\#§}$	117.33±3.53^{F}	20.67±2.03^{bx}	14.0±1.73	13.0±0.58^{by*}	7.33±1.20	218.67±2.91$^{cz\#p}$	138.67±3.93^{F}
4.0	257.33±6.6$^{cz\#iF}$	158.67±9.61$^{\Psi}$	32.0±2.65^{cya§}	22.67±2.96	21.0±1.53$^{cz\#p}$	14.67±1.45	310.33±5.5$^{cz\#iW}$	195.67±8.01$^{\Psi}$

Level of significance between CMC+IR and SCE+IR: $p < 0.01$; CMC = Carboxymethyl cellulose; SCE = *Syzygium cumini* extract; IR = Irradiation.

^{a}: $p<0.05$, ^{b}: $p<0.01$, ^{c}: $p<0.001$ V/s control.

^{x}: $p<0.05$, ^{y}: $p<0.01$, ^{z}: $p<0.001$ V/s 0.5Gy.

*: $p<0.05$, ^{a}: $p<0.01$, $^{\#}$: $p< 0.001$ V/s 1 Gy.

§: $p< 0.05$, ^{p}: $p< 0.01$, ^{i}: $p< 0.001$ V/s 2Gy.

^{F}: $p< 0.01$, ^{W}: $p< 0.001$ V/s 3 Gy.

Ψ: $p<0.001$ V/s 4 Gy.

The frequencies of BNC with one, two and multiple MN are represented separately in Figures 1.2–1.4. The frequency of BNC with one MN was elevated in a dose-dependent manner in the CMC+irradiation group and a maximum number of MNBNC were scored for 4 Gy the last irradiation dose studied (Figure 1.2). Administration of animals with 50 mg/kg body weight of SCE before exposure to different doses of γ-radiation significantly resulted in a significant inhibition of MNBNC bearing one MN when compared with the CMC + irradiation group. The protection factor for BNC with one MN was 2.7 for 0.5 Gy that declined with increasing

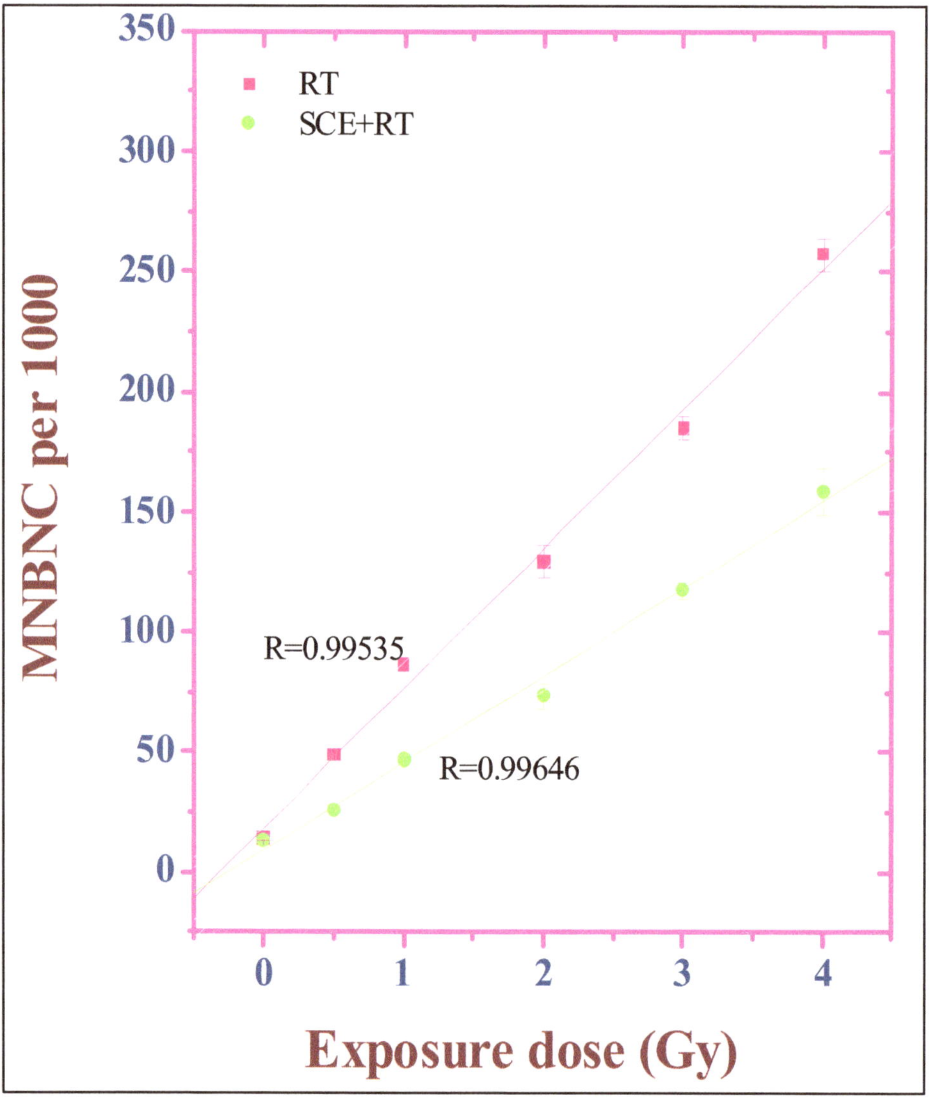

Figure 1.2: Effect of SCE on the Induction of BNC with One MN in Cultured Splenocytes Exposed to different Doses of γ-irradiation. ● CMC + irradiation; ■ SCE + irradiation.

dose of radiation (Table 1.2). The curve fitting of data revealed a linear dose-response relationship for both groups (Figure 1.1 and Table 1.1).

Table 1.2: Protection Factor of SCE against the Radiation-Induced Micronuclei in the Splenocytes of Mice Exposed to different Doses of γ-radiation

Exposure Dose (Gy)	*Protection Factor for Various Types of MNBNCs*			
	One	*Two*	*Multiple*	*Total*
0.5	2.7	1.8	0	2.6
1	2.2	1.6	1.6	2.0
2	1.9	1.4	1.7	1.8
3	1.6	1.5	1.7	1.6
4	1.7	1.4	1.4	1.6

The frequency of BNC with two MN showed a dose-dependent rise in the CMC + irradiation group (Figure 1.3). Oral administration of SCE before irradiation resulted in a significant reduction in the frequency of BNC with two MN when compared with the CMC + irradiation group (Table 1.1). The highest frequency of MNBNC with 2 MN was observed at 4 Gy irradiation. A highest protection factor of 1.8 was obtained for 0.5 Gy that declined steadily with increasing dose of radiation (Table 1.2). Both CMC and SCE+ irradiation group showed a linear dose response relationship (Table 1.1).

The BNCs with multiple (≥ 3) MN were conspicuous by their absence for CMC+ sham- irradiation group (Figure 1.4) and with a further increase in the irradiation dose, the frequency of BNC with multiple MN increased in a dose dependent fashion and the maximum number of multiple MNBNCs was observed for 4 Gy. Oral treatment of animals with SCE before irradiation resulted in a significant decline in the frequency of multiple MNBNCs at all the exposure doses and protection factor of approximately 1.6 was observed for 1-3 Gy, except 4 Gy where it was 1.4 (Table 1.2). The dose response was linear for CMC+irradiation and SCE+ irradiation groups.

Hydroxyl Radical-Scavenging

This assay shows the ability of SCE to inhibit hydroxyl radical-mediated deoxyribose degradation in a reaction mixture with EDTA. The relative extents of inhibition of deoxyribose degradation will give an indication of •OH scavenging activity. SCE inhibited the generation of hydroxyl radical in a dose-dependent manner, and a maximum inhibition of OH free radicals was observed at 350 μg (51 per cent); thereafter a steady state was attained (Figure 1.5).

Superoxide Anion Radical Scavenging

Superoxide anions are a precursor to active free radicals that have potential of reacting with biological macromolecules and thereby inducing tissue damage (Halliwell and Gutteridge, 1984). SCE inhibited the generation of $O_2^{\bullet -}$ radical in a dose-dependent manner, and a highest inhibition was observed at 250 μg (56.8 per cent); thereafter, a steady state was attained (Figure 1.6).

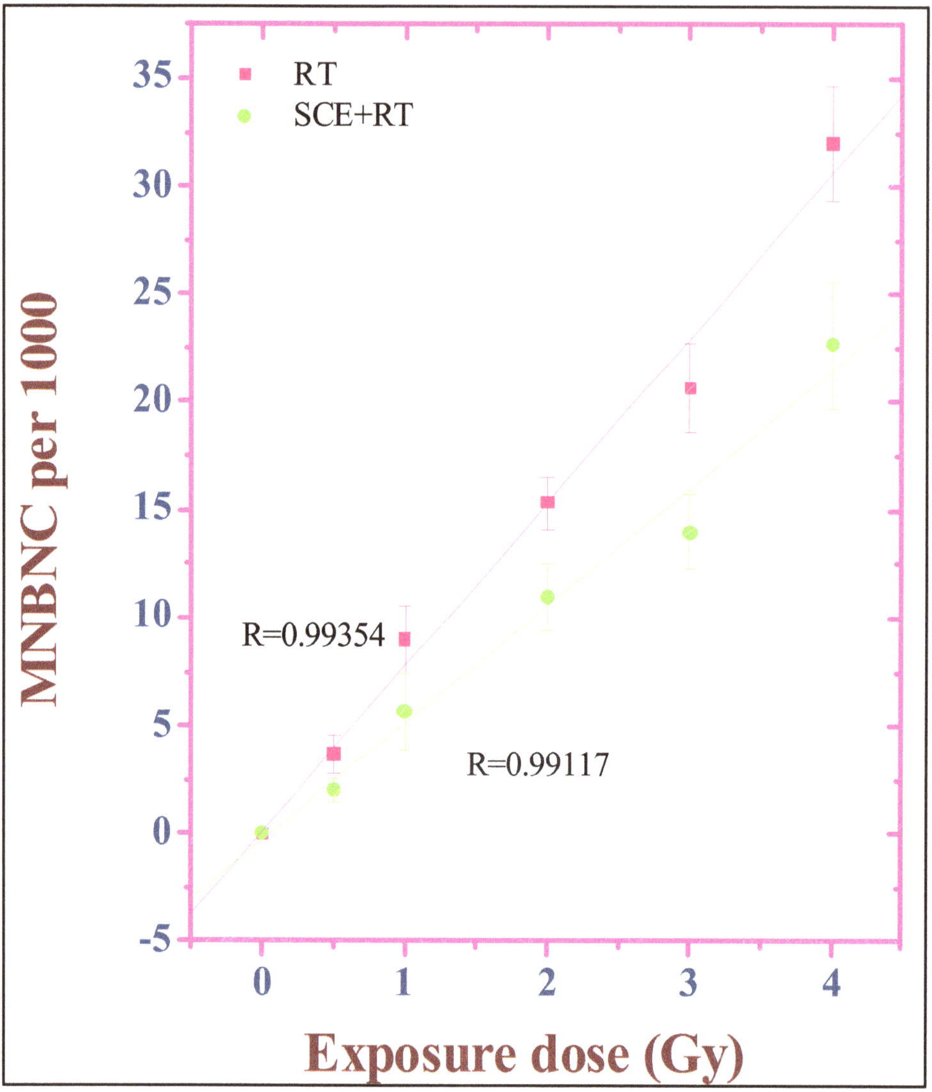

Figure 1.3: Effect of SCE on the Induction of BNC with Two MN in Cultured Splenocytes Exposed to different Doses of γ-irradiation. ● CMC + irradiation; ■ SCE + irradiation.

DPPH Free Radical Scavenging

DPPH has been widely used to evaluate the free radical scavenging effects of various antioxidants substances (Ozcelik *et al.*, 2003). In the DPPH assay, the antioxidants were able to reduce the stable radical DPPH to the yellow colored diphenyl-picrylhydrazine. The method is based on the reduction of alcoholic DPPH solution in the presence of a hydrogen-donating antioxidant due to the formation of the non-radical form DPPH-H by the reaction. With this method it is possible to determine the antiradical power of an antioxidant by measuring of a decrease in the

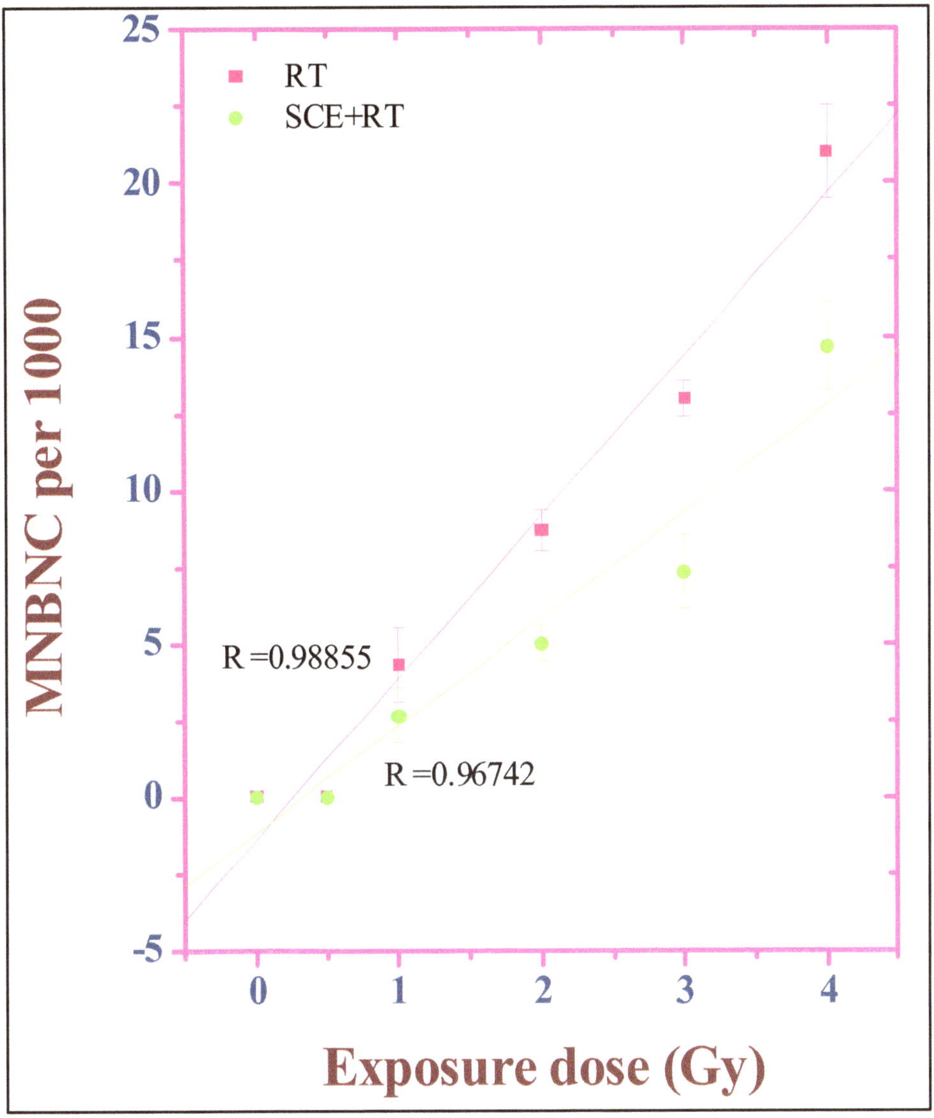

Figure 1.4: Effect of SCE on the Induction of BNC Bearing Multiple MN in Cultured Splenocytes Exposed to different Doses of γ-irradiation. ■ CMC + irradiation; ● SCE + irradiation.

absorbance of DPPH• at 517 nm. Becuase of a resulting color change from purple to yellow, the absorbance decreased when the DPPH• was scavenged by an antioxidant through donation of hydrogen to form a stable DPPH• molecule. In the radical form, this molecule had an absorbance at 517 nm which disappears after acceptance of an electron or hydrogen radical from an antioxidant compound to become a stable diamagnetic molecule (Matthaus, 2002). Figure 1.7 illustrates a decrease in the concentration of DPPH radical due to the scavenging ability of SCE in a concentration

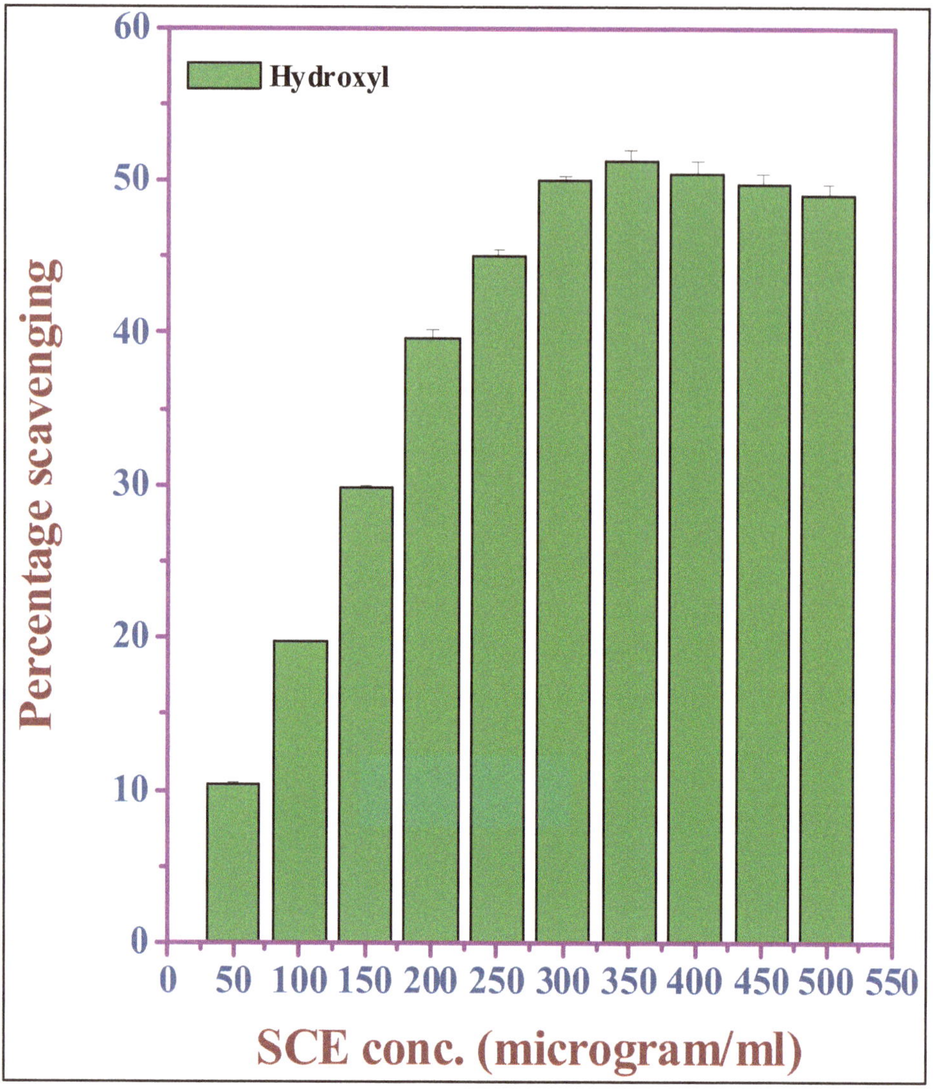

Figure 1.5: Effect of Various Concentrations of SCE on the Scavenging of Hydroxyl Free Radicals.

dependent manner and a maximum inhibition was observed at 80 µg (72.6 per cent); thereafter a steady state was attained.

ABTS Radical Scavenging

Generation of the ABTS radical cation forms the basis of one of the spectrophotometric methods that have been applied for the measurement of the total antioxidant activity of solutions of pure substances, aqueous mixtures and beverages (Miller *et al.*, 1996). A more appropriate format for the assay is a decolorization

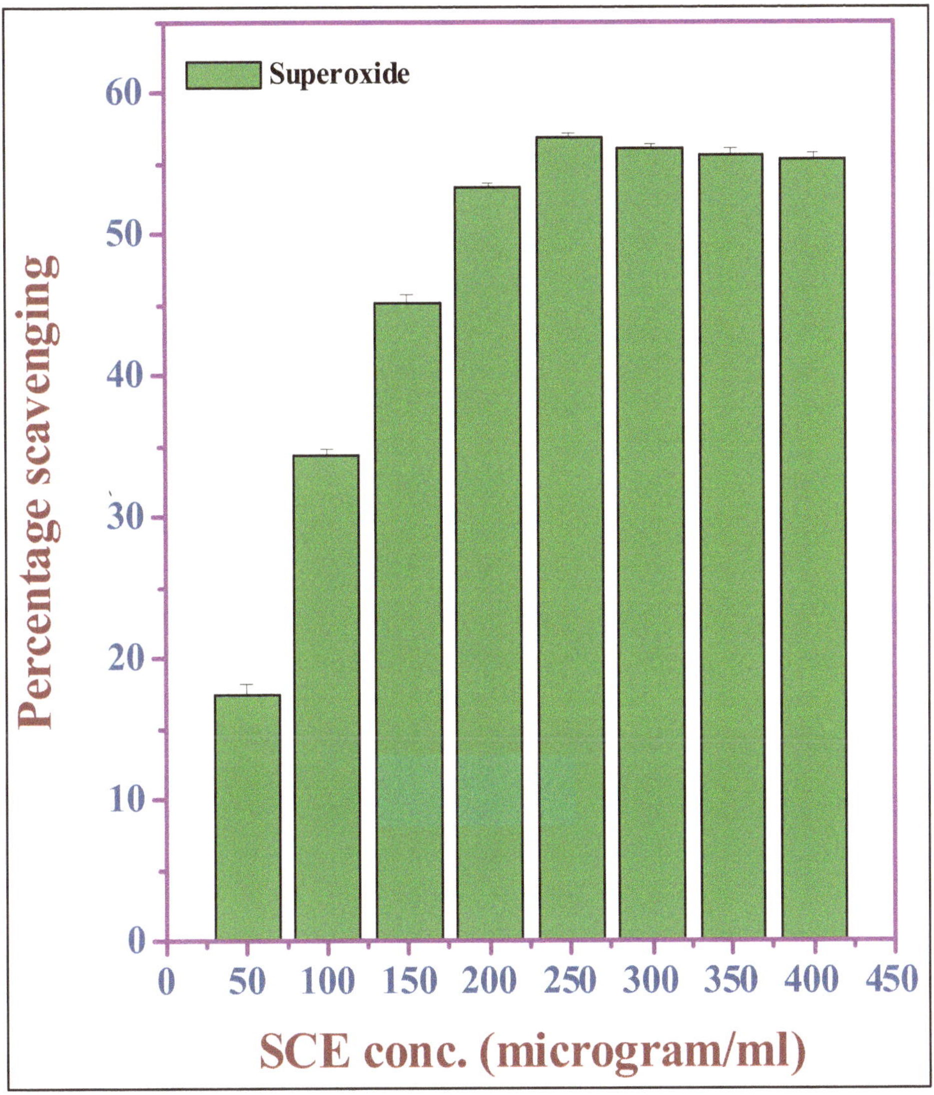

Figure 1.6: Effect of Various Concentrations of SCE on the Scavenging of Superoxide Anion Free Radicals.

technique where the radicals are generated directly in a stable form prior to reaction with putative antioxidants. The improved technique for the generation of $ABTS^{\bullet+}$ described here involves the direct production of the blue/green $ABTS^{\bullet+}$ chromophore through the reaction between ABTS and potassium persulphate. SCE inhibited the generation $ABTS^{\bullet+}$ radical in a concentration dependent manner (10-100 µg) and a maximum inhibition was observed at 80 µg (72.6 per cent), which declined steadily thereafter (Figure 1.8).

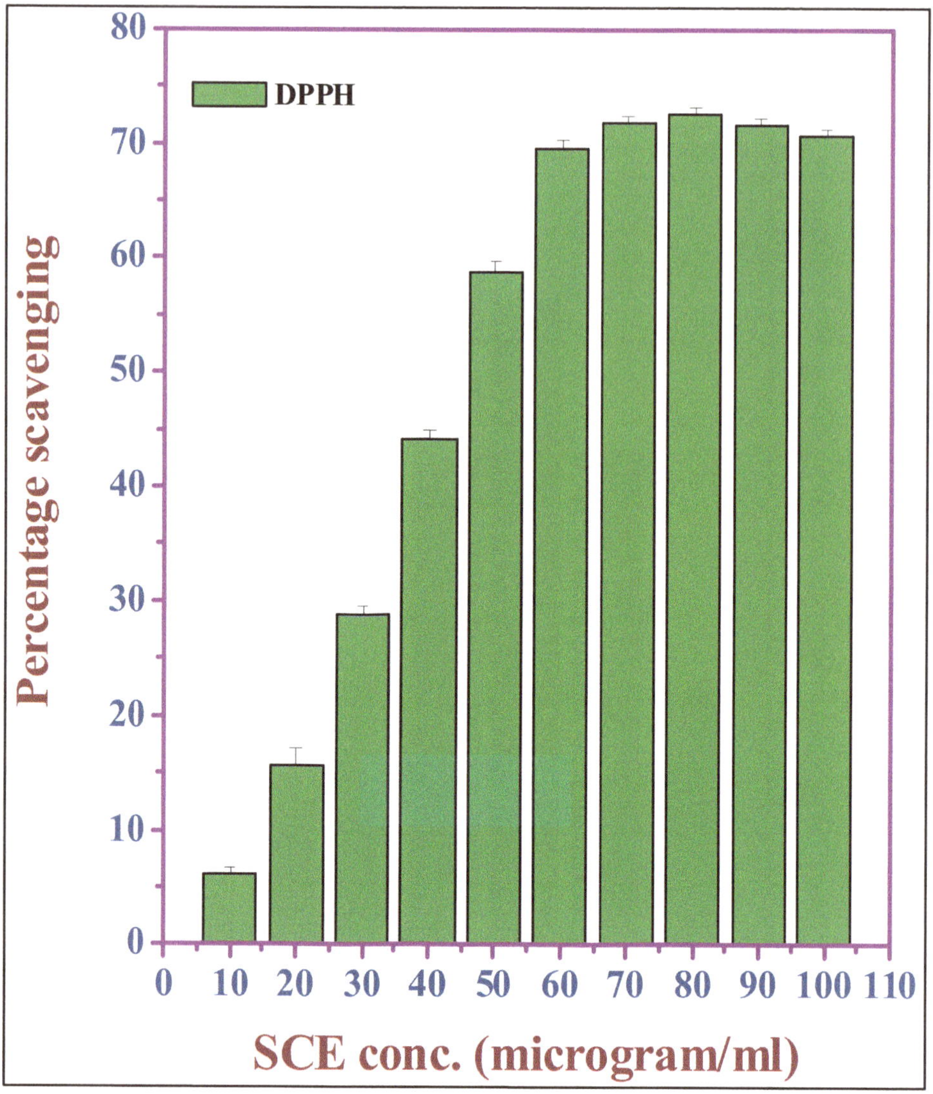

Figure 1.7: Effect of Various Concentrations of SCE on the Scavenging of DPPH Free Radicals.

Inhibition of Lipid Peroxidation-by SCE

Lipid peroxidation is an important pathophysiological event in a variety of diseases, drug toxicities and traumatic or ischemic injuries. In the central nervous system, for example the role of oxygen free radicals and lipid peroxidation results in the irreversible loss of neuronal tissue following brain injury (Brown and Hall, 1992) and degenerative neurological disorders (Halliwell, 1989). SCE inhibited the formation of TBARS in a dose dependent manner (50-500 μg) in brain homogenates and the maximum inhibition was observed at 350 μg (65.34 per cent), which declined steadily thereafter (Figure 1.9).

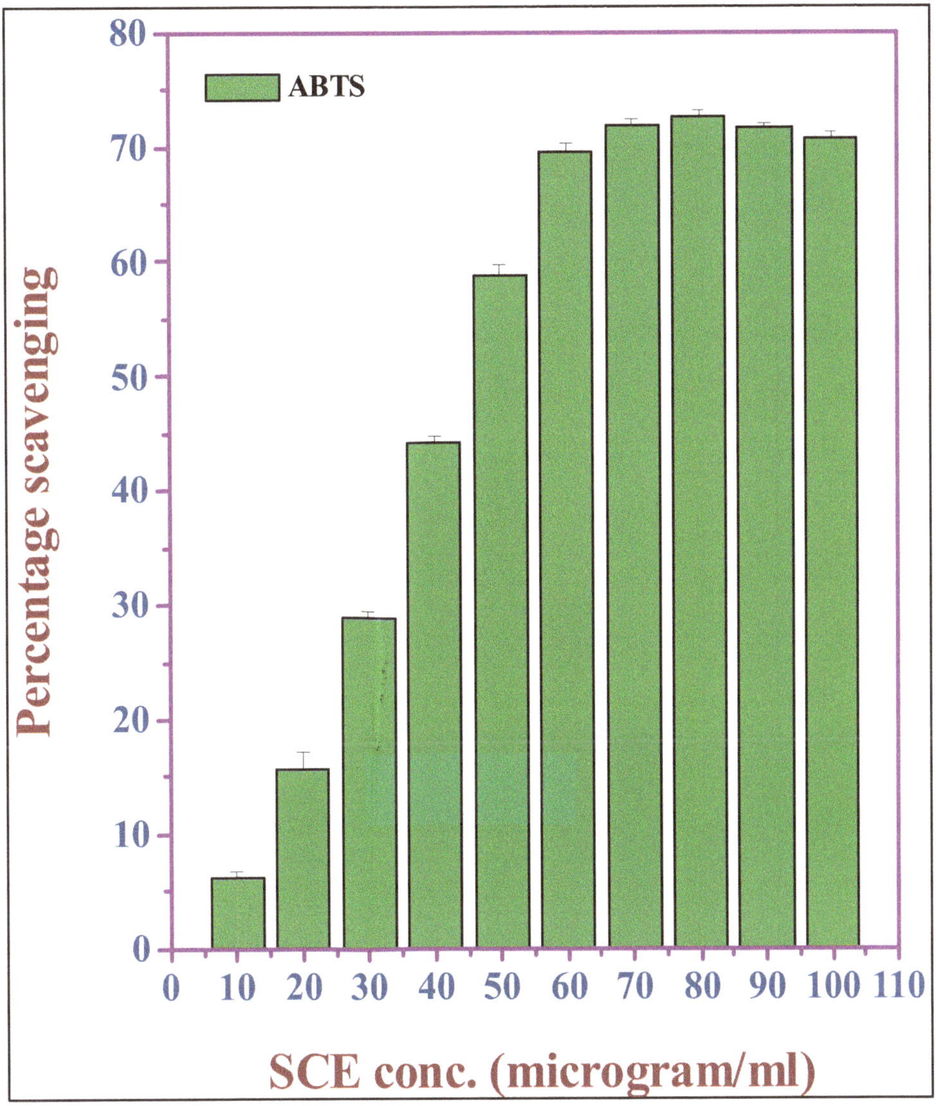

Figure 1.8: Effect of Various Concentrations of SCE on the Scavenging of $ABTS^+$ Cation Free Radicals.

Discussion

The DNA damage especially the double strand breaks produced by ionizing radiation are converted into chromosome aberrations, that are subsequently expressed as micronuclei after one cell division (Natarajan *et al.*, 1994; Ward, 1998; Fenech, 2006). Micronuclei can be assessed in dividing cells easily and they are the measure of DNA damage (Majer *et al.*, 2001). The micronucleus assay has been used to screen genotoxic agents and can serve as a short term test to screening radioprotective ability

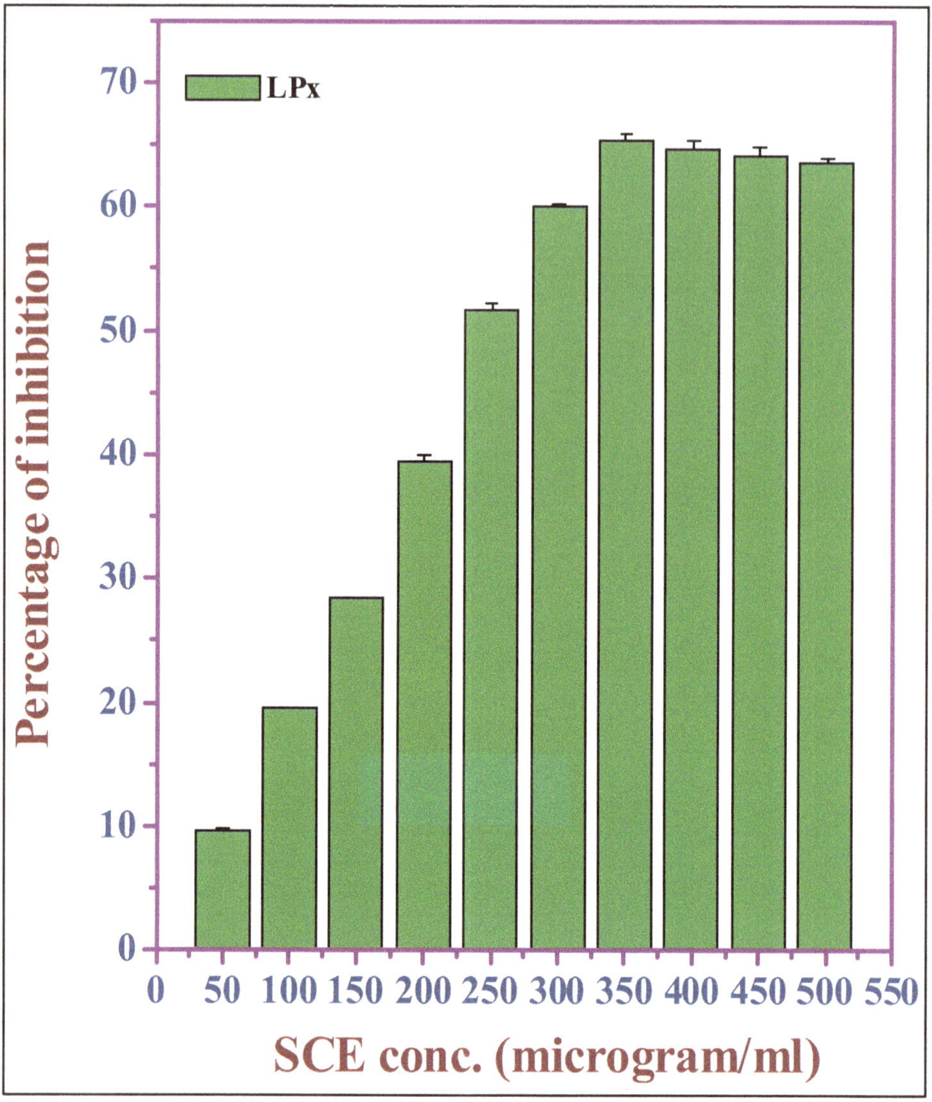

Figure 1.9: Effect of Various Concentrations of SCE on Lipid Peroxidation in Brain Homogenate.

of any pharmacological agent (Midander and Revesz, 1980; Heddle *et al.*, 1983; Jagetia and Venkatesha, 2005; Jagetia, 2007). Therefore, present study was undertaken to investigate the radioprotective potential of SCE in cultured splenocytes of mice exposed to different doses of γ-radiation.

The acentric chromosome or chromatid fragments or whole chromosomes or chromatids that lag behind in anaphase of cell division due to defective kinetochore or spindle fiber/s are not included in the daughter nuclei during telophase and are converted into small rounded or spherical extranuclear bodies that can be observed in the cytoplasm of daughter cell/s and are known as micronuclei (Heddle and Carrano, 1977). The assessment of micronuclei gives an indirect indication of DNA damage as micronuclei harboring chromosomal fragments may arise from direct double-strand DNA breakage, conversion of SSBs into DSBs after cell replication, or inhibition of DNA synthesis. Exposure of animals to different doses of gamma radiation resulted in a dose-dependent elevation in MNBNCs and the dose-response relationship was linear. A resembling effect has been observed earlier (Jagetia *et al.*, 1994; Jagetia and Aruna, 1997; Belyakov *et al.*, 1999; Jagetia and Adiga, 2000; Jagetia and Reddy, 2002; Jagetia *et al.*, 2003). However, a linear quadratic response has also been reported (Huber *et al.*, 1994; Vral *et al.*, 1998; Wuttke, 1998; Jagetia and Baliga, 2002). Irradiation not only increased the frequency of BNC with one MN but also with 2 or multiple MN indicating induction of complex multiple sites of DNA damage and production of these lesions was radiation dose dependent. Irradiation has been reported to increase the frequency of cells bearing one, two or multiple MN in lymphocytes and various cultured cell lines (Catena *et al.*, 1997; Belyakov *et al.*, 1999; Jagetia and Aruna, 1999, 2000; Jagetia and Adiga, 2000; Jagetia and Baliga, 2002; Jagetia *et al.*, 2003; Jagetia and Venkatesha, 2005).

Treatment of animals with 50 mg/kg body weight of SCE significantly reduced the frequency of MNBNCs when compared with CMC + irradiation group. An earlier study from this laboratory has reported that leaf extract of *Syzygium cumini* reduced the MNBNC frequency in HPBLs; however, no attempt was made to study the effect of SCE after different doses of radiation (Jagetia and Baliga, 2002). Similarly, *Aegle marmelos* leaf extract has been reported to protect human peripheral blood lymphocytes (HPBLs) against the radiation-induced MN formation (Jagetia *et al.*, 2003). The extracts of *Spirulina plantenis*, a blue-green alga have been reported to protect mice bone marrow against the radiation-induced micronucleus formation (Qishen *et al.*, 1989). The plant formulations like Liv.52 and abana have also been reported to protect mice bone marrow cells against the radiation-induced micronuclei formation (Jagetia and Ganapathi, 1989; Jagetia and Aruna, 1997; Jagetia *et al.*, 2006). Likewise, plant polyphenols like naringin and mangiferin, have been reported to protect mice and HPBLs against radiation induced MN formation (Jagetia and Reddy, 2002; Jagetia and Venkatesha, 2005). Certain other plant flavonoids like luteolin, quercetin, kaempferol, morin, rutin, fisetin, myricetin, epicatechin, genistein, eridictyol, phloretin, naringin and hesperidin have been found to protect mice bone marrow cells against the radiation-induced micronuclei (Shimoi *et al.*, 1994, 1996; Jagetia and Reddy, 2002; Hosseinmehr and Nemati, 2006). SCE not only protected against the formation of BNC with one MN but also with 2 or multiple MN in the cultured splenocytes after exposure to different doses of γ -radiation, indicating that SCE could efficiently protect complex multiple sites of DNA damage. An identical effect has been observed earlier with *Agele marmelos* extract and mangiferin in HPBLs (Jagetia *et al.*, 2003; Jagetia and Venkatesha, 2005). Likewise, a reduction in PCE bearing two or more MN has been observed in mice bone marrow treated with (E) 4-(4-N,N-dimethyl aminophenyl) but-3-en-2-one (Jagetia *et al.*, 1994), naringin a a citrus bioflavonoid (Jagetia and Reddy, 2002), the extracts of abana (Jagetia and Aruna, 1997), *S. cumini* (Jagetia and Baliga, 2002) and *A. marmelos* (Jagetia *et al.*, 2003) and also the mangiferin (Jagetia and Venkatesha, 2005).

The exact mechanism by which SCE protects against the genotoxic effects of radiation is not known. There may be several putative mechanisms by which SCE may have protected mouse splenocytes against the radiation-induced micronuclei formation. Ionizing radiations are potent inducer of free radicals and oxidative stress. Superoxide anion, arising either through metabolic processes or following oxygen "activation" by physical irradiation, is considered the "primary" ROS, whereas the hydroxyl radical, •OH, has a high reactivity, making it a very dangerous radical with a very short *in vivo* half life (Pastor *et al.*, 2000). These free radicals elicit a cascade of events that eventually interact with molecular DNA (Halliwell and Gutteridge, 1999). Hence scavenging of these radiation-induced free radicals by SCE may be one of the important mechanisms to reduce radiation induced DNA damage. This contention is supported by the *in vitro* results in this study, where SCE scavenged •OH, $O_2^{\bullet -}$, DPPH and $ABTS^{\bullet +}$ free radicals and inhibited lipid peroxidation in a concentration dependent manner. Further, some of the flavonoids including quercetin, kaempferol and myricetin, which are present in the SCE, have been also reported to scavenge free radicals like OH, $O_2^{\bullet -}$, and inhibit lipid peroxidation earlier (Maridonneau-parini *et al.*, 1986; Korina and Áfanas'ev, 1997; Abalea *et al.*, 1999). Similarly, kaempferol and quercetin have been reported to suppress the cytotoxicity of Superoxide ion and hydrogen peroxide in Chinese hamster V79 cells (Nakayama *et al.*, 1993; Nakayama, 1994). The polyphenol ellagic acid, which is also present in SCE, has been reported to be antimutagenic, chemopreventive (Tanaka, 1994), antioxidant and it has also been found to inhibit the radiation-induced lipid peroxidation in the liver of mice (Theresiamma *et al.*, 1996). The presence of flavonoids and ellagic acid in SCE extract might have been responsible for the observed radioprotection in mice splenocytes. The presence of SCE before irradiation may have also increased the antioxidant status of mice (data not shown) thus resulting in the protection of mice against radiation induced DNA damage. Thus, the radioprotective effect of SCE may be due to free radical scavenging, and increased antioxidant status. Although no attempt has been made to unfurl the molecular mechanisms however, there is no reason to believe that SCE may not have acted using these molecular mechanisms, since radiation has been reported to induce the transcriptional activation of NF-κB (Meng *et al.*, 2003), COX-II, and LOX (Tessner *et al.*, 2004), which have been associated with inflammation and oxidative stress (D'Acquisto *et al.*, 1998; Natoli, 2006) including DNA damage. SCE pretreatment may have blocked the activation of NF-κB, by reducing DNA damage, COX-II at mRNA level and LOX, thus protecting against the radiation-induced DNA damage. Quercetin, a flavonoid present in SCE has been reported to inhibit the activation of NF-κB and COX-II mRNA (Natarajan *et al.*, 1998; de Pascual-Teresa *et al.*, 2004). Myricetin, has been reported to increase the expression of DNA polymerase beta gene in a dose dependent manner, an enzyme responsible for the error-free DNA repair (Abalea *et al.*, 1999). Since SCE contains myricetin the reduction in the radiation-induced micronuclei by SCE may be also owing to the error-free DNA repair in the SCE + irradiation group.

The present study demonstrates that oral administration of jamun extract for five consecutive days before exposure to various doses of γ-radiation reduced the radiation-induced micronuclei formation and also protected splenocytes against the complex multiple sites of DNA damage. The radioprotective activity of jamun may not be due to single mechanism but may be due to interplay of several mechanisms. It may protect the cells against radiation-induced DNA damage by scavenging of free radicals and blocking radiation-induced lipid peroxidation. It may also up regulate antioxidant mechanisms of cell thereby reducing DNA damage. Jamun may have also inhibited the activation of NF-κB and COX-II mRNA. It may have also up regulated DNA polymerase and efficiently repaired the lesions induced by radiation in the cellular genome.

References

Abalea, V., Cillard, J., Dubos, M.P., Sergent, O., Cillard, P., and Morel, I. (1999). Repair of iron-induced DNA oxidation by the flavonoid myricetin in primary rat hepatocyte cultures. *Free Radical Biology and Medicine*, **26**: 1457-1466.

Belyakov, O.V., Prise, K.M., Trott, K.R., and Michael BD. (1999). Delayed lethality, apoptosis and micronucleus formation in human fibroblasts irradiated with X-rays or alpha-particles. *International Journal of Radiation Biology*, **75**: 985-993.

Braughler, J.M., Duncan, L.A., and Chase, R.L. (1986). The involvement of iron in lipid peroxidation. Importance of ferric to ferrous ratios in initiation. *Journal of Biological Chemistry*, **261**: 10282-10289.

Brown, S.A., and Hall, E.D. (1992). Role of oxygen-derived free radicals in the pathogenesis of shock and trauma with focus on central nervous system injuries. *Journal of the American Veterinary Medical Association*, **200**: 1849-59.

Catena, C., Parasacchi, P., Conti, D., Sgura, A., Trenta, G., Righi, E., Trinci, M.M., and Trinci, M. (1997). Peripheral blood lymphocyte decrease and micronucleus yields during radiotherapy. *International Journal of Radiation Biology*, **72**: 575-585.

Cole, R.J., Taylor, N., Cole, J., and Arlett, C.F. (1981). Short-term tests for transplacentally active carcinogens. I. Micronucleus formation in fetal and maternal mouse erythroblasts. *Mutation Research*, **80**: 141-57.

Cucinotta, F.A., and Wilson,.JW. (1994). Estimates of cellular mutagenesis from cosmic rays. NASA TP-3453. pp 1-10.

D'Acquisto, F., Sautebin, L., Iuvone, T., DiRosa, M., and Carnuccio, R. (1998). Prostaglandins prevent inducible nitric oxide synthase protein expression by inhibiting nuclear factor-kappaB activation in J774 macrophages. *FEBS Letters*, **440**: 76-80.

de Pascual-Teresa, S., Johnston, K.L., DuPont, M.S., O'Leary, K.A., Needs, P.W., Morgan, L.M., Clifford, M.N., Bao, Y., and Williamson, G. (2004). Quercetin metabolites downregulate cyclooxygenase-2 transcription in human lymphocytes ex vivo but not *in vivo*. *The Journal of Nutrition*, **134**: 552-557.

Fenech, M., and Morley, A.A. (1985). Measurement of micronuclei in lymphocytes.*Mutation Research*, **147**: 29-36.

Fenech, M. (2006). Cytokinesis-block micronucleus assay evolves into a "cytome" assay of chromosomal instability, mitotic dysfunction and cell death. *Mutation Research*, **600**: 58-66.

Goaz, P. (1987). Radiation Biology. In: Goaz, P., White, S., editors. Oral radiology principles and interpretation. 2nd ed. CV Mosby Co.

Halliwell, B. (1989). Oxidants and the central nervous system: some fundamental questions. Is oxidant damage relevant to Parkinson's disease, Alzheimer's disease, traumatic injury or stroke? *Acta neurologica Scandinavica. Supplementum,* **126**: 23-33.

Halliwell, B., Gutteridge, J.M.C., and Aruoma, O.I. (1987). The deoxyribose method: a simple "test-tube" assay for determination of rate constants for reactions of hydroxyl radicals. *Analytical Biochemistry*, **165**: 215-219.

Halliwell, B., and Gutteridge, J.M.C. (1984). Lipid peroxidation, oxygen radicals, cell damage, and antioxidant therapy. *Lancet*, **1**: 1396-7.

Halliwell, B., and Gutteridge, J.M.C. (1999). Free radicals in biology and medicine, 3rd ed. Oxford University Press.

Heddle, J.A., and Carrano, A.V. (1977). The DNA content of micronuclei induced in mouse bone marrow by gamma-irradiation: evidence that micronuclei arise from acentric chromosomal fragments. *Mutation Research*, **44**: 63-69.

Heddle, J.A., Hite, M., Kirkhart, B., Mavournin, K., MacGregor, J.T., Newell G.W., and Salamone, M.F. (1983). The induction of micronuclei as a measure of genotoxicity. A report of the U.S. Environmental Protection Agency Gene-Tox Program. *Mutation Research*, **123**: 61-118.

Hosseinimehr, S.J., and Nemati, A. (2006). Radioprotective effects of hesperidin against gamma irradiation in mouse bone marrow cells. *The British Journal of Radiology*, **79**: 415-418.

Huber, R., Schraube, H., Nahrstedt, U., Braselmann, H., and Bauchinger, M. (1994). Dose-response relationships of micronuclei in human lymphocytes induced by fission neutrons and by low LET radiations. *Mutation Research*, **306**: 135-141.

Hyland, K., Voisin, E., Banoun, H., and Auclair, C. (1983). Superoxide dismutase assay using alkaline dimethylsulfoxide as superoxide anion-generating system. *Analytical Biochemistry*, **135**: 280-7.

Jagetia, G.C. (2007). Radioprotective potential of plants and herbs against the effects of radiation. *Journal of Clinical Biochemistry and Nutrition*, **40**: 74-81.

Jagetia, G.C., and Adiga, S.K. (2000). Correlation between cell survival and micronuclei formation in V79 cells treated with vindesine before exposure to different doses of gamma-radiation. *Mutation Research*, **448**: 57-68.

Jagetia, GC., and Aruna, R. (1997). The herbal preparation abana protects against radiation-induced micronuclei in mouse bone marrow. *Mutation Research*, **393**: 157-163.

Jagetia, G.C., and Aruna, R. (1999). Effect of various concentrations of acyclovir on cell survival and micronuclei induction on cultured HeLa cells. *Mutation Research*, **446**: 155-165.

Jagetia GC., Aruna R. (2000). Correlation between cell survival and micronuclei-induction in HeLa cells treated with adriamycin after exposure to various doses of gamma-radiation. *Toxicology Letters*, **115**: 183-193.

Jagetia, G.C., Baliga, M.S., and Venkatesh, P. (2005). Influence of seed extract of *Syzygium Cumini* (Jamun) on mice exposed to different doses of gamma-radiation. *Journal of Radiation Research*, **46**: 59-65.

Jagetia, G.C., and Baliga, M.S. (2002). *Syzygium cumini* (Jamun) reduces the radiation-induced DNA damage in the cultured human peripheral blood lymphocytes: a preliminary study. *Toxicology Letters*, **132**: 19-25.

Jagetia, G.C., and Baliga, M.S. (2003). Evaluation of the radioprotective effect of the leaf extract of *Syzygium cumini* (Jamun) in mice exposed to a lethal dose of gamma-irradiation. *Nahrung*, **47**: 181-185.

Jagetia, G.C., Ganapathi, N.G., Venkatesh P., Rao N., and Baliga, M.S. (2006). Evaluation of the radioprotective effect of Liv 52 in mice. *Environmental and Molecular Mutagenesis*, **47**: 490-502.

Jagetia, G.C., and Ganapathi, N.G. (1989). Inhibition of clastogenic effect of radiation by Liv. 52 in the bone marrow of mice. *Mutation Research*, **224**: 507-510.

Jagetia, GC., Jacob, P.S., and Rao, M.N. (1994). (E)4-[4-N, N-dimethyl-aminophenyl]but-3-en-2-one (DMAP) treatment inhibits the radiation-induced micronucleus formation in bone marrow of BALB/c mice. *Mutation Research*, **306**: 71-80.

Jagetia, G.C., Menon, K.S., and Jain, V. (2001). Genotoxic effect of hydroquinone on the cultured mouse splenocytes. *Toxicology Letters*, **121**: 15-20.

Jagetia, G.C., and Reddy, T.K. (2002). The grapefruit flavanone naringin protects against the radiation-induced genomic instability in the mice bone marrow: a micronucleus study. *Mutation Research*, **519**: 37-48.

Jagetia, G.C., Venkatesh, P., and Baliga, M.S. (2003). Evaluation of the radioprotective effect of *Aegle marmelos* (L.) Correa in cultured human peripheral blood lymphocytes exposed to different doses of gamma-radiation: a micronucleus study. *Mutagenesis*, **18**: 387-393.

Jagetia, G.C., and Venkatesha, V.A. (2005). Effect of mangiferin on radiation-induced micronucleus formation in cultured human peripheral blood lymphocytes. *Environmental and Molecular Mutagenesis*, **46**: 12-21.

Jagetia, G.C., and Venkatesha, V.A. (2006a). Mangiferin protects human peripheral blood lymphocytes against γ-radiation–induced DNA strand breaks: a fluorescence analysis of DNA unwinding assay. *Nutrition Research*, **26**: 303–311.

Jagetia, G.C., and Venkatesha, V.A. (2006b). Treatment of mice with stem bark extract of Aphanamixis polystachya reduces radiation-induced chromosome damage. *International Journal of Radiation Biology*, **82**: 197-209.

Korin, L.G., and Afanas'ev, I.B. (1997). Antioxidant and chelating properties of flavonoids. *Advances in Pharmacology*, **38**: 151-63.

Krepinsky, A.B.,and Heddle, J.A. (1983). Micronuclei as a rapid and inexpensive measure of radiation induced chromosomal aberrations. In: Radiation induced chromosome damage in man. New York: Alan R. Liss Inc.

Littlefield, L.G., McFee, A.F., Sayer, A.M., O'Neill, J.P., Kleinerman, R.A., and Maor, MH. (2000). Induction and persistence of chromosome aberrations in human lymphocytes exposed to neutrons *in vitro* or *in vivo*: Implications of findings in 'retrospective' biological dosimetry. *Radiation Protection Dosimetry*, **88**: 59-68.

MacLean-Fletcher, S., and Pollard T.D. (1980). Mechanism of action of cytochalasin B on actin. *Cell*, **20**: 329-41.

Majer, B.J., Laky, B., Knasmuller, S., and Kassie, F. (2001). Use of the micronucleus assay with exfoliated epithelial cells as a biomarker for monitoring individuals at elevated risk of genetic damage and in chemoprevention trials. *Mutation Research*, **489**: 147-172.

Maridonneau-Parini, I., Braquet, P., and Garay, R.P. (1986). Heterogeneous effect of flavonoids on K^+ loss and lipid peroxidation induced by oxygen-free radicals in human red cells. *Pharmacological Research Communications*, **18**: 61-72.

Matthaus, B. (2002). Antioxidant activity of extracts obtained from residues of different oilseeds. *Journal of Agricultural and Food Chemistry*, **50**: 3444-52.

Meng A., Yu, T., Chen, G., Brown, S.A., Wang, Y., Thompson, J.S., and Zhou, D. (2003). Cellular origin of ionizing radiation-induced NF-kappaB activation *in vivo* and role of NF-kappaB in ionizing radiation-induced lymphocyte apoptosis. *International Journal of Radiation Biology*, **79**: 849-61.

Mensor, L.L., Menezes, F.S., Leitao, G.G., Reis, A.S., dos Santos, T.C., Coube, C.S., and Leitao, S.G. (2001). Screening of Brazilian plant extracts for antioxidant activity by the use of DPPH free radical method. *Phytotherapy Research*, **15**: 127-30.

Midander, J., and Revesz, L. (1980). The frequency of micronuclei as a measure of cell survival in irradiated cell populations. *International Journal of Radiation Biology and Related Studies in Physics, Chemistry, and Medicine*, **38**: 237-42.

Miller, N.J., Castelluccio, C., Tijburg, L., and Rice-Evan,s C. (1996). The antioxidant properties of theaflavins and their gallate esters–radical scavengers or metal chelators? *FEBS Letters*, **392**: 40-4.

Morgan, W.F., Day, J.P., Kaplan, M.I., McGhee, E.M., and Limoli, CL. (1996). Genomic instability induced by ionizing radiation. *Radiation Research*, **146**: 247-58.

Muller, H.J. (1927). Artificial transmutation of the gene. *Science*, **66**: 84-87.

Muruganandan, S., Pant, S., Srinivasan, K., Chandra, S., Tandan,S.K., Lal, J., and Prakash, R.V. (2002). Inhibitory role of *Syzygium cumini* on autacoid-induced inflammation in rats. *Indian Journal of Physiology and Pharmacology*, **46**: 482-6.

Nakayama, T., Yamada, M., Osawa, T., and Kawakishi, S. (1993). Suppression of active oxygen-induced cytotoxicity by flavonoids. *Biochemical Pharmacology*, **45**: 265-267.

Nakayama, T. (1994). Suppression of hydroperoxide-induced cytotoxicity by polyphenols. *Cancer Research*, **54**: 1991s-1993s.

Natarajan, A.T., Balajee, A.S., Boei, J.J., Chatterjee, S., Darroudi, F., Grigorova, M., Noditi, M., Oh, H.J., Slijepcevic, P., and Vermeulen, S. (1994). Recent developments in the assessment of chromosomal damage. *International Journal of Radiation Biology*, **66**: 615-23.

Natarajan, A.T., Darroudi, F., Mullenders, L.H., and Meijers, M. (1986). The nature and repair of DNA lesions that lead to chromosomal aberrations induced by ionizing radiations. *Mutation Research*, **160**: 231-236.

Natarajan, K., Manna, S.K., Chaturvedi, M.M., and Aggarwal, B.B. (1998). Protein tyrosine kinase inhibitors block tumor necrosis factor-induced activation of nuclear factor-kappaB, degradation of IkappaBalpha, nuclear translocation of p65, and subsequent gene expression. *Archives of Biochemistry and Biophysics*, **352**: 59-70.

Natoli, G. (2006). Tuning up inflammation: how DNA sequence and chromatin organization control the induction of inflammatory genes by NF-kappaB. *FEBS Letters*, **580**: 2843-2849.

Ozcelik, B., Lee, J.H., and Min, D.B. (2003). Effects of light, oxygen and pH on the 2,2-diphenyl-1-picrylhydrazyl (DPPH) method to evaluate antioxidants. *Journal of Food Science*, **68**: 487-490.

Pastor, N., Weinstein, H., Jamison, E., and Brenowitz, M. (2000). A detailed interpretation of OH radical footprints in a TBP-DNA complex reveals the role of dynamics in the mechanism of sequence-specific binding. *Journal of Molecular Biology*, **304**: 55-68.

Patt, H.M., Tyree, E.B., Straube, R.L., and Smith, D.E. (1949). Cysteine Protection against X Irradiation. *Science*, **110**: 213-214.

Peters, L.J., Brock, W.A., Johnson, T., Meyn, R.E., Tofilon, P.J., and Milas, L. (1986). Potential methods for predicting tumor radiocurability. *International Journal of Radiation Oncology, Biology, Physics*, **12**: 459-467.

Qishen, P., Guo, B.J., and Kolman, A. (1989). Radioprotective effect of extract from Spirulina platensis in mouse bone marrow cells studied by using the micronucleus test. *Toxicology Letters*, **48**: 165-169.

Ramirez, R.O., and Roa, C.C. Jr. (2003). The gastroprotective effect of tannins extracted from duhat (*Syzygium cumini* Skeels) bark on HCl/ethanol induced gastric mucosal injury in Sprague-Dawley rats. *Clinical Hemorheology and Microcirculation*, **29**: 253-61.

Schmid, W. (1975). The micronucleus test. *Mutation Research*, **31**: 9-15.

Shafi, P.M., Rosamma, M.K., Jamil, K., and Reddy, P.S. (2002). Antibacterial activity of *Syzygium cumini* and *Syzygium travancoricum* leaf essential oils. *Fitoterapia*, **73**: 414-6.

Sharma, O.P. (1976). Antioxidant activity of curcumin and related compounds. *Biochemical Pharmacology*, **25**: 1811-1812.

Shimoi, K., Masuda, S., Furugori, M., Esaki, S., and Kinae N. (1994). Radioprotective effect of antioxidative flavonoids in gamma-ray irradiated mice. *Carcinogenesis*, **15**: 2669-72.

Shimoi, K., Masuda, S., Shen, B., Furugori, M., and Kinae, N. (1996). Radioprotective effects of antioxidative plant flavonoids in mice. *Mutation Research*, **350**: 153-61.

Suffness, M., and Douros, J. (1979). Drugs of plant origin. In: Methods Cancer Research. pp. 73-126.

Sweeney, T.R. (1979). A survey of compounds from the antiradiation drug development program of the US Army Medical Research and Development Command. Washington DC: Government Printing Office. pp. 308-318.

Tanaka, T. (1994). Cancer chemoprevention by natural products. *Oncology Reports*, **1**: 1139-1155.

Tessner, T.G., Muhale, F., Schloemann, S., Cohn, S.M., Morrison, A.R., and Stenson, W.F. (2004). Ionizing radiation up-regulates cyclooxygenase-2 in I407 cells through p38 mitogen-activated protein kinase. *Carcinogenesis*, **25**: 37-45.

Thresiamma, K.C., George, J., and Kuttan, R. (1996). Protective effect of curcumin, ellagic acid and bixin on radiation induced toxicity. *Indian Journal of Experimental Biology*, **34**: 845-847.

Vral, A., Louagie, H., Thierens, H., Philippe, J., Cornelissen, M., andde Ridder L. (1998). Micronucleus frequencies in cytokinesis-blocked human B lymphocytes after low dose gamma-irradiation. *International Journal of Radiation Biology*, **73**: 549-55.

Wallace, S.S. (1994). DNA damages processed by base excision repair: biological consequences. *International Journal of Radiation Biology*, **66**: 579-89.

Ward, J.F. (1988). DNA damage produced by ionizing radiation in mammalian cells: identities, mechanisms of formation, and reparability. *Progress in Nucleic Acid Research and Molecular Biology*, **35**: 95-125.

Warrier, P.K., Nambiar, V.P.K., and Raman Kutty, C. (1996). Indian medicinal plants. Hyderabad India: Orient Longman Ltd.

Wuttke, K., Muller, W.U., AND Streffer, C. (1998). The sensitivity of the *in vitro* cytokinesis-blocked micronucleus assay in lymphocytes for different and combined radiation qualities. *Strahlentherapie und Onkologie*, **174**: 262-8.

Utilisation and Management of Medicinal Plants Vol. 3 (2015) *Pages* **27–52**
Editors: **V.K. Gupta and A. Kaul**
Published by: **DAYA PUBLISHING HOUSE, NEW DELHI**

2

Ginger Rhizome (*Zingiber officianale* Rosc.) Protects Mice Whole Body Exposed to Different Doses of γ-Radiation

Ganesh Chandra Jagetia[1*]

ABSTRACT

The radioprotective effect of hydroalcoholic extract of the rhizome of Zingiber officinale (ZOE) was studied in the mice exposed to various doses of gamma radiation. The preliminary screening of ZOE for radioprotection showed that 10 mg/kg ZOE was the optimum dose, when administered consecutively for 5 days by intraperitoneal route. The mice were injected with 10 mg/kg of ZOE or DDW intraperitoneally for five consecutive days before exposure to 6, 7, 8, 9, 10, 11 and 12 Gy of radiation. The animals from all the groups were observed for thirty days for the development of symptoms of radiation sickness and mortality. Pretreatment of mice with ZOE reduced the severity of symptoms of radiation sickness and mortality at all the exposure doses, and the number of survivors also increased in the ZOE + irradiation group when compared with the concurrent DDW + irradiation group. The ZOE treatment protected the mice against the GI death as well as bone marrow deaths. The $LD_{50/30}$ was found to be 8.2 Gy for the DDW + irradiation group, while it was 9.4 Gy for ZOE + irradiation groups, respectively. The administration of ZOE and OSE resulted in an increase of LD_{50} by 1.2 Gy and simultaneously the extract of Ocimum sanctum (OSE) and 2-mercaptopropioonyl glycine (MPG) were used as positive controls. Concurrently the dose reduction factor (DRF) was found to be 1.15 for ZOE + irradiation group. The optimum protective dose of 10 mg/kg was 1/50 of LD_{50} dose (500 mg/kg) of ZOE. The lipid peroxidation and glutathione status was evaluated in the surviving animals on day 31 post-irradiation. The irradiation of animals resulted in a dose dependent elevation in the lipid peroxidation. Conversely, the GSH declined in a dose dependent manner in the DDW + irradiation group. The treatment of animals with

---

1 Department of Zoology, Mizoram University, Aizawl – 796 004, Mizoram, India.
* *Author*: E-mail: gc.jagetia@gmail.com

ZOE before irradiation caused a significant depletion in the lipid peroxidation followed by a significant elevation in the GSH contents in the liver of mice at 31 days post-irradiation. The ZOE also showed a dose dependent antimicrobial activity in all the microorganisms studied. However, the best activity was observed for Pseudomonas aeruginosa, Salmonella typhimurium and Escherichia coli. The ZOE extract was also able to inhibit the proliferation of Candida albicans.

Keywords: *Zingiber officinale* (ginger), *Ocimum santum*, 2-mercaptopropionylglycine (MPG), Radiation, Mice survival, Acute toxicity, Radioprotection, Antibacterial, Antimicrobial, Lipid peroxidation, Glutathione.

Introduction

The search for radioprotectors started with the realization of the need for a safeguard against the military use of atomic weapons. With the recognition that normal tissue protection in radiotherapy is as important as the destruction of the cancer cells, the focus of radioprotection research became more therapy oriented. The first report of the use of chemicals to protect against the radiation-induced damage appeared in 1949, when Patt *et al.*, reported that cysteine protected mice and rats against the radiation-induced sickness and mortality. Thereafter, several chemical compounds and their analogues have been screened for their radioprotective ability. However, the practical applicability of the majority of these synthetic compounds remained limited, owing to their high toxicity at their optimum protective dose (Sweeny, 1979). To reduce the toxic side effects of synthetic compounds a need was felt to explore the alternative/s to the synthetic compounds, which could be less toxic and highly effective at non-toxic dose levels. The plants have been used to treat various ailments in human beings since the time immemorial and the herbal preparations have been usually considered safe and nontoxic in comparison to the synthetic compounds. Therefore, it is quite natural that the choice of alternative radioprotectors falls on plants/plant products.

The rhizome of *Zingiber officinale*, commonly known as ginger, is consumed worldwide as a spice and flavouring agent. The rhizome of ginger has been reported to possess diverse medicinal properties in the traditional Indian system of medicine, the Ayurveda, and it is widely used in several medicinal preparations. (Nadkarni, 1976; CHEMEXCIL, 1992; Warrier *et al.*, 1996). Depending on the season, type of climate, geographic region and the plant processing, the ginger has been reported to possess terpenes like ar-curcumene, farnesene, β-bisabolene, γ-selinene, β-elemene, β-sesquiphellandrene, phellandrene, camphene, α-pinene, β-pinene, cumene, myrecene, limonene, cymene, cineole, citral, zingiberene and basabolene. The ginger also contains the oxygenated monoterpenes like the 2-heptanol, 2-nonanol, n-nonanol, n-decanal, methyl heptenone, 1,8-cineole, borneol, bornyl acetate, linalool, geranial and neral. The oxymethyl phenols like gingerol, shogaol, zingerone and paradol were also found to be present in different proportions (Wealth of India, 1972; Rastogi and Mehrotra, 1990; CHEMEXCIL, 1992). Most of these compounds have been reported to possess antioxidant and free radical scavenging activities (Aeschbach *et al.*, 1994; Tanaka, 1994) and to increase the antioxidant enzymes (Kog Ah-ng *et al.*, 2000) and decrease lipid peroxidation (Ahmed *et al.*, 2000). Ginger has been reported to be antiemetic, anti-inflammatory, antioxidant, antidepressant, hypoglycemic, and antimicrobial (Phillips *et al.*, 1993; Jafri *et al.*, 2010; Auta *et al.*, 2011; Singh *et al.*, 2012; Mahluji *et al.*, 2013).

The lesson from the experience with radioprotectors world wide is that the animal studies with death as the end point is the most confirmatory, because the 30 days time period after lethal whole body irradiation clearly indicates the capacity of the drug, in test to modulate the recovery and regeneration of the gastrointestinal epithelium and the hemopoietic progenitor cells in the bone marrow, the two most radiosensitive organs that are essential for sustenance of life. The common usage, wide acceptability in human beings, diverse medicinal and antioxidative properties attributed to ginger rhizome, stimulated us to obtain an insight into the radioprotective effect of ginger (*Zingiber officinale*) in the mice exposed whole-body gamma radiation.

Materials and Methods

The animal care and handling were carried out according to the guidelines issued by the World Health Organization, Geneva, Switzerland and the INSA (Indian National Science Academy, New Delhi, India). Ten to twelve weeks old male Swiss albino mice weighing 30 to 36 g were selected from an inbred colony maintained under the controlled conditions of temperature (23± 2°C), humidity (50±5 per cent) and light (10 and 14 h of light and dark, respectively). The animals had free access to the sterile food and water. Four to six animals were housed in a polypropylene cage containing sterile paddy husk (procured locally) as bedding throughout the experiment. The institutional animal ethical committee had approved the study.

Preparation of the Extract

The ginger rhizomes (small local variety) were collected during the month of April, and the leaves of *Ocimum santum* in the month of September 2001. The plants were identified by Dr. G. K. Bhat, Department of Botany, Poorna Prajna College, Udupi, India). The ginger rhizomes were separated from the stems, freed of scales, cleaned, shade dried, powdered, and extracted. Briefly, one hundred grams of the ginger powder was extracted with 50 per cent aqueous ethanol at 50 to 60°C using a Soxhlet apparatus for 72 h. The cooled liquid extract was concentrated by evaporating its liquid contents. An approximate 22 per cent yield of the extract was obtained. Briefly, hundred grams of the powder was extracted in pure double distilled water at 50 to 60°C under reflux. The cooled liquid extract was centrifuged and the supernatant was collected and concentrated by evaporating its liquid contents in a lyophilizer to obtain a yield of 18 per cent. The extracts were stored in deep freeze at -70°C till use. Henceforth, the extract of *Zingiber officianale* will be called as ZOE.

Preparation of the Drug Solution

The drug 2-mercaptopropionylglycine (MPG) was procured from Santen Pharmaceuticals Limited, Osaka, Japan. MPG was freshly dissolved in the sterile double distilled water (DDW) so as to give a concentration of 2 mg/ml immediately before use. The pH of the solution was adjusted to 6.4 by the addition of 0.1 N sodium hydroxide before administration, while ZOE was dissolved at a concentration of 1 mg/ml in DDW.

Mode of Administration

The animals were administered with 0.01ml/g b. wt. of DDW or ZOE or MPG intraperitoneally, unless otherwise stated and the animals were divided into the following groups:

DDW+ Irradiation

The animals of this group received DDW as described above before exposure to different doses of γ irradiation.

MPG + Irradiation Group

The animals of this group were administered with 20 mg/kg b. wt. of MPG intraperitoneally thirty minutes before exposure to different doses of γ -radiation (Nagata *et al.*, 1972).

ZOE + Irradiation

This group of animals was injected with different doses or 10 mg/kg b. wt. of ZOE intraperitoneally once daily for 5 consecutive days before exposure to radiation (Jagetia and Baliga, 2002).

Irradiation

One hour after the last administration of DDW, or ZOE, and thirty minutes after the administration of MPG the prostrate and immobilized animals (achieved by inserting cotton plugs in the restrainer) were whole body exposed to ^{60}Co gamma radiation (Theratron, Atomic Energy Agency, Canada) in a specially designed well-ventilated acrylic box. A batch of ten animals was irradiated each time at a dose rate of 1.66 Gy/min at a source to animal distance (midpoint) of 70 cm. The following experiments were conducted:

Determination of Acute Drug Toxicity of ZOE

The acute toxicity of the ZOE was determined according to Prieur et al. (1973) and Ghosh (1984). Briefly, the animals were allowed to fast by withdrawing the food and water for 18 h. The fasted animals were divided into several groups of 10 each. Each group of animals was injected with various doses 125, 200, 250, 300, 400, 500, 600, 750, 800, 1000, 2000, 3000, and 4000 mg/kg body weight (b. wt.) of freshly prepared extract of ginger intraperitoneally. The animals were provided with food and water immediately after the drug administration. Mortality of the animals was observed up to 14 days post drug treatment.

Selection of Optimum ZOE Dose

The animals were divided into two groups DDW + irradiation and ZOE + irradiation group. The former group received 0.1 ml/g b. wt. of DDW, while the ZOE + irradiation group was administered with 5, 10, 20, 40 and 80 mg/kg b. wt. of ZOE intraperitoneally once daily for five consecutive days. One hour after the last administration on the fifth day the animals of both the groups were exposed to 10 Gy of γ-radiation (Jagetia and Baliga, 2002). This allowed the preliminary screening of the optimum drug dose and 10 mg/kg b. wt. of ZOE was found to be the best radioprotective dose and therefore, further studies were carried out using this dose of ZOE.

Selection of Route of ZOE Administration

An attempt has also been made to select the most effective route of administration of ZOE for radioprotection, where the animals were administered orally or intraperitoneally with 10 mg/kg b. wt. of ZOE once daily for five consecutive days before exposure to 10 Gy of γ-radiation.

Selection of Optimum Fraction Size of ZOE

To select the optimum fraction size of ZOE for radioprotection, the animals were administered with ZOE as a single dose of 50 mg or a daily dose of 25 mg for two days and 10 mg for five days intraperitoneally. The animals were exposed to 10 Gy of γ-radiation one hour after the last administration of ZOE.

The Radioprotective Effect of ZOE

To ascertain the radioprotective ability of ZOE the animals were divided into three groups as described above and one hour after the last administration of DDW or ZOE the animals were exposed to 6, 7, 8, 9, 10, 11 and 12 Gy of γ-radiation along with the positive controls.

The animals of all the experiments were monitored daily for the development of symptoms of radiation sickness, and mortality up to 30 days post-irradiation. The dose reduction factor (DRF) was calculated by the method of Miller and Tainter (1944).

$$DRF = \frac{LD_{50} \text{ of the ZOE + irradiation group}}{LD_{50} \text{ of DDW+ irradiation group}}$$

Biochemical Estimations

The animals exposed to different doses of radiation from both the ZOE + radiation and DDW + radiation groups, which survived up to 30 days were euthenised on the 31st day, after exposure. The livers were perfused with an ice-cold saline transcardially. The whole liver was removed, blot dried, weighed and a 10 per cent homogenate was prepared with an ice-cold 0.2 M sodium phosphate buffer pH 8.0 using a homogenizer (Yamato LSG LH-21, Japan).

Estimation of Protein

Total protein, were estimated by the method of Lowry, *et al.* (1954) using bovine serum albumin as standard.

Glutathione (GSH)

GSH contents were measured by the method of Moron *et al.* (1979). Briefly, proteins were precipitated by 25 per cent TCA, centrifuged and the supernatant was collected. The supernatant was mixed with 0.2 M sodium phosphate buffer pH 8.0 and 0.06 mM DTNB and incubated for 10 minutes at room temperature. The absorbance of the sample/s was read against the blank at 412 nm in a UV-Visible spectrophotometer (Shimadzu UV-260, Shimadzu Corp, Japan) and the GSH concentration was calculated from the standard curve.

Lipid Peroxidation (LPx)

LPx was measured by the method of Beuege and Aust (1984). Briefly, the tissue homogenate was mixed with TCA-TBA-HCl and the resultant mixture was heated for 15 min in a boiling water bath. After centrifugation the absorbance was recorded at 535 nm using a UV-Visible spectrophotometer (Shimadzu UV-260, Shimadzu Corp, Japan). The lipid peroxidation has been expressed as MDA in nM per mg protein.

Antimicrobial Activity

The antimicrobial activity of ZOE was evaluated on enteric pathogens such as *Escherichia coli, Klebsiella, Salmonella typhimurium, Shigella dysenteriae, Vibrio choleraoe, Pseudomonas aeruginosa* and *Staphylococcus aureus*. Antifungal activity of the ginger extract was also performed against *Candida albicans*, one of the etiological agents causing diarrhoea in the immunocomprimised hosts.

Antimicrobial activity of ginger extract was estimated by agar diffusion method (Cruickshank *et al.*, 1975; Treagan, 1982). Briefly, the bacterial and *Candida albicans* colonies were inoculated into peptone water and nutrient broth and incubated at 37°c for 4-6 hours. The turbidity was compared using Mc Farland's turbidity standard tube no: 5. The test strains of bacteria and *Candida albicans* were inoculated in the sterile petri plates of Muller-Hinton agar (MHA) and Sabourauds dextrose agar (SDA) by lawn culture method. The wells of 6 millimeter diameter were punched in the agar medium with the help of sterile borer. These wells were inoculated with 15, 22.5, 30, 37.5 and 45 mg of ZOE. The DDW and solvent control were run simultaneously. The plates were incubated at 37°C overnight and the diameter of zones of inhibition was measured in millimeters.

Analysis of Data

The statistical significance between the treatments was determined using the "Z" test for the survival studies, while the student's *t*-test was used for biochemical and antimicrobial activities. All the data are expressed as mean ± SEM (standard error of the mean).

Results

The results are expressed as percent survival after exposure to various doses of gamma radiation (Figure 2.1).

Determination of Acute Toxicity of ZOE

The administration of different doses of ZOE viz. 125, 200, 250 and 300 mg/kg b. wt. did not induce any mortality during the whole observation period. However, a further increase in the drug dose to 400 mg/kg b. wt. resulted in a 20 per cent reduction in the survival of mice. An increase in the dose of ZOE up to 500 mg/kg b. wt. resulted

Figure 2.1: Kapalan Meier's Estimate of Survival of Mice Treated with Ginger Extract (ZOE) before Whole Body γ-irradiation of Mice. a) selection of drug dose; b) selection of route of administration; c) selection of drug fractionation (a, b and c exposed to 10 Gy irradiation) and d) effect of various doses of radiation.

a

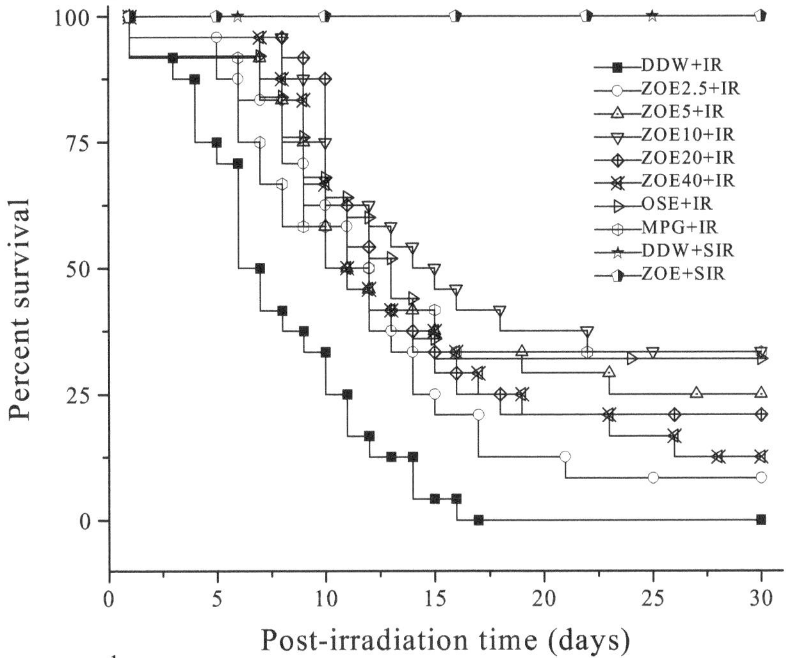

b

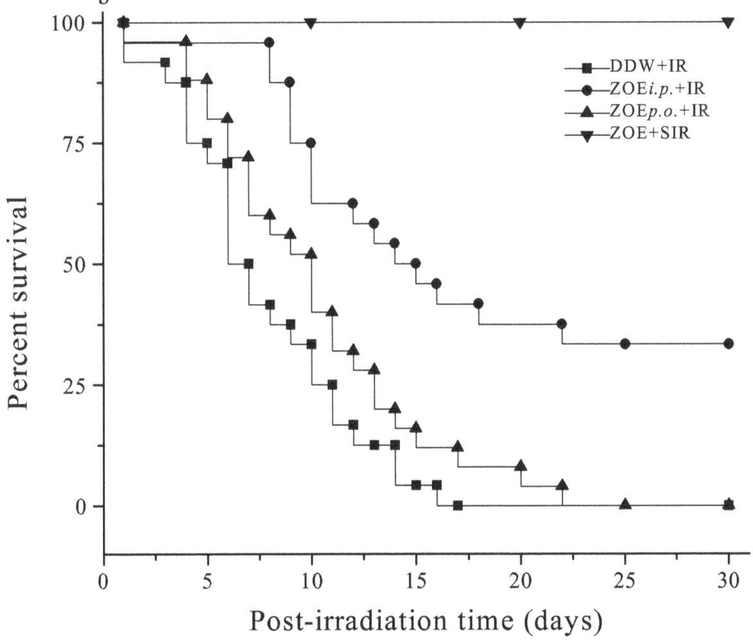

Contd...

Figure 2.1–*Contd...*

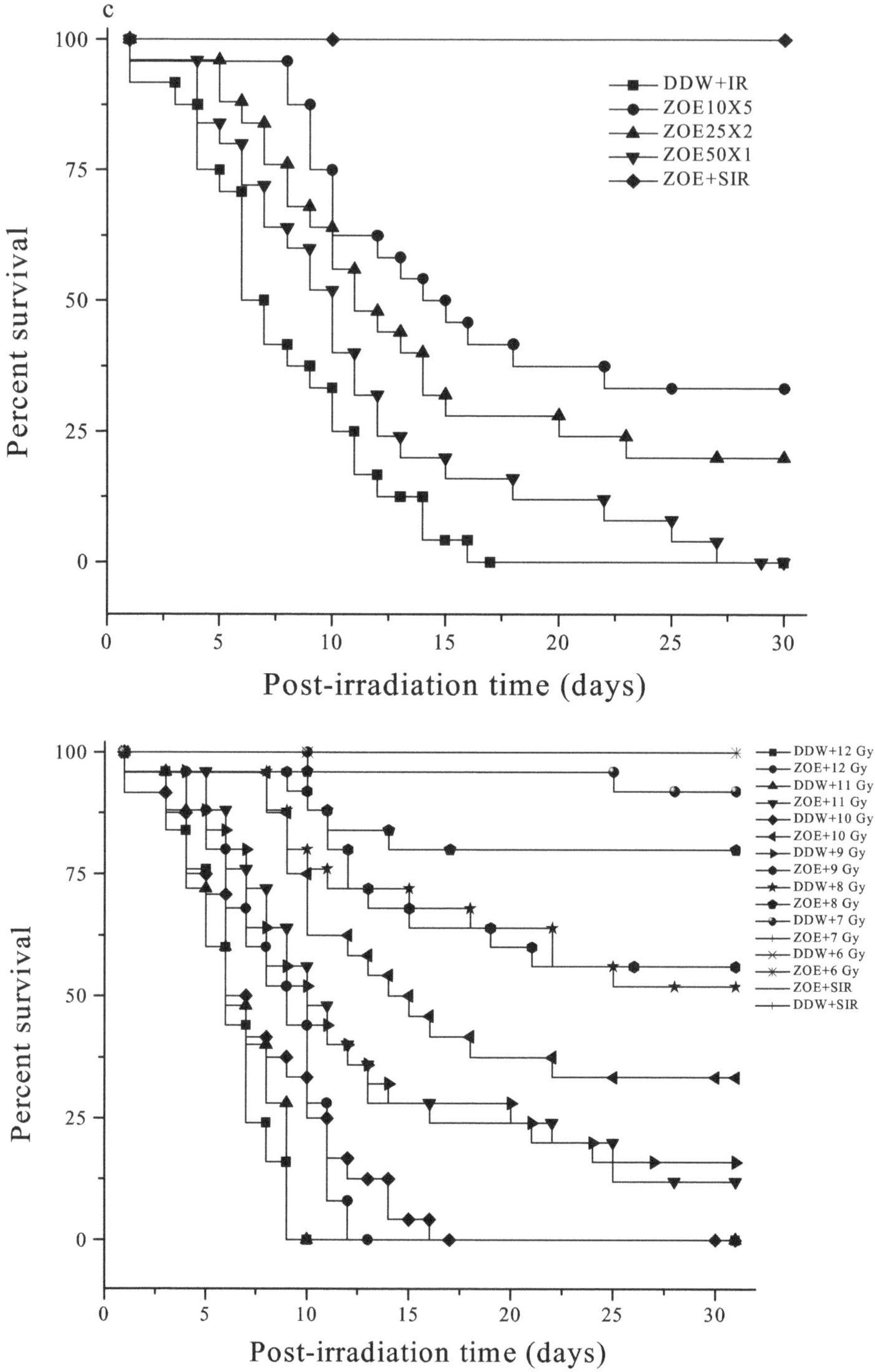

in a 50 per cent reduction in the survival of mice and 80 per cent of the animals died when the drug dose was increased to 600 mg/kg. 100 per cent mortality was observed at 600 mg/kg and thereafter up to a dose of 4000 mg/kg b. wt. ZOE.

Selection of Optimum Dose of ZOE

The optimum dose of ZOE for radiation protection, was selected by administrating the mice with 0, 2.5, 5, 10, 20 and 40 mg/kg b. wt. ZOE before whole body exposure to 10 Gy gamma radiation. The animals were monitored daily up to 30 days for the development of symptoms of radiation sickness, and mortality. The daily administration of different doses of ZOE for five consecutive days did not cause drug-induced mortality. The first mortality in the DDW + 10 Gy irradiation group was observed on day 3 and all the animals died by 16 days post-irradiation (Table 2.2). Exposure of mice to DDW + 10 Gy irradiation induced the symptoms of severe radiation sickness like reduction in the food and water intake, irritability, watering of eyes, epilation, weight loss, emaciation, lethargy, diarrhea, facial edema etc. The pretreatment of mice with various doses of ZOE either delayed or reduced the severity of radiation sickness. The onset of radiation-induced mortality was also delayed in the ZOE + irradiation group when compared with the DDW + 10 Gy irradiation group. The longest delay was observed for 10 and 20 mg/kg ZOE, where the first mortality was reported by day 8 post-irradiation (Table 2.2) while a shortest delay was observed for 2.5 mg/kg, where the first mortality occurred on day 5 post-irradiation. Mortality in the 5 and 40 mg/kg dose of ZOE was first observed at day 7 post irradiation, while it was on day 3 for DDW + 10 Gy irradiation.

Majority of the animals (66.66 per cent) of the DDW + 10 Gy irradiation group, died within 10 days after irradiation. However, the administration of MPG and resulted in 41.66 and 32 per cent mortality by day 10 and the survival was 1.75 and 2 fold greater than the DDW + 10 Gy irradiation, respectively. The treatment of mice with various doses of ZOE had an ameliorating effect on the gastrointestinal tract as evidenced by an increase in the 10 day survival of mice, where an increase of 1.87, 1.75, 2.25, 2.65 and 2.0 fold was observed for 2.5, 5, 10, 20 and 40 mg/kg ZOE, respectively. A lowest mortality of 25 per cent was observed in the animals treated with 20 mg/kg ZOE before irradiation. Other doses of ZOE also reduced the mortality, in comparison with DDW + 10 Gy irradiation, however, a significant elevation in the 10 day survival was observed only for 20 mg/kg ZOE ($p< 0.001$) treatment (Table 2.2).

Analysis of thirty day survival revealed a drug dose dependent increase in the survival of irradiated animals up to a dose of 10 mg/kg in ZOE + irradiation group, where a highest survival of 33.33 per cent was observed as compared to the DDW+irradiation group, where no survivors were reported (Figure 2.1). An increase in the drug dose to 20 and 40 mg resulted in a 13.5 per cent and 20.8 per cent reduction in the survival when compared with the 10 mg/kg ZOE + irradiation (Table 2.2). The lower doses of 2.5 and 5 mg/kg ZOE also increased the survival by 8.33 per cent and 25 per cent when compared to the DDW + irradiation group, where no survivors could be observed. A significant elevation in the survival was observed only in the animals that received 5, 10, and 20 mg/kg ZOE before exposure to 10 Gy. A maximum survival of 33.33 per cent was observed for 10 mg/kg ZOE and it was considered as an optimum dose for radioprotection (Table 2.2).

Selection of Route of Administration

The administration of 10 mg/kg ZOE, once daily, consecutively for five days by oral and intraperitoneal routes before irradiation to 10 Gy had an ameliorating effect on the gastrointestinal tract as evidenced by an increase in the 10 day survival of mice, where an increase of 1.57 and 2.25 fold was observed for oral and i. p. administration, respectively. The first mortality was observed on day three and only 33.33 per cent animals survived by day 10 in the DDW + 10 Gy irradiation group animals. The first mortality in the orally administered ZOE animals was observed on day four, where 52 per cent animals survived, while the first mortality in the i. p. injected ZOE animals was observed on day seven and 75 per cent survival was seen on day 10 post-irradiation. The survival after intraperitoneal administration of ZOE was 1.44 fold greater when compared with the oral administration.

The analysis of thirty-day survival revealed that the radioprotective activity was better when the drug was administered through i.p. rather than the oral route, as 33.33 per cent animals survived for the former after 30 days post irradiation than the latter where no survivors were observed. In spite of that, the oral treatment of ZOE had prolonged the life span and ameliorated the radiation sickness as the last mortality was observed on day 25 post irradiation in this group.

Selection of Optimum Fraction Size of ZOE

The cumulative dose of 50 mg/kg b. wt. of ZOE administration in five fractions for five days (10 mg/kg for 5 days) was found to be better than a single administration of 50 mg or 25 mg for two consecutive days as it afforded better protection against both the GI and hemopoetic syndromes.

The first mortality was observed at day 4, 5 and 8 in the animals administered with ZOE as single, two and five fractions, respectively, and 52, 64 and 75 per cent survival was observed at day 10 post irradiation after one, three and five fraction of ZOE. This increase in survival was 1.56, 1.92 and 2.25 fold greater than the DDW + 10 Gy irradiation group. The animal survival was 1.44 and 1.92 fold greater in the group that received 10 mg/kg ZOE once daily for 5 consecutive days, than that of two fractions of 25 mg and single fraction of 50 mg, respectively.

The analysis of thirty day survival revealed a fraction size dependent increase in the survival of irradiated animals as a maximum survival of 33.33 per cent was observed for the five fractions, 20 per cent in the two fraction while none of the animals, which received 50 mg/kg ZOE survived till 30 day (Figure 2.1). However there was an increase in the life span of animals receiving single dose of ZOE, as the last mortality was observed on day 28 post-irradiation. Since maximum survivors were observed after intraperitoneal administration of 10 mg/kg ZOE consecutively for five days before irradiation, this regimen of administration was considered superior to others and further studies were carried out adopting this regimen only.

Radioprotective Effect of ZOE

The radioprotective action of ZOE was evaluated by using optimum dose of 10 mg/kg, i.p. Administration for five consecutive days before exposure to 6, 7, 8, 9 and 10 Gy of γ-radiation. The irradiation of animals to different doses of radiation resulted in the development of symptoms of radiation sickness within 2-4 days after exposure, depending on the irradiation dose in the DDW + irradiation group. The exposure of mice to higher radiation doses resulted in the early onset of the symptoms of radiation sickness. The main symptoms included reduction in the food and water intake, irritability, watering of eyes, epilation, weight loss, emaciation, lethargy, diarrhea, and ruffling of hairs. Facial edema was also observed in a few animals between one and two weeks after exposure to 9 and 10 Gy. Some of the animals exposed to higher radiation doses exhibited paralysis and difficulty in locomotion during the second week after exposure. The severity of the symptoms increased with the increase in radiation dose.

The whole body irradiation of mice to 6 Gy did not induce mortality in both the groups (Table 3). However, with the increasing dose of radiation the survival declined in a dose dependent manner till a nadir in the survival was reached at 10 Gy, where no survivors were reported beyond 16 days post-irradiation. With the increase in the exposure dose the onset of mortality was also advanced (Table 2.3). The survival was plotted on to the log, while the exposure dose on the linear scale and the $LD_{50/30}$ was found to be 8.2 Gy for this group (Figure 2.2). The pretreatment of mice with 10 mg/kg ZOE delayed or reduced the severity of radiation sickness and also delayed the onset of radiation-induced mortality when compared with the concurrent DDW + irradiation group (Table 2.3). This delay in onset of mortality was almost by 4-5 days in the ZOE + irradiation group when compared with the DDW + irradiation group. The GI deaths were fewer compared to DDW + irradiation group for all exposure doses (Figure 2.2).

The pretreatment of mice with ZOE reduced the 10 day mortality by 1.28, and 2.25 folds for 9 and 10 Gy, respectively. The number of survivors increased significantly in the ZOE + irradiation group at 30 days post-irradiation ($p < 0.05$ for 9 and $p < 0.01$ for 10 Gy) when compared with the concurrent DDW + irradiation group (Table 2.3). The pretreatment of mice with ZOE reduced the 30 day mortality by 1.25, 1.8 and 3 folds for 7, 8 and 9 Gy, respectively. The $LD_{50/30}$ was found to be 9.4 Gy, resulting in an increase of 1.2 Gy (Figure 2.2) and the dose reduction factor was found to be 1.15.

Glutathione

The results are expressed as glutathione (GSH) µM/g tissue (Table 2.4). GSH contents remained unaltered in the DDW + sham-irradiation group. Similarly, the administration of ZOE alone before sham-irradiation did not alter the glutathione contents. The exposure of animals to different doses of gamma radiation resulted in a significant and dose dependent decline in the GSH contents in the DDW + irradiation group. The ZOE treatment before irradiation elevated the GSH contents significantly when compared with the concurrent DDW + irradiation group. The GSH contents reached to normal levels up to 7 Gy in ZOE +irradiation group. However, further

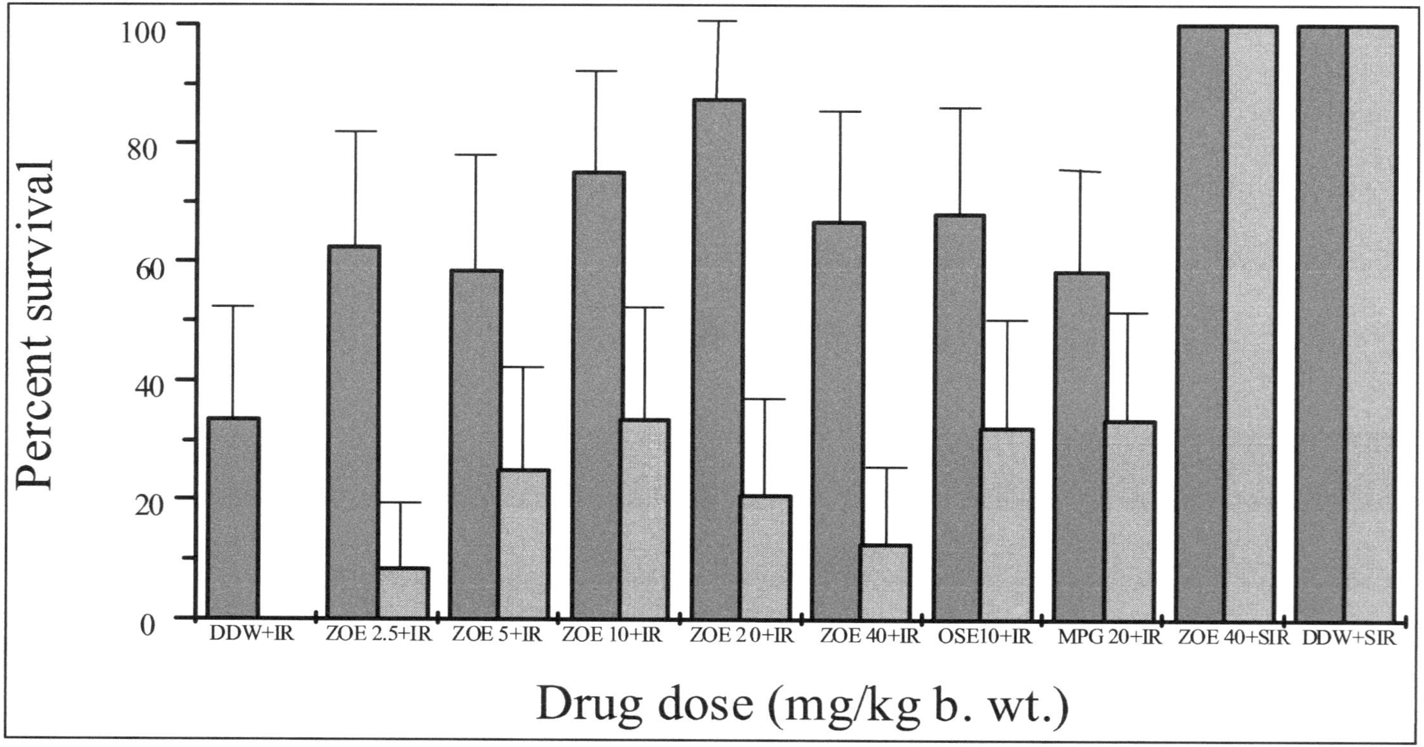

Figure 2.2: Effect of various doses of ZOE, OSE (10 mg/kg) and MPG (20 mg/kg) on the survival of mice exposed to 10 Gy of γ-radiation. Dark bars: day 10 and day Light bars: 30 survival. IR = irradiation, SIR = sham -irradiation. Error bars (95 per cent confidence limit).

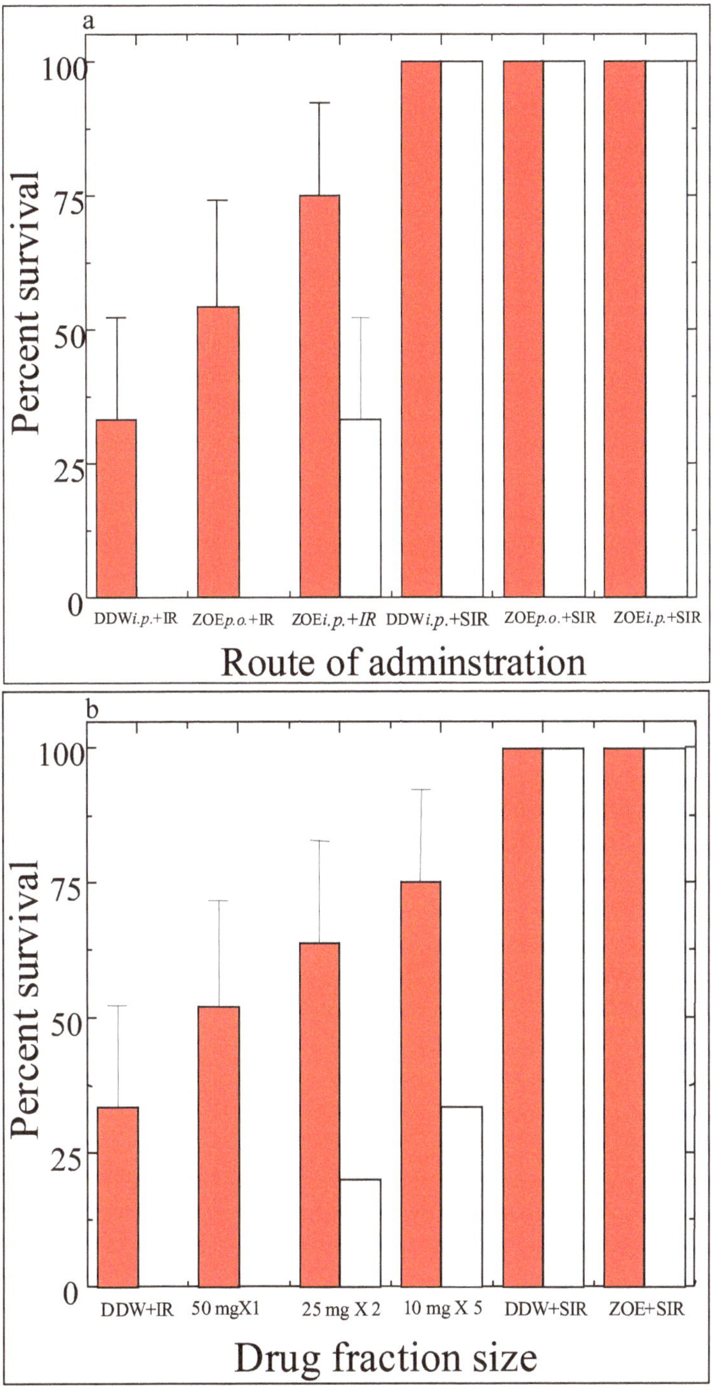

Figure 2.3: Effect of Route of Administration (a) of 10 mg/kg b. wt. of Ginger Extract (ZOE) and Fractionation of 50 mg/kg ZOE (b) on the Survival of Mice Whole-Body Exposed to 10 Gy of γ- radiation. ■ Survival 10 and 30 days. □ Error bars 95 per cent confidence limit.

increase in the irradiation dose resulted in the decline in GSH contents in ZOE + irradiation group when compared with the sham irradiation group (0 Gy).

Lipid Peroxidation (LPx)

The lipid peroxidation is expressed in terms of ηM MDA/mg protein (Table 2.4). LPx remained unaltered in DDW + sham-irradiation group at all post-exposure time periods. The administration of ZOE alone before sham-irradia-tion did not increase the MDA concentration, which was almost similar to that of the DDW + sham-irradiation group (Table 2.4). The irradiation of animals increased the lipid peroxidation in a dose dependent manner in both the DDW + irradiation and ZOE + irradiation groups and a peak level was observed at 10 Gy irradiation. The ZOE pretreatment significantly reduced the LPx-induction in the ZOE + irradiation group, thereby protecting against the radiation-induced lipid peroxidation at all the exposure doses studied. In spite of the reduction in the LPx by ZOE, the LPx levels were not regained to normal at day 31 post-irradiation (Figure 2.4).

Antimicrobial Activity

The zones of inhibition for alcoholic extract of ginger extract are shown in table. The ginger extract showed low to moderate inhibitory activity on enteric bacteria and *Candida albicans*. Among the bacteria, *Pseudomonas aeruginosa* is moderately sensitive followed by *Salmonella typhimurium*. The inhibitory effect of ZOE was best demonstrated at a concentration of 37.5 mg.

Discussion

Normal tissue tolerance is a major dose-limiting factor in radiotherapy and chemotherapy of cancer. In addition to its utility in the cancer treatment, an efficient, non-toxic radioprotector could also prove useful in occupational settings, where ionizing radiations are used (*e.g.*, defense, airline, military and research personnel, nurses, dental assistants, radiotherapy and nuclear medicine technicians, etc.) or in accidental exposures, which leave radioactivity in the environment and also during space travel to protect astronauts from the effects of high doses of radiation associated with solar flares.

With the escalation in the terrorist attack world wide, there is every likelihood that the committed terrorists may be able to procure the nuclear fissile material clandestinely and it is feared that some of the terrorist organizations may actually possess nuclear fissile material, which they may use clandestinely without hesitation to spread terror among the innocent people and the government establishments. As a result the threat of nuclear exposure looms large on the general public than ever before. Therefore, there is an urgent requirement to screen non-toxic radioprotector, which may be of great help in such situation/s for the benefit of general public and also for the population of the cancer patients undergoing radiation treatment. Dietary ingredients can be very useful, if they are found to protect against the deleterious effect of radiation, as it would not put extra burden to the body and they can be safely manipulated without toxic manifestations. Ginger is one of the dietary ingredients possessing several medicinal properties and is used in several ancient systems of

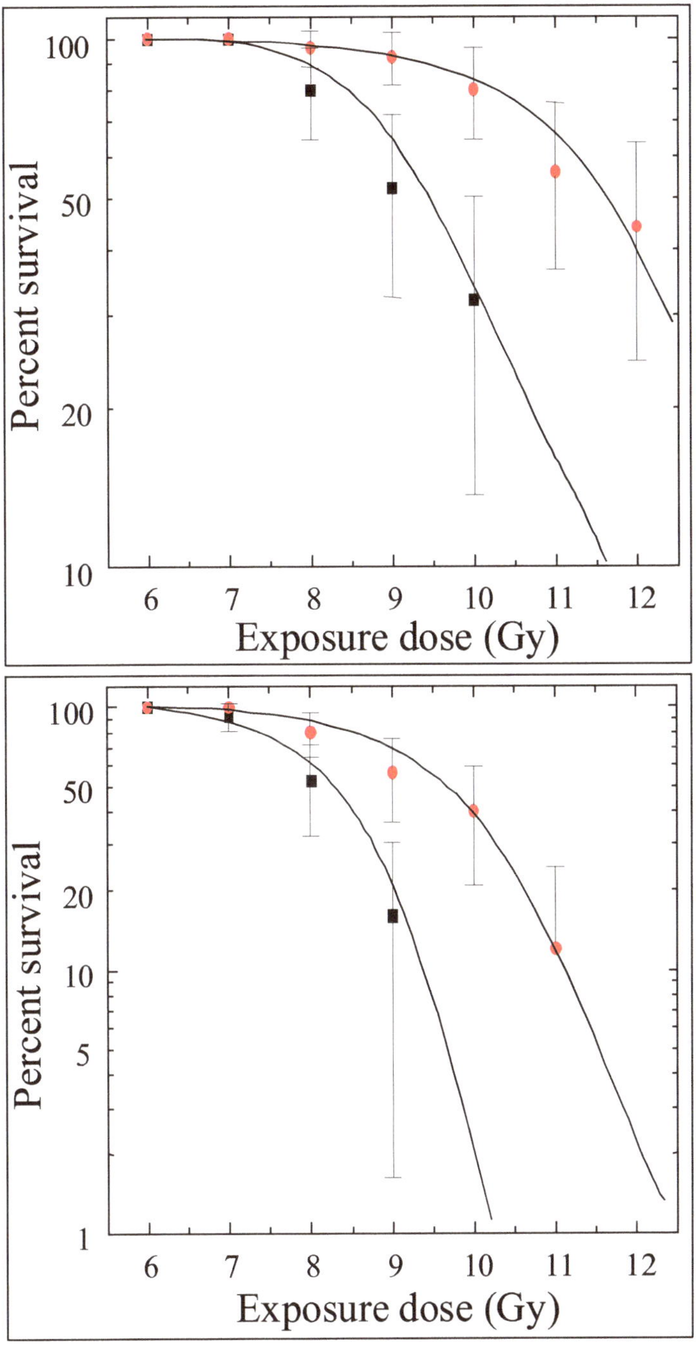

Figure 2.4: Effect of 10 mg/kg Body Weight of Ginger Extract on the Survival of Mice Exposed to different Doses of Gamma Radiation. DDW + irradiation (circles) and ZOE + irradiation (squares) groups, a) 10-day survival and b) 30-day survival. Error bars (95 per cent confidence limit).

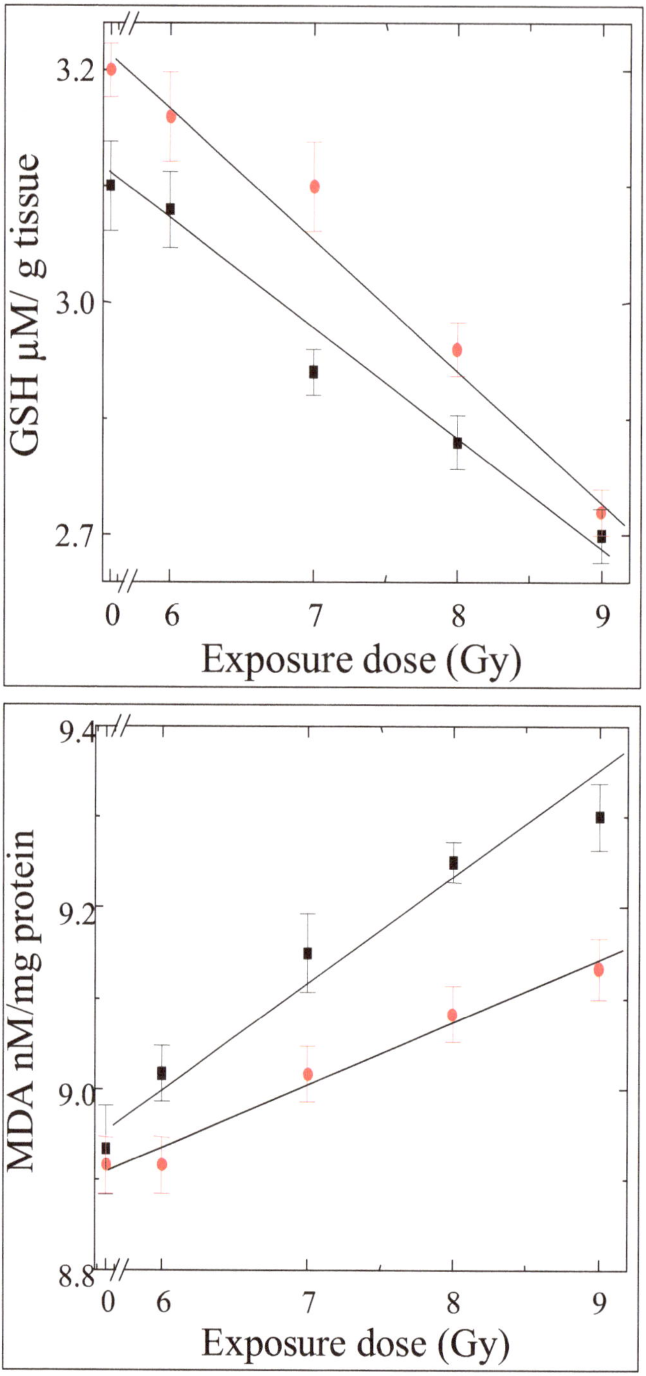

Figure 2.5: Alteration in the Radiation-Induced Free GSH Contents (a) and Lipid Peroxidation (b) in the Liver of Mice Whole Body Exposed to different Doses of γ-radiation. DDW + irradiation (squares) and ZOE + irradiation (circles).

medicine including Ayurveda, Greek, Roman, Siddha and Unani, (Nadkarni, 1976). Therefore, the aim of the present study was to evaluate the radioprotective effect of ginger in the mice exposed to different doses of whole body γ-irradiation.

The exposure of animals to 10 Gy resulted in the induction of radiation sickness within 2-4 days in the DDW + irradiation group. The death of 66.66 per cent of the animals exposed to 10 Gy of γ- radiation within 10 days is due to the functional failure of the gastrointestinal (GI) tract (Bond *et al.*, 1965; Jagetia and Baliga, 2002). Increase in the radiation dose advanced the radiation sickness and mortality and none of the animals survived beyond 11 and 9 days post-irradiation after exposure to 11 and 12 Gy, respectively. The reduction in the irradiation dose prolonged the appearance of the radiation sickness and the radiation-induced mortality, where 80 per cent and 52 per cent animals survived after exposure to 8 and 9 Gy, respectively. The animals that were exposed to 7 Gy irradiation exhibited classical hemopoetic syndrome and 8 per cent (2/25) animals died by 30 days post-irradiation. The $LD_{50/30}$ was found to be 8.2 Gy for the DDW + irradiation group. The above results are in contravention with the earlier studies carried out in our laboratory, where 100 per cent survival was observed for 7 Gy irradiation (Jagetia *et al.*, 1993; Uma Devi and Ganasoundari, 1995). This is due to the use of higher dose rate (1.66 Gy) in the present study, while lower dose rates (0.83 and 1 Gy) were used in the earlier studies. In the earlier study the $LD_{50/30}$ was 9.7 Gy, while it has been reduced to 8.2 in the present study owing to the use of higher dose rate exposure. A similar effect has been observed for the micronuclei induction, recently (Jagetia and Reddy, 2002).

A single whole-body exposure of mammals to ionizing radiation resulted in a complex set of symptoms whose onset, nature, and severity are a function of the radiation quality, radiation dose rate and total radiation dose. It is a well-established fact that the ionizing radiation at cellular level can induce damage in the biologically important macromolecules such as the DNA, proteins, lipids and carbohydrates in the various organs. While some damage may be expressed early, the other may be expressed over a period of time depending upon the cell kinetics and the radiation tolerance of the tissues. The proliferating cells are highly sensitive to the effect of radiation; therefore, the effect of whole body irradiation is mainly felt by the highly proliferating germinal epithelium, gastrointestinal epithelium and the bone marrow progenitor cells. Of these the germinal epithelium does not have a life supporting function to the exposed individual, while the gastrointestinal epithelium and the bone marrow progenitor cells are crucial for the sustenance of life and any damage to these cells will impair the normal physiological processes drastically causing adverse impact on the survival. The gastrointestinal epithelium is less sensitive than the bone marrow progenitor cells but as the cell transit time is quick, it is expressed earlier than the hemopoetic syndrome (Bond *et al.*, 1965). In mice death within 10 days post-irradiation is due to the gastrointestinal damage (Bond *et al.*, 1968; Jagetia and Baliga, 2002). The bone marrow stem cells are more sensitive to radiation damage than the intestinal crypt, but the peripheral blood cells have a longer transit time than the intestinal cells and hence the gastrointestinal syndrome appears earlier than the bone marrow syndrome and in mice death due to irradiation from 11 to 30 days post-irradiation is due to the hemopoetic damage inflicted by radiation (Bond *et al.*, 1965; Jagetia and Baliga, 2002).

Administration of mice with 2.5, 5, 10, 20 and 40 mg/kg b. wt. of ZOE by the intraperitoneal route for five consecutive days once daily before exposure to 10 Gy of radiation resulted in the reduction in the radiation-induced mortality when compared with the DDW + irradiation group. However, the optimum dose for radiation protection was 10 mg/kg (cumulative of 50 mg) above which the radioprotective effect declined. The earlier studies on radioprotection have shown that an agent in test (for radioprotective action) acts only at a particular dose and above which it may not be protective and in some cases it can even be toxic (Thomson 1959; Nagata *et al.*, 1972; Uma Devi and Ganasoundari, 1995; Jagetia and Baliga, 2002). The reason may be that after a particular concentration, a compound may start manifesting its toxic effects. The active principle of *Plumbago rosea*, the plumbagin, has been reported to stimulate the granulocytes *in vitro* at pico to femto gram range, while at higher doses it resulted in the immunosuppressive activity (Wagner *et al.*, 1988). A similar action cannot be ruled out for ZOE that has offered an optimum protection at 10 mg/kg (cumulative 50 mg), while the higher doses resulted in the decline in the protective action of ZOE. The same schedule when followed for the oral administration did not afford significant protection when compared to the intraperitoneal route and none of the animals survived beyond 25 days post-irradiation. However, the oral administration of ZOE delayed the radiation-induced sickness and mortality, when compared to the concurrent DDW + irradiation group. This may be due to the faster absorption of the ZOE through the i.p. route. A similar observation has also been reported for Ocimum sanctum and abana, a polyherbal formulation (Uma Devi and Ganasoundari, 1995, Jagetia and Aruna, 1997).

The intraperitoneal administration of mice with different fraction size *i.e.* 50 mg single administration, 25 mg consecutively for two days before irradiation to 10 Gy of gamma radiation although delayed the onset of mortality and reduced the symptoms of radiation sickness when compared with the non-drug treated irradiated controls, could not provide protection comparable to 10 mg administered daily, consecutively for five days before exposure to 10 Gy radiation. A similar observation has been reported earlier for *Moringa oleifera* (Rao *et al.*, 2001).

The ZOE treatment provided protection against the radiation sickness and mitigated the animal sufferings. The pattern of survival in ZOE + irradiation group was similar to that of the DDW + irradiation group except that the mortality was delayed. This clearly indicates the effectiveness of ZOE in arresting GI death, where the number of survivors was higher than that of the irradiated control. This reduction in GI death may be due to the protection of intestinal epithelium, which would have allowed proper absorption of the nutrients, when compared with the DDW + irradiation, and MPG + irradiation groups, the positive controls used in this study. The ZOE administration in all the groups rendered better protection against the radiation-induced gastrointestinal damage than and MPG + irradiation. Ginger has been reported to reduce the cisplatin induced nausea, vomiting and to reverse the cisplatin-induced emesis and delay in gastric emptying in dogs and rats, respectively (Sharma *et al.*, 1997, 1998). The ginger and its active constituents beta-sesquiphellendrene, beta-bisabolene, ar-curcumene, 6-shogaol, zingiberene, terpenoid and 6-gingerol have all also been observed to inhibit the experimentally induced gastric ulcerations in rats (Yamahara *et al.*, 1988; Al-Yahya *et al.*, 1989; Wu *et al.*, 1990; Yamahara *et al.*, 1992). Recently, ginger has been reported to stimulate/enhance the digestive enzymes of intestinal mucosa like lipase, sucrase, pancreatic amylase, trypsin and chymotrypsin in rats (Patel and Srinivasan, 1996; 2000), which would have helped the absorption of the food and nutrients in a better way resulting in the prolongation of life of animals in the ZOE + irradiation group than that of DDW + irradiation group.

The ZOE showed good antimicrobial activity on the enteric pathogens and the best activity was observed in *Pseudomonas aeruginosa, Staphylococcus aureus, Salmonella typhimurium, Shigella dysenteriae, Escherichia coli* and *Vibrio cholera*, while the least activity was observed in *Klebsiella aerogenes*. It was also found to be active against *the Candida albicans* and these results agree with the earlier reports on the antimicrobial activity of ginger (Gugnani and Ezenwanze, 1985; Warrier *et al.*, 1997; Chen *et al.*, 1985; Yin and Cheng, 1998). The antimicrobial activity of ZOE may be due to the presence of 1-8 cineol in it, which has also been reported to be antibacterial. The radioprotective plants like *Ocimum sanctum* and mint have also been reported to possess antimicrobial activity.

The hemopoietic syndrome is induced at low doses of irradiation and is manifested by hemopoietic stem cell depletion, and ultimately by the depletion of mature hemopoietic and immune cells. The pretreatment of mice with ZOE significantly reduced the bone marrow deaths in the ZOE + irradiation group, especially for 5, 10, 20 and 40 mg/kg, where a significant elevation in the survival has been observed. This increase in 30 day survival may be owing to the protection afforded by ZOE to the stem cell compartment of the bone marrow, which continued to supply the requisite number of cells in the survivors. The LD_{50} for this group was found to be 9.4 and ZOE treatment has increased it by 1.2 Gy over the LD_{50} of DDW + irradiation group, and the dose reduction factor was 1.15. A similar effect has been observed earlier for *Ocimum sanctum* and *Mentha averensis* (Uma Devi and Ganasoundari, 1995, Jagetia and Baliga, 2002).

Plants are complex mixtures of compounds and no single compound can provide the desired activity. It can be characterized as a polyvalent action and interpreted as additive or, in some cases, potentiating and the final formulation will have a concentrated desired activity and the undesired activities will be diluted or absent altogether (Kulkarni, 1997). As far as the authors are aware, this is probably the first report on the radioprotective effect of ginger. However, certain other plants like *Panax ginseng, Ocimum sanctum, Moringa oleifera* and *Mentha averensis* have been observed to protect mice against the radiation-induced mortality (Zhang *et al.*, 1987; Uma Devi and Ganasoundari, 1995; Rao *et al.*, 2001; Jagetia and Baliga, 2002).

The exact mechanism of action of ZOE is not known, however, it may scavenge free radicals produced by radiation and thus reduce the radiation-induced damage to the cellular genome. Our contention is supported by the observation, where ginger has been found to scavenge superoxide anions and H_2O_2 *in vitro* (Krishnakantha and Lokesh, 1993; Cao *et al.*, 1993). Similarly, the volatile oil of ginger has also been reported to scavenge hydrogen peroxide in chondrocytes induced by fulvic acid

(Guo *et al.*, 1997). The aqueous and the organic solvent extracts of ginger have been found to possess antioxidant property (Jitoe *et al.*, 1989). The curcuminoids; cassumunin A and cassumunin B isolated from the tropical ginger, *Zingiber cassumunar*, have been reported to reduce the oxidative stress induced by hydrogen peroxide in the rat thymocytes and it was better than curcumin (Nagano *et al.*, 1997).

The importance of the cellular membrane as a critical target in the enhancement of radiation-induced cell lethality has been emphasized (Alper, 1968). Lipid peroxidation induced by radiation is known to be due to the attack of free radicals on the fatty acid component of membrane lipids (Raleigh *et al.*, 1977). Lipid peroxidation is consid-ered to be an important effect of ioniz-ing radiation on biological membranes (Leyko and Bartosz, 1986). While DNA damage causes the radiation-induced reproductive cell death, membrane lipids are thought to be critical targets in the interphase cell death (Raleigh and Shum, 1983). Radiation-induced lipid peroxidation causes damage to the cellular membrane by altering the fluidity of the biological membranes, which progressively leads to cell degradation, and thereby affecting the biological defence mechanism. Noda *et al.* (1993) have reported that the product of lipid peroxidation, such as malonaldehyde (MDA), damages the enzyme system and DNA. Lipid peroxidation has been used as an end-point to study the action of oxidizing and free radical producing agents as well as to investigate the effects of intracellu-lar radical scavengers. The ZOE administration significantly reduced the amount of MDA than the concurrent control groups. This inhibition of lipid peroxidation by ginger extract may also have been responsible for the observed radioprotection. The ginger has been found to inhibit the induction of lipid peroxidation in the rat liver microsomes *in vitro* (Reddy and Lokesh, 1992) and in rats treated with the pesticide malathion (Ahmed *et al.*, 2000).

Several investigators have reported that lipid peroxidation start as soon as the supply of endogenous GSH is exhaust-ed, and that the addition of the GSH promptly stops further peroxidation (Christophersen, 1968). Glutathione (GSH) is involved in numerous cellular reductive reactions (Meister and Anderson, 1983). It is related to the repair of radiation-induced free radicals by hydrogen atom donation, rejoining of DNA strand breaks by participating in enzymatic reactions as a cofactor, and in the repair of DNA damage, resulting in protection (Révész and Malaise, 1983). The significant eleva-tion in the GSH level at all exposure doses, by ZOE may be responsible for the scavenging of radiation induced free radicals including LPx and thereby resulting in the protection against the radiation induced mortality. Zinger oil has been reported to elevate GSH levels in the liver of mice (Bharali *et al.*, 1998). Similarly, ginger has also been reported to increase the blood glutathione contents in the blood and reduce the lipid peroxidation by maintaining the activities of the antioxidant enzymes superoxide dismutase, catalase and glutathione peroxidase in rats exposed to malathion (Ahmed *et al.*, 2000). The ginger oil has been found to increase the acid-soluble sulphydryls, glutathione S-transferase and aryl hydrocarbon hydroxylase activity in the liver of mice (Banerjee *et al.*, 1994), a similar action of ginger in present study may not be ruled out.

Conclusions

From our study it is clear that ZOE, provided protection against the radiation-induced sickness and mortality and the optimum protective dose of 10 mg/kg is safe from the point of drug induced toxicity, as it is 1/50 of the LD_{50} dose of 500 mg/kg b. wt. The presence of gingerone, gingerols, curcuminoids and terpenes in the ZOE might have been responsible for the observed radioprotection in mice. The combination of factors like free radical scavenging by antioxidant components of ZOE, inhibition of lipid peroxidation, elevation of the GSH and the increased activity of antioxidative enzymes would have contributed to the observed radioprotective action of ZOE. Further studies are planned to explore the applicability of ZOE in cancer treatment by looking for the preferential protection to the normal tissues, and also to identify the active component(s) responsible for the radioprotective properties with emphasis on the study of their mechanism(s) of action.

Acknowledgements

We thank Prof. M. S. Vidyasagar, and Dr. J. Velumurugan, Department of Radiotherapy and Oncology, Kasturba Medical College, Manipal, India for providing the necessary irradiation facilities and radiation dosimetry. We also wish to thank Dr. Prof. P.G Shivananda and Dr. Mamatha Ballal, Miss Devaki and Miss Vinitha, Department of Microbiology for their assistance and help in carrying out antimicrobial studies. Thanks are also due to Prof. Gopalkrishna Bhat Poorna Prajna College Udupi, Karnataka, for identifying the plant material of Financial assistance provided by the Indian Council of Medical Research, Govt. Of India, New Delhi is gratefully acknowledged.

References

Aeschbach, R., Loliger, J., Scott, B.C., Murcia, A., Butler, J., Halliwell, B., and Aruoma, O.I. (1994). Antioxidant actions of thymol, carvacrol, 6-gingerol, zingerone and hydroxytyrosol. *Food and Chemical Toxicology*, **32**: 31-36.

Ahmed, R.S., Seth, V., and Banerjee B. D. (2000). Influence of dietary ginger (*Zingiber officinales* Rosc) on antioxidant defense system in rat: comparison with ascorbic acid. *Indian Journal of Experimental Biology*, **38**: 604-606.

Ahmed, R.S., Seth, V., Pasha, S.T., and Banerjee, B. D. (2000). Influence of dietary ginger (*Zingiber officinales* Rosc) on oxidative stress induced by malathion in rats. *Food and Chemical Toxicoogy*, **38**: 443-450.

Al-Yahya, M.A., Rafatullah, S., Mossa, J.S., Ageel, A.M., Parmar, N.S., and Tariq, M. (1989). Gastroprotective activity of ginger *Zingiber officinale* rosc., in albino rats. *American Journal of Chinese Medicine*, **17**: 51-56.

Auta, K.I., Galadima, A.A., Bassey, J.U., Olowoniyi, O.D., Moses,O.O., and Yako, A.B. (2011). Antimicrobial properties of the ethanolic extracts of *Zingiber officinale* (Ginger) on *Escherichia coli* and *Pseudomonas areuginosa*. *Annals of Biological Research*, **2(3)**: 307-311.

Beuege, J.A., and Aust, S.D. (1978). Microsomal lipid peroxidation. *Methods in Enzymology*, **30**: 302-310.

Bond, V.P., Fliedner, T.M., and Archambeau, J.O. (1965). Mammalian radiation lethality. *Academic press* New York. USA.

Cao, Z.F., Chen, Z.G., Guo, P., Zhang, S.M. Lian, L.X., Luo, L., and Hu, W.M. (1993).Scavenging effects of ginger on superoxide anion and hydroxyl radical. *Chung Kuo Chung Yao Tsa Chih.*, **18**: 750-751.

Chang, W.S., Chang, Y.H., Lu, F.J., and Chiang, H.C. (1994). Inhibitory effects of phenolics on xanthine oxidase. *Anticancer Research*, **14**: 501-506.

Chaumont, J.P., and Leger, D. (1992). Campaign against allergic moulds in dwellings. Inhibitor properties of essential oil geranium "Bourbon", citronellol, geraniol and citral. *Annals of Pharmacology*, **50**: 156-166.

CHEMEXCIL (1992). *Selected medicinal plants of India*. Basic Chemicals, Pharmaceutical and Cosmetic Export Promotion Council, Bombay 400 039, India. 205-207.

Chung, W.Y., Yow,C.M., and Benzie, I.F. (2003). Assessment of membrane protection by traditional Chinese medicines using a flow cytometric technique: preliminary findings. *Redox Reports*, **8**: 31-33.

Cruickshank,R., Duguid, J.P., Manoin, B.P., and Swain, R.H.A. (1975). *The practice of Medical Microbiology Vol. II*, pp. 196, London, Churchill Livingstone.

Geeta, D.M. Vasudevan, R., Kedlaya, S., Deepa, M. Ballal (2001). Activity of *Ocimum sanctum* (the traditional Indian medicinal plant) against the enteric pathogens. *Indian Journal of Medical Science*, **55**: 434-8.

Ghosh, M. N. (1984). Toxicity studies. In: *Fundamentals of experimental pharmacology*, edited by Ghosh MN Scientific Book Agency, Calcutta, India. pp. 153-158,.

Gong, Q.M., Wang, S. L., and Gan, C. (1989). A clinical study on the treatment of acute upper digestive tract hemorrhage with wen-she decoction. *Chung Hsi I Chieh Ho Tsa Chih.*, **9**: 272-273.

Gugnani, H.C., and Ezenwanze, E.C. (1985). Antibacterial activity of extracts of ginger and African oil bean seed. *Journal of Communicable Diseases*, **17(3)**: 233-226.

Guh, J.H., Ko, F.N. Jong, T.T., and Ting, C.M. (1995). Antiplatelet effect of gingerol isolated from *Zingiber officinale*. *Journal of Pharmacy and Pharmacolog.*, **47**: 329-32.

Guo, P., Xu, J., Xu, S., and Wang, K. (1997). Inhibition of hydrogen peroxide production on chondrocytes induced by fulvic acid by ginger volatile oil. *Chung Kuo Chung Yao Tsa Chih.*, **22**: 559-561.

Holden, C. (1997). Antibiotic basis for spice use. *Science*, **277**: 321.

Hori, Y., Miura, T., Hirai, Y., Fukumura, M., Nemoto, Y., Toriizuka, K., and Ida, Y. (2003). Pharmacognostic studies on ginger and related drugs-part 1: five sulfonated compounds from *Zingiberis rhizome* (Shokyo). *Phytochemistry*, **62**: 613-7.

Jafri, S.A., Abass, S., and Qasim, M. (2010). Hypoglycemic Effect of Ginger (*Zingiber officinale*) in Alloxan Induced Diabetic Rats (*Rattus norvagicus*). *Pakistan Veterinary Journal*, **31(2)**: 160-162.

Jagetia G. C., and Baliga, M. S. (2002). Influence of the leaf extract of *Mentha arvensis* Linn. (mint) on the survival of mice exposed to different doses of gamma radiation. *Strahlentherapie und Oncologie*, **178**: 91-98.

Jagetia, G.C., and Baliga, M.S. (2002). Cystone, an ayurvedic herbal drug imparts protection to the mice against the lethal effects of gamma-radiation: a preliminary study. *Nahrung*, **46**: 332-336.

Jagetia, G.C., and Reddy, T.K. (2002). The grapefruit flavanone naringin protects against the radiation-induced genomic instability in the mice bone marrow: a micronucleus study. *Mutation Research*, **519**: 27-48.

Jagetia, G.C., Baliga, M.S., Aruna, R., Rajanikant, G.K., and Jain, V. (2003). Effect of abana (a herbal preparation) on the radiation-induced mortality in mice. *Journal of Ethnopharmacology*, **86**: 1–7.

Jagetia, G.C., Baliga, M.S., Malagi, K.J. and Kamath, M.S. (2002). The evaluation of the radioprotective effect of Triphala (an ayurvedic rejuvenating drug) in the mice exposed to gamma-radiation. *Phytomedicine*, **9**: 99-108.

Jagetia, G.C., Ganapathi, N.G., Unnikrishnan, and M.K. (1993). Copperglycinate protects mice exposed to various doses of gamma radiation. *Strahlentherapie und Oncologie*, **5**: 323-328.

Jitoe, A., Masuda, T., Tengah, I.G.P., Supragupta, D.N., Gara, I.W., and Nakatani, N. (1989). Antioxidant activity of tropical ginger extracts and analysis of the contained curcuminoids. *Medical Hypotheses*, **29**: 25-28.

Kilic, D., Sayan, H., Gonul, B., and Egehan, I. (2000). The effect of granulocyte macrophage-colony stimulating factor on glutathione and lipid peroxidation in a rat model. *European Journal of Surgical Oncology*, **26**: 701-704.

Kong, T., Ah-Ng, Yu, R.,Chen, C., Mandlekar, S., and Primiano, T. (2000). Signal transduction events elicited by natural products: role of MAPK and caspase pathways in homeostatic response and induction of apoptosis. *Archives of Pharmacal Research*, **23**: 1-16.

Krishnakantha, T.P. and Lokesh, B.R. (1993). Scavenging of superoxide anions by spice principles. *Indian Journal of Biochemistry and Biophysics*, **30**: 133-134.

Leyko, W., and Bartosz, G. ((1986). Membrane effects of ionizing radiation and hyperthermia. *Inernational Journal of Radiation Biology*, **49**: 743-70.

Lowry, O. H. Roebrough, N.J., and Farr, A. L. (1951). Protein measurement with the folin phenol reagent. *Journal of Biological Chemistry*, **193**: 265-275.

Lucy Treagan (1982). *Medical microbiology*- lab procedures, pp.233, Philadelphia, W.B. Saunders Company.

Mahluji, S., Ostadrahimi, A., Mobasseri, M., Attari, V. E., and Payahoo, L. (2013). Anti-Inflammatory Effects of *Zingiber Officinale* in Type 2 Diabetic Patients. *Advanced Pharmaceutical Bulletin*, **3(2)**: 273-276.

Meyer, K., Schwartz, J., Crater, D., and Keyes, B. (1995). Zingiber officinale (ginger) used to prevent 8-Mop associated nausea. *Dermatology Nurse*, **7**: 242-244

Miller, L.C., and Tainter, M.L. (1944). Estimation of the ED_{50} and its error by means of logarithmic-probit graph paper. *Proc Soc Exp Biol Med.*, **57**: 261.

Moron, M.S., Depierre, J.W., and Mannervik, B. (1979). Levels of glutathione, glutathione reductase and glutathione-s-transferase activities in rat lung and liver. *Biochimica Biophysca Acta*, **582**: 67-78.

Nadkarni A,K. (1976). Indian Materia Medica, 3rd edition, Popular Press Ltd. Mumbai, India. pp 1308-1315.

Nagano, T., Oyama, Y., Kajita, N., Chikahisa, L., Nakata, M., Okazaki, E, and Masuda, T. (1997). New curcuminoids isolated from Zingiber cassumunar protect cells suffering from oxidative stress: a flow-cytometric study using rat thymocytes and H_2O_2. *Japanese Journal of Pharmacology*, **75**: 363-370.

Nagata, H., Sugahara, T., and Tanaka, T. (1972). Radiation protection by 2-mercaptopropionylglycine in mice. *Journal of Radiation Research*, **13**: 163-166.

National committee for Clinical Laboratory Standards (1990). Methods for dilution antimicrobial susceptibility tests for bacteria that grow aerobically. Approved Standard M7-A2. NCCLS.

Park, K.K., Chun, K.S., Lee, J.M. Lee, S.S. and Surh, Y.J. (1998). Inhibitory effects of [6]-gingerol, a major pungent principle of ginger, on phorbol ester-induced inflammation, epidermal ornithine decarboxylase activity and skin tumor promotion in ICR mice. *Cancer Letters*, **129**: 139-144.

Patro, B. S., Rele, S., Chintalwar, G. J., Chattopadhyay, S., Adhikari, S., and Mukherjee, T. (2002). Protective activities of some phenolic 1,3-diketones against lipid peroxidation: possible involvement of the 1,3-diketone moiety. *Chembiochem.*, **3**: 364-370.

Patt, H,M,, Tyree, E.B., Straube, R.L., and Smith, D.E. (1949). Cysteine protection against X-irradiation. *Science*, **110**: 213-214.

Pattnaik, S., Subramanyan, V.R., Bapaji, M., and Kole, C.R. (1997). Antibacterial and antifungal activity of aromatic constituents of essential oils. *Microbios.*, **89**: 39-46.

Pecoraro, A., Patel, J., Guthrie, T., and Ndubisi, B. (1988). Efficacy of ginger as an adjunctive anti-emetic in acute chemotherapy-induced nausea and vomiting. *ASHP Midyear Clinical Meeting*, **33**: 429E.

Phillips, S., Ruggier, R., and Hutchinson, S.E. (1993). *Zingiber oflcinale* (Ginger)-an antiemetic for day case surgery. *Anaesthesia*, **48**: 715-717.

Prieur, D.J., Young, D.M., Davis, R.D., Cooney, D.A., Homan, E.R., Dixon, R.L., and Guarino, A.M. (1973). Procedures for preclinical toxicologic evaluation of cancer chemotherapeutic agents: protocols of the laboratory of toxicology. *Cancer Chemotherapy Reports*, **4**: 1-28.

Raleigh, J. A., Kremers, W., and Gaboury, B. (1977). Dose-rate and oxygen effects in models of lipid membranes: linoleic acid. *Inernational Journal of Radiation Biology*, **31**: 203-213.

Rao, A. V., Devi, P. U., and Kamath, R. (2001). *In vivo* radioprotective effect of *Moringa oleifera* leaves. *Indian J Exp Biol.*, **39**: 858-63.

Rastogi R.M., and Mehrotra, B.N. (1990). Compendium of Indian medicinal plants. Central Drug Research Institute, Lucknow, India. **1**, 388-389.

Reddy, A. C., and Lokesh, B. R. (1992). Studies on spice principles as antioxidants in the inhibition of lipid peroxi-dation of rat liver microsomes. *Molecular Cell Biochemistry*, **111**: 117-124.

Singh, R.P., Jain, R., Mishra, R., and Tiwari, P. (2012). Antidepressant activity of hydroalcoholic extract of *Zinger officinale. International Research Journal of Pharmacy*, **3(20)**: 149-151.

Sweeney, T.R. (1979). A survey of compounds from the antiradiation drug development program of the U.S. Army Medical Research and development command. Government Printing office, Washington, D.C Publication. 308-318.

Tanaka, T. (1994). Cancer chemoprevention by natural products. *Oncology Reports*, **1**: 1139-1155.

Thomson, F. (1962). *Radiation protection in mammals*. Reinhold Publishing Corporation. USA.

Villanova, PA. Noda, Y., McGeer,P. L., McGeer, E. G., and Comporti, M. (1993). Lipid peroxidation. Biopathological significance. *Molecular Aspects of Medicine*, **14**: 199-207.

Warrier, P.K., Nambiar, V.P.K., and Ramankutty, C. (1996). *Indian medicinal plants*. Orient Longman Ltd., Hyderabad, India. **5**: 225-228.

Wealth of India (1972). *Raw materials*. Council of Scientific and Industrial Research, New Delhi, India. 43, 337-346.

Wu, H., Ye, D. Bai, Y., and Zhao, Y. (1990). Effect of dry ginger and roasted ginger on experimental gastric ulcers in rats. *Hongguo Zhong Yao Za Zhi.*, **15**: 278-280.

Yamahara, J., Hatakeyama, S., Taniguchi, K., Kawamura, M., and Yoshikawa, M. (1992). Stomachic principles in ginger. II. Pungent and anti-ulcer effects of low polar constituents isolated from ginger, the dried rhizoma of *Zingiber officinale* Rosc cultivated in Taiwan. The absolute stereostructure of a new diarylheptanoid] *Yakugaku Zasshi.*, **12**: 645-655.

Yamahara, J., Mochizuki, M., Rong, H.Q., Matsuda, H., and Fujimura, H. (1988). The anti-ulcer effect in rats of ginger constituents. *Journal of Ethnopharmacoogy*, **23**: 299-304.

Yoshikawa, M., Hatakeyama, S., Taniguchi, K., Matuda, H., and Yamahara,. J. (1992). 6-Gingesulfonic acid, a new anti-ulcer principle, and gingerglycolipids A, B, and C, three new monoacyldigalactosyl glycerols, from *Zingiberis rhizoma* originating in Taiwan. *Chemical and Pharmaceutical Bulletin* (*Tokyo*), **40**: 2239-2241.

Zhang, J.S., Sigdestad, C.P., Gemmell, M.A., and Grdina, D.J. (1987). Modification of radiation response in mice by fractionated extracts of *Panax ginseng*. *Radiation Research*, **112**: 156-163.

Utilisation and Management of Medicinal Plants Vol. 3 (2015) *Pages* **53–67**
Editors: **V.K. Gupta and A. Kaul**
Published by: **DAYA PUBLISHING HOUSE, NEW DELHI**

3

Optimization of Conditions for the Extraction of Oil from the Seeds of Castor and Soybean using Three Phase Partition (TPP) and Soxhlet Method

Mital J. Kaneria[1], Kalpna D. Rakholiya[1], Krunal V. Nagani[1], Neha M. Parmar[1] and Sumitra V. Chanda[1*]

ABSTRACT

In this study, a simple separation technique, Three Phase Partitioning (TPP) and soxhlet method, were used for the extraction of oil from seeds of castor and soybean. Three phase partitioning, a method generally used for protein separation, has been evaluated for extraction of oil with different optimising conditions like ratio of t-butanol (v/v), ammonium sulphate (w/v) addition to an aqueous slurry and the effect of various pH. Another method applied for the oil extraction was soxhlet extraction method using hexane solvent. In castor, the yield was almost same by both the extraction techniques while in soybean it was slightly less. Under optimized condition, best condition for the oil extraction of castor was the ratio of slurry to t-butanol was 1:2 (v/v), ammonium sulphate concentration was 50 per cent (w/v) and pH was 7, while for soybean, the ratio of slurry to t-butanol was 1:2 (v/v), ammonium sulphate concentration was 50 per cent (w/v) and pH was 4. TPP can be effective alternative approach for extraction of oil since it is economic, short processing time and simple.

Keywords: Castor, Soybean, Seeds, Oil extraction, Three Phase Partition (TPP) method, Soxhlet extraction method.

1 Phytochemical, Pharmacological and Microbiological Laboratory (PPML), Department of Biosciences (UGC-CAS), Saurashtra University, Rajkot – 360 005, Gujarat, India.

* *Corresponding author*: E-mail: svchanda@gmail.com

Introduction

India is a varietal emporium of medicinal plants and is one of the richest countries in the world in regard to genetic resources of medicinal plants. It exhibits a wide range in topography and climate, which has a bearing on its vegetation and floristic composition (Parekh and Chanda, 2007). Many Indian plants have been used from time immemorial to treat various diseases and infections in traditional medicinal systems of India such as Siddha, Ayuveda, Naturopathy etc. (Chanda *et al.*, 2013a).

Herbs are staging a comeback and herbal 'renaissance' is happening all over the globe. The herbal products today symbolize safety in contrast to the synthetics that are regarded as unsafe to human and environment. Although herbs had been priced for their medicinal, aromatic qualities for centuries, the synthetic products of the modern age surpassed their importance, for a while. It has been estimated that in developed countries such as United States, plant drugs constitute as much as 25 per cent of the total drugs, while in fast developing countries such as China and India, the contribution is as much as 80 per cent. Thus, the economic importance of medicinal plants is much more to countries such as India than to rest of the world (Suneetha and Chandrakanth, 2006; Nasab and Khosravi, 2014).

Medicinal plants have been well-known natural sources for the treatment of various diseases since antiquity. Search for eternal health and longevity and to seek remedy to relieve pain and discomfort promoted the early man to explore his immediate natural surroundings by trying many plants, animal products, minerals etc. for developing a variety of therapeutic agents. Medicinal plants continue to be important therapeutic aids for alleging ailments of human mankind (Coutinho *et al.*, 2008).

Historically, plants have provided a source of inspiration for novel drug compounds, as plant derived medicines have made large contributions to human health and well-being. Their role is twofold in the development of new drugs: (1) they may become the base for the development of a medicine, a natural blueprint for the development of new drugs, or; (2) a phytomedicine to be used for the treatment of disease. Several effective medicines used in their natural state such as aloe, opium were selected as therapeutics agents based on empirical evidence of their clinical application by traditional societies from different parts of the world (Dike *et al.*, 2012; Dhami, 2013; de Boer and Cotingting, 2014).

The primary benefits of using plant derived medicines are that they are relatively safer than synthetic alternatives, offering profound therapeutic benefits and more affordable treatment. Medicinal plants continue to be an important therapeutic aid for alleviating the ailments of humankind. Today, there is a renewed interest in traditional medicine and an increasing demand for more drugs from plant sources. This revival of interest in plant derived drugs is mainly due to the current widespread belief that "green medicine" is safe and more dependable than the costly synthetic drugs, many of which have adverse side effects (Nair and Chanda, 2007).

Medicinal plants are relied upon by 80 per cent of the world's population, and in India the use of plants as therapeutic agents remains an important component of the traditional medicinal system. A number of plants have been documented for their biological (Grover *et al.*, 2002; Gajera *et al.*, 2005; Patrick *et al.*, 2011; Kaneria and

Chanda, 2013; Umara *et al.*, 2014) and antimicrobial properties (Chanda *et al.*, 2010; Vijayarathna *et al.*, 2012; Chanda *et al.*, 2013b; Saleh and Otaibi 2013;Rakholiya *et al.*, 2014). Nature has been a source of medicinal agents for thousands of years. Extraction of bioactive compounds from medicinal plants permits the demonstration of their physiological activity. It also facilitates pharmacology studies leading to synthesizing of a more potent drug with reduced toxicity (Manna and Abalaka, 2000; Mary *et al.*, 2012).

Drug resistance is one of the most serious and widespread problem in all developing countries and it poses a very serious threat to public health (Levy and Marshall, 2004; Mandal *et al.*, 2009). In fact, infectious diseases are the major cause of death in the developing countries and accounts for almost 50 per cent of it. Antimicrobial agents are essentially important in reducing the global burden of infectious diseases. Therefore, there is a need to develop alternative antimicrobial drugs for the treatment of infection diseases from various sources such as medicinal plants (Cordell, 2000).

Medicinal plants used in folklore remedies have drawn interest of researchers in discovering solution to multiple drug resistance against commercial antimicrobial drugs. Many naturally occurring compounds found in plants have shown to possess antifungal, antibacterial and antiprotozoan activities and serve as a source of antimicrobial agents that could be used either systemically or locally (Heinrich *et al.*, 2004; Thabti *et al.*, 2014).

Moreover, the increasing use of plant extracts in food, cosmetic and pharmaceutical industries suggests that, in order to find active compounds, a systematic study of medicinal plants is very important (Nostro *et al.*, 2000).

Essential Oils

Essential oils are used in a wide variety of consumer goods such as detergents, soaps, toilet products, cosmetics, pharmaceuticals, perfumes, confectionery food products, soft drinks, distilled alcoholic beverages (hard drinks) and insecticides. The world production and consumption of essential oils and perfumes are increasing very fast. Production technology is an essential element to improve the overall yield and quality of essential oils. The traditional technologies pertaining to essential oil processing are of great significance and are still being used in many parts of the world. Water distillation, water and steam distillation, steam distillation, cohobation, maceration and enfleurage are the most traditional and commonly used methods. Maceration is adaptable when oil yield from distillation is poor. Distillation methods are good for powdered almonds, rose petals and rose blossoms, whereas solvent extraction is suitable for expensive, delicate and thermally unstable materials like jasmine, tuberose, and hyacinth. Water distillation is the most favored method of production of citronella oil from plant material (Abed, 2007). Essential oils are generally not only antibiotic and anticarcinogenic (Carson *et al.*, 1995; Mabona, and Vuuren, 2013), but also have a sedative effect on stress. It has been shown that these essential oils contain ketone, terpene and phenolic ether for sedation (Carvalho-Freitas and Costa, 2002).

Sources of Natural Essential Oils

Essential oils are generally derived from one or more plant parts, such as flowers (*e.g.* rose, jasmine, carnation, clove, mimosa, rosemary, lavander), leaves (*e.g.* mint, ocimum spp., lemongrass, jamrosa), leaves and stems (*e.g.* geranium, patchouli, petitgrain, verbena, cinnamon), bark (*e.g.* cinnamon, cassia, canella), wood (*e.g.* cedar, sandal, pine), roots (*e.g.* angelica, sassafras, vetiver, saussurea, valerian), seeds (*e.g.* castor, fennel, soybean, coriander, caraway, dill, nutmeg), fruits (bergamot, orange, lemon, juniper), rhizomes (*e.g.* ginger, calamus, curcuma, orris) and gums or oleoresin exudations (*e.g.* balsam of peru, balsam of tolu, storax, myrrh, benzoin) (Kuriakose and Joe, 2012; Singh *et al.*, 2013; Olmedo *et al.*, 2014).

Depending upon the plant family, essential oils may occur in specialized secretary structures such as glandular hairs (Labiatae, Verbenaceace, Geraniaceae), modified parenchymal cells (Piperaceae), resin canals (Conifers), oil tubes called vittae (Umbelliferae), lysigenous cavities (Rutaceae), schizogenous passages (Myrtaceae, Graminae, Compositae) or gum canals (Cistacae, Burseraceae). It is well known that when a geranium leaf is lightly touched, an odour is emitted because the long stalked oil glands are fragile. Similarly, the application of slight pressure on a peppermint leaf will rupture the oil gland and release oil. In contrast, pine needles and eucalyptus leaves do not release their oils until the epidermis of the leaf is broken. Hence, the types of structures in which oil is contained differ depending on the plant type and are plant-family specific. Unfortunately, not enough is known even today about these oil secretary structures to carefully categorize them. From the practical standpoint, they can be categorized into superficial and subcutaneous oils. Based on the currently available information, it may be inferred that oils of the Labiatae, Verbenaceae and Geraniaceae families are the only superficial oils known; consequently, the others are considered subcutaneous oils (Moller *et al.*, 2007; Seth *et al.*, 2012; Boukhris *et al.*, 2013).

Essential Oil Constituents

Major constituents of essential oils is clear that most essential oils consist of hydrocarbons, esters, terpenes, lactones, phenols, aldehydes, acids, alcohols, ketones, and esters. Among these, the oxygenated compounds (alcohols, esters, aldehydes, ketones, lactones, phenols) are the principal odour source. They are more stable against oxidizing and resinifying influences than other constituents (Turek and Stintzing, 2012; Devappa *et al.*, 2013; Ng *et al.*, 2014). On the other hand, unsaturated constituents like monoterpenes and sesquiterpenes have the tendency to oxidize or resinify in the presence of air and light. The knowledge of individual constituents and their physical characteristics, such as boiling point, thermal stability and vapor-pressure-temperature relationship is of paramount importance in technology development of oxygenated compounds (Abed, 2007). Biological activity of essential oils depends on their chemical composition which is determined by the genotype and influenced by environmental and agronomic conditions (Hosni *et al.*, 2013).

Materials and Methods

Collection of the Plants Material

The seeds of the *Glycine max* L. Merr (soybean) and *Ricinus communis* L. (castor) belonging to the families Fabaceae and Euphorbiaceae respectively, were purchased in the month of September, 2010 from Jamnagar, Gujarat, India. Plant seeds were cracked, the shells carefully removed and crushed in a homogenizer to fine powder and stored in air tight bottles.

Three Phase Partitioning Method

The slurry was prepared by grinding castor/soybean seed kernels (5 g/30 ml) in distilled water by gentle stirring with a magnetic stirrer. The pH of the slurry was adjusted to the desired value with 0.1 N NaOH or 0.1 N HCl. Appropriate amount of ammonium sulphate was added to this slurry and vortexed gently, followed by the addition of the appropriate amount of organic solvent. This was subsequently incubated at the desired temperature for specified period of time (1 h) for the three phase formation. The three phases formed were separated by centrifugation at 2000 g for 10 min. The upper organic phase, lower aqueous phase and interfacial precipitate layer were observed. The upper organic layer was collected and evaporated (at the particular boiling point) to obtain oil extracted in this phase. The amount of oil recovered were calculated as percentages of oil compare to extracted by soxhlet method (Sharma *et al.*, 2002).

Soxhlet (Hot Continuous Extraction) Method

In this method, the finely ground crude powder is placed in a porous bag or "thimble" made of strong filter paper, which is placed in chamber of the soxhlet apparatus. The extracting solvent in flask is heated and its vapors condense in condenser. The condensed extractant drips into the thimble containing the crude powder and extracts it by contact. When the level of liquid in chamber rises to the top of siphon tube, the liquid contents of chamber siphon into the flask. This process is continuous and is carried out until a drop of solvent from the siphon tube does not leave residue when evaporated. The advantage of this method, compared to previously described methods, is that large amounts of extract (oil) can be extracted with a much smaller quantity of solvent (Soxhlet, 1879). This affects tremendous economy in terms of time, energy and consequently financial inputs. At small scale, it is employed as a batch process only, but it becomes much more economical and viable when converted into a continuous extraction procedure on medium or large scale.

Results

In Three Phase Partitioning (TPP) method, ammonium sulphate and t-butanol was used for oil extraction. The dry powder of *Glycine max* and *Ricinus communis* seed kernels was added with different amount of ammonium sulphate (30 per cent, 40 per cent, 50 per cent, 60 per cent, etc. w/v) followed by the addition of various ratios of t-butanol. After 1 hour, the three phases were formed and the upper phase was collected which consisted of oil.

In three phase partitioning extraction (TPP), maximum oil in the upper phase (organic phase) is obtained but depends on slurry:butanol ratio. Ammonium sulphate is of critical importance and must be optimized (Dennison and Lovrein, 1997). The ratio (v/v) of organic solvent to the starting aqueous phase has been generally found critical in optimizing the TPP process (Sharma *et al.*, 2000). The yield of extraction experiments on *Glycine max* seed performed by three phase partition is shown in Figure 3.1a, with different ratios of t-butanol and slurry. The maximum oil yield obtained was 15 per cent at 1:2 ratios of t-butanol and slurry. Decreasing the t-butanol volume (1:1) decreased the oil yield whereas increasing the t-butanol ratio (1:3) to twice the volume did not lead to any significant increase. So, 1:2 (t-butanol:slurry) ratio was employed in further studies. The effect of various pH on oil extraction by TPP is shown in Figure 3.1b. The best pH for *Glycine max* was 4.0 at 50 per cent (w/v) ammonium sulphate concentration, giving an oil yield of 13 per cent (w/w). The oil yield decreased when pH was neutral (7.0) or alkaline (9.0). The pH of the medium is also a very important parameter in three phase partitioning. Hence, the effect of pH on the partitioning of slurry was also investigated. The slurry was saturated with 50 per cent (w/v) ammonium sulphate and then the pH of the medium was adjusted. The effect of varying amount of ammonium sulphate during TPP is shown in Figure 3.1c. Similar oil content was obtained by using 40 per cent and 50 per cent (w/v) ammonium sulphate. Decreasing the amount of ammonium sulphate decreased the oil yield while increasing the amount of ammonium sulphate did not make a significant difference to the oil yield. Thus, 50 per cent (w/v) gave best results and the same was employed in further studies.

In case of *Ricinus communis*, when the volume of slurry and t-butanol was in the ratio of 1:1 and 1:2, the oil yield was same (41 per cent) (Figure 3.2a) but when the ratio was increased to 1:3, the oil yield decreased to 33 per cent. On the other hand, when the ratio was 1:0.5, the oil yield was 38 per cent. Therefore, 1:2 ratio was taken as optimum ratio giving oil yield of 41 per cent. The effect of varying ammonium sulphate concentration during TPP extraction is shown in Figure 3.2b. The oil yield obtained was maximum (57 per cent) when ammonium sulphate concentration was 50 per cent (w/v). When ammonium sulphate concentration was decreased, oil yield decreased but when ammonium sulphate concentration was increased, it did not increase the oil yield. Therefore, 50 per cent (w/v) ammonium sulphate was employed in further studies. The effect of pH of slurry in TPP extraction is shown in Figure 3.2c. The best pH for *Ricinus communis* was 7.0 at 50 per cent ammonium sulphate concentration, giving an oil yield of 57 per cent (w/w). In both acidic (4.0) and alkaline (9.0) pH, the oil yield decreased.

The amount of oil recovered was calculated as percentages of oil obtained by Soxhlet extraction using hexane solvent. The oil yield of *Glycine max* was 20.78 per cent while that of *Ricinus communis* was 56.50 per cent. The comparison of oil obtained by soxhlet method with TPP method of *Glycine max* is shown in Figure 3.3. When the ratio of t-butanol and soybean slurry was varied, the maximum percentage of oil obtained was 71 per cent (w/w) at 1:2 ratio, using 50 per cent (w/v) ammonium sulphate (Figure 3.3a). When different concentrations of ammonium sulphate was used, almost similar percentage of oil (72 per cent w/w) was obtained at 40 per cent

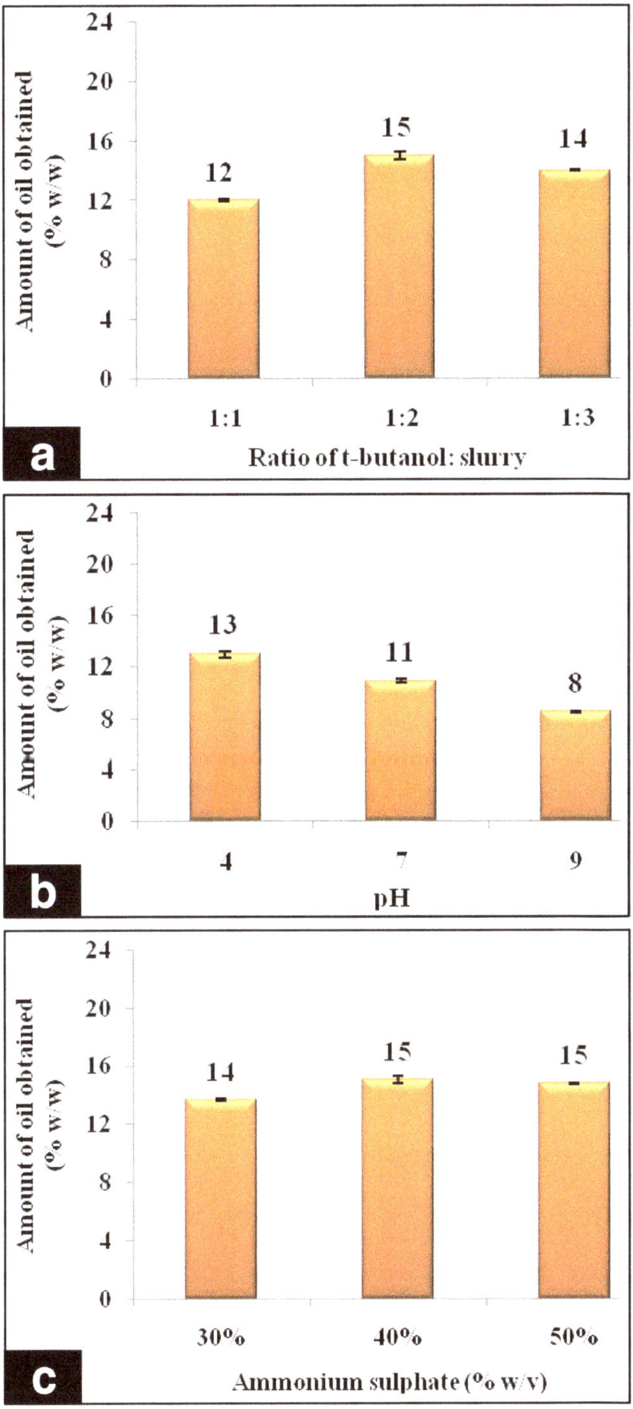

Figure 3.1a-c: Effect of different Parameter on Oil Extraction by TPP Method from *G. max* Seeds.

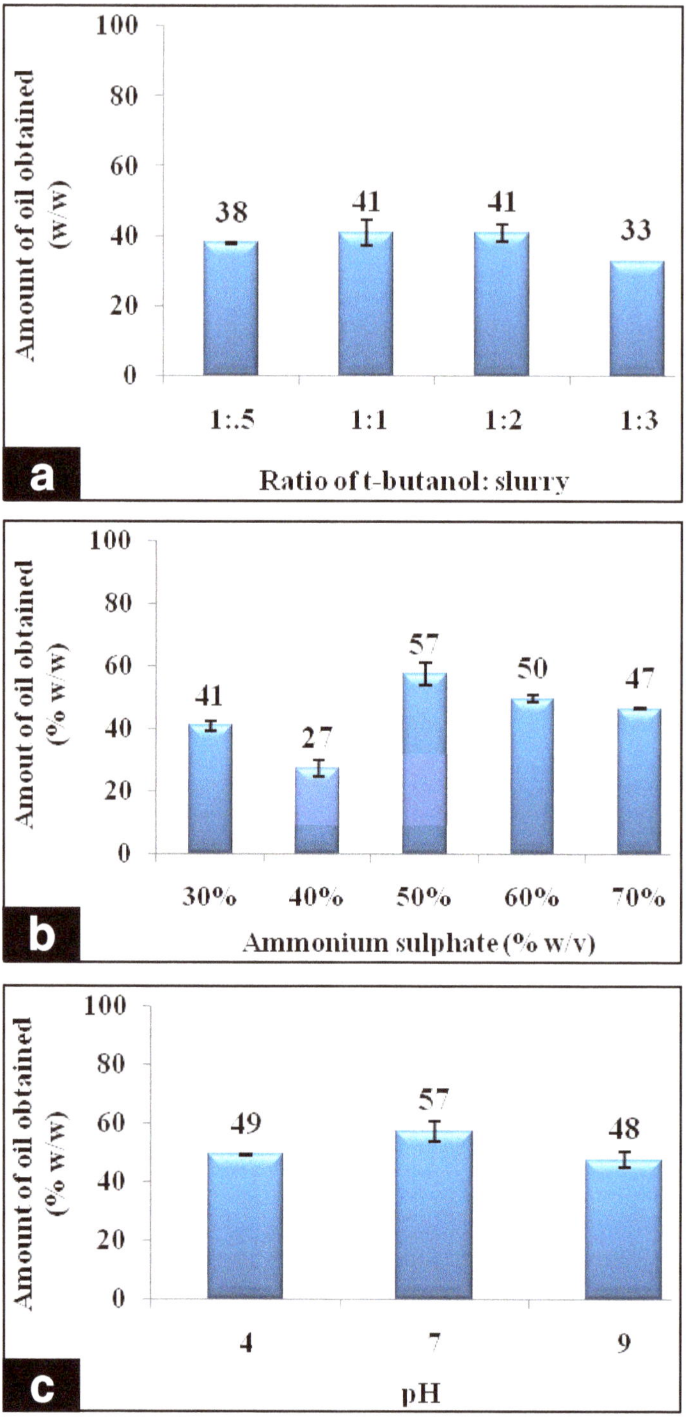

Figure 3.2a-c: Effect of different Parameter on Oil Extraction by TPP Method from *R. communis* Seeds.

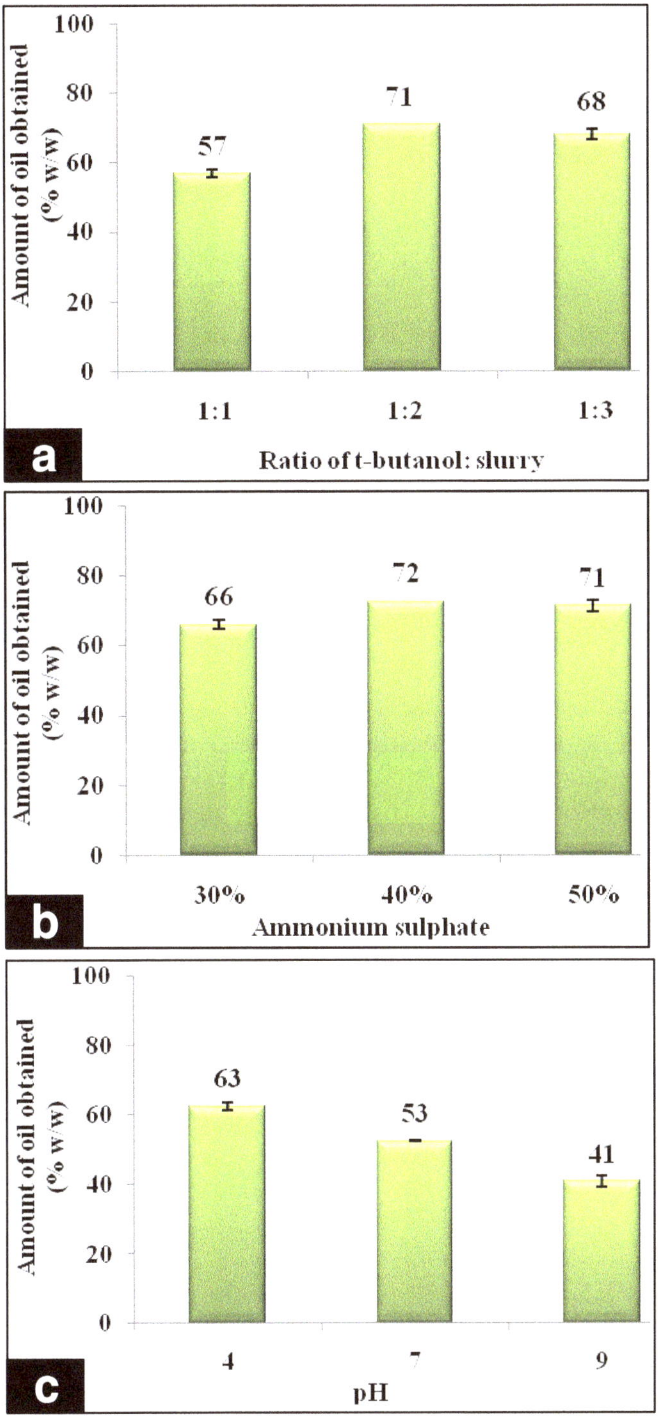

Figure 3.3a-c: Comparison of Oil Obtained by Soxhlet Method with TPP Methods from *G. max* Seeds.

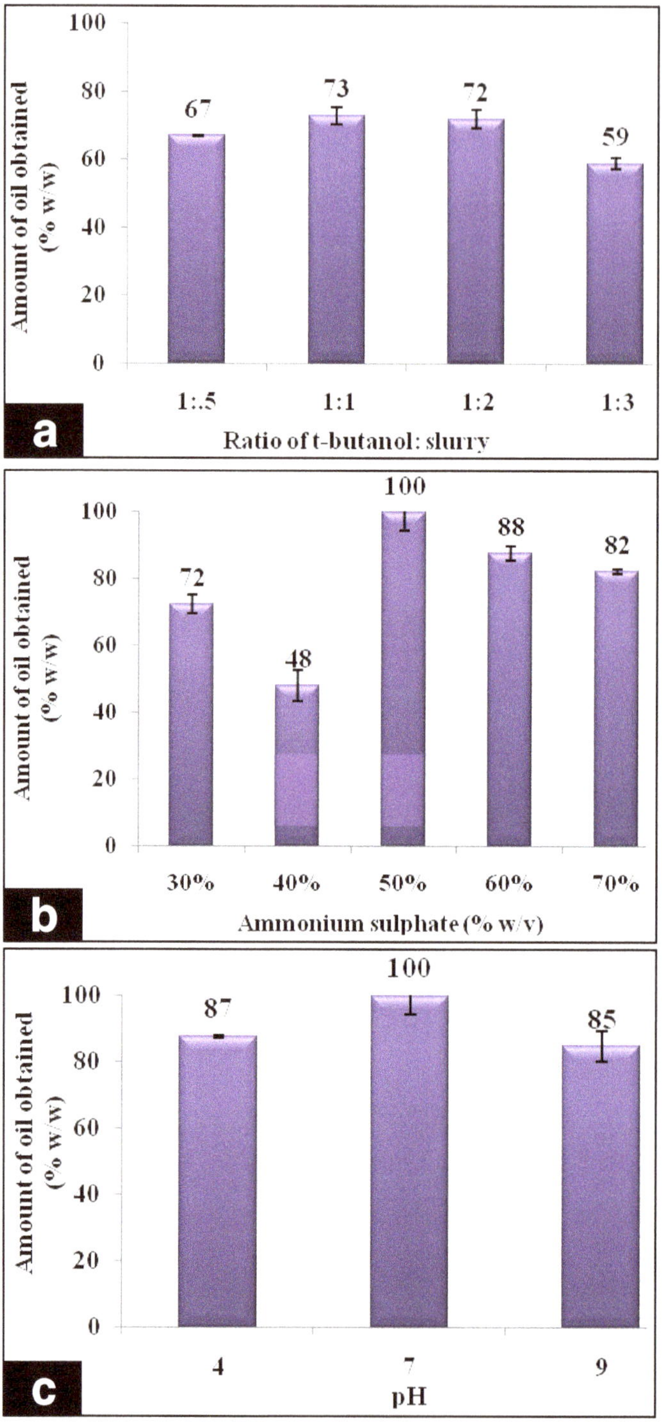

Figure 3.4a-c: Comparison of Oil Obtained by Soxhlet Method with TPP Methods from *R. communis* Seeds.

and 50 per cent (w/v) ammonium sulphate concentration, while the oil yield decreased to 66 per cent when the concentration of ammonium sulphate was decreased to 30 per cent (Figure 3.3b). Figure 3.3c shows the effect of varying pH on oil yield of *Glycine max*. Maximum yield was at pH 4 and as the pH increased, the oil content decreased. Therefore, it can be concluded that for best oil extraction of *Glycine max*, the ratio of slurry to butanol is 1:2, ammonium sulphate concentration is 50 per cent w/v and pH is 4.

In the case of *Ricinus communis*, different ratio of t-butanol to castor slurry were used in three phase partitioning experiments and the results are given in Figure 3.4a. Almost similar percentage of oil was obtained at 1:1 and 1:2 ratio of t-butanol. On decreasing or increasing the ratio of t-butanol, the percentage of oil yield decreased. The change in percentage of oil yield when ammonium sulphate concentration is changed is shown in Figure 3.4b. The percentage of oil yield was 72 per cent at 30 per cent ammonium sulphate concentration which decreased drastically to 48 per cent when ammonium sulphate concentration was 40 per cent. However, on increasing the concentration of ammonium sulphate to 50 per cent, the oil yield again increased greatly (100 per cent) which was maximum. Further increase in ammonium sulphate concentration, the oil yield decreased slightly (Figure 3.4b). The effect of varying pH on *Ricinus communis* percentage oil yield is shown in Figure 3.4c. Maximum oil yield was at pH 7 while at acidic and alkaline pH *i.e.* at pH 4 and pH 9, the percentage of oil yield slightly decreased. Therefore, it can be concluded that for best oil extraction of *Ricinus communis*, the ratio of slurry to butanol is 1:2, ammonium sulphate concentration is 50 per cent w/v and pH is 7.

Discussion

TPP is a simple and rapid procedure which is a combination of salting out and alcoholic precipitation with several advantages; it can be readily scaled up; it generally works at room temperature, the chemicals used can be recycled and it is cheap (Akardere *et al.*, 2010). t-butanol has a higher boiling point (84°C) than hexane (69 °C). Thus, addition of volatile organic compounds to the atmosphere (Rosenthal *et al.*, 1996) during the process will be much lower even if open systems are used. Simpler extraction designs will be possible. Lusas *et al.* (1997) have also suggested that possibility of separation of the organic solvents by chilling. This is considered desirable since chilling uses up much less energy than that required to evaporate the organic solvents by heating. t-butanol with a freezing point of 11 °C compares well with isopropanol (-89.5 °C) and hexane (-95 °C) (Abraham *et al.*, 1988).

In Soxhlet extraction method using hexane as solvent yielded almost similar amount of oil but TPP is simpler extraction designs, has short processing time of only about 2 h while the Soxhlet method takes more time (10-12 h) for oil extraction (Shah *et al.*, 2004). So, TPP can be an efficient alternative approach to oil extraction but more extensive modeling and scale-up studies are admittedly required before one can recommend it for use at the industrial level. Therefore, it can be stated that, TPP method is better than Soxhlet method for oil extraction.

Conclusions

The yield of oil from seeds of *Glycine max* and *Ricinus communis* by two different extraction techniques was investigated. In *R. communis*, the yield was almost same by both the extraction techniques, while in *G. max* it was slightly less. Under optimized condition, best condition for the oil extraction of *R. communis*, the ratio of slurry to t-butanol was 1:2 (v/v), ammonium sulphate concentration was 50 per cent (w/v) and pH was 7, while in case of *G. max*, the ratio of slurry to t-butanol was 1:2 (v/v), ammonium sulphate concentration was 50 per cent (w/v) and pH was 4. TPP can be effective alternative approach for extraction of oil since it is economic, short processing time and simple.

Acknowledgements

The authors thank Prof. S.P. Singh, Head, Department of Biosciences, Saurashtra University, Rajkot, Gujarat, India for providing excellent research facilities. The authors, Dr. Mital Kaneria, Dr. Kalpna Rakholiya and Dr. Krunal Nagani are also thankful to University Grants Commission, New Delhi, India for providing financial support.

References

Abed, K.F. (2007). Antimicrobial activity of essential oils of some medicinal plants from Saudi Arabia. *Saudi Journal of Biological Science*, **14**: 53-60.

Abraham, G., Hron, R.J., and Koltun, S.P. (1988). Modeling the solvent extraction of oil seeds. *Journal of American Oil Chemists Society*, **67**: 165-167.

Akardere, E., Ozer, B., Celem, E.B., and Onal, S. (2010). Three phase partitioning of invertase from baker's yeast. *Separation and Purification Technology*, **72**: 335-339.

Boukhris, M., Nasri-Ayachi, M., Mezghani, I., Bouaziz, M., Boukhris, M., and Sayadi, S. (2013). Trichomes morphology, structure and essential oils of *Pelargonium graveolens* L'Hér. (Geraniaceae). *Industrial Crops and Products*, **50**: 604-610.

Carson, C.F., Cookson, B.D., Farrelly, H.D., and Riley, T.V. (1995). Mechanism of action of *Melaleuca alternifolia* (Tea Tree) oil on *Staphylococcus aureus* determined by time-kill, lysis, leakage, and salt tolerance assays and electron microscopy. *Journal of Antimicrobiology Chemotheraphy*, **35**: 421-424.

Carvalho-Freitas, M.I., and Costa, M. (2002). Anxiolytic and sedative effects of extracts and essential oils from *Citrus aurantium* L. *Biology of Pharmacology Bulletin*, **25**: 1629-1633.

Chanda, S., Dudhatra, S., and Kaneria, M. (2010). Antioxidative and antibacterial effects of seeds and fruit rind of nutraceutical plants belonging to the family Fabaceae. *Food and Function*, **1**: 308-315.

Chanda, S., Rakholiya, K., and Parekh, J. (2013a). Indian medicinal herb: Antimicrobial efficacy of *Mesua ferrea* L. seed extracted in different solvents against infection causing pathogenic strains. *Journal of Acute Disease*, **2**: 277-281.

Chanda, S., Rakholiya, K., Dholakia, K., and Baravalia, Y. (2013b). Antimicrobial, antioxidant, and synergistic properties of two nutraceutical plants: *Terminalia catappa* L. and *Colocasia esculenta* L. *Turkish Journal of Biology*, **37**: 81-91.

Cordell, G.A. (2000). Biodiversity and drug discovery a symbiotic relationship. *Phytochemistry*, **55**: 463-480.

Coutinho, H.D., Lobo, K.M., Bezerra, D.A., and Lobo, I. (2008). Peptides and proteins with antimicrobial activity. *Indian Journal of Pharmacology*, **40**: 3-9.

de Boer, H.J., and Cotingting, C. (2014). Medicinal plants for women's healthcare in southeast Asia: A meta-analysis of their traditional use, chemical constituents, and pharmacology. *Journal of Ethnopharmacology*, **151**: 747-767.

Dennison, C., and Lovrein, R. (1997). Three phase partitioning: concentration and purification of proteins. *Protein Expression and Purification*, **11**: 149-161.

Devappa, R.K., Makkar, H.P.S., and Becker, K. (2013). Shelf-life of isolated phorbol esters from *Jatropha curcas* oil. *Industrial Crops and Products*, **49**: 454-461.

Dhami, N. (2013). Trends in Pharmacognosy: A modern science of natural medicines. *Journal of Herbal Medicine*, **3**: 123-131.

Dike, I.P., Obembe, O.O., and Adebiyi, F.E. (2012). Ethnobotanical survey for potential anti-malarial plants in south-western Nigeria. *Journal of Ethnopharmacology*, **144**: 618-626.

Gajera, H.P., Patel, S.V., and Golakiya, B.A. (2005). Antioxidant properties of some therapeutically active medicinal plants an overview. *Journal of Medicinal Aromatic Plant Science*, **27**: 91-100.

Grover, J.K., Yadav, S., and Vats, V. (2002). Medicinal plants of India with anti-diabetic potential. *Journal of Ethnopharmacology*, **81**: 81-100.

Heinrich, M., Barnes, J., Gibbons, S., and Williamson, E.M. (2004). Fundamentals of pharmacognosy and phytotherapy. *Churchill Livingstone, Edinburgh* 4–7.

Hosni, K., Hassen, I., Sebei, H., and Casabianca, H. (2013). Secondary metabolites from *Chrysanthemum coronarium* (Garland) flowerheads: Chemical composition and biological activities. *Industrial Crops and Products*, **44**: 263-271.

Kaneria, M.J., and Chanda, S.V. (2013). The effect of sequential fractionation technique on the various efficacies of pomegranate (*Punica granatum* L.). *Food Analytical Methods*, **6**: 164-175.

Kuriakose, S., and Joe, H. (2012). Qualitative and quantitative analysis in sandalwood oils using near infrared spectroscopy combined with chemometric techniques. *Food Chemistry*, **135**: 213-218.

Levy, S.B., and Marshall, B. (2004). Antibacterial resistance worldwide: causes, challenges and responses. *Nature Medicine*, **10**: S122 - S129.

Lusas, E.W., Watkins, L.R., Koseoglu, S.S., Rhee, K.C., Hernandez, E., Raiz, M.N., and Johnson, D. (1997). Final report: IPA as an extraction solvent. *INFORM* **8**: 280-305.

Mabona, U., and Vuuren, S.F.V. (2013). Southern African medicinal plants used to treat skin diseases. *South African Journal of Botany*, **87**: 175-193.

Mandal, S., Moudgil, M., and Mandal, S.K. (2009). Rational drug design. *European Journal of Pharmacology*, **625**: 90–100.

Manna, A., and Abalaka, M.E. (2000). Preliminary screening of the various extracts of *Physalis angulala* L. for antimicrobial activities. *Spectrum Journal*, **7**: 119-125.

Mary, H.P.A., Susheela, G.K., Jayasree, S., Nizzy, A.M., Rajagopal, B., and Jeeva, S. (2012). Phytochemical characterization and antimicrobial activity of *Curcuma xanthorrhiza* Roxb. *Asian Pacific Journal of Tropical Biomedicine*, **2**: S637-S640.

Moller, J.K.S., Catharino, R.R., and Eberlin, M.N. (2007). Electrospray ionization mass spectrometry fingerprinting of essential oils: Spices from the Labiatae family. *Food Chemistry*, **100**: 1283-1288.

Nair, R., and Chanda, S.V. (2007). Antibacterial activities of some medicinal plants of the western region of India. *Turkish Journal of Biology*, **31**: 231-236.

Nasab, F.K., and Khosravi, A.R. (2014). Ethnobotanical study of medicinal plants of Sirjan in Kerman Province, Iran. *Journal of Ethnopharmacology*, **154**: 190-197

Ng, C., Leong, X., Masbah, N., Adam, S.K., Kamisah, Y., and Jaarin, K. (2014). Heated vegetable oils and cardiovascular disease risk factors. *Vascular Pharmacology*, **61**: 1-9.

Nostro, A., Germano, M.P., Angelo, V.D., Marino, A., and Cannatelli, M.A. (2000). Extraction methods and bioautography for evaluation of medicinal plant antimicrobial activity. *Letters in Applied Microbiology*, **30**: 379.

Olmedo, R., Nepote, V., and Grosso, N.R. (2014). Antioxidant activity of fractions from oregano essential oils obtained by molecular distillation. *Food Chemistry*, **156**: 212-219.

Parekh, J., and Chanda, S. (2007). *In vitro* antimicrobial activity and phytochemical analysis of some Indian medicinal plants. *Turkish Journal of Biology*, **31**: 53-58.

Patrick, A., George, A.K., Kwame, S., and Kwabena, O.A. (2011). Free radical scavenging and anti-lipid peroxidative effects of a hydro-ethanolic extract of the whole plant of *Synedrella nodiflora* (L.) gaertn (Asteraceae). *Free Radicals and Antioxidants*, **1**: 70-78.

Rakholiya, K., Vaghela, P., Rathod, T., and Chanda, S. (2014). Comparative study of hydroalcoholic extracts of *Momordica charantia* L. against foodborne pathogens. *Indian Journal of Pharmaceutical Sciences*, **76**: 148-156.

Rosenthal, A., Pyle, D.L., and Niranjan, K. (1996). Aqueous enzymatic processes for edible oil extraction. *Enzyme Microbiology Technology*, **19**: 402-420.

Saleh, F.A., and Otaibi, M.M. (2013). Antibacterial activity of Date plam (*Pheonix dectylifera* L.) fruit at different riping stage. *Journal of Food Process Technology*, **4**: 1-6.

Seth, R., Mohan, M., Singh, P., Haider, S.Z., Gupta, S., Bajpai, I., Singh, D., and Dobhal, R. (2012). Chemical composition and antibacterial properties of the essential oil and extracts of *Lantana camara* Linn. from Uttarakhand (India). *Asian Pacific Journal of Tropical Biomedicine*, **2**: S1407-S1411.

Shah, S., Sharma, A., and Gupta, M.N. (2004). Extraction of oil from *Jatropha curcas* L. seed kernels by enzyme assisted three phase partitioning. *Industrial Crops and Products*, **20**: 275-279.

Sharma, A., Khare, S.K., and Gupta, M.N. (2002). Three phase partitioning for extraction of oil from soybean. *Bioresource Technology*, **85**: 327-329.

Sharma, A., Sharma, S., and Gupta, M.N. (2000). Purification of alkaline phosphatase from chicken intestine by three phase partitioning and use of phenyl-sepharose 6B in the batch mode. *Bioseparation*, **9**: 155-161.

Singh, P., Singh, S., Kapoor, I.P.S., Singh, G., Isidorov, V., and Szczepaniak, L. (2013). Chemical composition and antioxidant activities of essential oil and oleoresins from *Curcuma zedoaria* rhizomes. *Food Bioscience*, **3**: 42-48.

Soxhlet, F. (1879). Die gewichtsanalytische Bestimmung des Milchfettes. *Polytechnisches Journal*, **232**: 461-465.

Suneetha, M.S., and Chandrakanth, M.G. (2006). Establishing a multi-stakeholder value index in medicinal plants - an economic study on selected plants in Kerala and Tamilnadu States of India. *Ecological Economics*, **60**: 36-48.

Thabti, I., Elfalleh, W., Tlili, N., Ziadi, M., Campos, M.G., and Ali, F. (2014). Phenols, flavonoids, and antioxidant and antibacterial activity of leaves and stem bark of *Morus* species. *International Journal of Food Properties*, **17**: 842–854.

Turek, C., and Stintzing, F.C. (2012). Impact of different storage conditions on the quality of selected essential oils. *Food Research International*, **46**: 341-353.

Umara, S., Umarb, K., Hasnath, A., Md. Sarwarc, G., Khand, A., Ahmade, N., Ahmade, S., Katiyarf, C.K., Husaing, S.A., and Khana, H.A. (2014). *Boswellia serrata* extract attenuates inflammatory mediators andoxidative stress in collagen induced arthritis. *Phytomedicine*, (In Press)

Vijayarathna, S., Zakaria, Z., Chen, Y., Latha, L.Y., Kanwar, J.R., and Sasidharan, S. (2012). The antimicrobial efficacy of *Elaeis guineensis*: characterization, *in vitro* and *in vivo* studies. *Molecules*, **17**: 4860-4877.

Utilisation and Management of Medicinal Plants Vol. 3 (2015) *Pages* **69–76**
Editors: **V.K. Gupta and A. Kaul**
Published by: **DAYA PUBLISHING HOUSE, NEW DELHI**

4

Chemical Characterization of Important Aromatic Herbs Grown in Temperate Conditions of Central Himalayan Region of India

Archana P. Raina[1*] and K.S. Negi[2]

ABSTRACT

Himalayan region is known to be a rich repository of medicinal and aromatic plants. The agro-climatic condition prevailing in this region provides an ideal habitat for the natural growth of variety of plants and herbs, which provide raw materials for pharmaceutical, phyto-chemical, food flavoring and cosmetic industries. There is an urgent need to conserve the available species of medicinal and aromatic species and also to start their cultivation on commercial scale to meet domestic and global demands. A study was undertaken to collect wildly grown and important introduced aromatic plants of Rosmarinus officinalis, Salvia officinalis, Thymus serphyllum, Origanum vulgare, Origanum majorana and Satureja montana from temperate region of Central Himalayas. These aromatic plants are being maintained and conserved in the ex-situ Field Gene Bank of NBPGR Regional Station at Bhowali, district Nainital, Uttarakhand State in India. The essential oils isolated from these aromatic herbs have been chemically characterized by gas chromatography (GC) and gas chromatography-mass spectrometry (GC-MS). The essential oil content, aroma constituents

1 Germplasm Evaluation Division, National Bureau of Plant Genetic Resources, New Delhi – 110 012, India.

2 National Bureau of Plant Genetic Resources, Regional Station, Bhowali – 263 132, Uttarakhand, India

* *Corresponding author*: E-mail: aprraina@yahoo.co.in; raina@nbpgr.ernet.in

and different chemotypes identified in these plants have been reported in this paper. GC-MS profile of these essential oils revealed the presence of important aroma constituents that can be used in perfumery, food and flavor industry.

Keywords: Aromatic herbs, Essential oil composition, Gas chromatography-mass spectrometry, Marjoram, Oregano, Rosemary, Sage, Savory, Thyme.

Introduction

The increasing demand in developed countries for natural flavors offers tremendous potential for medicinal and aromatic crops as source of natural flavors. Aromatic plants are important in the food industry not only for their use as flavorings but also for their anti-oxidant properties (Piccaglia *et al.*, 1993). In the recent past, there has been increasing interest in the biological effects of the herbal spices. Many chemical compounds present in these herbs have therapeutic value such as antioxidant, antiseptic and antimicrobial activities. Due to heavy demand from industries leading to over exploitation, destruction of habitats and unsustainable harvesting practice expedited for an urgent and systematic approach to collect and cultivate many valuable herbs from temperate region of Central Himalayas. Some wildly grown and important introduced aromatic plants *viz., Rosmarinus officinalis* (Rosemary), *Salvia officinalis* (Sage), *Thymus serphyllum* (Wild Thyme), *Satureja montana* (Winter Savory), *Origanum majorana* (Marjoram) and *Origanum vulgare* (Oregano) were collected through several explorations from temperate regions of Uttarakhand. These plants were conserved in the *ex-situ* Field Gene Bank of NBPGR Regional Station at Bhowali, district Nainital, Uttarakhand State. The chemical composition of essential oils is determined in the first instance by its species and variety and also by agronomic conditions, harvest time and type of processing followed. The environmental and ecological characteristics of the particular growing region provide each aromatic plant of a definite species and variety with specific fingerprints that are demonstrated by its distinct organoleptic properties due to its composition. It is now possible with the use of powerful analytical techniques, to recognize these singularities in great detail and thus avoid adulterations. Thus, in this report the essential oil composition of these important aromatic crops cultivated in Uttarakhand region of India are given, and compared with those found in plants of different geographical origins. These plants are commercially important and this study allows not only an evaluation of the qualities of their essential oils and of their specific characteristics, but also may be useful in the future selection in order to find the best matching between environmental conditions and the composition of different species.

Materials and Methods

Plant Material

Aromatic plants of *Rosmarinus officinalis, Salvia officinalis, Thymus serphyllum, Satureja montana, Origanum majorana* and *Origanum vulgare* ssp. hirtum were collected from parts of Uttarakhand state located in Central Himalayan region. All the six plant species were identified and voucher specimens of each were deposited in

Herbarium of NBPGR, New Delhi. These plants are being maintained and conserved in the *ex-situ* Field Gene Bank of NBPGR Regional Station, Bhowali, district Nainital, Uttarakhand State, at 1600 m altitude, 79°30′ N latitude and 29° 20′ E longitude. It has a cold humid, sub-temperate climate with an annual rainfall of 1600 mm and temperature range from 2.2°C to 32° C. Aerial parts of above plants were harvested during the flowering period, when the crop was in full bloom stage. These were shade dried at room temperature and then subjected for hydro distillation process for 4 h for isolation of essential oils using a Clevenger-type apparatus. These oils were stored in a glass vial in the dark at 4°C for further chemical analysis.

Gas Chromatography and Gas Chromatography-Mass Spectrometry Analysis

Gas chromatography(GC) was carried out using Perkin Elmer Autosystem 8500 gas chromatograph, fitted with a flame ionization detector and a Carbowax 20M capillary column (50m × 0.25 mm i.d., 0.25µm film thickness). Nitrogen was used as carrier gas at the flow rate of 1 mL min^{-1}. The oven temperature was programmed from 60 °C to 240 °C at the rate of 3 °C/min with a hold of 10 min each at 60 °C and 240 °C. The injector and detector temperature was maintained at 220 °C and 250 °C, respectively. The sample (0.1 µL) was injected neat in split ratio (1:40) at 220 °C. Area percentage reports obtained by GC/FID was directly applied for quantification purpose. Gas chromatography-mass spectrometry (GC-MS) analyses were done on Agilent GC 7890 equipped with MSD 5975C and a HP-5MS capillary column (30m × 0.25 mm i.d., 0.25µm coating) and temperature programming was done as described above. Helium (1 mL min^{-1}) was used as carrier gas. The mass unit conditions were ion source 250°C, ionization energy 70 eV and acquisition mass range of m/z 40 to 400. The components were identified by comparing their relative retentions times with the retention times of authentic standards, and mass spectra with NIST, WILEY library data of the GC-MS system and literature data (Adams, 1995).

Results

The essential oil contents obtained in these plants were: *Rosmarinus officinalis* (0.29 per cent), *Salvia officinalis* (0.96 per cent), *Satureja montana* (0.88 per cent), *Thymus serphyllum* (0.60 per cent), *Origanum majorana* (0.37 per cent) and *Origanum vulgare* ssp. hirtum (0.24 per cent) on dry weight basis. Results of GC-MS analyses of chemical composition of the essential oils of six commercially important aromatic herbs of Uttarakhand are given in Table 4.1. Essential oil composition of these herbs is discussed here and compared with oil composition reported by various workers from different geographical regions of world.

Rosemary (*Rosmarinus officinalis*)

The chemical composition of *R. officinalis* oil showed high proportion of oxygenated monoterpenes (50.69 per cent), and monoterpene hydrocarbon (41.66 per cent) while sesquiterpenes hydrocarbons and oxygenated sesquiterpenes were present in very low proportions of 2.35 per cent and 0.48 per cent, respectively. The major components present in rosemary oil were α-pinene(21.24 per cent), camphene(7.67 per cent), myrcene (6.22 per cent), 1,8-cineole (20.79 per cent), camphor

Table 4.1: Chemical Composition (per cent) of Essential Oils of Aromatic Herbs Grown in Uttarakhand, India

Compound	*RI**	*Rosemary*	*Sage*	*Thymus*	*Savory*	*Marjoram*	*Oregano*
α-thujene	931	0.15	0.78	0.03	–	0.34	0.11
α-pinene	938	21.24	7.23	3.03	0.15	0.53	2.00
camphene	952	7.67	2.12	1.09	0.04	0.11	1.14
sabinene	976	0.74	–	–	–	6.91	–
β-pinene	980	4.20	1.83	0.25	–	0.34	0.96
3-octanol	984	–	–	–	0.09	–	0.65
myrcene	989	6.22	0.12	1.48	0.16	1.63	1.92
Δ-3-carene	1008	0.43	0.98	2.07	0.10	–	0.90
α-terpinene	1017	–	0.86	0.28	0.40	–	0.51
β-phellandrene	1025	–		–	–	0.34	0.34
p-cymene	1026	0.75	0.17	11.83	6.51	6.83	9.79
limonene	1031	0.75	–	–	–	–	–
1,8-cineole	1035	20.79	8.31	–	–	0.20	0.49
(E)-β-ocimene	1044	–	–	–	–	0.05	–
trans-sabinene hydrate	1048	–	–	–	1.13	3.86	1.77
γ-terpinene	1063	0.85	–	11.27	–	0.37	15.39
cis-sabinene hydrate	1065	–	–	–	–	15.76	0.54
terpinolene	1086	–	–	–	–	1.63	–
linalool	1099	1.11	4.38	0.26	1.76	0.88	1.22
α-thujone	1101	–	31.68	–	2.63	–	–
β-thujone	1112	–	3.67	–	–	–	–
cis-p-mentha-2-en-1-ol	1118	–	–	–	–	1.10	–
trans-p-mentha-2-en-1-ol	1136	–	–	–	–	1.88	–
camphor	1146	11.85	10.80	–	0.28	–	–
borneol	1172	4.30	–	0.96	0.43	0.63	2.64
terpinen-4-ol	1189	–	–	–	0.59	31.15	0.37
α-terpineol	1189	–	0.19	2.30	1.40	3.71	1.10
verbenone	1204	6.55		–	–	–	–
linalyl acetate	1254	1.11	–	1.53	1.16	2.41	0.85
geraniol	1254	0.10	–	0.63	0.48	–	–
bornyl acetate	1284	3.33	6.65	–	–	0.32	–
thymol	1289	–	0.21	54.86	0.60	–	33.94
carvacrol	1298	–	0.13	0.58	77.23	–	6.90
carvacrol acetate	1370	–	–	–	–	–	7.68
methyl cinnamate	1376	–	0.13	–	0.22	–	–

Contd...

Table 4.1–*Contd...*

Compound	*RI**	*Rosemary*	*Sage*	*Thymus*	*Savory*	*Marjoram*	*Oregano*
geranyl acetate	1379	0.61	–	–	–	–	–
β-caryophyllene	1408	2.35	4.38	4.07	0.83	2.49	1.02
α-humulene	1454	–	1.91	–	0.09	0.11	0.80
bicyclogermacrene	1500	–	–	–	–	1.90	–
β-bisabolene	1505	–	–	–	0.43	–	1.36
sesquiphellandrene	1521	–	0.38	1.68	–	–	–
nerolidol	1531	–	0.87	–	–	–	–
spathulenol	1577	–		–	–	–	0.34
caryophyllene oxide	1582	0.48	0.52	0.36	1.69	0.88	0.20
viridiflorol	1596	–	3.73	–	–	–	–
farnesol	1698	0.21	1.20	1.02	0.16	0.07	–
Total		95.79	93.23	99.58	98.56	86.43	94.93

*: RI: Retention Indices on HP-5MS capillary column.

(11.85 per cent), β-pinene(4.20 per cent), bornyl acetate (3.33 per cent), verbenone (6.55 per cent) and β-caryophyllene (2.35 per cent). From the present analysis, it is evident that the quality of rosemary oil of the plants under investigation is similar to chemotype rich in α-pinene, 1,8-cineole and camphor. Similar reports of rosemary oil composition have been reported from South India (Narayanan, 2003) and Uttarakhand (Kumar *et al.*, 2004). Variation in chemical composition of rosemary oil with geographical regions has been reviewed in detail by Lawrence, 2001. Recently rosemary oil has gained importance as one of the most potent spices as natural antioxidant. This oil also finds use in medicinal preparations such as carminative, diaphoretic, circulatory and cardiac stimulant and is used in treating headache.

Sage (*Salvia officinalis*)

S. officinalis is one of the most important species out of the Salvia species studied so far. The monoterpene fractions (78.91 per cent) were the most abundant in this oil and it was particularly rich in oxygenated derivatives (65.68 per cent); α-thujone (31.68 per cent), camphor (10.80 per cent), 1,8-cineole(8.31 per cent) and β-thujone (3.67 per cent) were found as the major compounds. In contrast, sesquiterpenes hydrocarbons were 6.29 per cent and oxygenated sesquiterpenes were 7.04 per cent. These included β-caryophyllene (4.38 per cent), α-humulene (1.91 per cent), viridiflorol (3.73 per cent). The high proportion of oxygenated monoterpenes is noteworthy as these exhibit antimicrobial, anti-inflammatory and antioxidant features (Boelens and Boelens, 1997). Quality of sage oil is determined on the basis of its thujone content; higher the thujone content better is the oil quality. Sage oil is used as a mild tonic, astringent, antioxidant, carminative and for flavouring sausages, canned meat seasoning and liquors.

***Thymus serphyllum* (Wild thyme)**

The chemical composition of the essential oils of *T. serphyllum* showed a total of 20 components accounting for 99.58 per cent of the oil. Thyme essential oil was characterized by the presence of phenolic monoterpene thymol (54.86 per cent) which gives it typical strong and spicy smell associated with thyme oil. A thymol rich chemotype of *T. serphyllum* has been reported earlier from Uttarakhand hills by Verma *et al.*, 2011. Other components identified were carvacrol (0.58 per cent) and biosynthetic precursors ρ-cymene (11.83 per cent), γ-terpinene (11.27 per cent). These four components composed 78.54 per cent of the total thyme oil. Other monoterpenes present were δ-3-carene (2.07 per cent), α-terpineole (2.30 per cent), myrcene (1.48 per cent), β-caryophyllene(4.07 per cent), linalool (1.53 per cent). This thymol rich chemotype is of commercial interest due to high antimicrobial activity attributed by phenols.

Winter Savory (*Satureja montana*)

Essential oil of *S. montana* showed carvacrol (77.23 per cent) as the major component out of total 23 components identified. Other components present in this oil were ρ-cymene (6.51 per cent), α-thujone (2.63 per cent), caryophyllene oxide (1.69), linalool (1.76), α-terpineole (1.40 per cent), trans-sabinene hydrate (1.13 per cent) and thymol (0.60 per cent). Phenol content (carvacrol) varies from 23.8 to 67.8 per cent (Lawrence, 1996). It is possible to use this carvacrol rich oil or herb itself as a natural product with antimicrobial and antioxidative effects. The herb is antibacterial, antispasmodic, astringent, carminative, digestive diuretic, expectorant, laxative, stomachic, sedative and antifungal.

Marjoram (*Origanum majorana*)

Chemical composition of *O. majorana* oil showed a total of 34 components which accounted for 90.85 per cent of oil. It showed terpinen-4-ol (31.15 per cent), cis-sabinene hydrate (15.76 per cent), p-cymene (6.83 per cent), sabinene (6.91 per cent), trans-sabinene hydrate (3.86 per cent), α-terpineole (3.71 per cent) as the main constituents. The amounts of oxygenated monoterpenes were more in *O. majorana*. Cis-sabinene hydrate is a compound of intensive spicy "marjoramy" aroma whereas trans-sabinene hydrate has no typical "marjoramy" properties. Baser *et al.*, 1993 reported two chemotypes of *O.majorana*, the cis-sabinene hydrate/terpinen-4-ol chemotype and the carvacrol/thymol chemotype. Sarer *et al.*, 1982 reported carvacrol (65 per cent) and thymol (4 per cent) as the two most prominent components in *O. majorana* oil whereas terpinene-4-ol (37.1 per cent), p-cymene (12.05 per cent), α-terpineole (7.15 per cent), carvacrol (3.60 per cent), trans-sabinene hydrate (2.41 per cent), cis-sabinene hydrate (1.43 per cent) and thymol (0.7 per cent) were reported in oil from Greece (Komaitis *et al.*, 1992). Present studies showed *O. majorana* oil composition was found close to the one reported from South India (Mallavararpu *et al.*, 1991) except some minor variations. The oil quality produced was close to oil produced in Europe and Southern India. *Origanum* species have been used in medicine and as spices since antiquity mainly because of their essential oils, which consist of a considerable amount of carvacrol and thymol. The plant possesses thyme like flavor and is used for flavoring foods and beverages.

Oregano (*Origanum vulgare*)

O. vulgare essential oil showed 28 compounds accounting 94.64 per cent of the total oil. The major components identified were thymol (33.92 per cent), carvacrol (6.90 per cent), γ-terpinene (17.67 per cent), p-cymene (11.99 per cent). Both thymol and carvacrol constituted 40.82 per cent, for phenolic character. A sum of the two phenols and their precursors constituted the bulk of *O. vulgare* oil (70.48 per cent). Malestic *et al.*, 2000 reported thymol content of 49.3 per cent along with 24.6 per cent carvacrol in the oil of Mediterranean region. Several chemo types of *O. vulgare* have been reported from different places (Pande and Mathela, 2000; Kaul *et al.*, 1996). The chemical composition of oregano essential oils is of great importance due to its medicinal and biological properties. A survey of the literature reveals that *O. vulgare* ssp. hirtum possess principal components phenols, ρ -cymene and γ-terpinene. The ratio of thymol or carvacrol varies and is correlated. Russo *et al.*, 1998 reported that thymol biosynthesis pathway is favoured with altitude and essential oils rich in phenolic compounds are reported to possess high levels of antimicrobial activity.

Discussion

Chemical composition of essential oils from six commercially important aromatic herbs of Uttarakhand revealed the presence of important aroma constituents that find an amazing wide application in many industries for scenting and flavorings of all types of consumers finished products. They can be used in perfumery, cosmetics, pharmaceutical, food and flavour industry. Both the oils and their active constituents are of great commercial value. The analysis of different essential oils from plants grown at Uttarakhand region of India suggested that the climate and altitude affect the components and their percentage. The current surge of interest for herbal products in preference over the synthetic aroma products has changed the trade scenario much in favour of natural essential oil and aroma chemical. During the past few years there has been a market increase in the interest shown on many herbal spices, which is used to enhance many kinds of foods. Due to high biological activity, these aromatic plants are widely used these days in food preservation as well as in the cosmetic and pharmaceutical industry. Chemical characteristics and yield of an essential oil depend particularly on season and location of cultivation, variety and the developmental stage of the plants at harvest date, respectively. Therefore, for different uses aromatic plants may be grown at different geographic for various uses.

References

Adams, R.P. (1995). *Identification of essential oils by Ion Trap Mass Spectroscopy.* Carol Stream, IL, Allured Publishing Corporation.

Baser, K.H.C., Ozek, T., Tumen, G., and Sezik, E. (1993). Composition of the essential oil of Turkish *Origanum* species with commercial importance. *J. Essent. Oil Res.,* **5**: 619-623.

Boelens, M.H., and Boelens, H. (1997). Chemical and sensory evaluation of three sage oils. *Perfum. Flavor.,* **22**: 19-40.

Kaul, V.K., Singh, B., and Sood, R.P. (1996). Essential oil of *Origanum vulgare* L. from North India. *J. Essent. Oil Res.*, **8**: 101-103.

Komaitis, M.E., Ifanti-Papotragianmi, N., and Melissaari Panagiotou, E. (1992). Composition of the essential oil of marjoram (*Origanum vulgare* L.). *Food Chem.*, **45**: 117-118.

Kumar, N., Lohani, H., Lehari, A., and Diwivedi, K. N.(2004). Composition of oil from rosemary *(Rosmarinus officinalis)* grown in Kumaon hills of Uttarakhand. *Indian Perfumer*, **48**: 411-414.

Lawrence, B.M. (1996). Progress in essential oils- Savory oil. *Perfum. Flavor.*, **21**: 55-57.

Lawrence, B.M. (2001). Progress in essential oils- Rosemary oil. *Perfum. Flavor.*, **26**: 50-62.

Malestic, J., Milas, M., and Jerkovic, I.(2000). Essential oil and glycosidically bound volatiles of *Origanum vulgare* L. ssp. *hirtum* (Linkj) Ietswaart. *Flav. Fragr. J.*, **15**: 190-194.

Mallavarapu, G.R., Kulkarni, R.N., and Ramesh,S. (1991). Essential oil of *Majorana hortensis. Indian Perfumer*, **37**: 81-86.

Narayanan, P. (2003). Comparison of quality of essential oils from South India. *FAFAI J.*, **5**: 47-56.

Pande, C., and Mathela, C.S. (2000). Essential oil composition of *Origanum vulgare* L. ssp. *vulgare* from Kumaon Himalayas. *J. Essent. Oil Res.*, **12**: 441-442.

Piccaglia, R., and Marotti, M. (1993). Characterization of several aromatic plants grown in Northern Italy. *Flav. Fragr. J.*, **8**: 115-122.

Russo, M., Galletti, G., Bocchini, P., and Carnacini, A. (1998). Essential oil composition of wild populations of Italian Oregano spice (*Origanum vulgare* ssp. hirtum): A preliminary evaluation of their use in chemotaxonomy by cluster analysis.1. *J. Agri. Food Chem.*, **46**: 3741-3746.

Sarer, E., Scheffer, I.I.C., and Svaendsen, A. B. (1982). Monoterpenes in the essential oil of *Origanum majorana. Planta Medica*, **46**: 236-239.

Verma, R.S., Verma, R. K., Chauhan, A., and Yadav, A. K. (2011). Seasonal variation in essential oil content and composition of thyme, *Thymus serphyllum* L. cultivated in Uttarakhand hills. *Ind. J. Pharma. Sci.*, **73**: 233-235.

Utilisation and Management of Medicinal Plants Vol. 3 (2015) *Pages* **77–84**
Editors: **V.K. Gupta and A. Kaul**
Published by: **DAYA PUBLISHING HOUSE, NEW DELHI**

5

Identification of Constituents of Essential Oils of *Zanthoxylum nitidum* (Roxb.) DC. and *Zanthoxylum rhetsoides* Drake from Vietnam

Tran D. Thang[1]*, Do N. Dai[2]
and Isiaka A. Ogunwande[3]*

ABSTRACT

The chemical compounds identified from the essential oils of the leaves of two Zanthoxylum species (Euphorbiaceae) cultivated in Vietnam are being reported. The oils were obtained by hydrodistillation while the chemical constituents were analysed by means of gas chromatography (GC-FID) and gas chromatography/mass spectrometry (GC-MS). The major constituents of Zanthoxylum nitidum (Roxb.) DC were β-caryophyllene (16.8 per cent), germacrene D (9.7 per cent) and (E)-nerolidol (9.3 per cent). The leaf oil of Zanthoxylum rhetsoides Drake comprised mainly of ρ-mentha-1(7),3-diene (15.0 per cent), α-phellandrene (11.3 per cent), sabinene (10.1 per cent), (E)-β-ocimene (7.3 per cent) and α-cadinol (7.6 per cent).

Keywords: *Zanthoxylum nitidum*, *Zanthoxylum rhetsoides*, Essential oils, Monoterpenes, Sesquiterpenes.

1 Faculty of Chemistry, Vinh University, 182-Le Duan, Vinh City, Nghean Province, Vietnam.

2 Natural Products Research Unit, Department of Chemistry, Faculty of Science, Lagos State University, Badagry Expressway Ojo, P. M. B. 0001, Lasu Post Office, Ojo, Lagos, Nigeria

3 Faculty of Agriculture, Forestry and Fishery, Nghe An Economics University, Vinh City, Nghe An Province, Vietnam.

* *Corresponding author*: E-mail: thangtd@vinhuni.edu.vn; isiaka.ogunwande@lasu.edu.ng

Introduction

Zanthoxylum is the largest and most widespread genus in the Rutaceae family of plants, containing about 250 species native to warm temperate and subtropical areas worldwide. These species includes small and large trees, deciduous and evergreen trees and shrubs, most of them with spiny trunks (Negi *et al.,* 2011). Many species have been investigated for their non-volatile chemical constituents as well as biological and pharmacological potentials. In this paper, we report on the volatile constituents identified from the leaf oils of relatively poorly studied *Zanthoxylum* species growing in Vietnam. Our findings into the chemical volatile constituents of some Vietnamese flora have been recently published (Hieu *et al.,* 2014).

Zanthoxylum nitidum (Roxb.) DC., is a morphologically variable prickly shrub. It is aromatic. It is sometimes a climbing plant. The leaves are made up of several leathery oval leaflets which are up to 12 by 8 cm in size. Flowers, which occur in the leaf axils, have yellow-green petals a few millimeters long. The fruit is a red-brown follicle. This species is used medicinally for many purposes, including the treatment of toothache, fever, cough, vomiting, and diarrhea. It is also used as an insecticide, as piscicide and used medicinally for pain management and other purposes (Patiño *et al.,* 2012). The fruits are used as a condiment. It can also protect the stomach from peptic ulcers (Bhattacharya and Zaman, 2012) and acts as an antibacterial action (Bhattacharya *et al.,* 2009). Biologically active compounds such as analgesic and anti-inflammatory active alkaloids (Jiang *et al.,* 2006; Jiang *et al.,* 2007; Di *et al.,* 2009), benzenoids (Jih-Jung *et al.,* 2001) and cytotoxic compounds (Cheng-Hui *et al.,* 2009) have been isolated from the plant. Concerning its volatile constituents, linalool (23.3 per cent), limonene (12.9 per cent), α-terpineol (8.3 per cent), α-pinene (7.9 per cent) were the main compounds identified in the fruit oil, while limonene (33.1 per cent), geraniol (10.6 per cent) and carvone (9.6 per cent) occurred in higher proportion in the leaf (Bhattacharya and Zaman, 2009). The Australian oil of *Z. nitidum* contained β-caryophyllene (11.0 per cent), α-humulene (7.2 per cent), 2-tridecanone (7.7 per cent) and an unidentified 2-alkanone (10.7 per cent) as principal components (Brophy *et al.,* 2009).

Zanthoxylum rhetsoides Drake (Vietnamese: Cay sen la'to or Hoang moc nhieu gai) is a tree up to 10 m high, flowering white, growing in China, India, Malaysia, Myanmar and in the mountain regions of Northern and Central Vietnam. The dried fruit (Vietnamese: hat sen) is a preferred spice for beef [14)]. Monoterpenes were the major compounds of its fruit essential oils. These were limonene (21.7 per cent), ρ-cymene (15.3 per cent), β-phellandrene (11.9 per cent), α-pinene (8.2 per cent), sabinene (6.9 per cent) and terpinen-4-ol (6.0 per cent) from Vietnam sample (Weyerstahl *et al.,* 1992), while ρ-mentha-1(7),3-diene (30 per cent), β-phellandrene (11 per cent), linalool (11 per cent) and myrcene (10 per cent) occurred in Chinese sample (Liangfeng *et al.,* 1992).

Materials and Methods

Plants Collection

Leaves of *Z. nitidum* and *Z. rhetsoides* were collected from Pu Huong Nature reserve, Nghean Province, Vietnam, in August 2011. Voucher specimens HDT 805

HDT 263 respectively have been deposited at the Botany Museum Vinh University, Vietnam, for future references.

Isolation of the Oils

Air-dried and pulverized samples of each plant material (0.5 kg each) were subjected to hydrodistillation separately for 4 h at normal pressure according to Vietnamese Pharmaocopeia (1997).

Gas Chromatography (GC) Analysis

Gas chromatography (GC) was performed on an Agilent Technologies HP 6890 Plus Gas chromatograph equipped with a Flame ionization detector (FID) and fitted with HP-5MS columns (both 30 m x 0.25 mm, film thickness 0.25 µm, Agilent Technology, Berkshire, United Kingdom). The analytical conditions were: carrier gas H_2 (1 mL/min), injector temperature (PTV, programmed temperature vaporization injection) 250 °C, detector temperature 260 °C, column temperature programmed from 40 °C (2 min hold) to 220 °C (10 min hold) at 4 °C/min. Samples were injected by splitting and the split ratio was 10:1. The volume injected was 1.0 µL. Inlet pressure was 6.1 kPa. Each sample was analyzed thrice.

Gas Chromatography/Mass Spectrometry (GC-MS) Analysis

An Agilent Technologies HP 6890N Plus Chromatograph fitted with a fused silica capillary HP-5MST column (30 m x 0.25 mm, film thickness 0.25 µm) and interfaced with a Mass spectrometer HP 5973 MSD was used for the GC-MS analysis. The conditions were the same as described above with He (2 mL/min) as carrier gas. The MS conditions were as follows: ionization voltage 70 eV; emission current 40 mA; acquisitions scan mass range of 35-350 amu at a sampling rate of 1.0 scan/s.

Identification of the Constituents

The identification of constituents was performed on the basis of retention indices (RI) determined by co-injection with reference to a homologous series of *n*-alkanes (C_9-C_{24}), under identical experimental conditions. Mass spectra of our test oils constituents were compared with known spectra of molecules in NIST 08 Libraries (on ChemStation HP) and Wiley 9th Version and the home-made MS library built up from pure substances and components of known essential oils. The retention indices of all the compounds of the essential oils were also compared with literature values (Adams, 2007; Joulain and Koenig, 1998). The relative amounts of individual components were calculated based on the GC peak area (FID response) without using correction factors.

Results

The plant samples yielded low contents of essential oils: 0.18 per cent and 0.32 per cent (v/w), respectively, calculated on a dry weight basis for *Z. nitidum* and *Z. rhetsoides*. All the oil samples were light yellowish in colour. Table 5.1 gives the chemical compositions of volatiles oils of the three *Zanthoxylum* species analyzed by GC-FID and GC-MS. The identified compounds represent 97.6 per cent and 92.1 per cent of the total oil contents of *Z. nitidum* and *Z. rhetsoides* respectively. It could be

Table 5.1: Percentage Compositions of *Zanthoxylum* Species

Compounds[a]	*RI*[b]	*RI*[c]	*Z.n*	*Z.r*
α-Thujene	931	924	–	1.0
α-Pinene	939	932	3.0	3.9
Camphene	953	946	0.1	1.1
Sabinene	976	969	Tr	10.1
β-Pinene	980	974	0.1	0.6
Myrcene	990	988	0.4	2.5
(3*E*)-Hexen-1-yl acetate	1001	1001	0.3	–
α-Phellandrene	1006	1002	0.1	11.3
Δ-3-Carene	1011	1008	–	0.3
α-Terpinene	1017	1014	Tr	3.2
p-Cymene	1028	1020	–	2.2
β-Phellandrene	1031	1025	1.3	–
β-Terpinene [ρ-mentha-1(7),3-diene]	1036	1036	–	15.0
(*Z*)-β-Ocimene	1042	1042	–	1.0
(*E*)-β-Ocimene	1052	1044	0.5	7.3
γ-Terpinene	1061	1054	Tr	4.6
(*E*)-4,8-Dimethyl-1,3,7-nonatriene	1080	1089	0.1	0.1
α-Terpinolene	1090	1086	Tr	1.3
Linalool	1100	1095	0.3	0.3
n-Nonanal	1102	1100	0.1	–
allo-Ocimene	1128	1128	0.1	0.9
Geijerene	1142	1138	0.1	–
Camphor	1143	1141	0.5	1.4
trans-5-methyl-2-(1-methylethyl)-Cyclohexanone	1150	1142	0.3	0.1
1,3-Dimethoxy Benzene	1169	1165	0.4	–
Terpinen-4-ol	1177	1174	Tr	1.1
α-Terpineol	1186	1186	0.1	Tr
2-Undecanone	1297	1293	2.2	Tr
Bicycloelemene	1337	1338	5.0	0.5
α-Cubebene	1351	1345	0.2	Tr
Cyclosativene	1368	1369	–	0.2
α-Ylangene	1370	1373	0.3	–
α-Copaene	1377	1374	1.0	Tr
β-Cubebene	1388	1387	–	0.3
β-Elemene	1391	1389	3.6	0.5

Contd...

Table 5.1–*Contd...*

Compounds[a]	*RI*[b]	*RI*[c]	*Z.n*	*Z.r*
α-Gurjunene	1412	1409	0.5	–
β-Caryophyllene	1419	1417	16.8	3.8
β-Gurjunene	1431	1431	2.2	–
γ-Elemene	1437	1434	–	0.1
α-Guaiene	1437	1437	–	0.1
Aromadendrene	1441	1439	0.8	–
Selina-4(15),6-diene	1451	1450	–	0.2
α-Humulene	1454	1452	5.6	0.7
Germacrene D	1480	1484	9.7	1.9
α-Amorphene	1485	1483	0.7	0.2
β-Selinene	1489	1489	1.6	0.8
cis-Cadina-1,4-diene	1496	1495	–	0.1
α-Selinene	1498	1498	–	1.1
Epizonarene	1502	1501	0.2	–
(*E,E*)-α-Farnesene	1510	1505	1.9	–
γ-Cadinene	1514	1513	0.1	–
δ-Cadinene	1525	1522	4.0	1.0
Elemol	1548	1548	–	0.1
Germacrene B	1557	1559	5.8	–
(*E*)-Nerolidol	1564	1561	9.3	–
Spathulenol	1578	1577	2.6	0.4
Caryophyllene oxide	1583	1582	–	0.6
Viridiflorol	1596	1592	3.6	0.2
Ledol	1599	1602	0.6	1.0
β-Oplopenone	1608	1607	–	–
epi-α-Muurolol	1646	1640	1.8	1.6
Aromadendrene epoxide	1648	1639	0.2	–
β-Eudesmol	1651	1649	0.2	–
α-Cadinol	1653	1652	1.6	7.6
epi-α-bisabolol	1688	1683	–	0.2
cis-*Z*-α-Bisabolene epoxide	1704	1704	0.4	0.2
(*Z,E*)-Farnesol	1723	1722	0.8	0.1
Mint sulfide	1738	1740	0.1	–
Benzyl benzoate	1760	1759	0.1	–
6,10,14-Trimethyl-2-pentadecanone	1844	1847	0.2	–
ent-Pimara-8(14),15-diene	1939	1939	0.1	–

Contd...

Table 5.1–*Contd...*

Compounds[a]	*RI*[b]	*RI*[c]	*Z.n*	*Z.r*
n-Hexadecanoic acid	1962	1960	0.1	0.1
Phytol	2125	1942	6.0	0.1
(*E*)-Asarone	1675	1679	0.1	–
Total	97.6	92.1		
Monoterpene hydrocarbons	5.7	65.3		
Oxygenated monoterpenes	0.9	2.8		
Sesquiterpene hydrocarbons	60.2	11.5		
Oxygenated sesquiterpenes	21.0	12.1		
Diterpenes	6.0	0.1		
Others	3.8	0.3		

a: Compounds identified by RI from column, co-injection, literature MS pattern and literature retention indices; b: Retention indices on HP-5 MS capillary column; c: Literature retention indices.

Tr: Trace amounts < 0.1 per cent; –: Not identified; *Z.n*: *Zanthoxylum nitidum*; *Z.r*. *Zanthoxylum rhetsoides*.

seen that sesquiterpene hydrocarbons (60.2 per cent) and oxygenated counterparts (21.2 per cent) represent the class of compounds present in *Z. nitidum*. The major constituents of *Z. nitidum* were β-caryophyllene (16.8 per cent), germacrene D (9.7 per cent) and (*E*)-nerolidol (9.3 per cent). There are significant quantities of phytol (6.0 per cent), germacrene B (5.6 per cent), α-humulene (5.6 per cent), and bicycloelemene (5.0 per cent). However, monoterpenes compounds (65.3 per cent) represent the dominant class of compound in the present in *Z. rhetsoides*. The leaf oil of *Z. rhetsoides* comprised mainly of β-terpinene (15.0 per cent), α-phellandrene (11.3 per cent), sabinene (10.1 per cent), α-cadinol (7.6 per cent) and (*E*)-β-ocimene (7.3 per cent).

Discussion

Previous reports on the volatile constituents revealed an abundance of β-caryophyllene (24.6 per cent), γ-elemene (14.7 per cent) and bicyclogermacrene (12.9 per cent) in the leaf of *Z. nitidum* (Luong *et al.*, 2003). However, limonene (44.1 per cent), geranial (12.1 per cent), neral (10.9 per cent) and linalool (6.8 per cent) were the main compounds of the seed oil (Dung *et al.*, 1992). The abundance of β-caryophyllene and α-humulene makes the composition of the studied leaf oil similar to those from previous report (Luong *et al.*, 2003; Brophy *et al.*, 2009), but differs due to the absence of limonene, geraniol and carvone (Bhattacharya and Zaman, 2009); 2-tridecanone (Brophy *et al.*, 2009), as well as γ-elemene and bicyclogermacrene (Luong *et al.*, 2003) among its constituents. However, while sesquiterpenes compounds predominates in this leaf oil and previous ones (Luong *et al.*, 2003; Brophy *et al.*, 2009), monoterpenoid compounds predominate in the other reports from the leaf analysed from India (Bhattacharya and Zaman, 2009).

Monoterpenes (68.1 per cent) were the dominant class of compound in the present investigation of the leaf oil of *Z. rhetsoides*, previously reported for the fruit oils

(Liangfeng *et al.*, 1992; Weyerstahl *et al.*, 1992). However, the oil samples differed in their quantitative composition of these monoterpene compounds. The high content of β-terpinene and sabinene was in agreement with previous result on the fruit oils (Liangfeng *et al.*, 1992; Weyerstahl *et al.*, 1992). Its contents of α-pinene, myrcene, *p*-cymene, linalool, and terpinen-4-ol were low when compared with the fruit oils. However, the absence of β-phellandrene and limonene among the leaf volatile makes it different from previous results on the fruit volatiles (Liangfeng *et al.*, 1992; Weyerstahl *et al.*, 1992).

It is well known that the content and composition of essential oils depends on the origin, environmental conditions, phenological stage, plant parts analyzed, harvesting time and processing methods. Some of these factors may have been responsible for the observed variations in the oil contents when compared with other species from other parts of the world.

References

Adams, R.P. (2007). Identification of Essential Oil Components by Gas Chromatography/Quadrupole Mass Spectrometry. 4th Edition, Carol Stream. IL: Allured Publishing.

Bhattacharya, S. and Zaman, K. (2012). Protective effect of *Zanthoxylum nitidum* bark in chemical and stress induced gastric mucosal lesions in male albino rats. *International Journal of Pharmacoogy*, **8**: 450-54.

Bhattacharya, S., Zaman, M.K. and Haldar, P.K. (2009). Antibacterial activity of stem bark and root of Indian *Zanthoxylum nitidum*. *Asian Journal of Pharmacy and Clinical Research*, **2**: 30-34.

Bhattacharya, S. and Zaman, K. (2009)/Essential oil composition of fruits and leaves of *Zanthoxylum nitidum* grown in upper Assam region of India. *Pharmacogosy Research*, **1**: 148-151.

Brophy, J.J., Goldsack, R.J., Forster, P.I. and Hutton, I. (2009). Composition of the leaf oils of the Australian and Lord Howe Island species of *Zanthoxylum* (Rutaceae). *Journal of Essential Oil Research*, **12**: 285-291.

Cheng-Hui, Y., Ming-Jen, C., Shiow-Ju, L., Cheng-Wei, Y., Hsun-Shuo, C. and Ih-Sheng, C. (2009). Secondary metabolites and cytotoxic activities from the stem bark of *Zanthoxylum nitidum*. *Chemistry and Biodiversity*, **6**: 846-857.

Di, G., Ding-Xiang, L., Yao, S.H.I., Jing-Yu, L. and Zhi-Da, M. (2009). A new benzophenanthridine alkaloid from *Zanthoxylum nitidum*. *Chinese Journal of Natural Medicine*, **7**: 274-277.

Dung, N.X,; Nga, T.H. and Leclercq, P.A. (1992). Essential oil from the seed of *Zanthoxylum nitidum* DC. *Journal of Pharmacy (Vietnamese)*, **4**: 21-24.

Hieu, L.D., Thang, T.D., Hoi, T.M. and Ogunwande, I.A. (2014). Chemical composition of essential oils from four Vietnamese species of *Piper* (Piperaceae). *Journal of Oleo Science*, **63**: 211-217.

Jiang, H., Wei-Dong, Z., Run-Hui, L., Chuan, Z., Yun-Heng, S., Hui-Liang, L., Ming-Jin, L. and Xi-Ke, X. (2006). Benzophenanthridine alkaloids from *Zanthoxylum nitidum* (Roxb.) DC, and their analgesic and anti-inflammatory activities. *Chemistry and Biodiversity*, **3**: 990-995.

Jiang, H., Wei-Dong, Z., Yun-Heng, S., Chuan, Z., Lei, X., Run-Hui, L., Bin, W. and Xi-Ke, X. (2007). Alkaloids from *Zanthoxylum nitidum* (Roxb.) DC. *Biochemical Systematics and Ecology*, **35**: 114-117.

Jih-Jung, C., Yu-Hsien, L., Shiow-Hwa, D., Tsong-Long, H. and Ih-Sheng, C. (2001). New benzenoids and anti-inflammatory constituents from *Zanthoxylum nitidum*. *Food Chemistry*, **125**: 282-287.

Joulain, D. and Koenig, W.A. (1998). The Atlas of Spectral Data of Sesquiterpene Hydrocarbons. E.B. Verlag, Hamburg, Germany.

Liangfeng, Z., Li, Y., Li, B., Lu, B. and Xia, N. (1992). Aromatic Plants and Essential Oil Constituents, Hai Feng Publishing Company, Hongkong, pp. 74,.

Luong, N.X., Hac, L.V., Thang, T.D., Tung, L.V. and Nguyen, N.D. (2003). Essential oil of the leaves of *Zanthoxylum nitidum* DC. In: *Proceeding, The Tenth Asian Chemical Congress*, Ed. Nguyen N.D., Hanoi, Vietnam, pp. 143.

Negi, J.S., Bisht, V.K., Bhandari, A.K., Singh, P. and Sundriyal, R.C. (2011). Chemical constituents and biological activities of the genus *Zanthoxylum*: A review. *African Journal of Pure Applied Chemistry*, **5**: 412-416.

Patiro, L.O.J., Prieto, R.J.A. and Cuca, S.L.E. (2012). *Zanthoxylum* genus as potential source of bioactive compounds. *In*: Bioactive Compounds in Phytomedicine, Vol. I. Ed. By Rasooli, A, Elsevier, London.

Vietnamese Pharmacopoeia. (1997). Medical Publishing House, Hanoi, Vietnam, pp. 1-134.

Weyerstahl, P., Marschall, H., Splittgerber, U., Son, P.T., Giang, P.M. and Kaul, V.K. (1992). Constituents of the essential oil from the fruits of *Zanthoxylum rhetsoides* Drake from Vietnam and from the aerial parts of *Zanthoxylum alatum* Roxb. from India. *Flavour Fragrance Journal*, **14**: 225-229.

Utilisation and Management of Medicinal Plants Vol. 3 (2015) *Pages* **85–92**
Editors: **V.K. Gupta and A. Kaul**
Published by: **DAYA PUBLISHING HOUSE, NEW DELHI**

6

Essential Oil Constituents of Vietnamese Species of *Fissistigma bracteolatum* Chatt. and *Fissistigma chloroneurum* (Hand.-Mazz.) Tsiang

Tran D. Thang[1]*, Do N. Dai[2] and
Isiaka A. Ogunwande[3]*

ABSTRACT

In the present investigation, we studied the essential oil compositions of the leaves and stem barks of two Fissistigma plants cultivated in Vietnam. The air-dried plant samples were hydrodistilled and the oils analysed by means of gas chromatography (GC-FID) and gas chromatography/mass spectrometry (GC-MS). The leaf of Fissistigma bracteolatum Chatt afforded oil whose major constituents were β-myrcene (17.0 per cent), β-caryophyllene (7.4 per cent), Δ-cadinene (6.9 per cent), bicyclogermacrene (6.2 per cent) and α-pinene (5.6 per cent). However, β-caryophyllene (20.0 per cent), Δ-cadinene (9.9 per cent), α-humulene (6.4 per cent), bicyclogermacrene (6.3 per cent) and linalool (6.2 per cent) were the principal components of the stem. On the other hand, while germacrene D (20.9 per cent), Δ-elemene (18.9 per cent), β-gurjunene (11.0 per cent), valencene (10.2 per cent) and β-elemene (9.3 per

---

1 Faculty of Chemistry, Vinh University, 182-Le Duan, Vinh City, Nghean Province, Vietnam.

2 Faculty of Agriculture, Forestry and Fishery, Nghe An Economics University, Vinh City, Nghe An Province, Vietnam.

3 Natural Products Research Unit, Department of Chemistry, Faculty of Science, Lagos State University, Badagry Expressway Ojo, P.M.B. 0001, Lasu Post Office, Ojo, Lagos, Nigeria.

* *Corresponding author*: E-mail: thangtd@vinhuni.edu.vn; isiaka.ogunwande@lasu.edu.ng

cent) were the main compounds identified in the leaf oil of Fissistigma chloroneurum (Hand.-Mazz.) Tsiang, the stem comprised mainly of valencene (34.0 per cent), β-selinene (10.6 per cent) and germacrene D (8.7 per cent).

Keywords: *Fissistigma bracteolatum, Fissistigma chloroneurum,* Essential oil composition, Monoterpenes, Sesquiterpenes.

Introduction

Fissistigma is a genus of plant in the family Annonaceae. It consists of about 80 species which are widely distributed from east India to north-east Australia (Kessler and van Heusden, 1993; Ban, 2000). As part of our continuing interest in the chemical analysis of Vietnamese flora (Thang *et al.*, 2014), this article reports the chemical compositions of essential oils from two *Fissistigma* species. *Fissistigma bracteolatum* Chatt., is a scandent evergreen shrub which grows up to 10 m long. The leaves are ovate-oblong in shape. Flowering takes place from March to June while fruiting occurs between August and November. This herb is used for the treatment of contusions and strains (Li and Gilbert, 2001). Phytochemical studies have revealed that *F. bracteolatum* contains chalcones and dihydrochalcones with potent inhibitory effects against superoxide anion production (Lien *et al.*, 2000; Yu *et al.*, 2005; Zhu *et al.*, 2010; Wu *et al.*, 2013), phenanthrene (Zhu *et al.*, 2010), flavonoids (Porzel *et al.*, 2000) and alkaloids (Deng *et al.*, 2002).

Fissistigma chloroneurum (Handel-Mazzetti) Tsiang, is a climber with papery and oblong leaves. The fruits are nearly spherical while the seeds are glabrous and hairless. Flowering occurs from March to November and fruiting takes place from July to December (Li and Gilbert, 2001). Phytochemical studies on the leaves of *F. chloroneurum* collected in Vietnam, yielded three compounds, stearic acid, geranylfamesol, β-sitosterol (stigmast-5-ene-3-ol) and stigmasterol (Thuy *et al.*, 2005). The hydrodistillation of the leaf of the plant yielded oil in which the major components were (*E*)-α-bergamotene (52.0 per cent), (*E*)-β-caryophyllene (9.0 per cent) and β-bisabolene (6.0 per cent), mainly sesquiterpenes (Höferl *et al.*, 2012).

Materials and Methods

Plants Collection

Mature leaves and stem barks of *F. bracteolatum* and *F. chloroneurum* collection were collected from the Natural Reserve at Dac Nông Province, Vietnam, in October 2013. Botanical identification was made by Curators at Herbarium of Institute Ecology and Biological Resources, Vietnam Academy of Science and Technology, Vietnam where voucher specimens BVT 129 and BVT 133, respectively were deposited.

Extraction of the Oils

0.5 Kg of air-dried and pulverized plant samples were subjected to separate hydrodistillation for 3 h at normal pressure, according to the Vietnamese Pharmacopoeia (1997). The yields obtained were 0.25 per cent and 0.21 per cent (v/w, *F. bracteolatum*; leaf and stem respectively) and 0.15 per cent and 0.1 per cent (v/w,

F. chloroneurum; leaf and stem respectively) calculated on a dry weight basis. Oil samples of *F. bracteolatum* were colourless while those of *F. chloroneurum* were light yellow coloured.

Gas Chromatography (GC) Analysis of the Oils

Gas chromatography (GC) analysis was performed on an Agilent Technologies HP 6890 Plus Gas chromatograph equipped with a FID and fitted with HP-5MS column (30 m x 0.25 mm, film thickness 0.25 µm, Agilent Technology). The analytical conditions were: carrier gas H_2 (1 mL/min), injector temperature (PTV) 250 °C, detector temperature 260°C, column temperature programmed from 40 °C (2 min hold) to 220°C (10 min hold) at 4°C/min. Samples were injected by splitting and the split ratio was 10:1. The volume injected was 1.0 µL. Inlet pressure was 6.1 kPa. Each sample was analyzed thrice.

Gas Chromatography-Mass Spectrometry (GC-MS) Analysis

An Agilent Technologies HP 6890N Plus Chromatograph fitted with a fused silica capillary HP-5 MS column (30 m x 0.25 mm, film thickness 0.25 µm) and interfaced with a mass spectrometer HP 5973 MSD was used for the GC/MS analysis, under the same conditions as those used for GC analysis. The conditions were the same as described above with He (1 mL/min) as carrier gas. The MS conditions were as follows: ionization voltage 70 eV; emission current 40 mA; acquisitions scan mass range of 35-350 amu at a sampling rate of 1.0 scan/s.

Identification of the Constituents

The identification of constituents was performed on the basis of retention indices (RI) determined by co-injection with reference to a homologous series of *n*-alkanes, under identical experimental conditions. Further identification was performed by comparison of their mass spectra with those from NIST 08 Libraries (on ChemStation HP) and Wiley 9th Version and the home-made MS library built up from pure substances and components of known essential oils, as well as by comparison of their retention indices with literature values (Adams 2007; Joulain and Koenig 1998).

Results

Table 6.1 displayed the identities of the compounds and the percentage composition present in both oil samples. The main constituents in the leaf oil of *F. bracteolatum* were β-myrcene (17.0 per cent), β-caryophyllene (7.4 per cent), Δ-cadinene (6.9 per cent) and bicyclogermacrene (6.2 per cent). There were significant amounts of α-pinene (5.6 per cent), β-elemene (4.7 per cent), Δ-elemene (4.4 per cent) and α-cadinol (4.3 per cent). On the other hand β-caryophyllene (20.0 per cent), Δ-cadinene (9.9 per cent), α-humulene (6.4 per cent), bicyclogermacrene (6.3 per cent) and linalool (6.2 per cent) were the major components of the stem oil. It also featured sizeable quantities of γ-elemene (5.0 per cent), bicycloelemene (4.7 per cent), α-pinene (4.2 per cent) and benzyl benzoate (4.0 per cent).

Monoterpene compounds were less common in both the leaf and stem oils of *F. chloroneurum*. Sesquiterpene compounds comprising of germacrene D (20.9 per cent), Δ-elemene (18.9 per cent), β-gurjunene (11.0 per cent), valencene (10.2 per cent) and β-

Table 6.1: Chemical Constituents of Essential Oils of *F. bracteolatum* and *F. Chloroneurum*[a]

Compounds[a]	*RI*[c]	*F. bracteolatum*		*F. chloroneurum*	
		Leaf	*Stem*	*Leaf*	*Stem*
α-Pinene	939	5.6	4.2	–	0.3
Camphene	953	0.5	0.3	–	–
Verbenene	962	0.3	–	–	–
Sabinene	976	–	–	–	0.2
β-Pinene	980	0.9	0.3	–	0.1
β-Myrcene	990	17.0	0.3	0.1	0.1
α-Phellandrene	1006	0.2	1.4	–	0.1
Δ-3-Carene	1011	0.1	–	–	0.1
α-Terpinene	1017	0.1	1.8	0.1	Tr
o-Cymene	1024	–	0.1	–	–
β-Phellandrene	1028	–	–	–	0.2
Limonene	1032	3.2	–	0.1	–
(*Z*)-β-Ocimene	1043	0.1	0.1	–	–
(*E*)-β-Ocimene	1052	0.3	1.3	0.1	0.1
γ-Terpinene	1061	0.1	0.1	–	Tr
α-Terpinolene	1090	0.3	0.2	–	Tr
Linalool	1100	0.5	6.2	0.1	Tr
(*E*)-4,8-Dimethyl-1,3,7-nonatriene	1110	–	0.1	–	–
Methyl salicylate	1197	–	0.1	–	–
Methyl chavicol	1203	0.3	–	–	0.2
Safrole	1287	–	1.5	–	–
Bornyl acetate	1289	0.4	–	–	–
Bicycloelemene	1327	0.6	4.7	1.2	0.2
Δ-Elemene	1340	4.4	–	18.9	4.3
α-Cubebene	1351	2.7	0.3	0.1	–
Neryl acetate	1362	0.3	–	–	–
Cyclosativene	1371	–	0.1	–	0.2
α-Ylangene	1375	–	1.2	–	–
α-Copaene	1377	1.5	3.2	0.7	1.4
Geranyl acetate	1381	0.1	–	–	–
β-Cubebene	1388	1.0	–	6.0	1.2
β-Elemene	1391	4.7	1.2	9.3	1.1
Cyperene	1399	3.7	–	–	0.1
α-Gurjunene	1412	0.2	0.2	0.2	–
trans-α-Bergamotene	1413	–	–3	–	0.2

Contd...

Table 6.1–*Contd...*

Compounds[a]	*RI*[c]	*F. bracteolatum*		*F. chloroneurum*	
		Leaf	*Stem*	*Leaf*	*Stem*
β-Caryophyllene	1419	7.4	20.0	–	3.5
β-Gurjunene	1434	0.4	–	11.0	0.9
γ-Elemene	1437	–	5.0	–	–
α-Guaiene	1440	–	0.4	–	–
Aromadendrene	1441	–	0.4	–	0.1
Isoeugenol	1451	–	0.7	–	–
α-Humulene	1454	2.9	6.4	5.0	3.2
3,7-Guaiadiene	1461	0.6	–	–	–
β-Agarofuran	1474	–	–	–	2.6
γ-Gurjunene	1477	1.7	–	–	–
γ-Selinene	1484	0.2	–	–	–
Germacrene D	1485	2.9	0.6	20.9	8.7
α-Amorphene	1485	–	1.3	–	–
β-Selinene	1486	0.9	0.9	–	10.6
epi-Bicyclosesquiphellandrene	1489	2.1	–	–	–
Eudesma-4,11-diene	1490	–	0.7	–	–
Valencene	1490	1.2	1.2	10.2	34.0
β-Guaiene	1491	1.3	–	–	–
Zingiberene	1494	0.4	1.7	–	–
Cadina-1,4-diene	1496	0.9	–	–	0.2
Bicyclogermacrene	1500	6.2	6.3	–	–
α-Muurolene	1500	–	–	0.3	5.0
Epizonarene	1502	–	–	–	0.2
β-Bisabolene	1506	–	–	0.5	0.3
γ-Cadinene	1514	1.1	1.3	–	–
α-Panasinsene	1518	–	–	–	6.0
Δ-Cadinene	1525	6.9	9.9	3.3	–
Eudesma-4 (15), 11-dien-9-one	1543	–	0.5	–	–
Calacorene	1546	–	–	–	0.1
Selina-3, 7(11)-diene	1547	0.5	0.9	–	–
Guaia-3,9-diene	1556	0.4	–	–	–
Germacrene B	1561	–	–	0.6	–
(*E*)-Nerolidol	1563	0.8	2.4	0.2	0.6
Ledol	1573	–	–	0.5	0.6
Spathulenol	1578	0.6	0.3	1.4	0.6

Contd...

Table 6.1–*Contd...*

Compounds[a]	*RI*[c]	*F. bracteolatum*		*F. chloroneurum*	
		Leaf	*Stem*	*Leaf*	*Stem*
Caryophyllene oxide	1583	–	1.6	–	1.1
Globulol	1585	–	–	–	1.2
Viridiflorol	1593	0.7	0.4	1.3	–
Longiborneol	1599	0.3	–	–	1.5
β-Oplopenone	1608	–	–	0.5	0.6
α-Copaene-8-ol	1633	–	–	0.8	–
Isospathulenol	1636	–	–	1.1	–
Hinesol	1640	–	–	–	1.6
α-Bisabolol	1645	–	–	0.2	0.2
t-Muurolol	1646	–	–	0.3	–
β-Eudesmol	1651	–	–	0.3	–
α-Cadinol	1654	4.3	1.9	0.6	2.7
1,6-dimethyl-4-(1-methylethyl)-Naphthalene	1671	0.5	–	–	–
Juniper camphor	1691	–	0.5	–	–
Mint sulfide	1741	–	–	–	0.6
Benzyl benzoate	1760	0.9	4.0	0.4	0.4
Azulone	1772	–	0.3	–	–
Phytol	1942	1.0	0.4	0.3	0.6
Total	95.2	98.9	96.6	98.2	
Monoterpene hydrocarbons	28.7	10.1	0.4	1.2	
Oxygenated monoterpenes	1.6	8.5	0.1	0.2	
Sesquiterpene hydrocarbons	55.8	68.4	88.2	85.1	
Oxygenated sesquiterpenes	6.7	7.1	7.2	10.7	
Diterpenes	1.0	0.4	0.3	0.6	
Others	1.4	4.4	0.4	0.4	

a: SD, Standard deviation, values were not significant and therefore excluded from the Table to avoid congestion; b: Elution order on HP-5MS column; c: Retention indices on HP-5MS column.

Tr: Trace amount, < 0.1 per cent; –: Not identified.

elemene (9.3 per cent) were the main compounds identified in the leaf oil of *F. chloroneurum*. Other significant ones were β-cubebene (6.0 per cent) and α-humulene (5.0 per cent). However, valencene (34.0 per cent), β-selinene (10.6 per cent) and germacrene D (8.7 per cent) were also the major sesquiterpene hydrocarbon compounds of the stem oil. Other notable compounds were α-panasinsene (6.0 per cent), α-muurolene (5.0 per cent) and Δ-elemene (4.3 per cent).

Discussion

Previous analysis of essential oils of *F. bracteolatum* afforded oils whose major constituents were *epi*-α-muurolol (17.8 per cent), caryophyllene oxide (13.1 per cent) and bicycloelemene (10.5 per cent) in the leaf, while β-bourbonene (13.6 per cent), (*E*)-β-ocimene (12.6 per cent) and (*E,E*)-α-farnesene (11.2 per cent) were present in the stem (Thang *et al.*, 2014). However, myrcene (83 per cent) was the main compound in another analysis (Thang and Dung, 2007). Although sesquiterpene compounds predominated in this result, both qualitative and quantitative variations were observed amongst the analyses. The oil of *F. bracteolatum* may be thought to exist in two chemical forms, one with high monoterpene content (Thang and Dung, 2007) and other with dominance of sesquiterpenes (Thang *et al.*, 2014; this study). Moreover, (*E*)-β-ocimene, bicycloelemene and caryophyllene oxide were identified in low amounts while β-bourbonene (13.6 per cent), (*E,E*)-α-farnesene and *epi*-α-muurolol were conspicuously absent in this investigated oil samples.

On the other hand, the main compounds of previously analysed oil of *F. chloroneurum* were either present in low quantity *i.e.* (*E*)-α-bergamotene and (*E*)-β-caryophyllene or could not be detected in this analysed oil (*i.e.* β-bisabolene). Most notable fact is that sesquiterpenes were the main class of compounds in this and previous study (Höferl *et al.*, 2012).

The observed compositional differences between the oil samples may due to factors such as the age of the plant, genetic, handling conditions, time of collection and environmental and climatic conditions between the various points of analysis.

Acknowledgements

The authors wish to thank the NAFOSTED (Vietnam) for the financial support of this study through the Project No. 104.01-2012.18.

References

Adams, R.P. (2007). Identification of Essential Oil Components by Gas Chromatography/Quadrupole Mass Spectrometry. 4th Edition, Carol Stream. IL: Allured Publishing.

Ban. T.N. (2000). *Annonaceae*. In: Ban, T.N., Editor. Flora of Vietnam. Vol. 1. Ho Na (Annonaceae). Hanoi: Publishing House of Science and Technology; p. 102.

Deng, Y., Chen, J., and Wu, F.E. (2002). Two new aporphine alkaliods from *Fissistigma bracteolatum*. *Chinese Chemistry Letter*, **13**: 862-864.

Hoeferl, M., Dai, D.N., Thang, T.D., Jirovetz, L., and Schmidt, E. (2013). Leaf essential oils of six Vietnamese species of *Fissistigma* (Annoanceae). *Natural Product Commununications*, **8**: 663-665.

Joulain, D., and Koenig, W.A. (1998). The Atlas of Spectral Data of Sesquiterpene Hydrocarbons. E.B. Verlag, Hamburg, Germany.

Kessler, P.J., and van Heusden, E.C.H. (1993). The Annonaceae of the Balikpapan Samarinda area, East Kalimantan, Indonesia. *Rheedea*, **3**: 63-65.

Li, B., and Gilbert, M.G. (2001). Annonaceae. In: Flora of China, Vol. Wu, Z., Raven, P.H. and Hong, D. (eds.) Science Press, Beijing and Missouri Botanical Garden Press, St Louis pp. 672-675, 708.

Lien, T.P., Porzel, A., Schmidt, J., Sung, T.V., and Adam, G. (2000). Chalconoids from *Fissistigma bracteolatum*. *Phytochemistry*, **53**: 991-995.

Porzel, A., Phuong, L.T., Schmidt, J., Drosihn, S., Wagner, C., Merzweiler, K., Van Sung, T., and Adam, G. (2000). Fissistigmatins A-D. Novel type natural products with flavonoid-sesquiterpene hybrid structure from *Fissistigma bracteolatum*. *Tetrahedron*, **56**: 865-872.

Thang, T.D., and Dung, NX. (2007). Progress in the study of some *Fissistigma* species from Vietnam. In: Jirovetz, L., Nguyen, X.D. and Varshney, V.K. editors. Aromatic plants from Asia, their chemistry and application in food and therapy. Dehradun, India: Har Krishan Bhalla and Sons; p. 119–126.

Thang, T.D., Luu, H.V., Dung, V.C., Tuan, N.N., Hung, N.H., Dai, D.N., and Ogunwande, I.A. (2014) Chemical constituents of essential oils from the leaves and stem barks of four Vietnamese species of *Fissistigma* (Annonaceae). *Natural Product Research*, **28**: 174-184.

Thuy, T.T., Tam, N.T.T., Anh, N.T., and Sung, T.V. (2005). First results of a study in chemical *Fissistigma chloroneurum* (Hand-Mazz) Tsiang. *Journal of Medicinal Materials Hanoi*, **10**: 49-51.

Vietnamese Pharmacopoeia. (1997). Medical Publishing House, Hanoi, Vietnam, pp. 1-134.

Wu, Y.C., Sureshbabu, M., Fang, Y.C., Wu, Y.H., Lan, Y.H., Chang, F.R., Chang, Y.W., and Hwang, T.L. (2013). Potent inhibition of human neutrophil activations by bractelactone, a novel chalcone from *Fissistigma bracteolatum*. *Applied Toxicology and Pharmacology*, **266**: 399-407.

Yu, H.L., Chen, C.Y., Rong, C.F., Long, H.T., Chaung, L.C., and Chang, W.Y. (2005). Potential antiinflammatory activities of bractelactone and other compounds isolated from *Fissistigma bracteolatum*. *Helvetica Chimica Acta*, **88**: 905-909.

Zhu, H., Xiaoling, L., Xiaohong, S., Qiangzhi, X., and Binghua, J. (2010). Dihydrochalcones and phenanthrene derivatives from *Fissistigma bracteolatum*. *Journal of Medical Colleg PLA.*, **25**: 226-234.

Utilisation and Management of Medicinal Plants Vol. 3 (2015) *Pages* **93–119**
Editors: **V.K. Gupta and A. Kaul**
Published by: **DAYA PUBLISHING HOUSE, NEW DELHI**

7

Ethnopharmacological Approaches to Wound Healing: Exploring Medicinal Weeds

Puja Ray[1*], Abhijit Dey[1], Souryadeep Mukherjee[1], A. Balakrishnan[2] and S.V. Lakshmi[2]

ABSTRACT

Over the centuries, societies around the world have developed their own traditional methods to make use of medicinal plants for their pain-relieving and healing abilities. Although some of those traditions and medicinal practices may seem unscientific, all of them attempted to overcome pain and suffering with an aim to enhance the quality of life. Many of the plant species contain active constituents that have a direct pharmacological action on the human systems. However, the human body is so much complex that only a very few drugs are approved by regulatory authorities for treating traumatic injuries. The ancient oriental system of medicine has identified the medical potency of some of the herbs generally consider unwanted (weeds) for wound healing purposes. These weeds which are easily traced and abundantly available are used both in herbal and conventional medicine and offer benefits that synthetic drugs lack. The focus of this review is on 12 different, widely researched weeds Portulaca oleracea L., Centella asiatica L. urban, Tridax procumbens L., Desmodium triflorum L., Cynodon dactylon L., Ageratum conyzoides L., Chromolaena odorata L., Heliotropium indicum L.,

1 Department of Biological Sciences, Presidency University, 86/1 College Street, Kolkata – 700 073, West Bengal, India.

2 Amirta Center for Nanomedicine and Molecular Medicine, AIMS, Ponnekkara-41, Kerala, India.

* *Corresponding author*: E-mail: puja.dbs@presiuniv.ac.in

Hemidesmus indicus R. Br., Tragia involucrata L., Hypericum patulum Thumb and Hypericum perforatum L., reputed for their tissue regeneration and wound healing ability. The objective of the present article is to keep abreast with the medicinal plants used in various aspects of wound healing.

Keywords: Medicinal weeds, Wound healing, Tissue regeneration, Drugs.

Introduction

Over the decades, investigators have been using a variety of approaches to study the fundamental biology of wound repair and tissue regeneration. Wound repair or healing, is the body's natural process of regenerating dermal and epidermal tissue comprising of a sequence of closely complex coordinated biochemical events (Phillips *et al.,* 1991). Wound healing is initiated by a stimulus of injury to the tissues and passes through four different phases (Table 7.1) which over lap in time (Baie and Sheikh, 2000a; Baie and Sheikh, 2000b; Bonner, 2000; Ross, 1980). Chronic wounds could be arrested in any of these phases but commonly disruption occurs in the inflammatory or proliferative phase with the accretion of excessive extracellular matrix and matrix metaloproteinases, such as collagenase and elastase, which results in premature degradation of collagen and growth factors. The abrasion could become more serious in absence of optimal healing environment. The presence of cytotoxic factors *i.e.*, infection, ischaemia, presence of foreign bodies, tissue maceration, etc can seriously delay the healing process (Phillips *et al.,* 1991). Hence, wound dressings play an important role in providing the required protection against these cyto-toxic factors. These dressings in combination with effective medicine modify the physiology of the wound environment thereby fastening the healing process (Udupa *et al.,* 1995).

Table 7.1: Phases of Wound Healing (Phillips *et al.*, 1991)

Phase of Healing	*Period Post Injury*	*Cells Involved in Phase*
Hemostasis	Immediate	Platelets
Inflammation		Neutrophis
Proliferation	1-4 days	Macrophages
Granulation		Lymphocytes
		Anglocytes
		Neurocytes
Contracture	4-21 days	Fibroblasts
		Keratinocytes
Remodeling	21days- 2 years	Fibrocytes

Oriental medicinal practice like Ayurveda has shown that many of the plant derived products have remarkable capability for wound healing and tissue regeneration (Dhar *et al.,* 2002; Padma, 2005) when applied as dressings. According to WHO (World Health Organization), over 80 per cent of the people in developing countries, who have limited access to the modern medicine depend on plant based medicines (Trivedi, 2006). The medicinal value of plants increases in the presence of certain intrinsic chemical substances which on interaction with human body, produce specific physiological action. Recent studies have shown that some of the important medicinal properties might be found in less exotic plants whose virtues have not yet been discovered and generally considered unwanted, commonly known as weeds (Kaushik., 1988; Nadkarni, 1976). Weeds have a controversial nature. They are considered to be competitive, persistent, pernicious, and interfere negatively with human interest (Stepp, 2004). To the agriculturist, they are plants out of place that need to be controlled, in an economical and practical way. In this context, the negative impacts of weeds directly affect all living beings. However due to their easy and abundant availability, some known weeds have been put to certain medicinal uses since ancient times (Izzo and Ernest, 2001; Jain, 1968), but medicinal virtues of such plants are overshadowed due to lack of literacy of common man and due to their weedy nature. It is ironical that even after possessing such rich heritage of medicinal fauna around us the vast majorities of currently available wound healing drugs are synthetic and are not derived from bio-active constituents of these medicinal plants (Geldenhuys, 2007).

In a systematic study published during 1997, it was evaluated that 157 of 520 drugs (30 per cent) approved by Food and Drug Administration (FDA) in the USA during 1983-94 (11 years) were natural products or their derivatives (Cragg *et al.*, 1997). This report revealed that when focused efforts are made to discover natural products for clinical use, the success level rises dramatically. Thus during the same period, 61 per cent of anticancer agents approved were natural products or their derivatives. In the absence of targeted studies, there were no analgesics, antidepressants, anxiolytics or other wound healing active drugs derived from natural products which were approved during the 11 year time period analyzed (Bajpai *et al.*, 2005; Hussain *et al.*, 2007, Salas *et al.*, 2005). The current scenario has not changed much. Although identification of hits and leads from secondary plant metabolites continue to be a major goal of many drug discovery projects, most such efforts do not target for the search of agents potentially useful for wound healing activity. In addition, few reports on wound healing activities of herbal extracts and their active constituents that do regularly get published are, in general, not subsequently evaluated adequately in terms of their potential for identifying structurally or functionally novel wound healing drugs. The main goals of such reports continue to be the evaluation of traditionally known wound healing herbal remedies or to identify their active constituents. In general, main objectives of these reports have been to generate more evidence based knowledge justifying their conventionally known therapeutic uses. However, in view of the facts that clinical efficacy of several therapeutically active herbal extract based remedies have been repeatedly demonstrated in properly controlled clinical trials and that their efficacy can not be easily interpreted in terms of current concepts of pharmacology, they deserve proper attention as readily available sources for structurally and functionally novel classes of wound healing drugs. The present review investigates 12 different reputed weeds used for wound treatment purposes. Table 7.2 briefly describes the pharmacodynamics of these weeds. Figure 7.1 depicts the active components responsible for wound healing isolated from the

Table 7.2: Medicinal Weeds and their Pharmacodynamics

Sl.No.	*Scientific Name*	*Common Name*	*Family*	*Pharmacodynamics*
1.	*Portulaca oleracea* (L.)	Purslane, Pigweed	Portulaceae	Anti-inflammatory, epithelization and antimicrobial
2.	*Centella asiatica* (L.) Urban	Asiatic Pennywort	Apiaceae	Antioxidant, epithelization and collagenization
3.	*Tridax procumbens*	Coatbuttons	Compositae	Burn wound epithelization
4.	*Desmodium triflorum* (L.)	Three-flower beggarweed	Fabaceae	Analgesic, anti-inflammatory, antifungal
5.	*Cynodon dactylon* (L.)	Dhub grass	Poaceae	Effective coagulant
6.	*Ageratum conyzoides* (L.)	Goatweed	Asteraceae	Antispasmotic and antibacterial
7.	*Chromolaena odorata* (L.)	Siam weed,	Asteraceae	Epithelization, antioxidant, antimicrobial and anticoagulation
8.	*Heliotropium indicum* (L.)	Heliotrope	Boraginaceae	Anti-inflammatory, epithelization and collagenization
9.	*Hemidesmus indicus* R. Br.	Indian Sarsaparilla	Asclepiadaceae	Antioxidant and antiulcerogenic
10.	*Tragia involucrata* (L.)	Indian stinging nettle	Euphorbiaceae	Antibacterial, neovascularization, granulation, diaphoretic, analgesic and anti-inflammatory
11.	*Hypericum patulum* Thumb	Goldencup St. John's wort	Hypericaceae	Epithelization and collagenization
12.	*Hypericum perforatum* (L.)	Perforate St. John's wort	Hypericaceae	Anti-depressant, epithelization and collagenization

Figure 9.1: Active Components for Wound Healing from Plants.

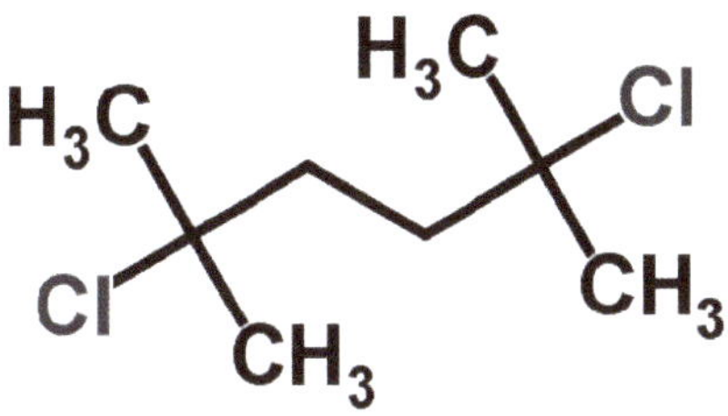

2, 4-dimethyl hexane

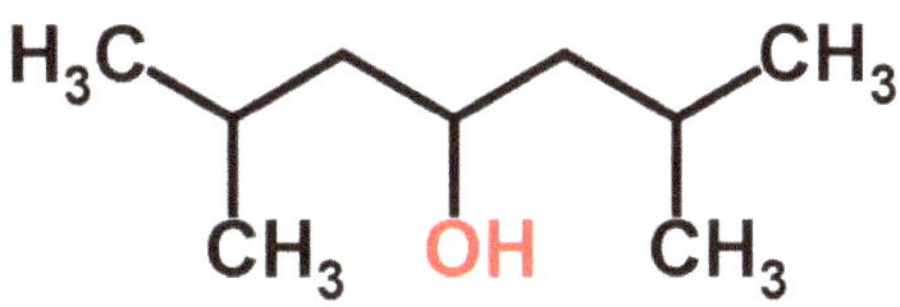

2, 6-dimethyl heptanes

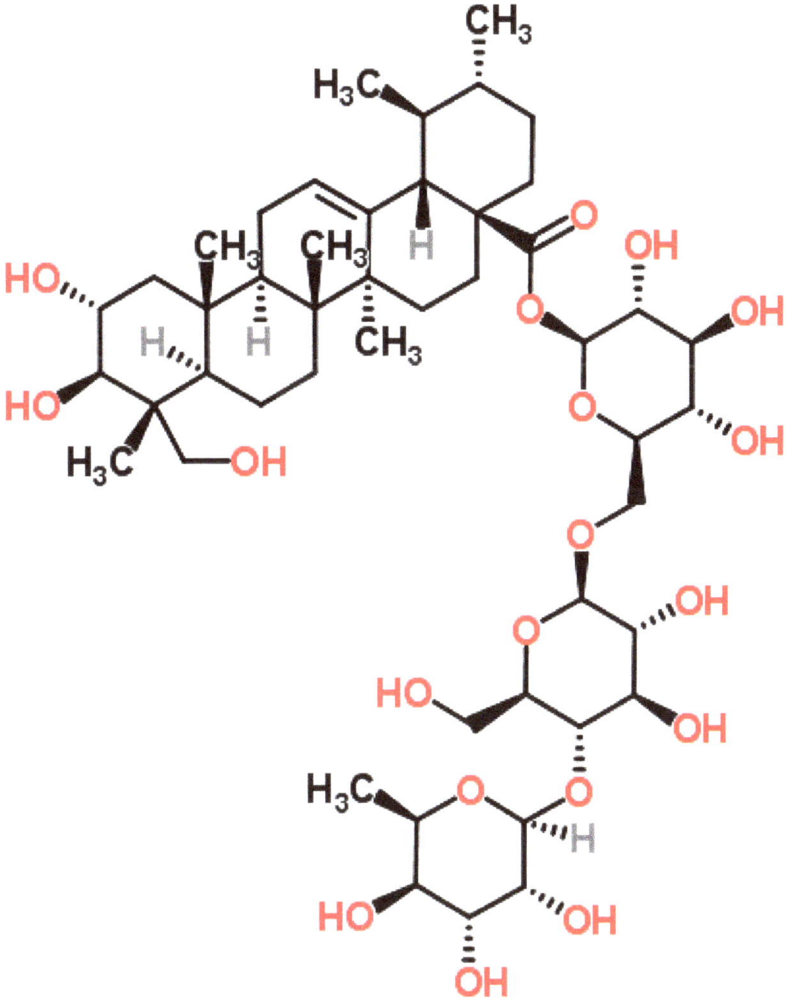

Asiatic acid

Contd...

Figure 9.1–*Contd...*

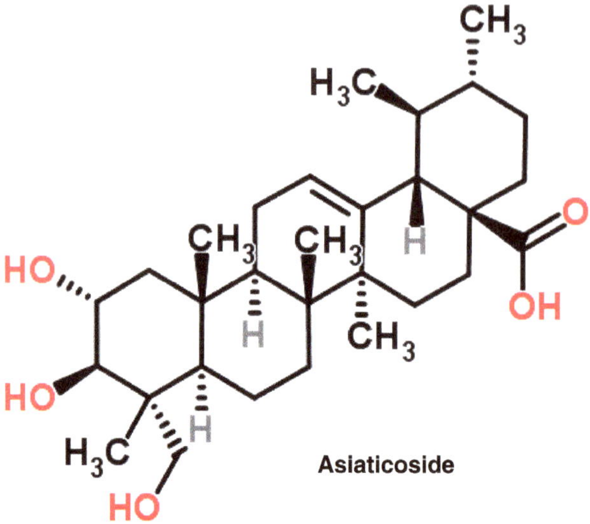

Asiaticoside

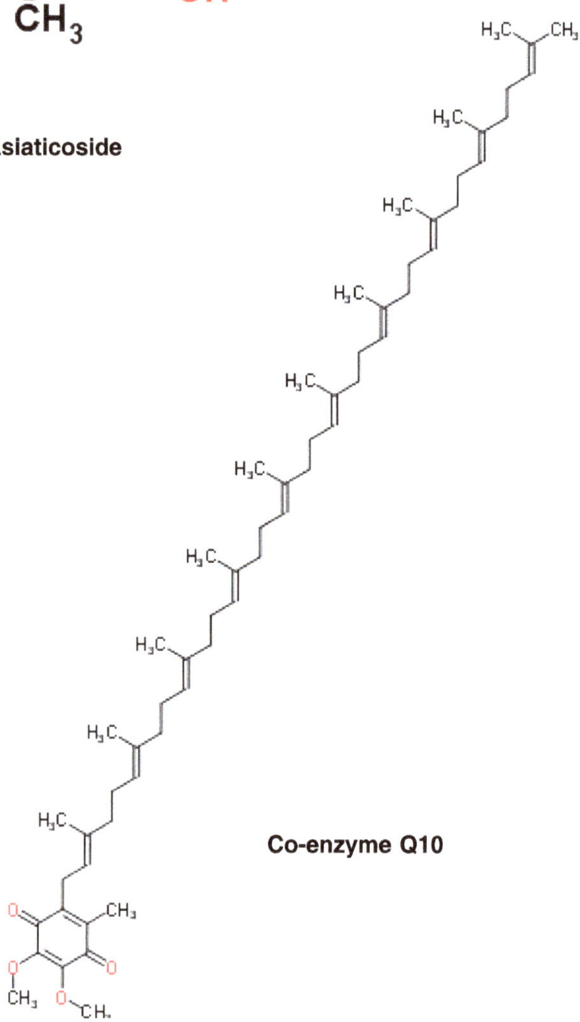

Co-enzyme Q10

Contd...

Figure 9.1–*Contd...*

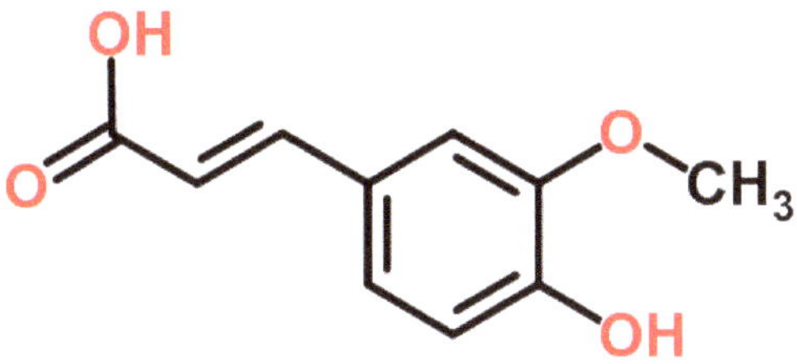

Ferulic acid

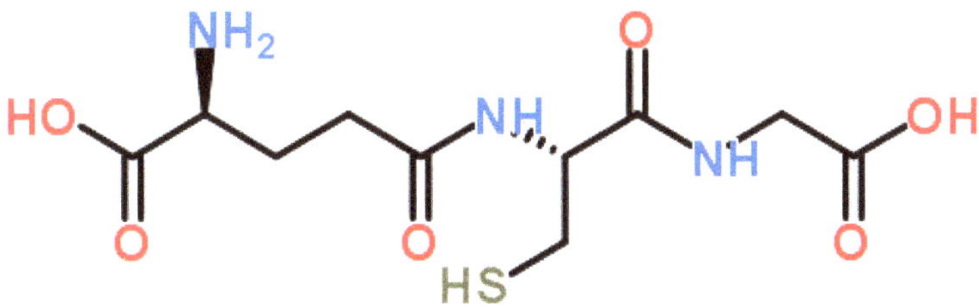

Glutathione

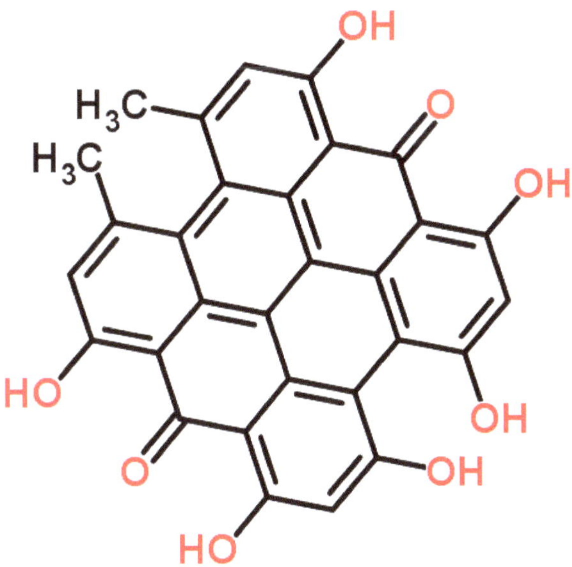

Hypericin

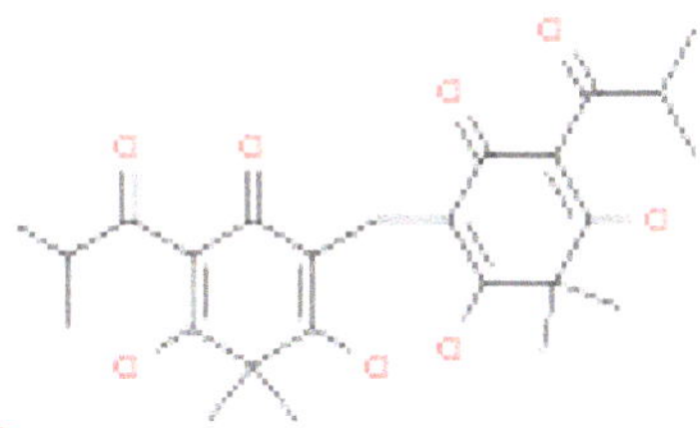

Japonicine-A

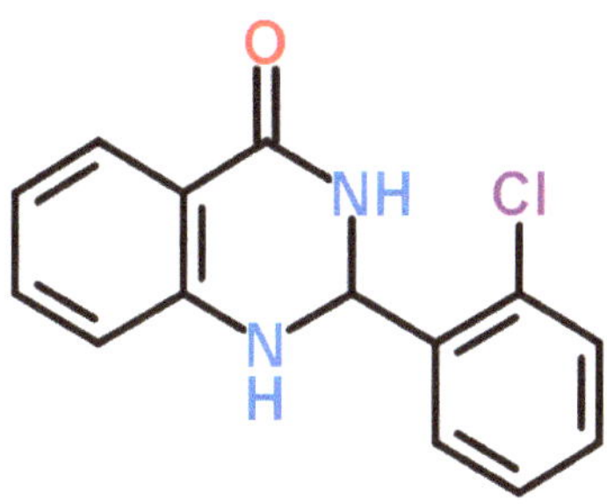

Madecassic acid

Contd...

Figure 9.1–*Contd...*

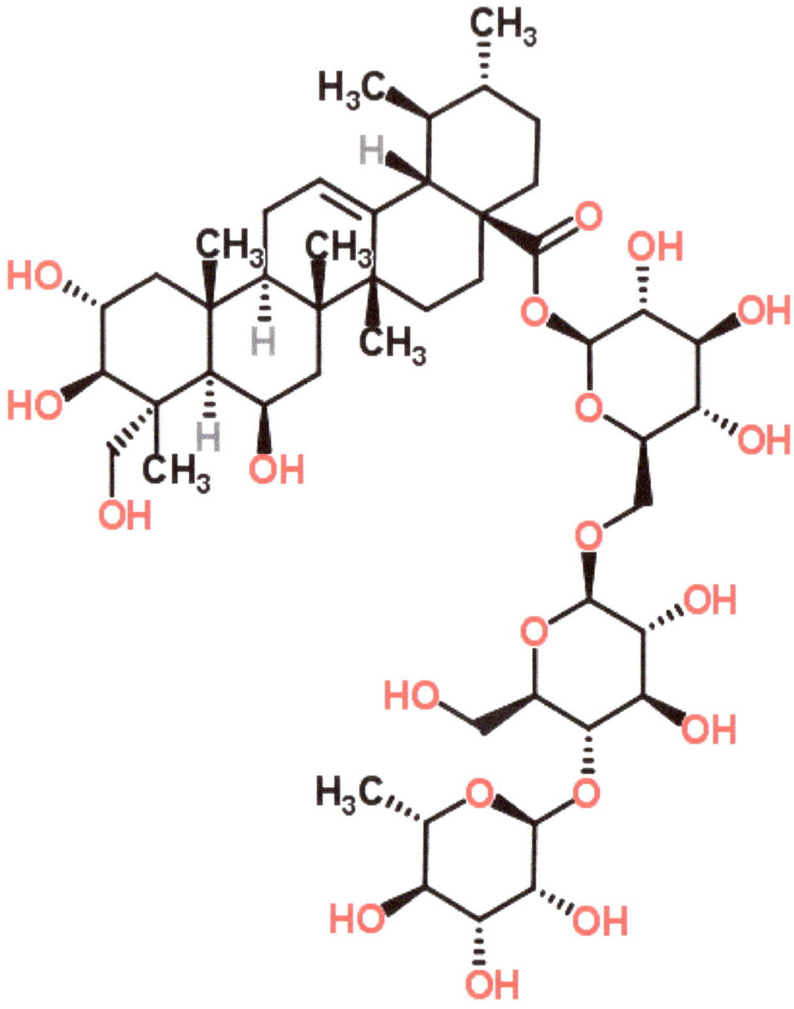

Madecassoside

Methylnonane

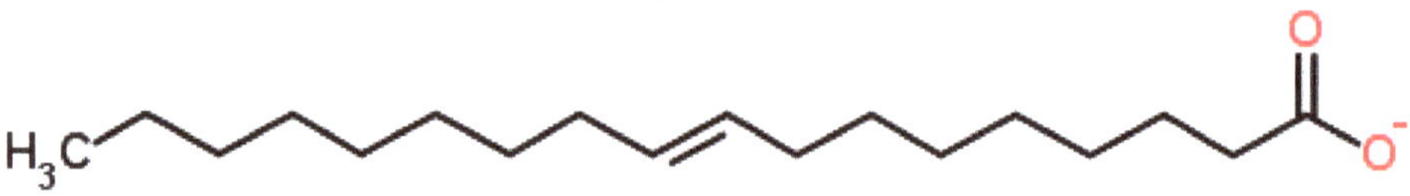

Omega 3 fatty acid

Contd...

Figure 9.1–*Contd...*

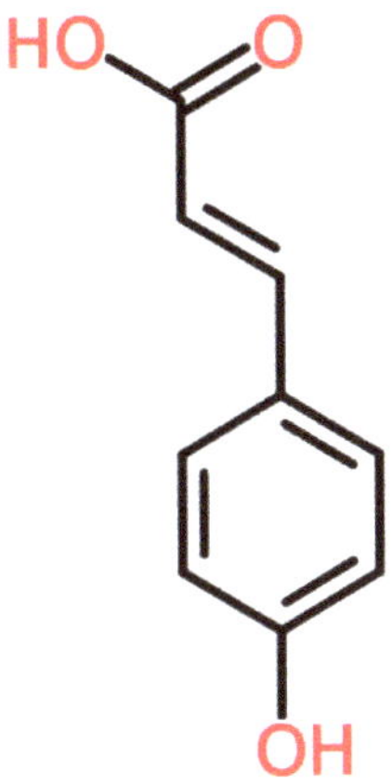

p-coumaric acid

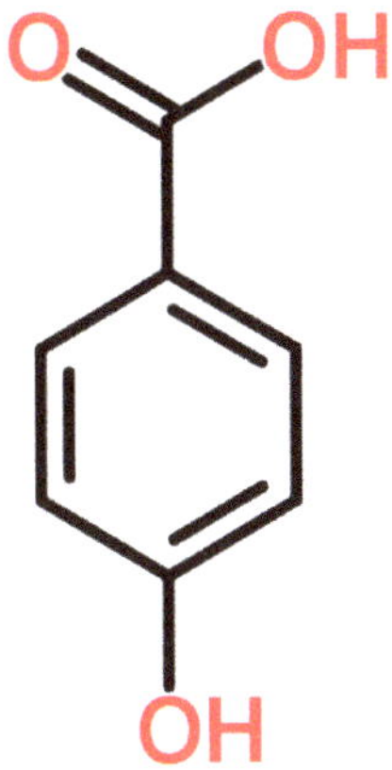

p-hydroxybenzoic acid

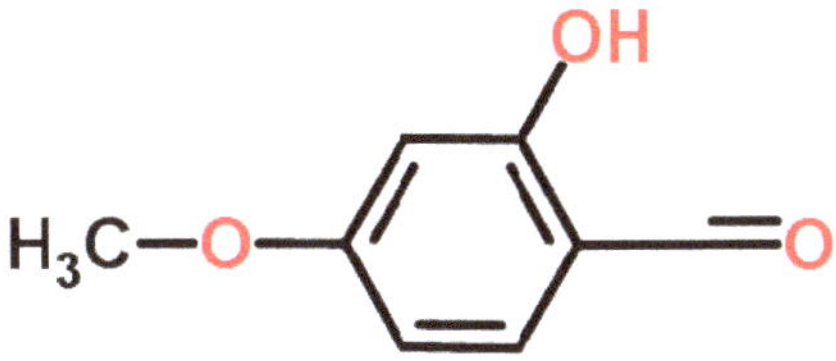

p-methoxy salicylic aldehyde

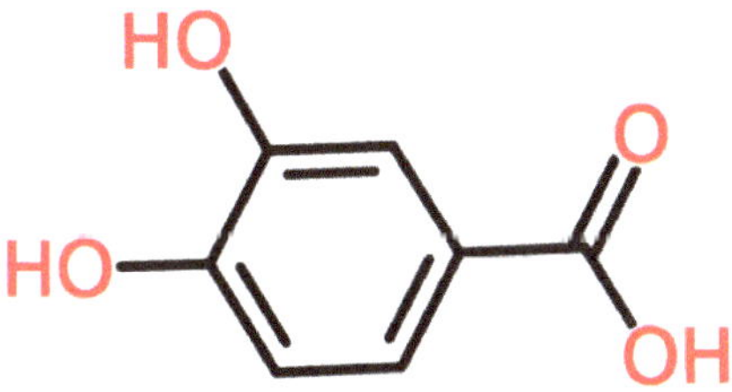

Protocatechuic acid

Shellsol

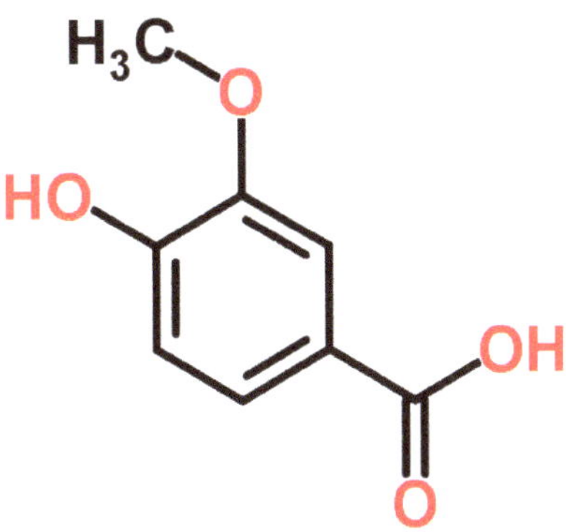

Vanillic acid

Vinyl hexylether

plants. This review combines the information included in previous publications carried out by different researchers and hopes to give an insight into past, present and future topical treatment strategies for wound healing.

Portulaca oleracea (PO)

PO is also known as common purslane, pigweed, little hogweed or pusley. It belongs in the family Portulaceae and is a native of India and the Middle East. It has smooth, reddish, mostly fleshy stems and alternate leaves clustered at stem joints and ends, which can reach upto 40 cm in height. The yellow flowers have five regular parts and are up to 6 mm wide. Seeds are formed in a tiny pod, which opens when the seeds are ready. It has a taproot with fibrous secondary roots and is able to tolerate poor, compacted soils and drought (Byrne and McAndrews, 1975). Ancients Greek used it as a remedy for constipation and inflammation of the urinary system (Megaloudi, 2005). In the Dominican Republic, all parts of this weed are used in treatments for internal parasites by mixing it with other plants (*e.g.*, *Chenopodium ambrosioides*) (Bairacli-Levy, 1991). WHO has listed this weed as one of the most used medicinal plants and given the term 'Global Pancreas' (Dweck, 2001; Samy *et al.*, 2004). The herb abounds in a milky juice and has been reported to be used in kidney diseases. The leaves are astringent, refrigerant, diuretic and emollient while seeds are described as demulcent and diuretic. (Samy *et al.*, 2004) They also relieve tormina and tenesus. The juice of the stem may be beneficial in prickly heat as well as when burning sensation is felt. The medicinal properties are attributed to the presence of large amounts of l-norepinephrine (l-noradrenaline; 0.25 per cent in fresh herb) in the ethanolic extract. This neurohormone extract has vasopressor and antihypotensive activities and reduces haemorrhage at the tissue level. It also has vitamins A, C and E and the minerals: calcium and potassium salts, phosphorus, manganese, silicon, copper and high levels of potassium and iron. It is also a good source of glutathione and Omega 3 fatty acids, Co-enzyme Q10 which makes the plant a great source of anti-oxidants (Dweck, 2001; Lim and Quah, 2007). These combinations of anti-oxidants speed up the healing process.

Studies on their anti-inflammatory activity demonstrated significant reduction in the increase in paw volume induced by carrageenan in wistar rats (Chan *et al.*, 2000; Parry *et al.*, 1993). Chan and co-workers (Chan *et al.*, 2000) found that the 400-mg/kg dose of the extract from these plants gave an earlier onset and higher and longer activity of anti-inflammation and analgesia than the synthetic drug diclofenac sodium (4-mg/kg). Rashed and team (Rashed *et al.*, 2003) found that PO also accelerates the wound healing process in white swiss mice by decreasing the surface area of the wound and increasing the tensile strength (*i.e.* resistance of the repaired tissue to break under tension and may indicate in part the quality of the repaired tissue). The greatest contraction was obtained at a single dose (SD) of 50 mg and the second greatest by two multiple doses (MD) of 25 mg. Analysis of the excision wound for scars after 20 days showed no scars on animals treated which was an indication for complete healing. They attributed this accelerated wound healing to the collagen fiber formations present in ethanolic extract. The results suggested that treatment with fresh homogenized crude extract of PO may have a beneficial influence on the various phases of wound healing like fibroplasias, collagen synthesis, and wound contraction, resulting in faster healing.

The plant also showed incredible antimicrobial activity. The methanol extract from its seed had high activity against both Gram-positive organisms, *Bacillus subtilis* (20mm), *Staphylococcus aureus* (15mm) and against one Gram-negative bacterium *Pseudomonas aeruginosa* (18mm). The methanol extract of seeds of PO at 100mg/ml concentration was as effective as 10µg/ml of Gentamicin and Tetracycline against *Bacillus subtilis,* 10µg/mlGentamicin against *Staphylococcus aureus* and 20µg/ml Gentamicin against *Pseudomonas aeruginosa* (Yagoub *et al.,* 2006). The antifungal activity of PO extracts against hyphal growth of filamentous *Trichophyton* fungi was evaluated by Bongoh *et al.,* in real time using an automatic single-cell bioassay system. The ethanolic extract showed a specific and marked activity against dermatophytes of the genera *Trichophyton* (Bongoh *et al.,* 2000). These studies showed PO to be an effective disinfectant during wound healing process.

Centella asiatica (CA)

CA of the Apiaceae (Umbelliferae) family is also known as Gotu kola, Indian Pennywort, Jal Brahmi and Mandookaparni. It is a native to Sri Lanka, northern Australia, Indonesia, Iran (Brinkhause and Linder, 2000; Bradwejn *et al.,* 2000; Maquart *et al.,* 1990; Maquart *et al.,* 1999), Malaysia, Melanesia, New Guinea, and other parts of Asia. It is a slender, herbaceous perennial plant with creeping stems bearing roots on their nodes. The leaves are borne on pericladial petioles of around 20 cm. Flowers are minute pinkish red in color, 3 to 5 in a clusters (umbels) near the surface of the soil. It grows along ditches and in low wet areas.

The literature reveal that CA has been used for wound healing, better circulation, memory enhancement, sedative, anti-stress, anti-anxiety, an aphrodisiac, adaptogen cancer, immune booster, respiratory ailments, treatment of skin disorders (such as psoriasis and eczema), periodontal disease, burn and scar treatment, revitalizing connective tissue, arthritis, treatment of liver and kidneys, detoxifying the body and high blood pressure etc. However, none of these claims have been evaluated by the FDA, but the research carried out by various groups and universities concluded that more research is needed to validate the medicinal value of this ancient herb (Hussain *et al.,* 2007).

The extract of the leaves has been reported to heal wounds, burns and ulcerous abnormalities of the skin (Kartnig, 1998) through stimulation of collagen and glycosaminoglycan synthesis. Titrated extract of CA has four major triterpenoid (having a C_{30} skeleton) components (*i.e.* asiatic acid, asiaticoside, madecassic acid and madecassoside). Among these components, asiaticoside (Sikareepaisan *et al.,* 2008), a trisaccharide triterpene, has been identified as the most active compound associated with the healing of wounds. This was evidenced by the observed increase in antioxidant levels at an initial stage of healing of excision-type cutaneous wounds in rats (Shukla *et al.,* 1999) and increased proliferation and production of types I and III pro-collagen mRNA and protein levels of human dermal fibroblasts (Lu *et al.,* 2004a) in response to the presence of this substance. However, the sugar portion of asiaticoside is not considered as an essential pharmacophore for biological activity, but enhances the wound healing effect of asiatic acid in topical application, and could be greatly simplified without loss of biological activity (Allegra *et al.*, 1981; Cesarone *et al.*, 1994; Cervenka and Jahodar *et al.*, 2006; Hussin *et al.*, 2007; Pointel *et al.*, 1987; Shim *et al.*, 1996). The study on wound-healing potential of the ethanolic extract of the plant on Wistar albino rats using incision, excision, and dead space wounds models showed significantly increase in the wound breaking strength in incision wound model compared to controls (Suguna *et al.*, 1996; Sunil *et al.*, 1998). The extract-treated wounds were found to epithelize faster, and the rate of wound contraction was significantly increased as compared to control wounds. The results indicated that the leaf extract promotes wound healing significantly and is able to overcome the wound-healing suppressing action of dexamethasone in a rat model (Shetty *et al.*, 2006). Another study showed that topical application of asiaticoside, isolated from CA, twice daily for 7 days to excision-type cutaneous wounds in rats led to increased enzymatic and non-enzymatic antioxidants, namely superoxide dismutase (35 per cent), catalase (67 per cent), glutathione peroxidase (49 per cent), vitamin E (77 per cent) and ascorbic acid (36 per cent) in newly formed tissues. This enhanced induction of antioxidant levels at an initial stage of healing which may have benefits on the healing properties of this substance (Shukla *et al.*, 1999).

The effect of CA ointment on wound healing was studied in bovine calves by evaluating their macroscopic and biochemical aspect. Macroscopically, the animals treated with CA ointment healed completely on 24th day post treatment. Lower values of alkaline phosphatase and zinc, and higher values of proteins, collagen, elastin, iron, copper and magnesium were recorded in CA ointment. There was no significant difference in respect of clinical and biochemical observations among the CA treated and commercially obtained canisep cream treated wounds (Gogoi *et al.*, 2006).

Asiaticoside was found to induce cell-cycle progression, proliferation and collagen synthesis in human dermal fibroblasts (Bonte *et al.*, 1994; Tenni *et al.*, 1988). Scientists thought, this extracellular matrix synthesis might be related to certain gene expression (Boiteau and Ratsimamanga *et al.*, 1956; Lu *et al.*, 2004b; Poizot and Dumez *et al.*, 1978). They found a close correlation among the gene profile, mRNA and protein production in the cells response to asiaticoside stimulation. In another study, the molecular pathways underlying the diverse biological activity of the triterpeniod compounds isolated from CA were studied with gene microarrays and real-time reverse transcription polymerase chain reaction (real-time RT-PCR) to quantify the expression of 1053 human genes in human fibroblasts. The results show that triterpenes from the plant evoke a gene-expression response consistent with their prevailing medical uses in the treatment of connective tissue disorders such as wound healing (Coldren *et al.*, 2003).

Tridax procumbens (TP)

TP is a species of flowering plant in the Compositae family and commonly known as coat buttons and tridax daisy (Saxena and Albert, 2005). It is native to the tropical Americas but it has been introduced to several tropical, subtropical, and mild temperate regions worldwide. It is a semi-prostrate annual or short-lived perennial, with stems up to 50 cm long. It is a weed of pastures and a wide range of annual and perennial crop types. The plant bears daisy-like yellow-centered white or yellow flowers with three-toothed ray florets. The leaves are toothed and generally arrowhead-shaped. Its fruit is a hard achene covered with stiff hairs and having a feathery, plumelike white pappus at one end. The plant is invasive in part because it produces so many of these achenes, up to 1500 per plant, and each achene can catch the wind in its pappus and be carried some distance. This weed can be found in fields, meadows, croplands, disturbed areas, lawns, and roadsides in areas with tropical or semi-tropical climates.

TP is widely known for its efficacy on burn wound healing when administered orally and topically. Studies by Babu *et al.* (Babu *et al.*, 2003) on epithelization period and wound contraction done on anaesthetized rats with partial wounds created by pouring hot molten wax at 80° C revealed that the percentage of burn wound contraction was significantly increased when the aqueous weed extract was administered in oral and topical group as compared to the control group on the 16th day. The period of burn wound epithelization significantly hastened in the topically administered group.

Studies by Margaret and team (Margaret *et al.*, 1998) showed that the aqueous extract of TP leaves was lyophilized and had a significant healing effect on the excision wound of rat skin fibroblast and paw oedema. Wound contraction was found to be comparable in the TP and ibuprofen (non-steroidal anti-inflammatory drug originally marketed as Nurofen) treated groups. Epithelialization was significant in the *Tridax* group. They also observed that in the carageenin induced oedema model, inhibition of oedema was comparable in 200 mg/kg TP treatment, and the specific activity of the enzyme gamma glutamyl transpeptidase was comparable with the TP extract.

Udupa and co workers (Udupa *et al.*, 1995) studied the effects of the whole plant extract (WPE), aqueous extract (AE), butanol fraction (BF) and petroleum ether fraction (PEF) of TP on dead space wound healing in albino rats. In this study granuloma tissue harvested from 10 day old wounds was used for estimation of lysyl oxidase activity, tensile strength and other biochemical parameters. Lysyl oxidase activity, protein and nucleic acid contents as well as the tensile strength were all increased significantly in animals treated with WPE and AE. On the other hand, BF- and PEF-treated animals showed a decrease in all these parameters except tensile strength. In these two groups the hexosamine levels were increased. Overall, WPE seems more active as compared to the other extracts in the healing process. They anticipated that TP may be having a dual role: one a stimulatory (direct) effect in the initial phase of wound healing and the other a depressant (indirect) effect in the later stage (Udupa *et al.*, 1991).

***Desmodium triflorum* (DT)**

DT, also known as threeflower ticktrefoil, or creeping tickfoil, is a small prostrate annual or perennial legume with a woody taproot. The stems are strongly branched to 50 cm frequently rooting at the nodes to form a mat. The leaves are trifoliate with leaflets up to 12 mm long and 10 mm wide. The flower cluster consists of 1–3 pink to purple flowers in leaf axils. The pods are flat, segmented, 6–18 mm long and 2–3.5

mm wide with 3–5 articles, and covered with minute hooked hairs. It is commonly known as creeping tick trefoil, three-flower beggarweed (Jain, 1968).

The plant decoction is also used as a mouthwash and as an expectorant. The roots are reputedly used in bilious (liver dysfunction) complaints, in fever and dysentery. The galactagogue leaves are ground with cow's milk to treat *diahhroea* due to indigestion and also in convulsions. Expressed juice from a well macerated plant, is applied to abscesses and wounds that do not heal readily (Singh and Panda, 2005). The paste of whole-plant is also applied externally to check bleeding due to cuts and heal the wound (Prusti and Behera 2007). The other medical actions include anti-spasmodic, sympathomimetic, central nervous system stimulation, curare-mimetic activity and diuretic and anthelmintic activity. In Indonesia, Malaysia, the Philippines, Laos and India the crushed plant or a poultice of the leaves is externally applied on wounds, ulcers, and for skin problems in general, apparently for its antiseptic properties. (Jain, 1968; Singh and Panda, 2005). However there are only few literatures available in respect with its wound healing activity.

In one of these studies, hexane, ethyl acetate and methanol extracts from DT were tested to evaluate the analgesic and anti-inflammatory activities in mice. The results showed both the hexane and methanol extracts at 300 mg/kg body weight showed statistically significant inhibition of rat paw oedema by 35.8 and 38.97 per cent, respectively, at the third hour after carrageenan injection. (Chowdhury *et al.*, 2005). However not much was reported in terms of the wound healing mechanism in this work.

***Cynodon dactylon* (CD)**

CD also known as durva grass, Bermuda Grass, Dog's Tooth Grass, Bahama Grass, Devil's Grass, Couch Grass, Indian Doab, Grama, and Scutch Grass, is a grass native to north Africa, Asia and Australia and southern Europe. It is a highly invasive, hardy perennial grass with creeping culms. The blades are a grey-green colour and are short, usually 4-15 cm long with rough edges. The erect stems can grow 1-30 cm (rarely to 90 cm) tall (Singh and Panda, 2005). The stems are slightly flattened, often tinged purple in color. The seed heads are produced in a cluster of 3-7 spikes (rarely 2) together at the top of the stem, each spike 3-6 cm long. It has a deep root system; in drought situations with penetrable soil, the root system can grow to over 2 m deep, though most of the root mass is less than 60 cm under the surface. The grass creeps along the ground and root wherever a node touches the ground, forming a dense mat (Jain, 1968).

CD is reputed as a remedy in epitaxis, haematuria and scabies. The juice of the plant has an astringent effect apart from being useful in cuts and wounds. Decoction of the root is used as diuretic, in dropsy, in hysteria, epilepsy, insanity, syphilis and as astringent. Infusion of root is useful in stopping the bleeding from piles (Sharma, 2003).The paste of the plant reportedly helps in the immediate stoppage of the blood from any cuts or injuries. The aqueous extract dried and powdered mixed with paraffin when applied as ointment produces wound healing effect. Treatment with the ethanolic extract in wistar rats has shown to promote wound contraction and reduced the time for closure showing healing potential comparable to 1 percent framycetin sulphate cream. (Charde *et al.*, 2003) Experimentally-induced wounds in rabbits healed in 12 days following application of aqueous leaf extracts in white soft paraffin (as 1 per cent ointment). Controls treated with paraffin alone healed in 16 days. (Subramanian and Nagarajan, 1988). However, fewer reports and inconclusive works do not give the needed scientific evidence based knowledge justifying their traditionally known therapeutic uses.

Ageratum conyzoides L. (AC)

AC is a native of the Central America and the Caribbean. (Baker, 1965; Correa, 1984; Cruz, 1985; Lorenzi, 1982;). It is commonly known as Billygoat-weed, Chick weed, Goatweed, Whiteweed. AC is a half hardy, erect, herbaceous annual, upto 80 cm tall; stems are covered with fine white hairs, leaves are opposite, pubescent with long petioles and include glandular trichomes (Achola *et al.*, 1994). The inflorescence contains 30 to 50 pink flowers arranged as corymbs and is self-incompatible (Jhansi and Ramanujam, 1987; Kaul, and Neelangini, 1989; Kleinschimidt, 1993; Ramanujam and Kalpana, 1992). The fruit is an achene with an aristate pappus and is easily dispersed by wind. Seeds are positively photoblastic, and viability is often lost within 12 months (Marks and Nwachuku, 1986; Ladeira *et al.*, 1987).

Traditionally the plant parts are boiled with oil and applied in rheumatism. The root extract is useful in kidney stone (Gravena *et al.*, 1993). Several pharmacological investigations have been conducted to determine its efficacy as an antimicrobial. For instance, Durodola (Durodola, 1977) verified inhibitory activities of ether and chloroform extracts against *in vitro* development of *Staphylococcus aureus*. Almagboul *et al.* (Almagboul *et al.*, 1985), using methanolic extract of the whole plant, verified the inhibitory action on the development of *Staphylococus aureus, Bacillus subtilis, Eschericichia coli,* and *Pseudomonas aeruginosa*. Studies initiated by Chah *et al.*, to evaluate the antibacterial and wound healing activities of methanolic extracts, showed it to be comparable with Cicatrin powder (a topical powder containing Neomycin sulphate 3,300 gm, Zinc bacitracin 250 gm, I-Cysteine 2 mg, Glycine 10 mg, dl-Threonine 1 mg) when treated on the excision wound in albino rats. (Chah *et al.*, 2006). It was also found that topical application of AC caused significant greater wound contraction than the controls used (Oladejo *et al.*, 2003) in wistar rats.

Okwori and co-workers (Okwori *et al.*, 2007) investigated *in vitro* antibacterial activities of the extracts on some selected bacterial pathogens. He reported that the hexane extracts of the leaf, stem and root had 100 per cent susceptibility to all the bacterial isolates used *viz., Staphylococcus aureus, Yersinia enterocolitica, Salmonella gallinarum* and *Escherichia coli*. The aqueous leaf extracts gave 75 per cent susceptibility, while methanolic leaf extract gave 50 per cent. The extracts inhibited the growth of the bacterial isolates in a concentration dependent manner. Phytochemical analysis of the dried leaves revealed the presence of Resins, Alkaloids, Tannins, Glycosides and Flavonoids which were all antibacterial in their activities. The dried stems revealed the presence of Resins Saponins, Tannins, Glycosides and Flavonoids while the dried roots contained Resins, Alkaloids Saponins and Flavonoids.

Bioka *et al.* (Bioka *et al.*, 1993) reported effective analgesic action in rats using aqueous extract of leaves of *A. conyzoides* (100 to 400 mg/kg). Assays released in

Kenia (African island), with aqueous extract of the whole plant, demonstrated muscle relaxing activities, confirming its popular use as an antispasmotic (Achola *et al.*, 1994).

These studies showed that the extracts carried the potentials of introducing a new template into modern medicine and would require further investigation in the empirical development of a new antibacterial for wound healing.

Chromolaena odorata (CO)

CO is a shrub of the Asteraceae family native to the new world tropics. It is an invasive weed of field crops in Africa and Asia where it has been introduced. It is now found all over tropical Asia, western Africa, and in parts of Australia (Ngono *et al.*, 2006). Other names for the plant include Siam weed and Christmas bush, and common floss flower. It is fast growing and forms a very dense canopy. CO is a perennial shrub that stands at least 3 m in the open and up to 10 m as climber. The shrub often forms dense thickets with their opposite, wide-spreading branches which prevent the establishment of other species. The leaves of CO are ovate, petiolate and tri-nerved from the base. The florets range from white to pale blue and are present in terminal oblong capitula. Seeds are brown to black, 4–5 mm long, with white parachute-like white hairs (pappus) at the top of the seed. It is used as a traditional medicine in Indonesia (Schmidt and Schilling, 2000) against burns and cuts.

Crude ethanol extracts from the leaves gives a compound called Eupolin. Eupolin ointment has been shown to promote the healing of soft tissue wounds and burns. The studies on wound re-epithelization of cultured human keratinocytes *in vitro* showed that the human keratinocyte proliferation was stimulated by low concentrations of the extract (from 0.1 to 5 μg/ml), cell differentiation by higher concentrations (50 to 300 μg/ml), and migration by intermediate concentrations (5 to 60 μg/ml) (Phan *et al.*, 2001a).

The Eupolin also demonstrated to be a powerful antioxidant which protected the fibroblasts and keratinocytes *in vitro*. The studies showed that this compound comprises of phenolic acids (protocatechuic, p-hydroxybenzoic, p-coumaric, ferulic and vanillic acids) and complex mixtures of lipophilic flavonoid aglycones (flavanones, flavonols, flavones and chalcones) which were powerful antioxidants to protect cultured skin cells against oxidative damage (Phan *et al.*, 2001b). Further studies on the anti-oxidant effects of both total ethanol and polyphenolic extracts from the plant leaves, on hydrogen peroxide and hypoxanthine-xanthine oxidase induced damage to human fibroblasts and keratinocytes, showing that the protection of cells against destruction by inflammatory mediators was one of the ways in which the extracts from the CO contribute to wound healing (Thang *et al.*, 2001).

Eupolin extract also increased expression of several components of the adhesion complex and fibronectin by human keratinocytes. Using indirect immuno-fluorescence it was found that increased expression (dose-dependent) of laminin 5, laminin 1, collagen IV, and fibronectin. The expression of the β1 and β4 integrins was upregulated by the extract at low concentrations (0.1 and 1 μg/ml), but the expression was decreased at higher doses of Eupolin (10 μg-150 μg/ml). Phan and coworker showed that Eupolin stimulated the expression of many proteins of the adhesion complex and fibronectin by human keratinocytes. It was concluded that the adhesion complex proteins are essential to stabilize epithelium, which contributed to the clinical efficacy of Eupolin in healing (Phan *et al.*, 2000). Clinical studies using this extract have shown antimicrobial and anticoagulation effects. However, the mechanism by which Eupolin affects cells involved in the wound healing process remains unknown (Phan *et al.*, 1998). Further studies by Ngono Ngane *et al.* (2006), have showed that an aqueous ethanol extract of leaves of CO and some of its fractions inhibit the *in vitro* growth of *Cryptococcus neoformans*, *Microsporum gypseum*, *Trichophyton mentagrophytes* and *Trichophyton rubrum*,which can play an important role in wound healing.

Heliotropium indicum Linn. (HIL)

It is commonly known as Erysipelas plant, Indian heliotrope, scorpion weed, turnsole and wild clary. This herb is native to pacific islands and usually associated with the moist rich soils of the lowland tropics near rivers and lakes, on dikes and roadsides, and in waste places, fallow rice fields, pools, ditches, or muddy soils which dry out periodically (Holm *et al.*, 1977). HIL is slightly woody at base, with a long taproot, stem erect, unbranched or branches few, 30-80 cm tall. It has large leaves (length 4.2 cm) which are alternate or sometimes subopposite. The stems are lightly hairy. It has small white flowers which turn yellow in throat and are arranged on a curled-up structure in two rows, unfurling as flowers open. (Stone, 1970)

Even though Ayurvedic practitioners in India have used HIL for centuries, there is no hard scientific evidence as to the positive effects of this herb, beside few Indian studies performed. Ethnic communities in India have been using the flowers as abortifacient and emmenagogue while whole plant has been used on boils, insect stings and sores. Several folklores report its tribal use in antidermatosis, skin pruritus, scabies sore throats and tonsillitis. The leaves are emollient and diuretic. Juice of leaves used for facial acne, gum-boils, sores and wounds. Decoction of roots or any plant part used for asthma. It is also useful in skin infections and wounds. Studies showed that when ethanolic extracts from the dried parts of the plants was topically applied on excision and incision wounds in rats, the results showed increased percentage of wound contraction and completed wound healing by 14th day indicating rapid epithelization and collagenization (Reddy *et al.*, 2002). Comparative study with medicinal plants like *Plumbago zeylanicum* and *Acalypha indica* showed HIL exhibiting better healing performance. A rapid biosynthetic activity was observed during initial phase of granulation and in remodeling phase. The maturation of collagen by the formation of inter and intra-molecular cross links improved wound contraction.

Srinivas *et al.* (Srinivas *et al.*, 2000) studies on the anti-inflammatory effect of HIL, on carrageenin induced hind paw oedema and cotton pellet granuloma in rats showed significant anti-inflammatory effect in both acute and subacute models of inflammation.

Hemidesmus indicus R. Br. (HIR)

HIR commonly known as Indian Sarsaparilla or false sarsaparilla is a native to South Asia. HIR is a slender, laticiferous, twining, sometimes prostrate or semi-erect shrub. Roots are woody and aromatic. The stem is numerous, slender, terete, thickened at the nodes. The leaves are opposite, short-petioled, very variable, elliptic-oblong to linear-lanceolate. The flowers are greenish outside, purplish inside, crowded in sub-sessile axillary cymes. It is used to make beverages and also used in traditional medicine. In Ayurveda it goes by the name of ananthamoola or Anantmula. The plant enjoys a status as tonic, alterative, demulcent, diaphoretic, diuretic and blood purifier. It is employed in nutritional disorders, syphilis, chronic rheumatism, gravel and other urinary diseases and skin affections. It is administered in the form of powder, infusion or decoction as syrup. It is also a component of several medicinal preparations (Jain, 1968).

Chemical analysis of the root showed the presence of coumarins, volatile oil the chief component of which is p-methoxy salicylic aldehyde, two sterols and a pregnane glycoside (Puri, 2003). Sharma (Sharma, 2003) reported the use of root decoction as blood purifier while Mary *et al.* (Mary *et al.*, 2003) observed that the methanolic extract of HIR also inhibited ADP-induced platelet aggregation *in vitro* (50–250 µg), which was comparable to commercial heparin.

The methanolic extract of the root bark has shown both the *in vitro* and *ex vivo* antioxidant potential (Ravishankara *et al.*, 2002). It was reported that the doses of 300 and 450 mg/kg of aqueous ethanolic extract are effective in treating ulcers in rats (Anoop and Jagdeesan, 2003). This antiulcerogenic (Mary *et al.*, 2003) effect was mainly because of its high mucoprotective activity, depicted by a selective increase in prostaglandin content. However in both the above studies the compounds associated with in these weeds extract was not explored for wound healing activity.

Tragia involucrata Linn. (TI)

TI is a perennial evergreen twining herb with hispid stinging hairs and is a native to countries of Indian subcontinent. It belongs to the family Euphorbiaceae and is a small annual herb. It is characterized for its leaf which causes itching on skin when come in contact with. There are certain small hair-like projections on the lower side of the leaf. These hairs are responsible for the itching effect. Traditionally the extracts from this plant has diaphoretic, analgesic and anti-inflammatory properties. The root-paste preparation is used to evict guinea worms. It is also used in the treatment of itchy boils of skin. Root decoction was found useful in relieving bronchitis and attendant fever. It also soothes scorpion sting (Singh and Panda, 2005).

Research carried out by Samy and group showed that the methanol extract of this weed comprises of nine compounds namely vinyl hexylether, shellsol, 2, 4-dimethyl hexane, 2-methylnonane and 2, 6-dimethyl heptane (Samy *et al.*, 2006a). These isolated compounds were studied against *Escherichia coli*, *Proteus vulgaris* and *Staphylococcus aureus* which revealed these compounds exhibiting bactericidal effects against all the tested bacteria. The compound shellsol showed effective killing of wound causing bacteria, *Staphylococcus aureus*. In rat wound model, rats that received shellsol showed complete healing after 24 days. Histological examination revealed an increase in the fibroblast, neovascularization, granulation and thickness of scar tissue after the treatment of shellsol with no toxic response as compared to control. Perumal and co workers (Samy *et al.*, 2006b) showed that the methanol extract of the roots of *T. involucrata* topically tested at doses of 100 and 200 mg/kg exerted significant wound healing effect in *Staphylococcus aureus*-induced excision wound in rats.

Hypericum patulum Thumb (HPT)

HPT is commonly known as Goldencup St. Johnswort. The natural range of the plant includes East Asia mainly Western China, Japan and Himalayas. It is a glabrous shrub with brownish bark upto 1 m height with opposite, simple, semi-evergreen, 1 - 2 ½ inches long, ovate-oblong leaves and produces 2 inches wide, bright yellow flowers in mid-summer. The plant extract is considered diuretic and spasmolytic. The seed extracts are useful as anthelminthic, astringent and diuretic (Kartnig, 1988).

Wound healing mechanisms of the methanol extract obtained from leaves were investigated on different experimental models of wounds in rats (Mukherjee *et al.*, 2000). The extract in the form of an ointment with two different concentrations (5 per cent and 10 per cent w/w ointment of leaf extract in simple ointment base) was applied on excision and an incision wound model in rats. Both concentrations of the ointment showed significant responses in both the wound types tested when compared with the control group. Studies revealed that the effect produced by the extract ointment, in terms of wound contracting ability, wound closure time, regeneration of tissues at wound site, tensile strength of the wound and histopathological characteristics were comparable to those of a standard drug nitrofurazone ointment. However detail studies on the compound causing wound healing was not included in the study.

Hypericum perforatum L. (HPL)

HPL is commonly known as St. John's wort and Klamath weed. It is a native to Eurasia and is a perennial herb with a stout creeping rhizome which bears clumps of erect, branched stems. The opposite sessile, ovate to oblong leaves have numerous reddish, translucent grandular dots. The showy flowers are arranged in terminal cymes (Bakshi *et al.*, 2003). The sepals and petals are black dotted, especially at the edges. The extract from this weed is known to contain upto 1 per cent ehereal oil, ~10 per cent catechin tannin, the flavonoglycoside hyperosid, rutin and querctrin and ~0.1 per cent of hypericin (a naphtodianthrone) and pseudohypericin (Hickey and King, 1981; Wichtl, 1986).

Other than being an important wound healer the herb is an astringent, aromatic detersive, resolutive, antihelmintic, diuretic, emmenagogues and purgative. It is used to treat chronic inflammation of internal organs and gynecological disorders. Japonicine-A extracted from the plant showed anti malarial activity in mice (Singh and Panda, 2005).

Studies (Santarem and Astarita, 2003) indicated that hypericin plays a major role in the wound healing activity while studies by Singh and Panda (Singh and Panda, 2005) reported an antibacterial preparation, novoimanin, made from the plant was effective against infections caused by *Staphylococcus aureus* in concentration of 0.1 ppm. Ozturk and co-workers compared the wound-healing effect of HPL extract with dexpanthenol and Titrated Extract of *C. asiatica* (TECA) on cultured chicken embryonic fibroblasts (Ozturk *et al.*, 2007). The chicken embryonic fibroblasts from fertilized eggs were incubated with the plant extract, dexpanthenol and TECA. Findings obtained in the study indicated that HPL extract exhibited a wound-healing mechanism similar to that of TECA. The HPL extract stimulated the fibroblast collagen production and the activation of fibroblast cells in polygonal shape, which plays a role in wound repair by closing the damaged area.

Conclusions

In the present review, information about popular medicinal weeds traditionally used for wound healing is surveyed. Taking into account the wound healing properties of the weeds mentioned here, it could be concluded that not all medicines need to be synthetic in origin, in fact the world's best chemist is probably Mother Nature. Medicinal compounds are not relegated to a certain family of plants and the extract from these weeds could be often beneficial to us. However, precaution should be taken as the extracts obtained may not always be suitable to the human system. But, perhaps the most important idea behind all of this is that significant research efforts into active plant compounds should continue. Not only have weeds provided us with some amazing medicines, but we have only scratched the surface of this category of chemicals. Hundreds of weed species remain to be analyzed for possible medicines. Thus with a global interest and expanding market of plant based raw material for the manufacture of drugs, pharmaceuticals etc, aholistic management action plan with proper documentation is necessary, specially when inclination towards chemical based drugs are decreasing because of health and environmental concerns. At the same time, there must be immediate efforts to ensure the conservation of diverse biological resources along with the preservation and application of local cultural knowledge on the use of these resources.

References

Achola, K.J., Munenge, R.W., and Mwaura. A.M. (1994). Pharmacological properties of root and aerial parts extracts of *Ageratum conyzoides* on isolated ileum and heart. *Fitoterapia*, **65**: 322–325.

Allegra, C., Pollari, G., Criscuolo, A., Bonifacio, M., and Tabassi, D. (1981). *Centella asiatica* extract in venous disorders of the lower limbs. Comparative clinicoinstrumental studies with a placebo. *Clin Ter.*, **99(5)**: 507-13.

Almagboul, A.Z, Farroq, A.A., and Tyagi. B.R. (1985). Antimicrobial activity of certain sudanese plants used in folkloric medicine: Screening for antibacterial activity, part II. *Fitoterapia*, **56**: 103–109.

Anoop, A., and Jegadeesan, M. (2003). Biochemical studies on the anti-ulcerogenic potential of *Hemidesmus indicus* R.Br. var. *indicus*. *J. Ethnopharmacol.*, **84**: 149-156.

Babu, G., Sanjeeva and Bairy, K.L. (2003). Effect of *Tridax procumbens* on burn wound healing. *Ind. Drugs*, **40**: 488-491.

Baie, S.Hj., and Sheikh, K.A. (2000a). The wound healing properties of *Channa striatus*-cetrimide cream-tensile strength measurment. *J. Ethnopharm.*, **71**: 93–100.

Baie, S.Hj., and Sheikh, K.A. (2000b). The wound healing properties of *Channa striatus*-cetrimide cream-wound contraction and glycosaminoglycan measurement. *J. Ethnopharm.*, **73**: 15–30.

Bairacli-Levy, J. (1991). The Complete Herbal Handbook for Farm and Stable. Faber, Boston, pp. 75-150.

Bajpai, M., Pande, A., Tewari, S.K., and Prakash, D. (2005). Phenolic contents and antioxidant activity of some food and medicinal plants. *Int J Food Sci Nutr.*, **56**: 287-291.

Baker, H.G. (1965). Characteristics and modes of origin of weeds. Academic Press, New York.

Bakshi, D.N., Sensarma, P., and Pal, D.C. (2003). A lexicon of medicinal plants in India (Vol 1). Naya Prokash, Kolkata, India, pp. 522-530.

Bioka, D., Banyikwa, F.F., and Choudhuri. M.A. (1993). Analgesic effects of a crude extract of *Ageratum conyzoides* in the rat. *Acta Hort.*, **332**: 171–176.

Bisht, M.K., Bhatt, K.C., and Gaur, R.D. (1988). Folk medicines of Arakot valley in district Uttarkashi: an ethnological study. In Purshottam Kaushik (Ed.) Indigenous Medicinal Plants. Today Tommorow Printers and Publishers, New Delhi, pp 163-165.

Boiteau, P., and Ratsimamanga, A.R. (1956). Asiaticoside extracted from *Centella asiatica* and its therapeutic uses in cicatrization of experimental and refractory wounds leprosy, cutaneous tuberculosis and lupus. *Therapie*, **11**: 125-149.

Bongoh, M., and Hwang, J. (2000). Detection of antifungal activity in *Portulaca oleracea* by a single cell bioassay system. *Phytotherapy research*, **14**: 329-332.

Bonner, J. (2000). Scar wars. *Chem. Ind. (london)*, **23**: 770–773.

Bonte, F., Dumas, M., Chaudagne, C., and Meybeck, A. (1994). Influence of asiatic acid, madecassic acid, and asiaticoside on human collagen I synthesis. *Planta Med.*, **60**: 133-135.

Bradwejn, J., Zhou, Y., Koszycki, D., and Shlik, J. (2000). A double-blind, placebo-controlled study on the effects of Gotu Kola (*Centella asiatica*) on acoustic startle response in healthy subjects. *J. Clin. Psychopharmacol.*, **20**: 680-684.

Brinkhause, B., and Lindner, M. (2000). Chemical, pharmacological and clinical profile of the east asian medical plant *Centella asiatica*. *Phytomedicine*, **5**: 427-435.

Byrne, R., and McAndrews, J. H. (1975). Pre-Columbian puslane (*Portulaca oleracea* L.) in the New World. *Nature*, **253**: 726-727.

Cervenka, F., and Jahodar, L. (2006). Plant metabolites as nootropics and cognitives. *Ceska Slov Farm*, **55**: 219-229.

Cesarone, M.R., Laurora, G., De Sanctis, M.T., Incandela, L., Grimaldi, R., Marelli, C., and Belcaro, G. (1994). The microcirculatory activity of *Centella asiatica* in venous insufficiency-a double-blind study. *Minerva Cardioangiol.*, **42**: 299-304.

Chah, K.F., Eze, C.A., Emuelosi, C.E., and Esimone, C.O. (2006). Antibacterial and wound healing properties of methanolic extracts of some Nigerian medicinal plants. *J Ethnopharm.*, **104**: 164-167.

Chan, K., Islam, M.W., Kamil, M., Radhakrishnan, R., Zakaria, M.N.M., and Habibullah, M. (2000). The analgesic and anti-inflammatory effects of *Portulaca oleracea* L. subsp. *sativa* (Haw.) *Celak. J. Ethnopharm.*, **73**: 445–451.

Charde, M.S., Fulzele, S.V., Sathurwar, P.M., and Dorle, A.K. (2003).Wound healing activity of Durva ghrita. *Ind. J. Pharma. Sci.*, **65**: 482-485.

Chowdhury, K. K., Achinto, S., Bachar, S. C., and Kundu, J. K. (2005). Analgesic and anti-inflammatory activities of *Desmodium triflorum* DC. *J Bio. Sci.*, **5**: 581-583.

Coldren, C.D., Hashim, P., Ali, J.M., Oh, S.-K., Sinskey, A.J., and Rha, C. (2003). Gene expression changes in the human fibroblast induced by *Centella asiatica* triterpenoids. *Planta Med.*, **69**: 725-732.

Correa, M.P. (1984). Dicionario das plantas úteis do Brasil e das exóticas cultivadas. *Ministério da Agricultura Rio de Janeiro, IBDF*, **2**: 139-155.

Cragg, G.M., Newman D.J., and Snader K.M. (1997). Natural products in drug discovery and development. *J Na. Prod.*, **60**: 52-60.

Cruz, G.L. (1985). Dicionário das plantas úteis do Brasil, 3 ed. Civilização Brasileira, Rio de Janeiro.

Dhar, U., Manjkhola, S., and Joshi, M. (2002). Current status and future strategy for development of medicinal plants sector in Uttaranchal, India. *Curr. Sci.*, **83**: 956-964.

Durodola, J.I. (1977). Antibacterial property of crude extracts from a herbal wound healing remedy - *Ageratum conyzoides*, L. *Planta Med.*, **32**: 388-390.

Dweck, A. C. (2001). Purslane – *Portulaca oleracea* The global panacea. *Personal Care Magazine*, **2**: 7-15.

Geldenhuys, C.J. (2007). Weeds or useful medicinal plants in the rural home garden? *Food Nutr. Bull.*, **28**: S392-S397.

Gogoi, J., Dutta, B., and Lekharu, J.C. (2006). Effect of canisep cream and free omental allograft on wound healing in bovine calves. *Ind. Veter. J.*, **83**: 385-388.

Gravena, S., Coletti, A., and Yamamoto, P.T. (1993). Influence of green cover with *Ageratum conyzoides* and *Eupatorium pauciflorum* on predatory and phytophagous mites in citrus. *Bul. OILB-SROP.*, **16**: 104–114.

Hickey, M., and King, C. (1981). 100 Families of flowering plants, Cambridge University Press, Cambridge, pp. 567-578.

Holm, L. G., Plucknett, D. L., Pancho, J. V., and Herberger, J. P. (1977). The world's worst weeds: distribution and biology. East-West Center, University Press of Hawaii. pp. 609-632.

Husain, G.M., Mishra, Deepa, Singh, P.N., Rao Ch. V., and Kumar Vikas (2007). Ethnopharmacological review of native traditional medicinal plants for brain disorders. *Phcog. Rev.*, **1**: 19-28.

Hussin, M., Abdul-Hamid, A., Mohamad, S., Nazami, S., Ismail, M., and Bejo, M. H. (2007). Protective effect of *Centella asiatica* extract and powder on oxidative stress in rats. *Food Chem.*, **100**: 535–541.

Izzo, A.A., and Ernst, E. (2001). Interactions between herbal medicines and prescribed drugs: a systematic review. *Drugs*, **61**: 2163–2175.

Jain, S.C. (1968). Medicinal Plants. National Book Trust of India, New Delhi. pp 176-187.

Jhansi, P., and Ramanujam, C.G.K. (1987). Pollen analysis of extracted and squeezed honey of Hyderabad, India. *Geophytology*, **17**: 237–240.

Kartnig, T. (1998). Herbs, spices and medicinal plants. In: L.E. Cracker, J.E. Simon eds. Vol. 3, Arizona, USA: Oryx Press; pp. 145-173.

Kaul, M.L.H., and Neelangini, S. (1989). Male sterility in diploid *Ageratum conyzoides* L. *Cytologia*, **54**: 445–448.

Kleinschmidt, G. (1993). Colony nutrition on the Atherton tableland. *Australasian Beekeeper*, **94**: 453–464.

Ladeira, A.M., Zaidan, L.B.P., and Figueiredo-Ribeiro, R.C.L. (1987). *Ageratum conyzoides* L. (Compositae): Germinação, floração e ocorrência de derivados fenólicos em diferentes estádios de desenvolvimento. *Hoehnea*, **15**: 53–62.

Lim, Y.Y., and Quah, E.P.L. (2007). Antioxidant properties of different cultivars of *Portulaca oleracea*. *Food Chem.*, **103**: 734-740.

Lorenzi, H. (1982). Plantas daninhas do Brasil. H. Lorenzi, Nova Odessa.

Lu, L., Ying, K., Wei, S., Liu, Y., Lin, H., and Mao, Y. (2004). Dermal fibroblast-associated gene induction by asiaticoside shown *in vitro* by DNA microarray analysis. *Br. J. Dermatol.*, **151**: 571–578.

Lu, L., Ying, K., Wei, S., Fang, Y., Liu, Y., Lin, H., Ma, L., and Mao, Y. (2004). Asiaticoside induction for cell-cycle progression, proliferation and collagen synthesis in human dermal fibroblasts. *Int. J Dermat.*, **43**: 801-807.

Malik, C.P. (2006). Recent approaches in the development and standardization of formulations of herbal drugs and the role of biotechnology in their upgradation. Herbal medicines: Traditional practices. (Trivedi P.C. ed) Aavishkar Publishers, distributors, Jaipur, India. pp. 322-330.

Maquart, F.X., Bellon, G., Gillery, P., Wegrowski, Y., and Borel, J.P. (1990). Stimulation of collagen synthesis in fibroblast cultures by a triterpene extracted from *Centella asiatica*. *Connect Tissue Res.*, **24**: 107-120.

Maquart, F.X., Chastang, F., Simeon, A., Birembaut, P., Gillery, P., and Wegrowski. Y. (1999). Triterpenes from *Centella asiatica* stimulate extracellular matrix accumulation in rat experimental wounds. *Eur. J. Dermatol.*, **9**: 289-296.

Margaret, I., Srinivasa, R.P., and Kaiser, J. (1998). Anti-inflammatory profile of *Tridax procumbens* in animal and fibroblast cell models. *Phytoth. Res.*, **12**: 285-287.

Marks, M.K., and Nwachuku. A.C. (1986). Seed-bank characteristics in a group of tropical weed. *Weed Res.*, **26**: 151–157.

Mary, N. K., Achuthan, C. R., Babu, B. H., and Padikkala, J. (2003). *In vitro* antioxidant and antithrombotic activity of *Hemidesmus indicus* (L.) R.Br. *J. Ethnopharmacol.*, **87**: 187-191.

Megaloudi, F. (2005). Wild and Cultivated Vegetables, Herbs and Spices in Greek Antiquity. *Environ Archaeol.*, **10**: 73-82.

Mukherjee, P.K., Verpoorte, R., and Suresh, B. (2000). Evaluation of *in-vivo* wound healing activity of *Hypericum patulum* (Family: Hypericaceae) leaf extract on different wound model in rats. *J. Ethnopharmacol.*, **70**: 315-321.

Nadkarni, K.M. (1976). The Indian Materia Medica Vol. I, (Popular Prakashan Pvt.) Ltd., Mumbai, India, 12-92.

Ngono, N. A., Ebelle E. R., Ndifor, F., Biyiti, L., Amvam P.H., and Bouchet, P. (2006). Activity of *Chromolaena odorata* (L.) King and Robinson (Asteraceae) of Cameroon. *Chemotherapy*, **52**: 103-106

Okwori, A. E. J., Dina, C.O., Junaid, S., Okeke, I. O., Adetunji, J. A., and Olabode, A. O. (2007). Antibacterial activities of *Ageratum conyzoides* extracts on selected bacterial pathogens. *Internet J Microbiol.*, **4**.

Oladejo, O.W., Imosemi, I.O., Osuagwu, F.C., Oyedele, O.O., Oluwadara, O.O., Ekpo, O.E., Aiku, A., and Akang, E.E. (2003). A comparative study of the wound healing properties of honey and *Ageratum conyzoides*. *Afr. J. Med. Med. Sci.*, **32**: 193-196.

Ozturk, N., Korkmaz, S., and Ozturk, Y. (2007). Wound-healing activity of St. John's Wort (*Hypericum perforatum* L.) on chicken embryonic fibroblasts. *J. Ethnopharmacol.*, **111**: 33-39.

Parry, O., Marks, J.A., and Okwuasaba, F.K. (1993). The skeletal muscle relaxant action of *Portulaca oleracea*: role of potassium ions. *J. Ethnopharm.*, **40**: 187–194.

Phan, T.T., Hughes, M.A., and Cherry, G.W. (1998). Enhanced proliferation of fibroblasts and endothelial cells treated with an extract of the leaves of *Chromolaena odorata* (Eupolin), an herbal remedy for treating wounds. *Plastic and Reconstructive Surgery*, **101**: 756-765.

Phan, T.T., Allen, J., Hughes, M.A., Cherry, G., and Wojnarowska, F. (2000). Upregulation of adhesion complex proteins and fibronectin by human keratinocytes treated with an aqueous extract from the leaves of *Chromolaena odorata* (Eupolin). *Eur. J. Dermatol.*, **10**: 522-527.

Phan, T.T., Hughes, M.A., and Cherry, G.W. (2001). Effects of an aqueous extract from the leaves of *Chromolaena odorata* (Eupolin) on the proliferation of human keratinocytes and on their migration in an *in vitro* model of reepithelialization. *Wound Repair Regen.*, **9**: 305-313.

Phan, T.-T., Wang, L., See, P., Grayer, R.J., Chan, S.-Y., and Lee, S.T. (2001). Phenolic compounds of *Chromolaena adorata* protect cultured skin cells from oxidative damage: Implication for cutaneous wound healing. *Biol. Pharmaceut. Bull.*, **24**: 1373-1379.

Phillips, G.D., Whitehe, R.A., and Kinghton, R. (1991). Initiation and pattern of angiogenesis in wound healing in the rat. *Am. J. Ana.*, **192**: 257–262.

Pointel, J.P., Boccalon, H., Cloarec, M., Ledevehat, C., and Joubert, M. (1987). Titrated extract of *Centella asiatica* (TECA) in the treatment of venous insufficiency of the lower limbs. *Angiology*, **38**: 46-50.

Poizot, A., and Dumez. D. (1978). Modification of the kinetics of healing after iterative exeresis in the rat. Action of a triterpenoid and its derivatives on the duration of healing. *C R Acad Sci Hebd Seances Acad Sci D.*, **286**: 789-792.

Prusti, A.B., and Behera, K.K. (2007). Ethnobotanical exploration of malkangiri district of Orissa, India. Ethnobotanical Leaflets edition.

Puri, H. S. (2003). Rasayana: Ayurvedic Herbs for Longevity and Rejuvenation by Dr published by Taylor and Francis, London, pp. 43-45.

Ramanujam, C.G.K., and Kalpana, T.P. (1992). *Tamarindus indica* L. an important forage plant for *Apis florea* F. in south central India. *Apidologie*, **23**: 403–413.

Rashed, A.N., Afifi, F.U., and Disi, A.M. (2003). Simple evaluation of the wound healing activity of a crude extract of *Portulaca oleracea* L. (growing in Jordan) in *Mus musculus* JVI-1. *J. Ethnopharm.*, **88**: 131–136.

Ravishankara, M. N., Shrivastava, N., Padh, H., and Rajani, M. (2002). Evaluation of antioxidant properties of root bark of *Hemidesmus indicus* R. Br. (Anantmul). *Phytomedicine*, **9**: 153-160.

Reddy, J. S., Rao, P. R., and Reddy, M. S. (2002).Wound healing effects of *Heliotropium indicum*, *Plumbago zeylanicum* and *Acalypha indica* in rats. *J Ethnopharmacol.*, **79**: 249-251.

Ross, R. (1980). Inflammation, cell proliferating and connective tissue formation in wound repair. In: Hunt, T.K. (Ed.), Wound Healing and Wound Infection. Appleton-Century-Crofts, New York, pp.1-45.

Salas, C.L., Fernández, M.M., and Chica, A.M.M. (2005). Topical chemotherapy for the treatment of burns. *Rev Enferm.*, **28**: 67-70.

Samy, J., Sugumaran, M., and Lee, K.L.W. (2004). Herbs of Malaysia: An Introduction to the medicinal, culinary, aromatic and cosmetic use of herbs. Times Edition, Kuala Lumpur.

Samy, R.P., Gopalakrishnakone, P. Houghton, P., and Ignacimuthu, S. (2006a). Purification of antibacterial agents from *Tragia involucrata*-A popular tribal medicine for wound healing. J. *Ethnopharmacol.*, **107**: 99-106.

Samy, R.P., Gopalakrishnakone, P., Sarumathi, M., and Ignacimuthu, S. (2006b). Wound healing potential of *Tragia involucrata* extract in rats. *Fitoterapia*, **77**: 300-302.

Santareìm, E.R., and Astarita, L.V. (2003). Multiple shoot formation in *Hypericum perforatum* L. and hypericin production. *Brazilian J Plant Physiol.*, **15**: 43-47.

Saxena, V. K., and Albert, S. (2005). β-Sitosterol-3-O-β-D-xylopyranoside from the flowers of *Tridax procumbens*. Linn. *J. Chem. Sci.*, 1**17(3)**: 263-266.

Schmidt, G.J., and Schilling, E.E. (2000). Phylogeny and Biogeography of *Eupatorium* (Asteraceae: Eupatorieae) Based on Nuclear ITS Sequence. *Amer. J. Bot.*, **87**: 716–726.

Sharma, R. (2003). Medicinal plants of India – An enclycopaedia. Daya Publ. House, Delhi. India. pp. 302-310.

Shetty, B.S., Udupa, S.L., Udupa, A.L., and Somayaji, S.N. (2006). Effect of *Centella asiatica* L (Umbelliferae) on normal and dexamethasone-suppressed wound healing in Wistar albino rats. *Int. J. Lower Extrem Wounds*, **5**: 137-143.

Shim, P. J., Park, J. H., Chang, M. S., Lim, M. J., Kim, D. H., Jung, Y. H., Jew, S. S., Park, E. H., and Kim, H. D. (1996). Asiaticoside mimetics as wound healing agent. *Bioorg. Med. Chem. Lett.*, **6**: 2937–2940.

Shukla, A., Rasik, A. M., and Dhawan, B. N. (1999). Asiaticoside-induced elevation of antioxidant levels in healing wounds. *Phytoth. Res.*, 1**3**: 50–54.

Sikareepaisan, P., Suksamrarn, A., and Supaphol, P. (2008). Electrospun gelatin fiber mats containing a herbal - *Centella asiatica* - Extract and release characteristic of asiaticoside Nanotechnol. 19: art. no. 015102 2008.

Singh, M.P., and Panda, H. (2005). Medicinal herbs with their formulations. Daya Publishing House, Delhi. Vol. 2. pp. 954-978.

Srinivas, K., Rao, M.E.B., and Rao, S.S. (2000). Anti-inflammatory activity of *Heliotropium indicum* Linn and *Leucas aspera* spreng. in albino rats. *Indian J. Pharmacol.*, **32**: 37-38.

Stepp, J.R. (2004). The role of weeds as sources of pharmaceuticals. *J. Ethnopharm.*, **92**: 163-166.

Stone, B. C. (1970). The flora of Guam. *Micronesica*, **6**: 1-65.

Subramanian, S., and Nagarajan, S. (1988). Wound healing activity of *Pongamia pinnata* and *Cynodon dactylon*. *Fitoterapia*, **59**: 43-44.

Suguna, L., Sivakumar, P., and Chandrakasan, G. (1996). Effects of *Centella asiatica* extract on dermal wound healing in rats. *Indian J Exp Biol.*, **34**: 1208-1211.

Sunil, K., Parameshwaraiah, S., and Shivakumar, H.G. (1998). Evaluation of topical formulations of aqueous extract of *Centella asiatica* on open wounds in rats. *Indian J Exp Biol.*, **36**: 569-572.

Tenni, R. Zanaboni, G., De Agostini, M.P., Rossi, A., Bendotti, C., and Cetta, G. (1988). Effect of the triterpenoid fraction of *Centella asiatica* on macromolecules of the connective matrix in human skin fibroblast cultures. *Ital J Biochem.*, **37**: 69-77.

Thang, P.T., Patrick, S., Teik, L.S., and Yung, C.S. (2001). Anti-oxidant effects of the extracts from the leaves of *Chromolaena odorata* on human dermal fibroblasts and epidermal keratinocytes against hydrogen peroxide and hypoxanthine-xanthine oxidase induced damage. *Burns*, **27**: 319-327.

Udupa, S.L., Udupa, A.L., and Kulkarni, D.R. (1991). Influence of *Tridax procumbens* on lysyl oxidase activity and wound healing. *Planta Med.*, **57**: 325-327.

Udupa, A.L., Kulkarni, D.R., and Udupa, S.L. (1995). Effect of *Tridax procumbens* extracts on wound healing. *Inter. J. Pharm.*, **33**: 37–40.

Wichtl, M. (1986). *Hypericum perforatum* L. -Das Johanniskraut. *Zeitschrift fur Phytotherapie*, **3**: 87-90.

Yagoub, S. O., Shami, H.A., S., Ahmed, B., and Zohier, A. (2006). Antimicrobial activity of some medicinal plants against some Gram positive, Gram negative and fungi. Almagboul Scientific Research Outlook and Technology Development in the Arab World (SRO4) Symposium MnE6 – 872.

Utilisation and Management of Medicinal Plants Vol. 3 (2015) *Pages* **121–133**
Editors: **V.K. Gupta and A. Kaul**
Published by: **DAYA PUBLISHING HOUSE, NEW DELHI**

8

Inter-specific Variation Studies on *Cissus* Species using Histochemical and SDS-PAGE

I. Revathy[1], M. Johnson[1*], T. Shibila[1], M. Narayani[1] and S. Ramalakshmi[1]

ABSTRACT

The present investigation was aimed to reveal the inter-specific variation among the selected Cissus species using histochemical and SDS-PAGE analysis. Using histochemical studies the presence of metabolites lignin and tannin was detected by standard method. The isolation and separation of protein using SDS-PAGE was performed by the standard methods. Tannin and lignin are profusely present in the outer layer of the all the studied Cissus species. The studied Cissus species expressed bands at different regions and with varied MW-Rf values. C. repanda expressed maximum number (10) of protein bands followed by C. xavieriensis showed 6 bands; C. quadrangularis represented 5 bands and Cissus quadrangularis var. rotundus displayed only four protein bands. The results of the present study revelaed the inter-specific variations among the studied Cissus species. These histochemical and protein profile can be used as pharmcognostical marker to identify the medicinal sources from its adulterants in the pharmaceutical industries and find solution for the taxonomical identity.

Keywords: *Cissus*, Histochemistry, Lignin, Tannin, SDS-PAGE, Pharmcognostical marker.

Introduction

Cissus is a genus of approximately 350 species of woody vines belonging to the grape family Vitaceae commonly known as Hadjod. It is cosmopolitan in distribution,

1 Centre for Plant Biotechnology, Department of Botany, St. Xavier's College (Autonomous), Palayamkottai - 627 002, Tamil Nadu, India.

* *Corresponding author*: E-mail: ptcjohnson@gmail.com

through a large number of *Cissus* species is found in the tropics. It has been reported to have antidiabetic, antioxidant, anti neurogenerative diseases, anti-flatulence, antifungal, anti-inflammatory, analgesic, antibacterial, protects from cardiovascular disease, increases longevity and cancer suppressive agents (Athar *et al.*, 2009; Joseph and Raj, 2011; Sun *et al.*, 2010; Szkudelski and Szkudelska, 2011). Histochemistry deals with the identification of chemical components of cells and tissues. The applications of histochemical characters are now a common practice for the identification and characterization of taxon to resolve taxonomic problems (Kadam *et al.*, 2013). In recent years, the importance of histochemistry in solving critical biosystematic problems is as popular as the use of cytological, biochemical and molecular markers.

Traditionally, genetic diversity is assessed based on morphological features such as plant height, reproductive features, day length sensitivity, local adaptation etc. A wide spectrum of simple and overlapping variations is now documented in plants (Rout *et al.*, 2010). Protein analysis (SDS-PAGE) is simple to use in a variety of applications in plant taxonomical research. The information on polymorphism using protein and isozyme profiles in a set of genotype is useful in tagging genes of interest and genetic mapping in long run to facilitate marker assisted selection where as the banding positions and regions of activity are variable based on the developments of plants (Johnson *et al.*, 2011). During the past 30 years, various techniques have been used for analysis of genetic variability at the molecular level in plants *viz.*, isozymes (Geetha and Wooten, 2003), RFLP (Mohapatra and Rout, 2005) and RAPD (Theerakulpisut *et al.*, 2008). Protein profiles are still powerful tools for determining genetic homology at the molecular level and for solving problems in systematic methodology.

A few phytochemical, pharmacological and pharmacognostical studies were carried out on selected taxa of *Cissus quadrangularis* (Shah, 2011; Austin *et al.*, 2009; Joseph *et al.*, 2011; Sathyaprabha *et al.*, 2010; Rosy and Rosakutty, 2012; Sathish Kumar *et al.*, 2012; Sivasubramanian and Brindha, 2013; Eswaran *et al.* 2012; Sumitra Chanda *et al.*, 2013) but there is no report on the chemosystematic studies on the *Cissus* taxa using histochemical and protein profiles. With this background, the present study was aimed to produce low cost biomarkers for the medicinally important *Cissus* species *viz.*, *Cissus quadrangularis*, *Cissus quadrangularis* var. *rotundus*, *Cissus xavieriensis* and *Cissus rependa* using histochemical and SDS-PAGE analysis.

Materials and Methods

Collection of Plant Materials

Cissus species *viz.*, *Cissus xavieriensis* Manickam *et al.* and *Cissus quadrangularis* var. *rotundus*, *Cissus quadrangularis* L. and *Cissus repanda* Vhal. were selected for the present investigation. *C. xavieriensis* and *C. quadrangularis* var. *rotundus* were collected from the natural habitats at St. Xavier's College, Palayamkottai, Tamil Nadu, India. *C. quadrangularis* was collected from Vadakankulam, Tamil Nadu, India. *C. repanda* was collected from Keeriparai, Kanyakumari District, Tamil Nadu, India.

Histochemical Analysis

Fresh free hand sections were made from the stem and leaves of selected *Cissus* species. Histochemical tests were made on the fresh sections of the selected plants treated with specific reagents to identify the presence or absence of metabolites like lignin and tannin. To detect lignin, Lugol's iodine and ammonium hydroxide were used whereas to detect tannin the mixture of Lugol's Iodine (60-70 per cent) and sulphuric acid were used.

Protein Analysis

For the electrophoretic studies, 500 mg of fresh young leaves were harvested from the mother plants and washed in de-ionized water and ground using pre-chilled mortar and pestle with 0.1M Tris buffer (pH 7.0). The resultant slurry was centrifuged at 10,000 rpm for 10 min at 4°C and the supernatant were collected and separated by slab gel electrophoresis. SDS-PAGE was carried out at 25°C in the air conditioned room (Anbalagan, 1999). After running the electrophoresis, the gel was stained with silver nitrate staining procedure have been developed for visualizing proteins that can be subsequently digested and recovered from the gel. The gel was observed using a Vilber Loubermat gel documentation system and banding profiles of protein was determined by the migration from the origin towards the anode. The similarity and variation between the bands were estimated by NTSsys software analysis and the cladogram was constructed.

Results

Histochemical Studies

The occurrence and distribution of various metabolites (lignin and tannins) in *C. quadrangularis*, *C. quadrangularis* var. *rotundus*, *C. xavieriensis* and *C. repanda* were illustrated in Figures 8.1 and 8.2. Tannin and lignin are profusely present in the epidermis of the all the studied *Cissus* species. The occurrence and distribution of metabolites lignin and tannin in selected *Cissus* species were illustrated in Tables 8.1 and 8.2.

Protein Analysis

For biosystematic analysis of *Cissus* species, SDS-PAGE was performed. Multiple regions of activity were obtained in the protein electrophoretic system of selected *Cissus* species *viz.*, *Cissus quadrangularis*, *Cissus quadrangularis* var. *rotundus*, *Cissus xavieriensis* and *Cissus repanda*. The studied plants expressed protein bands at different regions and with varied MW-Rf values (Table 8.3; Figure 8.3). *C. repanda* expressed maximum number (10) of protein bands followed by *C. xavieriensis* with 6 bands, *C. quadrangularis* represented 5 bands and *Cissus quadrangularis* var. *rotundus* displayed only four protein bands.

Based on the occurrence of protein bands in the gel system, the protein profiles of studied *Cissus* species gel electrophoresis system was classified into ten regions. Region 1 showed two bands, $PP1^{1}$ (MW- 263.02 kD) and $PP1^{2}$ (MW- 218.77 kD) expressed its unique presence in *C. repanda* and *C. xavieriensis* respectively. Region 2 depicted with single band, $PP2^{1}$ (MW- 199.5 kD) was shared by *C. quadrangularis* and

Table 8.1: Histochemical Localization of Lignin in Stem and Leaves of various *Cissus* Species

Plants	*Reagents*	*Colouration*	*Phytochemicals*	*Tissue Location*	
				Stem	*Leaf*
C. quadrangularis	Champerline	Yellowishorange	Lignin	Xylem, phloem, epidermis	Few epidermis and outerlayer of vascular bundles
C. quadrangularis var. *rotundus*				Epidermis, xylem, phloem	Few in vascular bundles
C. xaveriensis				Xylem, slightly in phloem	–
C. rependa				–	

Table 8.2: Histochemical Localization of Tannins in Stem and Leaves of various *Cissus* Species

Plants	*Reagents*	*Colouration*	*Phytochemicals*	*Tissue Location*	
				Stem	*Leaf*
C. quadrangularis	Lugol's	Brick Redyellow	Tannin	Absence of tannin	Very few in epidermis
C. quadrangularis var. *rotundus*				Epidermis, xylem, phloem	Scattered Epidermis, xylem, phloem
C. xaveriensis				Epidermis, few in cortex, phloem and xylem	–
C. rependa				–	Few in phloem cells

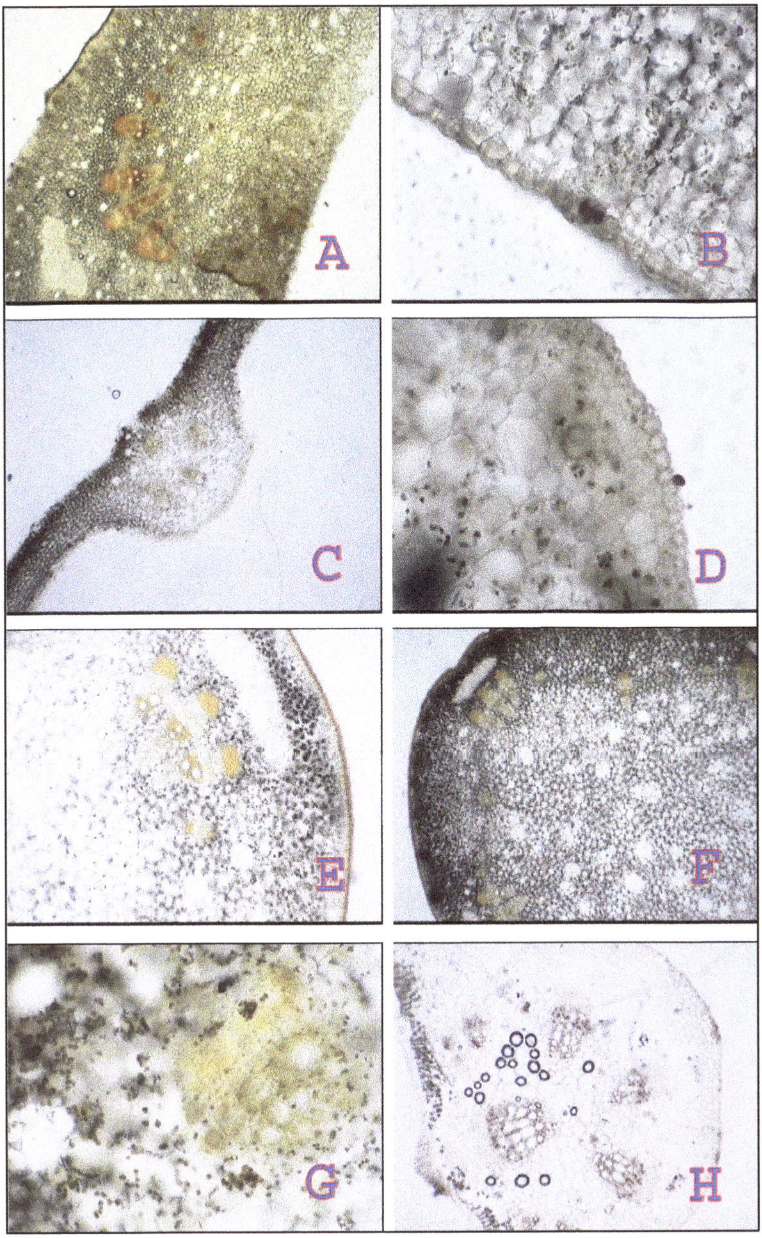

Figure 8.1: Histochemical Studies of *Cissus quadrangularis* L. and *Cissus quadrangularis var. rotundus.*

A: Lignin presence in *Cissus quadrangularis* L. stem; B: Tannin existence in *Cissus quadrangularis* L stem; C: Lignin occurence in *Cissus quadrangularis* L Leaves; D: Tannin incidence in *Cissus quadrangularis* L. Leaves; E: Lignin presence in *Cissus quadrangularis var. rotundus* stem; F: Tannin existence in *Cissus quadrangularis var. rotundus* stem; G: Lignin occurence in *Cissus quadrangularis var. rotundus* Leaves; H: Tannin incidence in *Cissus quadrangularis var. rotundus* Leaves.

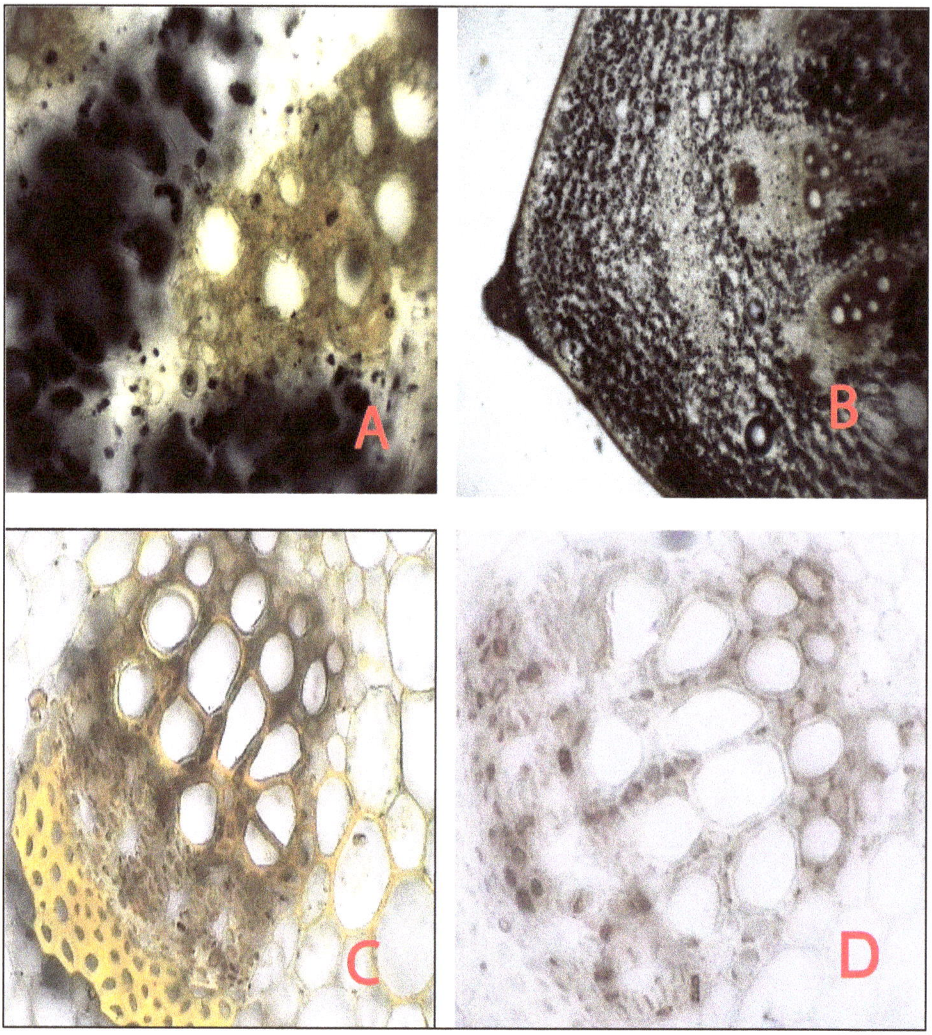

Figure 8.2: Histochemical Studies of *Cissus xavieriensis* and *Cissus repanda*.

A: Lignin occurence in *Cissus xaveriensis* Leaves; B: Lignin and Tannin presence in *Cissus xavieriensis* stem; E: Lignin presence in *Cissus repanda* Leaves; F: Tannin incidence in *Cissus repanda* Leaves.

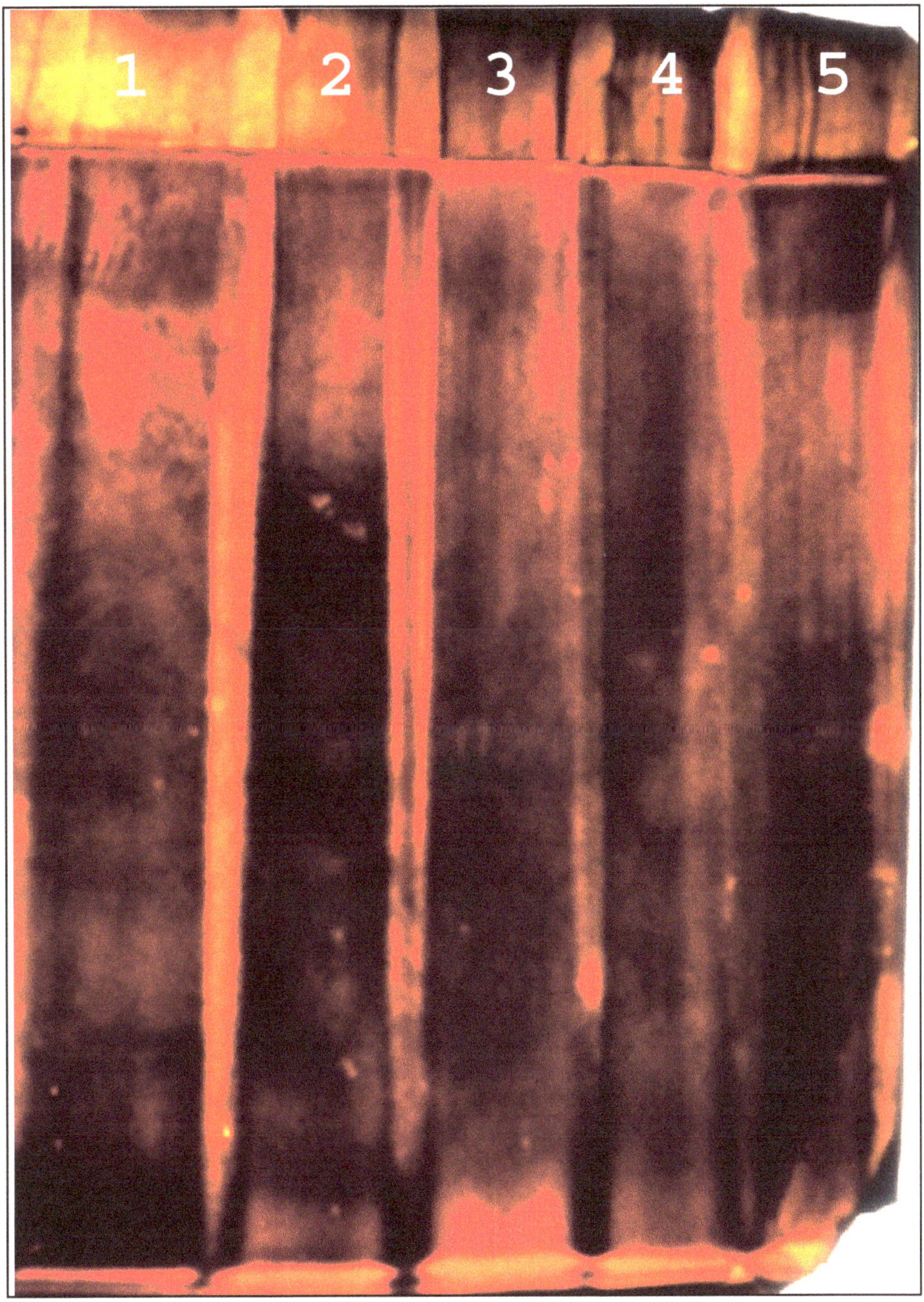

Figure 8.3: Proteomic Studies of *Cissus* Species.

L1: *C. quadrangularis*; L2: *C. quadrangularis* var. *rotundus*; L3: *C. xavieriensis*; L4: *C. repanda*; L5: Protein Marker.

Table 8.3: SDS-PAGE Profile of Studied *Cissus* Species

Rf	*Molecular Weight (kD)*	*Region*	*Position*	L_1	L_2	L_3	L_4
0.05	263.02	1	$PP\ 1^1$				+
0.08	218.77		$PP\ 1^2$			+	
0.1	199.5	2	$PP2^1$	+			+
0.32	95.49	4	$PP4^1$				+
0.37	72.44		$PP4^2$			+	
0.4	63.09	5	$PP5^1$	+			
0.42	57.54		$PP5^2$				+
0.47	41.68		$PP5^3$	+			+
0.49	39.81		$PP5^4$		+		
0.52	38.3	6	$PP6^1$			+	
0.54	36.3		$PP6^2$	+			
0.55	34.67		$PP6^3$				+
0.59	27.5		$PP6^4$		+		
0.62	21.8	7	$PP7^1$			+	
0.69	19.05		$PP7^2$		+		+
0.71	18.19	8	$PP8^1$	+		+	
0.76	16.59		$PP8^2$				+
0.81	14.12	9	$PP9^1$			+	
0.86	11.48		$PP9^2$				+
0.91	10	10	$PP10^1$		+		+
Total				5	4	6	10

L1: *C. quadrangularis*; L2: *C. quadrangularis* var. *rotundus*; L3: *C. xavieriensis*; L4: *C. repanda*.

C. repanda. Region 3 was failed to show active bands. Region 4 revealed two bands, $PP4^1$ (MW-95.49) and $PP4^2$ (MW-72.44 kD) was displayed its uniqueness by showing its existence in *C. repanda* and *C. xavieriensis* correspondingly. Region 5 displayed four bands. The protein $PP5^1$ (MW- 63.09 kD), $PP5^2$ (MW- 57.54 kD) and $PP5^4$ (MW- 39.81 kD) was dispalyed its distinctive occurrence in *C. quadrangularis*, *C. repanda* and *C. quadrangularis* var. *rotundus* respectively. $PP5^3$ (MW- 41.68 kD) was showed its jointly presence in *C. quadrangularis* and *C. repanda*. Region 6 illustrated with four active bands. $PP6^1$ (MW- 38.3 kD), $PP6^2$ (MW- 36.3 kD), $PP6^3$ (MW- 34.67 kD) and $PP6^4$ (MW- 27.5 kD) was showed its unique occurrence in *C. xavieriensis*, *C. quadrangularis*, *C. repanda* and *C. quadrangularis* var. *rotundus* correspondingly. The region 5 and 6 clearly distinguished the studied *Cissus* species. These proteins profile can be used as biochemical marker to distinguish the medicinally important plants. Region 7 displayed two bands with MW-Rf values 0.62 and 0.69. The protein $PP7^1$ (MW- 21.8 kD) was demonstrated its restricted occurrence only in *C. xavieriensis*.

PP7[2] with molecular weight 19.05 kD was shared by *C. quadrangularis* var. *rotundus* and *C. repanda*. Similarly, Region 8 expressed two active bands. PP8[1] (MW- 18.19 kD) was dispalyed the common existence in *C. quadrangularis* and *C. xavieriensis*. PP8[2] (MW-16.59 kD) showed its unique presence in *C. repanda*. Region 9 represented two protein bands at positions PP9[1] (MW- 14.12 kD) and PP9[2] (MW-11.48 kD) were observed in *C. xavieriensis* and *C. repanda*. Region 10 explained with single band, PP10[1] (MW- 10 kD) was common to *C. quadrangularis* var. *rotundus* and *C. repanda*.

Based on the SDS-PAGE profile of studied *Cissus* species, the cladogram was constructed. The cladogram shows two clusters C_1 and C_2 (Figure 8.4). The cluster C_1 is branched into two nodes, C_1N^1 and C_1N^2. C_1N^1 showed the distinct character of *C. quadrangularis*. C_1N^2 explained the similarity between *C. quadrangularis* var. *rotundus* and *C. repanda*. The cluster C_2 displayed the unique character of *C. xavieriensis*.

Discussion

Natural products research has an important role in the drug discovery process. Chemical and biological research on natural products over the past two centuries has not only provided drugs for the treatment of various human ailments, but has provided the stimulus for the development of modern synthetic organic chemistry and the emergence of medicinal chemistry as a major route for the discovery of novel and more effective therapeutic agents. Secondary metabolites produced by plants constitute a rich source of bioactive substances and nowadays the scientific interest has increased due to the search for new drugs from plant origin. The medicinal value of plants lies in some chemical substances. The most important of these bioactive compounds of plants are alkaloids, flavonoids, tannins and phenolic compounds (Siddique *et al.*, 2011; Renisheya *et al.*, 2012).

Proteins are direct gene products that reflect characteristic gene expression. Numerous researches revealing the full genome sequence of model plants. However, gene sequence does not give complete information about gene function, developmental and regulatory biology or biochemical kinetics (Qureshi *et al.*, 2007). Proteins are macromolecules directly responsible for most biological processes in a living cell, while protein function is directly influenced by post-translational modifications, which cannot be followed by genome studies. Therefore, it is necessary to conduct proteomic studies which elucidate protein presence under certain environmental conditions (Porubleva *et al.*, 2000). In the present study, SDS-PAGE was carried out to identify the biochemical similarity and variation among the studied *Cissus* species *viz.*, *C. quadrangularis*, *C. quadrangularis* var. *rotundus*, *C. xavieriensis* and *C. repanda* and the cladogram was constructed. The SDS-PAGE results revealed the proteomic difference among the studied *Cissus* speceis.

In recent years, limitations of numerical taxonomy *viz.* morphology, anatomy and cytology have been over come by biochemical and molecular markers (Johnson *et al.*, 2012). Babu *et al.* (2010) used the protein markers to classify the medicinally and economically important plants belonging to the genus *Amaranthus* using SDS-PAGE. Similar to the previous observations, in the present study also we observed the intra-generic variation among the selected *Cissus* species. The results of the present study confirm the closeness and divergence between the selected four species of *Cissus* and

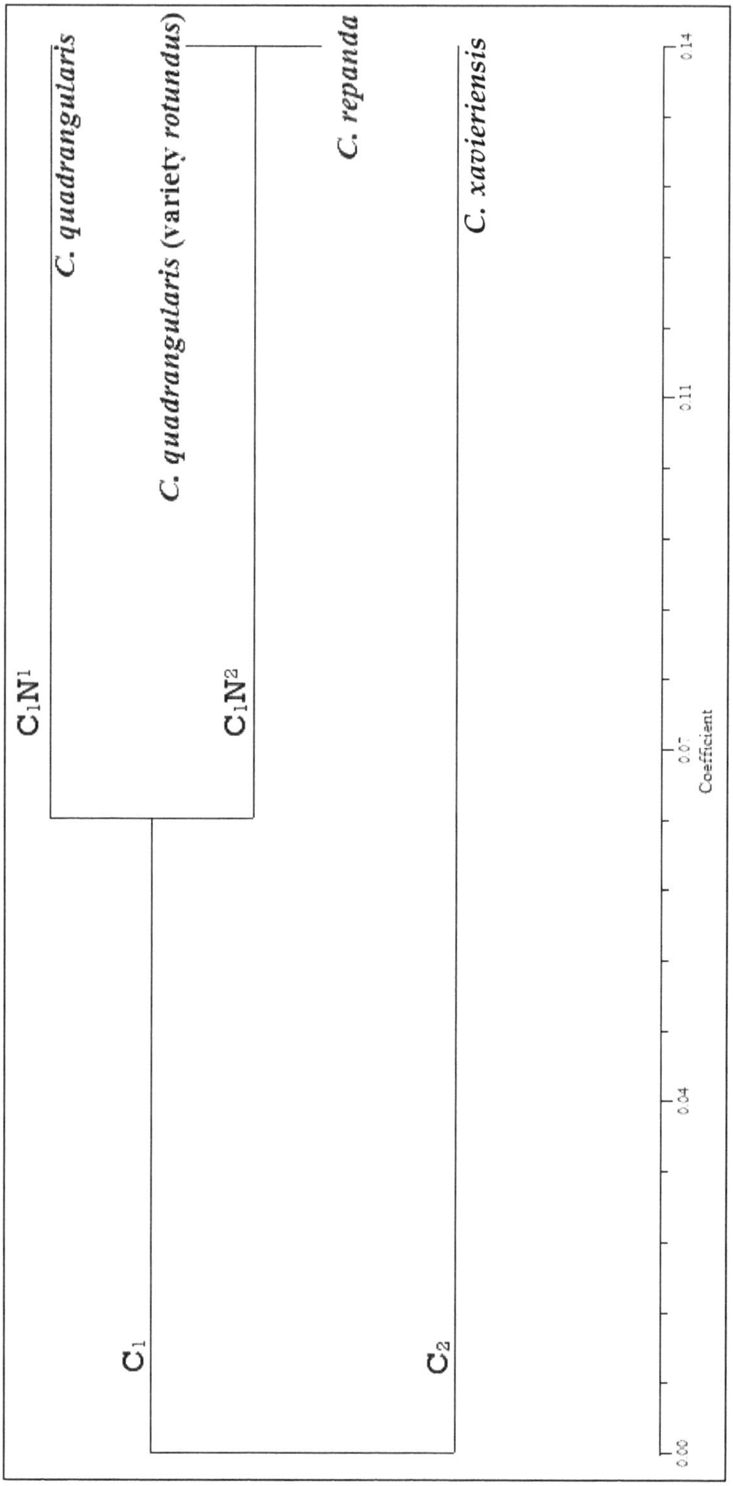

Figure 8.4: Cladogram of Studied *Cissus* Species.

these protein profiles may be used as a biochemical marker to distinguish the medicinally important species from its adulterant in the pharmaceutical industry.

Conclusion

Histochemical screening of the leaves and stem of various *Cissus* species revealed the presence of tannin and lignin in different tissues. It is found that more amount of chemical constituents' occurence in the vascular bundle is higher than other tissues. The SDS-PAGE banding profile revealed the variation among four *Cissus* species which may be used as a biochemical marker to distinguish the medicinally important species. It is necessary to study further the structural and functional roles of these reported proteins to use as a marker for the selected species of *Cissus*.

References

Anbalagan, K. (1999). An introduction to electrophoresis, Electrophoresis Institute, Yercaud, Tamil Nadu.

Athar, M., Back, J.H., Kopelovich, L., Bickers, D.R., and Kim, A.L. (2009). Multiple molecular targets of resveratrol: Anti-carcinogenic mechanisms. *Arch.Biochem. Biophys.*, **486**(2): 95-102.

Austin R. S., Danielle Vidaurre, George Stamatiou, Robert Breit, Nicholas, J. P., Dario Bonetta *et al.* (2011). Next-generation mapping of Arabidopsis genes. *The Plant Journal*, **67**: 715–725

Babu, A., Raja, D. P., Johnson M., and Sivasankari G. (2010). Variability studies of selected taxa of *Amaranthus*. *Journal of Basic and Applied Biology*, **4** (1 and 2): 125-136.

Eswaran, R., Anandan A., Doss A., Sangeetha G., and Anand, S. P. (2012). Analysis of chemical composition of *Cissus quadrangularis* Linn. By GC-MS, *Asian J Pharm Clin Res.*, **5**(2): 139-140.

Geetha, T., and Wooten, M.W. (2003). Association of the atypical protein kinase C-interacting protein p62/ZIP with nerve growth factor receptor TrkA regulates receptor trafficking and Erk5 signaling. *J Biol Chem.*, **278**: 4730–39.

Johnson, M., Dooslin Mary, D., and Babu, A. (2011). Salt stress responses on protein profile in *Vigna unguiculata* L. *Journal of Stress Physiology and Biochemistry*, 7 (4): 241-249

Johnson, M., Janakiraman, N. and Irudayaraj, V. (2012). Isozyme analysis on different varieties of sugarcane. *Journal of Stress Physiology and Biochemistry*, **8**(2): 22-31.

Joseph, B., and Justin Raj, S. (2011). Comparative study on various properties of five medicinally Important plants. *International journal of Pharmacology*, **7**(2): 206-211.

Sun, Z.L., Wang. Y.,Guo. M.L., and Li, Y.X. (2010). Two new hepaprotective saponins from semen celosiae. *Fitoterapia*, **81**(5): 375-380.

Joseph, L.N., Watson, J.E.M., Maloney, R.F., and Possingham, H.P. (2011). Securing non-flagship species from extinction. *Conserv Lett.*, doi: 10.1111/j.1755-263X.2011.00174.x.

Kadam, R., Bourne, D., Kompella, U., and Aquilante, C. (2013). Effect of cytochrome P450 2C8*3 on the population pharmacokinetics of pioglitazone in healthy caucasian volunteers. *Biol. Pharm. Bull.*, **36**: 245–251

Mohapatra, A., and Rout, G. (2005). Identification and analysis of genetic variation among rose cultivars using random amplified polymorphic DNA. *Naturforsh*, **60**: 611-617

Porubleva, L., and Chitnis, P.R. (2000). Proteomics: A powerful tool in the post-genomic era. *Indian J Biochem Biophys.*, **37**: 360-8.

Qureshi, M.I., Qadir, S., and Zolla, L. (2007). Proteomics based dissection of the stress responsive pathways in plants. *J. Plant Physiol.*, **164**: 1239-1260.

Renisheya, J.J.M.T., Johnson, M., Mary, U.M., and Arthy, A. (2011). Antibacterial activity of ethanolic extracts of selected medicinal plants against human pathogens. *Asian J Trop Biomed.*, S76-S78.

Rosy, A.B., and Rosakutty, P. J. (2012). GC-MS analysis of methanol wild plant and callus extracts from three *Cissus* species, Family Vitaceae. *Journal of Chemical and Pharmaceutical Research*, **4**(7): 3420-3426

Rout, J., Das, P., and Upreti, D. K. (2010). Epiphytic lichen diversity in a reserve forest in southern Assam, northeast India. *Tropical Ecology*, **51**: 281-288.

Sathish Kumar, T., Anandan, A., and Jegadeesan M. (2012). Identification of chemical compounds in *Cissus quadrangularis* L. Variant-I of different sample using GC-MS analysis. *Archives of Applied Science Research*, **4** (4): 1782-1787

Sathyaprabha, G., Kumaravel.S., Ruffina. D., and Praveenkumar. P. (2010). A comparative study on antioxidant, proximate analysis, antimicrobial activity and phytochemical analysis of *Aloe vera* and *Cissus quadrangularis* by GC-MS. *Journal of Pharmacy Research*, **3**(12): 2970-2973.

Shah, U. (2011). *Cissus quadrangularis* L: Phytochemicals, traditional uses and pharmacological activities – A review. *Int J Pharm Sci.*, **3(4)**: 4 - 44.

Sivasubramanian, R., and Brindha, P. (2013). *In- vitro* cytotoxic, antioxidant and GC-MS studies on *Centratherum punctatum* cass. *Int J Pharm Pharm Sci.*, **5 (3)**: 364-367.

Siddique, Y.H., Ara, G., Faisal, M., and Afzal, M. (2011). Protective role of *Plumbago zeylanica* extract against the toxic effects of ethinylestradiol in the third instar larvae of transgenic *Drosophila melanogaster* (hsp70-lacZ) Bg9 and cultured human peripheral blood lymphocytes. *Alter Med Studies*, **1**: 26-29.

Sumitra Chanda, Yogesh Baravalia, and Krunal Nagani (2013). Spectral analysis of methanol extract of *Cissus quadrangularis* L. stem and its fractions. *Journal of Pharmacognosy and Phytochemistry*, **2**(4): 132-140.

Szkudelski, T., and Szkudelska, K. (2011). Anti-diabetic effects of resveratrol. *Ann. N. Y. Acad. Sci.*, **1215**: 34-39.

Theerakulpisut, P., Kanawapee, N., Maensir, D., Bunnag, S. and Chantaranothai, P. (2008). Development of species-specific SCAR markers for identification of three medicinal species of *Phyllanthus*. *Journal of Systematics and Evolution*, **46** (4): 614–621.

Vijayalakshmi, A., Kumar, P.R., SakthiPriyadarsini, S., and Meenaxshi, C. (2013). Pharmacognostic and phytochemical investigation of the aerial parts of *Cissus quadrangularis* Linn. *Journal of Scientific and Innovative Research*, **2** (5): 872-879.

Utilisation and Management of Medicinal Plants Vol. 3 (2015) *Pages* **135–145**
Editors: **V.K. Gupta and A. Kaul**
Published by: **DAYA PUBLISHING HOUSE, NEW DELHI**

9

Antioxidant, Anti-proliferative Activities and Phenolic Content of *Ipomoea nil* (L.) Roth.

Madhulika Bhagat[1*] and A.K. Saxena[2]

ABSTRACT

Ipomoea nil is a species of morning glory known by several common names, including picotee morning glory, ivy morning glory, and Japanese morning glory. However, very little documentation is available for its antioxidant and anticancer properties. This study describes the antioxidant and anti-proliferative activity of Ipomoea nil.The antioxidant and anti-proliferative activities of aerial parts of I. nil were investigated.In the present study, an attempt has been made to evaluate in vitro antioxidant activity of different extracts of this plant by using the 1,1-diphenyl-2-picryl-hydrazyl (DPPH) radical scavenging assay, total antioxidant capacity, nitric oxide scavenging activity, ion chelating activity, hydroxyl radical scavenging activity and its correlation with total phenolic content. Anti-proliferative activity is evaluated against human cancer cell lines at different concentrations by using sulphorohadamine assay. Alcoholic extract of the Ipomoea nil aerial parts showed significant antioxidant and anti-proliferative properties when compared with hydro-alcoholic and aqueous extract presenting lower IC_{50} *values particularly in free radical scavenging activity, including DPPH scavenging activity (11.79±0.35µg/mL) and total antioxidant activity (30.17±1.8/mL). The nitric oxide radical scavenging, ion chelating and hydroxyl radical activity was also found maximum with* IC_{50} *=33.99±1.35µg/mL, 24.13±0.61µg/mL and 42.17±1.47µg/mL respectively. Dose dependent increase in the activity was observed in all the assays. The*

1 School of Biotechnology, University of Jammu, Jammu – 180 006, J&K, India.

2 Cancer Pharmacology Division, Indian Institute of Integrative Medicine, Canal Road, Jammu – 180 001, J&K, India.

* *Corresponding author*: E-mail: madhulikasbt@gmail.com

total phenolic and flavonoid contents were in the range of 243-189 mg gallic acid-equivalent (GAE)/g and 81-79 mg rutin-equivalent (RE)/g extract respectively. Significant anti-proliferatve activity was observed against Hep-2, liver cancer cell line (IC_{50}=9.2±1.35μg/mL) and SiHa, cervix cancer cell line (IC_{50}=12.23±5.41μg/mL) respectively. These findings demonstrate that Ipomea nil alcoholic extracts exhibit high phenolic content, antioxidant activity, and potent anti-proliferative activity and may prove to be an effective cancer chemopreventive agent.

Keywords: Anti-proliferative, Free radical scavenging, Human cancer cell lines, *Ipomea*, Medicinal plants, Sulphorohadamine blue.

Introduction

Free radicals are implicated for more than 80 diseases including diabetes mellitus, arthritis, cancer and several others. These free radicals are known to be scavenged by synthetic antioxidants, but due to their adverse side effects leading to carcinogenicity; search for effective and natural antioxidants has become crucial (Choi *et al.*, 2007; Adeolu *et al.*, 2009). Traditionally, Indian medicinal plants have been used from ages, as remedies for various human diseases as they are known to contain components of therapeutic value. Several isolated plant constituents as well as extracts have been acknowledged to possess antioxidant effects against free radicals in biological systems (Mervat *et al.*, 2009). Flavonoids and phenolic compounds are extensively distributed in plants and reported to exert multiple biological effects including antioxidant, free radical scavenging abilities, anti-inflammatory, anti-allergic, anti-microbial, anti-thrombotic, cardio-protective, anti-cancer and vasodilatory effects (Balasundram *et al.*, 2006).

Ipomea nil belongs to family Convolvulaceae. This genus consists of more than 200 species that are widely distributed in tropical and sub-tropical regions of the countries. Some of them are frequently used in folk medicine for the treatment of several diseases (Souza *et al.*, 2000). Traditionally, the seeds are thermogenic, purgative, vermifuge, depurative, anti-inflammatory, carminative. They are useful in vitiated condition of *kapha* and *vata*, inflammations, constipation, verminosis, skin diseases, leucoderma, scabies, dyspepsia, flatulence, bronchitis, gout, arthralgia, cephalgia, hepatopathy, splenopathy and fever (Warrier *et al.*, 1994). The seeds of this plant are acrid, sweet and contain diterpene glycosides, and they posses moderate to mild anticancer property against various cancer cell lines (Kim *et al.*, 2009). So, keeping the above medicinal values in mind, the present study was aimed to identify the active antioxidant and anticancer extracts of aerial parts of *Ipomea nil* that can be further fractionated and may exploited for use in food and pharmaceutical industry.

Materials and Methods

Plant Collection

Aerial parts of *Ipomea nil* were collected from local area of Jammu region in the month of December and were authenticated at source by taxonomist of the institute. A voucher specimen has been deposited at the herbarium of the Institute vide IIIM collection. The authenticated and freshly collected aerial parts were chopped and air dried under shade.

Extraction

Three extracts of the plant material (aerial parts) were made with 95 per cent alcohol, alcohol-water (1:1) and water by using repeated solvent extraction procedure. Dried powdered plant material was percolated in 95 per cent ethanol at ambient temperature for 16 h. The solvent was decanted and the process was repeated four times. The pooled solvent was evaporated under reduced pressure to yield alcoholic extract. Similarly, hydro-alcoholic extract was prepared. The dried plant material was soaked in alcohol-water (1:1, 1 L) and the extract was obtained. The dried powdered plant material was heated with distilled water (1.5 L) on steam bath for 2 h for the preparation of aqueous extract, the supernatant was decanted and filtered through celite powder and the process was repeated four times. The pooled extract was concentrated on rotavapour and dried in a lyophilizer.

DPPH Radical Scavenging Activity

DPPH scavenging activity was measured by spectrophotometric method (Sreejayan and Rao, 1996). To 1 mL of various concentrations of the extract, 1 mL of DPPH solution (0.1 mM) was added. An equal amount of methanol and DPPH served as control. After 20 min of incubation in dark, the absorbance was recorded at 517 nm. The experiments were performed in triplicate and the percentage of inhibition calculated (Prasanth *et al.*, 2000) by the formula:

Inhibition per cent = (Control- Test/Control)/control X 100

Total Antioxidant Capacity

Total antioxidant capacity was measured by spectrophotometric method (Shirwaikar *et al.*, 2003). 0.1 mL of the extract (10 mg/mL) dissolved in water was mixed in centrifuge tube with 1 mL of reagent solution (0.6 M sulphuric acid, 28 mM sodium phosphate and 4 mM ammonium molybdate).The tubes were capped and incubated in a thermal block at 95 °C for 90 min. After cooling to room temperature, the absorbance of aqueous solution was measured at 695 nm against blank. Ascorbic acid was used as the standard and total antioxidant capacity was expressed as the equivalents of ascorbic acid.

Nitric Oxide Radical Scavenging Activity

Nitric oxide was generated from sodium nitroprusside and measured by Griess' reaction (Green *et al.*, 1982). Sodium nitroprusside (5 mM) in standard phosphate buffer saline solution (0.025 M, pH 7.4) was incubated with different concentrations (2-1000 µg/mL) of the test extract dissolved in phosphate buffer saline (0.025 M, pH 7.4) and the tubes were incubated at 25°C for 5 h. Control experiments were conducted in the identical manner using the equivalent amounts of buffer. After 5 h, 0.5 mL of the sample was diluted with 0.5 mL of Griess' reagent (1 per cent sulphanilamide, 2 per cent *o*-phosphoric acid and 0.1 per cent naphthyl ethylenediamine dihydrochloride). The absorbance of the chromophore formed during diazotization of nitrite with suphanilamide and its subsequent coupling with napthyl ethylene diamine was read at 546 nm. The experiments were repeated in triplicate.

Ion Chelating Activity

Ion chelating activity was determined by colorimetric method (Benzie and Strain, 1996; Benzie and Szeto, 1999).The reaction mixture containing 1 mL 0.05 per cent *o*-phenathroline in methanol, 2 mL ferric chloride 200 µM and 2 mL various concentrations of the test compound was incubated at ambient temperature for 10 min and the absorbance of the sample was measured at 510 nm. The experiments were performed in triplicate.

Scavenging of Hydroxyl Radical

Extracts of different concentrations prepared in 2 per cent alcohol were taken in different test tubes and evaporated on a water bath. To these, 1 mL of Iron-EDTA solution (Iron was added as 2.0 mg Fe ($FeSO_4$) mixed with Na_2EDTA as an aqueous solution in a 1:1 molar ratio), 0.5 mL of 0.018 per cent EDTA and 1 mL of DMSO (0.85 per cent v/v in 0.1 M phosphate buffer, pH 7.4) were added and the reaction was initiated by adding 0.5 mL of 0.22 per cent ascorbic acid to each of the test tubes. Test tubes were tightly capped and heated on water bath at 80°-90°C for 15 min. The reaction was terminated by addition of 1 mL of ice-cold trichloroacetic acid (17.5 per cent w/v), kept aside for 2 min and the formaldehyde formed was determined by adding 3 mL of Nash reagent (75g of ammonium acetate, 3ml glacial acetic acid, 2ml acetyl acetone in to 1lt. distilled water) after 10-15 min for color development (Bhavani *et al.*, 1998). Intensity of yellow color formed was spectrophotometrically measured at 412 nm against reagent blank and the scavenging percentage of hydroxyl radical was calculated by comparing the results of the samples with that of the blank (Klein *et al.*, 1991).

Determination of Total Phenolic Content

The total phenolic content in the *I. nil* extracts was determined spectrophotometrically according to the Folin-Ciocalteu method (Singleton *et al.*, 1999) using galic acid as a standard (the concentration range: 0.025 to 0.5 mg/mL). The reaction mixture was prepared by mixing 1 mL of the methanolic solution (concentration 0.3 mg/mL), of the methanolic solution of the extract, 9 mL of distilled water, 1 mL of Folin-Ciocalteus reagent and 10 mL of 7 per cent sodium carbonate. After 90 minutes incubation at room temperature, the absorbance was determined spectrophotometrically at 765 nm. The total phenolic content was expressed as GAE in milligram per gram dry extract.

Determination of Total Flavonoid Content

The total flavonoid content was determined according to the aluminium chloride colorimetric method (Lin and Tang, 2007). Each plant extract (2 mL, 0.3 mg/mL) in methanol were mixed with 0.1 mL of 10 per cent aluminium chloride hexahydrate, 0.1 mL of 1 M potassium acetate and 2.8 mL of deionized water. After 40 minutes incubation at the room temperature, the absorbance of the reaction mixture was determined spectrophotometrically at 415 nm. Rutin was chosen as the standard (the concentration range: 0.005 to 0.1 mg/mL) and the total flavonoid content was expressed as milligram RE per g of dry extracts.

In vitro Anti-proliferative Activity

All the three extracts were evaluated for anti-proliferative activity against human cancer cell lines in 96 well microplates. Cell suspension (100 µl) containing 10^5 to 2 x10^5 cells/ml and 100 µl of tested sample were added into each well. After incubation in 5 per cent CO_2 at 37°C for 72 h, the cytotoxicity was determined by colorimetric method as described by Skehan and his coworkers (Skehan *et al.*, 1990). In brief, after 48 h incubation of cell lines, the medium was discarded and cells were fixed with 100 µl ice-cold 40 per cent trichloroacetic acid (TCA, Aldrich Chemical). Thereafter, the cells were incubated at 4 °C for 1 h and plates were washed five times with cold water. The excess water was drained off and the plates were left to dry, 50 µL of SRB stain (10 mg w/v in 1 per cent acetic acid) were added to each well for 30 min and subsequently washed with 1 per cent acetic acid to remove unbound stain, 10 mM tris base (pH: 10.5) was used to solubilize the dye. The plates were shaken vigorously, and the absorbance was measured at 540 nm using an ELISA reader (model 450, BIO-RAD, USA). Untreated cells were used as negative control. All determinations were performed in triplicate.

Statistical Analysis

Data was expressed as means SD Significant differences (p<0.05) between means of controls and treated cells were analyzed by ANOVA analysis.

Results

Antioxidant Activity

The antioxidant activity of alcoholic, 50 per cent hydro-alcoholic and aqueous extracts of *I.nil* were measured by various methods including DPPH scavenging activity, total antioxidant capacity, scavenging of nitric oxide radical, ion chelating activity, scavenging of hydroxyl radical. The results obtained by these studies were mentioned in Table 9.1.

DPPH is relatively stable free radical. The assay is based on measurement of scavenging ability of antioxidants towards stable DPPH radical. According to DPPH assay, alcoholic extract showed maximum activity with IC_{50} = 11.79 µg/mL and no significant change when compared with standard ascorbic acid having IC_{50} = 16.03 µg/mL. It revealed that alcoholic extract have maximum antioxidant activity followed by 50 per cent hydro-alcoholic extract with IC_{50} 42.17 µg/mL and least activity was demonstrated by aqueous extract when compared with standard antioxidant ascorbic acid in DPPH assay.

The total antioxidant capacities of extracts were calculated based on the formation of phosphomolybdenum complex. It was observed that alcoholic and 50 per cent hydro-alcoholic extracts showed more or less same maximum activity with IC_{50} values 30.17 µg/mL and 45.18µg/mL, respectively.

Table 9.1: IC_{50} (µg/ml) Values Obtained from different Antioxidant Assays of the Extracts of *Ipomea nil* Aerial Parts

Extracts	*DPPH Activity*	*Total Antioxidant Capacity*	*Nitric Oxide Activity*	*Ion Chelating Activity*	*Hydroxyl Radical Activity*
95 per cent alcoholic extract	11.79±0.35	30.17±1.8	33.99±1.35	24.13±0.61	42.17±1.47
50 per cent aqueous-alcoholic extract	42.17 ±1.47	45.18±2.03	41.43±1.86	85.5±2.74	104.4±2.6
AqueousExtract	115.5±6.93	126.2±7.86	104.4±2.6	199.5±2.97	221±5.38
Standard (Ascorbic acid)	16.03±0.22	14.18±0.85	11.56±0.9	7.9±0.93	5.63±0.67

Data presented as mean ± standard deviation of 3 parallel measurements. Different letters with in a column for a particular treatment represent significance at $P < 0.05$ applying Tukey's test

Similarly, the nitric oxide radical scavenging and ion chelating activity was also found maximum for alcoholic extract with IC_{50} 33.99 µg/mL and IC_{50} = 24.13 µg/mL respectively. Again when evaluated for hydroxyl radical activity the alcoholic extract showed maximum activity with IC_{50} = 42.17 µg/mL compared with 5.63 µg/mL ± 0.67 of ascorbic acid.

In general, extracts with a high antioxidant activity showed a high phenolic content. It is usually observed that plant extracts with a high phenolic content also contained a high flavonoid content (Maisuthisakul *et al.*, 2007). The amount of total phenolic and flavonoids for all extracts are demonstrated in Table 9.2. Alcoholic extract demonstrated significantly higher total phenolic and flavonoids when compared with hydro-alcoholic and aqueous extracts (Table 9.2).

Table 9.2: Total Phenolic Content and Total Flavonoids of *Ipomea nil* Aerial Parts

Extracts	*Total Phenolic Content, mgGAE/g*	*Total Flavonoids, mgRE/g*
95 per cent alcoholic extract	243.98 ± 2.14	81.52 ± 0.24
50 per cent aqueous-alcoholic extract	210.16 ± 2.15	82.18 ± 0.53
Aqueous extract	189.59 ± 2.03	79.13 ± 0.47

Anti-proliferative Activity

Alcoholic extract of *I.nil* was evaluated against four different human cancer cell lines *viz.*, A-549 (lung), HCT-15 (colon), Hep-2(liver), SiHa (cervix) along with standard drugs (Table 9.3). The alcoholic extract exhibited a dose-dependent growth inhibitory effect after a continuous exposure of 48 h period. Maximum cytotoxic effect of alcoholic extract was observed on Hep-2 and SiHa cancer cell lines. In terms of sensitivity, comparable responses were observed against all the four cancer cell lines and indicated a dose-response relationship by the extract.

Table 9.3: Antiproliferative Activity of *Ipomea nil* Extracts against Human Cancer Cell Lines

Plant Extracts	*Growth Inhibition IC $_{50}$ (µg/ml)*			
	A-549	*HCT-15*	*Hep-2*	*SiHa*
Alcoholic	43.5± 1.41	19.2±1.62	9.2±1.35	12.23±5.41
50 per cent hydro-alcoholic	56.12±2.31	33.6±2.10	48.3±3.12	53.14±2.87
Aqueous	76.9±2.95	68.3±1.31	55.02±2.51	69.5±1.75
5-Flurouracil	27.5±0.014	20.4±0.12	22.3±0.05	18.2±1.68

Discussion

Oxidative stress has been implicated in the pathology of many diseases and conditions including diabetes, cardiovascular diseases and inflammatory conditions cancer and ageing (Marx, 1987). Antioxidants may offer resistance against the oxidative stress by scavenging the free radicals, scavenging hydroxyl ions, nitric oxide and by many other mechanisms and thus prevention of disease (Braugghler, 1986).

DPPH method is commonly used to assess radical scavenging of any antioxidant substance because it is a quick, reliable and reproducible method to search *in vitro* general antioxidant of pure compounds as well as plant extracts (Koleva *et al.*, 2002). Antioxidant substances interact with DPPH, a stable free radical and transfer electrons or atoms of hydrogen to DPPH, thus neutralizing ("scavenging") the free radical (Marxen *et al.*, 2007). DPPH radical scavenging abilities of the extracts were significantly comparable to those of ascorbic acid, a well known antioxidant compound with health benefits for humans (Boots *et al.*, 2008). However, all the extracts in our study exhibited appreciable scavenging activity and there was a significant correlation between DPPH radical scavenging activity and polyphenolic content. IC_{50} value of alcoholic extracts indicates that it has a highest proton donating ability among all the three extracts (Table 9.1). Earlier it has been reported that extract with their proton-donating ability, could serve as free radical inhibitors or scavengers, and could act possibly as primary antioxidants (Marxen *et al.*, 2007). Similar results were observed for Total antioxidant activity, where the activity of alcoholic extract increased with increase in concentration of the extracts indicating the potential of alcoholic extract. Nitric oxide is a free radical produced in mammalian cells, involved in the regulation of various physiological processes. However, excess production of NO is associated with several diseases (Ialenti, 1993). The present studies also proved that alcoholic extract of *I.nil* have potential to reduce the nitric oxide radicals. This can be co-related with the antioxidant principles present in the extract which compete with oxygen to react with nitric oxide thereby inhibiting the generation of nitrite.

The results also implied that the extracts interfered with the formation of ferrous-*o*-phenanthroline complex. Here, ortho substituted phenolic compounds may exert pro-oxidant effect by interacting with iron. O-phenthroline quantitatively forms complex with Fe^{2+} which gets disrupted in the presence of chelating agent (Marcocci *et al.*, 1994), therefore suggesting that alcoholic extract has metal chelating activity.

Hydroxy radical scavenging activities of extracts were assayed by generating hydroxyl radicals using ascorbic acid-iron-EDTA (Blois, 1958). The hydroxyl radical formed by the oxidation react with DMSO to yield formaldehyde. The formaldehyde production from DMSO provides a convenient method to detect hydroxyl radicals formed during oxidation of DMSO by Fe^{3+}/ascorbic acid system. In comparison to all the three extracts, alcoholic extract again showed the prominent hydroxyl radical scavenging activity.

Flavonoids are the most ubiquitous groups of plant secondary metabolites (Havsteen, 2002). These metabolites have antioxidant activity through scavenging or chelating process and their effects on human nutrition and health are substantial (Cook and Samman, 1996). Flavonoids have been widely used in cancer treatments, coronary heart diseases, gastrointestinal ulcers and rheumatic diseases (Sreejayan and Rao, 1996). In the present study, antiradical efficiency in the methanolic extract was positively correlated with total phenolic contents (Table 9.1). Our data also indicates that polyphenols are important components in methanolic extracts which could be exploited for their use in free radical scavenging activity. Our results also indicate that *Ipomoea nil* contains significant amount of phenolic and flavonoid compounds and alcoholic extract of aerial parts of the plant possess higher levels of total phenol and flavonoid as compared to 50 per cent hydro-alcoholic and aqueous extracts. So, it can be predicted that the antioxidant activity may be due to the total phenolic and flavanoid content present in the plant. In most cases, total phenolic content as determined by Folin-Ciocalteu method was correlated with the antioxidant capacities (Roginsky and Lissi, 2005). The results obtained in the present studies may be attributed to several reasons *viz,* the inhibition of ferryl- perferyyl complex formation; scavenging of OH or superoxide radicals or by changing the ratio of Fe3+/Fe 2+; reducing the rate of conversions of ferrous to ferric or by chelation of the iron itself.

Anti-proliferative activity of three extracts has been analyzed against four different human cancer cell lines. Among them, alcoholic extract showed maximum activity with the lowest IC50 of 9.2±1.35 against Hep-2 cell line. 50 per cent hydro-alcoholic and aqueous extracts have shown moderate cytotoxic effect against all the cell lines. So, the results indicates that the methanolic extracts of *I. nil* possess significant anti-proliferative properties and bioactive molecules from the plant can be used as therapeutics. With the growing knowledge of mechanisms of human diseases, particularly metabolic diseases like cancer, diabetes, liver diseases and hypertension, understanding the role played by highly reactive oxygen species such as free radicals becomes increasingly relevant. Research on medicinal plants for natural antioxidants and anticancer agents is an attempt to find out newer means of fighting free radicals and cancer.

References

Adeolu, A.A., Florence, O.J., Srinivas, K., Masika, P., and Anthony, J.A. (2009). Asessment of the medicinal potential of the methanol extracts of the leaves and stems of *Buddleja saligna*. *Complem Altern Med.*, **9**: 21-28.

Amakura, Y., Tsutsumi, T., Sasaki, K., Nakamura, M., Yoshida, T., and Maitani T. (2008). Influence of food polyphenols on aryl hydrocarbon receptor-signaling pathway estimated by *in vitro* bioassay. *Phytochem.*, **69**: 3117-3130.

Balasundram, N., Sundram, K., and Sammar, S. (2006). Phenolic compounds in plants and agri-industrial by-products: Antioxidant activity, occurrence, and potential uses. *Food Chem.*, **68**: 191-203.

Blois, M.S. (1958). Antioxidants determinations by the use of stable free radical. *Nature*, **181**: 1199-1200.

Boots, A.W., Haenen, G.R.M.M., and Bast, A. (2008). Health effects of quercetin: From antioxidant to nutraceutical. *Eur. J. Pharmacol.*, **585**: 325-337.

Braugghler, J.M., Duncan, C.A., and Chase, L.R. (1986).The involvement of iron in lipid peroxidation: Importance of ferrous to ferric ratio in initiation. *J. Biol. Chem.*, **261**: 102-182.

Choi, Y., Jeong, H.S., and Lee, J. (2007). Antioxidant activity of methanolic extracts from some grains consumed in Korea. *Food Chem.*, **103**: 130-138.

Cook, N.C., and Samman, S. (1996). Flavonoids-chemistry, metabolism, cardioprotective effects, and dietary sources. *J Nutr Biochem.*, **7**: 66-76.

Gumul, D., Korus, J., and Achremowicz, B. (2007). The influence of extrusion on the content of polyphenols and antioxidant/antiradical activity of rye grains (secale cereale l.). *Acta Sci Pol Technol Aliment*, **6**: 103-111.

Green, L.C., Wagner, D.A., Glogowski, J., Skipper, P.L., Wishnok, J.S., and Tannenbaum, S.R. (1982). Analysis of nitrate and 15N in biological fluids. *Anal. Biochem.*, **239**: 131-138.

Han, S.J., Ryu, S.N., and Kang, S.S. (2004). A new 2-arylbenzofuran with antioxidant activity from the black colored rice (*Oryza sativa* L.) bran. *Chem Pharm Bull.*, **52**: 1365-1366.

Havsteen, B.H. (2002).The biochemistry and medical significance of the flavonoids. *Pharmacol Therapeut.*, **96**: 67-202.

Ialenti, A., Moncada, A., and Di Rosa, M. (1994). Modulation of perspective for the 1990s. *Nature*, **234**: 462.

Kim, K.H., Choi, S.U., and Lee, K.R. (2009). Diterpene glycosides from the seeds of *Pharbitis nil*. *J Nat Prod.*, **72**: 1121–1127.

Klein, S.M., Cohen, G., and Cederbaum, A.I. (1991). Production of formaldehyde during metabolism of dimethyl sulfoxide by hydroxyl radical generating system. *Biochem.*, **20**: 6006-6012.

Koleva, I.I., Van Beek, T.A., Linssen, J.P.H., de Groot, A., and Evstatieva, L.N. (2002). Screening of plant extracts for antioxidant activity: A comparative study on three testing methods. *Phytochem. Anal.*, **13**: 8-17.

Lee, Y.R., Woo, K.S., Kim, K.J., Son, J.R., and Jeong, H.S. (2007). Antioxidant activities of ethanol extracts from germinated specialty rough rice. *Food Sci Biotechnol.*, **16**: 765-770.

Lin, J.Y., and Tang, C.Y. (2007). Determination of total phenolic and flavonoid contents in selected fruitsand vegetables, as well as their stimulatory effects on mouse splenocyte proliferation. *Food Chem.*, **101**: 140–147.

Maisuthisakul, P., Suttajit, M., and Pongsawatmanit, R. (2007). Assessment of phenolic content and free radical-scavenging capacity of some Thai indigenous plants. *Food Chem.*, **100**: 1409–1418.

Marcocci, L., Packer, L., Droy-Lefaiz, M.T., Sekaki, A., and Gardes-Albert, M. (1994). Antioxidant action of *Gingko biloba* extract. *Meth. Enzymol.*, **234**: 462-465.

Marx, J.L. (1987). Oxygen free radicals linked too many diseases. *Science*, **235**: 529.

Marxen, K., Vanselow, K.H., Lippemeier, S., Hintze, R., Ruser, A., and Hansen, U.P. (2007). Determination of DPPH radical oxidation caused by methanolic extracts of some microalgal species by linear regression analysis of spectrophotometric measurements. *Sensors*, **7**: 2080-2095.

Mervat, M.M., El Far, Hanan, A.A., and Taie. (2009). Antioxidant activities, total anthrocynins, phenolics and flavonoids contents of some sweet potato genotypes under stress of different concentrations of sucrose and sorbitol. *Australian J Basic Applied Sci.*, **3**: 3609-3616.

Noh, K.S., Yang, M.O., and Cho, E.J. (2002). Nitrite scavenging effect of *Umbelligeraeceae*. *Korean J Food Cookery Sci.*, **18**: 8-12.

Petti, S. and Scully, C. (2009). Polyphenols, oral health and disease: a review. *J dentistry*, **37**: 413-423.

Prasain, J.K., Wang, C.C., and Barnes, S. (2004). Flavonoids and isoflavones (photoestrogens): absorption, metabolism, and bioactivity. *Free Radical Bio Med.*, **37**: 1324-1350.

Prasanth, K.V., Shasidhara, S., Kumar, M.M., and Sridhara, B.Y. (2000). Effect of *Luffa echinta* on lipid peroxidation and free radical scavenging activity. *J. Pharm. Pharmacol.*, **52**: 891-894.

Roginsky, V., and Lissi, E.A. (2005). Review of methods to determine chain breaking antioxidant activity in food. *Food Chem.*, **92**: 235-254.

Shirwaikar, A., Govindrajan, R., Rastogi, S., Vijaykumar, M., Rawat, A.K.S., Ehlotra, S.M., and Puspangandan, P. (2003). Studies on the antioxidant activities of *Desmodium gangeticum*. *Biol. Pharm. Bull.*, **26**: 1424-1427.

Singleton, V.L., Orthofer, R., and Lamuela-Raventos, R.M. (1999). Analysis of total phenols and other oxidation substrates and antioxidants by means of Folin–Ciocalteu reagent. *Meth. Enzymol.*, **299**: 152–178.

Singleton, V.L., and Rossi, J.A. (1965). Colorimetry of total phenolics with phosphomolybdic-phosphotungstic acid reagents. *American Journal of Enology and Viticulture*, **16**: 144–158.

Skehan, P., Storeng, R., Scudiero, D. *et al.* (1995). New colorimetric cytotoxicity assay for anticancer drug screening. *Journal of the National Cancer Institute*, **82**: 1107-1112.

Souza, M.M., Madeira, A., Berti, C., Krogh, R., Yunes, R.A., and Filho, C. (2000). Antinociceptive properties of the methanolic extract obtained from *Ipomea pescaprae* (L.) R. Br. *J. Ethnophaqrmacol.*, **69**: 85-90.

Sreejayan, N., and Rao, M.N.A. (1996). Free radical scavenging activity by curcuminoids. *Drug Res.*, **46**: 169-171.

Tylor, B.S., Kion, Y.M., Wang, Q.I., Sharpio, R.A., Billiar, T.R., and Geller, D.A. (1997). Nitric oxide down regulates hepatocyte-inducible nitric oxide synthase gene expression. *Arch Surg.*, **132**: 1177-1183

Warrier, P.K., Nambiar, V.P.K., and Ramankutty, C. (1994). Indian medicinal plants. *Orient Blackswan.*, **231**.

Yin, J.Y., Kwon, G.J., and Wang, M.H. (2007).The antioxidant and cytotoxic activities of *Sonchus oleraceus* L. extracts. *Nutr Res Pract.*, **1**: 189-194.

Utilisation and Management of Medicinal Plants Vol. 3 (2015) *Pages* **147–163**
Editors: **V.K. Gupta and A. Kaul**
Published by: **DAYA PUBLISHING HOUSE, NEW DELHI**

10

Antioxidant Mineral Elements in Plants

C.V. Lamarão[1,3], C.C. Silva[2], O. Canhoto[3], M.F. Ferreira[2], K.L.S. Rocha[2], R.G.A. Fontes[2] and V.F. Veiga Junior[3*]

ABSTRACT

The predominant elements found in all living organisms are carbon, oxygen, nitrogen and hydrogen. Since these are the essential elements in organic chemistry, for a long time it was considered that only the organic compounds and the chemical reactions associated with them were vital for the living beings. It was also believed that the elements so-called inorganic, had little or no meaning in life. At the time there were some well-known exceptions, like the presence of iron in blood, calcium in teeth and bones, and the presence of phosphorous in several metabolic processes. Several inorganic elements mainly metals are part of important processes for the human beings, such as their action as antioxidants which delay or inhibit the oxidation of enzymatic or non-enzymatic processes, these elements are zinc, copper, manganese and selenium. These elements are also essential for the activation of several metabolic reactions with enzymatic catalysis related with the reproductive system and also with growth process. It has also been proven their action on the protection against diseases, in particular in the treatment of diabetes mellitus, since it is associated with deficiencies in the metabolism of antioxidant elements as zinc, chrome and manganese. There are also applications of antioxidant elements for therapeutic purposes, however researchers are still trying to evaluate the benefits of their therapeutic applications and the health risks associate with their consumption. The aim of this chapter was to evaluate the main characteristics of antioxidant mineral elements such as zinc, copper, manganese and selenium that are present in plants and their direct effect in human metabolism and in human health.

Keywords: Antioxidant mineral elements, Zinc, Copper, Manganese, Selenium, Plants.

1 College of Agricultural Sciences, Amazonas Federal University, Manaus, Brazil.
2 Higher School of technology, Amazonas State University, Manaus, Brazil.
3 Chemistry Department, Amazonas Federal University, Manaus, Brazil.
* *Corresponding author*: E-mail: valdirveiga@ufam.edu.br

Introduction

According to the World Health Organization (WHO), medicinal plants are the ones that possess a therapeutic action from their roots, leafs, seeds, barks or other plant parts. The therapeutic action may be classified as purgative, stimulant or as any other pharmacologic activity when prescribed to higher animals.Usually, the main studies on medicinal plants are those describing the organic compounds in the search for the main active compound. These studies are mostly related with nutrition and their objective is to determine important macro or micronutrients from species with a specific commercial value (Belakova *et al.*, 1995; Majid *et al.*, 1995; Morais *et al.*, 1995).

Moreover its recognised nutritive and healing properties, many medicinal plants also have antioxidant characteristics. An antioxidant agent can be defined as a substance that when present in very low concentrations in comparison with the oxidative substrate, inhibits or delays significantly the substrate from the oxidation process. This definition is very general since it also includes macromolecules that can be affected by oxidation, for instance lipids, carbohydrates and also DNA. In this definition, from a biological point of view, any substance that is able to inhibit the action of a specific oxidative enzyme, can be considered an oxidative agent. Besides its use for medical purposes, these compounds have other industrial applications, such as food preservatives, in cosmetics, and as a protective agents against materials deterioration. They are also used as gasoline additives to prevent the formation of gums which interfere in the motor performance (Diaz *et al.*, 1996; Nogueira *et al.*, 1998).

The concentration of the antioxidant elements zinc, copper, manages and selenium in plants will vary according to several factors including the species, age, root distribution, the physical and chemical composition of soil, the amount and distribution of such elements, the grows methods and general climate conditions (Lopes *et al.*, 1996; Lovkova, 1999a).

The specific composition and concentration of the antioxidant elements in plants is related with the way they are absorbed.Since the absorption cells have the capability to exclude certain compounds, the composition and concentration of these elements is directly related with the protoplasm specificity (Jain *et al.*, 1992; Kar *et al.*, 1999).

Plants have specific needs of these antioxidant elements for their grows, this does not mean that such amount will be sufficient for the nutritional needs of those animals that eventually eat that plant (Lopes *et al.*, 1998; Lovkova, 1999b).

The scientific evolution of both chemistry and biology led to an increasing determination to have a better understanding on the chemical composition of the plants used for medicinal purposes throughout the ages. This curiosity is related with the fact that the molecular composition gives an indication of the chemical profile of the plant species, collected in a specific place at a given time, under the influence of certain environmental factors. All these factors are directly related in the biological activity of the plant. The search for natural compounds is increasing since they present very little harm to public health, nonetheless, the natural antioxidant compounds may not always be a solution due to their potential toxicity (Lopes *et al.*, 1996; Sathiyamoorthy, 1997; Lovkova, 1999a; Lovkova, 1999b; Wesolowsky, 2000).

The inorganic elements have a very important rule in human health, and their use in the prevention of certain diseases in on demand (Tian Xie *et al.*, 2003; Rajasekaran *et al.*, 2005). In this perspective, copper is a good antioxidant element that is a component of several enzymes involved in the production of cellular energy, and is also present in the development of cognitive tissue and in the production of melanin. Zinc, besides its activity as an antioxidant agent, is of extreme importance for the cellular metabolism and several enzymatic reactions are zinc-dependent, having a very important rule in the immune response. Manganese is also linked with the enzymatic activity and acts as an activator for many enzymes that are associated with the metabolic process of carbohydrates, amino acids and cholesterol. Moreover its activity as an antioxidant agent, it also participates in the cartilage and bones formation. Finally, selenium is considered as the "new" oligo-element that presents the following metabolic functions: is present in the synthesis of thyroid hormones, has antioxidant activity and also presents proven efficiency against peroxides (oxidant agents that can attack cells) since it is part of the enzyme glutathione peroxidase, an enzyme that attacks peroxides (Food Ingredients Brazil, 2008).

The following aspects will be discussed in this chapter, the relation between the antioxidant elements zinc, copper, manganese and selenium with the metabolism and health of plants and human beings, toxicity, their applications in the pharmacological and bio-cosmetic markets, and their implications on food with a special emphasis on the antioxidant activity of such elements.

Material and Methods

We performed a literature review of research published by Brazilian and international groups, using the research databases SCOPUS, ISI, and SCIELO and others Internet search tools.

A summary is given below of the main characteristics of antioxidant mineral elements in plants.

Results and Discussion

The Function of Antioxidant Elements in Plants

Zinc

This elements is very toxic for plants unless it is present in low concentrations. Nevertheless, the presence of zinc is required in residual amounts in order to have a good plant metabolic development. Its absence can cause structural abnormalities on the root-ends, deficiencies of grow in the aerial parts and a plant incapacity to form seeds (Terres *et al.*, 2001; Kawashima *et al.*, 2003; Catania *et al.*, 2009).

Zinc is part of the enzymes carbonic anhydrase and glutamate dehydrogenase, the later one is an activator for several enzymatic systems. Zinc is also required for the synthesis in dolacetic acid, an important plant hormone. It has been proven that a number of physiologic anomalies, mostly in tree species, is a consequence of a deficiency of this element (Miller, 1938; Scherz and Kirchhoff, 2006; Catania *et al.*, 2009).

Copper

In the same way as zinc, copper is also extremely toxic for plants unless it is present in very small amounts. One of the functions of this element yet to be confirmed, in the plant metabolism is the action on a number of redox reactions since it is part of certain redox enzymes such as tirosinase and ascorbato oxidase, and also plastocianin (Rajurkar *et al.*, 1998; Catania *et al.*, 2009).

Manganese

Plants usually use manganese in very small amounts. The manganese compounds are very toxic for plants except when present in low concentrations. Usually the concentrations of this element are higher in the more physiologically active plant parts, especially leafs. It is a very stable elements, *i.e.* there is very little variations in its distribution on the different plant parts (Sarmani *et al.*, 1999; Catania *et al.*, 2009).

Manganese acts in plants mostly as a catalyst. This element directly participates in the redox mechanisms, especially in those that are related with iron compounds (Miller, 1938; Vasconcelos *et al.*, 2000; Catania *et al.*, 2009). It is also an activator of certain enzymatic systems, including enzymes that are activated by magnesium. Manganese is also present in the synthesis of chlorophyll because the chloroplasts are rapidly affected by the deficiency of this element (Catania *et al.*, 2009).

Selenium

Many vegetables, including the majority of plants used for agricultural purposes, store large quantities of selenium, even when they are grown in selenium reach soils. This plants when in a hydroponic production exhibit toxicity symptoms due to selenium even when this element is present in relatively low concentrations. The species that normally are selenium accumulators, may intake about 100 to a 1000 times more selenium without showing any signs of toxicity, in comparison to those plants that do not show this capacity. There are indications that this element may be an essential compound for those species (Brown and Shrift, 1981; Terry *et al.*, 2000; Sors *et al.*, 2005; Kubachka *et al.*, 2007; Catania *et al.*, 2009).

Correlation between the Antioxidant Elements and Human Metabolism

It is known that a large number of mineral elements are essential to human nutrition and other mammals, having specific roles the in human metabolism (Harper *et al.*, 1982; Almeida *et al.*, 1992; Franco, 1998). It is possible to mention as an example, that calcium and magnesium, which act in the formation of bones, teeth and tissue. Potassium, when associated with sodium, regulates the functioning of the muscular system and heartbeat (Franco, 1998).Some elements are found in relatively low amounts in the human body and are called oligo-elements such as zinc, iron, cobalt, manganese, nickel, fluorine, etc. (Favier, 1991).

Zinc, manganese and copper function as essential activators in a series of metabolic reactions catalysed by enzymes, thus being very important elements for the reproductive system and growth (Roth and Kirchgaessner, 1991; WHO, 1996; Franco, 1998).

Zinc is an important trace element in human nutritionand is present in many biochemical functions in the human metabolism. It is the activation factor of several enzymes (*e.g.* carboanhydrase, alkaline phosphatase and of several enzymes of the nucleic acid synthesis). It stabilizes the structures of RNA, DNA and the ribosomes, and influences hormone metabolism, as for example for insulin or gonadotropin (Massey, 1998; Institute of Medicine, 2001).

Selenium is an essential trace element in the normal body functions and can be present in the inorganic form, metallic (SeO) or oxyanions such as selenite ($SeO(OH)_2$) and selenate ($SeO_2(OH)_2$), and also in organic form, selenocysteine (Sec) and selenomethionine (SeMet), analogues of sulfur amino acids cysteine and methionine, respectively (Suzuki,2005).

Since the selenium compounds are recognized as selenium types, they are transformed into selenite, an intermediate common metabolite, and then used for the synthesis of selenoproteins (selenoenzymes) or expelled after their gradual methylation. The SeMet is a unique selenium compound, because when it is recognized as selenium type compound, it can be transformed into selenocysteine by a trans-selenic reaction, and then being lysed by β-lyase or y-lyase to selenite. However, SeMet can at the same time be used for protein synthesis between selenomethionine and methionine by methionine-tRNA (Schrauzer, 2000; Suzuki, 2005).

Selenocysteineis present asan amino acid residue in selenoproteins in plants and animals, which is incorporated intoan amino acid sequence of seleno proteins by a specific residue codon. The selenomethionine is generally present as a residue of SeMetproteins. Therefore, proteins containing seleniumin the form Sec are called selenoproteins, while those containing seleniumin the form of SeMetare called simply selenium-containing proteins (Schrauzer, 2000; Suzuki, 2005).

Currently five human iso-enzymes glutathione peroxidase (GPx) are known to contain selenocysteine, exhibiting tissue-specific and different substrate specificities expressions. They are:GPx-1, foundin the cytosol; GPx-2 specific enzyme found in the gastrointestinal tract; GPx-3 secreted in proteins found in plasma GPx-4; acts on oxidized lipids, is called phospholipid hydroperoxide glutathione peroxidase and a GPx-sn, specific enzymeofthe sperms nucleus (Ottaviano *et al.*, 2008; Steinbrenner, 2009).

All these GPxiso-enzymes have in common a catalytic triad in its active center consisting of selenocysteine, glutamine and tryptophan residues. Its antioxidant capacity is due to the reduction of hydrogen peroxide, organic hydroperoxide and phospholipid hydroperoxide (only the GPx-4), usually requiring two glutathione (GSH) equivalents asco-substrate (Maiorino, 1995; Brigelius-Flohé, 1999).

Correlation between the Antioxidant Elements and Human Health

Mineral elements serve as structural components of tissues and as constituents of the body fluids and vital enzymes in major metabolic pathways and are essential for the function of all cells (Underwood, 1977; Lozak *et al.*, 2002; Rajasekaran *et al.*, 2005). Their concentrations in living tissue and the adult human needs are somewhat lower than those of the bulk elements.They were not easily quantified by early analytical methods, hence the name "trace" elements (Underwood, 1977; Al-Ahmed *et al.*, 1989; Rajasekaran *et al.*, 2005).

Growing concern with environmental factors in human health over the last few years has aroused renewed interest on the trace elements. Any abnormalities in their metabolism have been demonstrated consequences in many human diseases (Al-Ahmed *et al.*, 1989; Rajasekaran *et al.*, 2005). In particular, diabetes mellitus has been shown to be associated with abnormalities in the metabolism of zinc, chromium, magnesium, and manganese (Niewoehner *et al.*, 1986; Rajasekaran *et al.*, 2005).

The control of diabetes mellitus is considered a global problem and successful treatment is yet to be discovered (Palumbo *et al.*, 2001; Rajasekaran *et al.*, 2005). Many herbs have been shown to have hypoglycemic action in animals and humans (Singh, 2001).

Some of the inorganic elements such as zinc, calcium, potassium, manganese, chromium, and vanadium can reduce the blood sugar level and their indirect role in the controlling of diabetes mellitus is being recognized. These inorganic trace elements are nutritionally essential for human beings (Ghosh *et al.*, 2002). Chromium possibly influences glucose metabolism by helping in the binding of insulin to its receptors and potentiating its action. Zinc enhances the binding of insulin to its receptor and potentiates the effect of insulin (Tang *et al.*, 2001). Calcium and potassium play an important role in the release of insulin from β-cells of islets of Langerhans. Vanadium was shown to mimic all or most of the actions of insulin in intact cell systems. Magnesium improves glucose disposal, primarily by stimulating glycolysis (Rajasekaran *et al.*, 2005).

The purpose of some inorganic elements like potassium, zinc, calcium, manganese and traces of chromium in the improvement of impaired glucose tolerance and their indirect role in the management of diabetes mellitus are being more and more recognised (Gurson and Saner, 1971; Mertz, 1981; Kinlaw *et al.*, 1983; Niewoehner *et al.*, 1986; Kar *et al.*, 1999). Besides, zinc ions are necessary for crystallization of insulin and insulin is stored as zinc complex. A regular potassium concentration isalso necessary for optimal insulin secretion. Hypokalaemia or potassium depletion can result in reduced glucose tolerance. Calcium, potassium and traces of chromium play an important role in the release of insulin from β-cells of islets of Langerhans (Underwood and Mertz, 1986; Kar *et al.*, 1999).

As the genotoxic damage caused by the accumulation of oxidative modifications in DNA bases and breaking the single or double strand from the DNA backbone,are prominent characteristics in the development of many forms of cancer, selenium has been suggested to exert their anti-carcinogenic functions using detoxification selenoenzymes of ROS. In humans, indirect evidence have been provided by genetic data for the connection between the risk of cancer and an altered expression pattern of selenoproteins, as expressions of the proliferation of cancer tissues (Steinbrenner and Sies, 2009).

Studies in animals and in cell cultures support the role of various anticancer selenoenzymes. The accelerated prostate carcinogenesis was observed in transgenic rats showing a compromised biosynthesis of selenoproteins and the selenium status has been shown to be inversely correlated with the extent of DNA damage in the prostate of hunting dogs in advanced age (Waters *et al.*, 2005; Diwadkar-Navsariwala *et al.*, 2006).

The high expression of GPx-1 as well as the selenium supplementation protected the mammal cells against DNA damage caused by induction of UV41 rays. The high expression of GPx-4 suppressed the growth of a solid tumourof L929 tumorigenic fibrosarcoma cells and increased the sensitivity of melanoma B16BL6 tumorigenic cells in angiodestructive therapy (Heirman *et al.*, 2006).

The anti-carcinogenic role of GPx-1, GPx-4 and selenoprotein P seams to rely mostly on their function as hydroperoxide-degrading enzymes. In contrast, GPx-2 and TrxR-1 display both anti- and pro-carcinogenic properties. This characteristic of GPx-2 was recently highlighted using a small interfering RNA (siRNA) to weaken the expression of the GPx-2, which inhibits the migration and invasion of adenocarcinoma human cells but at the same time sustained the growth of tumours in rats (Banning *et al.*, 2008).

Other seleno-compounds affect the pathway of the transduction signals in tumour cells, as demonstrated by methyl-selenic acid (MSEA). In fibrosarcoma cells, MSEA inhibited the activation of metalloproteinase 2 (MMP-2) matrix by the stimulation of 12-O-tetra-decanoylphorbol-13-acetate (PMA) in the tumour by suppressing pathway of nuclear factor kappa B (NF-B) signals. The protein C kinase (PKC), which is activated by PMA and involved in carcinogenesis, has been shown to be inactivated by MSEA locally generated(Park *et al.*, 2007; Gundimeda *et al.*, 2008).

In the review by Navarro-Alcarón *et al.* (2008), 14 studies were found showing significantly lower levels of selenium in cancer patients compared to the control group. Low concentrations of serum selenium were also correlated with an increased cancer risk. Furthermore, Czeczot *et al.* (2006) found a reduction in GPx activity in the hepato cellular carcinoma compared to adjacent normal tissue.

This decrease may cause the intensification of lipid peroxidation and an increased in the peroxidation final products such as malondialdehyde (MDA). At the same time, increased levels of MDA carcinogenic tissues were also found. Therefore, some researchers recommend selenium as a nutritional prophylactic against cancer in a dosage of 50-100 mg Se/day. Selenium has also been recommended as a co-adjunct for chemotherapy in the treatment of cancer, in high doses of 200 mg Se/day(Reid *et al.*, 2008).

Studies in animals have shown the protective effect of various forms of selenium against many kinds of cancers including liver, skin, breast and colon. In particular, the development of breast cancer induced by 7,12-dimethylbenzanthracene (DMBA) and skin cancer induced by ultraviolet light, were significantly reduced in animals fed selenium (Rafferty *et al.*, 2002; Kocdor *et al.*, 2005).

In rat models, selenium supplement also reduced significantly the chemically induced carcinogenic process of liver cancer when administered during the advancement and progression stages (Bjorkhem-Bergman *et al.*, 2005).

In the review by Bardia *et al.* (2008), the authors concluded that Se supplement appears to reduce the risk of cancer, especially among men, but the results should be interpreted with caution because of the variations in the concentrations of Se according to gender and geographical location. Men had higher levels of serum selenium supplementthan women and the majority of studies were conducted in areas with individuals with nutritional deficiency of selenium.

A greater number of epidemiological intervention studies using selenium supplement in cancer treatment may provide support for a better definition of the benefits and risks of different types and doses of selenium to be used (Gromadzinska *et al.*, 2008).

In addition, studies assessing the importance of selenium in cancer treatment in humans relate more to the effect of this mineral in prostate and colorectal cancer and is less referenced in cases of bladder cancer (Sieja *et al.*, 2004; Brinkman *et al.*, 2006).

Antioxidant Elements and their Toxicity

The use of plant species for therapeutic purpose has been reported throughout the ages, but only recently scientists seek to assess the extent of the therapeutic benefits and safety with their use, mainly due to reports of serious side effects and fatalities, some of those associated with the presence of toxic metals (Tyler, 1987; Jain *et al.*, 1992; Matos, 1994; Kakosy *et al.*, 1996; Stewart *et al.*, 1999; Ernst, 2002).

Plants can accumulate these metals in all their tissues and can transfer them to the food chain. This accumulation is one of the environmental topics of interest nowadays, not only because of the toxicity in plants of manyof these metals, but also by potential of adverse effects on animal health and human (Tavares and Carvalho, 1992; Maiga *et al.*, 2005).

Although some metals are considered essential to the development of plants and animals, the limits of tolerance to these microelements are, in general, are very low. Others, such as lead (Pb), are considered toxic because it is known to induce various disorders in humans and laboratory animals, by changing the antioxidant activity via the inhibition of the SH functional group in many enzymes. However, all elements in high concentrations can be harmful to the organism due to their direct interaction with DNA(Maiga *et al.*, 2005; Klassen and Watkins, 2001; Hsu and Guo, 2002).

Below its highlighted some features on the toxicity of the antioxidant mineral elements zinc, manganese, copper and selenium.

Zinc

Consumption of large quantities of this metal, either by water, food or nutritional supplements, may affect health. Acute ingestion of high doses can cause stomach cramps, nausea and vomiting. The ingestion of high doses for several months may cause anaemia,effects on the pancreas and decreased HDL cholesterol. The regulator of food and medicine in the United States (Food and Drug Administration - FDA) does not recommend the addition of any zinc compound directly to the food (Azevedo*et al.*, 2003; Scherz and Kirchhoff, 2006; Klaassen, 2008) addition. An excessive zinc intake leads to acute adverse effects like diarrhea, vomiting, abdominal cramps and headache. Zinc toxicity is reflected also in adverse effects like chronic functional impairment in the immunological response, altered copper and iron function and the cholesterol metabolism (Institute of Medicine, 2001).

Manganese

Little is known about the toxicity of manganese compounds. However, evidence indicates that several of these compounds may induce neurological effects observed due to chronic exposure in humans via inhalation and the intermediate and chronic exposure of animals viaoral intake (Who, 1999; Martins, 2003).

Copper

Copper is an essential element for living organisms when present in small quantities. In general the population may be exposed by inhalation, ingestion of food goods and water or by skin contact, however, the main route of exposure apart from the occupational is oral intake. Ingestion of copper salts cause vomiting, lethargy, acute hemolytic anaemia, kidney and liver damage and, in some cases, death. The ingestion of water containing high amounts of the metal may cause nausea, vomiting, abdominal pain and diarrhoea. Children are more sensitive to the exposure effects of copper. Continued exposure to high concentrations of this metal in food or water can cause liver damage in children (Maiga *et al.*, 2005; Klassen and Watkins, 2001; Hsu and Guo, 2002).

Selenium

There is a moderate to high health risk on the consumption of selenium. High levels in blood of selenium can result in a condition called selenosis. Symptoms include gastrointestinal pain, hair loss, white marks in nails, and some nerve damage. The toxicity of selenium is associated with faintness, loss of hair and nails, irritability, fatigue, miscarriage and infertility. Selenium is toxic in high doses,a daily intake above 400 µg (Maiga *et al.*, 2005; Klassen and Watkins, 2001; Hsu and Guo, 2002; Cozzolino, 2007).

Applications of Antioxidant Elements in the Pharmaceutical and Biocosmetics Industries

Both the pharmaceutical and the cosmetics industries, has grown extensively in Brazil. According to a survey by the Brazilian Association of Toiletries and Cosmetics Industry (APIHPEC),in 2007 existed in Brazil 1,659 companies operating in the personal products, perfumery and cosmetics markets. This numbers makes Brazil the third country in sales in the global market sector with a turnover of billions of dollars in latest years (Cozzolino, 2007).

Brazil is also the first in the perfume and deodorant market and second in hair, oral, child care and sun protection products market. In the pharmaceutical market, Brazil is first in Latin America and ranks 7th in the world ranking (Cozzolino, 2007). Among these industries there are various applications of certain chemical elements of which we highpoint Zinc, Manganese, Selenium and Copper.

Zinc

It is mainly used in the form of zinc oxide, which is white, slightly soluble in water, but very soluble in acids. Zinc oxide is an astringent substance with antiseptic and anti-inflammatory topical action. It is an opaque substance that reflects light (physical filter), preventing radiation from reaching the skin.It is also available in a micro form for a better cosmetic acceptance either alone or in combination with photo-chemical protectors. It is mainly used by industry in the manufacture of skin products, or with other compounds used in the pharmaceutical industry for the manufacture of sunscreens, deodorants, preparations for the treatment of fungal infections, acne and anti-dandruff shampoo (Azevedo *et al.*, 2003; Scherz and Kirchhoff, 2006; Klaassen, 2008).

Manganese

Its derivative compound potassium permanganate is used as an oxidant, bleaching and in the treatment of skin diseases. Their organic forms are used in fungicides and smokeinhibitors, among other uses. The metal element is also used in small quantities in the mangafodipir trisodium (MnDPDP) drug as contrast agent in magnetic resonance imaging (MRI) (Who, 1999; Martins, 2003).

Copper

It is widely used in makeup products within the cosmetics industry. It is considered an essential metal to keep youth and elasticity of skin. Dyes usually have copper as a main ingredient (Maiga *et al.*, 2005; Klassen and Watkins, 2001; Hsu and Guo, 2002).

Selenium

It is used in cosmetics with the function of protecting the skin from environmental oxidative stress, preventing skin aging. It is associated with the metabolism of vitamin E, and also has anti-inflammatory action and immune-stimulant (Maiga *et al.*, 2005; Klassen and Watkins, 2001; Hsu and Guo, 2002; Cozzolino, 2007).

Use of Antioxidant Elements in Human Food

The general understanding about the way the antioxidant mineral elements affect the human body is increasing andnew supplement approaches have been developed. In particular the processes involved in the transport of metal cations, which usually have highly specific metal-ligand bonds and are highly stable, offering an excellent way of exploring new and more efficient forms of enrichment. One of the ways still under investigation, is the development of new products in which the release of the added element can be efficiently controlled (Halliwell, 2001; Droge, 2002).

Zinc

The second most abundant metal in the biological systems is this inorganic element that also participates in the anti-oxidative process. Up to date, were identified over 200 metalloenzymes containing this element. Zinc is essential for the integrity and function of cell membranes. Its concentration in the cell membrane may be quite high depending on the cell type and is influenced by nutritional levels of zinc in the body. It is a structural and catalytic component of the superoxide dismutase enzyme (SOD), present in the cytoplasm of all cells, which has in its active centreone copper ion and one zinc ion. This mineral is also part of the extracellular superoxide dismutase (EC-SOD) enzyme, present in plasma, lymph and synovial fluid. The action of SOD is to catalyse the conversion of two radicals:superoxide ion to hydrogen peroxide and molecular oxygen. Zinc deficiency has a major impact on growth, wound healing, the immune response and reproduction. The commonly used compounds employed to complement zinc are inorganic sulphate or chloride salts, or organic acetate, gluconate or stearate. Zinc oxides are still widely used as food additives. Zinc is distributed throughout the plant and animal kingdoms in abundance. Some of the main sources of this mineral are: bovine meat, fish, poultry, dairy products, oysters, seafood, grains, nuts and beans (Azevedo *et al.*, 2003; Scherz and Kirchhoff, 2006; Klaassen, 2008; Ingredients Food Brazil, 2008).

Copper

Copper has an essential antioxidant function through the action on superoxide dismutase against oxidative stress, helping to neutralize free radicals that could cause serious cell damage. Copper deficiency produces a higher oxidation of lipids and the oxidized cells increase the risk of becoming carcinogenic. Copper realizes a function in the complicate defence mechanism of the organism against oxidation. As part of the enzymes found in and around the cells, copper helps the body to neutralize free radicals to prevent cell destruction. The foods that are reach in copper are meats, seafood, seeds and oleaginous (WHO, 1999; Martins, 2003; Ingredients Food Brazil, 2008).

Manganese

One of the main roles of manganese is to act as an antioxidant agent and help to protect human health from oxygen toxic forms. Although its deficiency was not yet observed in humans, it is known that its function is linked with various enzymes such as mitochondrial superoxide dismutase, arginase, phosphoglucomutase, cholinesterase, beta-keto oxidative decarboxylases, certain peptidases and ATPase. His function in rheumatoid arthritis comes from the action of enzymes that are part of the synthesis of collagen, hyaluronic acid and other constituents of joint cartilage. The manganese deficiency in animals causes sterility and irreversible testicular degeneration, inability of females to breast feed, birth of weak, ataxic and small size animals. It is used as a supplement in treatments with an increased intake of calcium and phosphorus, since these elements decrease the absorption of manganese. In foods, it is found in greater amounts in whole grains, vegetable seeds, tubers, fruits and in smaller amounts in milk, poultry and fish (Klassen and Watkins, 2001; Hsu and Guo, 2002; Maiga *et al.*, 2005; Ingredients Food Brazil, 2008). **Selenium**: Also participates in the antioxidant processes. It is a mineral (trace element) related with the protection against damage caused by oxidative stress. Several scientific studies have shown that various types of dietary antioxidants, including selenium, may be effective in suppressing the activation of pro-inflammatory pathways through the chelation of free radical molecules. One of the most important biological functions of selenium is its function as a constituent of glutathione peroxidase (GPx), an antioxidant enzyme that attacks lipid peroxides and inorganic compounds. Low selenium concentrations were associated with the occurrence of atherosclerosis. Selenium deficiency is often related to heart disease, joint and bone structure problems. The selenium addition has aroused interest in many countries, especially those where the element deficiency causes diseases. Several simple compound such as SeO_2, Na_2SeO_3 or Na_2SeO_4 proved to be physiologically active. Grown in rich medium selenium yeast show beneficial health effects. It is preferably combined with organic compounds with lower toxicity. Grains are good sources of selenium, depending on the concentration of selenium in the soil and water where they grow. Other selenium sources are: seafood, beef and poultry (Klassen and Watkins, 2001; Hsu and Guo, 2002; Maiga *et al.*, 2005; Cozzolino, 2007; Ingredients Food, 2008).

References

Al-Ahmed, F. A., El-Denshary, E. S., Zaki, M. *et al.* (2002).Interaction between diazepam andherbs and their infusions, *Sci. Total Environ.*, **289**: 33–40.

Azevedo, F.A., and Chasin, A.A.M. (2003). (eds). Metais: Gerenciamento da toxicidade. São Paulo: Editora Atheneu, 554p.

Banning, A., Kipp, A., Schmitmeier, S., Löwinger, M., Florian, S., Krehl, S., *et al.*(2008).Glutathione peroxidase 2 inhibits cyclooxygenase-2-mediated migration and invasion of HT-29 adenocarcinoma cells but supports their growth as tumors in nude mice. *Cancer Res.*, **68**: 9746-53.

Bardia, A., Tleyjeh, I.M., Cerhan, J.R., Sood, A.K., Limburg, P.J., Erwin, P.J., *et al.*(2008). Efficacy of antioxidant supplementation in reducing primary cancer incidence and mortality: systematic review and meta-analysis. *Mayo Clin Proc.*, **83**: 23-34.

Belakova, M., Havranek, E., and Bumbalova, A. (1995). Heavy metals and some other elements in medicinal plants determined by x-ray fluorescence. *J. Radioanal. and Nucl. Chem.*, **201(5)**: 431-437.

Bjorkhem-Bergman, L., Torndal, U.B., Eken, S., Nystrom, C., Capitanio, A., Larsen, E.H., *et al.* (2005). Selenium prevents tumor development in a rat model for chemical carcinogenesis. *Carcinogenesis*, **26**: 125-31.

Brigelius-Flohé, R. (1999). Tissue-specic functions of individual glutathione peroxidases. *Free Radic Biol Med.*, **27**: 951-65.

Brinkman, M., Buntinx, F., Muls, E., and Zeegers, M.P. (2006). Use of selenium in chemoprevention of bladder cancer. *Lancet Oncol.*, **7**: 766-74.

Brown, T.A., and Shrift, A. (1981). Exclusion of selenium from proteins of selenium tolerant *Astragalus* species. *Plant Physiol.*, **67**: 1051–1053.

Czeczot, H., Scibior, D., Skrzycki, M., and Podsiad, M. (2006). Glutathione and GSH-dependant enzymes in patients with liver cirrhosis and hepatocellular carcinoma. *Acta Biochim Pol.*, **53**: 237-41.

Diaz, O.D., Portuondo, M.C., Valdes, M.G., Chang, R.R., and Handal, E.(1996). Comparison of various digestion methods for the determination of different metals in *Petiveria alliacea* Lynn (Anamu) by flame absorption spectrometry. *Quim. Anal.*, **15(2)**: 123-128.

Diwadkar-Navsariwala, V., Prinsm G.S., Swanson, S.M., Birch, L.A., Ray, V.H., Hedayat, S., *et al.* (2006). Selenoprotein deciency accelerates prostate carcinogenesis in a transgenic model. *Proc Natl Acad Sci USA*, **23**: 8179-84.

Droge, W. (2002). Free radicals in the physiological control of cell function. *Physil Ver.*, **82**: 47-95.

Ernst, E. (2002). Toxic heavy metals and undeclared drugs in Asian herbal medicines. *Trends Pharmacol Sci.*, **23**: 136-139.

Favier, A. (1991). Les oligoéléments en nutrition humaine. In: CHAPPUIS, P. (Ed) Les Oligoéléments en Medicine etBiologic.Paris: EditionsMédicalesInternacionales, Cap. 3, p. 41-74.

Franco, F. (1998). Tabela de Composição de Alimentos. 9ª ed., Atheneu, São Paulo.

Ghosh, D., Bhattacharya, B., Mukherjee, B. *et al.* (1986). Role of chromium supplementation of diabetes mellitus. *Am. J. Med.*, **81**: 63–68.

Gromadzinska, J., Reszka, E., Bruzelius, K., Wasowicz, W., and Akesson, B. (2008). Selenium and cancer: biomarkers of selenium status and molecular action of selenium supplements. *Eur J Nutr.*, **47 (Suppl 2)**: 29-50.

Gundimeda, U., Schiffman, J.E., Chhabra, D., Wong, J., Wu, A., and Gopalakrishna, R.(2008). Locally generated methylseleninic acid induces specic inactivation of protein kinase C isoenzymes: relevance to selenium-induced apoptosis in prostate cancer cells. *J Biol Chem.*, **283**: 34519-31.

Gurson, C.T., and Saner, G. (1971). Effect of chromium on glucoseutilization in marasmic protein calorie malnutrition. *American Journal of Clinical Nutrition*, **24**: 1313–1319.

Halliwell, B. (2001). Free radicals and other reactive species in disease. In: *Encyclopedia of Life Sciences*, Nature Publishing Group, p. 1 – 7.

Harper, H. A., Rodwelf, V.W., and Mayes, R.A. 1982. Manual de Química Fisiológica.5*a* ed., Atheneu, São Paulo.

Harris, H. H., Pickering, I. J., and Salt, D. E., 2004. Production of Se-methylselenocysteine in transgenic plants expressing selenocysteine methyltransferase. *BMC Plant Biol.*, **4**: 1.

Heirman, I., Ginneberge, D., Brigelius-Flohé, R., Hendrickx, N., Agostinis, P., Brouckaert, P. *et al.* (2006). Blocking tumor cell eicosanoid synthesis by GPx 4 impedes tumor growth and malignancy. *Free Radic Biol Med.*, **40**: 285-94.

Hsu, P.C., and Guo, Y.L. (2002). Antioxidant nutrients and lead toxicity. *Toxicology*, **180**: 33-44.

Institute of Medicine, Food and Nutrition Board, (2001). Dietary Reference Intakes for Vitamin A, Vitamin K, Arsenic, Boron, Chromium, Copper, Iodine, Iron, Manganese, Molybdenum, Nickel, Silicon, Vanadium, and Zinc. National Academy Press, Washington, DC.

Jain, N., Shahoo, R.K., and Sondhi, S.M. (1992). Analysis for mineral elements of some medicinal plants. *Indian Drugs*, **29(4)**: 187-190.

Kakosy, T., Hudak, A., and Naray, M. (1996). Lead intoxication epidemic caused by ingestion of contamined ground paprika. *J Toxicol-Clin Toxic.*, **34**: 507-511.

Kar, A., Choudhary, B.K., and Bandyopadhyay, N.G.(1999). Preliminary studies on the inorganic constituents of some indigenous hypoglycaemicherbs on oral glucose tolerance test. *J. Ethnopharmacology*, **64(2)**: 179-184.

Kawashima, L.M., and Valente Soares, L.M. (2003). Mineral profile of raw and cooked leafy vegetables consumed in Southern Brazil. *Journal of Food Composition and Analysis*, **16**: 605–611.

Kinlaw, W.B., Levine, A.S., Morley, J.E., Silvis, S.E., and Mc-Clain, C.J. (1983). Abnormal zinc metabolism in type II diabetes mellitus. *The American Journal of Medicine*, **75**: 273–277.

Klaassen, C. D. (2008). (ed). Casarett e Doull's Toxicology: The basic science of poisons.Texas: Mc Graw Hill Medical, 1310p.

Klassen, C.D., and Watkins, III J.B. (2001). *Toxicologia, a Ciência Básica dos Tóxicos, de Casarett e Doull*. Portugal: Editora McGraw-Hill.

Kocdor, H., Cehreli, R., Kocdor, M.A., Sis, B., Yilmaz, O., Canda, T. *et al.* (2005). Toxicity induced by the chemical carcinogen 7,12-dimethylbenz[a]anthracene and the protective effects of selenium in Wistar rats. *J Toxicol Environ Health A.*, **68**: 693-701.

Kubachka, K.M., Meija, J., LeDuc, D.L., Terry, N., and Caruso, J. (2007). Selenium volatiles asproxy to the metabolic pathways of selenium in genetically modified *Brassica juncea*. *Environ. Sci. Technol.*, **41**: 1863–1869.

Lokova, M.Y., Sokolova, S.M., Ponomareva, S.M., Buzuk, G.N., Bykhovsky, V.Y., and Kliment'eva, N.I. (1999b). Specificity of elemental compositions of medicinal plants synthesizing alkaloids. *Appl. Biochem. and Microbiol.*, **35(1)**: 68-75.

Lopes, M.F.G., Morais, N.M.T., Nogueira, C.M.D., Vasconcelos, N.M.S., Silva, W.C., and Gonzaga, M.L.C. (1996). Determinações Analíticas em Alfavaca, Quebra-Pedra, Sabugueiro, Eucalipto e Cajá-Umbu. *Anais Assoc. Bras.Quím.*, **45(4)**: 155-157.

Lopes, M.F.G., Vasconcelos, N.M.S., Almeida, M.M.B., Nogueira, C.M.D., Morais, N.M.T., and Sá, M.J.H.C. (1998). Caracterização Analítica de Plantas Medicinais. *Rev. Bras.Farm.*, **79(3-4)**: 88-89.

Lovkova, M.Y., Sokolova, S.M., Buzuk, G.N., Bykhovsky, V.Y., and Ponomareva, S.M. (1999a). Specific features of the elemental composition of medicinal plants synthesizing phenolic compounds. *Appl. Biochem. and Microbiol.*, **35(5)**: 514-524.

Lozak, A., Soltyk, K., Ostapczuk, P., *et al.* (1997). Determination of selected trace elements in and Health (part A), **32(8)**: 2111-2123.

Maiga, A., Diallo, D., Bye, R., and Paulsen, B.S. (2005). Determination of some toxic and essential metal ions in medicinal and edible plants from Mali. *J Agric Food Chem.*, **53**: 2321.

Maiorino, M., Aumann, K.D., Brigelius-Flohé, R., Doria, D., Van denHeuvel, J., McCarthy, J., *et al.* (1995). Probing the presumed catalytic triad of selenium-containing peroxidases by mutational analysis of phospholipid hydroperoxide glutathione peroxidase (PHGPx). *Biol Chem Hoppe Seyler.*, **376**: 651-60.

Majid, A.A., Sarmani, S., Yusoff N.I., Wei Y.K., and Hamzah F. (1995). Trace elements in Malaysian medicinal plants. *J.Radioanal. and Nucl. Chem.*, **195(1)**: 173-183.

Martins, I. (2003). Manganês. In: Azevedo, F.A., Chasin, A.A.M. (eds). Metais: gerenciamento da toxicidade.São Paulo: *Editora Atheneu*, 147-153.

Massey, R., and Taylor, S. (1991). Aluminium in Food and the Environment.Royal Society of Chemistry, London.

Matich, A.J., and Brummell, D.A. (2009). Accumulation of an organic anticancer selenium compound in a transgenic Solanaceous species shows wider applicability of the selenocyste inemethyltransferase transgene from selenium hyperaccumulators. *Transgenic Res.*, **18**: 407–424.

Matos, F. J.A. (1994). Farmácias Vivas, Edições UFC. Fortaleza.

McKenzie, M.J., Hunter, D.A., Pathirana, R., Watson, L.M., Joyce, N., Rowan, D.D., Morais, N.M.T., Nogueira, C.M.D., Lopes, M.F.G., Vasconcelos, N.M.S., and Sá, M.J.H.C. (1995). Estudo Inorgânico Analítico de Plantas Medicinais. *Anais Assoc. Bras.Quím.*, **44(4)**: 14-19.

Mertz, W. (1981). The essential trace elements, *Science, 213*: 1332–1338.

Navarro-Alcaron, M., and Cabrera-Vique, C.(2008). Selenium in food and the human body: A review. *Sci Total Environ.*, **400**: 115-41.

Niewoehner, C. B., Alen, J. I., Boosalis, M. *et al.* (1992). Role of zinc supplementation in type IIoral antidiabetic agents on serum glucose, insulin and chromium levels in rats, *Biol. Sci.*

Nogueira, C.M.D., Lopes, M.F.G., Morais, N.M.T., Almeida, M.M.B., Vasconcelos, N.M.S., and Sá, M.J.H.C.(1998). Determinações de Elementos Minerais em Plantas Medicinais. *Anais Assoc. Bras. Quím.*, **47(1)**: 22-24.

Ottaviano, F.G., Handy, D.E., and Loscalzo, J. (2008). Redox regulation in the extracellular environment. *Circ J.*, **72**: 1-16.

Palumbo, P. J. (2001). Glycemic control, maltime glucose excursions, and diabetic complicationsin type 2 diabetes mellitus. *Mayo Clin. Proc.*, **76**: 609–618

Pari, L., and Saravanan, G. (2002). Antidiabetic effect of Cogent db, a herbal drug in alloxan induced diabetes mellitus. *Comp. Biochem. Physiol.*, **131:** 19–25.

Park, J.M., Kim, A., Oh, J.H., and Chung, A.S. (2007). Methylseleninic acid inhibits PMA-stimulated pro-MMP-2 activation mediated byMT1-MMP expression and further tumor invasion through suppression of NF-kappaB activation. *Carcinogenesis*, **28**: 837-47.

Rafferty, T.S., Walker, C., Hunter, J.A., Beckett, G.J., and McKenzie, R.C.(2002). Inhibition of ultraviolet B radiation-induced interleukin 10 expression in murine keratinocytes by selenium compounds. *Br J Dermatol.*, **146**: 485-9.

Rajurkar, N.S., and Damame, M.M. (1998). Mineral content of medicinal plants used in the treatment of diseases resulting from urinary tract disorders. *Appl. Radiation and Isotopes*, **49(7)**: 773-776.

Reid, M.E., Duffield-Lillico, A.J., Slate, E., Natarajan, N., Turnbull, B., Jacobs, E. *et al.* (2008). The nutritional prevention of cancer: 400 mcg per day selenium treatment. *Nutr Cancer*, **60:** 155-63.

Roth, H., and Kirchgaessner, M. (1991). In: Holtmeier, H.J., Kruse-Jarres, J. (Eds.), Zink, Biochemie, Physiologie, Pathophysiology und Klinik des Stoffwechsels des Menschen. WissenschaftlicheVerlagsgesellschaftmbH, Stuttgart, pp. 67–101.

Sarmani, S.B., Abugassa, I., Hamzah, A., and Yahya, M.D. (1999). Elemental analysis of herbal preparations for traditional medicines by neutron activation analysis with the k(0) standardization method. *Biol. Trace Elem. Res.*, **71(2)**: 365-376.

Sathiyamoorthy, P., Van Damme, P., Oven, M., and Golangoldhirsh, A. (1989). Heavy metals in medicinal and fodder plants of the Negev desert. *J. Environ. Science Rep.*, **9**: 347–350.

Schrauzer, G.N. (2000). Selenomethionine: a review of its nutritional significance, metabolism and toxicity. *J Nutr.*, **130**: 1653-6.

Sieja, K., and Talerczyk, M. (2004). Selenium as an element in the treatment of ovarian cancer in woman receiving chemotherapy. *Gynecol Oncol.*, **93**: 320-7.

Singh, N. S., Vats, P., Suri, S. *et al.* (2001). Effect of an antidiabetic extract of *Catharanthus roseus* on enzymic activities in streptozotocin induced diabetic rats. *J. Ethnopharmacol.*, **76**: 269–277.

Sors, T.G., Ellis, D.R., and Salt, D.E. (2005). Selenium uptake, translocation, assimilation and metabolic fate in plants. *Photosynth. Res.*, **86**: 373–389.

Steinbrenner, H., and Sies, H. (2009). Protection against reactive oxygen species by selenoproteins. *Biochim Biophys Acta.*, **1790**: 1478-85.

Stewart, M.J., Moar, J.J., Steenkamp, P., and Kokot, M. (1999). Findings in fatal cases of poisoning attributed to traditional remedies in South Africa. *Forensic Sci Int.*, **101**: 177-183.

Suzuki, K.T. (2005). Metabolomics of selenium: Se metabolites based on speciation studies. *J Sci.*, **51**: 107-44.

Tang, X., and Shay, N. F. (2001). Zinc has an insulin like effect on glucose transport mediated byphosphoinositol-3-kinase and Akt in 3T3-L1 fibroblasts and adipocytes, *J. Nutr.*, **131**: 1414–1420.

Tavares, T.M., and Carvalho, F.M. (1992). Avaliação da exposição de populações humanas a metais pesados no ambiente: exemplos do Recôncavo Bahiano. *Quim Nova*, **15**:

Terres, C., Navarro, M., Martin-Lagos, F., Gimenez, R., Lopez, H., and Lopez, C.M. (2001). Zinc levels in foods from southeastern Spain: relation to dairy intake. *Food Additives and Contaminants*, **18**, 687–695.

Terry, N., Zayed, A.M., de Souza, M.P., and Tarun, A.S. (2000). Selenium in higher plants. *Annu. Rev. Plant Physiol.: Plant Mol. Biol.*, **51**: 401–432.

Tian, J., Xie, A., Wang, S., Mehendale, *et al.* (2003). Antidiabetic effects of *Gymnema yunnanese* extract, *Pharmacol. Res.*, **47**: 323–329.

Tyler, V.E. (1987). Herbal medicine in America. *Planta Med.*, **14**: 1-4.

Underwood, E. J. (1977). *Trace Elements in Human and Animal Nutrition,* 4th ed., Academic, New York.

Vasconcelos, N.M.S., Almeida, M.M.B., Lopes, M.F.G., Nogueira, C.M.D., and Morais, N.M.T. (2000). Plantas Medicinais: Um estudo analítico. *Anais Assoc. Bras. Quím.*, **49(3)**: 115-118.

Waters, D.J., Shen, S., Glickman, L.T., Cooley, D.M., Bostwick, D.G., Qian, J., *et al.* (2005). Prostate cancer risk and DNA damage: translational signicance of selenium supplementation in a canine model. *Carcinogenesis*, **26**: 1256-62.

Wesolowski, M., Konieczynski, P., and Ulewicz, B. Influence of the elemental composition of medicinal fruits on the results of their thermal analysis. *J. Thermal Anal and Calorimetry*, **60(1)**: 299-304.

World Health Organization (WHO). (1996). Trace Elements in Human Nutrition and Health, Geneva.

World Health Organization (WHO). (1999). Manganese and its compounds. Geneva: WHO, 46p. (Concise international chemical assessment document, 12).

Utilisation and Management of Medicinal Plants Vol. 3 (2015) *Pages* **165–180**
Editors: **V.K. Gupta and A. Kaul**
Published by: **DAYA PUBLISHING HOUSE, NEW DELHI**

11

Citrus: A Genus with Miraculous Pharmacological Potential

Rasleen Sudan[1] and Madhulika Bhagat[1*]

ABSTRACT

Citrus species are flowering plants belonging to family Rutaceae. Natural and cultivated origin hybrids include commercially important fruits such as the oranges, lemons, limes and grapefruits. Citrus fruits also hold importance in the Indian and Chinese traditional systems of medicine. They have been used since ages for the treatment of medical ailments like cold, cough, respiratory problems, indigestion, liver and hair tonic, against bacterial and fungal infections and also for treating skin diseases. Citrus fruits are high in vitamin C, and are good sources of folate and potassium. Vitamin C is a powerful antioxidant and protects the body from damaging free radicals. Citrus fruits have been extensively studied and they have been proven to constitute compounds of therapeutic importance like limonin, naringenin, hesperetin etc., that possess anticancer, antioxidant, anti-inflammatory, anti-obesity, antimicrobial and many other pharmacological properties. This review article is an attempt to gather information about the therapeutic and pharmacological value of some economically important Citrus species belonging to major groups of lemons, oranges and grapefruits viz. C. limon, C. aurantifolium, C. medica, C. sinensis, C. aurantia, C. reticulata and C. paradisi.

Keywords: Citrus fruits, Pharmacological potential, Vitamin C, Antioxidant.

1 School of Biotechnology, University of Jammu, J&K State, India.
* *Corresponding author*: E-mail: madhulikasbt@gmail.com

Introduction

Botanical Description of Citrus Fruits

Kingdom: Plantae

Division: Angiosperms

Class: Eudicots

Sub-class: Rosids

Order: Sapindales

Family: Rutaceae

Genus: *Citrus*

Figure 11.1: *Citrus* Fruits.

General Description

Commonly these plants are large shrubs or small trees with spiny shoots. The flowers are solitary or in small corymbs, with five white petals and are very strongly scented. The fruit (Figure 11.1) of any citrus tree is considered a hesperidium (a kind of modified berry) because it has numerous seeds, is fleshy and soft, derives from a single ovary and is covered by a rind (Spiegel-Roy and Goldschmidt, 1996). The citrus trees thrive in a consistently sunny, humid environment with fertile soil and adequate rainfall or irrigation.

Geographical Distribution

Citrus fruits are believed to have originated in the part of Southeast Asia bordered by Northeast India, Burma (Myanmar) and the Yunnan province of China and were eventually distributed all over the world (Rainer and Scora, 1975). Today, around 70 per cent of the world's total citrus production is grown in the Northern Hemisphere, in particular countries around the Mediterranean and the United States.

Nutritive Value of Citrus Fruits

Citrus fruits contain an impressive list of essential nutrients, including carbohydrates, potassium, folate, calcium, thiamin, niacin, vitamin B_6, phosphorus, magnesium, copper, riboflavin, pantothenic acid and a variety of phytochemicals (Table 11.1).

Major Chemical Constituents of Citrus Fruits

Lemons and limes have particularly high concentrations of citric acid, which can constitute as much as 8 per cent of the dry weight of these fruits. Constituents of lemon volatile oil (2.5 per cent of the peel) include limonene, α-terpinene, α-pinene, β-citral, coumarins, linalool, geraniol, mucilage, pectins, and bio-flavonoids (Miyake *et al.*, 2007; Fischer and Philips, 2006; Manners, 2007). Oranges contain 1.5 per cent essential oil. The main phytoconstituents are D-limonene, citral, citronellal, nootkaton, n-decanal, n-dodecanal, linalyl acetate, geranyl acetate, citronelyl acetate and anthranil acid methyl ester (Milind and Dev, 2012). Hesperetin, naringin, and

Table 11.1: Nutritional Value of different Citrus Fruits (Lemons, oranges and grapefruit) Gutherie and Picciano, 1995

Constituents	*Nut. Value/ 100g Lemon*	*Nut. Value/ 100g Orange*	*Nut. Value/ 100g Grapefruit*
Energy	121 kJ (29 kcal)	197 kJ (47 kcal)	138 kJ (33 kcal)
Carbohydrates	9.32 g	11.75 g	8.41 g
Sugars	2.5 g	9.35 g	7.31 g
Dietary fiber	2.8 g	2.4 g	1.1 g
Fat	0.3 g	0.12 g	0.10 g
Protein	1.1 g	0.94 g	0.69 g
Thiamine (vit. B_1)	0.04 mg (3%)	0.087 mg (8%)	0.037 mg (3%)
Riboflavin (vit. B_2)	0.02 mg (2%)	0.04 mg (3%)	0.020 mg (2%)
Niacin (vit. B_3)	0.1 mg (1%)	0.282 mg (2%)	0.269 mg (2%)
Pantothenic acid (B_5)	0.19 mg (4%)	0.25 mg (5%)	0.283 mg (6%)
Vitamin B_6	0.08 mg (6%)	0.06 mg (5%)	0.043 mg (3%)
Folate (vit. B_9)	11 μg (3%)	30 μg (8%)	10 μg (3%)
Choline	5.1 mg (1%)	8.4 mg (2%)	7.7 mg (2%)
Vitamin C	53 mg (64%)	53.2 mg (64%)	33.3 mg (40%)
Calcium	26 mg (3%)	40 mg (4%)	12 mg (1%)
Iron	0.6 mg (5%)	0.1 mg (1%)	0.06 mg (0%)
Magnesium	8 mg (2%)	10 mg (3%)	9 mg (3%)
Manganese	0.03 mg (1%)	0.025 mg (1%)	0.013 mg (1%)
Phosphorus	16 mg (2%)	14 mg (2%)	8 mg (1%)
Potassium	138 mg (3%)	181 mg (4%)	148 mg (3%)
Zinc	0.06 mg (1%)	0.07 mg (1%)	0.07

naringenin are flavonoids found in citrus fruits. Oranges also contain other flavonoid antioxidants such as alpha and beta-carotenes, beta-cryptoxanthin, zea-xanthin and lutein (Figure 11.2). The main constituents of grapefruits are similar to other citrus fruits. However, grapefruit contains a number of polyphenolic compounds, including lycopene which is a remarkable natural antioxidant, the flavanone glycoside naringin, alongside the two furanocoumarins bergamottin and dihydroxybergamottin. These inhibit the drug-metabolizing enzyme isoform CYP3A4 predominately in the small intestine, but at higher doses also inhibit hepatic CYP3A4 (Veronese *et al.*, 2003). It is via inhibition of this enzyme that grapefruit increases the effects of a variety of drugs by increasing their bioavailability (Bailley *et al.*, 1998; He *et al.*, 1998). Unlike other fruits, grapefruit contains a large amount of naringin. Grapefruit contains large quantities of a simple polyamine called spermidine, which may be related to aging (Eisenberg *et al.*, 2009).

Figure 11.2: Major Chemical Constituents of Citrus Fruits.

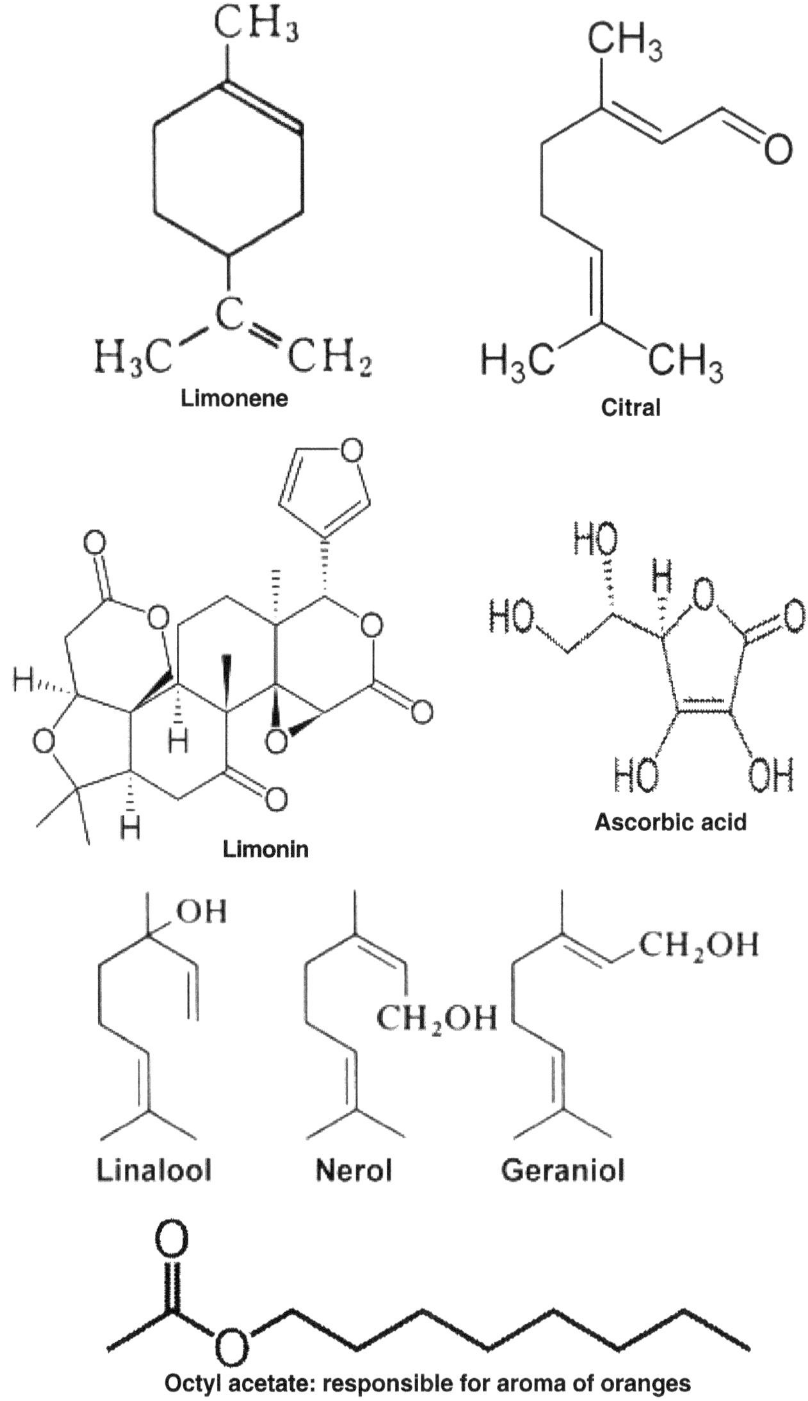

Contd...

Figure 11.2–*Contd...*

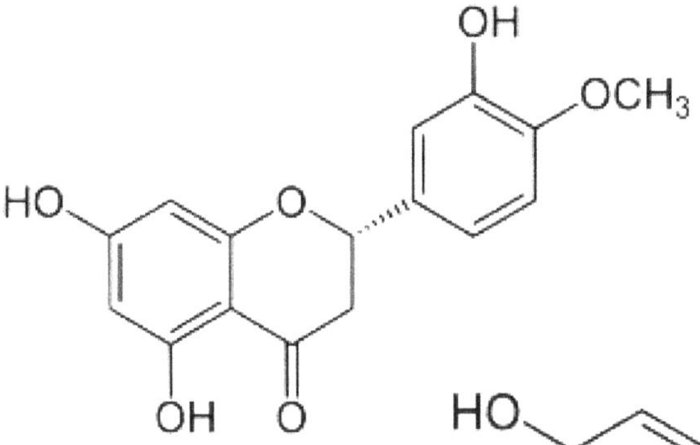

Hesperetin

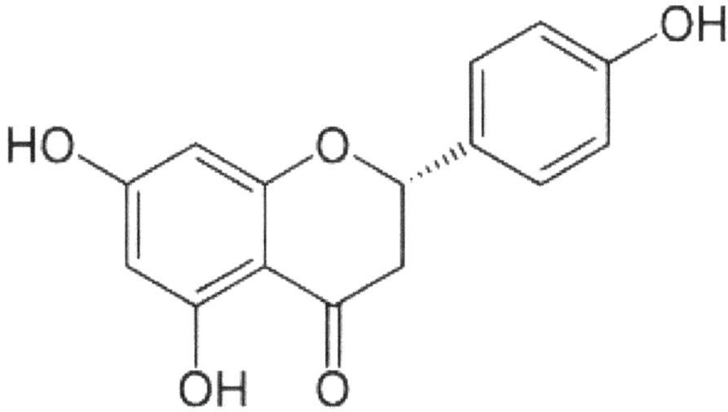

Naringenin

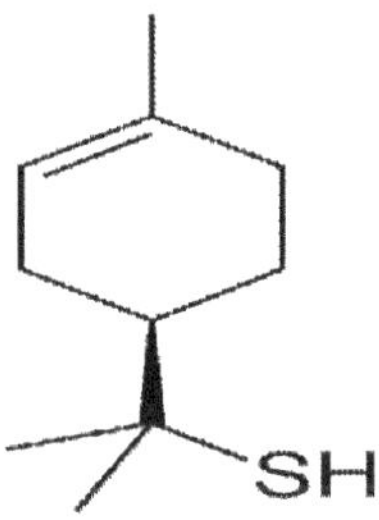

Grapefruit mercaptan

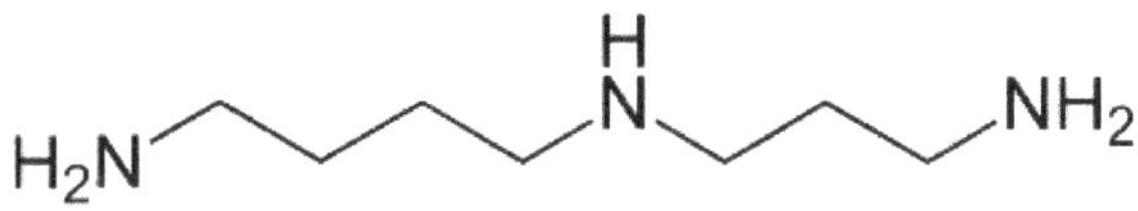

Spermidine

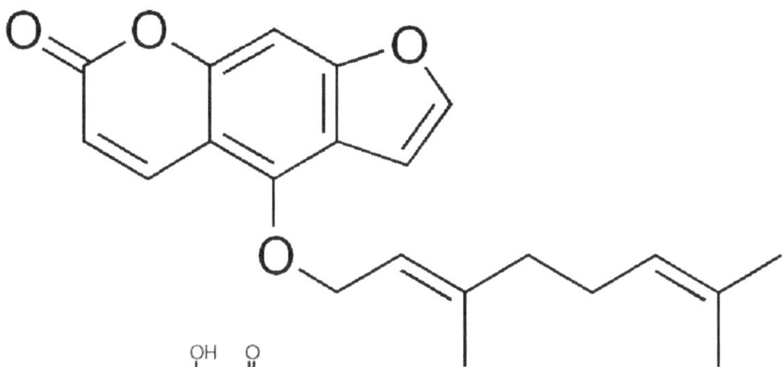

Bergamottin (furanocoumarin)

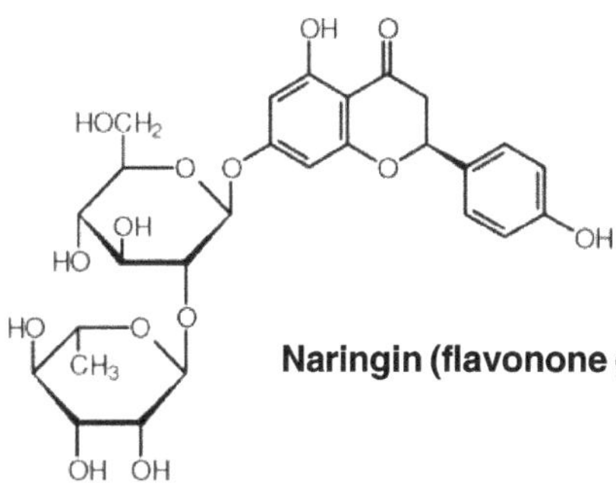

Naringin (flavonone glycoside)

Lemons and Limes

General Description

The lemon tree is an evergreen, tall tree. Its leaves are light green and the citrus fruit (lemon) is oval, small, and green to yellow. Unlike other citrus varieties, the lemon tree bears fruit continuously. Various species of lemons and limes are cultivated throughout the world. A few of them are described as under:

a) *Citrus limon*

General Description

The lemon (*Citrus limon;* commonly called as *nimbu* in Hindi) is a small evergreen tree native to Asia (Figure 11.3). The origin of the lemon is thought to be from Assam (Northeast India), northern Burma, and China. The lemon tree is tall and has stout spines. The leaves are dark green and evergreen. Flowers open to have 5 white petals. Fruits are globose to oblong and ripen to yellow (Morton, 1987).

Figure 11.3: *Citrus limon.*

Ethno-Medicinal Importance

Lemon juice has long been used as an astringent, diaphoretic, diuretic, gargle, lotion, tonic and laxative. It stimulates salivation and digestive juices in the stomach. Infusions prepared with the aerial (leaves) parts of *Citrus limon* are used in folk medicine for the treatment of obesity, diabetes, blood lipid lowering, cardiovascular diseases, brain disorders, and certain types of cancer (Campêlo *et al.*, 2011).

b) *Citrus aurantifolia*

General Description

Citrus aurantifolia (commonly called as Key lime; or *kaghzi nimbu* in Hindi), whose skin is thin and green in color and which have an oval or round shape (Figure 11.4). The flowers are yellowish white with a light purple tinge on the margins. Flowers and fruit are most abundant from May to September in the Northern Hemisphere (Golob, 1999).

Figure 11.4: *Citrus aurantifolia*.

Ethno-medicinal Importance

In the African sub-continent, key lime is used in treating malaria, diarrhoea, burns, gonorrhoea, stomach disorders and other infectious diseases (Ndukwe *et al.*, 2005). The mesocarp is also used as a very good facial scrub and helps in prevention of pimples due to its cleansing action on the skin (Onyeagba *et al.*, 2004).

c) *Citrus medica*

General Description

Citrus medica, commonly known as Citron, is a slow-growing shrub or small tree with evergreen leaves (Figure 11.5). The main content of a citron fruit is the thick white rind, which adheres to the segments. The citron tree is very vigorous with almost no dormancy, blooming several times a year, and is therefore fragile and extremely sensitive to frost (Morton, 1987). The citron is native to India where it borders on Burma, and Pakistan where it is found in valleys at the foot of the Himalayas, and in the Indian Western Ghats. Later on, it spread to Mediterranean countries and further to Europe (Hooker, 1882).

Ethno-medicinal Importance

From ancient through medieval times, the citron was used mainly for medical purposes: to combat seasickness, pulmonary troubles, intestinal ailments and other

Figure 11.5: *Citrus medica.*

disorders. The essential oil of the flavedo (the outermost, pigmented layer of rind) was also regarded as an antibiotic. Citron juice with wine was considered an effective antidote to poison. In Ayurvedic system of medicine, the fruit juice is used for treating conditions like nausea, vomiting, excessive thirst etc. (Arias and Laca, 2005).

Oranges

a) *Citrus sinensis*

General Description

Citrus sinensis is called sweet orange, which is a term derived from Sanskrit word "narang". The orange tree is an evergreen, flowering tree (Figure 11.6). It is subdivided into four classes with distinct characteristics: common oranges, blood or pigmented oranges, navel oranges, and acidless oranges (Kimball and Dan, 1999). Probably originating in Southeast Asia, oranges were already cultivated in China as far back as 2500 BC. Arabic people popularized sour citrus and oranges in Europe.

Ethno-medicinal Importance

The whole orange plant including fruits, leaves, flowers and peels are used as traditional medicine. Orange juice helps to eliminate toxins from the body and is also helpful in cases of anxiety disorders and stress. It is used in Mexican traditional medicines for treating tuberculosis. It helps in digestion, especially in cases of constipation. Chinese use it for treating cough, cold and respiratory disorders. In

Figure 11.6: *Citrus sinensis.*

France, it is used for curing angina, hypertension, diarrhea, menstrual disorders and palpitation (Milind and Dev, 2012).

b) *Citrus aurantium*

General Description

Commonly *Citrus aurantium* is referred as bitter orange, seville orange, sour orange, bigarade orange, or marmalade orange (Figure 11.7). *Citrus aurantium* is a spiny evergreen tree native to southern Vietnam, which is now grown throughout the Mediterranean region.

Figure 11.7: *Citrus aurantium.*

Ethnomedicinal Importance

Traditionally, essential oil extracted from bitter orange, (neroli oil) used to relieve tension, anxiety and also to increase circulation. It is cultivated for the essential oil expressed from the fruit, and for neroli oil and orange flower water, which are distilled from the flowers and used in perfume, as a flavoring and as a solvent.

c) *Citrus reticulata*

General Description

The Mandarin orange is a small citrus tree (*Citrus reticulata*) (Figure 11.8). The tree is drought-tolerant and is self-fertile. The mandarin is tender and is damaged easily by cold. It can be grown in tropical and subtropical areas. Mandarin oranges are native to China and Japan. They are now cultivated in all the countries and numerous other varieties have been developed from original mandarins like Satsuma, Kinnow, Clementines and Tangerines.

Figure 11.8: *Citrus reticulata.*

Ethno-medicinal Importance

Mandarins have been used in ayurveda (Chopra *et al.*, 1986). Traditional healers in China use mandarin orange as an active constituent in their remedies to treat the following ailments like poor digestion, constipation, colic, flatulence, heartburn, loss of appetite, dyspepsia, peptic ulcer, cirrhosis of the liver, inflammation of the liver and spleen, allergy, boost immunity, excessive phlegm, coughs, acute and chronic bronchitis, asthma, high cholesterol, arteriosclerosis, insomnia, stress, lumbago, skin problems, scurvy, hemorrhoid, mastitis, insulin resistance, cancer of the liver and pain in the breast (Jasim, 2012).

c) Grapefruits

General Description

The grapefruit (*Citrus × paradisi*) is a subtropical citrus tree known for its sour to semi-sweet fruit (Figure 11.9). These evergreen trees usually grow tall. The fruit is yellow-orange skinned and largely an oblate spheroid. The flesh is segmented and acidic, varying in colour depending on the cultivars, which include white, pink and red pulps of varying sweetness (Carrington and Fraser, 2003). This hybrid fruit was first developed in Barbados.

Figure 11.9: Grapefruit- *Citrus paradisi.*

Ethno-botanical Importance

Grapefruit juice is used for high cholesterol, atherosclerosis, cancer, a skin disease called psoriasis and for weight loss and obesity. Grapefruit seed extract is taken by mouth for bacterial, viral, and fungal infections including yeast infections. It is also used for the common cold and flu (influenza). It is also used as an ear or nasal rinse for preventing and treating infections; as a gargle for sore throats; and a dental rinse for preventing gingivitis and promoting healthy gums; and as a breath freshener. Some people inhale grapefruit vapours to help the body retain water, for headache, stress, and depression (Gupta *et al.*, 2011)

Pharmacological Importance of Citrus Fruits

Citrus fruits have long been valued for their wholesome nutritious and antioxidant properties. It is scientifically established that citrus fruits, especially oranges, by virtue of their richness in vitamins and minerals, have many proven health benefits.

Antioxidant Effects

Genus *Citrus* is highly regarded for showing its antioxidant properties and reported by numerous studies. Researchers have identified the naringenin, eriocitrin, hesperidin, and coumarins as antioxidants from this genus in *in vitro* and *in vivo* models (Miyake *et al.*, 1998; Miyake *et al.*, 1999). In addition to their unique phytonutrient properties, lemons and limes are an excellent source of vitamin C, one of the most important antioxidants in nature. Further, pectin fiber and lemon oil also possess antioxidant properties (Polunin, 1997). *Citrus limon* exhibits an antioxidant effect by preventing lipoperoxidation due to hydroxyl radical scavenging activity (Campelo *et al.*, 2011).

Antimicrobial and Anti-ulcer Effects

Lemon has been used externally for acne, fungus (ringworm and athlete's foot), sunburn, and warts (USDA, NRCS, 2008; Chevallier, 1996; Ensminger *et al.*, 1994). Lemon juice and lemon oil shows some bacteriostatic and antiviral action which is thought to be due to citral and linalool content. It inhibits the growth of *Aspergillus* mold (Ballot *et al.*, 1987) and has been used to disinfect drinking water (Alderman and Marth, 1976) and to inactivate rabies virus (D'Aquino and Teves, 1994). Grapefruit seed extract has been shown to have strong antimicrobial properties against fungi (Ignacio and Thai, 2005). It has also, been observed that regular intake of orange juice reduced the *Helicobacter pylori* infection which is a causative agent of ulcers (Simon *et al.*, 2003). Some citrus fruits like *C. sinensis*, *C. reticulata* and *C. maxima* have shown to possess antimicrobial activities against *S. aureus*, *S. pyogens*, *A. niger* and *C. albicans* (Mathur *et al.*, 2011).

Anti-cancer Effects

Several studies indicated the anticancer activity of the citrus genus at various levels. The flavonoids, eriocitrin, coumarins extracted from lemon fruit indicated the induction of apoptosis in acute myelomonocytic leukemia cells (Ogata *et al.*, 2000). The consumption of citrus fruits showed the protective effect against oral cancer (Palvia *et al.*, 2006). The structure of various chemical constituents of lemon and their relationship to cancer prevention has been investigated (Benavente-García *et al.*, 2007). Two citrus flavonoids, hesperetin and naringenin, are found to be effective inhibitors of human breast cancer cell proliferation *in vitro* (So *et al.*, 1996). In animal studies and laboratory tests with human cells, compounds in citrus fruits, including lemons and limes, called limonoids have been shown to help fight cancers of the mouth, skin, lung, breast, stomach and colon. New research adds to the growing body of evidence that modified citrus pectin (MCP), a natural supplement derived from the pith of citrus fruit, has a powerful effect against cancer. MCP enhanced the anti-cancer effects of two botanical formulas, one for breast cancer and one for prostate cancer. The combination treatment decreased cancer cell aggressiveness, reduced migration and inhibited metastasis in highly metastatic MDA-MB-231 human breast cancer cells and aggressive, PC-3 hormone refractory human prostate cancer cells (Jiang *et al.*, 2013).

Cardiovascular Disorders

Studies show that citrus flavonoids improve blood flow through coronary arteries, reduce the ability of arteries to form blood clots and prevent the oxidation of LDL ("bad") cholesterol, which is an initial step in the formation of artery plaques. Collectively, these studies add to the growing body of evidence that citrus flavonoids have marked lipid and lipoprotein lowering potential and demonstrate that naringenin and nobiletin reduce hepatic lipid accumulation and prevent lipoprotein overproduction, normalize glucose tolerance and insulin sensitivity and slow the progression of atherosclerosis (Mulvihill and Huff, 2012). Bitter orange contains the tyramine metabolites N-methyltyramine, octopamine and synephrine substances similar to epinephrine, which act on the α_1 adrenergic receptor to constrict blood vessels and increase blood pressure and heart rate (Gange *et al.*, 2006; Bui *et al.*, 2006). Frequent consumption of folate-rich foods, such as oranges and orange juice, tends to increase plasma folate levels and, thus, lower homocysteine levels.

Regulation of Fats and Lipid Accumulation

Nutrients in oranges are plentiful and diverse. Pectin, by its action as a bulk laxative, helps to protect the mucous membrane of the colon by decreasing its exposure time to toxic substances as well as by binding to cancer-causing chemicals in the colon. Pectin has also been shown to reduce blood cholesterol levels by decreasing its re-absorption in the colon by binding to bile acid in the colon. High intakes of vitamin

C (500 mg/day) obtained from the juice of freshly squeezed oranges, prevented a rise in the levels of oxidized LDL, even in the presence of a high-saturated fat diet (Harats *et al.*, 1998).

Anti-diabetic Effect

Citrus fruit peels contain naringin and hesperidin which regulate glucose regulatory enzymes and possess antihyperglycemic activity. They decrease the activity of glucose-6-phospatase and PEP. The mechanism of glucose regulation is believed to occur via anti peroxidation, inhibition of α- amylase enzyme which controls the conversion of carbohydrates to glucose (Milind and Dev, 2012; Kundusen *et al.*, 2011).

Anti-ageing

Grapefruit contains large quantities of a simple polyamine called spermidine, which may be related to aging. It is known to be necessary for cell growth and maturation, and as cells age their level of spermidine is known to fall. Scientists have shown that feeding spermidine to worms, fruit flies and yeast significantly prolongs their lifespan. In addition, adding spermidine to the diet of mice decreased molecular markers of aging, and when human immune cells were cultured in a medium containing spermidine, they also lived longer (Eisenberg *et al.*, 2009).

Other Uses

The sperm-immobilizing properties of lemon juice have been investigated as a potential topical vaginal contraceptive (Clarke, 2006). Lemon juice accelerated the transit of tetrofosmin through the liver parenchyma and improved myocardial imaging in a small study (Cherng, 2006). Researchers at The Ohio State University found that lemon oil aroma does not influence the human immune system, but enhances mood (Polunin, 1997). Orange has been proven to possess larvicidal activity against dengue vector *Aedes aegypti* (Warikoo *et al.*, 2012).

References

Alderman, G.G., and Marth, E.H. (1976). Inhibition of growth and aflatoxin production of *Aspergillus parasiticus* by citrus oils. *Z Leb. Unter. Forsch.*, **160(4)**: 353-358.

Arias, B.A.I., and Laca, L.R. (2005). Pharmacological properties of citrus and their ancient and medieval uses in the Mediterranean region. *J Ethnopharmacology*, **97**: 89–95.

Bailey, D.G., Malcolm, J., Arnold, O., and Spence, J.D. (1998). Grapefruit juice–drug interactions. *Br J Clin Pharmacol.*, **46(2)**: 101–10.

Ballot, D., Baynes, R.D., Bothwell, T.H., *et al.* (1987). The effects of fruit juices and fruits on the absorption of iron from a rice meal. *Br J Nutr.*, **57(3)**: 331-343.

Benavente-García, O., Castillo, J., Alcaraz, M., Vicente, V., Del Río, J.A., and Ortuño, A. (2007). Beneficial action of *Citrus* flavonoids on multiple cancer-related biological pathways. *Curr Cancer Drug Targets*, **7(8)**: 795-809.

Bui, L.T., Nguyen, D.T., and Ambrose, P.J. (2006). Blood pressure and heart rate effects following a single dose of bitter orange. *Annals of Pharmacotherapy*, **40 (1)**: 53–7.

Campêlo, L.M.L., Almeida, A.A.C., Freitas, R.L.M., Cerqueira, G.S., Sousa, G.F., Saldanha, G.B., Feitosa, C.M., and Freitas, R.M. (2011). Antioxidant and antinociceptive effects of *Citrus limon* essential oil in mice. *J Biomed Biotechnol.*, http: //dx.doi.org/10.1155/2011/678673.

Carrington, Sean., and Fraser, Henry C. (2003). "Grapefruit". *A~Z of Barbados Heritage.* Macmillan Caribbean. pp. 90–91.

Cherng, S.C, Chen, Y.H., Lee, M.S., Yang, S.P., Huang, W.S., and Cheng, C.Y. (2006). Acceleration of hepatobiliary excretion by lemon juice on 99mTc-tetrofosmin cardiac SPECT. *Nucl Med Commun.*, **27(11)**: 859-864.

Chevallier, A. (1996). Encyclopedia of Medicinal Plants. New York, NY: DK Pub; 81.

Chopra, R. N., Nayar, S. L., and Chopra, I. C. (1986). Glossary of Indian Medicinal Plants, New Delhi: Council of Scientific and Industrial Research.

Clarke, G.N., McCoombe, S.G. and Short, R.V. (2006). Sperm immobilizing properties of lemon juice. *Fertil Steril.*, **85(5)**: 1529-1530.

D'Aquino, M., and Teves, S.A. (1994). Lemon juice as a natural biocide for disinfecting drinking water. *Bull Pan Am Health Organ*, **28(4)**: 324-330.

Eisenberg, T., Knauer, H., Schauer, A., Büttner, S., Ruckenstuhl, C., Carmona-Gutierrez, D. *et al.*(2009). Induction of autophagy by spermidine promotes longevity. *Nat. Cell Biol.*, **11(11)**: 1305-14.

Ensminger, A., *et al.* (1994). Foods and Nutrition Encyclopedia. 2nd ed. Boca Raton, FL: CRC Press, pp. 1299-1302.

Fisher, K., and Phillips, C.A. (2006). The effect of lemon, orange and bergamot essential oils and their components on the survival of *Campylobacter jejuni, Escherichia coli O157, Listeria monocytogenes, Bacillus cereus* and *Staphylococcus aureus in vitro* and in food systems. *J Appl Microbiol.*, **101(6)**: 1232-1240.

Gange, C.A., Madias, C., Felix-Getzik, E.M., Weintraub, A.R., and Estes, N.A. (2006). Variant angina associated with bitter orange in a dietary supplement. *Mayo Clinic Proceedings*, **81 (4)**: 545–8.

Golob, P. (1999). Food and Agriculture Organization of the United Nations. Alphabetical List of Plant Families with Insecticidal and Fungicidal Properties. The use of spices and medicinals as bioactive protectants for grains. *Food and Agriculture Org.*, **13**.

Gupta, V., Kohli, K., Ghaiye, P., Bansal, P., and Lather, A. (2011). Pharmacological potentials of *Citrus paradisi*- an overview. *Int. Journal of Phytotherapy Research*, **1(1)**.

Gutherie, H., and Picciano, M. (1995). *Human nutrition.* St Louis, MO, USA, Mosby.

Harats, D., Chevion, S., Nahir, M., Norman, Y., Sagee, O., and Berry, B. (1998). *Citrus* fruit supplementation reduces lipoprotein oxidation in young men ingesting a diet high in saturated fat: presumptive evidence for an interaction between vitamins C and E *in vivo. American Journal of Clinical Nutrition*, **67**: 240-245.

He, K., Iyer, K.R., Hayes, R.N., Sinz, M.W., Woolf, T.F., and Hollenberg, P.F. (1998). Inactivation of cytochrome P450 3A4 by bergamottin, a component of grapefruit juice. *Chem. Res. Toxicol.*, **11 (4)**: 252–9.

Hooker, J.D. (1882). Flora of British India, 3, Authority for the secretary of state for India in council.

Jasim, A.R. (2012). Phytochemical Study of some flavonoids present in the fruit peels of *Citrus reticulata* grown in Iraq. *Kerbala Journal of Pharmaceutical Sciences*, **3**: 3-316.

Jiang, J., Eliaz, I., and Sliva, D. (2013). Synergistic and additive effects of modified citrus pectin with two polybotanical compounds, in the suppression of invasive behavior of human breast and prostate cancer cells. *Integr Cancer Ther.*, **12(2)**: 145-52.

Johnsen, S.P., Overvad, K., Stripp, C., Tjonneland, A., Husted, S.E., and Sørensen, H.T. (2003). Intake of fruit and vegetables and the risk of ischemic stroke in a cohort of Danish men and women. *Am. J Clin. Nutr.*, **78(1)**: 57-64.

Kimball, and Dan, A. (1999). Citrus processing: a complete guide (2nd ed). Springer, New York, pp. 450.

King, C.G.(1979).The isolation of vitamin C from lemon juice. *Fed Proc.*, **38(13)**: 2681-2683.

KunduSen, S., Haldar, P.K., Gupta M., Upal, K., Mazumder, Saha, P., Bala, A., Bhattacharya, S., and Kar, B. (2011). Evaluation of anti-hyperglycemic activity of *Citrus limetta* fruit peel in streptozotocin-induced diabetic rats. *Endocrinol.*, doi: 10.5402/2011/869273.

Manners, G.D. (2007). Citrus limonoids: analysis, bioactivity, and biomedical prospects. *J Agric Food Chem.*, **55(21)**: 8285-8294.

Mathur, A., Verma, S.K., Purohit, R., Gupta, V., Dua, V.K., Prasad, G.B.K.S., Mathur, D. Singh, S.K., and Sing, S. (2011). Evaluation of *in vitro* antimicrobial and antioxidant activities of peel and pulp of some citrus fruits. *IJPI's J. Biotech. And Biothe.*, **2.**

Milind, P., and Dev, C. (2012). Orange: range of benefits. *Int. Res. J. Pharm.*, **3(7).**

Miyake, Y., Mochizuki, M., Okada, M., Hiramitsu, M., Morimitsu, Y., and Osawa, T. (2007). Isolation of antioxidative phenolic glucosides from lemon juice and their suppressive effect on the expression of blood adhesion molecules. *Biosci. Biotechnol. Biochem.*, **71(8)**: 1911-1919.

Morton, J.F. (1987). In: Fruits of warm climates. Miami, FL, pp. 160–168

Murray, M. (1993). The Healing Power of Herbs. Rocklin, CA: Prima Publishing Co; pp. 143-366.

Mulvihill, E.E., and Huff, M.W. (2012). Protection from metabolic dysregulation, obesity, and atherosclerosis by citrus flavonoids: activation of hepatic PGC1α-mediated fatty acid oxidation. *PPAR Research*, 857142.

Murthy, C. K. N., Jayaprakasha, G. K., Kumar, V., Rathore, K. S., and Patil, B. S. (2011). Citrus limonin and its glucoside inhibit colon adenocarcinoma cell proliferation through apoptosis. *Journal of Agricultural and Food Chemistry*, **59 (6)**: 2314–2323.

Ndukwe, K.C., Okeke, I.N., Lamikanra, A., Adesina, S.K., and Aboderin, O. (2005). Antibacterial activity of aqueous extracts of selected chewing sticks. *J Contemp Dent Pract.*, **6(3)**: 86–94.

Ogata, S., Miyake, Y., Yamamoto, K., Okumura, K., and Taguchi, H. (2000). Apoptosis induced by the flavonoid from lemon fruit (*Citrus limon* BURM. f.) and its metabolites in HL-60 cells. *Biosci Biotechnol Biochem.*, **64(5)**: 1075-1078.

Onyeagba, R.A., Ugbogu, O.C., Okeke, C.U., and Iroakasi, O. (2004). Studies on the antimicrobial effects of garlic (*Allium sativum* Linn), ginger (*Zingiber officinale* Roscoe) and Lime (*Citrus aurantifolia* Linn). *Afri J Biotechnol.*, **3(10)**: 552–554.

Palvia, M., Pileggi, C., Nobile, C.G.,and Angelillo, I.F. (2006). Association between fruit and vegetable consumption and oral cancer: a meta-analysis of observational studies. *AmJClinNutr.*, **83(5)**: 1126-1134.

Polunin, M. (1997). Healing Foods. New York, NY: DK Publishing; pp. 64-65.

Rainer, Scora, R.W. (1975).On the history and origin of citrus. *Bulletin of the Torrey Botanical Club*, **102(6)**: 369–375.

Simon, J.A., Hudes, E.S., and Perez-Perez, G.I. (2003). Relation of serum ascorbic acid to *Helicobacter pylori* serology in US adults: the Third National Health and Nutrition Examination Survey. *J Am Coll Nutr.*, **22**: 283-289.

So, F.V., Guthrie, N., Chambers, A.F., Moussa, M., and Carroll, K.K. (1996). Inhibition of human breast cancer cell proliferation and delay of mammary tumorigenesis by flavonoids and citrus juices. *Nutr Cancer.*, **26**: 167–181.

Spiegel-Roy, Pinchas, and Eliezer E. Goldschmidt (1996). *Biology of Citrus*. Cambridge University Press, pp. 4.

USDA(L.) Burm. [medica × aurantifolia]. The PLANTS Database (http: // plants.usda.gov), National Plant Data Center, Baton Rouge, LA 70874-4490 USA.

Veronese, M.L., Gillen, L.P., Burke, J.P., Dorval, E.P., Hauck, W.W., Pequignot, E., Waldman, S.A., and Greenberg, H.E. (2003). Exposure-dependent inhibition of intestinal and hepatic CYP3A4 *in vivo* by grapefruit juice. *Journal of Clinical Pharmacology.*, **43(8)**: 831-9.

Warikoo, R., Ray, A., Sandhu, J.K., Samal, R., Wahab, N., and Kumar, S. (2012). Larvicidal and irritant activities of hexane leaf extracts of *Citrus sinensis* against dengue vector *Aedes aegypti* L. *Asian Pac J Trop Biomed.*, **2(2)**: 152–155.

Utilisation and Management of Medicinal Plants Vol. 3 (2015) *Pages* **181–197**
Editors: **V.K. Gupta and A. Kaul**
Published by: **DAYA PUBLISHING HOUSE, NEW DELHI**

12

Plumieride: An Immunopotentiating Candidate from *Plumieria acutifolia* Linn.

Anpurna Kaul[1*], Jasvinder Singh[1], N.K. Satti[1], Naseer Shah, A.K. Saxena[1] and V.K. Gupta[1]

ABSTRACT

Plumieride, an iridoid glucoside obtained from Plumieria acutifolia leaves was investigated for its effect on cell mediated and humoral immune responses and on cytokines levels in normal and immune suppressed mice. It stimulated the Delayed Type Hypersensitivity (DTH) reaction and humoral antibodies as well as Plaque Forming Cell (PFC) production in normal mice and restored to normal level, the responses in animals treated with Betamethasone (BMS) Azathioprine (AZA) and Cyclophosphamide (CY). In normal mice plumieride raised the intracellular level of interleukin -2, Tumor Necrosis Factor-α (TNF-α) and Interferon-γ (IFN-γ). It enhanced the release of IL-1α and TNF-α in AZA, CY and BMS treated mice, exhibited synergistic inhibitory effect on IL-2 production in AZA treated mice and restorative effect in CY and BMS treated mice. IFN- γ level was also restored to normal in AZA, CY and BMS treated mice respectively.

***Keywords*:** Plumieride, *Plumieria acutifolia*, Cell mediated immunity, Humoral immunity, Immunosuppressive, Immunorestorative, Cytokines.

Introduction

Plants containing iridoids have been used in folk medicine for centuries as bitter tonic, sedative, hypotensive, febrifuge, cough medicine, remedy for wound and skin diseases and for insecticidal properties (Atal *et al.*, 1986; Sticher, 1977). The plumerates,

1 CSIR-Indian Institute of Integrative Medicine, Canal Road, Jammu – 180 001, J&K State, India.

* *Corresponding author*: E-mail: akoul@iiim.ac.in

plumeric acid and Me plumerate, isolated from the leaves of *Plumieria acutifolia* have shown to posses strong antitumor activity (Fugimoto, 1988).

Plumeria acutifolia (Family: Apocynaceae) is a small tree with oblong, obovate leaves. Ethanolic extract of the leaves of *Plumieria acutifolia* was found to stimulate sheep erthrocyte (SRBC) induced DTH reaction and humoral antibody synthesis, hasten graft rejection and enhanced phagocytosis (Kaul, 1995) in mice. Iridoids, the bitter tasting monoterpenoid lactones, have recently gained increased recogninition because of their biological activities including choleretic (Das *et al.*, 1976), hepatoprotective (Ansari *et al.*, 1988), immunorestorative and immunostimulatory (Sharma, 1991; Sharma *et al.*, 1994), antiasthmatic (Dorsch *et al.*, 1991), anti-inflammatory and antiarthiritic (Singh *et al.*, 1993), laxative and diuretic (Harborne *et al.*, 1993), antibacterial (Rasool *et al.*, 2008) and antimicrobial (Surendra *et al.*, 2012) properties. This plant was also found biologically active as a diuretic, antipsychotic, antitumour agent and as an inhibitor of human immune deficiency virus type-1 (Mahabeer *et al.*, 2004). It was also reported that an endophytic colletotrichum gloeosporioides isolated from the same plant was found to produce a diterpenoid compound namely Taxol an anticancer drug (Nithya *et al.*, 2009).

A large variety of herbal drugs mentioned in Ayurveda, have been under study for their immunomodulating, adaptogenic and rejuvenating properties (Nayar, 1979) are in progress in medicinal and aromatic plants (Farnworth,1983; Labadie, 1993).

Bioactive natural products (Colegate *et al.*, 1993) are believed to promote positive health and maintain organic resistance against infections by re-establishing body equilibrium and conditioning the body tissues (Bhagwandas, 1978). Hence the drugs of plant origin are gaining increasing popularity and are being investigated for correction of immunological disorders (Aboolenein, 1980). Immunomodulation is a process, which alters the immune system of an organism by interfering with its functions. This interference results in either immunostimulation, an enhancement of immune reaction or immunosuppression that imply mainly to reduce resistance against infections and stress which may be because of environmental or chemotherapeutic factors (Patwardhan *et al.*, 1990). Immunostimulation and Immunosuppression both are needed to be tackled depending on the type of immunological disturbance. Recently, search for better moieties with these activities is becoming the field of major interest. Research focussed on the development of immunomodulators is directed towards activities that can be expressed in terms of stimulation or inhibition of immune factors and their integrated functions (Labadie, 1993). Recently the understanding of research on immuno-modulators has come up as a new field of immunopharmacology. Immunomodulation is an innovative strategy for overcoming incurable diseases involving cancer, AIDS, arthritis and allergies.

An in-depth study of the immune system is supposed to provide both the theoretical and therapeutic background of many chronic disorders. Keeping these factors in view, major efforts have to be directed towards the formulation of new strategies, to modulate the immune responses, to permit effective treatment of various ailments associated with immune system and thus development of a safe and effective immunomodulator for clinical use has become a major goal for many pharmaceutical investigators world over. In the present study, the immunostimulatory and immunorestorative effects of plumieride have been investigated being an iridoid glucoside on cell mediated and humoral immune responses in normal animals and in animals treated with most commonly used immune suppressive chemotherapeutic agents to ascertain its usefulness in immune suppressed conditions. Since cytokines have shown to be involved in initiation and regulation of immune responses (Georgiev *et al.*, 1993) the influence of plumieride on such cytokines as Interleukin-1α (IL-1α), Interleukin-2 (IL-2), Tumour Necrosis Factor-α (TNF-α) and Interferon γ (IFN-γ) production *in vivo* has also been studied.

Materials and Methods

Plumieride a white crystalline substance was received from Natural Product Chemistry Division of our Institute (IIIM, Jammu). Murine enzyme linked immunosorbent assay (ELISA) kits for IL-1α, IL-2, TNF-α and IFN-γ were procured from the Genzyme Corporation GC (Cambridge, M.A.,U.S.A), Cyclophosphamide (Endoxan Asta) from German remedies Limited (India), Azathioprine from Wellcome Laboratories (Paris, France), Betamethasone from Glaxo India, Ltd. (India), Levamisole from Biddle Sawyer Ltd.(India), Colloidal carbon (Indian ink) from market and RPMI 1640 medium from Gibco BRL Life Technologies Inc (Grand Island, N.Y., USA).

Preparation of Test Material

Dried leaves (500g) of *Plumeria acutifolia* were grounded and percolated with 50 per cent aqueous ethanol four times (4 x 2.5 Litres) for 14 hrs each. All the four extracts were pooled. The pooled aqueous extract was centrifuged, clear supernatant was evaporated to dryness on a wiped film evaporator at 50 ± 5°C, residue (extractive value 25 per cent) was fractionated with $CHCl_3$, EtOAc and *n* - BuOH successively.

The *n*- BuOH extract was chromatographed on a column of silica gel (60-120 mesh) and eluted with a gradient of MeOH in CHCl3. The $CHCl_3$: MeOH (5:1) eluate was rechromatographed on a silica gel (100-200 mesh) column using $CHCl_3$-MeOH: H_2O (6:1:0.1) as solvent. Fractions homogeneous on TLC were pooled, dried and charged on a sephadex LH-20 column. Column was eluted with MeOH to produce two fractions of 500ml each. Second fraction containing mainly the target compound Plumieride was subjected to further purification over on a sephadex LH-20. Column was eluted with MeOH: H_2O (3:2) to afford a fraction, which on crystallisation from EtOAc yielded a colourless amorphous powder soluble in MeOH.

HPLC

Standardisation of the extract was carried out on the basis of plumieride by HPLC using following conditions, where the per cent age of plumieride in the extract was found to be 1.25.

Column	RP-18 (E-Merck, 5μm, 4.6 x 250 mm temperature 30 degree.
Mobile Phase	MeOH: H_2O: 2:3
Flow rate	0.6ml/min.
λmax	225 nm

Plumieride amorphous powder, m.p. 229,$[\alpha]^{21}_{D}$-114 [c,0.5 per cent MeOH], MS: FABMS, $[M+Na]^+$ m/z 493was found to be a known iridoide *i.e.,* plumieride)-{1S-[1α,4aα,7α (R) 7aα]}-1-(β-D-glucopyranosyl)-4a,7a-dihydro-4'-(1-hydroxymethyl)-5'-oxospiro[Cyclo- penta (C) pyrano 7(H),2 '(5'H)-furan]-4-carboxylic acid methyl ester. Structure was finally confirmed by ^{1}H and ^{13}CNMR data similar to reported in literature.

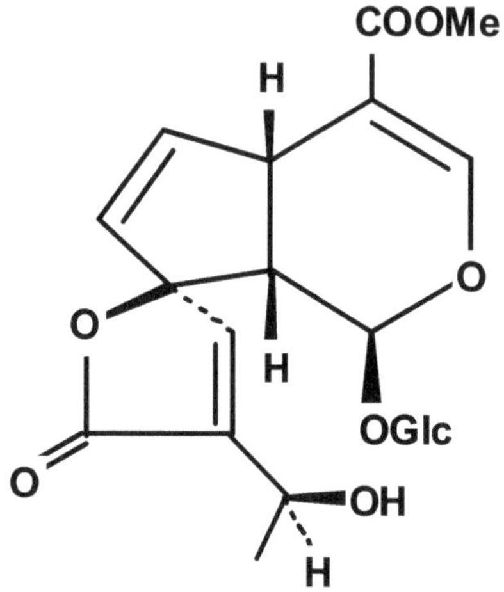

Plumieride *(Agoniadine)*

Keeping in view the high reputation of plumieride, 1- (β - D - Glucopyranosyloxy) - 49, 7a-di-hydro-4'-(1-hydroxyethyl)-5'-oxospiro[cyclopenta[c]pyran-7(1H),2'(5'H)-furan]-4-carboxyacid methyl ester, obtained from *Plumeria acutifolia* in traditional system of medicine, it was tempting to speculate that the restorative and rejuvenating power of this plant might be due to its action on immune system of the organism as no mention is made about its such effect anywhere in literature.

Acute Safety Study

An acute safety study was carried out in normal mice as per OECD guidelines no 423 (Martina *et al.,* 2013).

Antigen (SRBC)

Fresh sheep red blood cells (SRBC) collected aseptically from the jugular vein of sheep were stored in cold sterile Alsever's solution, washed three times with pyrogen free sterile normal saline (0.9 per cent NaCl w/v) and adjusted to a concentration of $5x10^9$ cells/ml for immunization and challenge at required time schedule.

Humoral Antibody Titre

Groups of 6 mice each were immunized by injecting 200µl of $5x10^9$ SRBC/ml intraperitoneally (i.p.) on day 0 and challenged 7 days later by injecting an equal volume of SRBC i.p. Blood samples were collected on day +7 (before challenge) for primary antibody titre and on day + 14 (7 days after challenge) for secondary antibody titre. Haemagglutination antibody titres were determined following the modified microtitration technique described by Nelson *et al.*(1967). The value of the highest serum dilution causing visible haemagglutination was taken as a titre. BSA - saline alone served as a control.

Delayed Type Hypersensitivity (DTH) Response

Mice were immunized by injecting 20 µl of 5x 10^9 SRBC/ml subcutaneously into the right hind foot pad. Seven days later the thickness of left hind foot was measured by volume differential meter considered as control. These mice were then challenged by injecting the same amount of SRBC intradermally into the left hind foot pad. The foot thickness was measured again at 24 h after challenge Doherty (1981).

Phagocytic Response

In vitro

The method of Lehrer, 1981 was followed. The percentage and average number of *Candida albicans* cells (heat killed) ingested by peritoneal murine macrophages was calculated.

In vivo

Carbon clearance assay to assess the effect of the drugs on phagocytic capacity of reticuleoendothelial system (RES) of the animals which constitutes initial defence mechanism against infectious microorganisms (Hudson *et al.*, 1980).

Skin Allograft Rejection

Fitted pinch of homologous grafts of skin were transplanted in groups of 10 inbred mice according to the method of Bellingham *et al.* (1951).Graft survival time was determined by daily observation, the criterion being epithelial survival. Extract was fed orally daily till graft rejection.

Plaque Forming Cell (PFC) Assay

Mice were immunized by an injection of 0.2 ml of 10 per cent (v/v) suspension of SRBC in normal saline. Five days later the animals were sacrificed and their spleen was removed aseptically. A single cell suspension of an individual spleen was prepared in RPMI 1640 medium and viability was checked by trypan blue dye exclusion test. Cell concentration was adjusted to 2x10^7cells/ml in RPMI1640. To 0.1 ml of this suspension, 0.3 ml of 5 per cent SRBC suspension, 0.1 ml of guinea pig serum and 0.5 ml of RPMI 1640 medium were added. A 150 µl quantity of this mixture was added to each Cunningham's chamber. The chambers were sealed with paraffin wax to prevent dehydration. These chambers were incubated for 60 to 90 min at 37°C in incubator. Plaques in SRBC lawn were enumerated using an inverted microscope. Each plaque represented a product of an individual antibody forming cell (Cunningham *et al.*, 1968).

Flowcytometric Studies

Flow cytometers take in a suspension of monodisperse (single, unclumped) cells and run them one at a time (single file) past a laser beam. As each cell passes through the laser beam, scattered and fluorescent light are quantitated. Sensitivity is limited by "autofluorescence", naturally fluorescent components of cells that set a background fluorescence intensity – fluorescent probes must emit substantially greater intensities in order for their signals to be quantitated accurately. Most flow cytometry is analytical: after the information is obtained as it passes through the cytometer, the sample is discarded. Some flow cytometry is preparative: living cells are sorted into separate containers based on the properties of each cell (Abbas, 1987).

Mycobacterium tuberculosis H37Rv Infection

The balb/c mice were infected intranasally with *Mycobacterium tuberculosis* H37Rv (10^3 CFU/mouse). The infected mice were divided in groups of 6 mice each. Treatment started one week after the infection. Mice from early control group were sacrificed, left lung was aseptically removed, homogenized and dilutions were plated on middle brook 7H10 agar to determine the CFU (Colony forming unit). Mice from late control and treatment group were scarified one week after the infection and lung CFU was determined.

Statistical analysis

The data was expressed as mean ± S.E. by student 't' test.

Results

Effect on Delayed Type Hypersensity Response and Humoral Antibody Titre (Primary and Secondary) in balb/c Mice

Plumieride was administered orally (1.56 - 25 mg/kg) for seven days post-immunization, an increase of 15.78 to 36.84 per cent in the expression of DTH reaction was observed in normal mice. The effect was dose dependent and highly significant ($P<0.001$) at 12.5 mg/kg dose (Table 12.1).

Plumieride (1.56 - 25mg/kg p.o.) produced a dose dependent increase in primary (7.69 - 30.76 per cent) and secondary (2.55-32.58 per cent) antibody synthesis in normal mice. The immunostimulatory effect of plumieride was highly significant at 12.5 mg/kg p.o. dose levels, and equally prominent on secondary antibody production (Table 12.1).

Effect on Phagocytic Function of Macrophages in Mice

The effect of plumieride was significant at 250 and 500 µg/ml dose and was highly significant at 1000 µg/ml. (Table 12.2).

Oral administration of plumieride (3.12 - 50 mg/kg) for consecutive seven days and thirty minutes prior to carbon injection enhanced the clearance rate of carbon particles from circulation in normal animals at the dose of 25 mg/kg p.o. The effect was highly significant at 25 and 50 mg/kg p.o. dose (Table 12.2).

Effect on Skin Allograft Rejection in Mice

Plumieride showed maximum and statistically significant effect (decrease in rejection time) that was 34.61 and 38.61 per cent at 12.50 and 25 mg/kg p.o. dose, respectively (Table 12.3). Cyclosporine a standard control at 5 mg/kg increased the rejection time by 39.69 per cent.

Table 12.1: Effect of Plumieride on Delayed Type Hypersensity Response and H umoral Antibody Titre (Primary and Secondary) in Balb/c Mice

Treatment Dose mg/kg P.O. per cent Change	*Cell Mediated response Humoral Antibody Titre*					
	Foot Thickness Mean±S.E.	*Per cent Activity*	*Pabt Mean± S.E.*	*Per cent Activity*	*Sabt Mean± S.E.*	*Per cent Activity*
ControlSRBC	0.95± 0.02	-	6.50± 0.22	–	6.66± 0.22	–
1.56	1.10 ± 0.02	15.78↑	7.00± 0.25	07.69↑	6.83 ± 0.16	02.55↑
3.12	1.18 ± 0.03 [a]	24.21↑	7.33± 0.21	12.76↑	7.16 ± 0.16	07.50↑
6.25	1.22 ± 0.08 [b]	28.42↑	7.50± 0.21[a]	15.38↑	8.00 ± 0.22[a]	20.12↑
12.5	1.30 ± 0.04 [c]	36.84↑	8.50± 0.25[c]	30.76↑	8.83 ± 0.22 [c]	32.58↑
25.0	1.24 ± 0.02 [b]	30.52↑	8.00± 0.21[b]	23.07↑	8.50 ± 0.21[c]	27.62↑
2.5	1.32 ± 0.03[c]	38.94!	8.33 ± 0.22	28.15↑	8.33±0.22	25.07↑

N = 6.

Data is represented as Mean ± S.E.

a: P<0.05, **b**: P<0.01, **c**: P<0.001; Pabt : Primary antibody titre; Sabt : Secondary antibody titre; ↑: increase.

Table 12.2: Effect of Plumieride on Phagocytic Function of Macrophages in Mice

Phagocytosis of heat killed Candida albicans by the murine macrophagesClearance rate of carbon particles from the circulation in normal animals

Treatment Dose μg/ml*	*Per cent Phagocytosis [in vitro study] Mean ± S.E.*	*Treatment** Dose mg/kg*	*Phagocytic Index [in vivo study] Mean ± S.E.*
Control	25.86 ± 2.02	Control	1.26 ± 0.12
10	27.00 ± 2.32(+4.40)	–	–
30	28.50 ± 1.95(+10.20)	3.12	1.34 ± 0.14(+6.34)
100	30.22 ± 2.11(+16.86)	6.25	1.42 ± 0.04(+12.69)
250	32.33 ± 1.42a(+25.01)	12.5	1.48 ± 0.12(+17.46)
500	38.00 ± 3.39 b(+ 46.94)	25.0	1.56 ± 0.08 a(+23.80)
1000	38.50 ± 2.92 b(+48.87)	50.0	1.59 ± 0.08 a(+26.19)
Levamisole10	40.23 ± 3.24 b(+ 55.56)	Levamisole2.5	1.68 ± 0.09 b(+33.33)

N=6.

Data is represented as mean± S.EF Figures in paranthesis represents percentage change.

+ : increase. **a**: P<0.01, **b**: P<0.001.

* : Number of observations for each dose 6.**: Number of animals in each group 6.

Table 12.3: Effect of Plumieride on Skin Allograft Rejection in Mice

Treatment	*Dose mg/kg p.o.*	*Rejection Time Days Mean ± S.E.*	*Graft Rejection (Per cent)*
Control	–	13.00 ± 0.25	–
Plumieride	3.12	11.12 ± 0.11	14.46↑
do	6.25	9.80 ± 0.21	24.60[a]↑
do	12.50	8.50 ± 0.22	34.61[b]↑
do	25.00	7.98 ± 0.23	38.61[b]↑
Cyclosporine	5.00	18.16 ± 0.16	39.69[b]↓

n = 6.

Data is represented as Mean ± S.E.

↑: Hastened; ↓: Delayed.

a: P<0.01, **b**: P<0.001; Student's 't' testNumber of animals in each group 6.

Effect on Plaque Forming Cell Assay (PFC) in Mice

Administration of plumieride (3.12 – 25 mg/kg) orally for 5 consecutive days beginning from day of immunization produced an increase in the number of antibody secreting cells (PFC) by 10.15 - 58.92 per cent in mouse spleen. The effect was highly significant at 12.5 - 25 mg/kg p.o. doses (Table 12.4).

Table 12.4: Effect of Plumieride on Plaque Forming Cell Assay (PFC) in Mice

Treatment	*Dose mg/kg p.o.*	*(B cell response) PFC/10⁶ Spleen Cells*	*Per cent Increase*
Control (SRBC)	–	6.50±0.22	–
Plumieridedododo	1.56	7.16±0.16	10.15↑
	3.12	8.00±0.25	23.07↑
	6.25	8.50±0.22[a]	30.76↑
	12.50	10.16±0.16[b]	56.30↑
	25	10.33±0.21[b]	58.92↑
Levamisole	2.5	11.33±0.21[b]	74.30↑

n=6.

Data is represented as mean± S.E.

↑: Increase.

a: P<0.01, **b**: P<0.001; Student's 't' test.

Number of animals in each group 6.

Effect on CD4⁺ and CD8⁺ T Cells Counts in Balb/c Mice

The value of the unsensitised group was 24.04 per cent of CD4⁺ T cells and 14.36 per cent per cent of CD8⁺ cells and in the sensitized group the value of CD4⁺ and CD8⁺ T cells was 30.60 per cent and 16.90 per cent respectively. This shows an increase in

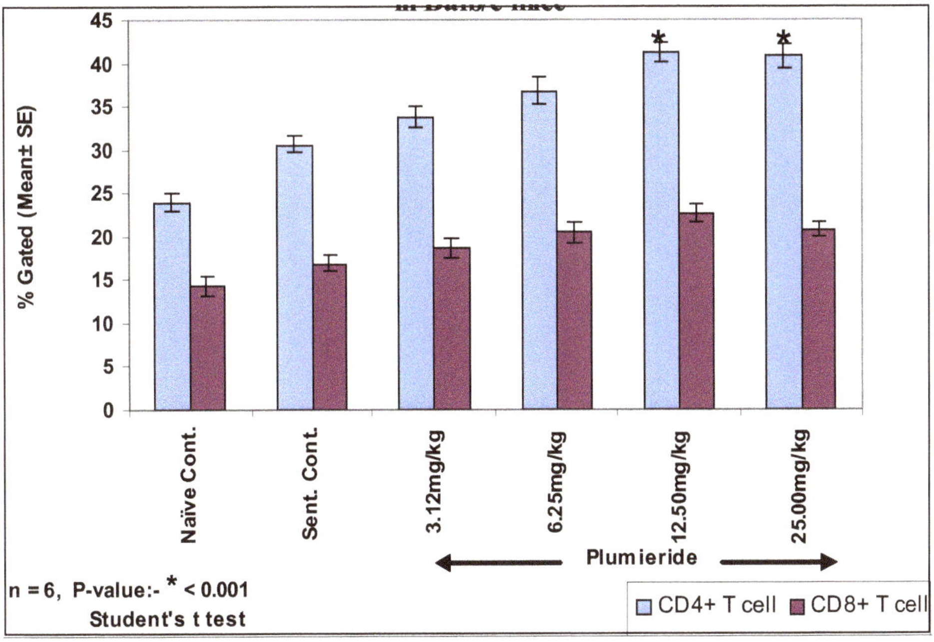

Figure 12.1: Effect on Plumeride on CD4⁺ and CD8⁺ T Cells Counts in Balb/c Mice.

$CD4^+$ and $CD8^+T$ cells in experimental animals treated with different doses of plumieride (Figure 12.1).

Flowcytometric Analysis of Plumieride on different Cytokines

The animals treated with plumieride at 3.12, 6.25, 12.50 and 25 mg/kg p.o. dose showed a maximum effect of 41.22 per cent of $CD4^+$ cells and 22.64 per cent of $CD8^+$ T cells at 12.50 mg/kg p.o. dose (Figure 12.1).

Animals treated with plumieride at graded oral doses showed increase in Intracellular

IL-2 expressions. The percentage of IL-2 secreted was maximum 19.44 per cent at 12.50 mg/kg per oral test drug dose (Figure 12.2).

The IFN-γ (Th1 signature cytokine) was enhanced in the groups treated with plumieride showing maximum expression (9.95 per cent) at 12.50 mg/kg oral dose (Figure 12.3).

The IL- 4 (Th2 signature cytokine) expression in sensitized control and non-sensitized control groups was 4.83 per cent and 3.80 per cent respectively (Figure 12.4).

Effect on IL-2 Expression in Balb/c Mice

Animals treated with plumeiride at graded oral doses showed increase in Intracellular IL - 2 expressions. The percentage of IL-2 secreted was 19.44 per cent at 12.50 mg/kg per oral test drug dose (maximum effect). Kaye *et al.* (1984). The percentage of IL-2 in the normal non-sensitized control group and in sensitized control group was 4.15 per cent and 6.06 per cent respectively (Figure 12.2).

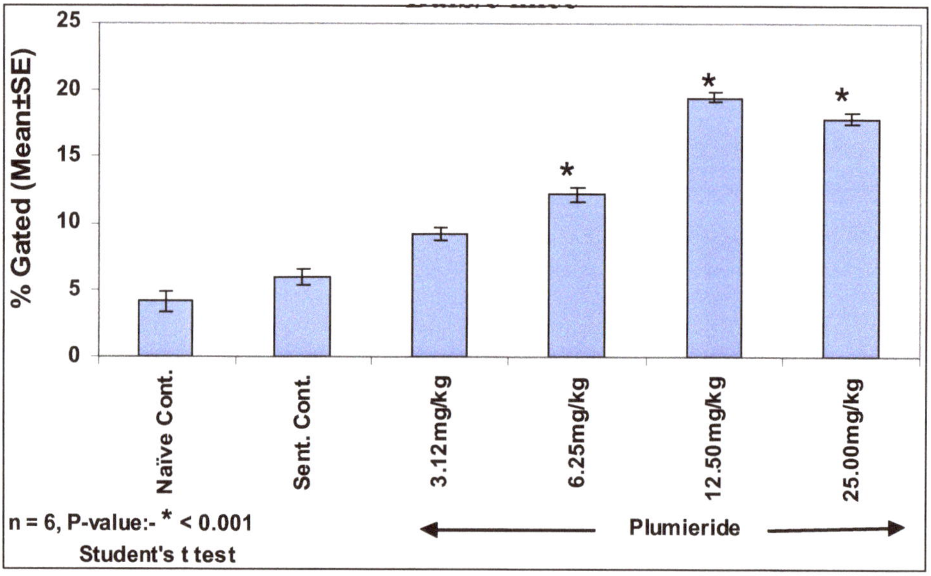

Figure 12.2: Effect on Plumeride on IL-2 Expression in Balb/c Mice.

Effect on IFN-gamma Expression in Balb/c Mice

The IFN-γ (Th1 signature cytokine) was enhanced in the groups treated with plumieride showing maximum expression (9.95 per cent) at 12.50 mg/kg oral dose. The sensitized control group showed 5.35 per cent and non-sensitized control showed 2.70 per cent of IFN-γ expression (Figure 12.3). There was an increase in all the test drug treated groups with maximum effect being observed at 12.50 mg/kg per oral dose.

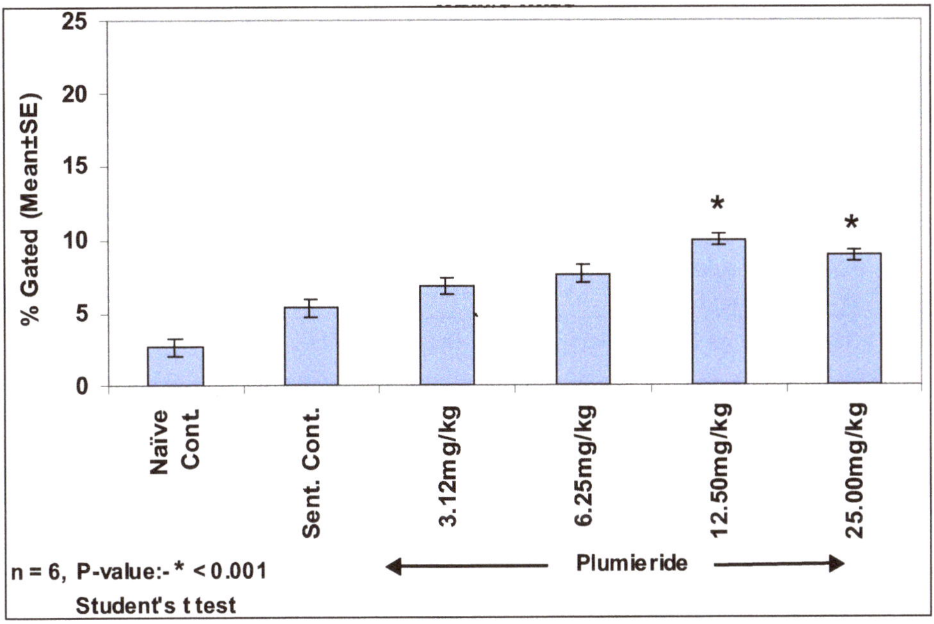

Figure 12.3: Effect on Plumeride on IFN-gamma Expression in Balb/c Mice.

Effect on Plumieride on IL-4 Expression in Balb/c Mice

The IL- 4 (Th2 signature cytokine) expression in sensitized control and non-sensitized control groups was 4.83 per cent and 3.80 per cent respectively.The maximum expression of IL- 4 in treated groups was 6.49 per cent at the dose of 12.50 mg/kg per oral dose. However, this was not statistically significant (Figure 12.4).

Efficacy on *Mycobacterium tuberculosis* Infection Model in Balb/c Mice

Treatment with plumieride at 12.5 mg/kg did not allow the bacteria to grow and the bacterial load was close to the early control. The combination of plumieride with rifampicin (anti-tuberculosis drug) was synergistic. The combination could reduce the CFU load to 3.0 log as compared to rifampicin alone that stood at 5.0 log. This combination reduced the initial CFU load (early control) by 1.7 log.

Figure 12.5 shows the CFU load in the left lung of the mice. One week after the infection the CFU load in the early control reached 4.7 log while in the late control

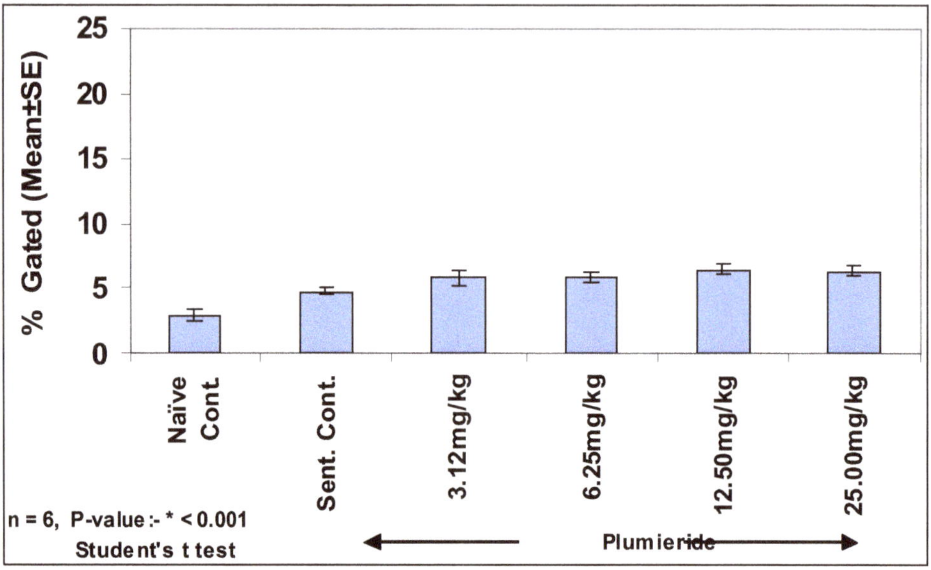

Figure 12.4: Effect on Plumieride on IL-4 Expression in Balb/c Mice.

was 6.8 log. The early control indicates the CFU load at the start of the treatment, whereas late control indicates the bacterial load of the untreated group.

Acute Safety Study

An acute safety study was carried out in female mice and no mortality or any change in normal general behaviour was observed upto the oral dose as high as 2000 mg/kg administered. as per OECD guidelines no. 423. No untoward symptom was observed over two week's period of observation in these experimental animals.

Discussion

Immunomodulators are a class of natural or synthetic agents which by modifying the immune system affects the therapeutics system. They have ability to augment, restore, inhibit or help to produce desired immune responses. Immunomodulatory drugs constitute thalidomide and its analogues, lenalidomide, pomalidomide and apremilast.

The present work describes that *Plumeria actifolia* on screening showed significant immunopotentiating activity. Plumeride an iridoid glucoside obtained from *Plumeria actifolia*. Iridoids are reported to possess a large number of biological activities. The immunepotentiating activity of iridoids were also reported by Ghule *et al.* (2012).

Plumieride produced a dose related increase in SRBC induced DTH reaction (cell mediated immunity) which shows the stimulatory effect on 'T' lymphocytes. An evidence that leads to support hypothesis of T-lymphocytes stimulation by the reduction in the homologous skin graft rejection time in mice treated with plumieride. The basic mechanism involved is reduction in graft rejection time by the stimulation of T- lymphocytes *i.e.*, CD^{4+} and CD^{8+} positive T cells.

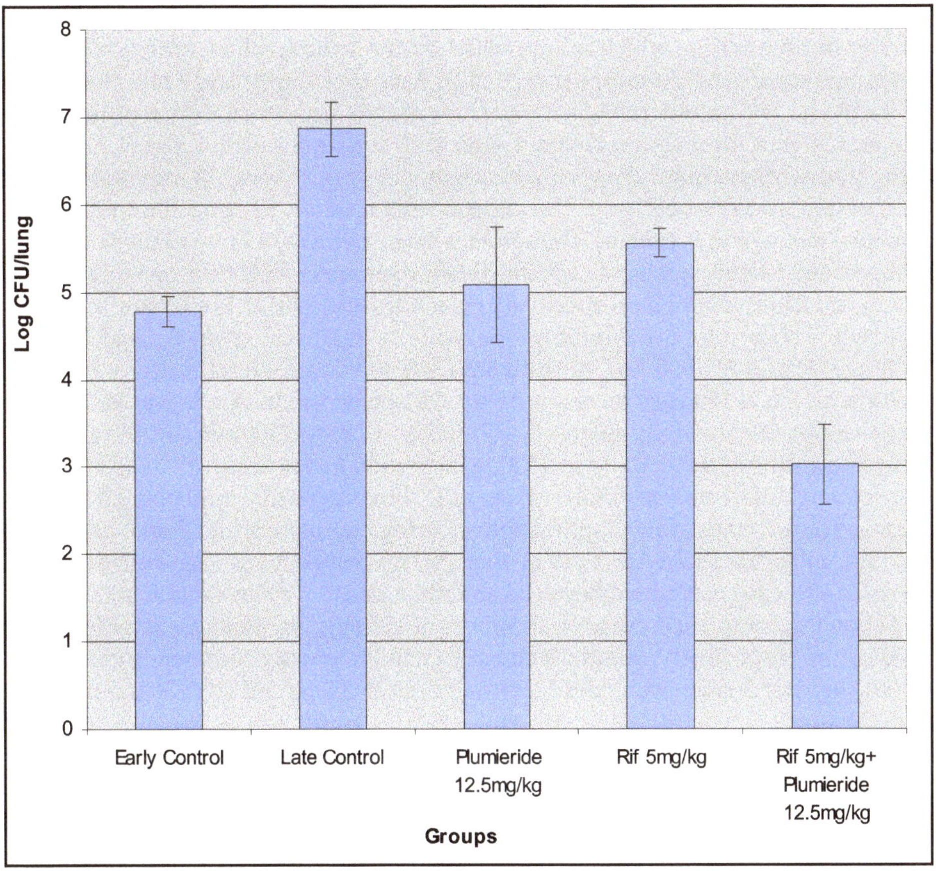

Figure 12.5: Efficacy of Plumieride in the *Mycobacterium tuberculosis* infection model in Balb/c Mice.

One of the major effector functions of CD^{4+} T cells is the activation of macrophages and this plays an important role in enhancing the activity at sites of insult. CD^{4+} T cell activation by plumieride may be one of the factors responsible for the increase in the functioning of the macrophages. The increased responsiveness of phagocytes is evident by the increase in the clearance of carbon particles from the reticuloendothelial system and also the hastening in the rate and subsets of T and B-lymphocytes, involved in phagocytosis *in vitro* by murine macrophages, thereby suggesting an increase in the functioning of macrophages (innate response) by causing stimulation of non-specific immune response.

The stimulation of humoral response against sheep red blood cells (SRBC) also indicates enhanced responsiveness of macrophages in antibody synthesis. Macrophages coordinate the processing and presentation of antigen to T and B cells and the increase of humoral response to SRBC reveals that plumieride may be acting as a source for the augmentation of these processes. The stimulation of the humoral response to SRBC by plumieride is also evident by the increase in the number of plaque forming cells. Indeed, these observations have shown the immunostimulatory activity of plumieride, which is supported by the antimicrobial, antibacterial and antifungal studies by Surendra *et al.* (2012), Ajay *et al.* (2010) and Laila *et al.*(2008) respectively. The antimicrobial activity showed a minimum inhibition of 16mm, 18 mm and 18 mm for methanol extract against *S. aureus*, *B. subtilus* and *E. coli* when compared with standard drug ciprofloxacin showing 26 mm, 28 mm and 24 mm zone of inhibition respectively. The aqueous extract showed 8mm 10mm diameter against *S. aureus* and *B. subtilus*. The aqueous extract showed 12 mm diameter against *A. niger* and 10 mm against *C. albicans* when compared with fluoconazole having 20mm and 22mm diameter respectively (Surendra *et al.*, 2012).The Ethanolic extract of leaves is showing partial antibacterial activity against *S. epidermidis* at 750 and 1000 µg/ml and at 1500 µg/ml showing complete antibacterial activity while in *Escherischia coli* at 1500 µg/ml respectively. Chloroform extract of leaves is showing partial antibacterial activity against *S. epidermidis* at 750 and 1000 µg/ml and showing complete antibacterial activity at 1500µg/ml while *Escherischia coli* at 1500 µg/ml respectively. Ethyl acetate extract of leaves is showing partial antibacterial activity against *S. epidermidis* at 1000 µg/ml and showing complete antibacterial activity at 1500 µg/ml while *Escherischia coli* at 1000 µg/ml respectively. Aqueous extract of leaves is showing partial antibacterial activity against *S. epidermidis* at 1000 µg/ml and showing complete antibacterial activity at 1500 µg/ml while *Escherischia coli* at 1500 µg/ml respectively. Standard ciprofloxacin is showing complete antibacterial activity against *S. epidermidis* and *Escherischia coli* at 500 µg/ml and 750 µg/ml (Ajay *et al.*, 2010).

Laila *et al.*(2008) observed that *P. rubra* inhibited the radial mycelial growth of all the test fungi at a concentration of 100 g/ml medium. The highest inhibition (62 per cent) of fungal radial mycelial growth was recorded against *A. ustus* with ethyl acetate extract at a concentration of 100 g/ml medium. Antifungal antibiotic nystatin (100g/ml medium) exhibited inhibitions of radial mycelial growth of all the four fungi respectively.

Th1/Th2 balance concept considers that immune response is usually polarized to give predominantly either a Th1 or Th2 response and fine-tuning of inhibitors, activators and regulatory signals ensures immunostasis. Much of the literature (Mossman *et al.*, 1989; Singh,1999; Giese, 2002) supports Th1/Th2 balance concept to the level of paradigm and many of T cell directed therapies have provided modest clinical benefits. Th1 response primarily promotes cytolytic T cells (CTLs), which are important in responding to infections. The results suggest that oral administration of plumieride selectively skewed the immune response towards Th1 response cells rather than Th2 cells by increased IFN-gamma, IL-2 versus IL- 4 cytokines levels, weak enhancement of the expression of IL-4, a Th2 signature cytokine may be due to complex cross talk between cytokines and their overlapping functions.

Plumieride at 12.5 mg/kg oral dose did not allow the bacteria (*Mycobacterium tuberculli*) to grow and the bacterial load was close to the early control in the experimental infected mice. The combination of plumieride with rifampicin showed synergism is being effective. The combination could reduce the CFU load to 3.0 log as compared to rifampicin alone that stood at 5.0 log. This combination reduced the initial CFU load (early control) by 1.7 log.

Our pragmatic study establishes unique immunostimulatory potential of test material and suggests its use in conditions where Th1 upregulation is required. Plumieride also showed stimulating immune responses comprising of administrating effective dose to experimental mice suffering from tuberculosis infection, thereof, capable of eliminating the internal pathogens. Such agents are being sought for the treatment of infectious diseases, immunodeficient diseases, or for generalized immunosuppression induced by drug treatment; for combination therapy with antibiotics; and as adjuvants for vaccines. All these findings suggest that plumieride has a significant immunostimulant activity suggestive of its possible usefulness as a therapeutic agent in immune compromised patients.

References

Abbas, A.K. (1987). Warner-Lambert Parke-Davis award lecture. Cellular interactions in the immuneresponse. The roles of B-lymphocytes and interleukin-4. *Am. J.Pathology,* **129**: 25- 33.

Aboolenein, A.A. (1980). Back to medicinal plants therapy. In: The History of Medicinal and Aromatic plants: *Proceedings of the second International Congress Alexandria, Egypt,* Dec.15 to **19.Ed**. Abdallah, A. (Hamdard Foundation Press, Pakistan): pp. 40-44.

Ajay, Singh, Baghel., Chanchal, Kumar, Mishra., Asha,Rani., D, Sasmal., and Rajesh, Kumar, Nema. (2010). Antibacterial activity of *Plumeria rubra* Linn. plant extract *J. Chem. Pharm. Res.,* **2(6)**: 435-440.

Ansari, R.A., Aswal, B.S., Chander, R., Dhawan, B.N., Garg, N.K., Kapoor, N.K., Kulshreshtha, D.K., Mehdi, H., Mehrotra, B.N., Patnaik, G.K., and Sharma, S.K.(1988). Hepatoprotective activity of kutkin, the iridoid glycoside mixture of *Picrorhiza kurroa. Indian J. Med. Res.,* **87**: 401-404.

Atal, C.K., Sharma, M.L., Kaul, A., and Khajuria, A. (1986). Immunomodulating agents of plant origin. Preliminary screening. *J. Ethnopharmac.,* **18**: 133-141.

Bhagwan Dash (1978). Fundamentals of Ayurvedic Medicine (Bansal Co., Delhi, India). p. ix-xvi.

Billingham, R.E., and Medawar, P.B. (1951). The technique of free skin grafting in mammals. *J.Exp.Biol.,* **28**: 385-485.

Das, P.K.,Tripathi, R.M., Agarwal,V.K., and Sanyal, A.K. (1976). Pharmacology of kutkin and its two organic acid constituents- Cinnamic acid and vanillic acid. *Indian J. Exp.Biol.,* **14**: 456- 458.

Cunningham,A.J., and Szenberg, A. (1968). Further improvement in plaque technique for detechting single antibody – forming cells. *Immunol.,* **14**: 599-601.

Colegate, M., Steven., and Russell, J.(1993). *Molyneux Bioactive Natural Products Detection, Isolation, and Structural Determination.* CRC Press 544 Pages.

Doherty, N.S. (1981). Selective effects of immunosuppressive agents against the delayed hypersensitivity response and humoral response to sheep red blood cells in mice. *Agents and Actions*, **11**: 237-242.

Dorsch, W., Stuppner, H., Wagner, H., Gropp, M.,Demoulin, S., and Ring, J.(1991). Antiasthmatic effects of *Picrorhiza kurroa*, Androsin prevents allergen- and PAV - induced bronchial obstruction in guinea pigs. *Int. Arch. Allergy and app. Immunology*, **95**: 128 - 133.

Fransworth, N. R. (1983). The role of medicinal plants in drug development. In Alfred Benson Symposium **20**. Natural products and Drug Development N.Copenhagen,17-30.

Fugimoto, Y.M.S.(1988). Plumerates, their antitumour activities and pharmaceuticals containing them. *Jpn. Kokai Tokyo Koho*., JP 6360, 949 (C.A. 109: 86332. P).

Georgiev, V.St., and Albright, J.F.(1993). Cytokines and their role as growth factors and in regulation of immune responses. In: Immunomodulating drugs Eds. V. St. Georgiev and H.Yamaguchi, The New York Academy of Sciences, *New York ANYAS*, **685**: 584-602.

Ghule, B.V., and Yeole, P.G.(2012). Immunomodulatory activities of iridoids fraction from *Barleria prionitis* Linn. *Journal of Ethnopharmacology*, **141**(1): 424–431

Giese, K. (2002). Unravelling novel intracellular pathways in cell-based assays. *Drug Discov. Today*, **7**: 179–186.

Harborne, J.B., and Baxter, H.(1993). (Eds) *Phytochemical Dictionary*, A Handbook of Bioactive compounds from plants. *Taylor and Francis, Ltd. London*. Washington, DC, pp. 569-578.

Hudson, L., and Hay, F.C.(1980). *Practical Immunology,2nd edition*, (Black Well Scientific Publication, London, U.K.): pp. 24-51.

Kaul, A. (1995).Effect of some important medicinal plants on immune system. Ph.D. Thesis University of Jammu.

Kaye, J., and Janeway, C.A. Jr.(1984).Induction of receptors for interleukin 2 requires T-cells, Ag: Ia receptor crosslinking and interleukin 1. *Lymphokine Res. J.*, **3**: 175-182.

Laila, Jarin., Shafiqur, M.D., and Anwar, M. N.(2008). Antibacterial and antifungal activity of crude extracts of *plumeria*. *J. B. Sci.*, **3**(1 and 2): 87-94.

Labadie, R.P. (1993). *Immunomodulatory compounds* In: Bioactive natural products eds. Steven M. Colegate, Russell, *J. Molyneux*, (*CRC Press Inc.)*: pp. 279-317.

Lehrer, R.I. (1981). Ingestion and Destruction of *Candida albicans*. In Methods for Studying Mononuclear Phagocytes Eds.Adms, D.O., Edelson, P.J. and Korem, H.S. pp.693-708.New York: Academic Press.

Mahabeer.P.Dobhal, Guolin Li, Any Gryshuk, Andrew Graham, Atul. K. Bhatanagar, Sirajud D. Khaja,Yogesh C. Joshi, M.C.Sharma, Allan Oseroff and Ravindra K. Pandey. (2004). Structural modifications of Plumeride isolated from *Plumeride bicolour* and the effect of these modifications on *in vitro* anticancer activity. *J. org. Chemistry*, **69**: 6165 - 6172.

Martina, Jonsson., Marika Jestoi., Alexis, V, Nathanail., Ulla-Maija, Kokkonen., Marjukka, Anttila.,Pertti, Koivisto., Pirkko, Karhunen., and Kimmo, Peltonen.(2013). Application of OECD Guideline 423 in assessing the acute oral toxicity of moniliformin *Food and Chemical Toxicology*, **53**: 27–32.

Mossman, T.R., and Coffman, R.L. (1989). TH1and TH2 cells: different patterns of lymphokine secretion lead to different functional properties. *Annu. Rev. Immunol.*, **7**: 145–173.

Nithya, K., and Muthumary, J. (2009). Growth studies of collectotrichum gloeosporiorides (Penz.) sacc. a Taxol producing Endophytic funus from *Plumeria actifolia*. Recent Research in *Science and Technology*, 2(**4**): 99-103. ISSN: 2076-5061.

Nayar, R.C. (1979). A review on the Ayurvedic drug "Sariva". *J. Res. Ind. Med. Yoga. Homeopath.*, **14(2)**: 69-79.

Nelson, D.S., and Mildenhall, P.(1967).Studies on cytophilic antibodies. The production by mice of macrophage cytophilic antibodies to sheep erythrocytes: relationship to the production of other antibodies and development of delayed type hypersensitivity. *Aust. J. Exp. Biol. Med. Sci.*, **45**: 113-130.

Patwardhan, B. Kulbag, D, Patki, P.S., and Nagsampagi, B. A. (1990). Search of *Immunomodulatory agents*: A review. *Indian Drugs*, **28(2)**: 56-53.

Rasool, S. N., Jaheerunnisa, S., Suresh Kumar, Chitta., and Jayaveera, K.N. (2008). Antimicrobial activities of *Plumeria acutifolia*. *Journal of Medicinal Plants Research*, **2(4)**: 077-080.

Sharma, M.L.(1991). Evaluation of immunomodulatory activity of some medicinal plants in mammals. Ph.D. Thesis University of Jammu.

Sharma, M.L., Rao, C.S., and Duda, P.L.(1994). Immunostimulatory activity of *Picrorhiza kurroa* leaf extract. *J. Ethnopharmacy*, **41**: 185-192.

Singh, G.B., Bani, S., Singh, S., Kaul, A., Khajuria, A., Sharma, M.L., Gupta, B.D., and Banerjee, S.K.(1993). Anti-inflammatory activity of the iridoids, Kutkin, Picroside-1and kutkoside from *Picrorhiza kurroa*. *Phytotherapy Res.*, **7**: 402-407.

Singh, V.K. (1999). The paradigm of Th1 and Th2 cytokines: its relevance to autoimmunity and allergy. *Immunol. Res.*, **20**: 147–161.

Sticher, O. (1977). Plant Mono, Di and Sesquiterpenoids with pharmacological or Therapeutical activity. In: New Natural Products and Plant Drugs with Pharmacological, Biological or Therapeutical activity, Eds H. Wagner and P.Wolf,Springer-Verlag., *New York, pp*. 137-176.

Surendra, Sharma., and Naresh. (2012). Antimicrobial potential of *Plumeria rubra* syn *Plumeria acutifolia* bark. *Der Pharma Chemica*, 4(**4**): 1591-1593

Utilisation and Management of Medicinal Plants Vol. 3 (2015) *Pages* **199–211**
Editors: **V.K. Gupta and A. Kaul**
Published by: **DAYA PUBLISHING HOUSE, NEW DELHI**

13

Potential Antidiabetic Medicinal Plants with Evidence of Preclinical and Clinical Investigations

Bhavesh M. Vyas[1], Dhara P. Mehta[2],
Shashipal P. Indoria[3] and Praful A. Talaviya[4*]

ABSTRACT

Diabetes mellitus is most common metabolic disorder in the world. Worldwide more than 366 million peoples are suffering from diabetes worldwide and day by day the prevalence of diabetes is due to rapid urbanization and lifestyle changes. Many of people with type 2 diabetes mellitus are using traditional plants as medicinal agents for diabetes therapy throughout the world. Though there are so many allopathic medicines are available for treatment of diabetes, herbal formulations are preferred and more accepted because of their having lesser side effects and also more economic as compared to allopathic medicine. Medicinal plants are having so many beneficial properties but there is lack of documentation of scientific research on plants. This chapter mainly focuses on most common antidiabetic herbs with preclinical and clinical evidences. This chapter clearly shows the antidiabetic properties and potential of medicinal plants with their effect on blood sugar and HbA1c level, which will provide a base and value added insight for further research on these plants.

Keywords: Blood sugar, Diabetes mellitus, HbA1c, Medicinal plants.

1 [1]Department of Pharmacology, AMC-MET Medical College, Ahmedabad India.
2 Department of Pharmacology, L.J Institute of Pharmacy, Ahmedabad, India.
3 Department of Pharmaceutics, RKDF College of Pharmacy, Bhopal, India.
4 Department of Pharmacology, Faculty of Pharmacy, Pacific University, Udaipur, India.
* *Corresponding author*: E-mail: talaviya.pr@gmail.com

Introduction

Diabetes Mellitus is a common metabolic disorder occurs due to insulin deficiency or insulin insufficiency (Uma *et al.*, 2006). The prevalence rate of diabetes is increasing rapidly worldwide which mainly affects the metabolic functions of body. The global prevalence of diabetes is estimated to be 8.3 per cent, representing 366 million people with diabetes and is expected to go up to 552 million people by 2025 (IDF diabetes atlas update 2012). One person out of four is suffering from diabetes worldwide. More than 80 per cent of world population is using traditional medicines for treatment of various diseases. Research done in Ayurveda and other systems of medicines has noted their use in diabetes mellitus (Yeh *et al.*, 2003). These medicines have lesser side effects and are economic and easily available. The active constituents of these plants are having antidiabetic property due to regeneration (repair) of B- cells of pancreas or increase insulin release and increase insulin sensitivity (Kavishankar *et al.*, 2011). Phytochemical constituents derived from plants present an exciting opportunity for the development of new therapeutic moiety. Phytochemicals are reported for having effect on onset and progression of chronic diseases, protection against oxidative stress and antiageing effect (Noor *et al.*, 2013).

There are huge numbers of bioactive compounds in the world which are renewable and non exhaustive. These medicines are also economic and readily available. From these wide varieties of species, many species have been reported to have antidiabetic activity. Thus, it is essential to preserve such medicinal plant species for benefit of humankind before it ceases to exist. There is also lack of documentation on antidiabetic activity of different medicinal plants (Kawalekar, 2011). There are many plants reported to exert antihyperglycemic activity, but active moiety which is having this effect is unknown. There is also lack of available information or scientific evidence for medicinal plants. This brief review mainly focused on most common medicinal plants used in diabetic patients and by CAM (Complementary and alternative medicine) practitioners.

Aloe – *Aloe vera* Linn. (Liliaceae)

Aloe vera is from Liliaceae family though it is a cactus like plant. There are 4 out of 240 species of Aloe are having nutritional value. According to preliminary research done on Aloe, gel like portion of leaf shown to having hypoglycemic effect at the dose of 15-30 ml per day (Klemens, 2006). Glucomannan, obtained from the gel of inner part of leaves, is having blood glucose lowering effect is noted by Shane-McWhorter (2001). Decrease in FBG (Fasting Blood Glucose) and HbA1c level was reported by Bunyapraphatsara *et al.* (1996) and Yongchaiyudha *et al.* (1996) in non randomized controlled trials on type 2 diabetes patients.

American ginseng and *Panax ginseng* – (Araliaceae)

Some species of ginseng are used for medicinal purpose. Research conducted by Kim *et al.* (2008) has noted that *Radix ginseng* is having antidiabetic activity in both animals and human. While Korean ginseng (0.1–1.0 g/mL) decreases blood glucose level by stimulation of insulin release from Beta-cells so having insulin mimetic activity. American ginseng (100 mg/kg) lowered levels of serum glucose and HbA_{1C}

in diabetic rats which is noted by Kim *et al.* (2008). American ginseng was administered for 8 weeks in diabetic patients (n=24); reported reduction in fasting blood glucose and HbA1c levels (Vuksan *et al.*, 2001). Decrease in blood sugar level and HbA1c was observed by Vuksan *et al.* (2001; 2008) in other randomized controlled trials on diabetic patients.

Garlic - *Allium sativum* L. (Liliaceae)

It is a perennial herb cultivated in all parts of India. Garlic extract significantly decreases serum glucose, lipid content, urea, uric acid, creatinine and liver enzymes levels, while increases serum insulin in diabetic rats but no effect was observed in normal rats. The antidiabetic effect of the garlic extract was more significant than glibenclamide (Eidi *et al.*, 2006). Precursor of allicin which is a sulphur containing amino acid (SACS) is reported to have anti diabetic activity on alloxan induced diabetes in rats. Garlic oil also has significant hypoglycemic effect. *In vitro* stimulation of isolated Beta-cells of rat pancreas by SACS was noted (Sheela and Augusti, 1992).

Onion - *Allium cepa* L. (Liliaceae)

Allium cepa wildly cultivated in Central Asia. Ether extract of dried onion powder have hypoglycemic activity in diabetic rabbits. Moreover, onion has antioxidant and hypolipidemic activity. Kumari *et al.* (1995) have noted that it restores liver hexokinase, glucose 6-phosphatase and HMG Co A (3-hydroxy-3-methylglutaryl-coenzyme A) reductase activities to normal level. It significantly controlled post prandial glucose levels in diabetic patients at the single dose of 50 gm of bulb juice. Mathew and Augusti (1975) have conducted one randomized controlled clinical trial of allyl propyl disulphide extract capsules in healthy volunteers (n=6) which was resulted reduction in FBG and increase in insulin concentration and insulin sensitivity.

Bael - *Aegle marmelose* (L.) Corr. (Rutaceae)

Aegle marmelose is native to India and also throughout Southeast Asia as a tree (Ponnachan *et al.*, 1993). It is cosmopolitan in distribution. It has antifungal, antiprotozoal, antimicrobial, antibacterial and antidiabetic activity. Leaf extract showed to reverse the decreased glycogen level, decrease blood urea and serum cholesterol level in diabetic rats by insulin mimetic effects (Ismail *et al.*, 2009). According to Ismail *et al.* (2009), bael leaves decoction 5mg daily for 16 weeks in Type 2 diabetic patients (n=20) showed decrease in FBG and PPG level significantly as compared to the other oral hypoglycemic agents. This effect was increased markedly when given in combination with Fenugreek powder 20 mg once daily.

Neem – *Azadirachta indica* A. Juss. (Meliaceae)

It is a tree obtained to India, Burma, Bangladesh, Sri Lanka, Malaysia and Pakistan, growing in tropical and semi-tropical regions. A 2g dose of powdered aqueous extract and alcoholic extract of neem when given three times in a day, it resulted in significant antihyperglycemic activity than oral hypoglycemic agents in uncontrolled type-2 diabetic patients (Waheed *et al.*, 2006). Leaf extract of neem has been observed to produce anti-hyperglycemic activity in streptozotocin diabetic rats without altered serum cortisol level (Gupta *et al.*, 2008). Kumari (2010) has reported that neem seed extract 2g three times a day for 40 days when administered to 23 non insulin dependent diabetic patients showed significant decrease in FBG and PPG level.

Ivy Gourd - *Coccinia indica* family (Curcubitaceae)

Coccinia indica (ivy gourd) wildly grows in India, is used to treat Madhuprameh in a traditional East Indian healing system (Yeh *et al.*, 2003). Combined extract *Coccinia indica* and Musa paradisiacal has β cell regeneration effect in STZ induced diabetic rats (Kumari, 2010). In diabetic patients dried extracts of *Coccinia indica* normalizes activities of Lipoprotein lipase, Glucose 6-Phosphatase and Lactate dehydrogenase (Mallick *et al.*, 2009). Kuriyan *et al.* (2008) has reported that alcoholic extract of herb have significant hypoglycemic effects on type 2 diabetes patients. In Diabetic patients 500mg/kg dried extract of *Coccinia indica* significantly increase insulin level by increasing insulin secretion or through glucose metabolic enzymes (Singh, 2011).

Fenugreek - *Trigonella foenum-graecum* L. (Fabaceae, Papilionaceae)

Fenugreek is an annual plant cultivated worldwide. The Fenugreek seed was a traditionally used by ancient Egyptians and spread to Asian countries such as China and India (Bawadi *et al.*, 2009; Kavishankar *et al.*, 2011). Its leaves and seeds are used for the treatment of diabetes in Ayurvedic and other ancient medical systems. 4-hydroxyisoleucine obtained from fenugreek seeds has α-D-glucosidase and α-amylase inhibitor activity. It also increases insulin secretion (Nahas and Moher, 2009). Significantly decreases plasma glucose levels in both normal and diabetic patients by Fenugreek extract was reported by Madar *et al.* (1988) and Sharma *et al.* (1990). Plant extract produces dose dependent hypoglycemic effect in both normal as well as diabetic rats (Khosla *et al.*, 1995). Fenugreek seeds improves glucose metabolism and restores function of creatinine kinase in heart, skeletal muscle and liver of diabetic rats. It also reduces hepatic and renal glucose-6-phosphalase and fructose -1, 6-biphosphatase activity (Gupta *et al.*, 1999).

Fig - *Ficus carica*- (Moraceae)

Diabetic patients of Spain and southwestern Europe are using *Ficus carica* leaf for the treatment of diabetes. Hypoglycemic effect in animals was observed but lack in human. In diabetic rats it also resulted in hypolipidemic activity (Perez *et al.*, 1996). Research done by Serraclara *et al.*, in 1995 on type 1 diabetic patients (n=10), in which administration of fig leaf tea for 28 days showed decrease in PPG and requirements of insulin. No adverse effects were reported (Serraclara *et al.*, 1998). Oral administration of *Opuntia ficus indica* Mill. (Indian fig) stems significantly decreases blood glucose in type 2 diabetic patients (n=8) after 2-3 hrs (Frati *et al.*, 1990).

Gaduchi - *Tinospora cordifolia* Miers. (Menispermaceae)

It is indigenous to the tropical parts of India, Myanmar and Sri Lanka. Aqueous extract of root shows significant antihyperglycemic effect in alloxan induced diabetic rats. A 400 mg/kg dose of aqueous extract which is equivalent to only one unit/kg of insulin, could elicit significant antihyperglycemic effect in different animal models (Kavishankar *et al.*, 2011; Dhaliwal, 1999).

Holy Basil - *Ocimum sanctum* L. (Lamiaceae)

It is well known as Tulsi, known for its medicinal properties. The aqueous extract of leaves reported for having hypoglycemic effect in both normal and diabetic rats (Vats *et al.*, 2002). Tulsi is having significant hypolipidemic effects in diabetic rats (Rai *et al.*, 1997). Decrease in blood glucose level and increase in renal, hepatic and muscle glycogen content was observed in 200 mg/kg dose when given for 30 days (Vats *et al.*, 2004). Agrawal *et al.* (1996) has noted in 40 type 2 diabetic patients decrease in glucose level in fasting and after meal when given 2.5g powder of fresh leaves of *Ocimum album* for 4 weeks (Agrawal *et al.*, 1996). In comparison with placebo *Ocimum sanctum* and *Ocimum album* Roxb leaves reported to have a significant hypoglycemic activity. Almost 17.6 per cent and 7.3 per cent decrease in fasting and postprandial blood glucose level respectively (Agrawal *et al.*, 1996).

Cinnamon

True cinnamon (*Cinnamomum verum*) is a small evergreen tree, the bark of which is a common spice. Cinnamon has been used traditionally to treat diabetes and other medical conditions. The aqueous extract enhances insulin receptor activity by multiple mechanisms. It also increases activity of glycogen synthase (Jarvill-Taylor *et al.*, 2001; Cao *et al.*, 2007). FBG level was decreased by 18 per cent - 29 per cent by administration of *Cinnamon cassia* in uncontrolled diabetic patients (n=60) was observed by Khan *et al.* (2003). In one trial, 43 patients with diabetes with an average HbA1c measurement of 7.1 per cent were treated with 1 g daily for 3 months. No change was reported in FBG or HbA1c levels (Blevins *et al.*, 2007).

Green tea - *Camellia sinensis* – (Theaceae)

Green tea, leaves of the *Camellia sinensis* plant, is heated to inactivate the enzymes to protect from the oxidation of fresh leaves. Polyphenol catechins, particularly epigallocatechin gallate are the main constituents responsible for the medicinal activity (Iso *et al.*, 2006; Zeimbekis and Vlachou, 2008). The catechins of tea increase insulin sensitivity and reduced β-cell damage *in vivo* and *in vitro* studies (Nahas and Moher, 2009). Daily consumption of tea decreases risk of type 2 diabetes by 48 per cent is proven in prospective and retrospective trials. However, there is no change in HbA1c level was observed in some trials of 1 to 2 months duration done by Ryu *et al.* (2006) and Fukino *et al.* (2005). Decrease in FBG levels by 30 per cent was observed when 1.5 L of partially oxidized tea is given to 20 patients for 28 days ($P<0.001$) (Chuang *et al.*, 1992).

Bitter melon - *Momordica charantia* – (Curcibitaceae)

Bitter melon fruits are used to treat diabetes in many traditional medicinal systems. It contains charantin, vicine and polypeptide-p (Noor *et al.*, 2013; Basch *et al.*, 2003), which may have effect on insulin secretion and hepatic glucose metabolism (Nahas and Moher, 2009). Two clinical trials have been done to evaluate the effects of bitter melon type 2 diabetic patients by Dans *et al.* (2007) and John *et al.* (2003). Both have reported to have no effect on HbA1c or FBG values. One Non-randomized open-label crossover short-term metabolic trial on 18 newly diagnosed type 2 diabetic patients was done by Welhinda *et al.* (1986), which resulted decrease in PPG (Postprandial Glucose) level.

Gurmar - *Gymnema sylvestre* - (Asclepiadaceae)

Gymnema sylvestre is also known as sugar destroyer in Hindi and Guajarati. Traditionally in ayurveda leaves of this plant are used to treat diabetes, cholesterol and obesity in India and part of Asia. Active constituent responsible for this action is Gymnemic acid but mechanism is unclear (Nahas and Moher, 2009; Bone, 2002). Ethanolic extract of plant reported to have significant hypoglycemic activity and also decrease HbA1c level when given in type 2 diabetic patients (n=22) at daily dose of 200mg ($P< 0.001$) (Baskaran *et al.*, 1990). According to Joffe (2001), 800mg extract to diabetic patients (n=65) which resulted in significant decrease in FBG and HbA1c level.

Indian gooseberry, Jamun - *Eugenia jambolana* – (Myrtaceae)

It is also known as *Syzygium cumini* (L.) Skeels, Jamun, Jambu, Black Plum or Black Berry (Wang *et al.*, 2013). Plants grow in different parts of India. Diabetic patients in India use decoction of kernels of *Eugenia jambolana* as household remedy for the treatment of diabetes. It is also noted for having antidiabetic effect varying with different diabetes level. In straptazocin induced diabetic mice administration of jamun pulp extract exhibited hypoglycemic activity within 30 min while the seed takes 1 day for its action. It was also noted that it increases insulin secretion in normal as well as diabetic animals. These extracts also inhibited insulinase activity from liver and kidney. *Syzygium cumini* extract also inhibit 5-glucosidase enzyme But these results did not get support from clinical trials results (Shinde *et al.*, 2008; Grover *et al.*, 2000; Teixeira *et al.*, 2006).

Ginkgo biloba – (Ginkgoacea)

The *Ginkgo biloba* native to China is available as a tree. It is also known as "living fossil" traditionally (Encyclopaedia Britannica 2012). Leaf extract of *G. biloba* contains flavonoid glycosides, terpene lactones and ginkgolic acids. Some positive and some negative results are obtained from clinical trials of *G. biloba* for anti diabetic activity. Though it increases plasma insulin level, it was noted by Kudolo (2000) that it has no effect in glucose level in healthy subjects. Kudolo (2001) also reported that 3 months consumption of *G. biloba* extract have significant hypoglycemic effect on NIDDM patients.

Plantago psyllium – (Plantaginaceae)

Plantago psyllium is a water-soluble fiber derived from husks of blonde Psyllium seeds. Rodriguez-Moran *et al.*(1998) reported that consumption of Psyllium 15 g/day for 6 weeks significantly decreases FBG level in type 2 diabetic patients (n=25). *Plantago psyllium* L. significantly decreases the incidence of increasing postprandial glucose by 14 per cent at breakfast and 20 per cent at dinner, 31 per cent after lunch compared to placebo was reported by Pastors *et al.* (1991).

Mango - *Mangifera indica* – (Anacardiaceae)

The tree is cultivated throughout India. Fruits and seeds are used for medicinal purpose traditionally. The aqueous extract of leaves shows throughout India and traditionally its seeds and fruits are used for treatment of various ailments (Mukesh and Namita, 2013). Leaf extract of mango showed antidiabetic activity in various animal models (Kemasari *et al.,* 2011; Luka and Mohammed, 2012). Waheed *et al.* (2006) registered that aqueous and alcoholic extract and powdered form of mango leaves with oral hypoglycemic agents can significantly restore the increased blood glucose value in type 2 diabetic patients uncontrolled with this hypoglycemic agents. While alcoholic extract alone has significant hypoglycemic activity in diabetic patients (Waheed *et al.,* 2006).

Indian kino tree – *Pterocarpus marsupium* (Fabaceae)

Plants grow throughout India and are use as hypoglycemic plant in folklore medicine. Aqueous bark extract lowered blood sugar and improved glucose tolerance of diabetics with no side effects observed (Gupta *et al.,* 2008). *Pterocarpus marsupium* contains Pterostilbene which is responsible for hypoglycemic activity. Flavonoids obtained from plant are proved for pancreatic β-cell regranulation activity and antihyperlipidemic activity. Also, the aqueous extract of the latex was reported for α-glucosidase inhibitory activity (Khan *et al.,* 2012). Aqueous extract (1g/kg, orally) of bark has been observed to produce anti-cataract activity in alloxan diabetic rats. Plant extract was prevented the hyper-triglyceridaemia and hyper-insulinaemia (insulin resistance) in type 2 diabetic patients (Gupta *et al.,* 2008).

Conclusion

Traditionally herbs are used as alternative medicinal agents all over the world. Most of them are now available as commercially for prevention and treatment of many diseases. Herbs are also improve health of humans and used to maintain well beingness. Similarly so many plants as above mentioned have been reported for their use in diabetes and also in its complications. However, less research was done on plants for their effect on cardiovascular risk factors and diabetes complications. None of this plant research has obtained advanced stage of development to replace the currently available synthetic medications. There is so many investigations are going on with the expectation to come up with a safer, effective and more economic compound for the prevention of constantly increasing diabetic population.

References

Agrawal, P., Rai, V., and Singh, R. (1996). Randomized placebo-controlled, single blind trial of holy leaves in patients with noninsulin-dependent diabetes mellitus. *International Journal of Clinical Pharmacology and Therapeutics*, **34**: 406-409.

Basch, E., Gabardi, S., and Ulbricht, C. (2003). Bitter melon (*Momordica charantia*): a review of efficacy and safety. *American Journal of Health-System Pharmacy*, **60**: 356-359.

Baskaran, K., Ahamath, B.K., Shanmugasundaram, K.R., and Shanmugasundaram, E. (1990). Antidiabetic effect of a leaf extract from *Gymnema sylvestre* in non-insulin-dependent diabetes mellitus patients. *Journal of Ethnopharmacol*, **30**: 295-305.

Bawadi, H.A., Maghaydah, S.N., and Tayyem, R.F. (2009). The postprandial hypoglycemic activity of fenugreek seed and seeds' extract in type 2 diabetics: a pilot study. *Pharmacognosy Magazine*, **5**: 134.

Blevins, S.M., Leyva, M.J., Brown, J., Wright, J., Scofield, R.H., and Aston, C.E. (2007). Effect of cinnamon on glucose and lipid levels in Non-insulin-dependent type 2 diabetes. *Diabetes Care*, **30**: 2236-2237.

Bone., K. Gymnema: a key herb in the management of diabetes. Port Townsend, WA: Townsend Letter for Doctors and Patients; 2002. Available from: www.townsendletter.com/Dec2002/phytotherapy1202.htm. Accessed on Apr 16 2014

Bunyapraphatsara, N., Yongchaiyudha, S., Rungpitarangsi, V., and Chokechaijaroenporn, O. (1996). Antidiabetic activity of *Aloe vera* L. juice II. Clinical trial in diabetes mellitus patients in combination with glibenclamide. *Phytomedicine*, **3**: 245-248.

Cao, H., Polansky, M.M., and Anderson, R.A. (2007). Cinnamon extract and polyphenols affect the expression of tristetraprolin, insulin receptor, and glucose transporter 4 in mouse 3T3-L1 adipocytes. *Archives of Biochemistry and Biophysics*, **459**: 214-222.

Chuang, L., Jou, T., Yang, W., Wu, H., Huang, S., and Tai, T. *et al.* (1992). Therapeutic effect of guar gum in patients with non-insulin-dependent diabetes mellitus. *Journal of the Formosan Medical Association*, **91**: 15-19.

Dans, A.M.L., Villarruz, M.V.C., Jimeno, C.A., Javelosa, M.A.U., Chua, J., and Bautista, R. *et al.* (2007). The effect of *Momordica charantia* capsule preparation on glycemic control in Type 2 Diabetes Mellitus needs further studies. *Journal of Clinical Epidemiology*, **60**: 554-559.

Dhaliwal, K.S. (1999). Method and composition for treatment of diabetes. 1999, US Patent 5886029.

Eidi, A., Eidi, M., and Esmaeili, E. (2006). Antidiabetic effect of garlic (*Allium sativum* L.) in normal and streptozotocin-induced diabetic rats. *Phytomedicine*, **13**: 624-629.

Frati, A.C., Jiménez, E., and Ariza, C.R. (1990). Hypoglycemic effect of *Opuntia ficus-indica* in non insulindependent diabetes mellitus patients. *Phytotherapy Research*, **4**: 195-197.

Fukino, Y., Shimbo, M., Aoki, N., Okubo, T., and Iso, H. (2005). Randomized controlled trial for an effect of green tea consumption on insulin resistance and inflammation markers. *Journal of Nutritional Science and Vitaminology*, **51**: 335-342.

Ginkgo, in *Encyclopaedia Britannica Encyclopaedia Britannica Online*, Encyclopaedia Britannica, London, UK, 2012 http: / /www.britannica.com/EBchecked/topic/ 233931/ginkgo. assessed on Apr 19 2014.

Grover, J., Vats, V., and Rathi, S. (2000). Anti-hyperglycemic effect of *Eugenia jambolana* and *Tinospora cordifolia* in experimental diabetes and their effects on key metabolic enzymes involved in carbohydrate metabolism. *Journal of Ethnopharmacol*, **73**: 461-470.

Gupta, D., Raju, J., and Baquer, N.Z. (1999). Modulation of some gluconeogenic enzyme activities in diabetic rat liver and kidney: effect of antidiabetic compounds. *Indian Journal of Experimental Biology*, **37**: 196-199.

Gupta, R., Bajpai, K.G., Johri, S., and Saxena, A. (2008). An overview of Indian novel traditional medicinal plants with anti-diabetic potentials. *African Journal of Traditional, Complementar and Alternative Medicines*, **5**: 1-17.

International diabetes federation, IDF DIABETES ATLAS UPDATE 2012, http: // www.idf.org/diabetesatlas/5e/Update2012. accessed on 01 Apr 2014.

Ismail, M.Y.M. (2009). Clinical evaluation of antidiabetic activity of *Trigonella* seeds and *Aegle marmelos* leaves. *World Applied Sciences Journal*, **7**: 1231-1234.

Iso, H., Date, C., Wakai, K., Fukui, M., and Tamakoshi, A. (2006). The relationship between green tea and total caffeine intake and risk for self-reported type 2 diabetes among Japanese adults. *Annals of Internal Medicine*, **144**: 554-562.

Jarvill-Taylor, K.J., Anderson, R.A., and Graves, D.J. (2001). A hydroxychalcone derived from cinnamon functions as a mimetic for insulin in 3T3-L1 adipocytes. *Journal of the American College of Nutrition*, **20**: 327-336.

Joffe, D. (2001). Effect of Extended Release Gymnema Sylvestre Leaf Extract (Beta Fast GXR) Alone or In Combination With Oral Hypoglycemics or Insulin Regimens for Type 1 and Type 2 Diabetes. *Diabetes Control Newsletter* 2001;76 www.diabetesincontrol.com/studies/gymnema1_2.shtml.

John, A.J. Cherien, R., Subhash, H.S., and Cherian, A.M. (2003). Evaluation of the efficacy of bitter gourd (*Momordica charantia*) as an oral hypoglycemic agent a randomized controlled clinical trial. *Indian Journal of Physiology and Pharmacology*, **47**: 363-365.

Kavishankar, G., Lakshmidevi, N., Murthy, S.M., Prakash, H., and Niranjana, S. (2011). Diabetes and medicinal plants-A review. *International Journal of Pharmaceutical and Biomedical Research*, **2**: 65-80.

Kawalekar, J. (2011). Survey and documentation of some of antidiabetic plants used by local vaidyas in gavanal village, at hukkeri taluk, belgaum district. *Plant Sciences Feed*, **1**: 34-38.

Kemasari, P., Sangeetha, S., and Venkatalakshmi, P. (2011). Antihyperglycemic activity of *Mangifera indica* Linn. in alloxan induced diabetic rats. *Journal of Chemical and Pharmaceutical Research*, **3**: 653-659.

Khan, A., Safdar, M., Khan, M.M.A., Khattak, K.N., and Anderson, R.A. (2003). Cinnamon improves glucose and lipids of people with type 2 diabetes. *Diabetes care*, **26**: 3215-3218.

Khan, V., Najmi, A.K., Akhtar, M., Aqil, M., Mujeeb, M., and Pillai K. (2012). A pharmacological appraisal of medicinal plants with antidiabetic potential. *Journal of pharmacy and bioallied sciences*, **4**: 27-42.

Khosla, P., Gupta, D., and Nagpal, R. (1995). Effect of *Trigonella foenum graecum* (Fenugreek) on blood glucose in normal and diabetic rats. *Indian Journal of Physiology and Pharmacology*, **39**: 173-173.

Kim, E.K., Song, M.Y., Hwang, T-O., Kim, H.J., Moon, W.S., and Ryu, D-G, *et al.* (2008). *Radix clematidis* extract protects against cytokine-and streptozotocin-induced β-cell damage by suppressing the NF-κB pathway. *International journal of molecular medicine*, **22**: 349-356.

Klemens, J. (2006). Herbs that Lower Blood Sugar. *Journal for the American Association of Integrative Medicine*, 1-5.

Kudolo, G.B. (2000). The effect of 3-month ingestion of Ginkgo biloba extract on pancreatic beta-cell function in response to glucose loading in normal glucose tolerant individuals. *The Journal of Clinical Pharmacology*, **40**: 647-654.

Kudolo, G.B. (2001). The effect of 3-month ingestion of Ginkgo biloba extract (EGb 761) on pancreatic beta-cell function in response to glucose loading in individuals with non-insulin-dependent diabetes mellitus. *The Journal of Clinical Pharmacology*, **41**: 600-611.

Kumari, D.J. (2010). Hypoglycaemic Effect of *Moringa Oleifera* And *Azadirachta Indica* in Type 2 Diabetes Mellitus. *The Bioscan*, **5**: 211-214.

Kumari, K., Mathew, B.C., and Augusti, K. (1995). Antidiabetic and hypolipidemic effects of S-methyl cysteine sulfoxide isolated from *Allium cepa* Linn. *Indian Journal of Biochemistry and Biophysics*, **32**: 49.

Kuriyan, R., Rajendran, R., Bantwal, G., and Kurpad, A.V. (2008). Effect of supplementation of *Coccinia cordifolia* extract on newly detected diabetic patients. *Diabetes Care*, **31**: 216-220.

Luka, C., and Mohammed, A. (2012). Evaluation of the antidiabetic property of aqueous extract of *Mangifera indica* leaf on normal and alloxan-induced diabetic rats. *Journal of Natural Product and Plant Resources*, **2**: 239-243.

Madar, Z., Abel, R., Samish, S., and Arad, J. (1988). Glucose-lowering effect of fenugreek in non-insulin dependent diabetics. *European Journal of Clinical Nutrition*, **42**: 51-54.

Mallick, C., De, D., and Ghosh, D. (2009). Correction of protein metabolic disorders by composite extract of *Musa paradisiaca* and *Coccinia indica* in streptozotocin-induced diabetic albino rat: An approach through the pancreas. *Pancreas*, **38**: 322-329.

Mathew, P., and Augusti, K. (1975). Hypoglycaemic effects of onion, *Allium cepa* Linn. on diabetes mellitus-a preliminary report. *Indian Journal of Physiology and Pharmacology*, **19**: 213.

Modak, M., Dixit, P., Londhe, J., Ghaskadbi, S., and Devasagayam, T.P.A. (2007). Indian herbs and herbal drugs used for the treatment of diabetes. *Journal of Clinical Biochemistry and Nutrition*, **40**: 163-173.

Mukesh, R., and Namita, P. (2013). Medicinal plants with antidiabetic potential: A review. *American-Eurasian journal of agricultural and environmental sciences*, **13**: 81-94.

Nahas, R., and Moher, M. (2009). Complementary and alternative medicine for the treatment of type 2 diabetes. *Canadian Family Physician*, **55**: 591-596.

Noor, A., Bansal, V.S., and Vijayalakshmi, M. (2013). Current update on anti-diabetic biomolecules from key traditional Indian medicinal plants. *Current science*, **104**: 721-727.

Pastors, J.G., Blaisdell, P.W., Balm, T.K., Asplin, C.M., and Pohl, S.L. (1991). Psyllium fiber reduces rise in postprandial glucose and insulin concentrations in patients with non-insulin-dependent diabetes. *The American Journal of Clinical Nutrition*, **53**: 1431-1435.

Perez, C., Dominguez, E., Ramiro, J., Romero, A., Campillo, J., and Torres M. (1996). A study on the glycaemic balance in streptozotocindiabetic rats treated with an aqueous extract of *Ficus carica* (fig tree) leaves. *Phytotherapy Research*, **10**: 82-83.

Ponnachan, P., Paulose, C., and Panikkar, K. (1993). Effect of leaf extract of *Aegle marmelose* in diabetic rats. *Indian Journal of Experimental biology*, **31**: 345-345.

Rai, V., Iyer, U., and Mani, U. (1997). Effect of Tulasi (*Ocimum sanctum*) leaf powder supplementation on blood sugar levels, serum lipids and tissues lipids in diabetic rats. *Plant foods for human nutrition*, **50**: 9-16.

Rodriguez-Morán, M., Guerrero-Romero, F., and Lazcano-Burciaga, G. (1998). Lipid- and glucose-lowering efficacy of *Plantago psyllium* in type II diabetes. *Journal of Diabetes and its Complications*, **12**: 273-278.

Ryu, O., Lee, J., Lee, K., Kim, H., Seo, J., and Kim, S. *et al.* (2006). Effects of green tea consumption on inflammation, insulin resistance and pulse wave velocity in type 2 diabetes patients. *Diabetes Research and Clinical Practice*, **71**: 356-358.

Serraclara, A., Hawkins, F., Perez, C., Domiìnguez, E., Campillo, J.E., and Torres, M.D. (1998). Hypoglycemic action of an oral fig-leaf decoction in type-I diabetic patients. *Diabetes Research and Clinical Practice*, **39**: 19-22.

Shane-McWhorter, L. (2001). Biological complementary therapies: a focus on botanical products in diabetes. *Diabetes Spectrum*, **14**: 199-208.

Sharma, R., Raghuram, T., and Rao, N.S. (1990). Effect of fenugreek seeds on blood glucose and serum lipids in type I diabetes. *European Journal of Clinical Nutrition*, **44**: 301-306.

Sheela, C., and Augusti, K. (1992). Antidiabetic effects of S-allyl cysteine sulphoxide isolated from garlic *Allium sativum* Linn. *Indian Journal of Experimental biology*, **30**: 523-526.

Shinde, J., Taldone, T., Barletta, M., Kunaparaju, N., Hu, B., and Kumar, S. *et al.* (2008). α-Glucosidase inhibitory activity of *Syzygium cumini* (Linn.) Skeels seed kernel *in vitro* and in Goto–Kakizaki (GK) rats. *Carbohydrate Research*, **343**: 1278-1281.

Singh, L.W. (2011). Traditional medicinal plants of Manipur as anti-diabetics. *Journal of Medicinal Plants Research*, **5**: 677-687.

Teixeira, C., Fuchs, F., Weinert, L., and Esteves, J. (2006). The efficacy of folk medicines in the management of type 2 diabetes mellitus: results of a randomized controlled trial of *Syzygium cumini* (L.) Skeels. *Journal of clinical pharmacy and therapeutics*, **31**: 1-5.

Torres, W.D. (2004). *Momordica charantia* Linn. (family: Cucurbitaceae)–chemistry and pharmacology. Las Vegas, NV: American Academy of Anti- Aging Medicine; 2004. Available from: http: //www.charantia.com/ampalaya_archive/ studies/1.doc Accessed Apr 17, 2014.

Uma, D.P., Selvi, S,. Suja, S., Selvam, K., and Chinnaswamy, P. (2006). Antidiabetic and hypolipidemic effect of *Cassia auriculata* in alloxan induced diabetic rats. *International Journal of Pharmacology*, **2**: 601-607.

Vats, V., Grover, J., and Rathi, S. (2002). Evaluation of anti-hyperglycemic and hypoglycemic effect of *Trigonella foenum-graecum* Linn, *Ocimum sanctum* Linn and *Pterocarpus marsupium* Linn in normal and alloxanized diabetic rats. *Journal of Ethnopharmacol.*, **79**: 95-100.

Vats, V., Yadav, S., and Grover, J. (2004). Ethanolic extract of *Ocimum sanctum* leaves partially attenuates streptozotocin-induced alterations in glycogen content and carbohydrate metabolism in rats. *Journal of Ethnopharmacol.*, **90**: 155-160.

Vuksan, V., Sievenpiper, J.L., Wong, J., Xu, Z., Beljan-Zdravkovic, U., and Arnason, J.T. *et al.* (2001). American ginseng (*Panax quinquefolius* L.) attenuates postprandial glycemia in a time-dependent but not dose-dependent manner in healthy individuals. *The American Journal of Clinical Nutrition*, **73**: 753-758.

Vuksan, V., Sievenpiper, J.L., Xu, Z., Wong, E.Y., Jenkins, A.L., and Beljan-Zdravkovic, U., *et al.* (2001). Konjac-Mannan and American ginseng: emerging alternative therapies for type 2 diabetes mellitus. *Journal of the American College of Nutrition*, **20**: 370-380.

Vuksan, V., Sung, M-K., Sievenpiper, J.L., Stavro, P.M., Jenkins, A.L., and *et al.* (2008). Korean red ginseng (*Panax ginseng*) improves glucose and insulin regulation in well-controlled, type 2 diabetes: Results of a randomized, double-blind, placebo-controlled study of efficacy and safety. *Nutrition, Metabolism and Cardiovascular Diseases*, **18**: 46-56.

Waheed, A., Miana, G., and Ahmad, S. (2006). Clinical investigation of hypoglycemic effect of seeds of *Azadirachta-inidca* in type-2 (NIDDM) diabetes mellitus. *Pakistan Journal of Pharmaceutical Sciences*, **19**: 322-325.

Welihinda, J., Karunanayake, E., Sheriff, M., and Jayasinghe, K. (1986). Effect of *Momordica charantia* on the glucose tolerance in maturity onset diabetes. *Journal of Ethnopharmacol.*, **17**: 277-282.

Yeh, G.Y., Eisenberg, D.M., Kaptchuk, T.J., and Phillips, R.S. (2003). Systematic review of herbs and dietary supplements for glycemic control in diabetes. *Diabetes care*, **26**: 1277-1294.

Yongchaiyudha, S., Rungpitarangsi, V., Bunyapraphatsara, N., and Chokechaijaroenporn, O. (1996). Antidiabetic activity of *Aloe vera* L. juice. I. Clinical trial in new cases of diabetes mellitus. *Phytomedicine*, **3**: 241-243.

Zeimbekis, A., and Vlachou, I. (2008). Effects of black and green tea consumption on blood glucose levels in non-obese elderly men and women from Mediterranean Islands (MEDIS epidemiological study). *European Journal of Nutrition*, **47**: 10-16.

Utilisation and Management of Medicinal Plants Vol. 3 (2015) *Pages* **213–222**
Editors: **V.K. Gupta and A. Kaul**
Published by: **DAYA PUBLISHING HOUSE, NEW DELHI**

14

Taraxacum officinale Exhibited Diuretic and Anti-inflammatory Effects in the Rat

M.E. Meléndez-Camargo[1*] and R. Silva-Torres[1]

ABSTRACT

Taraxacum officinale Weber (Asteraceae), commonly known as dandelion, is widely used in the Mexican Traditional Medicine to treat a variety of diseases. The aim of this study is to measure the anti-inflammatory and diuretic activities of the extract from T. officinale in the rat. Phytochemical analysis of this extract was made by the conventional tests. Wistar adult rats of 180 to 220 g of body weight (b wt) were used. The anti-inflammatory activity of lyophilized (100 y 200 mg/kg, b.wt.) of this plant on cotton pellet-induced granuloma model was evaluated and compare it with indomethacin (5 mg/kg b.wt.). The diuretic response to aqueous extract at doses of 100 and 200 mg/kg was carried out and compare it with that induced by furosemide at 4 mg/kg. Aqueous extracts, furosemide or vehicle were administered orally to rats and the effects on sodium, potassium and water balance were assessed. The phytochemical analysis revealed the presence of alkaloids, flavonoids, tannins, reducing sugars, and quinones. Lyophilized showed a high anti-inflammatory effect similar to indomethacin without affect stomach and duodenum epitheliums. Aqueous extracts at both doses and furosemide produced important and significant increments in urinary excretion of sodium, potassium and water in rats with respect to control group. This increase was dose dependent and aqueous extract at the highest dose induced sodium and water excretion similar to furosemide, potassium excretion increased but this was less than the one induced by furosemide.

---

1 Department of Pharmacy, Escuela Nacional de Ciencias Biológicas-Zacatenco del Instituto Politécnico Nacional, Av. Wilfrido Massieu s/n, Unidad Profesional Adolfo López Mateos, Col. Lindavista, CP 07738, México, DF, México.

* *Corresponding author*: E-mail: emelendezc@hotmail.com; mcamargo@ipn.mx

T. officinale exhibited anti-inflammatory and diuretic effects and support the evidence to substantiate the traditional use of T. officinale in inflammatory processes and in renal disorders.

Keywords: Anti-inflammatory effect, Dandelion, Diuretic activity, Flavonoids, Furosemide, Indomethacin, *T. officinale*, Water excretion.

Introduction

Taraxacum officinale Weber, (*Asteraceae* family) is a plant widely used in the Mexican Traditional Medicine to treat a variety of disease condition like renal and vesicle calculi, diarrhea, gout, abdominal inflammation, and rheumatism. This plant also possesses antitumor and antidiabetic activities and it prevents influenza and bronchitis (Rzedowski and Rzedowski, 2001; Schütz *et al.*, 2006). Recently has been demonstrated that is effective in the treatment of acute experimental hepatitis induced with CCl_4 (Favari *et al.*, 2013) *T. officinale*, commonly known as dandelion, contains many natural substances such as flavonoids, tannins and reducing sugars (Favari *et al.*, 2013; Hu and Kitts, 2003).

The phenolic compounds exhibited an important antioxidant effect mainly due to their redox properties and chemical structures such as hydroxyl groups, which can scavenge free radicals, chelate metal ions, quench singlet and triplet oxygen, or decompose peroxides and these metabolites are present in this plant (Zheng and Wang, 2001). Oxidative stress is one of the mechanisms involved in inflammatory process, then antioxidant action of the *T. officinale* extract must to protect against this disease. Marked anti-oxidative activity of *T. officinale* were demonstrated by aerial parts (Favari *et al.*, 2013) and flower crude extract (Zheng and Wang, 2001) *in vitro* model systems.

Diuresis is a critical for excretion of numerous compounds, maintaining the body´s water/electrolyte equilibrium and eliminating toxic substances. There are a great number of plant species that present the opportunity to discover potent diuretics. One of these is dandelion and no detailed pharmacological studies that could support the anti-inflammatory and diuretic activities of this plant have been reported (Schütz *et al.*, 2006; Zheng and Wang, 2001).

The aim of this study was to evaluate the anti-inflammatory and diuretic activities of *T. officinale* in rats to validate its use in the Mexican Traditional Medicine.

Materials and Methods

The plant was collected in April (2005) at the National School of Biological Sciences, Mexico, D.F. and botanically authenticated by the biologist Delfina Ramos through comparison with identified specimens deposited in the herbarium at the Botany Department, National School of Biological Sciences (Voucher # 222). The material was shade-dried and powdered.

Preparation of Aqueous Extract

The plant extract was prepared with 250 g of complete plant and was suspended in 1.5 liters of boiled distilled water for 5 minutes. After extraction, the mixture was filtered and concentrated in a vacuum evaporator to final volume of 150 ml and then lyophilized it. The yield obtained was 31.8 g (12.5 per cent). The lyophilized powder was used for all the experiments.

Phytochemical Analysis

Phytochemical analysis of aqueous extract was made by the conventional tests (Dominguez, 1978).

Animals

Adult female Wistar rats were used. They were housed and maintained in the animal house at room temperature (22 - 24°C) and 50 - 55 per cent relative humidity, with a 12/12 hour day/night cycle. Rats were fed with standard rodent diet and water *ad libitum*. Care and handling of animals followed internationally accepted procedures according to the Institute for Laboratory Animal Research's Guide for Care and Use of Laboratory Animals.

Animal Treatment, Induction of Inflammatory Process and Anti-inflammatory Activity

Cotton Pellet-Induced Granuloma Formation in Rats

The experiments were carried out in un-anesthetized adult female Wistar rats. A sterilized cotton pellet granuloma weighing 50 ± 5 mg was introduced subcutaneously under sterile conditions into the axilla of anesthetized rats (Márquez-Flores *et al.*, 2009). The animals were randomly allocated distributed by means of a table of random numbers into to four groups of 8 rats each. The test groups of animals were treated orally with lyophilized *T. officinale* (100 y 200 mg/kg b wt), daily for 7 days. Animals in the control and reference groups received 5 per cent sodium bicarbonate solution (1ml/kg b wt) and indomethacin (5 mg/kg b wt), respectively.

The animals were sacrificed on the eighth day. The pellets surrounded by granuloma tissue were dissected out carefully and recorded for wet weight, then, the pellets were dried at 60 °C to a constant weight. The difference between the weight of the pellet after drying and its initial weight was taken as the final granuloma weight. Mean weight of the granuloma tissue formed in each group was obtained and the results were expressed as wet dry and final granuloma weight.

Gastroduodenal Damage Study

The abdomen of all groups of the rats from the cotton pellet-induced granuloma formation were opened. The stomach and duodenum were removed and opened then the tissue was examined for macroscopic lesions.

Diuretic Activity

Effects of *T. officinale* on Water Balance in the Rat

The experiments were carried out in non anesthetized adult female Wistar rats. These were distributed by means of a table of random numbers into four groups: one control group received vehicle (distilled water, 1 ml/kg b wt), one furosemide-treated group (positive-control group, 4 mg/kg b wt) and two treated- groups with the lyophilized at the doses of 100 y 200 mg/kg b wt. In all groups, the administration was oral and there were eight animals per group. At the beginning of the experiment, the urinary bladder was emptied by gentle compression of the abdomen (t=0), as previously described (Meléndez *et al.*, 2004; Meléndez *et al.*, 2006).

Once the bladder was emptied, either the vehicle or furosemide or aqueous extract were administered. The animals were kept during 6 h in metabolic cages in a quiet and warm environment (21-23 °C) under fluid and food deprivation. At the end of this period the sample of urine was obtained (t= 6 h) and collected into calibrated pipettes (Dade, Miami, Fla.), also by abdomen compression to ensure complete emptiness of the bladder. The sample was added to the spontaneously voided urine that was collected under mineral oil and ice-cooled to prevent evaporation. Blood samples were obtained and centrifuged at 6000 g (Eppendorf centrifuge 5804 R, Germany) for 20 min to separate blood cells from serum. Osmolalities of plasma and urine were measured in a vapor pressure osmometer (Wescor, Logan, Utah) in triplicate. Urine volume was also measured and then osmolal and free water clearances were calculated.

Effect of Aqueous Extract on Electrolyte Balance

To study the effect of *T.officinale*, sodium and potassium concentrations were measured in the urine and plasma samples (Flamephotometer Corning 400, Corning Medical and Scientific, England) from rats treated with the lyophilized. Urinary excretions and clearances of sodium and potassium were calculated (Meléndez-Camargo *et al.*, 2014).

Effect of aqueous extract on glomerular filtration rate

Glomerular filtration rate (GFR) was estimated by the clearance of endogenous creatinine. To avoid the error due to tubular secretion of creatinine, only female animals, in which no secretion occurs, were used. Plasma and urinary creatinine were determined by the method of Jaffe alkaline picrate modified by Melendez *et al.* (2004). The following parameters, creatinine clearance (ml/min) and fractional excretion of sodium (FE_{Na}) and potassium (FE_K), were calculated using the conventional equation:

$$FE_{Na} = (C_{Na}/C_{creat}) \text{ X } 100,$$

where,

$$C_{Na} = U_{Na} \text{ X } V/P_{Na}, \text{ and } C_{creat} = U_{creat} \text{ X } V/P_{creat}.$$

Filtered sodium and potassium loads were also calculated.

Amount of Sodium and Potassium in Plant Extract

The content of Na^+ and K^+ were determined in lyophilized of the plant as described previously.

Statistical Analysis

Data were expressed as mean ± SD from 8 animals per group. Data were analyzed by one way ANOVA followed by Newman's test for comparison between control and treated groups with a significance of less than 0.05.

Results

Phytochemical Analysis

The phytochemical analysis of the present study revealed the presence of flavonoids, tannins, reducing sugars, quinones and alkaloids.

Anti-inflammatory Activities

Cotton Pellet-Induced Granuloma Formation in Rats

As shown in Table 14.1, lyophilized at both doses decrease the wet, dry and final granuloma weight, the difference between test groups and control group was statistically significant ($p<0.05$). *T. officinale* showed a high anti-inflammatory effect in all granuloma weights, being this result very similar to the one obtained with indomethacin.

Table 14.1: Anti-inflammatory Effect of *T. officinale* Lyophilized on Cotton Pellet-Induced Granuloma in Rats

Group	*Granuloma Wet Weight (g)*	*Granuloma Dry Weight (g)*	*Granuloma Final Weight (mg)*
5 per cent Sodium bicarbonate solution (1 ml/kg)	0.414 ± 0.038	0.107 ± 0.014	57.92 ± 13.0
Indomethacin (5 mg/kg)	0.360 ± 0.047*	0.077 ± 0.009*	27.26 ± 0.56*
Lyophilized (100 mg/kg)	0.324 ± 0.029*	0.079 ± 0.004*	29.63 ± 4.40*
Lyophilized (200 mg/kg)	0.310 ± 0.021*	0.068 ± 0.003*	26.30 ± 3.10*

Mean ± SD are shown. (n = 9). *Significantly different from control group ($p<0.05$).

Gastroduodenal Damage Study

The stomach and duodenum of the groups treated with either 5 per cent sodium bicarbonate solution or *T. officinale*, presented a normal epithelium, there were no signs of irritation or ulceration in the gastroduodenal tract; in comparison with the group treated with indomethacin, in which the presence of irritation was observed along the duodenum and in some of the cases gastrointestinal sores were present.

Diuretic Activity

Effect of *T. officinale* on the Water Balance in the Rat

An increase in the urinary volume was observed in lyophilized- treated group and in furosemide-treated group ($p < 0.05$; Figure 14.1). Osmolal and free water clearances increased in the treated groups either with furosemide or aqueous extract (Table 14.2).

Table 14.2: Effect of Aqueous Extract of *Taraxacum officinale* on Osmolal and Free Water Clearances

Groups	*Osmolal Clearance (µl/min)*	*Free Water Clearance (µl/min/)*
Control (vehicle)	5.8 ± 0.5(8)	–4.4 ± 0.8(8)
Furosemide (4 mg/kg)	14.0 ± 4.4*(8)	–9.6 ± 3.6 **(8)
Aqueous extract (100 mg/kg)	9.9 ± 2.3*(8)	–7.0 ± 1.4 *(8)
Aqueous extract (200 mg/kg)	12.8 ± 2.0*(8)	–8.3 ± 1.7 *(8)

Mean ± SD are shown. Figures in parenthesis indicate number of animals.

* $p < 0.05$, ** $p < 0.01$

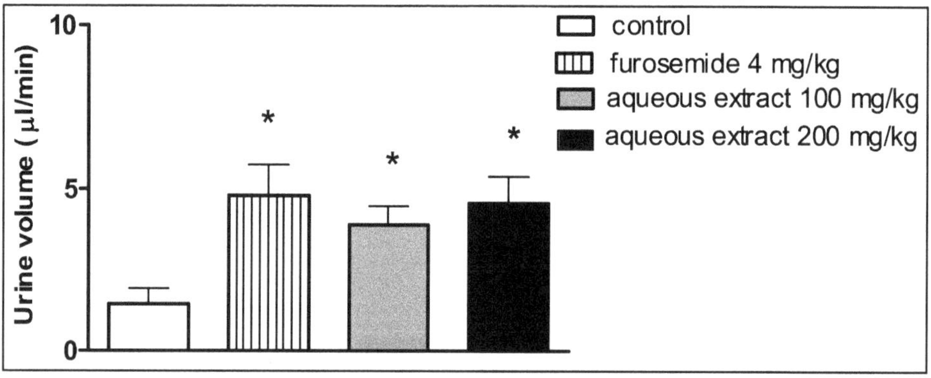

Figure 14.1: Urine Volume in Rats Treated with Aqueous Extract of *Taraxacum officinale*. Furosemide and aqueous extract (100 and 200 mg/kg) of *T. officinale* induced an increment in the urinary flow. Mean ± SE are depicted by the heights of the bars and brackets. *$p<0.05$.

Effects of *T. officinale* on the Electrolyte Balance in the Rat

Sodium clearance increased in the groups treated with either aqueous extract or furosemide (Figure 14.2A). The extract increases the potassium clearance in both treated groups, in contrast to furosemide, the plant does not affect the excretion of potassium in the same degree, the extract induce a lower effect than furosemide (Figure 14.2B). The basal values of plasma electrolytes were similar in all the groups. There were no significant changes in the plasma osmolality and hematocrit among the groups (data not shown).

Effect of *Taraxacum officinale* on the GFR and Filtered Sodium and Potassium Loads

In this study, the GFR remained unchanged in all treated-group and changes in the filtered sodium load were not significant. In the filtered potassium load similar results was obtained (data not shown).

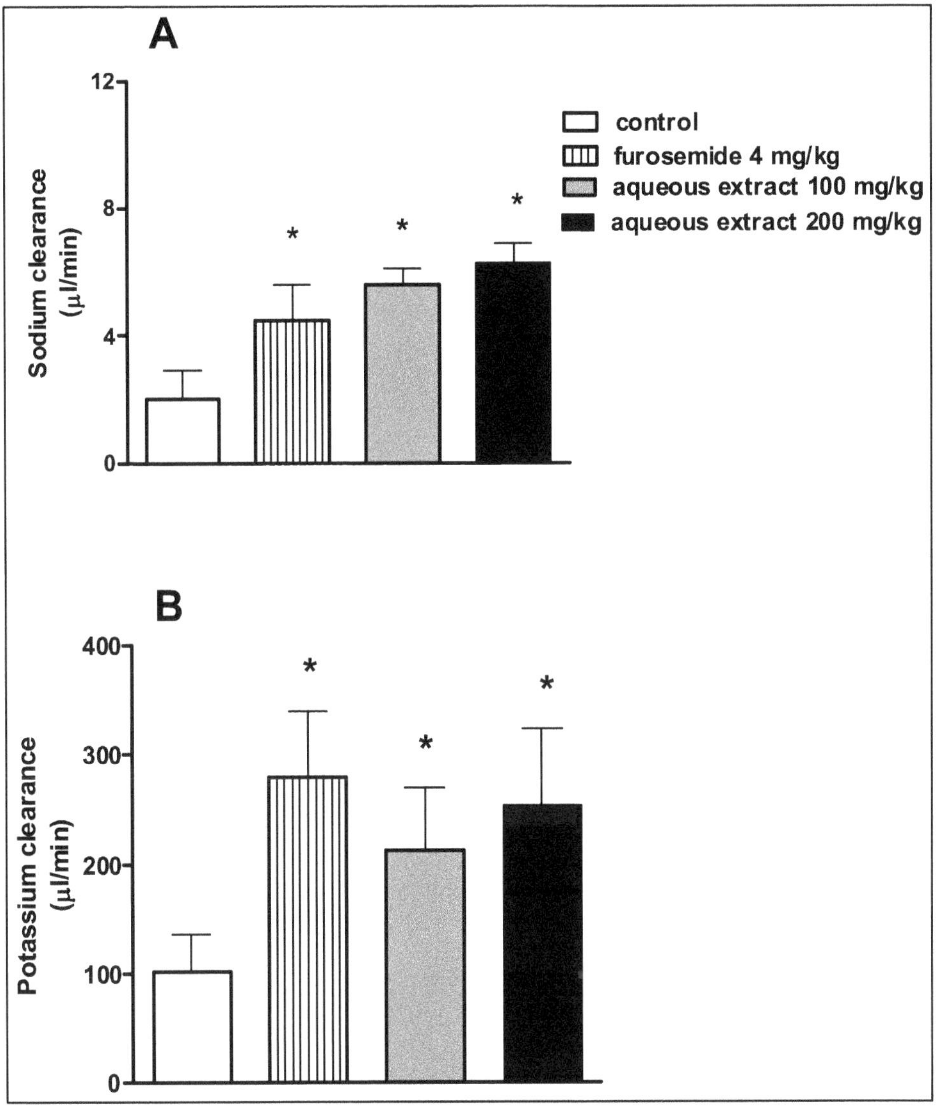

Figure 14.2: Sodium and Potassium Clearances in Rats Treated with Aqueous Extract of *Taraxacum officinale*. In panel 2A furosemide and aqueous extract (100 and 200 mg/kg) of *T. officinale* elicited significant increments in the sodium clearance. Mean ± SE are depicted by the heights of the bars and brackets, * $p<0.05$. In figure 2B the changes in potassium clearances were measured, furosemide and aqueous extract (100 and 200 mg/kg) of *T. officinale* increment potassium excretion, the extract induces less effect to furosemide. Mean ± SE are shown, * $p < 0.05$.

Sodium and Potassium Fractional Excretions (FE_{Na} and FE_K)

Significant increments in FE_{Na} and FE_K were seen in the experimental groups, the increment observed in FE_{Na} for the treated-groups with the aqueous extract of

T. officinale, at the dose 200 mg/kg was higher than furosemide, this was not observed for FE_K where the increment was less than furosemide (Table 14.3).

Table 14.3: Effect of *Taraxacum officinale* Aqueous Extract on Sodium and Potassium Fractional Excretions

Groups	*Sodium Fractional Excretion (FE_{Na} per cent)*	*Potassium Fractional Excretion (FE_K per cent)*
Control (vehicle)	0.154 ± 0.04(8)	5.80 ± 1.28(8)
Furosemide (4 mg/kg)	0.286 ± 0.08*(8)	11.20 ± 1.9*(8)
Aqueous extract (100 mg/kg)	0.322. ± 0.115*(10)	9.90 ± 1.3*(10)
Aqueous extract (200 mg/kg)	0.356. ± 0.15*(10)	9.56 ± 1.8*(10)

Mean ± SD are shown. Figures in parenthesis indicate number of animals.

* $p < 0.05$.

The Amount of Sodium and Potassium in Lyophilized of *T. officinale*

In the aqueous extracts of dandelion at the dose of 100 mg/kg, the potassium content was 22 mmol/l and the content of sodium was 19.4 mmol/l.

Discussion

The anti-inflammatory effect of some species of plant has been attributed to the presence of several secondary metabolites like the flavonoids, which have polyphenolics groups. Their aptitude to modify eicosanoid biosynthesis (antiprostanoid and anti-inflammatory responses), as well as prevent the plaquet aggregation has been described (Martínez-Flores *et al.*, 2002). Due to the fact that this class of metabolites was demonstrated in *T. officinale* lyophilized, it is suggested that the same ones should be responsible for the anti-inflammatory activity of this plant.

T. officinale contains several flavonoids such as: quercetin, luteolin and luteolin-7-0-glucoside (Rodríguez-Fragoso *et al.*, 2008), these phytoconstituent and tannins could inhibit lipid peroxidation.

The Nonsteroidal Antiinflamatory Drugs (NSAIDs), like indometacin, act on the araquidonic acid receptors, the two cyclooxygenase isoforms: cyclooxygenase-1 (COX-1) and cyclooxygenase-2 (COX-2), inhibiting prostaglandin biosynthesis. The gastrointestinal damage induced by these drugs is not selective, since besides inhibiting COX-2 present only in damaged or inflamed tissues, they inhibit COX-1, a protective isoform. When COX-1 is inhibited, the gastrointestinal tract becomes capable of being hurt by the gastric juice (Scheiman, 2005).

The results demonstrate that the groups treated with lyophilized, as well as the control groups, did not produce gastroduodenal irritation. Possibly the reason that irritation is not observed in these groups, is that the anti-inflammatory mechanism of this specie consists of the selective inhibition of COX-2 (Blobaum and Marnett, 2007).

Diuretics are drugs that increase the rate of urine flow and sodium excretion and are used to adjust the volume and/or composition of body fluids in a variety of clinical situations, including hypertension, heart failure, nephritic syndrome and cirrhosis. Loop diuretics such as furosemide can increase the urinary flow rate, also they are strongly saluretic inasmuch as they increase urinary sodium and chloride excretion (Jackson, 2001), that is why in this study furosemide was used as positive control. The most important consideration is that a large number of these plants are rich in potassium (Ranju *et al.*, 2011), which would not lead to potassium depletion, thus giving the benefits of potassium sparing diuretic such as spironolactone, triamterene, etc. (Jackson, 2001).

Through this work, the diuretic activity of the aqueous extract of the dandelion, as described in Traditional Medicine, for the treatment of renal and urinary tract diseases was demonstrated. *T. officinale* induced an increment in the urine output in the electrolyte excretion at 100 and 200 mg/kg of the liophilized in sodium clearance major to the one obtained with positive control. In contrast to furosemide, aqueous extract does not affect in the same degree the potassium clearance, this response is lower in FE_K that in FE_{Na}, this potassium-sparing effect should be viewed as a favorable feature of the extract in regards to electrolyte excretion (Meléndez *et al.*, 2004).

It is possible that the flavonoids present in the aqueous extract exerted its diuretic effect by inhibiting tubular reabsorption of water and electrolytes as such action has been suggested for some other plants (Jayakody *et al.*, 2011).

In conclusions, the present study demonstrate that *T. officinale* induces significant effects on urinary excretion of water and electrolytes and support the claims of diuretic efficacy and anti-inflammatory activity in Mexican Traditional Medicine.

Acknowledgements

This work was partially supported by the Secretaría de Investigación y Posgrado-IPN (SIP 20150511 and 20140838).

References

Blobaum, A.L., and Marnett, L.J. (2007). Structural and functional basis of cyclooxygenase inhibition. *J Med Chem.*, **50(7)**: 1425-1441.

Domínguez, X. A. (1978). Métodos de investigación fitoquímica, Limusa, México, D.F.

Favari, L., Arce-Diaz, C., Ortiz-Martínez, J., Pablo-Pérez, S., Soto C., and Meléndez-Camargo, M.E. (2013). Hepatoprotective and antioxidan effects of *Taracacum officinale* against carbón tetrachloride induced acute hepatic damage in the rat. *Rev Mex Cienc Farm.*, **44(4)**: 53-61.

Hu, C., and Kitts, D.D. (2003). Antioxidant, prooxidant and cytotoxic activities of solvent-fractionated dandelion (*Taraxacum officinale)* flower extracts *in vitro. J Agric Food Chem.*, **51**: 301-310.

Jackson, E.K. (2001). Diuretics. *In*: Goodman and Gilman's The pharmacological basis of therapeutics, Ed. By Hardman, J.G., Limbird, L.E., and Goodman, G.A., McGraw-Hill, New York, pp.755-788.

Jayakody, J.R.A.C., Ratnasooriya, W. D., Fernando, W.A.N.A., and Weerasekera, K.R. (2011). Diuretic activity of leaves extract of hot water infusion of *Ruta graveolens* L. in rats. *J Pharmacol Toxicol.*, **6(5)**: 525-532.

Márquez-Flores, Y.K., Montellano-Rosales, H., Campos-Aldrete, M.E., and Meléndez-Camargo, M.E. (2009) Anti-inflammatory activity of aqueous and methanol extracts of *Oenothera rosea* L´Hér. ex Ait in the rat. *Rev Mex Cienc Farm.*, **40(3)**: 11-16.

Martínez-Flórez, S., González-Gallego, J., Culebras, J.M., and Muñón, M.J. (2002). Los flavonoides: propiedades y acciones antioxidantes. *Nutr Hosp.*, **XVII (6)**: 271-278.

Meléndez, C.M.E., Berdeja, B., and Miranda, G. (2004). Diuretic effect of the aqueous extract of *Bidens odorata* in the rat. *J Ethnopharmacol.*, **95**: 363-366.

Meléndez, E., Cooley, O., and Ramírez, M. (2006). Evaluation of diuretic activity of six plants used for the treatment of urinary diseases in Mexico. *In*: Recent Progress in Medicinal Plants, Vol. 13, Ed. By Govil, J.N., Singh, V. K. and Arunachalam, C, Studium Press LLC, USA, pp. 323-336.

Meléndez-Camargo, M.E., Contreras-León, I., and Silva-Torres, R. (2014). Diuretic effect of alkaloids fraction extracted from *Selaginella lepidophylla* (Hook. et Grev.) Spring. *BLACPMA*, **13 (1)**: 92 – 99.

Ranju, G., Niranjan, S., Pradip, Y., Kumar, P.M., and Kumar, J.G. (2011). Evaluation of diuretic activity of *Platycladus oriental* leaves extract. *Inter Res J Pharm.*, **2(3)**: 117-120.

Rodríguez-Fragoso, L., Reyes-Esparza, J., Burchiel, S., Herrera-Ruiz, D., and Torres, E. (2008). Risks and benefits of commonly used herbal medicines in Mexico. *Toxicol Appl Pharmacol.*, **227**: 125-135.

Rzedowski, J., and Rzedowski, G. (2001). Flora fanerogámica del Valle de México. Instituto de Ecología A.C., México.

Scheiman, J.M. (2005). Effects of nonsteroidal antiinflammatory drugs, including COX-2 specific inhibitors, on the GI tract. *Am Soc Gastrointestinal Endosc.*, **12 (3)**: 1-4.

Schütz, K., Carle, R., and Schieber A. (2006). *Taraxacum*- A review on its phytochemical and pharmacological profile. *J Ethnopharmacol.*, **107**: 313-323.

Zheng, W., and Wang, S.Y. (2001). Antioxidant activity and phenolic compounds in selected herbs. *J Agric Food Chem.*, **49**: 5165-5170.

Utilisation and Management of Medicinal Plants Vol. 3 (2015) *Pages* **223–230**
Editors: **V.K. Gupta and A. Kaul**
Published by: **DAYA PUBLISHING HOUSE, NEW DELHI**

15

Trigonella foenum graecum (Methi Dana): An Effective Vaginal Contraceptive

Radha Sharma[1*], Leena Vaidya[2]
and Suman Jain[3]

ABSTRACT

In India from the ancient times Trigonella foenum graecum seeds as spice are used to flavor and improve the taste of food recipes. They are also used in cosmetics and medicinal preparations. After almost two decades of scientific eclipse, vaginal spermicidal contraceptive methods are gaining interest and importance due to safety and acceptability reasons. At the same time, increasing concern about the side effects of oral contraceptives, IUDs and other methods. This has resulted a shift in attention of investigators and users from these methods to traditional methods of fertility regulation. For this reason a study was carried out on seeds of Trigonella foenum-graecum (Fenugreek), to explore its contraceptive efficacy regarding to its spermicidal action. The Fenugreek extract showed immense potential as a spermicidal drug. An ideal contraceptive formulation can be developed for controlling the population explosion to some extent. Irritation studies showed that upto 2 per cent concentration range the extracts are safe while earlier studies showed that the minimum concentration of fenugreek extract for spermicidal action is 0.3 per cent. Further the irritation potential of fenugreek extract can be minimized by developing a formulation with suitable inert base so that a safe and effective vaginal contraceptive of herbal origin may be developed.

Keywords: Fenugreek, Intra-uterine Devices, Spermicidal, *Trigonella foenum-graecum*, Vaginal contraceptive.

1 [1]SriRam College of Pharmacy, Banmore, Morena, M.P., India.

2 Dr. H. S. Gaur University, Sagar, M.P., India.

3 SOS in Pharmaceutical Sciences,Jiwaji University, Gwalior, M.P., India.

* *Corresponding author*: E-mail: sharma.radha@rediffmail.com

Introduction

Trigonella foenum (Fenugreek) is an annual plant belongs to Fabaceae family. The plant has small round leaves, is cultivated worldwide as a semi-arid crop (Amin *et al.*, 2005; Balaraman *et al.*, 2006). Some other properties of trigonella are carminative, anticarcinogenic, hypoglycemic, hypocholestrolemic, antibacterial, laxative, diuretic, antiulcerative, antihypertensive, antithrombotic, antioxidant (Basch *et al.*, 2003; Eckstein *et al.*, 1969; Hamden *et al.*, 2010). The daily use of fenugreek seeds as dietary supplement is safe. It has good beneficial effects to raise blood hemoglobin by easy means. This is helpful in preventing as well as cure anemia (Kassaian *et al.*, 2009). *Trigonella foenum-graecum* (Fenugreek/methi) have great medicinal value in Indian homes and ancient medical system. Fenugreek seeds are thought to be a galactagogue that is often used to increase milk supply in lactating women (Kaviarsan *et al.*, 2007). Various herbal spermicidal agents such as Neem oil, Sapindus saponins, etc. have been reported yet, there is need to develop a safe, vaginal contraceptive having local actions so that to minimize systemic toxicities. Fenugreek seeds were shown to protect against breast and colon cancer (Kaviarasan *et al.*, 2007; Kaviarasan *et al.*, 2006; Kaviarasan *et al.*, 2004). The fenugreek seeds also have hepatoprotective property (Pandian *et al.*, 2002). Fenugreek seeds provide polysaccharide galactomannan and gitogenin with tigogenin and neotigogens. Some other bioactive constituents of fenugreek include mucilage, volatile oils and alkaloids such as choline and trigonelline, sotolone and pyrazines. The bitter taste of fenugreek seeds is mainly due to steroidal saponins and alkaloids. All the active constituents are non toxic on consumption (Pape *et al.*, 1991).

There is no ideal spermicidal agent investigated till date. Available spermicidal contraceptives also possess some serious physical and physiological side effects. At the same time, increasing concern about the side effects of oral contraceptives, intra uterine devices (IUDs) and other methods have shifted attention of investigators and users from these methods to traditional methods of fertility regulation. For this reason a study was carried out on seeds of *Trigonella foenum-graecum* (Fenugreek), to explore its contraceptive efficacy regarding to its spermicidal action.

Materials and Methods

Fenugreek extracts are rich source of saponins, which may be the main reason of its spermicidal activity and saponins shows hemolysis, *in vitro* hemolytic potency of the extracts was determined. *In vitro* assessment of local irritation by spermicidal agents is based on their hemolytic potency (Pape *et al.*, 1987; Petit *et al.*, 1993). Vaginal irritation potential was accessed by *in vivo* vaginal irritation studies on female rats as animal models. All the experiments were performed as per CPCSEA guidelines.

Preparation of the Extract

Dried seeds of Fenugreek were collected, powdered and extracted successively with various solvents. Extracts were concentrated and dried. Each extract was dissolved in physiological saline to produce various concentrations.

Hemolytic Study (RBC Hemolytic Assay)

Blood samples in heparinized tubes were obtained from local pathology lab. Erythrocytes were separated by centrifugation at 1,000 rpm for 15 min. Washed and resuspended in phosphate buffered saline (PBS). Test samples were prepared by combining 0.5 ml of RBC suspension with 3 ml of PBS containing various concentrations of alcohol extract (0.01 per cent - 2 per cent). min. The samples were than centrifuged at 1500 rpm for 3 min. Supernatant of each sample was collected and absorbance (A) was measured at 450 nm. Zero and 100 per cent hemolysis were determined with isotonic PBS and 0.1 per cent Triton X100 respectively (Petit *et al.*, 1993).

Vaginal Irritation Test in Rats

Eighteen female rats used were divided into 6 groups of three animals in each group. Group 1 was kept as control and 1 ml of saline (0.9 per cent) was applied vaginally to the animals. Group 2 to group 5 animals were treated intra vaginally with 1 ml of different concentrations of Fenugreek extract once per day for 10 consecutive days. Group 6, animals were treated with 1 ml of 5 per cent of N-9 was applied for 10 consecutive days. All the animals were sacrificed on day 11 and the reproductive tract was examined grossly and microscopically. Vaginal tissues were rapidly removed and fixed in 10 per cent neutral-buffered formalin. Vaginas were examined for epithelial ulceration, edema, leukocyte infiltration, and vascular congestion. Irritation scores were assigned based on the scoring system (Raju *et al.*, 2004). Total scoring of 0 to 8 is acceptable, scores of 9 to 11 indicate borderline irritation potential, and scores greater than 12 are potentially irritating.

Results and Discussion

Earlier studies proved that, the fenugreek seeds rich in proteins with essential amino acids, iron, ascorbate and folate content (Sharma, 1986). This helps to prevent and cure anemia and maintain good health. The fenugreek seeds are helpful in the improvement of the digestibility of food, which increases good absorbing capacity of food constituents in blood for best metabolic use in the body. It is commendable that till now no side effects were noted in subjects given with medicinal dosage of fenugreek seeds during study. *Trigonella foenum graecum* is a very common ingredient in Indian dishes. In India it is commonly known as methi Dana. Its use as bitter tonic, antipyretic anthelmintic and antiinflammatory has been reported earlier.

Hemolytic Study (RBC Hemolytic Assay)

Hemolytic potency was determined to access the local irritation of the Fenugreek extracts as compared to N-9. The per cent hemolysis was calculated as-

$$\% \text{ Hemolysis} = \frac{(\text{A450 in the extract samples} - \text{A450 in PBS})}{(\text{A450 in 0.1 per cent Triton X 100} - \text{A450 in PBS}) \text{ X } 100}$$

The results are summarized in Table 15.1. Study showed that ethanol extract of Fenugreek showed 50 per cent and 100 per cent hemolysis at above 0.6 per cent and

2.5 per cent concentrations respectively, while N-9 showed 50 per cent and 100 per cent hemolysis at 0.1 per cent and above 2 per cent levels, respectively.

Table 15.1: Results of Hemolytic Study

Concentration	*Per cent Hemolysis*	
	Alcohol Extract	*N-9*
2 per cent	86.0 per cent	98.0 per cent
1 per cent	63.0 per cent	92.0 per cent
0.5 per cent	42.0 per cent	60.0 per cent
0.1 per cent	39.0 per cent	40.0 per cent
0.05 per cent	17.0 per cent	35.0 per cent
0.01 per cent	9.0 per cent	17.0 per cent

* Values are mean of three readings; * Control PBS = 0 per cent.

* Standard 0.1 per cent Triton X100 = 100 per cent Hemolysis.

Vaginal Irritation Test in Rats

Group 1

No irritation reaction observed in vaginal epithelium of control group animals as the inflammatory score of the animals was in acceptable range *i.e.* 0–4 (Table 15.2).

Table 15.2: Results of Group 1 of Vaginal Irritation Test

Group 1		*Calculation of Score in Submucosal Layer Epithelial Lining*				
		Ulceration	*Oedema*	*Leukocytic Infiltration*	*Vascular Congestion*	*Total Score*
Control	**A**	0	1	1	0	2
Saline	**B**	0	0	1	0	1
	C	1	1	1	1	4

Group 2

No irritation reaction observed in vaginal epithelium of 0.3 per cent Fenugreek extract treated group animals as the inflammatory score of the animals was in acceptable range *i.e.*0-4 (Table 15.3).

Table 15.3: Results of Group 2 of Vaginal Irritation Test

Group 2		*Calculation of Score in Submucosal Layer Epithelial Lining*				
		Ulceration	*Oedema*	*Leukocytic Infiltration*	*Vascular Congestion*	*Total Score*
0.3 per cent	**A**	0	1	1	1	3
Fenugreek	**B**	0	2	0	1	3
Extract	**C**	0	0	1	1	2

Group 3

No irritation reaction observed in vaginal epithelium of 1.0 per cent Fenugreek extract treated group animals as the inflammatory score of the animals was in acceptable range *i.e.* 0-4 (Table 15.4).

Table 15.4: Results of Group 3 of Vaginal Irritation Test

Group 3		*Calculation of Score in Submucosal Layer Epithelial Lining*				
		Ulceration	*Oedema*	*Leukocytic Infiltration*	*Vascular Congestion*	*Total Score*
1 per cent	**A**	0	1	2	1	4
Fenugreek	**B**	0	0	1	0	1
Extract	**C**	1	0	1	1	3

Group 4

Mild irritation reactions observed in vaginal epithelium of 2.0 per cent Fenugreek extract treated group animals as the inflammatory score of the animals was in range 4–8 (Table 15.5).

Table 15.5: Results of Group 4 of Vaginal Irritation Test

Group 4		*Calculation of Score in Submucosal Layer Epithelial Lining*				
		Ulceration	*Oedema*	*Leukocytic Infiltration*	*Vascular Congestion*	*Total Score*
2 per cent	**A**	0	2	1	2	1
Fenugreek	**B**	0	1	1	0	2
Extract	**C**	1	1	2	1	5

Group 5

Irritation reactions observed in vaginal epithelium of 5.0 per cent Fenugreek extract treated group animals as the inflammatory score of the animals was in range 4–8 (Table 15.6).

Table 15.6: Results of Group 5 of Vaginal Irritation Test

Group 5		*Calculation of Score in Submucosal Layer Epithelial Lining*				
		Ulceration	*Oedema*	*Leukocytic Infiltration*	*Vascular Congestion*	*Total Score*
5 per cent	**A**	3	2	1	0	6
Fenugreek	**B**	2	2	2	3	9
Extract	**C**	1	2	3	2	8

Group 6

Irritation reaction observed in vaginal epithelium of N-9 treated group animals as the inflammatory score of the animals was in range above 9–10 (Table 15.7).

Table 15.7: Results of Group 6 of Vaginal Irritation Test

Group 6		*Calculation of Score in Submucosal Layer Epithelial Lining*				
		Ulceration	*Oedema*	*Leukocytic Infiltration*	*Vascular Congestion*	*Total Score*
5 per cent	**A**	3	2	2	2	9
N-9	**B**	4	3	2	1	10
	C	4	3	2	2	11

Previous animal experiments and human intervention research demonstrated that the antidiabetic effect of fenugreek seeds ameliorate most metabolic symptoms associated with diabetes (Sharma *et al.*, 1996; Srinivasan *et al.*, 2006;Udayasekhran *et at.*, 1996). It can be concluded from the earlier research work that *Trigonella foenum graecum* extract can lower kidney or body weight ratio, blood glucose, blood lipids and improve hematological properties. Fenugreek Supplements decreases serum cholesterol, triacylglycerol low density lipoprotein in human patients. This is also been tested in experimental models of hypercholesterolemia and hypertriglyceridemia. Hemolytic studies of extracts were performed to access local irritation potential as compared to the N-9 samples. The extract showed 50 per cent and 100 per cent hemolysis at 0.6 per cent and 2.5 per cent concentrations respectively. While N-9 showed 50 per cent and 100 per cent hemolytic at 0.1 per cent and 2 per cent concentrations. Alcoholic extract does not induce any pathological changes in the rat vagina. Results showed no significant difference between the treated and control groups of animals. Inflammatory score caused by Fenugreek extract at concentration 0.1-1 per cent was in acceptable range. At 2 per cent concentration the extract showed mild irritation reactions while, at 5 per cent concentration the score is in 8-10 range *i.e.* showing irritation reactions. Results were compared to the irritation scores in control (0.9 per cent saline treated animals) and N-9 treated animals group. The Fenugreek extract showed immense potential as a spermicidal drug and an ideal contraceptive formulation can be developed for controlling the population explosion to some extent. As irritation studies showed that upto 2 per cent concentration range the extracts are safe while our earlier studies showed that the minimum concentration of fenugreek extract for spermicidal action is 0.3 per cent. Further the irritation potential of fenugreek extract can be minimized by developing a formulation with suitable inert base so that a safe and effective vaginal contraceptive of herbal origin may be developed so far.

In Indian homes fenugreek as spice and as herb have great medicinal value and proved to reduce, and terminate most of the diseases by the use of medicines prepared out of it. On the basis of various ethno pharmacological literatures it can be concluded that fenugreek seeds can be used in simple remedies for treating a variety of ailments. Till date there is no ideal spermicidal agent investigated. The existing spermicidal contraceptives also possess some serious physical and physiological side effects. At the same time, increasing concern about the side effects of oral contraceptives, IUDs and other methods have shifted attention of investigators and users from these methods to traditional methods of fertility regulation. For this reason a study was carried out on seeds of *Trigonella foenum-graecum* (Fenugreek), to explore its contraceptive efficacy regarding to its spermicidal action. Thus the present study concludes that fenugreek have contraceptives property along with restorative and nutritive properties. From the literature the daily use of fenugreek seeds as dietary supplement is safe. It has good beneficial effects to raise blood hemoglobin.

Acknowledgement

One of author wish to express their gratitude and thankfulness to Union Grant Commission for providing grants for the completion of the above research work.

References

Amin, A., and Ayasha, A. (2005). Hemopreventive activities of *Trigonella foenum graecum* (Fenugreek) against breast cancer. *Cell Biol International*, **29**: 687-94.

Balaraman, R., and Dangwal, S.M. (2006). Antihypertensive effect of *Trigonella foenum –graecum* seeds in experimeninduced ypertension in rats. *Pharm Biology*, **44**: 568-575.

Basch, E., and Catherne, U. (2003).Therapeutic applications of fenugreek. *Altern Med Review*, **8**: 20-27.

Eckstein, P., and Jackson, M.,C.(1969). Comparison of vaginal tolerance tests of spermicidal preparations in rabbits and monkeys. *Reprod. Fertil.*, **20**: 85-93.

Hamden, K., and Jaouadi, B. (2010). Inhibitory effect of Fenugreek Galactomannan on digestive enzymes related to Diabetes, Hyperlipidemia, and liver-Kidney dysfunctions. *Biotech Biopro Engineering*, **15**: 407-413.

Kassaian, N., and Azadbakht, L. (2009). Effect of Fenugreek seeds on blood glucose and lipid profiles in type 2 diabetic patients. *Int J Vitamin Nutr Research*, **17**: 737-743.

Kaviarasan, S., and Naik, G. (2007). *In vitro* studies on antiradical and antioxidant activities of Fenugreek (*Trigonella foenum-graecum*) seeds. *Food Chemistry*, **103**: 31-37.

Kaviarasan, S., and Nalini, Ramamurty. (2006). Fenugreek (*Trigonella foenum-graecum*) seed extract prevents ethanol-induced toxicity and apoptosis in Chang liver cells. *Alcohol Alcoholism*, **41**: 267-273.

Kaviarasan, S., and Vijayalakshmi, K. (2004). Polyphenol-Rich extract of fenugreek seeds protect erythrocytes from oxidative damage. *Plant Foods for Human Nutrition*, **59**: 143-147.

Pandian, R. S., and Anuradha, C.V. (2002). Gastro-protective effect of fenugreek seeds (*Trigonella foenum–graecum*) on experimental gastric ulcer in rats. *J Ethnopharmacology*, **81**: 393-397.

Pape, W., and Hoppe, U. (1991). *In vitro* methods **for** the assessment of primary local effects of topically applied preparations. *Skin Pharmacol.*, **4(3)**: 205-212.

Pape, W., and Pfannanbecka, U. (1987). Validation of the red blood cell test system as *in vitro* assay for the rapid screening of irritation potential of surfactants. *Mol. Toxicol.*, 1**(4)**: 525-536.

Petit, P, and Sauvaire, Y. (1993). Effects of Fenugreek seed extract on feeding behavior in the rat: metabolic–endocrine correlates. *Pharmacol Biochem and Behavior,* **45**: 369-374.

Raju, J., and Patlolla, J. M. (2004). Diosgenin, a steroid saponin of *Trigonella foenum graecum* (Fenugreek), inhibits azoxymethane-induced aberrant crypt foci formation in F344 rats and induces apoptosis in HT-29 human colon cancer cells. *Cancer Epidemiol Biomarkers Prev.*, **13**: 1392-1398.

Sharma, R., D. (1986). Effect of fenugreek seeds and leaves on blood glucose and serum insulin responses in human subjects. *Nutrition Research*, **6**: 1353-1364.

Sharma, R, D., and Sarkar, A. (1996). Use of fenugreek seed powder in the management of non-insulin dependent diabetes mellitus. *Nutrition Research*, **16**: 1331-1339.

Srinivasan, K. (2006). Fenugreek (*Trigonella foenum–graecum*) A review of health beneficial physiological effects. *Food reviews International*, **22**: 203-224.

Udayasekhra, P. R., and Sesikiran, B. (1996). Short term nutritional and safety evaluation of fenugreek. *Nutrition Research*, **16**: 1495-15.

Utilisation and Management of Medicinal Plants Vol. 3 (2015) *Pages* **231–243**
Editors: **V.K. Gupta and A. Kaul**
Published by: **DAYA PUBLISHING HOUSE, NEW DELHI**

16

Evaluation and Studies of Medicinal Euphorbious Plants *Croton tiglium*: A Review

Ram P. Yadav[1] and Ajay Singh[1*]

ABSTRACT

Medicinal plant preparations are chemically complex and may contain one or many structurally related active compounds that produce a combined effect. Phytochemical studies help in standardizing the herbal preparations so as to get the optimal concentrations of these active constituents, as well as in preserving their activities.

Croton tiglium (Linn.) is a commonly occurring plant in garden of North-Eastern, U.P, India belonging to family euphorbiaceae. In view of extensive medicinal importance of this plant in review is an effect to compile all the information reported on its phytochemical and pharmacological activities as there information will be helpful in scientific use of the plant and formulation development.

The present work is an extensive review of the phytochemical and pharmacological reports of Croton tiglium to give comprehensive information in an attempt to provide direction of further research.

Keywords: Euphorbiaceae, Phytochemical, Phorbol esters, Flavonoids, Alkaloids.

Introduction

Many Euphorbiaceae are well known in different parts of the world as toxic and/or medicinal. The high diversity of the described effects is a reflex of the high

1 Natural Product Laboratory, Department of Zoology, D.D.U, Gorakhpur University, Gorakhpur, U.P., India.

* *Corresponding author*: E-mail: rampratapy@rediffmail.com

chemical diversity of this plant group. The genus *Croton* has about 1300 species of trees, shrubs and herbs and is found in the tropical and subtropical regions of North and South Hemispheres. A number of the *Croton* species are known for their medicinal qualities especially in Africa, Asia and South America. *Croton* has been found to possess secondary metabolites such as alkaloids, terpenoids, flavanoids and compounds such as diterpenoids. *Croton* spp. are commonly used for the treatment of non communicable diseases such as diabetes, cancers and other ailments such as digestive problems, dysentery, wounds, fevers, constipation, diarrhea, intestinal worms, malaria, pain ulcers, inflammation. The parts that are used for the treatments of the different kinds of disease are the leaves, the roots, the stem barks, the fruit and the seeds (Yibralign, 2007).

We have a number of herbaceous species in this country, but none of any economic importance. The croton plant is a native of India and is grown all through the East Indies. It is a small tree 15 to 20 feet height. The leaves are ovate, petiolate, acuminate, alternate, the margins faintly serrate. The flowers are borne in loose terminal spike like racemes, and are monocious, the male flowers being at the top of the raceme, the females below. The male flowers have five sepals, each sepal bearing a yellow gland, five petals, and from ten to twenty stamens with slender filaments. The female flowers have floral envelopes similar to those of the male, and a large sessile, three-celled ovary, thickly covered with stellate hairs and bearing three slender styles, each style dividing into two linear stigmas. The fruit is a three-celled capsule, each cell having a single seed which yields the croton oil of commerce. *Croton tiglium* is considered indigenous to Malabar, Ceylon, Amboina (of the Molucca islands), the Philippines and Java.

India is a land of immense biodiversity in which two (Eastern Himalayas and the Western Ghats of India) out of 25 hot spots of the World are located. This country is perhaps the largest producer of medicinal herbs and is rightly called the botanical garden of the World. It is generally estimated that in India over 6000 plants are used in traditional folk and herbal medicine representing about 75 per cent of the medicinal needs of the third World countries (Rashekharan, 2002). Medicinal plants are important for pharmacological research and drug development, not only when plant constituents are used directly as therapeutic agents, but also as a starting material for the synthesis of drugs or as models for pharmacologically active compounds (Mukherjee, 2003).

The medicinal plants research is an interdisciplinary and international collaboration exists between different scientific disciplines including botany, pharmacognosy, pharmacology, phytochemistry, medicine and toxicology. Some of the opportunities for natural products discovery and development as in pharmaceutical, agrochemical, cosmetic, fine chemical and neutraceutical (Said, 1970).

Scientific Classification of *Croton tiglium*

Kingdom: Plantae

(unranked): Angiosperms

(unranked):Eudicots

(unranked): Rosids

Order: Malpighiales

Family: Euphorbiaceae

Subfamily: Crotoneae

Genus: Croton

Species: *Croton tiglium*

Botany

Figure 16.1: *Croton tiglium.*

It is a small evergreen tree, 4.5-6.0 m in height with ash coloured smooth bark and young shoots sprinkled with stellate hairs. Leaves are oblong to ovate-lanceolate, obtuse or rounded at the 2-glanded box, acuminate, membraneous, yellowish green and minutely toothed. Flowers are small, unisexual, males on slender pedicels, females larger and on short thick pedicels. Fruits are ovoid or oblong trigonous capsules. Seeds are smooth, test a black and enclosing reddish brown oily endosperm (Warrier *et al.*, 1994). Other species belonging to the genus *Croton* are as *C. aromaticus* Linn. *C. caudatus* Geisel *C. jouera* Roxb. *C. malabaricus* Bedd. *C. oblongifolius* Roxb. *C. polyandrus* Roxb. syn. *Baliospermum montanum* Muell-Arg. *C. reticulates* (Chopra *et al.*, 1980).

Properties and Activity

Oil contains phorbol myristate acetate (Husain *et al.*, 1992). Seeds contain up to 20 per cent protein and 30-50 of Croton oil, liquid-liquid distribution procedures proved to be the separation to lipids. Iso-guanine-D-ribose (crotoniside) and saccharose were isolated from the seeds. In fractionation is of choice. The pre hydrogenated parent hydrocarbon of phorbol is a perhydrocyclopropabenzulene called tigliane. Phorbol, a tetracylic diterpene with a 5, 7, 6 and 3- member ring has 6 oxygen functions. Phorbol accounts for 3.4 per cent and 4- deoxy- 4- phorbol for 0.29 per cent of the weight of croton oil. Twenty five phorbol-12, 13-diesters have been detected (Hecker *et al.*, 1974).

Phorbol myristate acetate activates nitroblue tetrzolium reduction in human polymorphs. Seed and oil is purgative, rubefacient and anti-dote for snakebite. The seeds and oil are acrid, bitter, thermogenic, emollient, drastic purgative, digestive, carminative, anthelmintic, antiinflammatory, vermifuge, diaphoretic, expectorant, vesicant, irritant and rubefacient.

The identification and structural determination of the cathartic, skin irritant and tumor promoting constituents of Croton oil, the seed oil of *Croton tiglium* L.

(Euphorbiaceae), proved to be a daunting challenge to natural products chemicals that took from the mid 19th centaury until the 1960s to resolve, when it was determined that the toxicity of the species result from the presence of a complex mixture of esters of the tetra cyclic diterpene, phorbol (Hecker and Schmidt, 1974). Intoxication of human by Croton oil has resulted in bloody diarrhea, cyanosis, delirium, gastroenteritis, and respiratory disturbances and the ingestion of as fatal (Hecker and Schmidt, 1974). Croton seeds however are used in Chinese traditional medicine for a number of purposes, including the treatment of apoplexy, cancer colds delayed menstruation, edema, fever, paralysis, scabies, throat afflictions and toothache (Duke and Ayensus, 1985).

Chemistry of *Croton tiglium*

Croton chemistry is considerably diverse. Terpenoids are the predominant secondary metabolite constituents in the genus, chiefly diterpenoids, which may belong to the cembranoid, clerodane, neoclerodane, halimane, isopimarane, kaurane, secokaurane, labdane, phorbol and trachylobane skeletal types. Triterpenoids, either pentacyclic or steroidal, have frequently been reported for *Croton* species. Volatile oils containing mono and sesquiterpenoids, and sometimes also shikimate-derived compounds are not rare in the genus. Several species have been reported as sources of different classes of alkaloids, a fact that enhances considerably the importance of the genus from the medicinal point of view. Phenolic substances have frequently been reported, among which flavonoids, lignoids and proanthocyanidins predominate.

Antidiahorreal Activity

Phytochemicals isolated from medicinal plants are known to be effective in treating bacterial infections. Species under *Croton* genus are found in the different parts of the world and are widely used for the treatment of bacterial infections. Organic solvents (methanol, ethanol, acetone and hexane), inorganic solvents (water) and hydro distillation were described for extraction. From different *Croton* species different methods were used to extract the active contents from roots, leaves, stem-bark and seeds. The resulting extracts and fractions of the extracts were tested against nine Gram negative and eleven Gram positive bacteria in addition to sensitive and resistant strains of *Mycobacterium tuberculosis* using different antibacterial tests. The antibacterial properties were quantitatively evaluated by the minimum inhibitory concentration (MIC), the minimum inhibitory dose (MID) and the zone of inhibition (ZH).

Antibacterial Activity

Antibacterial properties of *Croton* species: the efficacy of the extracts, pharmacology and phytochemistry. Some species of *Croton* are known to contain proanthocyanidins or alkaloids. These alkaloids may be in the form of taspine or some of the several benzylisoquinoline-like compounds. Some of these compounds present in *Croton* species have been shown to have antibacterial properties (Peres *et al.*, 1997; Abo *et al.*, 1999). Evidence from *in vitro* studies point to the use of essential oils as antibacterial agents for wide range of pathogenic bacteria strains such as *Listeria monocytogenes, Listeria innocua, Salmonella typhimurium, E. coli, Shigella dysenteria, Bacillus cereus, Staphylococcus aureus* and *Salmonella typhimurium* (Burt, 2004, Nguefack *et al.*, 2004, Schmidt *et al.*, 2005). Biologically active compounds such as sonderianin, korberin A and B isolated from the genus *Croton tiglium* with specific antibacterial activities against *Mycobacterium smegmatis, Staphylococcus aureus, Bacillus subtilis, E. coli* (Mc Chesney *et al.*, 1991) are however well known.

Anti-flammatory

A Great part of pharmacological assays with *Croton* substances dealt with the clerodane *trans*-dehydrocrotonin, a wide diversity of effects having been noticed, including hypolipidemic, hypoglycaemic, antioestrogen and anti-cancer. Cytotoxic effects also have been observed in assays with alkaloids (taspine) and with secokaurene, labdane and cembranoid diterpenes. Several other effects of *Croton* substances have been registered, including anti-hypertensive, anti-inflammatory, antimalarial, antimicrobial, antispasmodic, antiulcer, antiviral and myorelaxant (Antonio *et al.*, 2007).

Anthelmic Activity

An anti-emetic is a drug that is widely used to control motion sickness, severe nausea and vomiting. Various types of anti-emetic drugs are used to control nausea and vomiting. Some of these works on the brain by preventing the activation of the medullar chemoreceptor trigger zone or vomiting centre and other drugs work on the gut by speeding up the rate at which the stomach empties and so help to move food through the intestines more quickly (Zachary *et al.*, 2004). The family Euphorbiaceae has been selected because many of the members of family Euphorbiaceae has been used as folk medicine in the treatment of different diseases like *Croton tiglium* used as a pain reliever and dry cough (Koche *et al.*, 2010).

Antifertility Activity

Antifertility activity of ses observed in the uterus at 15 days gestation. This is supported by the average number of corpus lutueum were also found very little, so the suspected the folliculogenesis not occurred, although ovulation occurs, the ovum cell quality is not perfect. It was thought to be caused by the active compounds in the extract of *B. balsamifera*, which interfere the growth of egg follicles. *C. tiglium* contains steroids and terpenoids which could be developed into medicine to help people achieve the goal of the national family planning program (Arifin *et al.*, 2007).

Antifungal Activity

Antifungal activity of crude extracts and supernatants may have been due to the presence of lipophilic compounds that may bind within or internal to the cytoplasmic membrane (Boyd and Beveridge, 1979; 1981), quinines (Mahoney *et al.*, 2000) or thionine, which affect growth of filamentous fungi mainly by causing membrane permeablization (Huang *et al.*, 2001). The selected plants *Croton tiglium* showed antifungal activity against the *Mucor mucedo* and *Aspergillus niger* more strongly than for *Aspergillus tamarii* and *Rhizopus solani*.

Haemolytic Activity

Although the presence of haemagglutinating and haemolytic proteins in the seed extract of *Croton tiglium* (Fam. Euphorbiaceae) has been known for a long time (Lin and Huang, 1976), the activities have been purified only recently. Banerjee and Sen (1981) reported the purification of a lectin from *C. tiglium* seeds with haemagglutinating activity towards erythrocytes of sheep, cow and a few other animals and haemagglutinating as well as haemolytic activity towards rabbit erythrocytes.

Molluscicidal Activity

Molluscicides of plant origin have gained greater importance since it is believed that natural products are ecologically sound and culturally than synthetic ones. A large number of plant families which possess natural molluscicidal activity have been identified. The plant phytochemical derived from plant resources can be used as an alternative to the synthetic molluscicides, The Euphorbious plants *Croton tiglium* have strong molluscicidal activity against freshwater snail *Lymnaea acuminata* (Yadav and Singh, 2001; 2002; 2007; 2010; 2012; 2013)

Anticancer Activity

The cancer-causing substances are often considered to be exclusively synthetic; there are numerous natural carcinogens, chemical compounds that occur in environment, and in food plants (Concon, 1988). Isaac Berenblum (1903-2000) discovered the potent inflammatory agent, croton oil extracted from *Croton tiglium* L. native or cultivated in Asia (India, Ceylon, China), Malay Argipelago and Africa (Zanzibar, Tanzania), and its most active ingredient, 12-O-tetradecanoylphorbol-13-acetate (TPA) in 1941 (Berenblum). Both agents now belong to classic tumour promoters. Phorbol esters extracted from the seeds of *Croton tiglium* showed inhibitory effect on HIV-induced cytopahtic effect (CPE) on MT-4 cells and a potent inhibitory effect on the proliferation of HIV-1 (EI-Mekkawy *et al.*, 2000).

Larvicidal Activity

As part of continued search for bioactive natural plant products screened larvicidal activity of crude oils from seeds of *Croton tiglium*. The plant is traditionally being used among the various tribal communities of this part of India for their medicinal and bio-insecticidal potency. *Croton tiglium* also had moderate larvicidal potency against both the targeted mosquitos larvae of *Aedes aegypti* and *Culex quinquefasciatus*. LC_{50} values of the crude oil extracts from seeds of *C. tiglium* against *Aedes aegypti* was found 23.64 ppm and that for *Culex quinquefasciatus* was found 54.77 ppm. (Rituparna *et al.*, 2012).

Antioxidant Activity

Use of *Croton tiglium* leaves as natural antiinfectant and natural blood coagulatory is of immense importance among the ethnic people. Natural antioxidants of plant origin have greater application and they also find use as nutraceuticals and phytoceuticals as they have significant infect on the skin of human health and diseases prevention (Baruah and Baruah, 2010).

Active Compound

Terpenes

Terpenes, terpenoids or isoprenoids are dimmers, trimers or polymers of isoprene units, which are usually jointed in a head to tail fashion. In plants the activated form of the isoprene unit (isopentyl pyrophosphate) which is the building block of each type of terpenoid is synthesized either by the mevalonic acid pathway (*e.g.*, sesquiternenoids) or the methylerythritolphosphate pathway (*e.g.*, mono- and diterpenoids) (Taiz and Zeiger, 2006). Isoprene units usually condense to form linear chain or ring compounds commonly containing carbon atom numbers of 10 (the monoterpenoids), 15 (the sesquiterpenoids, 20 (the diterpenoids), or 30 (the triterpenoids). Terpenoids with 25 carbons are rarely found.

Monoterpenes

Monoterpenes are commonly found in oils of *Croton tiglium*. Iridoids and pyrethrins are included in this group. Examples of monoterpenes commonly found in essential oils. They are widely used as insecticides and their pharmacological properties range from analgesic to anti-inflammatory (Taiz and Zeiger, 2006).

Sesquiterpenes

Sesquiterpenes are also constituents of essential oils of plants *Croton tiglium*, *e.g.*, bisabolol, humulene and caryophyllene. Sesquiterpenes lactones are well known as bitter principles and occur in families like the Euphorbiaceae. These compounds possess a broad range of activities due to the Ü-methylene-y-lactone moiety and epoxides. Their pharmacological activities are antibacterial, antifungal, anthelmintic, antimalarial and molluscicidal. An example is santonin, which is used as anthelmintic and antimalarial (Gurib-Fakim, 2006).

Diterpenes

Diterpenes are present in animals and plants and have therapeutic applications, for example, the famous taxol and its derivatives are anticancer drugs. Other examples are forskolin, which has hypertensive activity zoapatanol is an abortifacient while stevoside is a sweetening agent (Gurib-Fakim, 2006). The diterpenoids from a group of compounds having general molecular formula $C_{20}H_{32}$. They are not steam volatile, and are usually obtained from plants. Numerous species of Euphorbiaceae, which contain diterpenes, show molluscicidal properties. Diterpenes a common compound in *Croton* may be represented as clerodanes, cembranoid, halimanes, kauranes, labdanes, phorbol esters, trachylobanes and sarcapetalanes. Volatile oils are present in some of the *Croton* species (Salatino *et al.*, 2007).

Triterpenes

Triterpenes are C_{30} compounds arising from the cyclization of squalene. They are comprised of a variety of structurally diverse compounds, which include steroids. Tetracyclic terpenes and steroids have similar structures but have different biosynthetic pathways (Taiz and Zeiger, 2006).

Steroids contain a ring system of three six member and one five member ring. Because of the profound biological activities encountered, many natural steroids together with a considerable number of synthetic and semi-synthetic steroidal compounds are employed in medicine (*e.g.*, steroidal saponins, cardioactive glycosides, corticosteroid hormones and mammalian sex hormones). The pharmaceutical applications of triterpenes and steroids are considerable (Gurib-Fakim, 2006).

Phenolic Compounds

All phenolic compounds have an aromatic ring that contains various attached substitute groups such as hydroxyl, and methoxy (-O-CH_3) groups, and often other non aromatic ring structures. They range from simple structures with one aromatic ring to complex polymers such as tannins and lignins. Phenolics differ from lipids in being more soluble in water and less soluble in non-polar organic solvents. Some phenolics, however are rather soluble in ether, especially when the pH is low enough to prevent ionization of any carboxyl and hydroxyl group present. These properties greatly aid separation of phenolic from one another and from other compounds. Other classes of phenolic compounds include coumarines, quinones and flavonoids (Taiz and Zeiger, 2006).

Flavonoids

Flavonoids are 15-carbon compounds generally distributed throughout the plant kingdom. More than 2000 have been identified from plants (Taiz and Zeiger, 2006). They are responsible for the colour of flowers, fruits and some times leaves. Some may contribute to the colour by acting as a copigment. The name flavonoid refers to the latin word 'flavus' meaning yellow. Flavonoids protect the plant from UV-damaging effects and play a role in pollination by attracting animals with their colours (Gurib-Fakim, 2006). The basic structure of flavonoids is 2-phenyl chromane or an Ar-C_3-Ar skeleton. Biosynthetically they are derived from a combination of the shikimic acid and the acetate pathway.

Nitrogen Containing Compound

A large variety of plant secondary metabolites have nitrogen in their structures. Included in this category are such well known anti herbivore compounds such as alkaloids and cyanogenic glycosides, which are of considerable interest because of their toxicity to humans and their medicinal properties. Most nitrogenous secondary metabolites are biosynthesized from common amino acids (Taiz and Zeiger, 2006).

Alkaloids

The term 'alkaloids' has been defined as a cyclic organic compound containing nitrogen in a negative oxidation state, which has limited distribution in living organisms (Taiz and Zeiger, 2006). Based on their structures, alkaloids are divided in to several subgroups: non-heterocyclic alkaloids are divided in to 12 major groups according to their basic ring structure. Mescaline is an example of a non-heterocyclic or pseudo-alkaloid, tetrandrine is an example of a bisbenzylisoquinoline alkaloid while solasodine is a tripterpene alkaloid (Gurib-Fakim, 2006).

Free alkaloids are soluble in organic solvents and react with acids to form water-soluble salts. There are exceptions like berberine, which is quartenary ammonium

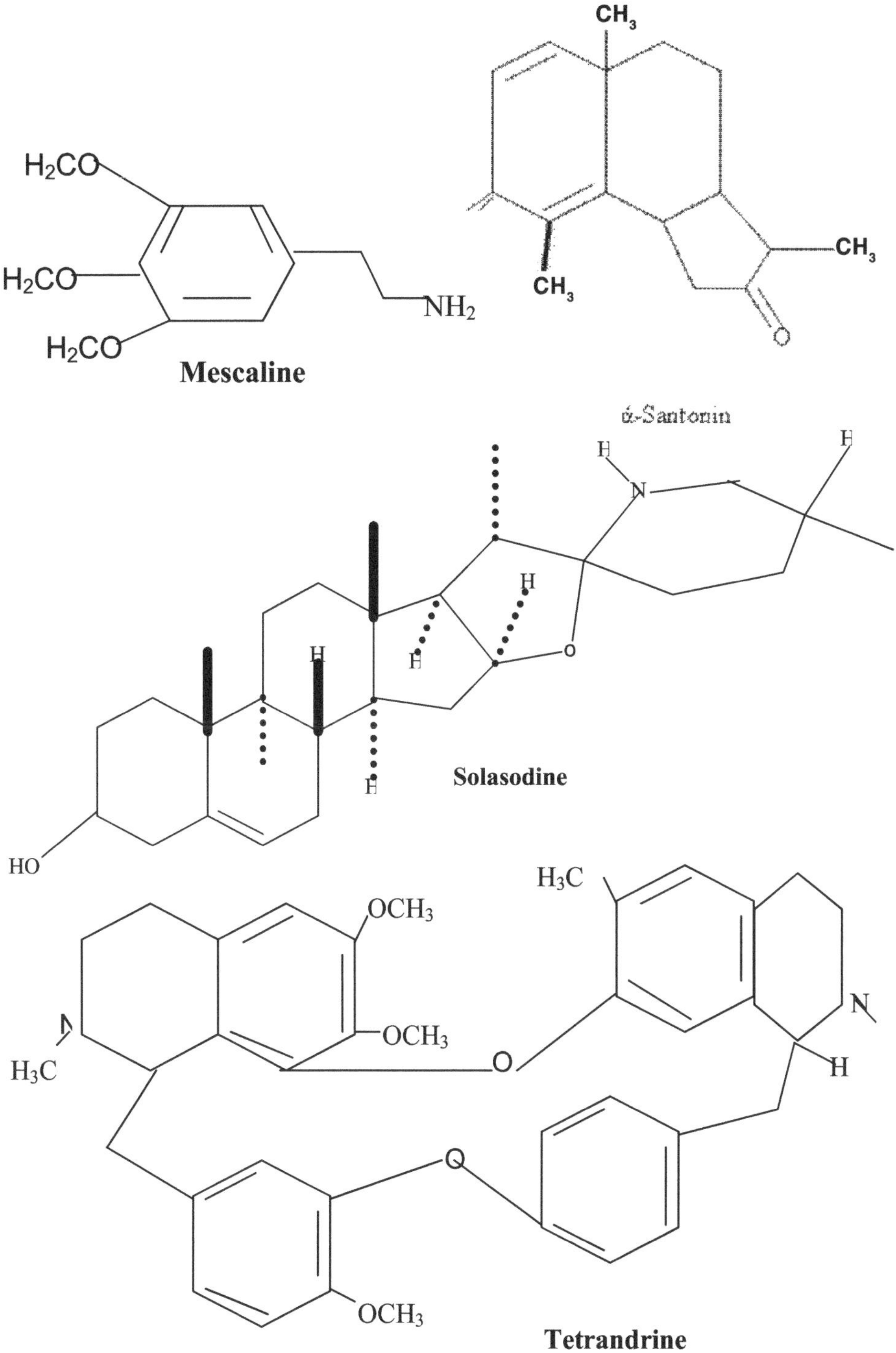

Figure 16.2: Chemical Compounds.

alkaloid. Most alkaloids are solids except for nicotine, which is a liquid. Alkaloids usually have a marked physiological action on humans or animals, and are some times believed to be waste products and a nitrogen source. They are thought to play an important role in plant protection, germination and plant growth stimulation.

Alkaloids are pharmaceutically significant, *e.g.* morphine as a narcotic, analgesic, codeine in the treatment of coughs and pain, colchicines in the treatment of gout, quinine as an antimalarial, quinidine as an antiarrythmic and L-hyoscyamine (in the form of its racemic mixtures known as atropine) as antispasmodic and for pupil dilation (Gurib-Fakim, 2006).

Cyanogenic Glycosides

Perhaps the most obvious defense-related secondary metabolites cyanogenic glucosides (Bennett and Wallsgrove, 1994). They are not in themselves toxic but are readily broken down to give off volatile poisons when the plant is crushed. Cyanogenic glycosides release the well known respiratory poisonous gas, hydrogen cyanide (Taiz and Zeiger, 2006).

Conclusion

It has been realized that many of the modern tools used for the cure of diseases are blunt, brutal, cumbersome and costly. What is needed is one that is safe, effective and less expensive. Billions of people on earth depend mainly on herbal medicines that have been significantly contributing to their primary health care needs. Therefore, the role of medicinal plants in providing effective health care services in most rural parts of the world is likely to continue far beyond the recently emerged 21st century.

The area of medicinal plant research is fast developing. Both pre clinical and clinical testing are integral components of medicinal plant research. Preclinical testing of plants for medicinal properties is of vital importance, not only to provide a scientific basis for their usage but also validate their historical utilization by traditional healers and herbalists, and thus provides the society with sources of new, effective and safe drugs. With the opening of newer vistas in the field of medicine including the modern molecular biology tools, high output automated bioassays and newer technologies for rapid structure determination in the area of medicinal plant research, the field of preclinical testing seems to have a bright future.

References

Abo, K.A., Ogunleye, V.O., and Ashidi, J.S. (1999). Antimicrobial potential of *Spondias mombin, Croton zambesicus* and *Zygotritonia crocea*. *Phytotherapy Research,* **13**(6): 494-7.

Antonio, S., Maria, Salatino, L.F., and Giuseppina, N. (2007). Traditional uses, Chemistry and Pharmacology of *Croton* species (Euphorbiaceae). *Journal Brazilian Chemical Society,* **18** (1): 11-33, 2007.

Arifin, H., Widianingsih, I., and Marusin, N. (2007) Pengaruh pemberian akut ekstrak etanol daun Capo (*Blumea basamifera* D.C.) terhadap gambaran morfolofis dan histologi hati mencit putih jantan. *Journal Sains dan Teknologi Farmasi,* **12** (2): 82-88.

Banerjee, K. K., and Sen, A. (1981). *Archives Biochemical Biophysics*, 212, 740.

Baruah, K., and Baruah, M. (2010). Nonconventional food plants of medicine use in Northeast India In: Medicinal plants of North east India edited by Bharthakur SK Ahmed A, Gogoi P, Dutta DN and Ahmed GU North Eastern development corporation Ltd Guwhati Assam 318

Bennett, R.N., and Wallsgrove, R. (1994). Secondary metabolites in plant defense mechanisms. *New Phytologist*, **127**(4): 617-633

Berenblum, I. (1941). The mechanism of carcinogenesis. A study of significance of cocarcinogenic action and related phenomena. *Cancer Research*, **1**: 807-814.

Boyd, I., and Beveridge, *E.G.* (1979). Relationship between the antibacterial activity towards *E. coli* NCTC 5933 and the physicochemical properties of some esters of 3,4,5-trihydrobenzoic acid (gallic acid). *Microbios*, **24**: 173-184.

Boyd, I., and Beveridge, *E.G.* (1981). Antimicrobial activity of some alkyl esters of gallic acid of 3,4,5-trihydrobenzoic acid against *E. coli* NCTC 5933 with particular reference to n-propil gallate. *Microbios*, **30**: 73-85.

Burt, S. (2004). Essential oils: their antibacterial properties and potential applications in foods– A review. *International Journal of Food Microbiology*, **94**: 223-53.

Chopra, I., Howe,T. G. B. Lintona, H. Linton, K. B. Rrichmondm, H., and Spellerd, C. E. (1981). The tetracyclines: prospects at the beginning of the 1980s. *Journal of Antimicrobial Chemotherapy*, **8**: 5-2 1.

Concon, J.M. (1988). Food Toxicology, Parts A and B. Marcel Dekker, New York.

Duke, J.A., and Ayensus, E.S. (1985). Medicinal plants of China 2. Algonac. MI reference publications Inc.

EI-Mekkawy, S., Meselhy, M.R. Nakamura, N. Hattori, M. Kawahata, T., and Otake, T. (2000). Anti-HIV phorbol esters from the seeds of *Croton tiglium*. *Phytochemistry*, **53**: 457-464

Gurib-Fakim, A. (2006). Medicinal plants: Tradition of yesterday and drugs of tomorrow. *Molecular Aspects Medical*, **27**(1): 1-93.

Hecker, E., and Schmidt, R. (1974). Phorbol esters- the irritants and cocarcinogens of *Croton tiglium* L. *Fortsch Chem Org Naturst.*, **39**: 377-467.

Hecker, H., Brun, R., and Reinhordt, C. (1974). Morphometric analysis of the midgut of female *Aedes aegypti* (L) (Insecta: Diptera) under various physiological conditions. *Cell Tissue Research*, 152: 37-49.

Hussain, R., Poindexter, R.W., and Ottesen, E.A. (1992). Control of allergic reactivity in human filariasis. Predominant localization of blocking antibody to the IgG4 subclass. *Journal Immunology*, **2**: 148: 2731-7.

Koche, D., Shirsat, R. Imran, S., and Bhadange, D.G. (2010). Phytochemical Screening of Eight Traditionally Used Ethnomedicinal Plants from Akola District (MS) India. **1**: 3.

Lin, J., and Huang, M. (1976). Taiwan Pharm. Assoc., **28**: 104.

Mahoney, N., Molyneux, R.J., and Campbell, B.C. (2000). Regulation of aflatoxin production by napthoquinones of walnut (*Juglans regia*). *Journal Agriculture Food Chemistry*, **48**: 4418-21.

Mc Chesney, J.D. Clark, A.M., and Silveira, E.R. (1991). Antimicrobial diterpenes of *Croton sonderianus*. II. ent-Beyer-15-en-18-oic acid. *Pharmacology Research*, **8**(10): 1243-7.

Mukherjee, P.K. (2003). GMP for Indian systems of medicine. In: P.K. Mukherjee, R. Verpoorte. Eds. GMP for Botanicals: Regulatory and quality issues on phytomedicines. New Delhi PA: Business Horizons; 99-112.

Nguefack, J., Budde, B., and Jakobsen, M. (2004). Five essential oils from aromatic plants of Cameroon: their antibacterial activity and ability to permeabilize the cytoplasmic membrane of *Listeria innocua* examined by flow cytometry. Lett. *Applied Microbiology*, **39**: 395-400

Peres, M.T. Monache, D.F. Cruz, A.B. Pizzolatti, M.G., and Yunes, R.A. (1997). Chemical composition and antimicrobial activity of *Croton urucurana* Baillon (Euphorbiaceae). *Journal Ethnopharmacology*, **56**(3): 223-6.

Qian Huang, Dongyu Liu, Paul Majewski, Leah C. Schulte, Joshua M. Korn, Richard A. Young, Eric S. Lander, and Nir Hacohen (2001). The Plasticity of Dendritic Cell Responses to Pathogens and their Components. *Science*, **294** (5543): 870-875.

Rajshekharan, P.E. (2002). Herbal Medicine (World of Science, Employment News, 21-27 November 2002), pp. 3.

Rituparna, B., Kalita, M.C. Goswami R.C., and Talukdar A.K. (2012). Larvicidal Efficacy of Crude Seed Extracts of Six Important Oil Yielding Plants of North East India against the Mosquitoes *Aedes aegypti* and *Culex quinquefasciatus*. *Journal of Biofertilizer and Bio-pesticides*, **3**: 2-4

Said, H.M. (1970). Hamdard Pharmacopoeia of Eastern medicine Hamdard National foundation, Karachi, 7.

Salatino, A., Salatino, M.L., and Fegri, G. (2007). Traditional uses, chemistry and pharmacology of Croton species (Euphorbiaceae). *Journal Brazilian Chemical Society*, **18**(1): 11–33.

Schmidt, E., Jirovetz, L., Buchbauer, G., Denkova, Z., Stoyanova, A., and Murgov, I. (2005). Antimicrobial testing and gas chromatographic analysis of aroma chemicals. *Journal Essential Oil Bearing Plants*,**8**: 99-106.

Taiz, L., and Zeiger, E. (2006). Plant physiology, 4th edn. Ch. 13.m Pp. 315-344. Sinauer Associates, Inc. Publishers, Massachussetts

Warrier, P.K., Nambiar, V.P.K., and Ramankutty, C. (1994). Indian medicinal plants– a compendium of 500 species. Vol. 2. Madras: Orient Longman; pp. 47.

Yadav, R.P., and Singh, A. (2007). Toxic Effects of Euphorbiales on freshwater snail *Lymnaea acuminata* in ponds. *Journal of Herbs Spices and Medicinal Plants,* **13** (2): 87-94.

Yadav, R.P., and Singh, A. (2013). Toxic effects of selected plant pesticides against freshwater snail *Lymnaea acuminata*. *International Journal of Traditional and Natural Medicines,* **2**(3): 149-163.

Yadav, R.P., and Singh, A. (2001). Environmentally safe molluscicide from two common Euphorbiales. *Iberus,* **19** (1): 65-73.

Yadav, R.P., and Singh, A. (2002). Toxic effects of latex of C*roton tiglium* on *Lymnaea acuminata* and *Channa punctatus. Iberus,* **20** (2): 31-44.

Yadav, R.P., and Singh, A. (2010). Toxic effects of Crotocaudin extracted from medicinal plant *Croton tiglium* stem bark on the vector snail and freshwater fish. *Zeitschrift fr Naturforschung* C, **65**c: 327-336.

Yadav, R.P., and Singh, A. (2012). Bio-pesticides used as snail control. Lambert Academic Publishing Germany, pp. 1-80.

Yibralign, Z. (2007). Masters thesis on Phytochemical Investigation on the stem bark of *Croton macrostachyus* (Bisana). Addis Ababa University.

Zachary, A.F., Robert, D.S., and Austin, G.B. (2004). Practical Selection of Antiemetics. *American Famacology Physiology,* **69**(5): 1169-1174.

Utilisation and Management of Medicinal Plants Vol. 3 (2015) *Pages* **245–270**
Editors: **V.K. Gupta and A. Kaul**
Published by: **DAYA PUBLISHING HOUSE, NEW DELHI**

17

Evaluation of Sustainable and Eco-friendly Approaches of Indian Seaweeds against Coastal Malarial Vector, *Anopheles sundaicus* (Diptera: Culicidae)

J. Subramaniam[1*], K. Murugan[1], K. Kovendan, P. Mahesh Kumar[1], D. Amerasan[1], C. Panneer Selvam[1], T. Nataraj[1], D. Dinesh[1], B. Chandramohan[1], R. Chandrasekar[2] and Jiang-Shiou Hwang[3]

ABSTRACT

Seaweeds are the extraordinary sustainable resources in the marine ecosystem which have been used as a source of food, feed, medicine and bio pesticides. The present study was carried out to establish the properties of major seaweeds (Dictyota dichotoma, Padina pavonica, Valoniopsis pachynema) extract on larvicidal, pupicidal and adulticidal activity against the coastal malarial vector, Anopheles sundaicus. The sea weeds were collected from the area around southeast coast of India. Seaweeds were washed with fresh water and shade-dried at room temperature. An electrical blender powdered the dried materials (seaweeds). The powder (800 g) of the leaf was extracted with 2.5 of organic solvents of methanol for 8 h using a soxhlet apparatus and then filtered. The crude seaweed extracts were evaporated to dryness in a rotary vacuum evaporator. The sea weed extracts are observed larvicidal, pupicidal and adulticidal effects after 24 h of exposure; however, the highest larval and pupal mortality was

1 Division of Entomology, Department of Zoology, School of Life Sciences, Bharathiar University, Coimbatore – 641 046, Tamil Nadu, India.

2 Department of Biochemistry and Molecular Biophysics, 238 Burt Hall, Kansas State University, Manhattan 66506, KS, USA.

3 Institute of Marine Biology, National Taiwan Ocean University, Keelung 202-24, Taiwan.

* *Corresponding author*: E-mail: jayapalsubramaniam@gmail.com

found in the methanol extract of D. dichotoma followed by V. pachynema P. pavonica respectively. V. Pachynema showed good aulticidal activity against coastal malarial vector, An. sundaicus. No mortality was observed in the control. The results of the sea weeds, D. dichotoma, P. pavonica, V. Pachynema is promising as good larvicidal and pupicidal, adulticidal properties of against coastal malarial vector, An. sundaicus. This is a new eco-friendly approach for the control of An. sundaicus mosquito as target species of vector control programs.

Keywords: Seaweeds, *D. dichotoma, P. pavonica, V. pachynema, An. sundaicus,* Larvicidal, Pupicidal, Adulticidal.

Introduction

Mosquitoes are the principal vector of many vector-borne diseases affecting human beings and animals, in addition to nuisance. Vector-borne diseases in India, *e.g.*, malaria, dengue, chikungunya, filariasis and Japanese encephalitis, cause thousands of deaths per year. India reports 1.48 million malarial cases and about 1,173 deaths, 1.4 million suspected and 1,985 confirmed chikungunya cases, 5,000 Japanese encephalitis cases and approximately 1,000 deaths, and 383 dengue cases and 6 deaths during 2006 and 2007 (Kumar *et al.*, 2007; WHO 2007; Gopalan and Das, 2009; Dhiman *et al.*, 2010).

Anopheles mosquitoes is the primary vector of malaria in India and other West Asian countries, and improved methods of control are urgently needed (Burfield and Reekie, 2005; Mittal *et al.*, 2005). Malaria infects more than 500million humans each year, killing approximately 1.2 to 2.7 million per year. About 90 per cent of all malaria cases occur in Africa, as does approximately 90 per cent of the world's malaria-related deaths (Breman *et al.*, 2004). Malaria, caused by *Plasmodium falciparum*, is one of the leading causes of human morbidity and mortality from infectious diseases, predominantly in tropical and subtropical countries (Snow *et al.*, 2005). Mosquito bites may also cause allergic responses including local skin reactions and systemic reactions such as *urticaria* and *angioedema* (Peng *et al.*, 2004). Botanical and microbial insecticides have been increasingly used for mosquito control because of their efficacy and documented non-toxic effects on non-target organisms (Ascher *et al.*, 1995). The highest number of malaria, *P. falciparum* cases, and malaria-related deaths is recorded from the state of Orissa located in the eastern part of India (Sharma *et al.*, 2010).

The predominantly brackish water breeding *Anopheles sundaicus* is a major vector of malaria in coastal areas of Southeast Asian countries (Dusfour *et al.*, 2007; Manguin *et al.*, 2008). *An. sundaicus.* was also an important vector of malaria along the east coast of India, including the state of Tamil Nadu until DDT spraying largely eliminated it from mainland India, confining it more to the Andaman and Nicobar islands (Manguin *et al.*, 2008; Das *et al.*, 2004).

Mosquito control, in view of their medical importance, assumes global importance. In the context of ever increasing trend to use more powerful synthetic insecticides to achieve immediate results in the control of mosquitoes, an alarming increase of physiological resistance in the vectors, its increased toxicity to non-target organism and high costs are noteworthy (WHO, 1975). Most of synthetic chemicals are expensive and destructive to the environment and also toxic to humans, animals and other non-target organisms. Besides, they are non-selective and harmful to other beneficial organisms. Some of the insecticides act as carcinogenic agents and are even carried through food chain which in turn affects the non-target organism. Therefore alternative vector control strategies, especially effective and low cost are extremely imperative (Piyarat *et al.*, 1974; Kalyanasundaram and Das, 1985, Subramaniam *et al.*, 2011).

Marine organisms are a rich source of structurally novel and biologically active metabolites many chemically unique compounds of marine origin with various biologically activity have been isolated have being developed as new pharmaceuticals (Blunt *et al.*, 2009; Lin *et al.*, 2010; Gorietal, 2011; Ravikumar *et al.*, 2009c, 2011). Varity of medicines and chemicals are prepared from seaweeds and their associates (Spavieri *et al.*, 2010; Ravikumar *et al.*, 2009b).

Seaweeds are macrophytic algae, aprimitive type of plants lacking true roots, stems and leaves. Most seaweeds belong to one of three divisions - the Chlorophyta (green algae), the Phaeophyta (brown algae) and the Rhodophyta (red algae). There are about 900 species of green seaweed, 4000 red species and 1500 brown species found in nature (Dawes, 1981).

Dictyota dichotoma is a marine brown alga distributed worldwide from temperate to subtropical regions. Species of brown algae belonging to family Dictyotaceae have been shown to be a rich source of interesting biologically active diterpenoids of versatile chemical classes (Vashishta, 1984; Duran *et al.*, 1997).

Taxonomy of *Dictyota dichotoma*

Kingdom: Harosa

Phylum: Phaeista (Phaeophyta)

Class:Dictyotales (Phaeophyceae)

Order: Dictyotaceae

Family: Dictyoteae

Genus: *Dictyota*

Species: *Dictyota dichotoma*

(Hudson) (J.V. Lamouroux 1809).

Padina pavonica (L.), a brown alga, was widespread in southeast coast of India. *P. pavonica* is an alga that exhibits calciotropic properties. This brown alga belongs to the pheophycae family and grows abundantly in the Mediterranean from the surface down to 60m below sea level.

Taxonomy of *Padina pavonica*

Kingdom: Harosa

Phylum: Phaeista (Ochrophyta)

Class: Dictyotales (Phaeophyceae)

Order: Dictyotaceae

Family: Zonarieae

Genus: *Padina*

Species: *Padina pavonica*

(Linnaeus) Thivy, 1960.

V. pachynema whereas used in many medicinal purpose worldwide. Cytotoxicity activities highly effective showed considerable action and their values LD_{50} was below 250µ g/ml (Manilal *et al.*, 2009)

Taxonomy of *Valoniopsis pachynema*

Kingdom: Viridaeplantae

Phylum: Ulvophyceae (Chlorophyta)

Class: Siphonocladales

Order: Valoniaceae

Family: *Valoniopsis*

Genus: *Valoniopsis*

Species: *Valoniopsis pachynema*

(G.Martens) Børgesen, 1934.

In the South-west coast of India, a group of scientists studied 13 groups of marine algae to evaluate the cytotoxic, larvicide, nematicide and ichthyotoxic activities on *Artemia salina* larvae. This Indian region is the only marine habitat with great marine algae diversity. 13 algae extracts between them *Dictyota dichotoma* and *Hypnea pannosa* showed lethal effect against the root nematode *Meloidogyne javanica*. *D. dichotoma* and *Valoniopsis pachynema* showed an ichthyotoxic activity. *A. orientalis, Padina tetrastromatica* and *Centeroceras clavulatum* showed activity against the urban mosquito larvae *Culex quinquefasiatus* (Manilal *et al.*, 2009).

Being rich in minerals, vitamins, trace elements and bioactive substances, seaweeds are called medical food of the 21st century. *Digenea* (Rhodophyta) produces an effective vermifuge (kainic acid). *Laminaria* and *Sargassum* species have been used in China for the treatment of cancer. Anti-viral compounds from Undaria have been found to inhibit the Herpes simplex virus, which are now sold in capsule form. Research is now being carried out into using *Undaria* extract to treat breast cancer and HIV. Another red alga *Ptilota* sp. produces a protein (a lectin) that preferentially agglutinates human B-type erythrocytes *in vitro*. Some calcareous species of *Corallina* have been used in bone-replacement therapy (Stein and Borden 1984). *Asparago psistaxiformes* and *Sarconema* sp. are used to control and cure goiter while heparin, a seaweed extract, is used in cardiovascular surgery (Sajid and Satam, 2003).

Recent findings evidenced that seaweeds contained antibacterial (Tuney *et al.*, 2006), antiviral (Garg *et al.*, 1992), antifungal (Tang *et al.*, 2002; Aliya *et al.*, 1999), cytotoxic (Tang *et al.*, 2002) and larvicidal potentials (Thangam and Kathiresan, 1991). The secondary metabolites synthesized by seaweeds demonstrate a broad spectrum of bioactivity varying from neurologically active in humans to algicidal, nematicidal, insecticidal and ichthyotoxicity in lower form of animals (smith *et al.*, 2004; Manilal *et al.*, 2009).

In the present paper we report the screening for larvicidal, pupicidal, repellency and adulticidal activities of 3extracts obtained from seaweeds *D. dichotoma, P. pavonica, V. pachynema* against coastal malarial vector, *An. sundaicus.*

Materials and Methods

Collection of Eggs and Maintenance of Larvae

The eggs and larvae of coastal malarial vector, *Anopheles sundaicus* were collected from Mandapam, Ramanathapuram District, Tamilnadu, India, using an "O"-type brush. These eggs and larvae were brought to the laboratory and transferred to 18×13×4-cm enamel trays containing 500 mL of water for hatching. The mosquito larvae were fed with pedigree dog biscuits and yeast at 3:1 ratio. The feeding was continued until the larvae transformed into the pupal stage.

Maintenance of Pupae and Adults

The pupae were collected from the culture trays and transferred to plastic containers (12×12 cm) containing 500 mL of water with the help of a dipper. The plastic jars were kept in a 90×90×90-cm mosquito cage for adult emergence. Mosquito larvae were maintained at 27+2°C, 75–85 per cent relative humidity, under a photoperiod of 14:10 L/D. A 10 per cent sugar solution was provided for a period of 3 days before blood feeding.

Blood Feeding of Adult Mosquito Vectors

The adult female mosquitoes were allowed to feed on the blood of a rabbit (a rabbit per day, exposed on the dorsal side) for 2 days, to ensure adequate blood feeding for 5 days. After blood feeding, enamel trays with water from the culture trays were placed in the cage as oviposition substrates.

Collection of Plant and Preparation of Extract

The three seaweeds (*D. dichotoma, P. pavonica, V. pachynema*) were collected from in and around southeast coast of India, Ramanathapuram, Tamil Nadu. The three seaweeds were washed with tap water and shade-dried at room temperature (27±2°C). Dried material was ground into powder using a electrical blender. From the powder, 500 g of the seaweed materials was extracted with 1.5 L of organic solvents of ethanol for using a Soxhlet apparatus boiling point ranging 60–80°C for 8 h (Vogel, 1978). The extracts were filtered through a Buchner funnel with Whatman number 1 filter paper. The crude plant extracts were evaporated to dryness in rotary vacuum evaporator. One gram of the plant residue was dissolved in 100 mL of acetone (stock solution) and considered as 1 per cent stock solution. From this stock solution, different concentrations were prepared ranging from 50, 150, 250, 350 to 450 ppm, respectively.

Larval/Pupal Toxicity Test

Laboratory colonies of mosquito larvae/pupae were used for the larvicidal/pupicidal activity. Twenty-five numbers of I- to IV-instar larvae and pupae were introduced into 500-mL glass beaker containing 249 mL of dechlorinated water and 1 mL of desired concentrations of different seaweeds extract was added. Larval food was given for the test larvae. At each tested concentration, two to five trials were made and each trial consisted of five replicates. The control was set up by mixing 1 mL of acetone with 249 mL of dechlorinated water. The larvae and pupae were exposed to dechlorinated water without acetone served as control. The control mortalities were corrected by using (Abbott's, 1925) formula.

$$\text{Corrected mortality} = \frac{\text{Observed mortality in treatment} - \text{Observed mortality in control}}{100 - \text{Control mortality}} \times 100$$

$$\text{Percentage mortality} = \frac{\text{Number of dead larvae/pupae}}{\text{Number of larvae/pupae introduced}} \times 100$$

The LC_{50} and LC_{90} were calculated from toxicity data by using probit analysis (Finney, 1971).

Repellent Bioassay

The stock solutions of the extracts were diluted with acetone, polysorbate 80, and distilled water to obtain test solutions of 50, 100, 150, 200 and 250 ppm. For repellent experiment, 50 laboratory-reared bloods starved adult female mosquitoes that were between 3 and 10 days old were placed into separate laboratory cages (45×45×40 cm). Before each test, the forearm and hand of a human subject were washed with unscented neutral soap, thoroughly rinsed, and allowed to dry 10 min before extracts application. The different plant extracts being tested were applied from the elbow to the fingertips. The arm was left undisturbed. An arm treated with acetone and polysorbate 80 served as control. The control and treated arms were introduced simultaneously into the cage. The numbers of bites were counted over 5 min, every 30 min, from 1800 to 0600 hours. Protection time was recorded as the time elapsed between repellent application and the observation period immediately preceding that in which a confirmed bite was obtained. If no bites were confirmed at 180 min, tests were discontinued, and protection time was recorded as 180 min. An attempt of the mosquito to insert its stylets was considered a bite. No mosquito attempted to bite the control arm during the observation period; that trial was discarded, and the test was repeated with a new batch of mosquitoes to ensure that lack of bites was due to repellence and not to mosquitoes not being predisposed to get a blood meal at the time. The experiments were conducted five times in separate cages, and in each replicate, different volunteers were used to nullify any effect of skin differences on repellency. It was observed that there was no skin irritation from the plant extract. The percentage protection was calculated using the following formula (Fradin and Day, 2002; Venkatachalam and Jebanesan, 2001).

$$\text{Protection} = \frac{[\{\text{No. of bitres received by control arm}\} - \{\text{No. of bites received by treated arm}\}]}{\{\text{No. of bites received by control arm}\}} \times 100$$

Adulticidal Bioassay

Sugar-fed adult female mosquitoes (5–6 days old) were used. The three seaweeds extract (*DDME, PPME* and *VPME*) were diluted with acetone to make different concentrations. The diluted plant extracts were impregnated on filter papers (140×120 mm). A blank paper consisting of only ethanol was used as control. The papers were left to dry at room temperature to evaporate off the ethanol overnight. Impregnated papers were prepared fresh prior to testing. The bioassay was conducted in an experimental kit consisting of two cylindrical plastic tubes both measuring 125×44 mm following the method in (WHO, 1981). One tube served to expose the mosquitoes to the plant extract, and another tube was used to hold the mosquitoes before and after the exposure periods. The impregnated papers were rolled and placed in the exposure tube. Each tube was closed at one end with a 16 mesh size wire screen. Sucrose-fed and blood starved mosquitoes (20) were released into the tube, and the mortality effects of the extracts were observed every 10 min for 3 h exposure period. At the end of 1, 2, and 3 h exposure periods, the mosquitoes were placed in the holding tube. Cottonpads soaked in 10 per cent sugar solution with vitamin B complex was placed in the tube during the holding period of 24 h. Mortality of the mosquitoes was recorded after 24 h. The above procedure was carried out in triplicate for plant extract of each concentration.

Statistical Analysis

The average larval mortality data were subjected to probit analysis for calculating LC_{50} and LC_{90}, and other statistics at 95 per cent confidence limits of upper confidence limit (UCL) and lower confidence limit (LCL) and chi-square values were calculated using the SPSS 16.0 version (Statistical software package) to find the regression equation values. Results with $P<0.05$ were considered to be statistically significant.

Results

The bio-efficacy of different seaweeds crude extracts was evaluated for the potential larvicidal and pupicidal activity of (*D. dichotoma, P. pavonica, V. pachynema*) marine plants used for this purpose. The effects of methanol extracts of the seaweeds of *D. dichotoma, P. pavonica, V. pachynema* were tested at 50–450 ppm and against the first to fourth instar larvae and pupae of malarial vector, *An. sundaicus*. Table 17.1 (Figure 17.1) provides the larvicidal and pupicidal activity of *D. dichotoma* against the coastal malarial vector, *An. sundaicus*. 31.6 per cent mortality was observed in first instar larvae at 50 ppm, whereas it has been increased to 97.4 per cent at 450 ppm concentration of *D. dichotoma* methonalic extract (*DDME*) treatment. Similar trend has been noted for all the instars and pupae of *An. sundaicus* at different concentration of *DDME* treatment. The LC_{50} and LC_{90} values were represented as follows; LC_{50} value of first instar was 170.34 ppm, second instar was 188.24 ppm, third instar was 213.65

Table 17.1: Larvicidal and Pupicidal Properties of Seaweed *D. dichotoma* Methonalic Extract against the Coastal Malarial Vector, *An. sundaicus*

Larval and Pupal Stages	*Per cent Larval and Pupal Mortality*					*Regression Equation*	*LC_{50} and (LC_{90})*	*95 per cent Confidence Limit*		*χ^2 (df-4)*
	Concentration of D. dichotoma (DDME) (ppm)									
	50	*150*	*250*	*350*	*450*			*LC_{50} LFL (UFL)*	*LC_{90} LFL (UFL)*	
I instar	31.6±1.07[a]	44.8±1.77[ab]	59.6±2.10[a]	78.6±1.40[a]	97.4±2.10[a]	Y=-0.872+ 0.005=X	170.34 (420.64)	80.6 (231.33)	338.06 (619.34)	7.31*
II instar	29.8±2.78[ab]	41.7±1.18[b]	56.4±1.17[b]	74.8±1.12[b]	93.2±2.08[ab]	Y=-0.882+ 0.005=X	188.24 (461.79)	159.19 (214.01)	417.26 (524.57)	4.16*
III instar	27.2±1.30[c]	38.6±1.81[c]	51.7±3.83[c]	70.2±2.68[c]	87.8±1.08[b]	Y=-0.915+ 0.004=X	213.65 (512.99)	183.94 (241.11)	460.11 (589.83)	2.36*
IV instar	25.0±2.08[cd]	36.8±2.16[d]	48.2±1.52[d]	66.7±2.34[d]	80.2±1.86[c]	Y=-0.864+ 0.004=X	235.4 (584.50)	202.41 (267.17)	515.14 (691.68)	1.14*
Pupa	22.8±2.14[d]	35.0±2.83[de]	44.4±1.27[e]	59.4±2.70[e]	70.8±3.14[d]	Y=-0.899+ 0.003=X	280.38 (679.91)	244.7 (319.78)	586.84 (834.00)	0.20*

Control-Nil mortality, *LFL:* Lower Fiducidal Limit; *UFL:* Upper Fiducidal Limit; χ^2: *Chi*-square value; *df:* Degrees of freedom.

Within a column means followed by the same letter(s) are not significantly different at 5 per cent level by DMRT. *Significant at $P < 0.05$ level.

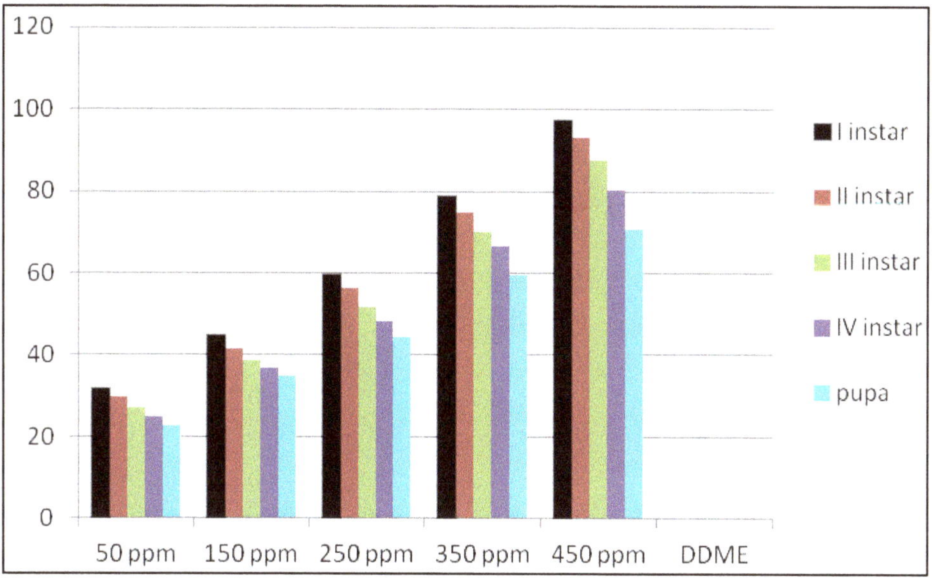

Figure 17.1: Larvicidal and Pupicidal Properties of Seaweed *D. dichotoma* Methonalic Extract against the Coastal Malarial Vector, *An. sundaicus.*

ppm, fourth instar was 235.40 ppm, and pupa was 280.38 ppm, respectively. The LC_{90} value of first instar was 420.64 ppm, second instar was 461.79 ppm, third instar was 512.99 ppm, fourth instar was 584.50 ppm, and pupa was 679.91 ppm, respectively.

Table 17.2 (Figure 17.2) shows the results of larval and pupal mortality of *An. sundaicus* (first to fourth instar larvae and pupae) after the treatment of *P. pavonica,* methonalic extract (*PPME*) at different concentrations (50 to 450 ppm). Mortality (27.8 per cent) was noted at first instar larvae by the treatment of *PPME* at 50 ppm, whereas it has been increased to 88.4 per cent at 450 ppm of *PPME* treatment, and 19.3 per cent mortality was noted at pupae by the treatment of *PPME* at 50 ppm, and it has been increased to 64.5 per cent at 450 ppm, respectively. Similar trend has been noted for all the larval instars and pupae of *An. sundaicus* at different concentrations of *PPME* treatment. The LC_{50} and LC_{90} values were represented as follows: LC_{50} value of first instar was 208.03 ppm, second instar was 231.63 ppm, third instar was 259.90 ppm, fourth instar was 293.35 ppm, and pupa was 327.26ppm, respectively. The LC_{90} value of first instar was 503.22 ppm, second instar was 540.61 ppm, third instar was 601.85 ppm, fourth instar was 668.44 ppm, and pupa was 735.76 ppm, respectively.

Table 17.3 (Figure 17.3) provides the considerable larval and pupal mortality after the treatment of *V. Pachynema* methonalic extract (*VPME*) against all the larval instars and pupae of coastal malarial vector, *An. sundaicus* at different concentrations of (50-450 ppm) *VPME*. The first instar larval mortality and pupal mortality was 29.4 per cent -92.4 per cent and 20.8 per cent -67.7 per cent, respectively. The LC_{50} and LC_{90} values were represented as follows: The LC_{50}values of first instar were192.53 ppm,

Table 17.2: Larvicidal and Pupicidal Properties of Seaweed *P. pavonica,* Methonalic Extract against the Coastal Malarial Vector, *An. sundaicus.*

Larval and Pupal Stages	*Per cent Larval and Pupal Mortality*					*Regression Equation*	*LC_{50} and (LC_{90})*	*95 per cent Confidence Limit*		*χ^2 (df-4)*
	Concentration of P. pavonica (PPME) (ppm)									
	50	*150*	*250*	*350*	*450*			*LC_{50} LFL (UFL)*	*LC_{90} LFL (UFL)*	
I instar	27.8±1.23a	38.8±1.57a	53.2±1.17a	71.6±1.10a	88.4±2.70a	Y=-0.903+ 0.004=X	208.03 (503.22)	178.31 (235.19)	451.97 (577.28)	2.14*
II instar	25.0±2.78ab	35.9±1.28b	49.6±1.21ab	67.8±1.12ab	84.2±1.18ab	Y=-0.961+ 0.004=X	231.63 (540.61)	202.11 (259.98)	483.35 (624.85)	1.29*
III instar	23.2±1.30bc	33.5±1.81c	45.8±3.83cd	63.2±2.68c	77.6±1.71cd	Y=-0.974+ 0.004=X	259.9 (601.85)	228.66 (292.07)	531.41 (710.05)	0.57*
IV instar	21.3±2.18d	30.2±2.16d	42.7±1.52d	59.2±2.14d	70.2±1.16d	Y=-1.002+ 0.003=X	293.35 (668.44)	259.71 (331.66)	581.94 (807.59)	0.28*
Pupa	19.3±2.14de	28.6±2.13e	40.3±1.17de	53.7±2.30e	64.5±2.14e	Y=-1.027+ 0.003=X	327.26 (735.76)	289.83 (374.83)	630.7 (913.24)	0.46*

Control-Nil mortality, *LFL:* Lower Fiducidal Limit; *UFL:* Upper Fiducidal Limit; χ^2: *Chi*-square value; *df:* Degrees of freedom.

Within a column means followed by the same letter(s) are not significantly different at 5 per cent level by DMRT. *Significant at $P < 0.05$ level.

Table 17.3: Larvicidal and Pupicidal Properties of Seaweed *V. Pachynema* Methonalic Extract against the Voastal Malarial Vector, *An. sundaicus.*

Larval and Pupal Stages	*Per cent Larval and Pupal Mortality*					*Regression Equation*	*LC_{50} and (LC_{90})*	*95 per cent Confidence Limit*		*χ^2 (df-4)*
	Concentration of V. pachynema (VPME) (ppm)									
	50	*150*	*250*	*350*	*450*			*LC_{50} LFL (UFL)*	*LC_{90} LFL (UFL)*	
I instar	29.4±1.77a	40.8±1.37--a	55.3±2.40a	74.6±1.70a	92.4±1.30a	Y=-0.895+ 0.005=X	192.53 (468.27)	163.61 (218.38)	422.92 (532.33)	3.88*
II instar	27.2±2.78ab	39.7±1.18ab	51.4±1.17b	70.8±1.12b	89.2±1.18b	Y=-0.921+ 0.004=X	210.22 (502.81)	180.93 (237.16)	451.93 (576.17)	3.19*
III instar	24.7±1.30cd	36.8±1.81bc	48.3±3.83c	67.4±2.68c	83.2±1.28c	Y=-0.951+ 0.004=X	234.43 (550.51)	204.41 (263.42)	490.96 (638.92)	1.29*
IV instar	23.0±2.08d	32.7±2.16d	44.3±1.52d	62.2±2.14d	75.2±1.16de	Y=-0.972+ 0.004=X	269.55 (624.79)	237.42 (303.58)	548.73 (743.70)	0.57*
Pupa	20.8±2.14e	30.3±2.13e	42.4±1.17e	56.2±2.30e	67.7±2.14e	Y=-0.987+ 0.003=X	306.28 (704.04)	270.57 (348.90)	607.04 (864.98)	0.38*

Control-Nil mortality, *LFL:* Lower Fiducidal Limit; *UFL:* Upper Fiducidal Limit; χ^2: *Chi*-square value; *df:* Degrees of freedom.

Within a column means followed by the same letter(s) are not significantly different at 5 per cent level by DMRT. *Significant at $P < 0.05$ level.

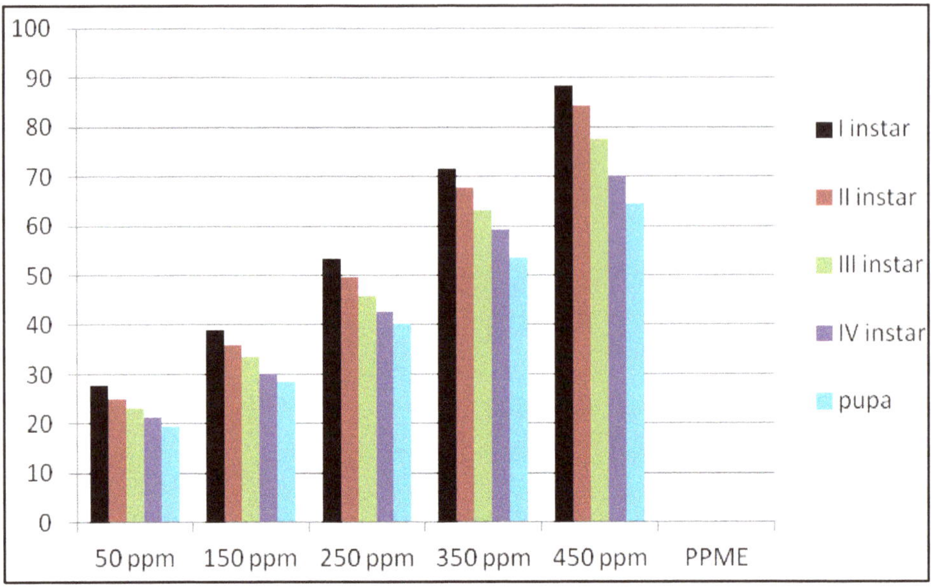

Figure 17.2: Larvicidal and Pupicidal Properties of Seaweed *P. pavonica*, Methonalic Extract against the Coastal Malarial Vector, *An. sundaicus.*

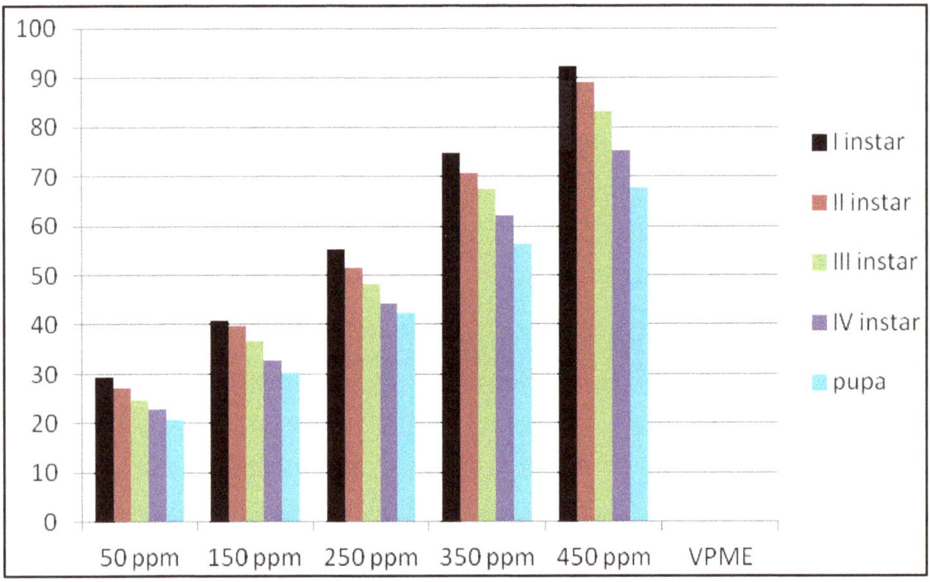

Figure 17.3: Larvicidal and Pupicidal Properties of Seaweed *V. Pachynema* Methonalic Extract against the Coastal Malarial Vector, *An. sundaicus.*

second instar was 210.22 ppm, third instar was 234.43 ppm, fourth instar was 269.55 ppm,and pupa was 306.28 ppm, respectively. The LC_{90}values of first instar were468.27 ppm, second instar was502.81 ppm, third instar was 550.51 ppm, fourth instar was 624.79, and pupa was 704.04 ppm, respectively. The χ^2 values are significant at

$P<0.05$ level. The 95 per cent confidence limits LC_{50}, LC_{90} (LFL–UFL) values were also calculated. Larval and pupal mortality was observed after24 h exposure. No mortality was observed in the control group. Regression equation and chi-square values were significant at $P<0.05$ level.

In the present observation, the results from the skin repellent activity of methanol extract of different seaweeds (*DDME, PPME* and *VPME*) against blood starved adult female of *An. sundaicus* are given in Table 17.4. The methanol extracts of three seaweeds show significant repellency against *An. sundaicus*. In this observation, this seaweed crude extract gave protection against mosquito bites without any allergic reaction to the test person, and also, the repellent activity is dependent on the strength of the seaweeds extract. The highest repellency of 180 min was observed in methanol extract of *D. dichotoma* (*DDME*) against *An. sundaicus* followed by *PPME* and *VPME* respectively. The results clearly show that repellent activity was dose dependent.

The results of the adulticidal activity of methanol extracts of seaweeds *D. dichotoma, P. pavonica, V. pachynema* (*DDME, PPME* and *VPME*) against the adult of coastal malarial vector, *An. sundaucus* are presentedin Tables 17.5 (Figure 17.4). Among the coastal malarial vector tested, the highest adulticidal activity was observed in the highest mortality followed by *V. pachynema, D. dichotoma* and *P. pavonicaat* higher concentrations, the adult showed restless movement for some time with abnormal wagging and died. The rates of mortality were directly proportional to concentration. The LC_{50} and LC_{90} values of methanol extracts against adulticidal activity of (*D. dichotoma, P. pavonica, V. pachynema*) *An. sundaicus* were the following: *D. dichotoma* (*DDME*) LC_{50} values were 147.18 ppm and LC90 values were 346.42 ppm; *P. pavonica* (*PPME*) LC_{50} values were 161.94 ppm and LC_{90} values were 372.88

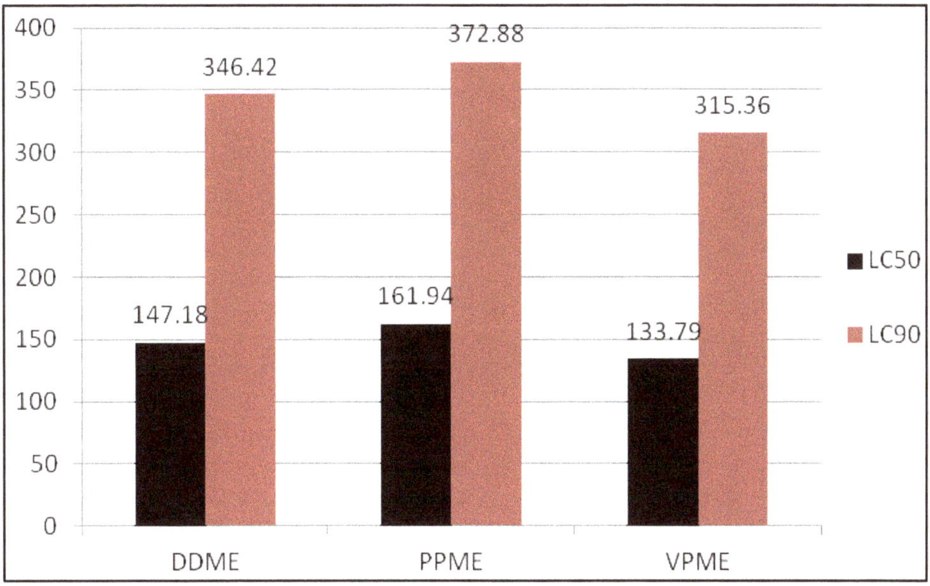

Figure 17.4: Adulticidal Activity of Methanol Extract of different Seaweeds against Coastal Malarial Vector, *An.sundaicus.*

Table 17.4: Repellent Activity of Three different Seaweed Methanol Extracts against Coastal Malarial Vector, *Anopheles sundaicus*

Mosquito Species	*Concentration (ppm)*	*Per cent of Repellency ± SD*					
		Time Post Application of Repellent (mins)					
		30	*60*	*90*	*120*	*150*	*180*
D. dichotoma (*DDME*)	50	98.2±2.16	91.2±2.28	86.4±2.70	79.8±3.34	73.4±3.20	70.2±3.70
	100	100.0±0.0	99.4±0.89	92.6±3.57	88.2±4.76	84.6±3.71	79.6±1.78
	150	100.0±0.0	100.0±0.0	100.0±0.0	97.8±1.30	93.4±2.70	87.8±2.28
	200	100.0±0.0	100.0±0.0	100.0±0.0	100.0±0.0	100.0±0.0	95.4±2.88
	250	100.0±0.0	100.0±0.0	100.0±0.0	100.0±0.0	100.0±0.0	100.0±0.0
P. pavonica (*PPME*)	50	89.2±1.78	84.2±1.30	80.4±1.14	76.4±1.51	72.8±1.30	69.2±0.83
	100	97.4±1.14	99.4±0.54	88.8±1.48	83.2±0.83	80.4±0.89	77.6±1.67
	150	100.0±0.0	100.0±0.0	97.6±1.51	93.4±0.89	89.4±1.51	85.8±1.64
	200	100.0±0.0	100.0±0.0	100.0±0.0	100.0±0.0	96.0±1.41	92.6±1.34
V. pachynema (*VPME*)	250	100.0±0.0	100.0±0.0	100.0±0.0	100.0±0.0	100.0±0.0	99.0±1.00
	50	82.6±1.51	79.8±1.09	74.6±0.54	71.8±1.30	68.8±1.14	63.2±1.34
	100	89.2±1.92	85.2±1.46	81.2±1.92	84.2±0.44	76.2±2.28	72.6±1.14
	150	97.6±1.14	94.6±1.01	91.6±1.14	90.0±2.34	84.4±1.78	80.8±1.30
	200	100.0±0.0	100.0±0.0	99.2±0.83	97.8±0.83	91.0±1.00	86.0±2.70
	250	100.0±0.0	100.0±0.0	100.0±0.0	100.0±0.0	96.4±0.54	93.6±1.92

Table 17.5: Adulticidal Activities of Methanol Extract of different Seaweeds against Coastal Malarial Vector, *An.sundaicus*

Seaweeds Species	*Concentration (ppm)*	*Mortality (per cent) (mean±SD)*	*LC_{50}, ppm (LFL-UFL)*	*LC_{90}, ppm (LFL-UFL)*	χ^2
D. dichotoma (*DDME*)	60	29.8±2.16e			
	120	43.2±2.58de	147.18	346.42	
	180	55.8±2.94cd	(126.28-165.38)	(310.76-400.22)	0.38
	240	72.8±2.28b			
	300	84.6±2.30a			
P. pavonica (*PPME*)	60	26.4±2.07ef			
	120	41.6±2.30de	161.94	372.88	
	180	53.2±2.77c	(141.25-180.96)	(332.08-436.05)	0.27
	240	67.2±3.27bc			
	300	81.6±1.34a			
V. pachynema (*VPME*)	60	31.6±2.65de			
	120	46.6±1.81d	133.79	315.36	
	180	60.4±3.57bc	(113.39-151.00)	(285.49-358.86)	1.37
	240	74.8±1.72b			
	300	90.6±2.70a			

ppm; and *V. Pachynema* (*VPME*) LC_{50}values were 133.79 ppm and LC_{90} values were 315.36 ppm, respectively.

Discussion

Mosquitoes are the principal vector of many vector-borne diseases affecting human beings and animals, in addition to nuisance. *Anopheles* species is the primary vector of malaria in India and other West Asian countries, and improved methods of control are urgently needed (Burfield and Reekie, 2005; Mittal *et al.*, 2005). Malaria infects more than 500 million humans each year, killing approximately 1.2–2.7 million per year. About 90 per cent of all malaria cases occur in Africa, as does approximately 90 per cent of the world's malaria-related deaths (Breman *et al.*, 2004). However, a high level of insecticide resistance has developed through chemical control of the vector and pests threatening the control strategies. To overcome these problems, it is necessary to search for alternative methods of vector control. The failure of chemical insecticides to control the insect and growing public concern for safe food and a healthy environment have catalyzed the search for more environmentally benign control methods for the management of the vectors. Bio pesticides provide an alternative to synthetic pesticides because of their generally low environmental pollution, low toxicity to humans, and other advantages (Liu *et al.*, 2000). Natural products of plant origin are widely used as control of agricultural pests and vectors. Essential oils from plants and algal extracts may be the rich alternative sources of mosquito larval control agents, as they constitute a rich source of bioactive compounds that are biodegradable into nontoxic products and potentially suitable for use in control of mosquito larvae. Many plant derived natural compounds were tested for mosquito control (Consoli *et al.*, 1994).

Marine macro algae or seaweeds are used in traditional medicine by coastal peoples in many areas of the world, particularly in Asia and the Caribbean, where parasitic diseases commonly occur. In particular, Japanese and Chinese cultures have used seaweed decoctions since ancient times to treat parasitic infections, as well as other medical problems (Moo-Puc *et al.*, 2008). Although marine macro algae are among the richest sources of chemically diverse natural products (Mayer *et al.*, 2009).

Seaweed contains components with larvicidal properties against mosquitoes. The larvicidal efficacy of the acetone extract of the thalli of *Caulerpascal pelliformis* var. *denticulata* was determined against late 2nd to early 3rd instars of *Culex pipiens* at concentrations ranging from 100 to 2,000 parts per million (ppm). At 1,200 ppm, the extract caused >70 per cent larval mortality at 24, 48, and 72 h exposure (Cetin *et al.*, 2010). *Oscillatoria* challenged with zinc having the most effective larvicidal property against *Culex quinquefasciatus* (say) has already been studied by Indira Vincent *et al.* (2011).

Earliar, (Manilal *et al.*, 2011) *Padina minor* showed significantly greater larvicidal activity compared to *D. linearis* at LC_{50}. Greater larvicidal activity of *Padina* than *Dictyota* was also observed in *P. tetrastromatica* and *D. dichotoma* from southwest Coast of India. The LD50 of *P. tetrastromatica* on the II instar larvae of *Cx. ouinquefasciatus* is 0.96 mg/ml (96 µg/ml) while *D. dichotoma* is 0.25 mg/ml (226 µg/

ml). Both *P. minor* and *D. linearis* showed greater potency compared to *Ulva fasciata* and *Grateloupia lithophila* collected from Kovalam, near Chennai (Poonguzhali and Nizha, 2012). In the present results of our research findings *D. dichotoma* was showed higher larvicidal and pupicidal activity against coastal malarial vector. *An. sundaicus* larvae having highest mortality (97.4 per cent) was observed in first instar larvae at 450 ppm concentration of *D. dichotoma* methonalic extract (*DDME*) treatment (Table 17.1). The LC_{50} and LC_{90} values were represented as follows; LC_{50} value of 1st- 4th instar and pupae were 170.34 ppm, 188.24 ppm, 213.65 ppm, 235.40 ppm, and pupa was 280.38 ppm, respectively. The LC_{90} value of 1st- 4th instar and pupae were 420.64 ppm, 461.79 ppm, 512.99 ppm, 584.50 ppm, and pupa was 679.91 ppm, respectively. Approximately 150,000 natural products are currently known, from which 22,000 are from marine organisms. In the last three decades, the number of metabolites isolated from marine species was significantly increased and many compounds have been reported among terpenes and polyphenols (Blunt *et al.*, 2011).We believe that marine seaweed products may serve as suitable synthetic alternatives to the future of insecticidal agents, as they are relatively safe, biodegradable and easily available worldwide (Manilal *et al.*, 2011).

In the previous report envisaged that marine plant extracts contain promising larvicidal activity (Thangam *et al.*, 1993). Polyhalogenatedmonoterpenes, aplysiaterpenoid A and telfairine, isolated from *Plocamium* sp. Exhibited insecticidal activity against mosquito larvae (Watanabe *et al.*, 1989). (Thangam *et al.*, 1993; Rao *et al.*, 1995) envisaged that the Indian marine plant extracts possessed potential larvicidal activity. Selvin and Lipton (2004) and Alarif *et al.* (2010) their studies also showed that secondary metabolites isolated from algae showed significant larvicidal activity against mosquitoes. In our results also confirmed that the seaweeds also very effective against *An. sundaicus*. We recorded, *PPME* against *An. sundaicus* (Table 17.2) had values of LC_{50} value of first instar was 208.03 ppm, second instar was 231.63 ppm, third instar was 259.90 ppm, fourth instar was 293.35 ppm, and pupa was 327.26 ppm, respectively. The LC_{90} value of first instar was 503.22 ppm, second instar was 540.61 ppm, third instar was 601.85 ppm, fourth instar was 668.44 ppm, and pupa was 735.76 ppm, respectively. Karmegam *et al.* (1997) was noted a very interesting observation their study was the prolonged larval stage for those which were not killed by the extracts. In the control groups, the larvae became pupae in 2–3 days, however, larvae of the same stage in *P. minor* and *D. linearis* extracts became pupae in 5–7 days. Prolonged larval stage was also observed in *Cx. quinquefasciatus* treated with 31.25 ppm of *Argenome mexicana* extract. This was attributed to hormonal imbalance which delayed larval metamorphosis.

Many authors have reported the marine plant extracts have a good larvicidal activity against mosquito vector control programmes (Thangam and Kathiresan, 1991; Thangam *et al.*, 1993). Manilal *et al.* (2009) have reported the larvicidal activity of seaweed extracts at the end of 24 h suggested that instars of larvae were highly sensitive and most effective against instars of Culex mosquito larvae. The LD_{50} value of *D. dichotoma* was observed 226µg/ml. Highest toxicity was observed in *D dichotoma,* was eco-friendly ichthyocides tests and considered as extremely toxic since 100 per cent mortality was observed even at lowest concentration 50 µg/ml. *D. dichotoma* has been early reported to have antifeedant activity (Valerie and Hay, 1986). Poonguzhali and Nisha (2012) have reported the larvicidal activity of three different extracts (methanol, acetone, and benzene) of *U. fasciata* and *G. lithophila* against the larvae of *Culex* was performed under laboratory evaluation. in terms of lethal concentration for 50 per cent and 90 per cent mortality (LC_{50} AND LC_{90}) values were represented as follows: for *U. fasciata* LC_{50} value of the methanol extract was 515.88 (mg/L) while for *G. Lithophila* LC_{50} value of the methanol extract was 431.90 (mg/L), LC_{90} value of the methanol extract of *U. fasciata* was 867.10(mg/L), while for *G. lithophila* LC_{90} value of the methanol extract was 860.60 (mg/L).

Osman *et al.* (2010) demonstrated that, the post coital contraceptive activity from a crude extract in marine red algae *Gelidiella acerosa* is due to the presence of various phytochemical components such as alkaloids, flavonoids, phenols, amino acid, steroids, tannins and carbohydrates. Chapagain *et al.* (2008) reported that, saponins can also serve as natural larvicidal compounds. Cetin *et al.* (2010) reported that, maximum larvicidal activity was observed in the case of *D. dichotoma* which caused 50 percent mortality of the larvae of *Anopheles stephensi* and *Culex quinqefasciatus* at the concentrations of 22 and 34 ppm and caused total mortality at 80 and 100 ppm. Ravikumar *et al.* (2011) have reported that, the seaweed *Chaetomorpha antennina* showed excellent *in vitro* and *in vivo* antiplasmodial activity. In our present result findgs (Table 17.3) (Figure 17.3) provides the considerable larval and pupal mortality after the treatment of *V. pachynema* methonalic extract (*VPME*) against all the larval instars and pupae of coastal malarial vector, *An. sundaicus*. at different concentrations of (50-450 ppm) *VPME*. The first instar larval mortality and pupal mortality was 29.4 per cent -92.4 per cent and 20.8 per cent -67.7 per cent, respectively. The LC_{50} and LC_{90} values were represented as follows: The LC_{50} values of first instar were 192.53 ppm, second instar was 210.22 ppm, third instar was 234.43 ppm, fourth instar was 269.55 ppm, and pupa was 306.28 ppm, respectively. The LC_{90} values of first instar were 468.27 ppm, second instar was 502.81 ppm, third instar was 550.51 ppm, fourth instar was 624.79, and pupa was 704.04 ppm, respectively. A number of other components of algal extracts have been shown to act as larvicidal agents against mosquito vectors. Thus, acetone extracts of the marine seaweeds *Caulerpa scapelliformis, D. dichotoma, Enteromorpha clathrata, E. intestinalis* and *Ulva lactuca* were active against fourth instar larvae of *A. aegypti* with LC_{50} values of 53.70, 61.65, 85.11, 67.70 and 91.20 ppm, respectively (Thangam and Kathiresan, 1991a). Extracts of *C. scapelli formis* exerted a weak larvicidal effect against *Culex pippens* with a reported $LC_{50}$and LC_{90} values were showed 338.9 ppm and 1,891.31 ppm (Cetin *et al.*, 2010). Methanolic extracts of *Westiellopsis* sp. (blue–green alga/cyanobacterium) are also reportedly active against fourth instar larvae of *A. aegypti* with an LC_{50} of 55.84 ppm (Rao *et al.*, 1999).

There are few published studies using essential oils or plant derivatives for repellency against mosquitoes. Many different techniques have been used to measure mosquito repellency. Besides different techniques, there are differences in biting pressure in a mosquito population which is another factor affecting repellency testing (Barnard *et al.*, 1998). *R. officinalis* was effective in terms of repellence time against *A. aegypti* in concentrations as low as 12.5 per cent. These results are consistent with reports of insect repellent properties of rosemary, including *C. pipiens pallens* (Choi *et al.*, 2002). The mosquito sensitivity to repellents varies among *Aedes*, *Anopheles*, and *Culex* mosquitoes (Amer and Mehlhorn, 2006). (Venkatachalam and Jebanesan, 2001a, b) have also reported that the repellent activity of methanol extract of *Ferronia elephantum* leaves against *A. aegypti* activity at 1.0 and 2.5 mg cm"2 concentrations gave 100 per cent protection up to 2.14±0.16 h and 4.00±0.24 h, respectively, and the total percentage protection was 45.8 per cent at 1.0 mg cm"2 and 59.0 per cent at 2.5 mg cm"2 for 10 h. The essential oil of *Tagetes minuta*, providing a repellency of 90 per cent protection for 2 h against *A. stephensi, C. quinquefasciatus* and *A. aegypti*, was reported by (Tyagi *et al.*, 1994). It was found that a CO_2 extract of the seeds of the Mediterranean plant *Vitex agnus castus* can be used as a spray to keep away especially *Ixodes ricinus* and *Rhipicephalus sanguineus* ticks from animals and humans for at least 6 h. In addition, mosquitoes, biting flies and fleas are also repelled for about 6 h (Mehlhorn *et al.*, 2005). In this observation, this seaweed crude extract gave protection against mosquito bites without any allergic reaction to the test person, and also, the repellent activity is dependent on the strength of the seaweeds extract. The highest repellency of 180 min was observed in methanol extract of *D. dichotoma* methonalic extract (*DDME*) against *An. sundaicus* followed by *PPME* and *VPME* respectively. The results clearly show that repellent activity was dose dependent. (Vimaladevi *et al.*, 2012) that the protection period of insoluble bound phenolic acid against *A. aegypti* was 120 min and that of soluble conjugated phenolic acid 90 min. The free phenolic acid compounds provided only a 30 min protection. In contrast, treatment of an arm with ethanol provided less than 30 min protection against the bite of *A. aegypti* mosquitoes. The percentage of repellency for *A. aegypti* ranged between 100 per cent for insoluble bound phenolic acid and 91.2 ± 1.17 per cent for soluble conjugated phenolic acid after 2 h. The free phenolic acid exhibited only 66.6 ± 0.71 per cent repellency after 2 h. A diluted ethanolic solution of phenolic acids applied to skin of the forearm was shown to be very eective in protection against *A. aegypti*. The repellent properties of natural products to mosquitoes and other pest insects were well known before the advent of synthetic repellents (Curtis *et al.*, 1989).

Benzene and methanol extracts of *Artemisia vulgaris* have been reported to have repellent activity against *A. aegypti* 17 (Yit *et al.*, 1985). Quelling, the insect repellent produced in China, derived from the extract of the lemon grass and eucalyptus plants were evaluated against mosquitoes. Essential oil obtained from *Vitex negundo* was used as repellent against *A. aegypti* (Hebbalkar *et al.*, 1992). Neem products are good mosquito repellents showing 90 per cent to 100 per cent protection against malaria vectors and about 70 per cent against *C. quinquefasciatus* (Ansari and Razdan, 1994).

The effects of the tested extract, adult emergence and adulticidal activity of the mosquitoes are remarkably greater than those reported for other plant extracts in the literature. For example at the highest concentration, 50 per cent inhibition of the emergence of the adult mosquitoes was observed by the use of the ethyl acetate fractions of *Calophyllum inophyllum* seed and leaf, *Solanum suratense* and *Samadera indica* leaf extracts and the petrol ether fraction of *Rhinocanthus nasutus* leaf extract on *C. quinquefasciatus, A. stephensi* and *A. aegypti* (Muthukrishnan and Puspalatha, 2001).

Nathan *et al.* (2005) considered pure limonoids of neem seed, testing for biological, larvicidal, pupicidal, adulticidal and antiovipositional activity, *A. stephensi* and the larval mortality was dose-dependent with the highest dose of 1 ppm azardirachtin evoking almost 100 per cent mortality, affecting pupicidal and adulticidal activity and significantly decreased fecundity and longevity of *A. stephensi*. The highest adulticidal effect was established from *Piper sarmentosum*, followed by *Piper ribesoides* and *Piper longum*, with LD_{50} values of 0.14, 0.15 and 0.26 µg/female, respectively (Choochote *et al.*, 2006). This result is also comparable to earlier reports of (Dua *et al.*, 2010) who observed that the adulticidal activity of the essential oil of *Lantana camara* was evaluated against different mosquitoes species on 0.208 mg/cm2 impregnated papers, and the KDT_{50} and KDT_{90} values of the essential oil were 20, 18, 15, 12 and 14 min, and 35, 28, 25, 18 and 23 min against *A. aegypti*, *C. quinquefasciatus*, *A. culicifacies*, *Anopheles fluvialitis* and *A. stephensi* with their percent mortality of 93.3 per cent, 95.2 per cent, 100 per cent, 100 per cent and 100 per cent, respectively. In the present adulticidal results, the LC_{50} and LC_{90} values of methanol extracts of (*D. dichotoma*, *P. pavonica*, *V. pachynema*) against *An. sundaicus* were the following: *D. dichotoma* (*DDME*) LC_{50} value was 147.18 ppm and LC_{90} value was 346.42 ppm; *P. pavonica* (*PPME*) LC_{50} value was 161.94 ppm and LC_{90} value was 372.88 ppm; and *V. pachynema* (*VPME*) LC_{50} value was 133.79 ppm and LC_{90} value was 315.36 ppm, respectively (Figure 17.4). Subramanian and Murugan (2103) have reported that seed extract of *Myristica fragrans* have high larvicidal, pupicidal, adulticidal and repllency activities against the malarial vector *An. stephensi*.

In conclusion, *D. dichotoma*, *P. pavonica* and *V. pachynema* oers a promising and potential bio control agent against *An. sundaicus*. Furthermore, safety considerations and the emergence of mosquito resistance to conventional insecticides make larvicide of natural origin preferable to synthetic compounds. Natural insecticides, especially those derived from plants that are more selective and easily degradable, are more promising in this aspect. The isolated bioactive phytochemical from the seaweed could be used in stagnant water bodies which are known to be the breeding grounds for mosquitoes. However, the exact mechanism and the compound responsible for the larvicidal activities are currently unclear. Therefore, further studies on the identification of the active compounds involved and their mode of action in field trials are needed to recommend *D. dichotoma*, *P. pavonica* and *V. pachynema* as an anti-mosquito product used to combat and protect from mosquitoes in a vector control program.

References

Abbott, W. S. (1925). A method of computing the effectiveness of insecticides. *Journal of Economic Entomology*, **18**: 267–269.

Alarif, W. M., Elnaga, Z. S. A., Ayyad, S. E. N., and Al-Lihaibi, S.S. (2010). Insecticidal metabolites from the green algae *Caulerpa racemosa*. *Clean Soil Air Water*, **38**: 548 – 557.

Aliya, R., and Shamaeel, M. (1999). Phytochemical evaluation of four coenocytic green seaweeds from the coast of Karachi. *Pakistan Journal of Marine Biology*, **5**: 65-76.

Amer, A., and Mehlhorn, H. (2006). Repellency effect of forty-one essential oils against *Aedes, Anopheles,* and *Culex* mosquitoes. *Parasitology Research,* **99(4)**: 478–490.

Ansari, M.A., and Razdan, R. K. (1994). Repellent action of *Cymbopogon martini* staf var. Sofia [sic] oil against mosquitoes. *Indian journal of malariology,* **31(3)**: 95–102.

Ascher, K. R. S., Schmutterer, H., Zebitz, C. P. W., and Naqvi, S. N. H. (1995). The Persian lilac or chinaberry tree: *Melia azedarach* L. In: Schmutterer H (ed) The *Neem* Tree: source of unique natural products for integrated pest management, medicine, industry and other purposes. Weinheim, VCH, pp 605–642.

Barnard, D. R., Posey, K. H., Smith, D., and Shreck, C. E. (1998). Mosquito density, biting cycle, and cage size effects on repellent test. *Medical and Veterinary Entomology,* **12**: 39-45.

Blunt, J.W., Copp, B. R., Hu, W. P., Munro, M. H. G., Northcote, P. T., and Prinsep, M. R. (2009). Marine natural products. *Natural Products and Reports,* **26**: 170–244.

Blunt, J.W., Copp, B. R., Munro, M. H. G., Northcote, P. T., and Prinsep, M. R. (2011). Marine natural products. *Natural Product Reports,* **28**: 196–268.

Breman, J.G., Martin, A. S., and Mills, A. (2004). Conquering the intolerable burden of malaria: what's new, what's needed: a summary. *American Journal of Tropical Medicine and Hygiene,* **71(2)**: 1–15.

Burfield T., and Reekie, S.L. (2005). Mosquitoes, malaria and essential oils. *International Journal of Aromatherapy,* **15**: 30–41.

Cetin, H., Gokoglu, M., and Emre, O. (2010). Larvicidal activity of the extract of seaweed, *Caulerpa scalpelliformis,* against *Culex pipiens. Journal of the American Mosquito Control Association,* **26(4)**: 433-435.

Chapagain, B. P., Saharan, V., and Wiesman, Z. (2008). Larvicidal activity of saponins from *Balanites aegyptiaca* callus against *Aedes aegypti* mosquito. *Bioresour Technol,* **99(5)**: 1165-1168.

Choi, W. S., Park, B. S., Ku, S. K., and Lee, S. E. (2002). Repellent activities of essential oils and monoterpenes against *Culex pipiens* pallens. *Journal of the American Mosquito Control Association,* **18**: 348–351.

Choochote, W., Chaithong, U., Kamsuk, K., Rattanachanpichai, E., Jitpakdi, A., Tippawangkosol, P., Chaiyasit, D., Champakaew, D., Tuetun, B., and Pitasawat, B. (2006). Adulticidal activity against *Stegomyia aegypti* (Diptera: culicidae) of three Piper spp. *Revista do Instituto de Medicina Tropical de São Paulo,* **48(1)**: 33–37.

Consoli, R. A., and Oliveira, R.L. (1994). Principais mosquitos de importancia sanitaria no Brasil, Fiocruz, Rio de Janeiro, pp. 225.

Curtis, C.F., Lines, J. D., Baolin, L., and Renz, A. (1989). Natural and synthetic repellents, pp.75–92. In: Curtis C.F. (ed.), Appropriate Technology in Vector Control, CRC, New York, 344 pp. ISBN-10: 0849347556.

Das, M. K., Wattal, S., Nanda, N., and Adak, T. (2004). Laboratory colonization of *Anopheles sundaicus*. *Current Science*, **86**: 1069–1070.

Dawes, C. J. (1981). Marine Botany. John Wiley and Sons, New York.

Dhiman, C. R., Pahwa, S., Dhillon, G. P. S., and Dash, P. A. (2010). Climate change and threat of vector-borne diseases in India: are we prepared? *Parasitology Research*, **106**: 763–773.

Dua, V. K., Pandey, A. C., and Dash, A. P. (2010). Adulticidal activity of essential oil of *Lantana camara* leaves against mosquitoes. *Indian Journal of Medical Research*, **131**: 434–439.

Duran R., Zubia, E., Ortega, M. J., and Salva, J. (1997). New diterpenoids from the alga *Dictyota dichotoma*. *Tetrahedron*, **53**: 8675-8688.

Dusfour, I., Blondeau, J., Vythilingam, I., Baimai, V., Trung, H., Sochanta, T., Bangs, M. J., and Manguin, S. (2007). Polymerase chain reaction identification of three members of the *Anopheles sundaicus* (Diptera: Culicidae) complex, malaria vectors in Southeast Asia. *Journal Medical Entomology*, **44**: 723-731.

Finney, D. J. (1971). Probit Analysis. Cambridge University, London: 68–78.

Fradin M. S., and Day, J. F. (2002). Comparative efficacy of insect repellents against mosquito bites. *New English Journal of Medical*, **13**: 347–418.

Garg, H. S., Sharma, T., Bhakuni, D. S., Pramanik, B. N., and Bose, A. K. (1992). An antiviral sphingosine derivative from green alga *Ulva fasiata*. *Tetrahedron Letters*, **33**: 1641–1644.

Gopalan, S.S., and Das, A. (2009). Household economic impact of an emerging disease in terms of catastrophic out-of-pocket health care expenditure and loss of productivity: investigation of an outbreak of chikungunya in Orissa, India. *Journal of Vector Borne Diseases*, **46**: 57–64.

Hebbalkar, D.S., Hebbalkar, G.D., Sharma, R.N., Joshi, V.S., and Bhat, V. S. (1992). Mosquito repellent activity of oils from *Vitex negundo* Linn. leaves. *Indian Journal of Medical Research*, **95**: 200–203.

Kalyanasundaram, M., and Das, P. K. (1985). Larvicidal and synergistic activity of plant extracts for mosquito control. *Indian Journal of Medical Research*, **82**: 19–23.

Karmegam, N., Sakthivadivel, M., Anuradha, V., and Daniel, T. (1997). Indigenous-plant extracts as larvicidal agents against *Culex quinquefasciatus* Say. *Bioresource Technology*, **59** : 137–40.

Kumar, A., Valecha, N., Jain, T., and Aditya, P. (2007). Dash burden of malaria in India: retrospective and prospective view. *American Journal of Tropical Medicine and Hygiene*, **77**: 69–78.

Lin, A. S., Stout, E. P., Prudhomme, J., Le Roch, K., Fairchild, C.R., and Franzblau, S. G. (2010). Bioactive bromophycolides R–U from the Fijian red alga *Callophycus serratus*. *Journal of Natural Products*,**73**: 275–278.

Liu, S. Q., Shi, J. J., Cao, H., Jia, F. B., Liu, X.Q., and Shi, G.L. (2000). Survey of pesticidal component in plant. In Entomology in China in 21st century, proceedings of conference of Chinese Entomological Society ed. Dianmo, Li Beijing, China: Science and Technique, pp. 1098-1104.

Manguin, S., Garros, C., Dusfour, I., Harbach, R. E., and Coosemans, M. (2008). Bionomics, taxonomy, and distribution of the major vector taxa of *Anopheles subgenus* Cellia in Southeast Asia: An updated review. *Infection, Genetics and Evolution*, **8**: 489-503.

Manilal, A., Sujith, S., Kiran, G. S., Selvin, J., Shakir, C., Gandhimathi, R., and Panikkar, M. V. N. (2009). Biopotentials of seaweeds collected from southwest coast of India. *Journal of Marine Science and Technology*, **17**: 67–73.

Manilal, A., Thajuudeein, N., Selvin, J., Idahyadulla, A., Surendra kumar, R., and Sujith, S. (2011). *In vitro* mosquito larvicidal activity of marine algae against the human vectors, *Culex quinequefaciatus* and *Aedes aegypti*. *Intrnational Journal of Zoological Research*, **7(3)**: 272-278.

Mayer, A. M., Rodríguez, A.D., Berlinck, R. G., and Hamann, M. T. (2009). Marine pharmacology in 2005–6: Marine compounds with anthelmintic, antibacterial, anticoagulant, antifungal, antiinfl ammatory, antimalarial, antiprotozoal, antituberculosis, and antiviral activities; affecting the cardiovascular, immune and nervous systems, and other miscellaneous mechanisms of action. *Biochimica et Biophysica Acta*, **1790**: 283–308.

Mehlhorn, H., Schmahl, G., and Schmidt, J. (2005). Extract of the seeds of the plant *Vitex agnus* castus proven to be highly efficacious as a repellent against ticks, fleas, mosquitoes and biting flies. *Parasitology Research*, **95**: 363–365.

Mittal, P. K, Adak, T., and Subbarao, S.K. (2005). Inheritance of resistance to *Bacillus sphaericus*toxins in a laboratory selected strain of *A. stephensi* (Diptera: Culicidae) and its response to *Bacillus thuringiensis* var. *israelensis*. *Current Sciences*, **89**: 442–443.

Moo-Puc, R., Robledo, D., and Freile-Pelegrin, Y. (2008). Evaluation of selected tropical seaweeds for *in vitro* anti-trichomonal activity. *Journal of Ethnopharmacology*, **120**: 92–97.

Muthukrishnan, J., and Puspalatha, E. (2001). Effects of plant extracts on fecundity and fertility of mosquitoes. *Journal of Applied Entomology*, **125**: 31–35.

Nathan, S.S., Kalaivani, K., Murugan, K., and Chung, P. G. (2005). Effects of neem limonoids on malarial vector *Anopheles stephensi* Liston (Diptera: Culicidae). *Acta Tropica*, **96**: 47–55.

Osman, M. E. H., Abushady, A. M., and Elshobary, M. E. (2010). *In vitro* screening of antimicrobial activity of extracts of some macroalgae collected from Abu-Qir bay Alexandria Egypt. *The African Journal of Biotechnology*, **9(12)**: 7203-7208.

Peng, Z., Beckett A.N., Engler, R. J., Hoffman, D. R., Ott, N. L., and Simons, F. E. R. (2004). Immune responses to mosquito saliva in 14 individuals with acute systemic allergic reactions to mosquito bites. *Journal of Allergy and Clinical Immunology*, **114**: 1189–1194.

Piyarat, S.W.K., Freed, M. and Roy, S. (1974). Biologically active plant extract for the control of mosquito larvae. *Mosque News*, 34–398.

Poonguzhali, T. V., and Nisha, L. J. (2012). Larvicidal activity of two seaweeds, *Ulva fasciata* and *Grateloupia lithophila* against mosquito vector, *Culex quinquefasciatus*. *International Journal of Current Science*, **4**: 163–8.

Rao, D. R., Mani, T. R., Rajendran, R., Joesph, A. S., Gjanana, A., and Reuben, R. (1995). Development of high level resistance to *Bacillus sphaericus* in a field population of *C.quinquefasciatus* from Kochi, India. *Journal of the American Mosquito Control Association*, **11**: 1 – 5.

Rao D. S., Girijavallabhan, K. G., Muthusamy, S., Chandrika, V., Gopinathan, C. P., Kalimuthu, S., and Najamuddin, M. (1999). Bioactive compounds from marine organisms with emphasis on the Indian Ocean. New Delhi, India, Oxford and IBH Publishing, 373-377.

Ravikumar, S., Ramanathan, G., Subhakaran, M., and Inbaneson S. (2009b). Antimicrobial compounds from marine halophytes for silkworm disease treatment. *International Journal of Medical Sciences*, **1(5)**: 184–191.

Ravikumar, S., Thajudeen, T., Suganthi, P., Inbeneson, S. J., and Vinothkumar, T. (2009c). Bioactive potential of seagrass bacteria against human bacterial pathogens. *Journal of Environmental Biology*, **31**: 387–389.

Ravikumar, S., Ramanathan, G., Inbaneson, S. J., and Ramu, A. (2011). Antiplasmodial activity of two marine polyherbal preparations from *Chaetomorpha antennina* and *Aegiceras corniculatum* against *Plasmodium falciparum*. *Parasitology Research*, **108**: 107-113.

Selvin, J., and Lipton, A. P. (2004). Biopotentials of *Ulva fasciata* and *Hypnea musiformis* collected from the peninsular coast of India. *Journal of Marine Science and Technology*, **12**: 1-6.

Sharma, S. K., Upadhyay, A. K., Haque, M. A., Tyagi, P. K., Raghavendra, K., and Dash, A. P. (2010). Wash-resistance and field evaluation of alphacypermethrin treated long-lasting insecticidal net (Interceptor) against malaria vectors *Anopheles culicifacies* and *Anopheles fluviatilis* in a tribal area of Odisha, India. *Acta Tropica*, **116(1)**: 24–30.

Smit, A. J. (2004). Medicinal and pharmaceutical uses of seaweed natural products: A review. *Journal of Applied Phycology*, **16**: 245–262.

Snow, R. W., Guerra, C. A., Noor, A. M., Myint, H. Y., and Hay, S. I. (2005). The global distribution of clinical episodes *of Plasmodium falciparum* malaria. *Nature*, **434**: 214– 217.

Spavieri, J., Kaiser, M., Casey, R., Hingley-Wilson, S., Lalvani, A., Blunden, G., *et al.*(2010). Antiprotozoal, antimycobacterial and cytotoxic potential of some British green algae. *Phytother Res.*, **24**: 1095

Stein, J. R., and Borden, C. A. (1984). Causative and beneficial algae in human disease conditions: a review. *Phycologia*, **23**: 485-501.

Subramaniam J., Murugan, K., kovendan, K., Mahesh kumar, P., panneerselvam, C., Nataraj, T., Amerasan, D., and Kalimuthu, K. (2011). Bio–efficacy of plant extracts against the malarial vector, *Anopheles stephensi* (Diptera: Culicidae). *Indian Journal of Applied Entomology*, **25(2)**: 131–135

Subramanian, J., and Murugan, K. (2103). Evaluation of larvicidal, pupicidal, repellent, and adulticidal activity of *Myristica fragrans* (Family: *Myristicaceae*) against malarial vector *Anopheles stephensi*. Proceedings of the national conference on national conference on insect diversity and systematics: special emphasis on molecular approaches (NCIDS-2013) p. 1-6.

Tang, H. F., Yi, Y. H., Yao, X. S., Xu, Q. Z., Zhang, S. Y., and Lin, H. W. (2002). Bioactive steroids from the brown algae *Sargassum carpophyllum*. *Journal of Asian Natural Product Research*, **4**: 95–105.

Thangam, T. S., and Kathiresan, K. (1991). Mosquito larvicidal activity of marine plant extracts with synthetic insecticides. *Botanica Marina*, **34**: 537-539.

Thangam, T. S., Kathiresan, K., and Mabbett, T. (1993). Mosquito larvicidal activity of seaweed extract against *Aedes aegypti* and *Culex quinquefasciatus*. *International Pest Control*, **35**: 94-95.

Tuney, I., Cadirci, B. H., Unal, D., and Sukatar, A. (2006). Antimicrobial activities of the extracts of marine algae from the coast of Urla (Izmir, Turkey). *Turkish Journal of Biology*, **30**: 171-175.

Tyagi, B. K., Ramnath, T., and Shahi, A. K. (1994). Evaluation of repellency effect of *Tagetus minuta*(Family: Compositae) against the vector mosquitoes *Anopheles stephensi* Liston, *Culex quinquefasciatus* Say and *Aedes aegypti* L. *International Pest Control*, 39–48.

Valerie, J. P., and Hay, M. E. (1986). Seaweed susceptibility to herbivory, Chemical and Morphology Correlates, *Marine Ecology-Progress Series*, **33**: 255–264.

Vashishta, B. R. (1984). Botany for degree students, Part 1 Algae, 7th. Ed., S. Chand and Company LTD, New Delhi, p. 5.

Venkatachalam, M. R., and Jebanesan, A. (2001a). Larvicidal activity of *Hydrocotyle javanica* Thunb. (Apiaceae) extract against *Culex quinquefasciatus*. *Journal of Experimental Zoology*, **4(1)**: 99–101.

Venkatachalam M. R., and Jebanesan, A. (2001b). Repellent activity of *Ferronia elephantum* Corr. (Rutaceae) leaf extract against *Aedes aegypti*. *Bioresource Technology*, **76(3)**: 287–288.

Vimaladevi S., Mahesh, A., Balaji, N., Dhayanithi, and Karthikeyan, N. (2012). Mosquito larvicidal ecacy of phenolic acids of seaweed *Chaetomorpha antennina* (Bory) Kuetz. against *Aedes aegypti*. *Biologia*, **67/1**: 212–216, 2012, DOI: 10.2478/s11756-011-0152-9.

Vogel, (1978).Textbook of practical organic chemistry. The English Language Book Society and Longman, London, p 1368.

Watanabe, K., Miyakado, M., Ohno, N., Okada, A., Yanagi, K., and Moriguchi, K. (1989). A polyhalogenated insecticidal monoterpene from the red alga, *Plocamium telfairiae*. *Phytochemistry*, **28**: 77–78.

World Health Organization (1975). Manual on Practical Entomology in Malaria Part I.Who division of malaria and other parasitic diseases, pp. 160.

World Health Organization (1981). Instructions for determining the susceptibility or resistance of adult mosquitoes to organochlorine, organophosphate and carbamate insecticides: diagnostic test. HO/VBC/81-807, Geneva.

World Health Organization (2007). Combating waterborne diseases at the household level. Part1.

Yit H. S., Ku-Hua, W. V., Kumamoto, J. H., and Axelrod, M. M. S. (1985). Isolation and identification of mosquito repellent in *Artemesia vulgaris*. *Journal of Chemical Ecology*, **11**: 1297–1306.

Utilisation and Management of Medicinal Plants Vol. 3 (2015) *Pages* **271–282**
Editors: **V.K. Gupta and A. Kaul**
Published by: **DAYA PUBLISHING HOUSE, NEW DELHI**

18

Larvicidal Activity of Indigenous Plant Extracts *against* Diamondback Moth *Plutella xylostella* (Lep: Plutellidae)

D. Amerasan[1*], K. Murugan[1], C. Panneerselvam[1], K. Kovendan[1], P. Mahesh Kumar[1], J. Subramaniam[1], T. Nataraj, D. Dinesh [1], B. Chandramohan[1] and R. Chandrasekar[2]

ABSTRACT

The present study explored the effects of Andrographis lineate (Acanthaceae), Cassia auriculata (Cesalpinaceae) and Solanum nigrum (Solanaceae) tested against third instar larvae of diamondback moth, Plutella xylostella. The medicinal plants were collected from Maruthamalai Hills, near the Bharathiar University campus in Coimbatore, Tamil Nadu, India. The plant extracts (chloroform, acetone and methanol) showed larvicidal effects after 48 h of exposure. Among three plants tested, the highest larvicidal activity was observed in methanol extract of A. lineate followed by C. auriculata and S. nigrum against P. xylostella. The rates of mortality were directly proportional to concentration of extract. The chloroform, acetone and methanol extract of A. lineate with LC_{50} values of 3.724, 2.424 and 2.190 per cent; C. auriculata with LC_{50} values of 4.238, 3.198 and 2.693 per cent; S. nigrum with LC_{50} values of 4.921, 3.573 and 3.217 per cent respectively, against third instar larvae of P.xylostella. The Chi-square values are significant at $p<0.05$ level. The Chi-square values in the bioassays indicated probably the heterogeneity of the test population. The 95 per cent confidence limits LC_{50} (LCL–UCL) and LC_{90} (LCL–UCL) were also calculated. No mortality

1 Division of Entomology, Department of Zoology, School of Life Sciences, Bharathiar University, Coimbatore – 641 046, Tamil Nadu, India.

2 Department of Biochemistry and Molecular Biophysics, 238 Burt Hall, Kansas State University, Manhattan 66506, KS, USA.

* *Corresponding author*: E-mail: ezhil_amar@yahoo.co.in

was recorded in the control.These results suggest that the leaf extracts of A. lineate, C. auriculata and S. nigrum have the potential to be used as an ideal ecofriendly approach for the control of the P. xylostella. Further detailed research is needed to identify the active ingredient in the extracts and implement the effective agricultural pest management program.

Keywords: *Plutella xylostella, Andrographis lineate, Cassia auriculata, Solanum nigrum,* Larvicide.

Introduction

Diamondback moth, *Plutella xylostella* L. is a serious pest of cauliflower, cabbage, broccoli, mustard, radish and turnip (Atwal, 1976; Eusebion and Rajesus, 1996; Capinera, 2001; Mohan and Gujar, 2003). Outbreaks of *P. xylostella* in Southeast Asia often caused more than 90 per cent crop loss (Verkerk and Wright, 1996). *P. xylostella* attacks the crop from the nursery stage onwards causing up to 52 per cent loss in marketable yield in cabbage (Krishnakumar *et al.*, 1984; Liang *et al.*, 2003; Shelton, 2004). Lingappa *et al.* (2000) reported that losses vary from 30 to 100 per cent. In India, the losses due to *P. xylostella* infestation is estimated to be US$ 16million annually in cultivated area of cabbage of 501,700 ha (Mohan and Gujar, 2003). Continuous use of chemical insecticides has led to the emergence and spread of resistance in agricultural pests and vectors of human diseases (Georghiou, 1990). and environmental degradation. Hence, new safe alternatives botanical products are useful and desirable tools in most pest management programs because they are effective and often non toxic to natural enemies having low environmental impact (Schmutterer, 1995; Haseeb *et al.*, 2004; Xu *et al.*, 2004). Botanicals have been shown to be effective larvicides and adulticides and in some cases are more eco-friendly against non-target animals and have been successfully tested for various biocontrol programmes (Markouk *et al.*, 2000).

Andrographis lineata Nees (Family: Acanthaceae) is a medicinally important in the traditional system of medicine in India and have been used extensively in snake bite and as anti-pyretic (Alagesaboopathi, 1999). It is also used as blood purifier and also in veterinary medicine. Hari Kishore *et al.* (2003) reported structure elucidation of three new flavonoids, 5,7,2´,3´,4´-pentamethoxyflavone (**1**), 2´-hydroxy-2,4´,6´-trimethoxychalcone (**2**) and dihydroskull capflavone I (**3**). The methanol and aqueous extracts of *A. lineata* established pharmacological evidence to support the folklore claim that it is used traditionally as a hepatoprotective agent (Sangameswaran *et al.*, 2008).

Cassia auriculata (Family: Cesalpinaceae) is an evergreen shrub that grows in many parts of India and in other parts of Asia. The flower, leaves, stem, root, and unripe fruit are used for treatment, especially in Ayurvedic medicine. Phytochemical analysis from root and leaf sample of *C. auriculata* which exhibited the presence of anthroquinones, alkaloids, flavonoids, steroids, tannins and phenolic compounds (Kalaichelvi *et al.*, 2009). *Cassia* species are rich sources of polyphenols, anthroquinone derivatives, flavonoids and polysaccharides and tannins. (Ayo *et al.*, 2007). The plant has been reported to antidiabetic, antiperoxidative and antihyperglyceamic (Manickam *et al.*, 2002) and microbicidal activity (Pari and Latha, 2003). The flowers are used to treat urinary discharges, nocturnal emissions, diabetes and throat irritation (Prakash, 2006). Cassia leaf extract affected the conidial germination of fungi *Microsporum gypsum* (Phongpaichit *et al.*, 2004).

Solanum nigrum (Family: Solanaceae) commonly known as black nightshade, grows as a weed, found in the dry parts of India and other parts of the world. It has a long history of medicinal usage and has been used as a traditional folk medicine for treating various ailments such as pain, inflammation, fever and liver disorders (Ong, 2003; Latiff, 2002). In India *S. nigrum* has been chosen as the important ingredient for herbal formulations, namely Liv 52, which is mainly used for treating liver diseases (Sandhir and Gill, 1999). Several photochemicals have been identified and isolated from *S. nigrum*, which contain alkaloids, flavonoids, saponins, tannins, phytic acid and hydrocyanic acid (Potawel *et al.*, 2008).

The present study aimed to explore the possibility of using the leaf extracts of some commonly available and ecofriendly plant extract such as *Andrographis lineate* (Acanthaceae), *Cassia auriculata* (Cesalpinaceae) and *Solanum nigrum* (Solanaceae), for *P. xylostella* management.

Materials and Methods

Plant Collection

The fully developed fresh leaves of *A. lineata*, *C. auriculata* and *S. nigrum* were collected from the Maruthamalai Hills, near the Bharathiar University campus in Coimbatore. It was authenticated by a plant taxonomist from the Department of Botany, Bharathiar University. A voucher specimen is deposited at the herbarium of the Entomology Division, Bharathiar University.

Extraction

The leaves were washed with tap water, shade-dried at room temperature (28±2°C) for 5 to 10 days. The air-dried materials were powdered separately using a commercial electrical blender. The finely ground plant material (500 g/solvent) was loaded in Soxhlet apparatus and was extracted with three different solvents, namely, of chloroform, acetone and methanol individually. The solvent from the extract was removed using a rotary vacuum evaporator to collect the crude extract. The crude residue of these plants varies with the solvents used. The *A. lineata*, *C. auriculata* and *S. nigrum* with three different solvents yielded 35.14, 42.23, 30.10, 40.20, 60.18g, 24.10, 31.25, 22.30, 34.12, 53.40 g and 14.40, 26.28, 13.10, 29.30, 40.20g of crude residue, respectively. Standard stock solutions were prepared at 1 per cent by dissolving the residues in acetone. From this stock solution, different concentrations were prepared ranging from 1.5, 3.0, 4.5, 6.0, and 7.5 per cent, respectively.

Insect Rearing

P. xylostella L. larvae were obtained from a culture that was maintained on cabbage leaf at laboratory conditions (26±1°C; 11±0.5 h photoperiod; 65-70 per cent R.H.). Newly emerged third instar larvae were used in bioassay studies.

Larvicidal Bioassay

Different concentrations of plant extracts were applied using leaf dip method. The treated leaves were exposed to the larvae. After 24 h of treatment, the larvae were continuously maintained on the non-treated fresh cabbage leaves. Fresh cabbage leaves were provided at every 24 h. larval mortality was recorded after 48 h of treatment. Five replicates were maintained for each treatment with 25 larvae per replicate. Percent mortality was calculated by (Abbott, 1925). The experiment was conducted at laboratory temperature of 27 ± 2°C with 14:10 light photoperiod and 75 ±5 per cent relative humidity. The LC_{50}, LC_{90} were calculated from toxicity data by using probit analysis (Finney, 1971).

$$\text{Abbott's Corrected Mortality} = \frac{\%\ \text{mortality in treatment} - \%\ \text{mortality in control}}{100 - \%\ \text{mortality in Control}} \times 100$$

Results

The results obtained from the present study confirm the larvicidal activity of chloroform, acetone and methanol extract of *A. lineata*, *C. auriculata* and *S. nigrum* against the third instar larvae of *Plutella xylostella* are presented in Tables 18.1–18.3 (Figure 18.1). Among three plants tested, the highest larvicidal activity was observed in methanol extract of *A. lineata* then *C. auriculata* and *S. nigrum* against *P. xylostella*. The rates of mortality were directly proportional to concentration of extracts. The LC_{50} and LC_{90} values were 2.190, 2.693, 3.217 per cent and 5.258, 7.302, 8.403 per cent, respectively. The Chi-square values are significant at $p<0.05$ level. The Chi-square values in the bioassays indicated probably the heterogeneity of the test population. The 95 per cent confidence limits LC_{50} (LCL–UCL) and LC_{90} (LCL–UCL) were also calculated. Each test included a control group with five replicates for each individual concentration. All the tested components that showed lethal effect and mortality were positively dose-dependent.The leaf extract of *A. lineata* was found to be most effective than *C. auriculata* and *S. nigrum* against larvae of *P. xylostella*.

Discussion

Plants have been used as source of medicine in virtually all cultures. In recent years, interest in screening medicinal plants for insect control has increased significantly. The advantage of using medicinal plants is that they are easily available and they have been used extensively for medicinal purposes, implying that they have low or no toxicity to humans. In addition, they can be applied to insect pests in the same way as conventional insecticides (Regnault-Roger, 1997; Schmutterer, 1997; Liang *et al.*, 2003; Charleston *et al.*, 2005; Liu *et al.*, 2007).

In the present study the larvicidal activity of chloroform, acetone and methanol extract of indigenous plants (*A. lineata*, *Cassia auriculata* and *Solanum nigrum*) against the third instar larvae of *P. xylostella*. All extracts including the methanol leaf extract of *A. lineata* were toxic to *P. xylostella*. Lee *et al.* (2001) reported that the methanol extract of *Curcuma longa* rhizome had strong insecticidal activity against *P. xylostella* larvae with 100 per cent mortality. Vanichpakorn and Vanichpakorn (2007)

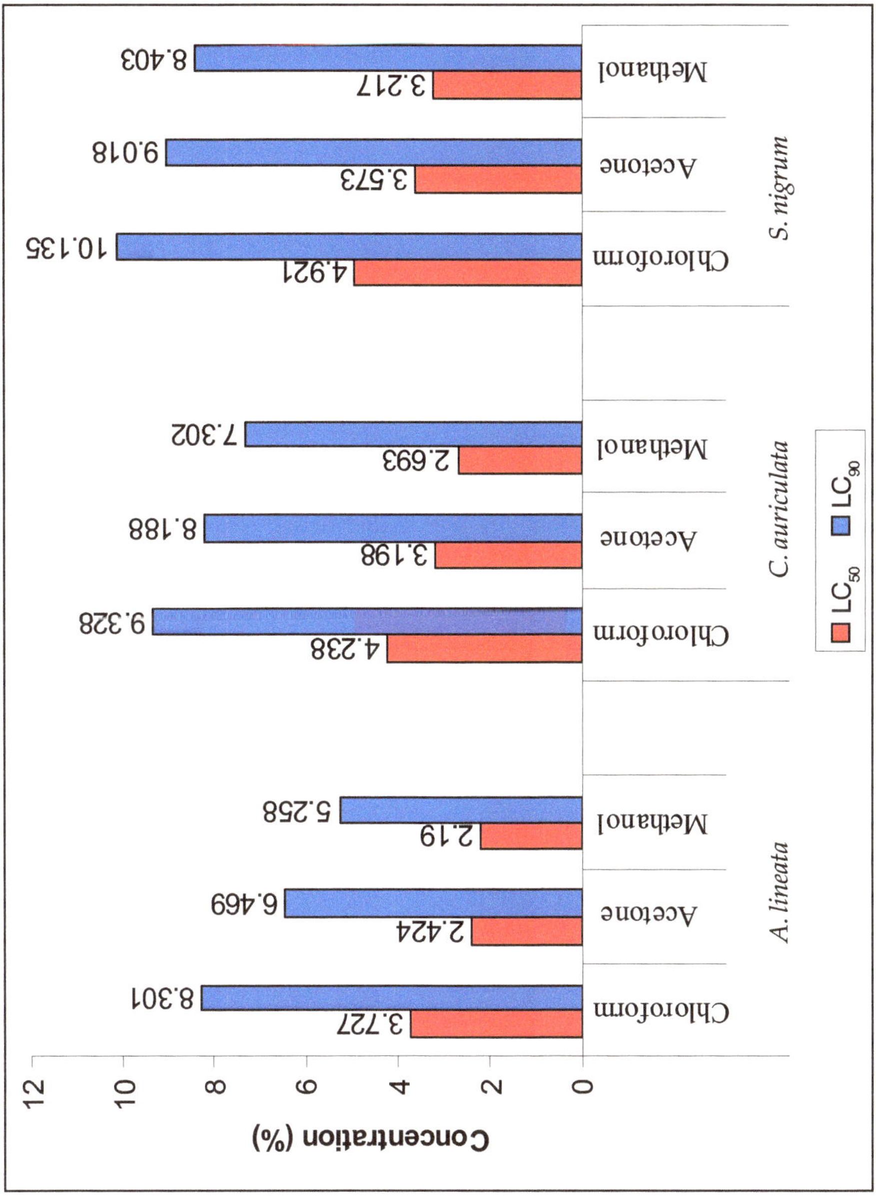

Figure 18.1: Graph Showing the LC_{50} and LC_{90} Values of *P. xylostella*.

Table 18.1: Larvicidal Activity of different Solvent Extract of *Andrographis lineate* against III Instars Larvae of *P. xylostella*

Name of the Extract	*Concentration (Per cent)*	*Per cent Mortality ± SD*	*LC_{50}, per cent (LFL–UFL)*	*LC_{90}, per cent (LFL–UFL)*	*χ^2*
Chloroform	Control	0.0±0.0	3.724	8.301	1.471*
	1.5	27.46 ± 2.02	(3.249-4.146)	(7.514-9.448)	
	3	38.37 ± 1.53			
	4.5	63.04 ± 2.33			
	6	72.48 ± 1.60			
	7.5	85.24 ±1.30			
Acetone	Control	0.0±0.0	2.424	6.469	0.244*
	1.5	39.46 ±1.76	(1.847-2.872)	(5.893-7.275)	
	3	56.26 ±1.75			
	4.5	74.44 ± 0.72			
	6	86.32 ±1.22			
	7.5	95 .34 ±1.70			
Methanol	Control	0.0±0.0	2.190	5.258	4.696*
	1.5	43.36 ± 1.09	(1.726-2.559)	(4.827-5.836)	
	3	59.88 ± 1.91			
	4.5	78.56 ± 2.01			
	6	95.76 ± 0.67			
	7.5	100 ±0.00			

LFL: Lower fiducidal limits; UFL: Upper fiducidal limits; χ^2: Chi square value.

* Significant at P< 0.05 level.

investigated the efficacy of ethanol extracts of 5 plant species for *P. xylostella* control. Chloroform extract was identified as the most toxic treatment against third instar larvae of *P. xylostella* at 5 per cent concentration. Chloroform extract recorded 69.6 per cent larvicidal activity at 96 h (Lingathurai *et al.*, 2011). Neto Bandeira *et al.* (2012) studied the ethanolic extracts of the flowers and fruits of *Muntingia calabura* were the most toxic against first instar *P. xylostella* larvae with LC_{50} values of 0.61 lg mL^{-1} and 1.63 lg mL^{-1} respectively, followed by the hexane extract of the fruits (LC_{50}= 5.5 lg mL^{-1}) and flowers (LC_{50}= 18.9 lg mL^{-1}). In our study the highest larvicidal activity was observed in methanol extract of *A. lineata* with LC_{50} and LC_{90} values of 2.190 and 5.258 per cent followed by *C. auriculata* (LC_{50}=2.693 per cent; LC_{90}=7.302 per cent) and *S. nigrum* (LC_{50}=3.217 per cent; LC_{90}=8.403 per cent) *respectively*. Similarly Pavela (2004) reported that leaf extract of *Marrubium vulgare* at 5.0 per cent concentration exhibited 42.2 per cent of larval mortality. The extracts might have arrested the various metabolic activities of the larvae. Ultimately the larvae failed to feed and the development was arrested in various instar stages. Ethyl acetate extract of *A. lineate*, methanol extracts of *Cocculus hirsutus*, and *Tagetes erecta* (EI_{50}=128.14, 79.39, 143.97, and 92.82 ppm; EI_{90}

Table 18.2: Larvicidal Activity of different Solvent Extract of *Cassia auriculata* against III Instars Larvae of *P. xylostella*

Name of the Extract	*Concentration (Per cent)*	*Per cent Mortality ± SD*	*LC_{50}, per cent (LFL–UFL)*	*LC_{90}, per cent (LFL–UFL)*	χ^2
Chloroform	Control	0.0±0.0	4.238	9.328	0.482*
	1.5	23.56 ± 1.98	(3.752-4.700)	(8.342-0.833)	
	3	37.22 ± 0.98			
	4.5	55 .38 ± 2.29			
	6	67.44 ± 1.73			
	7.5	78.08 ± 1.68			
Acetone	Control	0.0±0.0	3.198	8.188	1.849*
	1.5	35.54 ± 1.00	(2.612-3.673)	(7.343-9.464)	
	3	46.84 ± 1.44			
	4.5	61.66 ±1.72			
	6	73.12 ±1.95			
	7.5	89.68 ± 2.09			
Methanol	Control	0.0±0.0	2.693	7.302	3.694*
	1.5	40.72 ± 1.32	(2.078-3.170)	(6.592-8.343)	
	3	51.52 ± 1.00			
	4.5	66.56 ± 1.94			
	6	78.62 ± 1.16			
	7.5	94.78 ± 1.18			

LFL: Lower fiducidal limits; UFL: Upper fiducidal limits; χ^2: Chi square value.

* Significant at P< 0.05 level.

= 713.53, 293.70, 682.72, and 582.59 ppm), respectively, against *A. subpictus* (Gandhi Elango *et al.*, 2011).The mode of action and site of effect for larvicidal activities have been studied by various authors (Lewis *et al.*, 1993; Zafra-Polo, 1996). The biological activity of the experimental plant extracts may be due to various compounds, including phenolics, terpenoides, flavonoids, and alkaloids present in plants of these compounds may jointly or independently contribute to produce larvicidal effect against *P.xylostella*. Presence of more than one major phytocompound inhibited the growth of microbes and insects (Park *et al.*, 2002; Mansour *et al.*, 2004; Ahmad and Aqil, 2007). It has been observed that botanical insecticides affect insect physiology in many different ways (Senthil-Nathan *et al.*, 2004).

According to Pipithsangchan *et al.* (2004) it was shown that the methanol extract of *Azadirachta excelsa* (Jack) Jacobs. Seed showed insecticidal activity against *P. xylostella*. The LC_{50} values at 72 h after treatment of second and third instar larvae using topical application were 16,298.4 and 28,225.4 mg/l, respectively. Ling *et al.* (2008) also reported that *Mimordica charantia* ethanol extract had both feeding deterrent and larvicidal activities against *P. xylostella*. Baskar *et al.* (2009) reported the hexane fraction of *Atalantia monophylla* against *H. armigera*. Aggarwal *et al.* (2006) reported

Table 18.3: Larvicidal Activity of different Solvent Extract of *Solanum nigrum* against III Instars Larvae of *P. xylostella*

Name of the Extract	*Concentration (Per cent)*	*Per cent Mortality ± SD*	*LC_{50}, per cent (LFL–UFL)*	*LC_{90}, per cent (LFL–UFL)*	χ^2
Chloroform	Control	0.0±0.0	4.921	10.135	0.863*
	1.5	18.66 ± 1.38	(4.450-5.428)	(9.015-1.874)	
	3	31.34 ± 1.43			
	4.5	49 .76 ± 1.39			
	6	60.26 ± 2.03			
	7.5	72 .12 ± 1.69			
Acetone	Control	0.0±0.0	3.573	9.018	0.191*
	1.5	32.36 ± 1.44	(2.984-4.068)	(8.008-0.601)	
	3	43.62 ± 1.73			
	4.5	58.38 ± 1.98			
	6	70.78 ± 2.41			
	7.5	83.16 ± 1.74			
Methanol	Control	0.0±0.0	3.217	8.403	0.775*
	1.5	35.58 ± 1.32	(2.609-3.708)	(7.507-9.779)	
	3	46.34 ± 2.00			
	4.5	61.52 ± 1.70			
	6	73.68 ±2.01			
	7.5	87.44 ± 1.92			

LFL: Lower fiducidal limits; UFL: Upper fiducidal limits; χ^2: Chi square value.

* Significant at P< 0.05 level.

that *Bacillus thuringiensis aizawai* exhibited larvicidal activity against *H. armigera* and *P. xylostella* (Sengonca *et al.*, 2006). Elango *et al.* (2009a) reported that the leaf acetone, chloroform, ethyl acetate, hexane, and methanol extracts of *Andrographis lineata* were tested against fourth instar larvae of malaria vector, *Anopheles subpictus* and Japanese encephalitis vector, *C. tritaeniorhynchus* and the antifeedant activity of leaf aqueous extract was evaluated against the fourth instar larvae of gram pod borer, *Helicoverpa armigera* (Ramya and Jayakumararaj, 2009). Our results corroborated with the abovefindings. The high Chi-square values in the bioassays probably indicated the heterogeneity of the test population. Shi *et al.* (2007) reported that the ethanol, chloroform, and n-butyl alcohol extracts of *Veratrum nigrum* had insecticidal effects against *Mythimna separata* Walker, *Aphis craccivora* Koch, *Culex pipiens* pallens Coquillett, and *Tetranychus cinnabarinus* Boisduval.

In conclusion, the present study revealed the larvicidal activity of indigenous plants (*A. lineata*, *C.auriculata* and *S.nigrum*). These plant extracts have potential for the development of new and safe control products for *P. xylostella*.The present study plants are easily available, accessible, and affordable therefore the usage of traditional plants should be promoted among the local residents in order to reduce the crop damage as well as pest population. Further work is in progress to isolate and identify the insecticidal constituents of these plants and to develop a new botanical formulation for the management of *P. xylostella*.

References

Abbott, W.S. (1925). A method of computing the effectiveness of an insecticide. *J. Eco. Entomol.*, **18**: 265-266.

Aggarwal, N., Holaschke, M., and Basedow, T. (2006). Evaluation of biorational insecticides to control *Helicoverpa armigera* (Hubner) and *Spodoptera exigua* (Hubner) (Lepidoptera: Noctuidae) fed on Vicia faba L. *Giessen. Mitt. Dtsch. Ges. Allg. Angew. Ent.*, **15**: 245-250.

Ahmad, I., and Aqil, F. (2007). *In vitro* efficacy of bioactivie extracts of 15 medicinal plants against ESbL producing multi drug resistant bacteria. *Micobiol. Res.*, 162- 264.

Alagesaboopathi C., and Balu S. (1999). A*ndrographis wallich* ex and *Andrographis lineata* Nees is used as an herbal drug in the treatment of snake bite. *Indian J. economic and taxonomi Botany*, 23: 29-32.

Atwal, A.S. (1976). Agricultural pests of the India and Southeast Asia. Second Ed. Kalyani Publisher, Delhi, 502.

Ayo, R.G., Amupitan J.O., and Zhao Y. (2007). Cytotoxicity and antimicrobial studies of 1, 66 8 tri - hydroxyl- 3 – methyl – anthroquinones (Emodin) isolated from the leaves of *Cassia nigricans* Vahl. *Afr. J. Biotcchnology*, 6 **(11)**. 276.

Baskar, K., Kingsleys., Vendan, S.E., Paulraj, M.G., Duraipandiyan, V., and Ignacimuthu, S. (2009). Antifeedant Larvicidal and pupicidal activities of *Atalantia monphylla* (L.) Correa against *Helicoverpa armigera* (Hubner) Lepidoptera: Noctuidae). *Chemosphere*, **75**: 345-349.

Capinera, J.L. (2001). Handbook of vegetable pest. Academic press, San Diego, (pp.729).

Charleston, D.S., Kfir, R., Dicke, M., and Vet L.E.M. (2005). Impact of botanical pesticide derived from *Melia azedarach* and *Azadirachta indica* on the biology of two parasitoid species of the diamondback Moth. *Bio. Control*, **3**: 131–142.

Elango, G., Rahuman, A. A., Kamaraj, C., Bagavan, A., and Abduz Zahir, A. (2011). Efficacy of medicinal plant extracts against malarial vector, *Anopheles subpictus Grassi. Parasitol. Res.*, **108**: 1437-1445.

Elango, G., Rahuman, A.A., Bagavan, A., Kamaraj, C., Zahir, A.A., and Venkatesan, C. (2009a). Laboratory study on larvicidal activity of indigenous plant extracts against *Anopheles subpictus* and *Culex tritaeniorhynchus*. *Parasitol. Res.*, **104**: 1381–1388.

Eusebion, E.J., and Rejsus, B.M.M. (1996). Integrated pest management of DBM: the "Philippines Highland" eupelience. Proceedings of 3rd International Workshop, Kula Lumpur, Malaysis, 253-256.

Finney, D.J. (1971). Probit Analysis, third ed. Cambridge University Press, London, (pp: 383).

Georghiou G.P. (1990). Managing resistance to agrochemicals, from fundamental research to practical strategies. In: Green M.B., Le baron H.M., Mobreg W.K. (Eds) ACS Symposium series, 421. American Chemical Society.

Hari Kishore, P., Vijaya Bhaskar Reddy, N., and Kesava Reddy, N. (2003). Flavonoids isolated from *Andrographis lineata*. *J. of Phytochem*., **63**: 457- 461.

Haseeb, M., Liu, T.X., and Jones, W.A. (2004). Effects of selected insecticides on *Cotesia plutellae* (Hymenoptera: Braconidae), an endolarval parasitoid of *Plutella xylostella* (Lepidoptera: Plutellidae). *Biolo. Cont*., **49**: 33-46.

Kalaichelvi, K., Manian, S., and Chitravadivu, C. (2009). Qualitative Analysis of selected medicinal plants of Tamilnadu, India. Middle. *East J. of.Scientific. Res*., **4 (3)**: 144-146.

Krishnakumar Srinivasan, K.K., Ramachander, P.K., and Suman, C.L. (1984). Optimum control strategy of cabbage pests from a chemical control trial. *Singapore J. of Primary Industries*, **25**: 85-87.

Latiff K.M. (2002). Tumbuhan Ubatan Malayasia, Universiti Kebangsaan Malaysia in collaboration with ministry of science, Technology and Innovation. (pp. 589).

Lee, H.S., Shin, W.K., Song, C., Cho, K.Y., and Ahn, Y.J. (2001). Insecticidal activities of arturmerone identified in *Curcuma longa* rhizome against *Nilaparvata lugens* (Homoptera: Delphacidae) and *Plutella xylostella* (Lepidoptera: Yponomeutidae). *J. Asia-Pac. Entomol*., **4**: 181-185.

Lewis, M.A., Arnason, J.T., Philogene, J.R., Rupprecht, J.K., and Mclaughlin, J.L. (1993). Inhibition of respiration at site I by asimicin, an insecticidal acetogenin of the paw. *Asimina triloba* (Annonaceae). *Pesticide Biochem and Physiol*., **45**: 15-23.

Liang, G.M., Chen, W., and Liu, T.X. (2003). Effect of three neem-based insecticides on diamondback moth (Lepidoptera: Plutellidae). *Crop. Prot*., **22**: 333-340.

Ling, B., Wang, G., Y J., Zhang, M., and Liang, G. (2008). Antifeedant activity and active Ingredients against *Plutella xylostella* from *Momordica charantia* leaves. *Agric. Sci. China*, **7**: 466 -1473.

Lingappa, S., Basavanagoud, K., Kulkarni, K.A., Roopa, S.P., and Kambrekar, D.N. (2000). Threat to vegetable production by diamondback moth and its management strategies.- In: IPM System in Agriculture. (pp: 235-248).

Lingathurai, S., Ezhil Vendan, M., Paulraj, G., and Ignacimuthu, S. (2011). Antifeedant and larvicidal activities of *Acalypha fruticosa* Forssk. (Euphorbiaceae) against *Plutella xylostella* L. (Lepidoptera: Yponomeutidae) larvae. *J. of King. Saud. University. (Science)*, **23**: 11-16.

Liu, Z.L., Goh, S.H., and Ho, S.H. (2007). Screening of Chinese medicinal herbs for bioactivity against *Sitophilus zeamais* Motschulsky and *Tribolium castaneaum* (Herbst). *J. Stored. Prod. Res*., **43**: 290-296.

Manickam, P., Namasivaqyam, N., Periyasamy, V., and Rajagopal, S.K. (2002). Asia Pacific. *J. Cli. Nut.*, **11**: 157-163.

Mansour, F., Azaizeh, H., Saad, B., Tadmor, Y., Abo-Moch, F., and Said, O. (2004). The potential of middle eastern flora as a source of new safe bio-acaricides to control *Tetranychus cinnabarinus*, the carmine spider mite. *Phytoparasitica*, **32**: 66-72.

Markouk, M., Bekkouche, K., Larhsini, M., Bousaid, M., Lazrek, H.B., and Jana, M. (2000). Evaluation of some Moroccan medicinal plant extracts for larvicidal activity. *J of Ethanopharmacol.*, **73**: 293-297.

Mohan M., and Gujar, G.T. (20030. Local variation in susceptibility of the diamondback moth, *Plutella xylostella* (Linnaeus) to insecticides and role of detoxification enzymes. *Crop Protec.*, **22**: 495-504.

Neto Bandeira, G., Augusto Gomes Da Camara, C., Martins De Oraes, M., Barros, R., Muhammad, S., and Akhtar, Y. (2012). Insecticidal activity of *Muntingia calabura* extracts against larvae and pupae of diamondback, *Plutella xylostella* (Lepidoptera, Plutellidae). *J. King. Saud. University. (Science)*, **25(1)**: 83-89.

Ong, H.C. (2003). Sayuran Khasiat Makanan Dan Ubatan. Utusan Publication Distributors 2nd Bhd. Fakulti Sains: Universiti Malaya. (pp: 92-3).

Pari, L., and Latha, M. (2003). Antihyperglycaemic effect of *Cassia auriculata* in experimental diabetes and its effects on key metabolic enzymes involved in carbohydrate metabolism. *Cli. Exp. Pharmacol. Physiol.*, **30**: 38-43.

Park, B., Lee, S., Choi, W., Jeong, C., Song, C., and Cho, K. (2002). Insecticidal and acaricidal activity of pipernonaline and piperoctadecalidine derived from dried fruits of *Piper longum* L. *Crop. Prot.*, **21**: 249-25.

Pavela R. (2004). Insecticidal activity of certain medicinal plants. *Fitoterpia*, **75**: 745-749.

Phongpaichit, S., Pujenjob, N., Rukachaisirikul, V., and Ongsakul, M. (2004). Antifungal activity from leaf extracts of *Cassia alata* L., *Cassia fistula* L. and *Cassia tora* L. *J. Sci. Technol.*, **26(5)**: 741-748.

Pipithsangchan, S., Subhadhirasakul, S., Butpha, T., Phadoongsombat, N., and Chantrapromma, K. (2004). Effects of extracts from thiem seeds on diamondback moth (*Plutella xyllostella* Linn.). Songklanakarin. *J. Sci. Technol.*, **26**: 221-232.

Potawel, S.E., Sinha, S.D., Shroff, K.K., Dhalawat, H.J., Boraste, S.S., and Gandhi, S.P. (2008). *Solanum nigrum* Linn: A phytochemical Review. *Pharmacology*, **3**: 140-63.

Prakash, S.K. (2006). Effects of Herbal extracts towards microbicidal activity against pathogenic *Escherichia coli* in poultry. *Int. J. Poultry Sci.*, **5**: 259-261.

Ramya, S., and Jayakumararaj, R. (2009). Antifeedant activity of selected ethno-botanicals used by tribals of Vattal hills on *Helicoverpa armigera* (Hübner). *J. Pharm. Res.*, **2(8)**: 1414-1418.

Regnault-Roger, C. (1997). The potential of botanical essential oils for insect pest control. *Int. Pest. Manage. Rev.*, **2**: 25-34.

Sandhir, R., and Gill, K.D. (1999). Hepatoprotective effects of Liv-52 on ethanol induced liver damage in rats. *Indian. J. Exp. Biol.*, **37**: 762-6.

Sangameswaran, B., Reddy, T.C., and Jayakar, B. (2008). Hepatoprotective effect of leaf extracts of *Andrographis lineata* nees on liver damage caused by carbon tetrachloride in rats. *Phytother Res.*, **22**: 124-126.

Schmutterer, H. (1995). The Neem Tree: Source of Unique Natural Products for Integrated Pest Management, Medicine, Industry and Other Purposes.- VCH Publishers Inc., New York, (pp: 695).

Schmutterer, H. (1997). Side-effects of neem (*Azadirachta indica*) products on insect pathogens and natural enemies of spider mites and insect. *J. Appl. Entomol.*, **12**: 121-128.

Sengonca, C., Liu, B., and Zhu, Y.J. (2006). Insecticidal activity and antifeedant effect of a new type biocide GCSC-BtA against *Plutella xylostella* L. (Lep, Plutellidae). *J. Pest. Sci.*, **79**: 3-8.

Senthil-Nathan, S., Chung, P.G., and Murugan, K. (2004). Effect of botanicals and bacterial toxin on the gut enzyme of *Cnaphalocrocis medinalis*. *Phytoparasitica*, **32**: 433-43.

Shelton A.M. (2004). Management of the diamondback moth: Déjà vu all over again In: Endersby N., Ridland P.M. (Eds.) the Management of Diamondback Moth and Other Crucifer Pests. (pp. 3-8).- Proceedings of the Fourth International Workshop on Diamondback Moth. 26 29 November 2001, Melbourne, Australia.

Shi, Q.L., Wang, H.T., Li, P., and Tian, Z.J. (2007). Preliminary studies on the insecticidal activity of extract from *Veratrum nigrum*. *Nat. Prod. Res. Dev.*, **19**: 266-269.

Vanichpakorn, P., and Vanichpakorn, Y. (2007). Diamondback moth (*Plutella xylostella* L.) control on Chinese kale by application of plant extracts. *Khon. Kaen. Agric. J.*, **35**: 487- 495.

Verkerk, R.H.J., and Wright, D.J. (1996). multitrophic interactions and management of the diamondback moth: a review. *Bull. Entomol. Res.*, **86**: 205-216.

Xu, Y., Liu, T., Leibee, G.L., and Jones, W.A. (2004). Effects of selected insecticides on *Diadegma insulare* (Hymenoptera: Ichneumonidae), a parasitoid of *Plutella xylostella* (Lepidoptera: Plutellidae). *Bioconl. Sci. and Technol.*, **14**: 713-723.

Zafra-Polo, M.C., Gonzales, M.C., Estornell, E., Sahpaz, S., and Cortes, D. (1996). Acetogenins from Annonaceae: inhibitors of mitochond complex 1. *Phytochem.*, **42**: 253-271.

Utilisation and Management of Medicinal Plants Vol. 3 (2015) *Pages* **283–291**
Editors: **V.K. Gupta and A. Kaul**
Published by: **DAYA PUBLISHING HOUSE, NEW DELHI**

19

Antitrypanosomal Activity of *Xanthium strumarium* Leaves against *Trypanosoma evansi*

P. Shaba[1*], S. Dey[1] and R.K. Singh[2]

ABSTRACT

Without any new drug insight despite ever increasing threat of trypanosomosis with resistance of available trypanocides, there is a need to search for new drug. As a result of this, Xanthium strumarium leaf sample was extracted with methanolic solvent. The test extract of X. strumanium leaves at different concentrations (250-1000 µg ml^{-1})were screened against Trypanosoma evansi. It was carried out on Vero cells grown in Dulbecco's Modified Eagle Medium (DMEM) and supplemented with fetal calf serum (20-40 per cent) at appropriate conditions. In vitro cytotoxicity test of X. strumarium leaves extract at concentrations (1.56-100 µg ml^{-1}) was done on Vero cells grown in Dulbecco's Modified Eagle Medium (DMEM)but without foetal calf serum at appropriate conditions. Moderate antitrypanosmal activity was observed at concentrations of 750-1000 µg ml^{-1}. Extract of X. strumarium and diminazineaceturate were cytotoxic to Vero cells in all concentrations except at concentrations of 1.56-6. 25 µg ml^{-1}.

Keywords: Medicinal plant, *Xanthium strumarium* leaves, *Trypanosoma evansi, In vitro* trypanocidal activity, Cytotoxicity test.

1 Division of Medicine, Indian Veterinary Research Institute, Izatnagar – 243 122, Uttar Pradesh, India.

2 Indian Veterinary Research Institute, Regional Station, Mukteswar – 263 138, Uttranchal, India.

* *Corresponding author*: E-mail: shabamine@gmail.com

Present Address: College of Agriculture, P.M.B. 109, Mokwa, Niger State, Nigeria.

Materials and Methods

Chemicals

Silica gel-G for thin layer chromatography (TLC), solvents (hexane, chloroform, methanol, acetic acid and ethyl acetate) for extraction of plant material and development/analysis of TLC plates, vanillin for spray, and iodine for detection of bioactive constituents were used which were purchased from E. Merck, India.

Plant Materials

Fresh leaves of *Xanthium strumarium* (Asteraceae) were obtained from the premises of Indian Veterinary Research Institute, Izatnagar-243122, Bareilly, Uttar Pradesh. The leaves were subsequently identified at Institute of Himalayan Biosource and Technology, Palampur, Himachal Pradesh, India.

Preparation of Extract

The extraction was carried out according to the method of Stahl (1969). 20 g of dried *X. strumarium* leaves were powdered using laboratory pestle and mortar, and cold extracted with 200 ml of methanol (analytical grade). Residues obtained were extracted twice in the same medium. The filtrates were combined, dried at 37°C and stored at 4°C until used.

Solvent Systems

The following solvent systems were tested to develop the TLC plates according to the method of Stahl (1969).

Chloroform/hexane/acetic acid (50:50:1)

Chloroform/ethyl acetate/acetic acid (50:50:1)

Methanol and chloroform (20: 80)

Thin Layer Chromatography (TLC) Plates

Aliquot (0.2 ml) of extract was applied on TLC plates, dried under room temperature and immersed inside the appropriate solvent systems in a glass jar. It was done to detect the presence of bioactive constituents in applied extract. This was done in accordance with the method of Stahl (1969).

Animals

Swiss albino mice (20-30 g) of either sex were obtained from Animal Research Laboratory Section of Indian Veterinary Research Institute (IVRI) Izatnagar, maintained in standard environmental conditions and fed on a standard diet prepared by the institute with water *ad libitum*. Usage of mice in the experiment was strictly guided by laid down rules of committee on Ethics and Cruelty to Animals of the institute.

Test Organism

T. evansi was obtained from the Division of Parasitology, Indian Veterinary Research Institute (IVRI), Izatnagar and was maintained in the laboratory by serial sub-passages in Swiss albino mice. The strain was routinely tested for virulence as per the method of Williamson *et al.*(1982).

Trypanosomes Count

Counting of trypanosomes was carried out following the method of Lumsden *et al.*(1973). A number of fields (10-15) of each drop of blood or incubated media and parasites in triplicate were counted using glass slides under inverted microscope (400X). An average mean trypanosomes count was taken as number of trypanosomes per field.

In vitro Trypanocidal Activity

In vitro trypanocidal activity was carried out with modified method of Oliveira *et al.* (2004). A Vero cell line (SIGMA) was grown in Dulbecco's Modified Eagle Medium (DMEM) supplemented with 20-40 per cent foetal calf serum (FCS), GIBCO USA and antibiotics (100 iu penicillin, 100 µg streptomycin and 40 µg gentamycin) in 96-wells flat bottom microculture plates (NUNC, Denmark). Each well received 100 µl of DMEM containing $5x10^5$ cells ml^{-1}. Plates were incubated at 37°C under 5 per cent CO_2 for 12 h. After the formation of confluent monolayer, the medium was discarded and replaced with a fresh one. Finally, a high parasitaemic blood from mouse was diluted with DMEM to obtain $1x10^6$ parasites ml^{-1}. Suspension (100 ml of medium with trypanosomes) was added at the rate of 1:1 to test MPE of *X. strumarium* and the plates were incubated under the same conditions mentioned above. The test was repeated at least thrice.

Stock of test MPE of *X. Strumarium* (leaves) was solubilized in 1 per cent dimethylsuphoxide (DMSO). The concentration in the experiment had no deleterious effect by itself on host cells or parasites. 1 per cent DMSO in distilled water was used as control (Young *et al.*, 2000).

In vivo Infectivity Assessment

After incubation for anti-trypanosomal activity was completed, contents of ELISA plate wells with reduced and apparently killed trypanosomes from MPE of *X. strumarim* (leaves) was inoculated (0.1ml mouse-1) into two groups of mice (six group-1) intra-peritoneal, and observed for more than 30 days for parasitemia (Woo, 1970; lgweh *et al.*, 2002).

In vitro Cytotoxicity Test

It was done according to the method of Sidwell and Hoffman(1997). Vero cell line (SIGMA) was grown in DMEM in 96-wells microculture plates. Each well was seeded with 500,000 cells ml^{-1} and plates were incubated at 37°C with 5 per cent CO_2 for 48 h. After the formation of confluent monolayer, the supernatant was discarded and replaced with fresh medium. Confluent monolayer of Vero cell lines was treated with serial dilutions (1.56-100 µg ml^{-1}) of MPE of *X. strumarium* (leaves) in triplicate and incubated for 72 h consecutively under the same conditions described previously. At 24 h interval, plate was observed under inverted microscope for cytotoxic effects as compared to untreated normal cells that served as control. In each case, after 72 h of incubation, the culture media of the incubated Vero cells was discarded. Adhered cells were stained with a drop of crystal violet in phosphate buffered solution. Plates were then incubated for 24 h at 37°C in ordinary incubator. Plates were later observed under inverted microscope for cytotoxic effects. It was repeated thrice.

Statistical Analysis

Results of trypanocidal activity were expressed as mean ± SEM. Statistical analysis was done using Sigma stat (Jandel, USA).

Results and Discussion

Extraction

The solvent, methanol, was suitable in extraction of bioactive constituents as observed on TLC plates (plates not shown). Presence of bioactive constituents from MPE of *X. strumarium* leaves were detected on TLC plates. The solvent used in extraction of *X. strumarium* leaves is comparable to extraction of MPE of *Camellia sinensis* leaves on TLC plates (Shaba *et al.*, 2011c).

Thin Layer Chromatography Plates Analysis

Solvent system, methanol/chloroform (20:80), was more suitable than other solvent systems tested in the analysis of thin layer chromatography (TLC) plates with applied aliquots of plant extracts. TLC plates (plates not shown) showed different patterns of bioactive constituents of *X. strumarium* leaves, which were subsequently responsible for anti trypanosomal activity. This method is comparable to that used by Freiburghaus *et al.* (1998) in bioassay-guided isolation of a diasterolisomer of kolavenol from *Entada abyssinia* active on *T. brucei rhodesiense* and Shaba *et al.* (2011c) in extraction of *Camellia sinenesis* leaves.

In vitro Trypanocidal Activity

Anti-trypanosomal activity of *X. strumarium* leaves are contained in Table 19.1. Anti-trypanosomal activity varied from immobilization, reduction and killing of trypanosomes at different concentrations used. At a concentration of 250 μg ml^{-1} of MPE of *X. strumarium* leaves, there was no significant reduction of trypanosomes (40.00±0.0 to 35.67.±0.33). But at 750 μg ml^{-1} concentration of the extract, there was a marked reduction of trypanosomes count (40.00±0.0 to 1.33±0.33). At 1000 μg ml^{-1} of MPE of *X. strumarium*, trypanosomes were not detectable in the corresponding ELISA plate wells (40.00±0.0 to 0.0±0.00), which are comparable to 4 h of diminazine aceturate (Berenil, a standard drug at 50 μg ml^{-1}). Anti-trypanosomal activity is in line with *in vitro* trypanocidal activity of MPES of medicinal plants used in treatment of trypanosomosis in northern Nigeria at an effective concentration of 8.3 mg ml^{-1} (Wurochekke and Nok, 2004) and *in vitro* trypanocidal activity of methanolic extracts of *Quercus borealis* leaves and *Zingiber officinale* roots with complete killing of trypanosomes at 500 and 750 μg/ml (Shaba *et al.*, 2011b) and trypanocidal potential of *Camellia sinensis* leaves where trypanosomes were not detected in the corresponding ELISA plate wells at 250 μg/ml of the test extract at 4 h of incubation (Shaba *et al.*, 2011c). An average mean parasite count of 37.67±0.58 is statistically critical value. Average mean parasite count from 37.67±0.58 and below was significant between

Table 19.1: *In vitro* Trypanocidal Activity of Methanolic Extract of *Xanthium strumarium* Leaves against *Trypanosmae vansi* on Vero Cell Line

Concentration of Plant Extract in µg/ml	*1 h*	*2 h*	*3 h*	*4 h*	*5h*	*6 h*	*7 h*	*8 h*	*9 h*
250	39.33±0.33	39.67±0.88	38.67±0.33	38.33±C.33	38.67±0.33	37.00±00	35.67±0.33	35.33±0.33	35.67.±0.33
500	38.00±0.58	38.38±0.33	37.33±0.33	37.67±C,33	36.000±0.58	35.33±0.33	34.33±0.33	33.33±0.33	32.67±0.33
750	37.67±0.33	36.33±0.33	35.33±0.67	33.00±C.58	25.67±0.33	18.33±0.33	14.33±0.33	6.67±0.33	1.33±0.33
1000	34.33±0.33	28.00±o.58	22.67±0.33	13.67±C.67	7.00±0.58	0.00±0.00	0.00±0.00	0.00±0.00	0.00±0.00
Diminazine-aceturate (50) Positive control	23.33±0.88	9.6±0.67	1.67±0.67	0.00±00	0.00±00	0.00±00	0.00±00	0.00±00	0 . 0 0 ± 0 0
Control (Negative control)	40,00±0.00	40,00±0.00	40,00±0.00	40,00±0.00	40.00±0.00	40.00±0.00	40.00±0.00	40.00±0.00	40.00±0.00

Bioassay status: significant reduction of parasites counts from concentration of 750 µg/ml and complete killing of parasites at 1000 µg/ml at 6th hour of observation. An average mean parasites count of 37.67± 0.58 is statistically critical value. Average mean parasites counts from 37.67± 0.58 and below is significant between the treatment groups and negative control ($P \leq 0.05$ to 0.01).

Table 19.2: Cytotoxic Effects of Methanolic Extract of *Xanthium strumarium* Leaves on Vero Cell Line Compared to Diminazineaceturate (Berenil)

Concentration of Test Material in µg/ml	*Cytotoxic Effects of Extract and DA at Various Time Intervals of Incubation*						*Control*
	24 h		*48 h*		*72 h*		
	X. strumariumebula	*DA*	*X. strumarium*	*DA*	*X. strumarium*	*DA*	
100	100 per cent	66.6 per cent	100 per cent	100 per cent	100 per cent	100 per cent	0
50	100 per cent	33.3 per cent	100 per cent	100 per cent	100 per cent	100 per cent	0
25	100 per cent	0	100 per cent	33.3	100 per cent	66.6	0
12.5	100 per cent	0	100 per cent	0	100 per cent	33.3 per cent	0
6.25	0	0	66.6 per cent	0	100 per cent	0	0
3.13	0	0	0	0	33.3 per cent	0	0
1.56	0	0	0	0	0	0	0

The same concentrations were used for both MPE of *X. Strumarium* leaves diminazineaceturate(DA). Cytotoxicity effects increased with period of incubation depending on the concentrations of the extract and DA.

the treatment groups and negative control (*p* d" 0.05). Previously, xanthatin and xanthinosin were isolated from *X. strumarium* (Ramírez-Erosa *et al.*, 2007). Xanthatin has been isolated from *X. strumarium* and was found to be the major and most active compound against *T. b. brucei* with an IC(50) value of 2.63µg/mL and a selectivity index of 20 (Nilbret *et al.*, 2011). The mechanism of action of xanthatin according to Nilbret *et al.* (2011) was by xanthatin inducing apoptosis and inhibiting both PGE(2) synthesis and 5-lipoxygenase activity thereby avoiding unwanted inflammation. This mechanism of action may explain the antitrypanosomal activity of MPE of *X. strumarium* of leaves in this current report.

In vivo Infectivity Test

Group of mice inoculated with contents of ELISA plate wells with completely killed trypanosomes (40.00±0.0 to 0.0±0.00)by MPE of *X. strumarium* leaves survived for more than 30 days, while other group inoculated with contents of ELISA plate wells with reduced trypanosomes count(40.00±0.0 to 1.33±0.33)died of parasitaemia. *In vivo* Infectivity assessment of anti-trypanosomal activity is comparable to anti-trypanosomal effect of the aqueous extract of *Brassica oleracea* and MPE of *Terminalia chebula* dried fruits where inoculated in mice with contents of wells with apparently killed trypanosomes survived (Igweh *et al.*, 2002; Shaba *et al.*, 2007).

In vitro Cytotoxicity Test

In vitro cytotoxic effects of MPE of *X. strumarium* leaves, and di-minazineaceturate at the same concentrations on Vero cells depicted different effects such as distortion, swelling, sloughing and death of Vero cells compared to negative normal cells in control wells (Table 19.2). MPE of *X. strumarium* and diminazineaceturate were cytotoxic to Vero cells at all concentrations except at 1.56 and 1.56-6.25 µg ml^{-1}, respectively. These results are comparable to *in vitro* cytotoxicity tests of comparative extraction of *Camellia sinensis* leaves and methanolic extract of *Vitex negundo* leaves, in which similar cytotoxic effects were observed (Shaba *et al.*, 2011c; Shaba *et al.*, 2008).

It can be concluded from this current report that MPE of *X. strumarium* leaves at different concentrations demonstrated moderate degrees of trypanocidal activities in this report. Xanthatin and xanthinosinhave been previously isolated from *X. strumarium* leaves, and xanthatin was found to be the major and most active compound with its trypanocidal activity. Bioassay-guided isolation of active constituents and *in vivo* test in mice are required to fully ascertain the trypanocidal status of the plant materials.

Acknowledgements

Financial contributions towards research by India and Nigeria governments are highly acknowledged.

References

Asres, E., Bucar, F., Knauder, E., Yardley, V., Kendrick, H., and Croft, S.L. (2001). *In vitro* antiprotozoal activity of extract and compounds from the stem bark of *Combretum molle. Phytotherapy Research*, **15**: 6113-6117.

Cragg, G.M., Newman, D.J., and Snader, K.M. (1997). Natural products in drug discovery and development. *Journal of Natural Products*, **60**: 52-60.

Freiburghaus, F., Kaminsky, R., Nkuna, M.H.N., and Brun, R, (1996). Evaluation of African medicinal for their *in vitro* trypanocidal activity. *Journal of Ethnopharmacology*, **55**: 1-11.

Freiburghaus, F., Steck, A., Ptander, H., and Brun, R. (1998). Bioassay guided isolation of a diastereoisomer of kolavenol from *Entada absysinica* active on *Trypanosoma brucei rhodesiense. J.Ethnopharmacol.*, **61**: 179-183.

Gutteridge, W.E. (1985). Existing chemotherapy and its limitations. *British Medical Bulletin*, **41**: 162-168.

Hursey, B.S. (2000). The programme against African Trypanosomiasis. *Trends parasitol.*, **17**: 99-100.

Igweh, A.C., Aguiyi, J.C., and Okwuaasaba, F.K. (2002). Antitrypanosomal effect of the aqueous extract of *Barssica oleracea. J. Fitotera.*, **71**: 17-21.

Kalba, C.P., Farooquee, N.A. and Dhar, U. (2005). Traditional uses and conservation of timur (*Zanthoxylum armatum* DC.) through Social Institutions in Uttaranchal Himalaya, India. **3**: 224-230.

Lumsden, W.H.R., Herbert, W.J., and McNeilage, G.J.C. (1973). Techniques with trypanosomes. Churchil Livingstone, London.

Nok and Nock (2002). Transferin coupled azanthraquinone enhances the killing effects on trypanosomes. The role of lysosomalmannosidase. *J. Patasites*, **9**: 375-379.

Nibret, E.1., Youns, M., Krauth-Siegel, R.L., and Wink, M. (2011).Biological activities of xanthatin from *Xanthium strumarium* leaves. *Phytotherapy Reearch*, **25**: 1883-90.

Mehta, B., Kharya, M.D., Srivastava, R., and Verma, K.C. (1981). Antimicrobial and anthelmintic activities of the essential oil of *Zanthoxylum alatum* Roxb. *Indian Perfumer*, **25**: 19-21.

Ramírez-Erosa, I., Huang, Y., Hickie, R.A., Sutherland, R.G., Barl, B., and Can, J. (2007). Xanthatin and xanthinosin from the burs of *Xanthium strumarium* L. as potential anticancer agents. *Journal of Physiology and Pharmacology*, **85**: 1160-72.

Sarveswaran, S., Marati, R. V., and Marathaiveeran, P.B. (2006). Effects of *Terminalia arjuna* bark extract on opoptosis of human hepatoma cell line (HEPG2). *Intl J. Gastroenterol.*, **12**: 1015-1024.

Shaba, P., Pandey, N.N., Sharma, O.P., Rao, J.R., and Singh, R.K. (2006). Antitrypanosomal and cytotoxicity of methanolic *Plumbago zeylanica* root back against *Trypanosoma evensi. Indian J.Vet.Pub.Hlth.*, **4**: 31-36.

Shaba, P., Pandey, N.N., Sharma, O.P., Rao, J.R., and Singh, R.K. (2008). *In vitro* trypanocidal and cytotoxicity effects of methanolic extract of *Vitex negundo* leaves against *Trypanosoma evansi.* The 13th Congress of the Federation of Asian

Veterinary Associations. Pages 43-44.FAVA-OIE Joint Symposium on Emerging Diseases, Bangkok, Thailand.

Shaba, P., Pandey, N.N., Sharma, O.P., Rao, J.R., and Singh, R.K. (2011a). Anti-trypanosomal potential of methanolic extract of *Calotropis gigantean* leaves against *Trypanosoma evansi* and its cytotoxicity. *Intl. J. Biosource and Stress Mgt.*, **73**: 121-124.

Shaba, P., Pandey, N.N., Sharma, O.P., Rao, J.R., and Singh, R.K. (2011b). *In vitro* trypanocidal activity of methanolic extracts of *Quercus borealis* leaves and *Zingiber officinale* roots. *Greener J. AgricSci.*, **1**: 41-47.

Shaba, P., Pandey, N.N., Sharma, O.P., Rao, J.R., and Singh, R.K. (2007). Comparative antitrypanosomal activity of *Terminalia chebula* dried fruits against *Trypanosoma evansi*. *J. Planta Medic.*, **73**: 997-1034.

Sidwell, R.W., and Huffman, J.H. (1997). Antiviral drug resistance. *Res. Virol.*, **148**: 353-365.

Soejarta, D.D. (1996). Biodiversity prospects and benefit sharing. Perspective from medicinal plants from the field. *J. Ethnopharmacol.*, **51**: 1-15.

Seed, J.R. (2001). African trypanosomosis research in: 100 years of progress but questions remain. *Intl. J. parasitol.*, **31**: 434-442.

Stahl, E. (1969). Thin layer chromatography. A Laboratory Handbook. Springer, New York.

Tiwary, S.N., Naik, D.K., Tewary, P., Mittal, K., and Yadav, S. (2007).Chemical composition and larvicidal activities of thee essential oil of *Zanthoxylum armatam* DC (Rutaceae) against three mosquito vectors. *J. Vect. Borne Dis.*, **44**: 198-204.

Upadhyaya, K., and Ashok, P.K. (2010). Concentration dependent antioxidant activity of *Zanthoxylum armatum*. *Journal of Pharmacy Research*, **3**: 1581-1582.

Wiliamson, J., March, J.C., Scott-Finning, J.J. (1982). Drug synergy in experimental African trypanosomiasis. *Tropennmedizin und Parasitologie*, **33**: 76-82.

World Health Organization Health. (2001). Pan African tsetse and trypanosomiasis eradication campaign. Fifty-fifth World Health Assembly, WHO, Geneva.

World Health Organization. (2004). Anon; 2004 communicable Disease Surveillance and Rsponse. WHO, Geneva.

Woo, P.T.K. (1970). Evaluation of the haemataocrit centrifuge and other techniques for the field diagnosis of human trypanosomiasis and filariasis. *J. Acta Trop.*, **28**: 298-303.

Wurochekke, A.U., and Nok, A.J. (2004). *In vitro* anti trypanosomal activity of some medicinal plants used in the treatment of trypanosomosis in Northern Nigeria. *Afri. J. Biotech.*, **3**: 481-483

Young, V., Schmitz, V., Vanner-Santos, M.A., Lima, A.P.C.A., Lalmanach, G., Juliano, L., Gauther, F., and Scharfstein, J. (2000).Altered expression of cruzipain and a cathepsin B-like target in a *Trypanosoma cruzi* cell line displaying resistance to synthetic inhibitors of cysteine-proteinases. *J. Mol. Biochem. Parasitol.*, **109**: 47-59.

Utilisation and Management of Medicinal Plants Vol. 3 (2015) *Pages* **293–305**
Editors: **V.K. Gupta and A. Kaul**
Published by: **DAYA PUBLISHING HOUSE, NEW DELHI**

20

In vitro Trypanocidal Activity of Comparative Extractions of *Eupatorium adenophorum* Leaves and Flowers with Solvents of Different Polarities against *Trypanosma evansi*

P. Shaba[1*], N.N. Pandey,[1] and R.K. Singh[2]

ABSTRACT

Trypanosomosisis a zoonotic disease and a menace that is on the increase in recent years causing lots of loss in production and huge cultivatable fertile land invaded by Glossina spp, one of the of the vectors, that transmit the disease. Eupatorium adenophorum leaves and flowers were comparatively extracted with solvents of different polarity (hexane, chloroform, methanol and aqueous). The extracts obtained from the solvents were dried and tested against Trypanosoma evansi at different concentrations (250-1000 μg ml^{-1}) on Alsever's medium for antitrypanosomal activity. Trypanosomes (1x10^{6}/ml) were suspended in Alsever's solution with inactivated bovine serum at 58°C for 1 h. 180 μl of the medium was added to the test extracts of each solvent (20 μl)and incubated at 37°C with 5 per cent carbon dioxide for 5 h. On hourly basis, drops of the incubated mixture were observed under inverted microscope for antitrypanosomal activity. In vitro cytotoxicity test of methanolic extracts (MES) of both leaves and flowers at concentrations (1.56-100 g ml^{-1}) was done on Vero cells grown in

1 Division of Medicine, Indian Veterinary Research Institute, Izatnagar – 243 122, Uttar Pradesh, India.

2 Indian Veterinary Research Institute, Regional Station, Mukteswar – 263 138, Uttranchal, India.

* *Corresponding author*: E-mail: shabamine@gmail.com

Present Address: College of Agriculture, P.M.B. 109, Mokwa, Niger State, Nigeria.

Dulbecco's Modified Eagle Medium (DMEM) but without foetal calf serum at appropriate conditions. Results showed the order of antitrypanosomal activity of the leaves extracts was methanol, hexane, chloroform and aqueous. While that of flowers was methanol, aqueous, hexane and chloroform.Extracts of E. adenophorum flowers demonstrated more antitrypanosomal activity with maximal activity at 1000 µg ml^{-1} *(40.00±0.0 to 15.00±0.58) than leavesextracts (40.00±0.0 to 23.33±0.33). Moderate antitrypanosmal activity with significant difference (*$P \leq 0.05$ *to 0.01) was observed. Both methanolic extracts of E. adenophorum flowers, leaves and diminazineaceturate were cytotoxic to Vero cells in all concentrations except at concentrations of 1.56 and 1.56-6. 25 µg ml*$^{-1}$*, respectively.*

Keywords: *In vitro*, Trypanocidal activity, *Eupatorium adenophorum*, *Trypanosma evansi*, Vero cells.

Introduction

Trypanosomosis has becoming a menace in different parts of Africa and other parts of the world where the disease exist for more than a century (Abiodun *et al.*, 2012; WHO, 2007). It affects both animals and humans at distinct strata with lots of devastations in terms of morbidity and mortality rates (WHO, 2004). In recent years, trypanosomosis is on the increase and the animal's vectors, (*e.g.*, *Glossina* spp) are capturing more fertile lands thereby rendering the affected huge fertile cultivable land inhabitable for both animals and humans (WHO, 2004). *Trypanosoma brucei rhodesiense* in Africa have been incriminated in the cause of trypanosomosisin humans (Nok and Nock, 2002; WHO, 2004).

Direct losses due to trypanosomiasis are estimated to amount to between US$ 1-1.2 billion each year whereas theindirect impact of African Animal Trypanosomosis (AAT) on agriculture in sub-Saharan Africa exceeds this amount. Apondered evaluation extrapolated for the total tsetse-infested lands values the total losses, interms of agricultural Gross Domestic Product, at US$ 4.75 billion per year (FAO, 2004).The overall impact extends to the restricted access to fertile and cultivable areas, imbalancesin land use and exploitation of natural resources and compromised growth and diversificationof crop-livestock production systems (Mattioli *et al.*, 2004).

Resistance to limited classes of drugs available in the market have been reported at different parts of Africa and elsewhere the disease thrives (Nok and Nock, 2002; WHO, 2004).

Local usages of herbs to treat trypanosomosis have been validate and documented (Nok and Nock, 2002).

Anti-inflammatory and antifeedant of *E. adenophprum* leaves have been reported (Paraj *et.al.*, 2001; Ashim *et.al.*, 2011).

Evidence of trypaocidal activities of medicinal plants either extracts or isolated compounds have been validated by many researchers. These extracts/compounds may likely be a potential targets for antitrypanosomal drug development (WHO 2004; Shaba *et al.*, 2008, 2012, 2014).

On this premise, this research was designed to investigate possibility of extracting bioactive constituents using solvents of different polarities and possible anti-trypanosomal activity.

Materials and Methods

Chemicals

Silica gel-G for thin layer chromatography (TLC), solvents (hexane, chloroform, methanol, acetic acid and ethyl acetate) for extraction of plant material and development/analysis of TLC plates, vanillin for spray, and iodine for detection of bioactive constituents were used which were purchased from E. Merck, India.

Plant Materials

Fresh *Eupatorium adenophorum* leaves and flowers (Spreng, Crofton by weed) were obtained from Palampur, in Himachal Pradesh. The medicinal plant materials wereidentified at Institute of Himalayan Biosource and Technology, Palampur, Himachal Pradesh, India.

Preparation of Extract

The extraction was carried out according to the method of Stahl (1969). 20 g each of *Eupatorium adenophorum* leaves and flowers were powdered using laboratory pestle and mortar, and cold extracted with 200 ml of methanol (analytical grade). Residues obtained were extracted twice in the same medium. The filtrates were combined, dried at 37°C and stored at 4°C until used.

Solvent Systems

The following solvent systems were tested to develop the TLC plates according to the method of Stahl (1969).

Chloroform/hexane/acetic acid (50:50:1)

Chloroform/ethyl acetate/acetic acid (50:50:1)

Methanol and chloroform (20: 80)

Thin Layer Chromatography (TLC) Plates

Aliquot (0.2 ml) of extract was applied on TLC plates, dried under room temperature and immersed inside the appropriate solvent systems in a glass jar. It was done to detect the presence of bioactive constituents in applied extracts. This was done in accordance with the method of Stahl (1969).

Animals

Swiss albino mice (20-30 g) of either sex were obtained from Animal Research Laboratory Section of Indian Veterinary Research Institute (IVRI) Izatnagar, maintained in standard environmental conditions and fed on a standard diet prepared by the institute with water *ad libitum*. Usage of mice in the experiment was strictly guided by laid down rules of committee on Ethics and Cruelty to Animals of the institute.

Test Organism

*T. evansi*was obtained from the Division of Parasitology, Indian Veterinary Research Institute (IVRI), Izatnagar and was maintained in the laboratory by serial sub-passages in Swiss albino mice. The strain was routinely tested for virulence as per the method of Williamson *et al.*(1982).

Trypanosomes Count

Counting of trypanosomes was carried out following the method of Lumsden *et al.*(1973). A number of fields (10-15) of each drop of blood or incubated media and parasites in triplicate were counted using glass slides under inverted microscope (400X). An average mean trypanosomes count was taken as number of trypanosomes per field.

In vitro Trypanocidal Activity

In vitro trypanocidal activity was carried out according to the method of Talakal *et al.*(1995). *Eupatorium adenophorum* leaves and flowers were comparatively extracted with solvents of different polarities (hexane, chloroform, methanol and aqueous) and screened against *Trypanosoma evansi*. A high parasitemic blood of a mouse was diluted with Alsever's solution to obtain a final parasite concentration of $1x10^6$ parasites/ml. The medium consist of Alsever's solution and inactivated bovine serum at 58°C for 1 h. Suspension (180 µl of medium with trypanosomes) was added to 20 µl of extracts to test E. *adenophorum* (leaves and flowers) for antitrypanosomal activity and the plates were incubated at 37°C under 5 per cent $CO_{2.}$ The test was repeated at least thrice.

Stock of each extract of *E. adenophorum* leaves and flowers was solubilized in 1 per cent dimethylsuphoxide (DMSO). The concentration in the experiment had no deleterious effect by itself on host cells or parasites. 1 per cent DMSO in distilled water was used as control (Young *et al.*, 2000).

In vivo Infectivity Assessment

After incubation for anti-trypanosomal activity was completed, contents of ELISA plate wells with reduced and apparently killed trypanosomes from methanolic extracts (MES) of *E. adenophorum* (leaves and flowers) were inoculated (0.1ml mouse-1) into two groups of mice (six group-1) intra-peritoneal, and observed for more than 30 days for parasitemia (Woo, 1970; lgweh *et al.*, 2002).

In vitro Cytotoxicity Test

It was done according to the method of Sidwell and Hoffman. (1997). Vero cell line (SIGMA) was grown in DMEM in 96-wells microculture plates. Each well was seeded with 500,000 cells ml^{-1} and plates were incubated at 37oC with 5 per cent CO_2 for 48 h. After the formation of confluent monolayer, the supernatant was discarded and replaced with fresh medium. Confluent monolayer of Vero cell lines was treated with serial dilutions (1.56-100 µg ml^{-1}) of MES of *E. adenophorum* (leaves and flowers) in triplicate and incubated for 72 h consecutively under the same conditions described previously. At 24 h interval, plate was observed under inverted microscope for cytotoxic effects as compared to untreated normal cells that served as control. In each case, after 72 h of incubation, the culture media of the incubated Vero cells was discarded. Adhered cells were stained with a drop of crystal violet in phosphate buffered solution. Plate was then incubated for 24 h at 37°C in ordinary incubator. Plates were later observed under inverted microscope for cytotoxic effects.

Statistical Analysis

Results of trypanocidal activity were expressed as mean ± SEM. Statistical analysis was done using Sigma stat (Jandel, USA).

Results and Discussion

Extraction

In this extractions with solvents of different polarities, (hexane, chloroform, methanol and aqueous), the extraction of bioactive constituents by each solvent was observed on TLC plates. In this investigation, it appeared that methanol was more efficient in extraction of bioactive constituents presence in extracts of *E. adenophorum* (leaves and flowers) extracts as observed on TLC plates (Plates not shown). This method of extraction is in line with that of Freiburghaus *et al.* (1998) in bioassay-guided isolation of a diasterolisomer of kolavenol from *Entada abyssinia* active on *T. brucei rhodesiense* and comparative extraction of *Terminalia belirica* dried fruits where different solvents of polarities extracted bioactive constituents therein at different rates (Shaba *et al.*, 2008).

Thin Layer Chromatography Plates Analysis

Solvent system, methanol/chloroform (20:80), was more suitable than other solvent systems tested in the analysis of thin layer chromatography (TLC) plates with applied aliquots of *E. adenophorum* extracts (leaves and flowers). TLC plates (plates not shown) showed different patterns of bioactive constituents from extracts of *E. adenphorum* (leaves and flowers), whichwere subsequently responsible for anti-trypanosomal activity. This method is comparable to that used by Freiburghaus *et al.* (1998) in bioassay-guided isolation of a diasterolisomer of kolavenol from *Entada abyssinia* active on *T. brucei rhodesiense* and comparative extraction of *Terminalia belirica* dried fruits where different solvents of polarities extracted bioactive constituents therein at different rates (Shaba *et al.*, 2008).

In vitro Trypanocidal Activity

Anti-trypanosomal activates of each extract of *E. adenophorum* (leaves and flowers) are presented in Tables 20.1–20.8. Anti-trypanosomal activity varied from immobilization, reduction and the killing of trypanosomes at different concentrations used. Antitrypanosomal activity order of *E. adenophprum* leaves extracts (methanol, hexane, chloroform and aqueous) was the same as that of extraction of bioactive constituents. While that of *E. adenophorum* (flowers) was methanol, aqueous, hexane and chloroform. Extracts of *E. adenophorum* flowers demonstrated more antitrypanosomal activity with maximal activity at 1000 μg ml^{-1} (40.00±0.0 to 15.00±0.58) than leaves extracts (40.00±0.0 to 23.33 ±0.33). MES extracted bioactive

Table 20.1: *In vitro* Trypanocidal Activity of Comparative Extraction of *Eupatorium adenophorum* Flowers with Hexane Extract of different Polarities against *Trypanosma evansi* on Alsever's Medium

Concentration of Plant Extract in μg/ml	*1 h*	*2 h*	*3 h*	*4 h*	*5h*
250	40.00±0.0	40.00±0.0	40.00±0.0	38.33±0.33	36.67±0.33
500	40.00±0.0	40.00±0.0	40.00±0.0	37.33±0.33	36.33±0.33
750	40.00±0.0	40.00±0.0	40.00±0.0	34.67±0.33	33.00±0.0
1000	40.00±0.0	40.00±0.0	40.00±0.0	31.00±0.0	26.67±0.33
Control (Negative control)	40.00±0.0	40.00±0.0	40.00±0.0	40.00±0.0	40.00±0.0
Diminazineaceturate (50) Positive control					

Bioassay status: Reduction of trypanosomes counts at concentration of 1000 μg/ml and but no complete killing of trypanosomes at hours of observation. An average mean parasites count of 37.67± 0.58 is statistically critical value. Average mean trypanosomes counts from 37.67± 0.58 and below is significant between the treatment groups and negative control. ($P \leq 0.05$ to 0.01).

Table 20.2: *In vitro* Trypanocidal Activity of Comparative Extraction of *Eupatorium adenophorum* Flowers with Chloroform Extract of different Polarities against *Trypanosma evansi* on Alsever's Medium

Concentration of Plant Extract in μg/ml	*1 h*	*2 h*	*3 h*	*4 h*	*5h*
250	40.00±0.0	40.00±0.0	40.00±0.0	38.33±0.33	36.67±0.33
500	40.00±0.0	40.00±0.0	40.00±0.0	38.67±0.33	36.67±0.33
750	40.00±0.0	40.00±0.0	40.00±0.0	36.00±0.0	33.67±0.33
1000	40.00±0.0	40.00±0.0	40.00±0.0	33.33±0.33	29.67±0.33
Control (Negative control)	40.00±0.0	40.00±0.0	40.00±0.0	40.00±0.0	40.00±0.0
Diminazineaceturate (50) Positive control	22.33±0.33	9.000±0.58	1.333±0.33	0.0±0.0	0.0±0.0

Bioassay status: Reduction of trypanosomes counts from concentration of 1000 μg/ml butno complete killing of trypanosomes at hours of observation. An average mean trypanosomes count of 37.67± 0.58 is statistically critical value. Average mean trypanosomes counts from 37.67± 0.58 and below is significant between the treatment groups and negative control. ($P \leq 0.05$ to 0.01).

Table 20.3: *In vitro* Trypanocidal Activity of Comparative Extraction of *Eupatorium adenophorum* Flowers with Methanlic Extract of different Polarities against *Trypanosma evansi* on Alsever's Medium

Concentration of Plant Extract in μg/ml	*1 h*	*2 h*	*3 h*	*4 h*	*5h*
250	40.00±0.0	40.00±0.0	37.67±0.33	35.33±0.33	32.33±0.33
500	40.00±0.0	40.00±0.0	36.00±0.0	31.67±0.33	28.00±0.0
750	40.00±0.0	37.33±0.33	30.33±0.33	26.33±0.33	19.33±0.33
1000	40.00±0.0	33.33±0.33	28.33±0.33	22.00±0.0	15.00±0.58
Control (Negative control)	40.00±0.0	40.00±0.0	40.00±0.0	40.00±0.0	40.00±0.0
Diminazineaceturate (50) Positive control	22.33±0.33	9.000±0.58	1.333±0.33	0.0±0.0	0.0±0.0

Bioassay status: significant reduction of trypanosomes counts from concentration of 750 to 1000 μg/ml but no complete killing of parasites at any concentration/hour during the period of observation. An average mean trypanosomes count of 37.67± 0.58 is statistically critical value. Average mean trypanosomes count from 37.67± 0.58 and below is significant between the treatment groups and negative control. ($P \leq 0.05$ to 0.01).

Table 20.4: *In vitro* Trypanocidal Activity of Comparative Extraction of *Eupatorium adenophorum* Flowers with Aqueous Extract of different Polarities against *Trypanosma evansi* on Alsever's Medium

Concentration of Plant Extract in μg/ml	*1 h*	*2 h*	*3 h*	*4 h*	*5h*
250	40.00±0.0	40.00±0.0	38.00±0.0	35.00±0.0	30.67±0.33
500	40.00±0.0	36.33±0.33	30.33±0.33	26.33±0.33	23.67±0.33
750	40.00±0.0	37.33±0.33	28.67±2.8	27.00±0.0	20.33±0.33
100	40.00±0.0	35.00±0.0	28.00±0.0	23.00±0.0	16.33±0.33
Control (Negative control)	40.00±0.0	40.00±0.0	40.00±0.0	40.00±0.0	40.00±0.0
Diminazineaceturate (50) Positive control	22.33±0.33	9.000±0.58	1.333±0.33	0.0±0.0	0.0±0.0

Bioassay status: significant reduction of trypanosomes counts from concentration of 750 to 1000 μg/ml but no complete killing of trypanosomes at any concentration/hour during the period of observation. An average mean trypanosomes count of 37.67± 0.58 is statistically critical value. Average mean trypanosomes counts from 37.67± 0.58 and below is significant between the treatment groups and negative control ($P \leq 0.05$ to 0.01).

Table 20.5: *In vitro* Trypanocidal Activity of Comparative Extraction of *Eupatorium adenophorum* Leaves with Hexane Extraction of different Polarities against *Trypanosma evansi* on Alsever's Medium

Concentration of Test Material in µg/ml	*1 h*	*2 h*	*3 h*	*4 h*	*5h*
250	40.00 ±0.00	40.00 ±0.00	40.00 ± 0.00	40.00 ± 0.00	40.00 ± 0.00
500	40.00 ± 0.00	40.00 ± 0.00	40.00 ± 0.00	40.00 ± 0.00	37. 67 ± 0/33
750	40.00 ± 0.00	40.00 ± 0.00	40.00 ± 0.00	37.33 ± 0.67	31.67 ± 0.33
1000	40.00 ± 0.00	40.00 ± 0.00	40.00 ± 0.00	34.00 ± 0.00	27. 33 ± 0.67
Diminazineaceturate (50) Positive control	22.33±0.33	9.000±0.58	1.333±0.33	0.0±0.0	0.0±0.0
Control (Negative control)	40.00±0.00	40.00±0.00	40.00±0.00	40.00±0.00	40.00±0.00

Bioassay status: significant reduction of trypanosomes counts from concentration of 1000 g/ml but no complete killing of trypanosomes throughout hours of observations in any concentration. An average mean trypanosomes count of 37.67± 0.58 is statistically critical value. Average mean trypanosomes counts from 37.67± 0.58 and below is significant between the treatment groups and negative control. ($P \leq 0.05$ to 0.01).

Table 20.6: *In vitro* Trypanocidal Activity of Comparative Extraction of *Eupatorium adenophorum* Leaves with Chloroform Extract of different Polarities against *Trypanosma evansi* on Alsever's Medium

Concentration of Test Material in µg/ml	*1 h*	*2 h*	*3 h*	*4 h*	*5h*
250	40.00 ± 0.00	40.00 ± 0.00	40.00 ± 0.00	40.00 ± 0.00	40.00 ± 0.00
500	40.00 ± 0.00	40.00 ± 0.00	40.00 ± 0.00	40.00 ± 0.00	35.33 ± 0.67
750	40.00 ± 0.00	40.00 ± 0.00	40.00 ± 0.00	37. 00 ± 0.00	30 .00 ± 0.00
1000	40.00 ± 0.00	40.00 ± 0.00	40.00 ± 0.00	33. 67 ± 0.57	27.33 ± 0.88
Diminazineaceturate (50) Positive control	22.33±0.33	9.000±0.58	1.333±0.33	0.0±0.0	0.0±0.0
Control (Negative)	40.00±0.0	40.00±0.0	40.00±0.0	40.00±0.0	40.00±0.0

Bioassay status: significant reduction of trypanosomes counts at 1000µg/ml but no complete killing of trypanosomes throughout hours of observation. An average mean trypanosomes count of 37.67± 0.58 is statistically critical value. Average mean trypanosomes counts from 37.67± 0.58 and below is significant between the treatment groups and negative control. ($P \leq 0.05$ to 0.01).

Table 20.7: *In vitro* Trypanocidal Activity of Comparative Extraction of *Eupatorium adenophorum* Leaves with Methanolic Extract of different Polarities against *Trypanosma evansi* on Alsever's Medium

Concentration of Test Material in µg/ml	*1 h*	*2 h*	*3 h*	*4 h*	*5h*
250	40.00 ± 0.00	40.00 ± 0.00	37.67 ± 0,33	37.00 ± 0.00	34. 67 ± 0.33
500	40.00 ± 0.00	40.00 ± 0.00	37.00 ± .058	36 33 ± 0.33	30.00 ± 00
750	40.00 ± 0.00	40.00 ± 0.00	35.33 ± 0.33	33.67 ± 0.33	24.67 ± 0.67
1000	39.33 ± 0.58	40.00 ± 0.00	34.67 ± 0.33	31 00 ± 0.00	23.33 ± 0.33
Diminazineaceturate (50) Positive control	22.33±0.33	9.000±0.58	1.333±0.33	0.0±0.0	0.0±0.0
Control (Negative)	40.00±0.0	40.00±0.0	40.00±0.0	40.00±0.0	40.00±0.0

Bioassay status: significant reduction of trypanosomes counts from concentration of 750 µg/ml but no complete killing of parasites throughout hours of observation. An average mean trypanosomes count of 37.67± 0.58 is statistically critical value. Average mean trypanosomes counts from 37.67± 0.58 and below is significant between the treatment groups and negative control. (F ≤ 0.05 to 0.01).

Table 20.8: *In vitro* Trypanocidal Activity of Comparative Extraction of *Eupatorium adenophorum* Leaves with Water Extraction of different Polarities against *Trypanosma evansi* on Alsever's Medium

Concentration of Test Material in µg/ml	*1 h*	*2 h*	*3 h*	*4 h*	*5h*
250	40.00 ± 0.00	40.00 ± 0.00	40.00 ± 0.00	37.67 ± 0.33	35.67 ± 0.33
500	40.00 ± 0.00	40.00 ± 0.00	40.00 ± 0.00	37.33 ± 33	35.33 ± 0.67
750	40.00 ± 0.00	40.00 ± 0.00	37.33 ± 0.58	35.00 ± 55	34.00 ± 0.00
1000	40.00 ± 0.00	37.33 40.00 ± 0.000.38	34.67 ± 0.58	31.00 ± 0,55	29.67 ± 0.00
Diminazineaceturate (50)	22.33±0.33	9.000±0.58	1.333±0.33	0.0±0.0	0.0±0.0
Control (Negative) Positive control	40.00±0.0	40.00±0.0	40.00±0.0	40.00±0.0	40.00±0.0

Bioassay status: significant reduction of trypanosomes counts from concentration of 1000 µg/ml but no complete killing of trypanosomes throughout the period of observation. An average mean trypanosomes count of 37.67± 0.58 is statistically critical value. Average mean trypanosomes counts from 37.67± 0.58 and below is significant between the treatment groups and negative control. (P ≤ 0.05 to 0.01).

Table 20.9: Cytotoxic Effect of Methanolic Extract of *Eupatorium adenophorum* Leaves on Vero Cell Line Compared to Diminazineaceturate (Berenil)

Concentration of Test Material in µg/ml	*.Effects of Test Extract at Various Periods of Incubation (24 h, 48 h, 72 h)*						
	Eupatorium sp.	*Berenil*	*Eupatorium sp.*	*Berenil*	*Eupatorium sp.*	*Berenil*	*Control*
100	100 per cent	66.6 per cent	100 per cent	100 per cent	100 per cent	100 per cent	0
50	100 per cent	33.3 per cent	100 per cent	100 per cent	100 per cent	100 per cent	0
25	100 per cent	0	100 per cent	100 per cent	100 per cent	100 per cent	0
12.5	100 per cent	0	100 per cent	0	100 per cent	33.3 per cent	0
6.25	0	0	66.6 per cent	0	100 per cent	0	0
3.13	0	0	0	0	100 per cent	0	0
1.56	0	0	0	0	33.3 per cent	0	0

Eupatorium sp. was toxic to Vero cell line at all concentrations while, diminazine aceturate was toxic except at concentrations range 6.25-1.56 µg/ml. Same concentrations were used for diminazine aceturate (Berenil).

Table 20.10: Cytotoxic Effect of Methanolic Extract of *Eupatorium adenophorump* Flowers on Vero Cell Line Compared to Diminazineaceturate (Berenil)

Concentration of Test Material in µg/ml	*.Effects of Test Extract at Various Periods of Incubation (24 h, 48 h, 72 h)*						
	Eupatorium sp.	*Berenil*	*Eupatorium sp.*	*Berenil*	*Eupatorium sp.*	*Berenil*	*Control*
100	100 per cent	66.6 per cent	100 per cent	100 per cent	100 per cent	100 per cent	0
50	100 per cent	33.3 per cent	100 per cent	100 per cent	100 per cent	100 per cent	0
25	100 per cent	0	100 per cent	100 per cent	100 per cent	100 per cent	0
12.5	33.3 per cent	0	100 per cent	0	100 per cent	33.3 per cent	0
6.25	0	0	66.6 per cent	0	100 per cent	0	0
3.13	0	0	0	0	100 per cent	0	0
1.56	0	0	0	0	0	0	0

Eupatorium sp. and dimainazine aceturate were toxic to Vero cell line except at concentrations range 6.25-1.56 µg/ml. Same concentrations were used for diminazine aceturate (Berenil).

constituents more than other solvents system used as shown in their bioactivities. Moderate antitrypanosmal activity with significant difference (P d" 0.05 to 0.01) was observed. These antitrypanosomal activities are in line with that of comparative extraction of *Terminalia belirica* dried fruits where different solvents of polarities extracted bioactive constituents therein at different rates with corresponding bioactivities (Shaba *et al.*, 2008).

In vivo Infectivity Test

Two groups of mice were inoculated with MES contents of ELISA plate wells with moderate anti-trypanosomal activity. Both groups of mice died due to parasitaemia during the period of observation.This is because none of ELISA plate wells was devoid of trypanosomes. *In vivo* Infectivity assessment of anti-trypanosomal activity is comparable to anti-trypanosomal effects of the aqueous extract of *Brassica oleracea* buds (fruits), MPES of *Ageratum houstonionum* flowers and *Terminalia chebula* dried fruits where inoculated mice with contents of ELISA plate wells with apparently killed trypanosomes survived. And the other groups with reduced trypanosomes count, like MES of *E. adenophorum*, died of parasitaemia (Igweh *et al.*, 2002; Shaba *et al.*, 2011c; Shaba *et al.*, 2008; Shaba *et al.*, 2007).

In vitro Cytotoxicity Test

During the process of *in vitro* cytotoxic effects of MES of *E. adenophorum* (leaves and flowers) and diminazineaceturate at the same concentrations on Vero cells, distortion, swelling, sloughing and death of Vero cells compared to negative normal cells in control wells (Tables 20.9 and 20.10) were noted. Both MES of *E. adenophorum* flowers, leaves and diminazineaceturate were cytotoxic to Vero cells in all concentrations except at concentrations of 1.56 and 1.56-6. 25 µg ml^{-1}. These results are in agreement with *in vitro* cytotoxicity tests of extraction of *Ageratum houstonionum* flowers and methanolic extract of *Khaya senegalensis* root bark in which similar cytotoxic effects were observed (Shaba *et al.*, 2011c; Shaba *et.al.*, 2011b).

From this investigation, MES of *E. adenophorum* leaves and flowers possess moderate anti-trypanosomal activity. This is the first attempt to extract *E. adenophrum* leaves and flowers with solvent of different polarities, and to determine which of them will extract more of the bioactive constituents responsible for antitrypanosomal activity. MES of *E. adenophorum* leaves and flowers extracted more of the bioactive constituents than other solvents as seen on the TLC plates. Anti-trypanosomal activity of *E. adenophorum* leaves and flowers may be due to the presence of alkaloids and flavonoids already identified.

Acknowledgements

Financial contributions toward the research by India and Nigeria governments are highly acknowledged.

References

Abiodun, O.O., Gbotosho, G,O., Ajaiyeoba, E,O., Brun, R., and Oduola, A.M.(2012). Antitrypanosomal activity of some medicinal plants from Nigerian ethnomedicine. *Parasitol Res.*, **2**: 521-526.

Freiburghaus, F., Steck, A., Ptander, H., and Brun, R. (1998). Bioassay guided lsolation of a diastereoisomer of kolavenol from *Entadaabsys sinica* active on *Trypanosoma bruceirdesiense*. *Journal of Ethnopharmacology*, **61**: 179-183.

Igweh, A.C., Aguiyi, J.C., and Okwuaasaba, F.K., 2002. Antitrypanosomal effect of the aqueous extract of *Brassisca oleracea*. *Journal of Fitoterapia*, **71**: 17-21.

Lumsden, W.H.R., Herbert, W.J., and McNeilage, G.J.C. (1973). *In Vitro* Techniques with Trypanosomes. Churchil Livingstone, London.

Nok, A.J., and Nock, C. (2002). Transferin coupled azanthraquinone enhances the killing effects on trypanosomes. The role of lysosomalmannosidase. *Journal Patasites*, **9**: 375-379.

Oliveira, B.A., Pereira, D.G., Fernandes, A.M.A.P., DeCastro, S.L., Souza, A.R.M., Brito, A.O., and DeSouza Duran, N. (2004). Trypanocidal activity of 2-propen-1-amine derivatives on trypomastigotes culture and in animal model. *Parasitol Res.*, **9**: 11-25.

Shaba, P., Pandey, N.N., Sharma, O.P., Rao, J.R., and Singh, R.K. (2007). Comparative antitrypanosomal activity of *Terminalia chebula* dried fruits against *Trypanosoma evansi*. *Planta Medica*, **73**: 997-1034.

Sarveswaran, S., Marati, R.V., and Marathaiveeran, P.B. (2006). Effects of *Terminalia arjuna* bark extract on opoptosis of Human Hepatoma cell line (HEPG2). *Indian J. Gastrol.*, **12**: 1015-1024.

Shaba, P., Pandey, N.N., Sharma, O.P., Rao, J.R., and Singh, R.K. (2008). I*n Vitro* trypanocidal and cytotoxicity effects of methanolic extract of *Vitex negundo* leaves against *Trypanosoma evansi*. The 13th Congress of the Federation of Asian Veterinary Associations.Pages 43-44.FAVA-OIE Joint Symposium on Emerging Diseases, Bangkok, Thailand.

Shaba, P., Pandey, N.N., Sharma, O.P., Rao, J.R., and Singh, R.K. (2011a). *In vitro* trypanocidal activity of methanolic extracts of *Quercus borealis* leaves and *Zingiber officinale* roots. *Greener J. Agric. Sci.*, **1(1)**: 41-47.

Shaba, P., Pandey, N.N., Sharma, O.P., Rao, J.R., and Singh, R.K. (2012). Anti-trypanosomal activity of *Piper nigrum* L. (Black Pepper) against *Trypanosoma evansi*. *J. Vet. Adv.*, **2(4)**: 161-167.

Soulsby, E.J.J. (1982). *Helminths, Arthropods and Protozoa of Domesticated Animals*, 7th edn.Bailliere Tindall and Casell Ltd., London.

Shaba, P., Pandey, N.N., Sharma, O.P., Rao, J.R., and Singh, R.K. (2008). *In vitro* trypanocidal activity of comparative extraction of *Terminalia bellirica* (Combretaceae) dried fruits with solvents of different polarities against T*rypanosma evansi*. *Internet Journal of Veterinary Medicine*, **6(1)**:

Shaba, P., Pandey, N.N., Sharma, O.P., Rao, J.R., and Singh, R.K. (2011b). Antitrypanosomal activity of methanolic extract of *Khaya senegalensis* root bark against *Trypanosoma evansi*. *Intl. J. Food, Agric.Vet. Sci.*, **1(1)**: 21-26.

Shaba, P., Pandey, N.N., Sharma, O.P., Rao, J.R., and Singh, R.K. (2011c). Antitrypanosomal potential of methanolic extract of *Ageratum houstonionum* flowers against *Trypanosoma evansi*. *Intl. J. Basic appl. Medical Sci.*, **1(1)**: 149-154.

Sidwell, R.W., and Huffman, J.H. (1997). Antiviral drug resistance. *Res. Virol.*, **148**: 353-365.

Stahl, E. (1969). Thin layer chromatography. A Laboratory Handbook Springer, New York.

Paraj, S., Polana, S. P. V., Vidyasagar, S. A., and Al dosari, M. A. (2012). Antifeedant activity of three essential oils againstthe red palm weevil, *Rhynchophorus ferrugineus*. *Bulletin of Insectology*, **65**: *71-76.*

Williamson, J., March, J.C., and Scott-Finning, J.J. (1982). Drug synergy in experimental african trypanosomiasis. *Tropennmedizin und Parasitologie*, **33**: 76-82.

World Health Organization (2004). Anon; 2004 communicable Disease Surveillance and Rsponse.WHO, Geneva.

Woo, P.T.K. (1970). Evaluation of the haemataocrit centrifuge and other techniques for the field diagnosis of human trypanosomiasis and filariasis. *J. Acta Trop.*, **28**: 298-303.

Wurochekke, A.U., and Nok, A.J. (2004). *In vitro* anti trypanosomal activity of some medicinal plants used in the treatment of trypanosomosis in Northern Nigeria. *Afri. J. Biotech.*, **3**: 481-483.

Young, V., Schmitz, V., Vanner-Santos, M.A., Lima, A.P.C.A., Lalmanach, G., Juliano, L., Gauther, F., and Scharfstein, J. (2000). Altered expression of cruzipain and a cathepsin b-like target in a *Trypanosoma cruzi* cell line displaying resistance to synthetic inhibitors of cysteine-proteinases. *Mol. Biochem. Parasitol.*, **109**: 47-59.

Utilisation and Management of Medicinal Plants Vol. 3 (2015) *Pages* **307–313**
Editors: **V.K. Gupta and A. Kaul**
Published by: **DAYA PUBLISHING HOUSE, NEW DELHI**

21

Antibacterial and Antifungal Efficacy of Different Varieties of *Dicranopteris linearis*

M. Johnson[1]*, D. Patric Raja[1], V. Irudayaraj[1] and N. Janakiraman[1]

ABSTRACT

The present study was aimed to investigate the antimicrobial properties of ethanolic extracts of Dicranopteris linearis var. sebastiana, Dicranopteris linearis var. tenuis and Dicranopteris linearis var. brevis for their ability to inhibit the growth of microorganisms. The antimicrobial activity was evaluated using paper disc diffusion method against three pathogenic bacteria and two pathogenic fungi using three different concentrations viz., 25, 50 and 100 µg/ml. The maximum degree of antibacterial and antifungal activity was observed in D. linearis var. sebastiana followed by D. linearis var. tenuis. D. linearis var. brevis showed comparatively low degree of activity. All the five pathogens showed the maximum inhibition in the highest concentration (100 µg/ml). The results revealed that the crude extracts are proved to be good sources of potential antimicrobial compounds.

Keywords: *Dicranopteris linearis*, Antimicrobial, Disc diffusion.

Introduction

The value of medicinal plants to the mankind is very well proven. It is estimated that 70 to 80 per cent of the people worldwide rely chiefly on traditional health care system and largely on herbal medicines (Shanley and Luz, 2003). India harbours about 15 percent (3000 - 3500) medicinal plants, out of 20,000 medicinal plants of the world. About 90 percent of these are found growing wild in different climatic regions

1 Centre for Plant Biotechnology, Department of Botany, St. Xavier's College (Autonomous), Palayamkottai, Tirunelveli – 627 002, Tamil Nadu, India.

* *Corresponding author*: E-mail: ptcjohnson@gmail.com

of the country (Farombi, 2003). Medicinal plants represent a rich source of antimicrobial agents (Bhimba *et al.*, 2010; Darabpour *et al.*, 2010; Irudayaraj *et al.*, 2010; Jarrar *et al.*, 2010; Bhattacharjee *et al.*, 2010). Different parts of medicinal plants were used as extract for raw drugs and they possess varied medicinal properties. Some of these raw drugs are collected in smaller quantities by local use while many other raw drugs are collected in larger quantities and traded in market as raw material for many herbal industries (Uniyal *et al.*, 2006). The increasing failure of chemotherapeutics and antibiotic resistance exhibited by pathogenic microbial infectious agents has led to the screening of several medicinal plants for their potential antimicrobial activity (Elizabeth, 2005).

Though economic and medicinal values of higher plants, especially the angiosperms have been largely investigated (Pugazharasi *et al.*, 2009; Rajan and Jeevagangai, 2009, Patience *et al.*, 2010; Bagepalli *et al.*, 2010), pteridophytes have been unfortunately ignored. Pteridophytes make as important contribution to the earth's plant diversity. Being the second largest group of vascular plants, they form a significant, dominant component of many plant communities. Many ferns and fern allies are growing luxuriantly on the slopes of Western Ghats (Caius, 1935; Nair, 1959).

Materials and Methods

Healthy, disease free endemic ferns of *Dicranopteris linearis* var. *sebastiana*, *Dicranopteris linearis* var. *tenuis* and *Dicranopteris linearis* var. *brevis* were collected from Kothayar and Palni hills of Western Ghats, South India. The whole plants were air and shade dried for two weeks at room temperature and pulverized to powder using mortar and pestle. The powered materials (50 g) were extracted successively with 200 ml of ethanol by using Soxhlet extractor for 8 h at a temperature not exceeding the boiling point of the solvent. The ethanolic extracts were filtered using Whattman filter paper (No.1) and then concentrated in vacuum at 40°C using Rotary evaporator. The residues obtained were stored in a freezer -70° C until further tests (Aliero and Afolayan, 2006). Antibacterial and Antifungal studies were carried out by disc diffusion method (Paul Raj *et al.*, 2011; Balakumar *et al.*, 2011) against the pathogens *viz.*, *Proteus aureus*, *Klebsiella pneumoniae*, *Streptococcus* sp., *Aspergillus niger* and *Fusarium* sps. Different concentrations of extracts ranging from 25 to100 µg/ml were used for bacterial and fungal sensitivity test. The inhibition zone against the pathogenic bacteria and fungi were recorded. The experiments were repeated in triplicates and the results were documented.

Results

The ethanolic extracts were tested for antibacterial and antifungal activity against five human pathogens by using three different concentrations *viz.*, 25, 50 and 100 µg/ml. The maximum degree of antibacterial and antifungal activity was observed in *D. linearis* var. *sebastiana* followed by *D. linearis* var. *tenuis* and *D. linearis* var. *brevis* showed comparatively low degree of activity. The results of antibacterial and antifungal activity of three different varieties of *D. linearis* were shown in Table 21.1.

Table 21.1: Antibacterial and Antifungal Activity of different Varieties of *D. linearis*

Species	*Zone of Inhibition in cm*														
	P. aureus			*K. pneumoniae*			*Streptococcus sp.*			*A. niger*			*Fusarium sp.*		
Conc. in µg/ml	*25*	*50*	*100*	*25*	*50*	*100*	*25*	*50*	*100*	*25*	*50*	*100*	*25*	*50*	*100*
D. linearis var. *sebastiana*	1.2	2.4	3.1	1	1.8	2.2	0	1.2	1.8	0	1.1	2	0	1	1.6
D. linearis var. *tenuis*	0	1.2	1.8	0	0	1.6	0	0	1.2	0	0	1.2	0	0	0
D. linearis var. *brevis*	0	0	1.2	0	0	0	0	0	0	0	0	1.4	0	0	0

The ethanolic extract of *D. linearis* var. *sebastiana* showed the antibacterial activity against three pathogens *viz., P. aureus, K. pneumoniae* and *Streptococcus* sp. with the inhibition zones 3.1 cm, 2.2 cm and 1.8 cm and antifungal activity against two selected pathogens *viz., A. niger* and *Fusarium* sp. with 2 cm and 1.6 cm of inhibition zones respectively. All the five pathogens showed the maximum inhibition in the highest concentration (100 µg/ml). *D. linearis* var. *tenuis* exhibited the antibacterial activity against the selected bacteria *P. aureus, K. pneumoniae* and *Streptococcus* sp. with the inhibition zones 1.8 cm, 1.6 cm and 1.2 cm respectively and antifungal activity in *A. niger* with the inhibition zone 1.2 cm in the highest concentration (100 µg/ml). *Fusarium* sp. is resistant to the ethanolic extract of *D. linearis* var. *tenuis*. The antibacterial activity of *D. linearis* var. *brevis* was observed only against *P. aureus* with the maximum zone of inhibition 1.2 cm and antifungal activity only against *A. niger* with 1.8 cm of inhibition zones in 100 µg/ml concentration. *K. pneumonia, Streptococcus* sp. and *Fusarium* sp. are resistant to the ethanolic extracts of *D. linearis* var. *brevis*.

Discussion

The *in vitro* antibacterial and antifungal activity of the ethanolic extracts of three different varieties of *D. linearis* was quantitatively assessed on the basis of zone of inhibition. All the plants studied in the present investigation exhibited varying degree of zone of inhibition against the selected bacterial and fungal pathogens. The present study revealed that the antibacterial and antifungal efficacy of three different varieties of *D. linearis* against the selected strains and the degree of inhibition varied depending upon the concentration of the extract. The ethanolic extract of *D. linearis* var. *sebastiana* showed highest activity (5/5) against the pathogens followed by *D. linearis* var. *tenuis* (4/5) and *D. linearis* var. *brevis* (2/5). The presence of antimicrobial activity in a particular species may be due to the presence of one or more bioactive compounds such as alkaloids, glycosides, flavonoids, steroids, phenols, saponins, tannins etc. (Khan *et al.*, 2010). Recently, a number of plants have been reported for antimicrobial properties of pteridophytes across the world (Maridass and Ghanthikumar, 2008; Singh *et al.*, 2008; Maridass, 2009; Maridass and Ravichandran, 2009). The present observations confirmed the anti-microbial property and supplemented the previous observations on antimicrobial activity of pteridophytes.

Occurrence of bacterial and fungal diseases is a serious problem of the present world. This is because of the development of anti-microbial drug resistance of the pathogens and side effects exhibited by the drugs used for microbial diseases. Hence there is a great demand for safer, alternative and effective chemotherapeutic agents (Irobi and Darambolo, 1993). The results of present study showed diverse in effectiveness (zone of inhibition) which may be attributed due to the presence of the secondary metabolites. *Proteus* sp. can cause urinary tract infections and hospital acquired infections (URL, 2014). The ethanolic extracts of all the three varieties of *D. linearis* showed inhibition zone against *Proteus* sp. which illustrates that these three plant extracts can be used to cure urinary tract infections and hospital acquired infections. *K. pneumoniae* is an important cause of human infections and several diseases *viz.*, urinary tract infections, nosocomial infections, pneumonia, septicemias and soft tissue infections. The diseases caused by *K. pneumoniae* can result in death of patients who are immunodeficient (URL, 2014 a). In the present study, the ethanolic extracts of *D. linearis* var. *sebastiana* and *D. linearis* var. *tenuis* showed antibacterial activity against *K. pneumonia*. It suggested that the plants can be used to treat urinary tract infections, nosocomial infections, pneumonia, septicemias and soft tissue infections. Streptococcal infections cause a variety of health problems which includes skin diseases, blood infections, pneumonia and meningitis in newborns (URL, 2014 b). The ethanolic extracts of *D. linearis* var. *sebastiana* and *D. linearis* var. *tenuis* exhibited antibacterial activity against *Streptococcus* sp. which shows that these two plants can be used to cure skin diseases, blood infections, pneumonia and meningitis. *A. niger* is one of the most common causes of otomycosis (fungal ear infections), which can cause pain, temporary hearing loss, and in severe cases, damage the ear canal and tympanic membrane (URL, 2014 c). The ethanolic extracts of all the three varieties of *D. linearis* showed the antifungal activity against *A. niger*. It confirms that the plant extracts can be used to treat otomycosis and other causes. *Fusarium* infections in humans may occur in the nails and in the cornea (URL, 2014 d). The ethanolic extracts of *D. linearis* var. *sebastiana* only demonstrated the activity against the *Fusarium* sps. The results of the present study advised that the extracts of *D. linearis* var. *sebastiana* can be used to treat fusarial infections.

Conclusion

This research work gives further scope to screen the chemical constituents of the ethanolic extracts of different varieties of *D. linearis*. By isolating and identifying these bioactive compounds new drugs can be formulated which will be very useful to combat the various infectious diseases caused by pathogenic bacteria.

References

Aliero, A.A., and Afolayan, A.J. (2006). Antimicrobial activity of *Solanum tomentosu*. *Afr J Biotechnol.*, **5**(4): 369-72.

Bagepalli, S.A.K., Kuruba, L., Jayaveera, K.N., Devangam, S.S., Avalakondarayappa, A.K., and Bachappa, M. (2010). Antioxidant and antipyretic properties of methanolic extract of *Amaranthus spinosus* leaves. *Asian Pac J Trop Med.*, **3(9)**: 702-706.

Balakumar, S., Rajan, S., Thirunalasundari, T., and Jeeva. S. (2011). Antifungal activity of *Aegle marmelos* (L.) Correa (Rutaceae) leaf extract on dermatophytes. *Asian Pacific Journal of Tropical Biomedicine*, **4**: 309-312.

Bhattacharjee, I., Chatterjee, S.K., and Chandra, G. (2010). Isolation and identification of antibacterial components in seed extracts of *Argemone mexicana* L. (Papaveraceae). *Asian Pac J Trop Med.*, **3**(7): 547-551.

Bhimba, B.V., Meenupriya, J., Joel, E.L., Naveena, D.E., Kumar, S., and Thangaraj, M. (2010). Antibacterial activity and characterization of secondary metabolites isolated from mangrove plant *Avicennia officinalis*. *Asian Pac J Trop Med.*, **3**(7): 544-546.

Caius, J.F. (1935). The medicinal and poisonous ferns of India, *J Bombay Nat Hist Soc.*, 341-361.

Darabpour, E., Motamedi, H., and Nejad, S.M.S. (2010). Antimicrobial properties of *Teucrium polium* against some clinical pathogens. *Asian Pac J Trop Med.,* **3**(2): 124-127.

Elizabeth, K.M. (2005). Antimicrobial activity of *Terminalia bellerica. Indian J Clin Biochem.,* **20**: 150-153.

Farombi, E.O. (2003). African indigenous plants with chemotherapeutic potentials and biotechnological approach to the production of bioactive prophylactic agents. *African J Biotech.,* **2**: 662 - 671.

Irobi, O.N., and Darambolo, S.O. (1993). Antifungal activity of crude extracts of *Mitracarpus villosus* (Rubiaceae). *J Ethnopharmacol* **40**, 137-140.

Irudayaraj, V., Janaky, M., Johnson, M., and Selvan N. (2010). Preliminary phytochemical and antimicrobial studies on a spike-moss *Selaginella inaequalifolia* (hook. and grev.) Spring. *Asian Pac J Trop Med.,* **3(12)**: 957-960.

Jarrar, N., Abu-Hijleh, A., and Adwan, K. (2010). Antibacterial activity of *Rosmarinus officinalis* L. alone and in combination with cefuroxime against methicillin-resistant *Staphylococcus aureus*. *Asian Pac J Trop Med.,* **3**(2): 121-123.

Khan, Z.S., Shinde, V.N., Bhosle, N.O., and Sahera Nasreen. (2010). Chemical composition and Antimicrobial activity of Angiospermic plants. *Middle- East Journal of Scientific Research,* **6**(1): 56-61.

Maridass, M., and Ghanthikumar, S. (2008). Antibacterial activity of leaves of *Blechnum orientale* L., *Pharmacologyonline,* **3**: 58-60.

Maridass, M., and Ravichandran, B. (2009). The Screening of phytochemical and antioxidant activity of two endemic ferns, *Mycodium excertum* (Wall. Ex Hook) Copel and *Tectaria zeylanica* (Holtt) Sledge. *Pharmacologyonline,* **1**: 56-63.

Maridass, M. (2009). Antibacterial Activity of *Mecodium exsertum* (Wall.ex Hook) Copel –A Rare Fern. *Pharmacologyonline,* **1**: 1-7.

Nair, BK. (1959). Medicinal ferns of India. *Bull Nat Bot Gard, Lucknow,* **29**: 1-36.

Patience, O.O., Edwin, O.O., Philip, F.U, Ernest, K.D., and Obiorah. D.C. (2010). Seasonal variation for the antidiabetic activity of *Loranthus micranthus* methanol extract. *Asian Pac J Trop Med.,* **3**(3): 196-199.

Paul Raj, K., Irudayaraj, V., Johnson, M., and Patric Raja, D. (2011). Phytochemical and antibacterial activity of epidermal glands extract of *Christella parasitica* (L.) H. Lev. *Asian Pacific Journal of Tropical Biomedicine,* **1**(1): 8-11.

Pugazharasi, G., Meenakshi, S.A., Ramesh, K.N., Bastin, C.M., and Natarajan, E. (2009). Screening of antimicrobial activity of *Phyllanthus maderaspatensis* L. *J Basic Appl Biol.,* **3**(3 and 4): 43-49.

Rajan, S., and Jeevagangai, T.J. (2009). Studies on the antibacterial activity of Aegle marmelos-fruit pulp and its preliminary phytochemistry. *J Basic Appl Biol.,* **3**(1 and 2): 76-81.

Shanley, P., and Luz, L. (2003). The impacts of forest degradation on medicinal plant use and implication for health care in Eastern Amazonia. *BioScience*, **53** (6): 573 - 584.

Singh, M., Singh, N., Khare, P.B., and Rawat, A.K.S. (2008). Antimicrobial activity of some important *Adiantum* species used traditionally in indigenous systems of medicine. *J. Ethnopharmacol.*, **115**(2): 327-329.

Uniyal, S.K., Singh, K.N., Jomwal, P., and Lal, B. (2006). Traditional use of medicinal plants among the tribal communities of Chota Bengal, Western Himalaya. *J Ethnobiol Ethnomed.*, **2**: 1-14.

URL, 2014 a http: / /microbewiki.kenyon.edu/index.php/Klebsiella_pneumoniae

URL, 2014 b http: / /www.nlm.nih.gov/medlineplus/streptococcalinfections.html

URL, 2014 c http: / /en.wikipedia.org/wiki/Aspergillus_niger

URL, 2014 d http: / /en.wikipedia.org/wiki/Fusarium

URL, 2014 http: / /www.phageinternational.com/pathogens/proteus.html

Utilisation and Management of Medicinal Plants Vol. 3 (2015) *Pages* **315–320**
Editors: **V.K. Gupta and A. Kaul**
Published by: **DAYA PUBLISHING HOUSE, NEW DELHI**

22

Antipyretic Effect of Honey in Albino Rats

B.V. Owoyele[1*] and E.O. Adeyeye[1]

ABSTRACT

Honey was evaluated for its antipyretic effect in albino rats using the Brewers' yeast induced hyperpyrexia test. Pyrexia was induced by subcutaneously injecting 20 per cent (w/v) Brewers' yeast suspension (10ml/kg) into the dorsum of the rats. A total of twenty five albino rats were used for the study and the animals were randomly divided into five groups with each group containing five rats. A group received 10ml/kg of normal saline; another group received 10mg/kg of indomethacin; and the remaining three groups received 25, 50, 100ml/kg body weight of honey orally. All the animals were injected with Brewers' yeast and 17 hours later the rectal temperatures of each rat were measured with digital thermometer. The results showed that all the doses of honey administered significantly ($P<0.05$) reduced the rectal temperature of the hyperthermic animals in a dose dependant manner. The results confirmed that the honey used has antipyretic effect.

Keywords: Antipyretic, Brewers' yeast, Honey, Pyrexia, Rats, Temperature.

Introduction

Fever also known as pyrexia or controlled hyperthermia (Axelrod and Diringer, 2008) is a common medical sign characterized by an elevation of temperature above the normal range of 36.5–37.5 °C (98–100 °F) due to an increase in the body temperature regulatory set-point (Karakitsos and Karabinis, 2008). This increase in set-point triggers increased muscle tone and shivering. A fever is usually one of the body's immune responses that attempt to neutralize a bacterial or viral infection. Fever is different from an uncontrolled hyperthermia because hyperthermia is an increase in

1 Department of Physiology, Faculty of Basic Medical Sciences, College of Health Sciences, University of Ilorin, P. M. B. 1515,Ilorin, Nigeria.

* *Corresponding author*: E-mail: deleyele@yahoo.com or deleyele@gmail.com

body temperature over the body's thermoregulatory set-point and this is usually due to excessive heat production and/or insufficient thermoregulation (Axelrod and Diringer, 2008). In healthy adults, the range of normal, healthy temperatures for oral temperature is 33.2–38.2 °C (92–101 °F), for rectal it is 34.4–37.8 °C (94–100 °F), for tympanic membrane it is 35.4 – 37.8 °C (96–100 °F), and for axillary it is 35.5–37.0 °C (96–99 °F) (Sund-Levander *et al.*, 2002).

Honey was an important food for Homo sapiens from his very beginnings and the relationship of man with the honey bees dates back to the Stone age (Crane, 1983). According to Crane (1975), the first written reference to honey, a Sumerian tablet writing, dating back to 2100–2000 BC, mentions honey's use as a drug and an ointment. In most ancient cultures honey has been used for both nutritional and medical purposes (Crane, 1975; Jones, 2000; Crane, 1999; Allsop and Miller, 1996). For a long time in human history it was an important carbohydrate source and the only largely available sweetener until industrial sugar production began to replace it after 1800 (Crane, 1975). Generally, honey has been used as both nutrient and medicine since ancient time (Owoyele *et al.*, 2011). An alternative medicine branch, called apitherapy, has developed in recent years, offering treatments based on honey and other bee products against many diseases. In contemporary time important uses of honey are in the treatment of wounds, burns and infections (Molan, 1999; Owoyele *et al.*, 2011).

The present study aims at investigating the effect of honey produced from the apiary of University of Ilorin, Nigeria on hyperpyrexia using Brewer's yeast. This is because previous studies on the University of Ilorin type of honey (Owoyele *et al.*, 2011) showed that the honey had anti-inflammatory activity. Therefore honey might also have an antipyretic effect as part of its biological activities.

Materials and methods

Animals

Twenty five male albino rats (weight 116±3.2g) were used for the study. They were breed in the Faculty of Basic Medical Sciences Animal House of the University of Ilorin. The animals were housed in 5 groups in cages with free access to food and water. They were fed with muse cubes from Bendel Feeds, Yoruba road, Ilorin. The study was conducted in accordance with the ethical rules on animal experimentation approved by Physiology Department of the University of Ilorin. The honey used for the study was purchased from the apiary of University of Ilorin.

Antipyrexic Test

Hyperpyrexia was induced in the rats by subcutaneous injection of 20 per cent suspension of Brewer's yeast in the dorsum of all the rats according to the method described by Loux *et al.* (1972) and Al-Shabanah and Raza (2002). The rats were randomly divided into 5 groups of 5 rats each. The dose of the Brewer's yeast injected in the animals was 10mg/kg.

The rectal temperature of all the rats was measured before yeast injection and 17 hours (Pre-drug temperature) after the injection using digital thermometer.

The rats in group A were given normal saline (control); those in group B were given indomethacin (reference) while animals in groups C-E (the test groups) were given 25, 50, 100 ml/kg of the honey orally. The rectal temperature was measured at 60, 90, and 120 minutes after the oral administration of the respective of honey, saline and indomethacin. The mean temperature of each group were also determined and compared with the respective pre drug temperatures.

Statistical Analysis

All values were expressed as Mean ± Standard Error of Mean. Statistical significance was determined using AVOVA and Duncan Test. Values with $P < 0.05$ compared with control were considered significant.

Results

The results show a significant ($p<0.05$) decrease in rectal temperature at all time intervals in the animals treated with low dose of honey (25ml/kg). There was a significant ($p<0.05$) decrease in rectal temperature at 60 and 90 minutes in animals treated with 50ml/kg of honey. Also, there was a significant ($p<0.05$) decrease in rectal temperature at 90 minutes in animals treated with 100ml/kg of honey. The result is presented in Table 22.1.

Discussion and conclusion

In this present study, it was found that the low dose of honey (25ml/kg) induced hypothermia which was persistent to the end of the experiment, being maximal at120 minutes after honey administration. The same trend of the result was observed after the administration of other doses of honey. The effect of honey on rectal temperature was dose-dependant. Pyrexia of endogenous origin usually result from the stimulatory effect of some leucocyte (cytokines) on the preoptic region of the anterior hypothalamus and cells of organum vasculosum which leads to the release of prostaglandin E_2 (PGE_2) which in turn stimulates the hypothalamic thermostat to adjust the body temperature to a higher set point (Scammell *et al.*, 1996). The antipyretic activity of honey may be attributed to inhibition of prostaglandin release because honey has been found to have the capacity to reduce blood prostaglandins level (Al-Waili and Nader, 2004).

The antipyretic effect of honey may also not be unconnected with some of its constituents. Since some literature has shown that honey contain some active agents which has the capacity to influence hyperpyrexia and such constituents include; flavoniods, saponins and alkaloids. Flavonoids, saponins (Agbaje *et al.*, 2006). Some alkaloids and flavonoids have been implicated to have the ability to block and inhibit the synthesis of prostaglandin E2 (Backhouse *et al.*, 1994) which eventually reduce elevated body temperature in animals. Like wise some flavonoids have been credited with ability to suppress TNFα which may lead to anti pyresis (Chang *et al.*, 2007; Adesokan *et al.*, 2008).

In conclusion, this study has shown that honey has antipyretic effects in addition to several therapeutic effects that have been reported in several literatures. Nevertheless

Table 22.1: Antipyretic Effect of Honey in Albino Rats

Group	*Doses*	*Temperature Before Yeast (ºC)*	*17 hours after Yeast*	*Post Drug Temperature (ºC)*		
				60 mins	*90 mins*	*120 mins*
Control	10ml/kg	36.88±0.49	38.26±0.50	37.92±0.54	37.74±0.54	37.93±0.23
Indomethacin	10mg/kg	37.02±0.44	38.14±0.40*	36.82±0.20*	36.48±0.24*	36.44±0.38*
Honey	25ml/kg	37.32±0.10	38.02±0.26	37.20±0.19*	36.92±0.20*	36.90±0.24*
Honey	50ml/kg	36.94±0.43	38.62±0.40	36.72±0.34*	36.96±0.35*	37.62±0.35
Honey	100ml/kg	36.70±0.50	37.46±0.11	36.56±0.22*	36.32±0.38*	36.62±0.24*

Each value is Mean ± Standard Error of Mean. n= 5 and * $p<0.05$.

more investigations are required to completely unveil the specific active agents and the mechanism actions.

References

Adesokan, A.A., Yakubu, M.T. Owoyele, B.V., Akanji, M. A., Soladoye, A. O., and Lawal, O.K. (2008): Effects of administration of aqueous extract of aqueous and ethanolic extract of *Enantia chlorantha* stem bark on brewer's yeast –induced pyresis in rats. *African Journal of Biochemistry Research,* **2(7)**: 165 – 169.

Agbaje, E. O., Ogunsanya, T., and Aiwerioba, O. I. R. (2006). Conventional use of honey as antibacterial agent. *Annals of African Medicine,* **5(2)**: 78 – 81.

Allsop, K.A., and Miller, J.B. (1996). Honey revisited: A reappraisal of honey in pre-industrial diets. *Br J Nutr.,* **75**: 513–520.

Al-Shabanah, O.A., and Raza, M. (2002). Vigabatrin induced hypothermia and antipyrexia: Involvement of $GABA_A$ receptors. *Pharmacology Reviews and Communications,* **12**: 25-37.

Al-Waili, N. S., and Nader, S. B. (2003). Natural honey lowers plasma prostaglandin concentrations in normal individuals. Journal of Medicinal Food, **6(2)**: 129-133.

Axelrod, Y.K., and Diringer, M.N. (2008).Temperature management in acute neurologic disorders. *Neurol Clin.,* **26 (2)**: 585–603.

Backhouse, N., Delporte, C., Givernau, M., Cassels, B.K., Valenzuela, A., and Speisky, H. (1994). Anti-inflammatory and antipyretic effects of boldine. *Inflammation Research,* **42(3-4)**: 114-117.

Chang, C.P., Huang, W.T., Cheng, B.C., Hsu, C.C., and Lin, M.T. (2007). The flavonoid baicalin protects against cerebrovascular dysfunction and brain inflammation in experimental heatstroke. *Neuropharmacology,* **52(3)**: 1024-1033.

Crane, E. (1975). History of honey. In Crane E (ed): Honey, A Comprehensive Survey. London: William Heinemann, pp 439–488.

Crane, E. (1983). The Archaeology of Beekeeping. London: Gerald Duckworth and Co.

Crane, E. (1999). The World History of Beekeeping and Honey Hunting." London: Gerald Duckworth and Co.

Jones, R. (2000). Honey and healing through the ages. In Munn P, Jones R (eds): Honey and Healing. Cardiff: International Bee Research Association IBRA, pp. 1–4.

Karakitsos, D., and Karabinis, A. (2008). Hypothermia therapy after traumatic brain injury in children. *N. Engl. J. Med.,* **359** : 1179–80.

Loux, J.J., Depalma, P.D., and Yankell, S.L. (1972). Antipyretic testing of aspirin in rats. *Toxicol. Appl. Pharmacol.,* **22**: 672-675.

Molan, P. (1999). Why honey is effective as a medicine. 1. Its use in modern medicine. *Bee World,* **80**: 79–92.

Owoyele, B.V., Oladimeji, A., David, A.J., and Soladoye, A.O. (2007). Antipyretic activities of the leaf extract of *Chromolaena odorata*. *The Tropical Journal of Health Sciences*, **14**: 1-3.

Owoyele, B. V., Adenekan, O. T., and Soladoye, A.O. (2011). Effects of honey on inflammation and nitric oxide production in Wistar rats. *Journal of Chinese Integrative Medicine*, **9**: 447- 452

Sund-Levander, M., Forsberg, C., and Wahren, L.K. (2002). Normal oral, rectal, tympanic and axillary body temperature in adult men and women: a systematic literature review. *Scand J Caring Sci.*, **16**: 122–128.

Scammell, T. E., Elmquist, J. K., Grifn, J. D., and Saper, C. B. (1996). Ventromedial Preoptic Prostaglandin E2 Activates Fever-Producing Autonomic Pathways. *J. Neurosci.*, **16**: 6246–6254.

Utilisation and Management of Medicinal Plants Vol. 3 (2015) *Pages* **321–328**
Editors: **V.K. Gupta and A. Kaul**
Published by: **DAYA PUBLISHING HOUSE, NEW DELHI**

23

A Comparative Study of Five Medicinally Important Plant Extracts against Two Multi Drug Resistant Human Pathogenic Bacteria

Sunayana Saha[1], Abhijit Dey[1], Souryadeep Mukherjee[1] and Rabindranath Bhattacharyya[1*]

ABSTRACT

Medicinal plants are generating an ever increasing interest due to their effectiveness, low cost and minimum side effects associated with drugs derived from them. Different medicinal plants, viz., Piper beetle, Psidium guajava, Lantana camara, Euphorbia hirta and Cassia fistula were selected to check their antibacterial activity against two multi drug resistant urinary tract infecting bacteria – E.coli and Klebsiella pneumonae. To check the antibacterial effect, the leaves of above plants were collected, washed properly in tap water and then dried in shade for seven days. The leaves were finally dried in hot air oven at 60pC for two hours and then grinded in a motorized grinder. One gram of powdered leaves were suspended in different solvents viz. 70 per cent methanol, ethanol, chloroform, acetone and distilled water and kept in rotary shaker for 24 hours at 35pC. Finally the extracts were filtered and Whatmann filter paper discs of 5 mm diameter was dipped into the filtrate, dried and checked for antimicrobial activity following standard disc diffusion method. The methanolic extract of P. guajava was found to be most effective against K. pneumonae than E. coli, whereas methanolic extracts of P. beetle was found to be more active against E. coli. The methanolic extract of L. camara showed moderate activity against these bacteria. But the extract of C. fistula and E. hirta showed complete resistance against these pathogenic strains. Minimum Inhibitory Concentrations (MIC) were measured accordingly. Further purification of the extracts will be

1 Microbiology Research laboratory, Department of Biological Sciences, Presidency University, 86/1 College Street, Kolkata – 700 073, West Bengal, India.

* *Corresponding author*: E-mail: rabindranathbpc@yahoo.co.in

needed for the isolation of the pure compounds which can be helpful for the better understanding of their pharmacological potentialities.

Keywords: Medicinal plants, *Piper beetle, Psidium guajava, Lantana camara, Euphorbia hirta, Cassia fistula,* Antibacterial activity, *E. coli, Klebsiella pneumonae.*

Introduction

The increasing emergence of bacterial resistance to a large number of antibiotics causes major health concern. Nowadays, infections from common pathogens are mostly resistant to several antibiotics which undermine the ability of antibiotics to control the infections(Nerurkar *et al.*, 2012; Sood and Gupta, 2012; Bahadin *et al.*, 2011). Urinary Tract Infection (UTI), either hospital acquired or community acquired is one of the most common infections presently threatened of. An estimate of sufferings due to UTI across the globe may rise to 75 per cent of female population, and 15-25 per cent of this group will suffer from a relapse of this disease (Pouwels *et.al*, 2013; Hunstad, 2010). Studies on UTI in India are rare (Zhao and Stevens, 1998; Slawson *et al.*, 1992). In eastern India, UTI is a common infection found among all ages from infants to elderly persons. However in Eastern India studies on UTI and the susceptibility pattern of antibiotics along with the choice of correct antibiotics are still undergoing. Although empirical antimicrobial treatment for UTI is accepted clinically, bacteria are developing resistance to antibiotics faster than the development of new classes of antibiotics.

The (recurrent) empirical antimicrobial treatment in patients exerts significant resistance pressure on the uropathogens. Very often clinicians prescribe broad spectrum antibiotics instead of specific antibiotic during empirical treatment, in view of resistance of the causative organism against the antibiotic. On the individual patient level, this is a clinical problem in patients with urinary tract infections (UTI), in particular in women with recurrent UTI (rUTI). Antibiotic misuse and the patients' who are non compliant or who do not complete the course of antibiotic therapy cause an increase of antibiotic resistant bacteria.

A epidemiological survey was done and the most predominant causative agents of UTI was determined. The antibiotic resistance pattern of two principal UTI pathogens, *E. coli* and *Klebsiella* spp., was also studied. It was seen that the scenario of Eastern India appears in general to be similar to that found in other parts of India as well as other parts of the globe. Greater than 50 per cent resistance was observed for penicillin combinations.

The use of plants as the source of remedies for the treatment of many diseases dates back to prehistory and people of all continents have this old tradition. India is a land of its own diversity in flora with a rich of natural resources representing the natural biological heritage. Exploration of different medicinal plants and assessments of their potent antimicrobial activity can provide an alternative source of treatment. Utilization of the specific component of the medicinal plant in drug designing or drug remodeling can also prove to be very useful in combatting drug resistant pathogens. Nanoparticles produced from these plant extracts can also be utilized as an alternative treatment.

In herbal medicines, the active components almost always act in the presence of many other complementary substances that are generally harmless. Such components give better stability and effectiveness to the active component than that of its isolated and pure form (Shariff, 2001). The isolation and determination of the pharmacological properties of the active components of herbal remediesallow for better formulations with reduced toxicity, and synthesis of more potent drugs on a commercial scale (Manna and Ablaka, 2000; Ebana *et al.*, 1991). The studied medicinal plants are the single source of many beneficial components of herbal remedies, which can be used without any known detrimental effect (Gutiearrez *et al.*, 2008). The wide availability of these common medicinal plants in West Bengal andtheir effectiveness against the studied uropathogens, make these medicinal plants a very promising alternative to commercial antibiotics, that are losing efficacy in the treatment of UTI.

Materials and Methods

Collection of Plant Leaves and Leaf Powder Preparation

Fresh leaves of *Piper beetle, Psidium guajava, Lantana camara, Euphorbia hirta* and *Cassia fistula* were collected from different southern parts of, Kolkata. Fresh leaves were washed in tap water, weighed and kept for shed drying for 14 days. The leaves were finally dried in hot air oven at 60°C for 2 hours. The dried leaves were weighed again and crushed in motorized grinder. The leaf powder thus obtained were stored in an air tight container for further use.

Extract Preparation

Two gm of each of the dried leaf powder was weighed and suspended in 10 ml of each solvent, thus making a final concentration of 200mg/ml. The solvent chosen for the experiments were 70 per cent methanol, absolute ethanol, acetone and water.

The suspension was kept in a shaker-incubator system for 24 hours at 37 p C with continuous shaking for successful extraction of the leaf powder. After 24 hrs, the extract was filtered and the filtrate was assayed for its antimicrobial activity.

Antimicrobial Disc Preparation

Whatmann filter paper was punched using a clean punching machine to produce filter paper discs of 5 mm diameter. These discs were sterilized in autoclave and dipped in the plant extract filtrate and the surface moisture is dried.

Microorganism Selection

Clinical isolates of 2 common and predominant uropathogens of West Bengal, namely *E.coli* and *Klebsiella pneumonae* was selected as the test organisms. The organisms were collected from the BELLE VUE CLINIC, Kolkata. One loopful culture of both the organism was inoculated in 10 ml nutrient broth and incubated for 24 hrs at 37 p C in a rotary shaker at 180 rpm for obtaining a 24 hour young culture.

Kirby – Bauer Method of Disc Diffusion Assay

The 24 hrs young culture were spreaded on sterile nutrient agar plates. The plates were kept for 30 minutes for absorption of the inoculum on the plates. Discs containing leaf extract filtrate of each plant, control of each solvent and standard antibiotic(streptomycin) of concentration 5mg/mlwere impregnated on the inoculated agar plates.

After 24 hrs. the plates were observed for clear zone diameter, around the discs.

Minimum Inhibitory Concentration

The leaf extract filtrates of *Psidium guajava* and *Piper betel* were dried and diluted to make concentrations, as 250 mg/ml, 200mg/ml, 150 mg/ml, 100 mg/ml, 50mg/ml and 25 mg/ml. To each concentrations, sterile filter paper discs were dipped, dried and used against the isolated uropathogens for determining the MIC, following standard Disc Diffusion Method. A control disc was prepared by dipping the disc in the solvent solution without any plant dust. The discs were impregnated on nutrient agar plates overlaid with 24 hours broth culture of the clinical isolates. The plates were incubated for 24 hours at 37°C and observed for clear zone around the discs.

Results and Discussion

From the set of experiments done, it was seen that, among all the plant extracts used, the extracts of the leaves of *Psidium guajava* and *Piper betel* were most effective than any other plant extracts assayed (Table 23.1).

Table 23.1: Inhibition Zone of different Plant Extracts

Sl.No.	*Name of Medicinal Plant*	*Zone Diameter of K. pneumonae*	*Zone Diameter of E. coli*
1.	*Psidium guajava*	12mm	–
2.	*Piper betel*	–	11 mm
3.	*Lantana camara*	9 mm	6 mm
4.	*Euphorbia hirta*	–	6 mm
5.	*Cassia fistula*	–	–
6.	Standard antibiotic(streptomycin 5mg/ml)	–	–

P. guajava extracts in different solvents were assayed (Table 23.2). Only 70 per cent methanolic extract produced a significant clear zone against *K. pneumonae*.

Table 23.2: Inhibition Zone of *Psidium guajava* in different Solvent Extract

Sl.No.	*Name of Solvents*	*Zone Diameter of K. pneumonae*	*Zone Diameter of E. coli*
1.	70 per cent methanol	12mm	–
2.	Ethanol	–	–
3.	Acetone	–	–
4.	Hot water	–	–

P. betel extracts of different solvents were assayed (Table 23.3). Both methanolic and ethanolic extract showed significant antimicrobial activity against isolated *E. coli* strains.

Table 23.3: Inhibition Zone of *Piper betel* in different Solvent Extract

Sl.No.	*Name of Solvents*	*Zone Diameter of K. pneumonae*	*Zone Diameter of E. coll*
1.	70 per cent methanol	-	11
2.	Ethanol	–	6
3.	Acetone	–	–
4.	Hot water	–	–

The *L. camara* extracts with different solvents were assayed against both the strains (Table 23.4). For both the strains moderate activity was showed by methanolic and ethanolic extracts of *L. camara*.

Table 23.4: Inhibition Zone of *Lantana camara* in different Solvent Extract

Sl.No.	*Name of Solvents*	*Zone Diameter of K. pneumonae*	*Zone Diameter of E. coli*
1.	70 per cent methanol	9	6
2.	Ethanol	6	6
3.	Acetone	–	–
4.	Hot water	–	–

Different solvent extracts of *E.hirta* were assayed (Table 23.5). *E.coli* showed moderate sensitivity against the methanolic extract of the plants.

Table 23.5: Inhibition Zone of *Euphorbia hirta* in different Solvent Extract

Sl.No.	*Name of Solvents*	*Zone Diameter of K. pneumonae*	*Zone Diameter of E. coli*
1.	70 per cent methanol	–	6
2.	Ethanol	–	–
3.	Acetone	–	–
4.	Hot water	–	–

Different extracts of *C. fistula* were assayed (Table 23.6). Both the isolates showed complete resistance against all the extracts.

Scrutinizing results from Table 23.2–23.6 it can be stated that, the methanolic extract of *Psidium guajava* gave the highest zone diameter of 12mm against *Klebsiella pneumonae*. Whereas methanolic extract of *Piper betel* gave a zone diameter of 11 mm against *E.coli*.

Table 23.6: Inhibition Zone of *Cassia fistula* in different Solvent Extract

Sl.No.	*Name of Solvents*	*Zone Diameter of K. pneumonae*	*Zone Diameter of E. coli*
1.	70 per cent methanol	-	–
2.	Ethanol	–	–
3.	Acetone	–	–
4.	Hot water	–	–

Lantana camara extract in methanol solvent produced a clear zone of diameter 9mm and 6mm against *E.coli* and *Klebsiella pneumonae* respectively.

Whereas, methanolic extract of *Euphorbia hirta* showed a zone diameter of 6 mm against *E.coli* but failed to produce any clear zone against *Klebsiella pneumonae.*

But both the uropathogens found to be completely resistant against all the extracts of *Cassia fistula.*

The MIC result of *Psidium guajava* against *K. pneumonae* (Table 23.7) depicted that all the leaf extract concentrations, except, 50mcg/ml and 25 mcg/ml showed clear zone of increasing diameter with increasing concentrations of leaf extracts.

Table 23.7: MIC of *Psidium guajava* against *Klebsiella pneuminae*

Sl.No.	*Different Concentrations*	*Zone diameter of Klebsiella pneumonae in mm*
1.	Control	–
2.	250 mg/ml	15
3.	200mg/ml	13
4.	150 mg/ml	11
5.	100mg/ml	10
6.	50 mg/ml	–
7.	25mg/ml	–

Table 23.8: MIC of *Piper betel* against *E. coli*

Sl.No.	*Different Concentrations*	*Zone diameter of E. coli in mm*
1.	Control	–
2.	250 mg/ml	14
3.	200mg/ml	11
4.	150 mg/ml	10
5.	100mg/ml	08
6.	50 mg/ml	–
7.	25mg/ml	–

The MIC result of *Piper betel* against *E. coli* (Table 23.8) depicted that all the leaf extract concentrations, except, 50mcg/ml and 25 mcg/ml showed clear zone of increasing diameter with increasing concentrations of leaf extracts.

Conclusion

Now, as the rampant emergence of Multi Drug Resistant bacteria is making the future management of infectious diseases uncertain, the World Health Organization (WHO) is putting emphasis on plants as a source of large number of potential drugs with live-saving implications (Gutiearrez *et al.*, 2008). The use of leaf extracts might be sometimes more effective in reducing toxicity than the antibiotics because of buffering might take place between different constituents. From this study, the methanolic extract of *Psidium guajava* and *Piper betel* showed satisfactory potential as antibacterial agents effective against Multi Drug Resistant *Klebsiella pneumonae* and *E. coli*. Respectively, which are common uropathogens in Eastern India.

In MIC experiments the results revealed that the pathogens failed to grow in the presence of the effective dose of the *P. guajava* and *P. betel* active ingredients, confirming that the nature of the antibacterial activity was bactericidal but not bacteriostatic. However, further research is needed to identify active principle of the extract for better drug designing.

References

Bahadin, J., Teo, S,S,H., and Mathew, S. (2011). Aetiology of community-acquired urinary tract infection and antimicrobial susceptibility patterns of uropathogens isolated. *Singapore Med J.*, **52**: 415- 420.

Ebana, R.U.-B., Madunagu, B.-E., Ekpe, E.-D., *et al.* (1991). Microbiological exploitation of cardiac glycosides and alkaloids from *Garcinia kola, Borreriao cymoides, Kola nitida* and *Citrus aurantifolia. J. Appl. Bacteriol.*, **71**: 398–401.

Hunstad, D.A. (2010). Intracellular lifestyles and immune evasion strategies of uropathogenic *Escherichia coli. Annu Rev Microbiol.*, **64**: 203–221.

Manna, A., and Abalaka, M.-E. (2000). Preliminary screening of the various extracts of *Physalis angulala* (L.) for antimicrobial activities. *Spectr. J.*, **7**: 119–125.

Nerurkar, A., Solanky, P., and Naik. S.S. (2012). Bacterial pathogens in urinary tract infection and antibiotic susceptibility pattern. *J Pharm Biomed Sci.*, **21**: 1-3.

Pouwels, K.B., Visser, S.T., and Hak, E. (2013). Effect of pravastatin and fosinopril on recurrent urinary tract infections. *J Antimicrob Chemother.*, **68**: 708-714.

Shariff, Z.-U. (2001). Modern Herbal Therapy for Common Ailments. p. 9–84. Nature.

Pharmacy Series. vol. 1. Spectrum Books Ltd., Ibadan, Nigeria in association with Safari, Books (Export) Ltd., United Kingdom.

Slawson, R.M., Trevors, J.T. and Lee, H. (1992). Silver accumulation and resistance in *Pseudomonas stutzeri. Arch. Microbiol.*, **158**: 398–404.

Sood, S., and Gupta, R. (2012). Antibiotic resistance pattern of community acquired uropathogens at a tertiary care hospital in Jaipur, Rajasthan. *Indian J Community Med.*, **37**: 39-44.

Gutieárrez, R.-M., Mitchell, S., and Solis, R.-V. (2008). *Psidium guajava*: a review of its traditional uses, phytochemistry and pharmacology. *J. Ethnopharmacol.*, **117**: 1–27.

Zhao, G.J. and Stevens, S.E.(1998). Multiple parameters for the comprehensive evaluation of the susceptibility of *Escherichia coli* to the silver ion. *Biometals*, **11**: 27–32.

Utilisation and Management of Medicinal Plants Vol. 3 (2015) *Pages* **329–346**
Editors: **V.K. Gupta and A. Kaul**
Published by: **DAYA PUBLISHING HOUSE, NEW DELHI**

24

In vitro Antibacterial Properties and Dyeing Ability of *Punica granutum* L. Rind Extract for Healthcare Textile

R. Rathinamoorthy[1*]

ABSTRACT

*This study focused on the development of single stage dyeing cum functional finishing process of textile material. In this present work, Punica granutum L peel (Pomegranate rind) identified for their high content of coloured substances. The coloured substances were extracted using simple solvent extraction method and they were applied on the textile. To improve the colour fixation different mordanting methods were used with tannic acid as a mordent. The dyed cotton materials were evaluated for their colour strength Value (K/S) and CIE L*a*b* values. Based on the K/S values, it is observed that pre mordanting gave higher colour value. The treated fabrics were also tested for their antibacterial activity against bacterial strains like Staphylococcus aureus, Escherichia coli, Klebsiella pneumoniae, Proteus vulgaris and Salmonella typhi. The result indicates that the treated cotton fabric shows a clear microbial resistance with 27-32 mm zone of inhibition in the agar diffusion test against all the above mentioned strains. Similarly, to measure the microbe resistance quantitatively, the AATCC Test Method 100-2004 was performed. The result indicates that, the treated samples show 99 per cent of reduction against Staphylococcus aureus and 90 per cent reduction against Escherichia coli.*

Further, the finished samples were evaluated for the fastness properties of colour like fastness to washing, perspiration and light. The result indicates that, the wash fastness value in the range of 3, fastness to perspiration is 3-4 and light is 2. This suggests that, the potentiality of the selected plant source as a dye and as well as an antibacterial agent. The comfort properties

1 Assistant Professor, Department of Fashion Technology, PSG College of Technology, Coimbatore, India.

* *Corresponding author*: E-mail: r.rathinamoorthy@gmail.com

of textile like water absorbency, air permeability and water vapour permeability were indicates that, the influence of the process on those properties were meager. Hence, this study suggests possible diversified application of Punica granutum L extracts in the textile era.

Keywords: *Punica granutum* L., Mordanting, Colour strength, Agar diffusion method, Challenge test, Antimicrobial efficacy.

Introduction

Textile materials (natural and synthetic) used to be coloured for value addition, look and desire of the customers. Today there are more than ten thousand dyes available commercially and seven lakh tons of dyes are produced annually (Zolinger, 1987). Dyes can be of many different structural varieties like acidic, basic, disperse, azo, anthraquinone based and metal complex dyes among others. The textile industry is the largest consumer of dye stuffs. During the coloration process a large percentage of the synthetic dye does not bind and is lost to the waste stream (Weber and Adams, 1995). The textile industry consumes a substantial amount of water in its manufacturing processes used mainly in the dyeing and finishing operations of the plants. The wastewater from textile plants is classified as the most polluting of all the industrial sectors, considering the volume generated as well as the effluent composition (Sen and Demirer, 2003). Water pollution due to discharge of non biodegradable coloured effluents from textile dye manufacturing and textile dyeing still is one of the major environmental concerns in the world today (Shahid adeel *et al.*, 2009).

The interest towards the eco friendly dyeing, natural dye material is keeps on increasing in the developed and developing countries. Natural dyes of comprise those colourants that are obtained from animal or vegetable matter without chemical processing. Colourants agents of plants are derived from roots, leafs, bark and trunk or fruit. The introduction of natural dyes in textile dye houses is coupled to several requirements which have to be fulfilled *i.e.*, adaptation of traditional processes on modern equipment, appropriate supply of plant materials to dye houses and selection of material leading to product with acceptable fastness properties (Bechtold *et al.*, 2002; Bechtold and Turcanu, 2003).

Hence, in this work, *Punica granutum* L. peel (Pomegranate rind) was chosen based on the previous knowledge on the plant material, as representative sources for a plant-based yellow dye. The selection is also based on their potential importance is health care sector. The growth of micro organism in clothing causes unpleasant odor, staining, loss of mechanical strength and etc., and also cause health related problem to the wearer (Raja *et al.*, 2010). Hence it is important to provide necessary protection to the wearer from the micro organism; the fabric must have the bacterial resistant properties. There are several antimicrobial agents used to improve the functional ability of the clothing material (Thilagavathi *et al.*, 2005). But recently there are lot of attraction towards natural based herbs as an antimicrobial agent because of its eco friendly and health hazardless ness. *Punica granatum* L. has been widely used by traditional medicine in America, Asia, Africa and Europe for the treatment of different types of diseases (Gracious *et al.*, 2001; Kim *et al.*, 2002; Murthy *et al.*, 2001; Lamar *et al.*, 2008). It has been highlighted in many studies as having antimicrobial activity against a range of both Gram positive and negative bacterial (Navarro *et al.*, 1996). The fruits of *Punica granatum* (pomegranate) have been used to treat acidosis, dysentery, microbial infections, diarrhoea, helminthiasis, haemorrhage, and respiratory pathologies (Fuentes and Exposito, 1995). Melendez and Capriles (2006) have also reported that extracts from *Punica granatum* fruits possess strong *in vitro* antibacterial activity against many bacterial strains tested. Many studies have shown that the pomegranate peel extract has wound healing properties (Chidambara *et al.*, 2002). Antibacterial activity (Prashanth *et al.*, 2001), antifungal activity (Dutta *et al.*, 1998) and antimicrobial effect (Navarro *et al.*, 1998).

The dyeing ability of the selected plant material was evaluated using colour strength (K/S) value of dyed textile objectively by spectrophotometer. The dying process was carried out along with different mordanting process namely pre, post and simultaneous mordanting. The tannic acid is selected as the mordant based on their eco friendliness. The best colour yielding process and dye source were selected and evaluated for the fastness and physical properties on textile for performance analysis. The dyed textile materials were also evaluated for their antibacterial efficacy against most common pathogens.

Materials and Methods

Materials

The Herbs used in this research work is *Punica granutum* L. peel, are sourced from the retail stores of Coimbatore, Tamil nadu. The knitted interlock fabric with the specification of 36 CPI, 46 WPI and 157 GSM was sourced from Tirupur. Chemicals such as Methanol and Tannic acid were sourced from Prasad Associates, Coimbatore.

Plant Material Extraction

Fresh rinds of *Punica granatum* L. were shadow dried and made into fine powder. The methanolic and water extract of the powders were obtained. 10 gm of powder of each is soaked in methanol for 24 hours to obtain 30 per cent concentrated solution, resulting in active substances being dissolved in methanol. The extract were filtered and used for antimicrobial finishing (Rathinamoorthy and Thilagavathi, 2013).

Microorganisms

Bacterial cultures used in the present studies were obtained from Microbial Type Culture Collection (MTCC) IMTECH, Chandigarh. The bacterial strains *Staphylococcus aureus* (*MTCC 737*), *Escherichia coli* (*MTCC 1687*), *Klebsiella pneumoniae* (*MTCC 6644*), *Proteus vulgaris* (*MTCC 742*), *Salmonella typhi* (*MTCC 733*) were used.

Dyeing of Cotton Material

Mordanting can be achieved by pre-mordanting (before dyeing), simultaneously mordanting and dyeing or it may be a post mordanting system (after dyeing). Different types of mordants can be applied on the textile to increase the dye uptake of natural dyes. Extensive work has been reported (Paliwal, 2001; Sengupta, 2001; Prabu and

Premraj, 2001; Paul *et al.*, 2002) for dyeing of textiles with natural dyes adopting specific mordanting system for a particular textile material.

In pre-mordanting method, the textile substrate is first treated in aqueous solution of mordant for optimized time (30 - 60 minutes) and temperature (70 – 100 °C) with a ML ratio of 1:5 to 1:20 and then dried with or without washing. The mordanted textile material is then dyed following optimized dyeing conditions. For simultaneous mordanting and dyeing system, the textile substrate is immersed in a dye bath solution containing both mordant and dye in a definite quantity and dyeing may be started at the pre-determined optimum condition. In case of post-mordanting method of natural dyeing, the dyeing process is carried out for bleached textiles in the absence of mordant at pre-determined dyeing condition and the dyed fabric is treated in a separate bath called saturator containing suitable mordanting solution. After dyeing, the textile material is washed properly and soaping is carried out by 2g/L industrial soap solution (Ashis Kumar Samanta and Adwaita Konar, 2011). The methanol extract of herbal material was directly applied on the textile material by simple exhaust method with a material to liquor ratio of 1:10 and then Pad-dry-Cured. The fabric was then dried at 80°C for ten minutes.

Characterisation

Analyses of Colour Value Using Spectrophotometer

The *K*/*S* values of the dyed fabric was determined by measuring their surface reflectance using a computer-aided Macbeth 2020 plus reflectance spectrophotometer followed by calculating the *K/S* values using following Kubelka Munk equation with the help of relevant software (Prathiba Devi *et al.*, 2013).

$$K/S = \frac{(1 - R_{\lambda max})^2}{2R\lambda_{max}} = \alpha C_d$$

where α is the constant for a particular textile substrate and dye system; *K*, the coefficient of absorption; *S*, the coefficient of scattering; C_d, the concentration of dye; and *R*λmax, the surface reflectance value of the sample at a particular wavelength where maximum absorption occurs for a particular colourant. From the measured *K/ S* values for 100 per cent individual dye applied on a fabric, the calculated *K/S* values were obtained for samples dyed with a specific properties of a selected pair of dyes (dye A and dye B) using following equation (Prathiba Devi *et al.*, 2013).:

$$\text{Calculated } K/S^a = \left(\frac{m}{100} K/S^b\right) + \left(\frac{n}{100} K/S^c\right)$$

where (a) is observed for *m*:*n* proportion of A and B dyes; (b) is observed for 100 per cent dye A; and (c) is observed for 100 per cent dye B. Also, the coefficient of variation percentage (CV per cent) of *K*/*S* values was determined from the 10 point *K*/*S* data taken at 10 different points of the corresponding dyed fabric samples. The total colour difference (Δ*E*) values were observed by measuring *L**, *a**, *b** values before and after treatment using a computer-aided Macbeth 2020-plus reflectance spectrophotometer along with associated Colour-Lab plus software using the following CIELab equations:

$$\Delta E = [(\Delta L^*)^2 + (\Delta a^*)^2 + (\Delta b^*)^2]^{1/2}$$

where,

$$L^* = 116\,(Y/Y_o)^{1/3} - 16$$
$$a^* = 500\,[(X/X_o)^{1/3} - (Y/Y_o)^{1/3}]$$
$$b^* = 200\,[(Y/Y_o)^{1/3} - (Z/Z_o)^{1/3}]$$

Chroma (psychometricchroma) values in CIE Lab color space were calculated as follows

$$C_{(ab)}^* = (a^2 + b^2)^{1/2}\,,\ \Delta C^* = C^*_{1(ab)} - C^*_{2(ab)}$$

whereC*1(ab) and C*2(ab) are the chroma values for standard sample and produced sample. CIE 1976 metric Hue-Difference (ΔH) can be given for CIELab system using the following relationship:

$$\Delta H_{ab} = [(\Delta E_{ab}^*)^2 - (\Delta L^*)^2 - (C_{ab}^*)^2]^{1/2}$$

General metamerism index (MI) was calculated employing Nimeroff and Yurow's equation as follows (Prathiba Devi *et al.*, 2013):

$$MI = \frac{\Sigma(\Delta R\bar{x})^2}{X^2} + \frac{\Sigma(\Delta R\bar{y})^2}{Y^2} + \frac{\Sigma(\Delta R\bar{z})^2}{Z^2}$$

where, ΔR values refer to the difference in reflectance values between two pieces of the same fabric sample measured at two different wavelengths (under two viewing conditions i.e wavelength conditions of viewing illuminant used in the measurement); and x, y and z are the standard CIE-observer function for three primary colours, while X, Y and Z are the CIE tristrimulus values (average) of the respective samples chosen for determining MI values as mentioned above.

Agar Diffusion Method (SN 195920)

The treated and untreated fabric samples were placed in the AATCC bacteriostasis agar (Rathinamoorthy *et al.*, 2011), which has been previously inoculated (Mat culture) with a test organism. After incubation, a clear area of uninterrupted growth underneath and along the side of the test material indicates the antibacterial effectiveness of the fabric. The area of the inhibition zone is a measure of antibacterial effectiveness of the material.

Hohensteins Modified Challenge Test (JIS L 1902)

Specimens of the test material were shaken in a known concentration of bacterial suspension and the reduction in bacterial activity in standard time was measured. The efficiency of the antimicrobial treatment was determined by comparing the reduction in bacterial concentration of the treated sample with that of control sample expressed as a percentage reduction of standard time (Rathinamoorthy *et al.*, 2011).

The percentage of reduction of bacteria by the specimen treatments by the following formula

Per cent of reduction = 100 (B-A)/B

Here A= the number of bacteria recovered from the inoculated treated test specimen swatches in the jar incubated over 24 hr contact time.

B= the number of bacteria recovered from the inoculated treated test sample in jar immediately after inoculation (at 0 contact time)

Analysis of Fastness Properties

The dyed materials were analysed for the following fastness properties as per the standards. Wash fastness - AATCC 61, Light fastness - AATCC 16 E, Perspiration fastness - AATCC 15 (AATCC Technical Manual, 2007).

Physical Properties

The effect of natural dyeing process on the physical properties of dyed material was evaluated using the following test as per standards mentioned. Bursting strength – BS 4768, Water absorption - BS 4554, Water vapor permeability – BS 1970, Air permeability – BS 5636 (Saville, 1999).

Results and Discussion

Analyses of Colour Strength Value (K/S)

The colour strength properties of different combination of Jute/cotton combination were evaluated, which was dyed with hot brand reactive dyes. CIE L*a*b* (CIELAB) color scales based on the Opponent-Color Theory.

- L scale: Light vs. dark where a low number (0-50) indicates dark and a high number (51-100) indicates light.
- A scale: Red vs. green where a positive number indicates red and a negative number indicates green.
- B scale: Yellow vs. blue where a positive number indicates yellow and a negative number indicates blue.

The L value for each scale therefore indicates the level of light or dark, the a value redness or greenness, and the b value yellowness or blueness. All three values are required to completely describe an object's color (Hunter Labs, 1996). The Figure 24.1 represents the K/S Value of the natural plant extract finished textile at different treatments.

From the K/S values of the plant extract finished textiles, it can be noted that, For *Punica granutum* rinds, the higher colour strength value is noticed in the case of pre mordanting process. Hence, based on the higher colour depth, the above stated samples were processed for further characterization. The L, a, b, C and H values were discussed in Table 24.1.

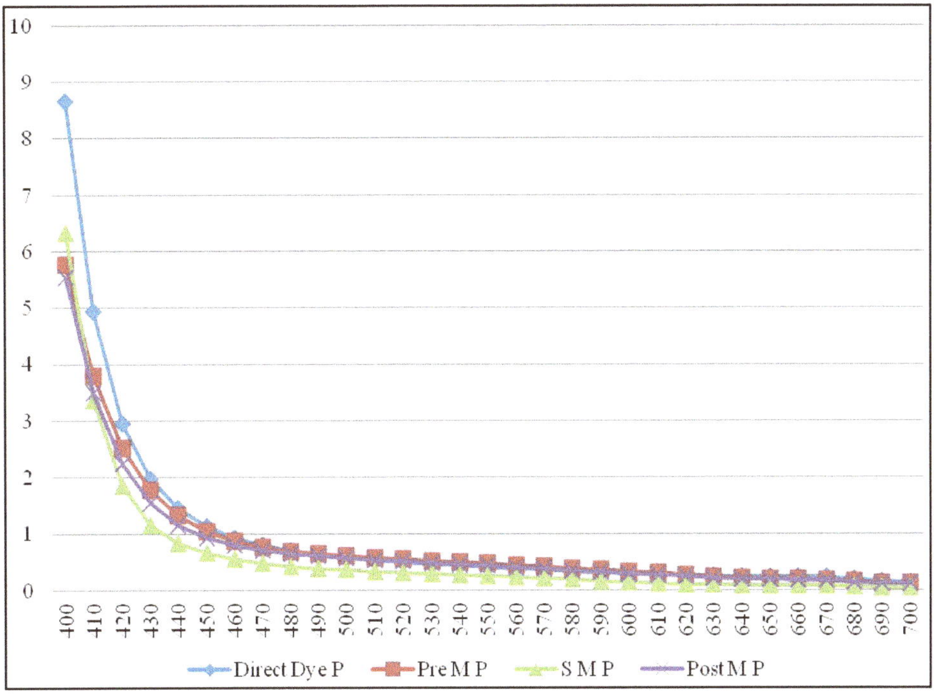

Figure 24.1: Colour Strength Value of *Punica granutum* Rind Extract Finished Textile.

Table 24.1: CIE L, a, b C and H Values for Selected Samples

Samples	*L**	*a**	*b**	*C*	*H*
Pre Mordanted *Punica Granutum* peels	69.359	2.509	20.266	20.421	82.909

The C axis represents Chroma or 'saturation'. This ranges from 0 at the centre of the circle, which is completely unsaturated (*i.e.* a neutral grey, black or white) to 100 or more at the edge of the circle for very high Chroma (saturation) or 'colour purity' (http://www.colourphil.co.uk/lab_lch_colour_space.shtml). The value indicated in Table 26.1 shows that, the dyed material were in the range 20. Which indicate the values of Chroma. The H axis represents Hue. The horizontal slice through the centre, cutting the 'sphere' in half, shows a coloured circle. Around the edge of the circle we see every possible saturated colour, or Hue. This circular axis is known as h° for Hue. The units are in the form of degrees° (or angles), ranging from 0° (red) through 90° (yellow), 180° (green), 270° (blue) and back to 0°. The result indicates that, the hue values is 82.9, its all yellow based colours (http://www.colourphil.co.uk/lab_lch_colour_space.shtml). The dyed fabrics are shown in Figure 24.2.

Agar Diffusion Test (SN 195920)

The result of Agar Diffusion Test for antimicrobial effectiveness of both water and Methanol extracts against standard test cultures *viz., Staphylococcus aureus* (*MTCC*

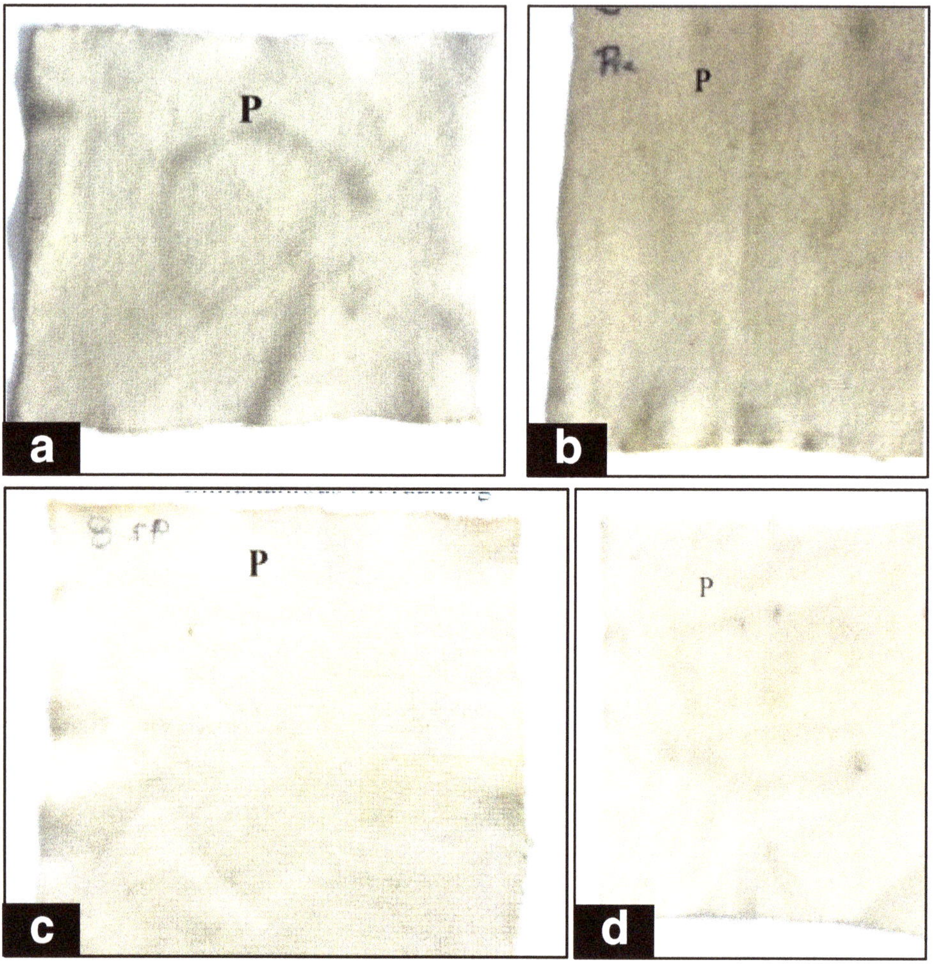

Figure 24.2: Plant Extract Dyed Textile Sample (a) Direct dyeing; (b) Pre- Mordanting; (c) Simultanious Mordanting and (d) Post Mordanting.

Table 24.2: Zone of Inhibition of Treated Textile Material for different Strains

Sl.No.	*Bacteria*	*Zone of Inhibition (in mm)*	
		Control	*Punica granatum L. Rind Extract*
1.	*Staphylococcus aures*	–	28
2.	*Escherichia coli*	–	26
3.	*Klebsiella pneumonia*	–	25
4.	*Proteus vulgaris*	–	32
5.	*Salmonella typhi*	–	28

–: No zone formation.

737) (gram positive), *Escherichia coli* (*MTCC 1687*), *Klebsiella pneumoniae* (*MTCC 6644*), *Proteus vulgaris* (*MTCC 742*) and *Salmonella typhi* (*MTCC 733*) (gram negative) organisms shows that the activity of treated samples is stronger for all the above mentioned strains. The methanol and water extract of the *Punica granatum* L Peel was applied on the textile material. The results indicates the presence of clear zone of inhibition of 27-38 mm diameter for methanol extract treated fabrics against all the five selected micro organism namely *Staphylococcus aureus* (gram positive), *Escherichia coli*, *Klebsiella pneumoniae*, *Proteus vulgaris* and *Salmonella typhi* (gram negative) Table 24.2 and Figure 24.3.

Hohensteins Modified Challenge Test

The Table 24.3 indicates that the finished material show high percentage of reduction against both *Staphylococcus aureus* and *Escherichia coli.* As a high *Punica granatum L* shows 99.5 per cent of reduction against *Staphylococcus aures.* And in the case of *Escherichia coli* 85 per cent of reduction was achieved by treated fabric. The results show that, in both agar diffusion and quantitative test, the methanol extract treated samples shows high activity against all the human pathogenic bacterial strains.

Table 24.3: Percentage of Bacterial Reduction by Quantitative Measurement

Sl.No.	*Bacterial Strains*	*Sample*	*At "0" Hour Contact*	*At "24" Hour Contact*	*Per cent of Reduction*
1.	*S. areaus*	*Punica granatum* L. rind	$220X10^{-4}$	$110X10^{-2}$	99.5 per cent
2.	*E. coli*	*Punica granatum* L. rind	140X10-3	21X10-3	85 per cent

Durability Evaluation

The treated samples were evaluated for washing durability by AATCC Test Method 61. The results of wash durability of methanolic extract treated fabric are given in Tables 24.4 and 24.5. The antimicrobial activity of the samples was assessed after 5 washes and 10 washes by using both agar diffusion and Hohensteins modified challenge test. There is a reduction in activity after subsequent washes.

Agar Diffusion Test

The agar diffusion test results shows, that there is a reduction in the inhibition zone after subsequent washes. The *Punica granatum* L. rind treated samples shows inhibition zone of 22 and 26 mm respectively for *Escherichia coli* and *Klebsiella pneumonia* after 5 washes. The 10 cycles washed samples shows clear inhibition zone only against *Klebsiella pneumonia* (24mm).

Hohensteins Modified Challenge Test

Decreases in the percentage of reduction were obtained as the washing frequencies increased. Similarly Samples that had been washed 5 times were Shown very low percentage of bacterial reduction in the case of *Escherichia coli*, around 74.52 per cent of activity. But the sample that had been washed 10 cycles shows no activity against *Escherichia coli.* However, while considering the *Staphylococcus aureus,* these

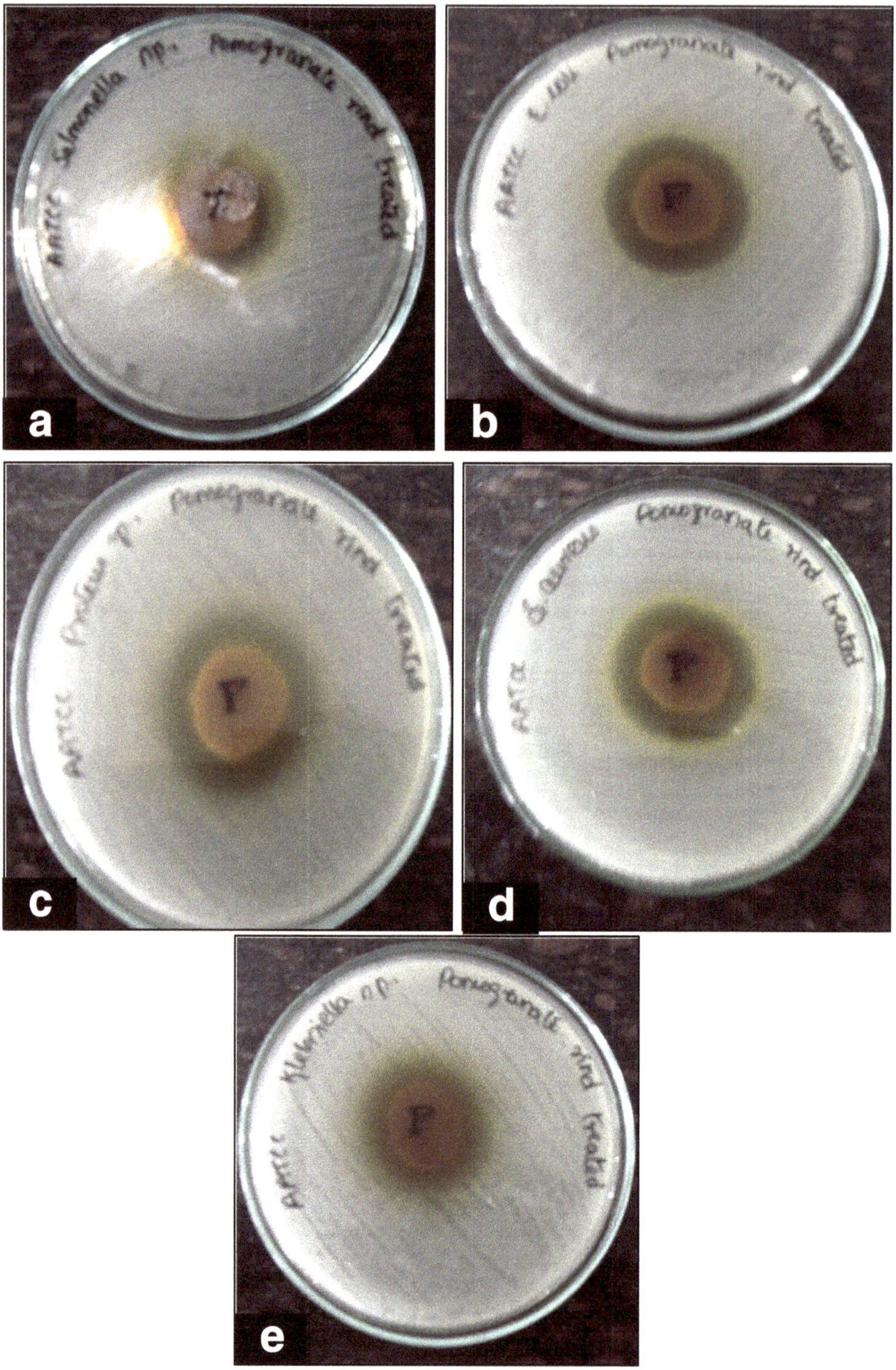

Figure 24.3: Agar Diffusion for Finished Textiles against (a) *Salmonella typhi;* (b) *E.coli;* (c) *Proteus vulgaris;* (d) *S. areaus;* (e) *Klebsiella pneumonia.*

Table 24.4: Durability Test by Quantitative Measurement

Sl.No.	*Bacterial Strains*	*Sample*		*At "0" Hour Contact*	*At "24" Hour Contact*	*Per cent of Reduction*
1	*S.Areaus*	*Punica granatum* L. rind	Control fabric	–	–	Nil
			After 5 Wash	–	–	Nil
			After 10 wash	–	–	Nil
2	*E.coli*	*Punica granatum* L. rind	Control fabric	–	–	Nil
			After 5 Wash	$212X10^{-4}$	$540X10^{-3}$	74.52 per cent
			After 10 wash	–	–	Nil

above stated samples were noted no activity for both after 5 and 10 cycles of wash. It is noted that the treated material having poor durability after commercial laundering. The poor durability of the antimicrobial agent in the textile material may be due to the weak Vander Val's or hydrogen bonds, between antimicrobial agent and cellulose material (Thilagavathi and Kannaian, 2008).

Table 24.5: Durability Evaluation of Treated Samples by Agar Diffusion Method

Sl.No.	*Sample*	*Bacterial Stain*	*After 5 wash Inhibition Zone (in mm)*	*After 10 wash Inhibition Zone (in mm)*
1.	*Punica granatum* L. rind	*Staphylococcus aures*	–	–
		Escherichia coli	22	–
		Klebsiella pneumonia	26	24
		Proteus vulgaris	–	–
		Salmonella typhi	–	–

–: No zone formation.

Analysis of Colour Fastness Properties

Colour Fastness to Washing

The Colour fastness to washing process is evaluated as per standard AATCC 61. The principle of the AATCC washing fastness test is to test the samples under appropriate conditions of temperature, detergent solution, bleaching and abrasive action so that the colour change is similar to that occurring after five hand, domestic or commercial launderings. The colour change can be obtained in a conveniently short time. The abrasive action is due to the result of the frictional effects of fabric against a container, the low liquor ratio and the impact of steel balls on the fabric. The result indicates that, the developed dyes were with the moderate rating of 3-4. The results are given in Table 24.6.

Table 24.6: Fastness to Washing

Sl.No.	*Staining*	*Pre Mordanted Punica granutum*
1.	Change in color	3
2.	Wool	3-4
3.	Acrylic	3
4.	Polyester	2-3
5.	Nylon	2
6.	Cotton	3-4
7.	Acetate	3

Colour Fastness to Perspiration

The color fastness to perspiration of the selected samples was tested As per AATCC standard. The average color fastness to perspiration is shown in the Table 24.7.

Table 24.7: Color Fastness to Perspiration

Test	*Sl.No.*	*Pre Mordanted Punica granutum*	
		Alkaline	*ACIDIC*
Change in color	1.	3	3
Wool	2.	3	3
Acrylic	3.	3-4	3-4
Polyester	4.	3-4	3-4
Nylon	5.	3-4	3-4
Cotton	6.	3	3
acetate	7.	3	3

The color fastness properties of dyed material were tested for acidic and alkaline Ph perspiration for the selected dyed fabrics. The results were indicates that the fastness rating of 3-4 was observed for all the dyed fabrics. However the dyes has considerable strength values around 3.

Colour Fastness to Light

The color fastness to light of the dyed fabric were tested against standard light source as per AATCC 16E, the average color fastness to light after 10 hours was shown in Table 24.7. The result shows that, the value of the sample is with rating of 2. It indicates the poor light fastness.

Effect of Dyeing Process on the Physical Properties

Since, the dyeing of cotton material was carried out with the methanol extract of selected plant based material. The effect of the extract treatment on the basic comfort related physical properties of textile were evaluated.

Water Absorbency

The water absorbency results are given in Table 24.8. From the results it is found that the natural plant extract dyed fabrics were capable of absorbing water quicker than the control sample. This is because of the natural dye particle deposition in and on the surface of the fabric. The average absorbency of the dyed material is very nearer to the control fabric. However, there is reduction in the absorbency noted. The reason for the quick absorption time in the case of dyed fabric may be attributed to the mordant pretreatments which lead the fabric structure and yarn more capable of absorbing water.

Table 24.8: Evaluation of Water Absorbency for Dyed Fabric

Sl.No.	*Control*	*Pre Mordanted Punica granutum*
1.	3.9	3.2
2.	3.5	3.3
3.	4.0	3.2
Avg	**3.8**	**3.2**

Water Vapour Permeability

The water vapour permeability is calculated for *Punica granutum* L. peel, dyed fabrics before and after dying process. The values were taken as an average value of five repeated measurements of water vapour permeability value as given in the Figure 24.4.

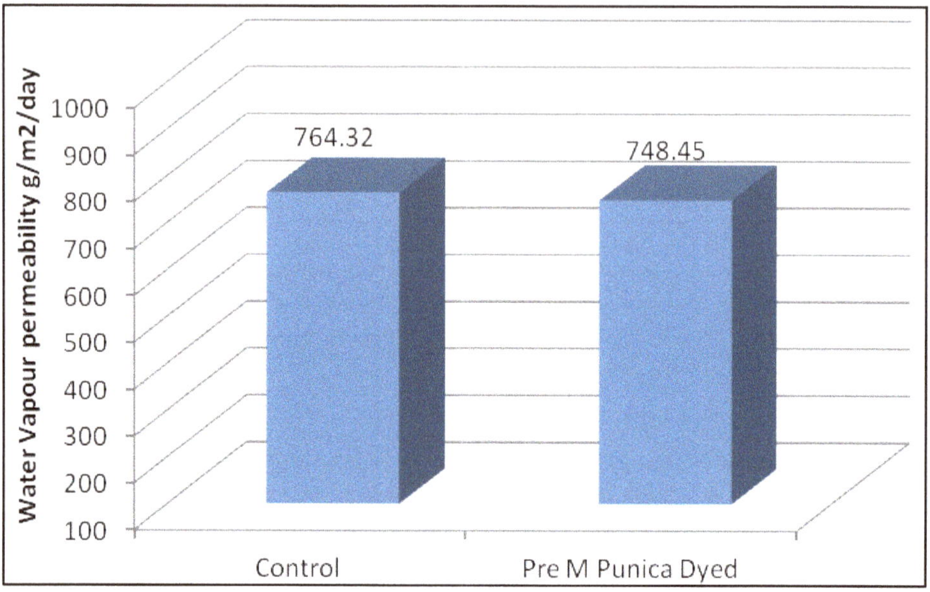

Figure 24.4: Water Vapour Permeability.

Water Vapour Permeability test results were shown in Figure 24.4. From the results it can be noticed that, the dyeing process basically reduced the water vapour permeability of the textile material. However, there is no significant amount of differences observed. This is may be because of the more amounts of deposits of plant extract particle on the surface of the dyed sample. This will close the space between inter and intra space between the yarn.

Air Permeability

The air permeability values of the natural dyed fabric were evaluated as per BS 5636. The results are shown in Figure 24.5. It can be noted that, the air permeability values were not affected much. However, there is a small amount of reduction is observed in all the dyed material. This may be because of the deposition and blocking of natural bye particle in inter and intra yarn spaces. But the permeability values are in the range of acceptance.

Bursting Strength

To study the effect of dyeing process on the strength of the fabric, the bursting strength of the fabric were evaluated as given in Table 24.9. The average strength of the dyed material was comparatively reduced than the control sample. This is because

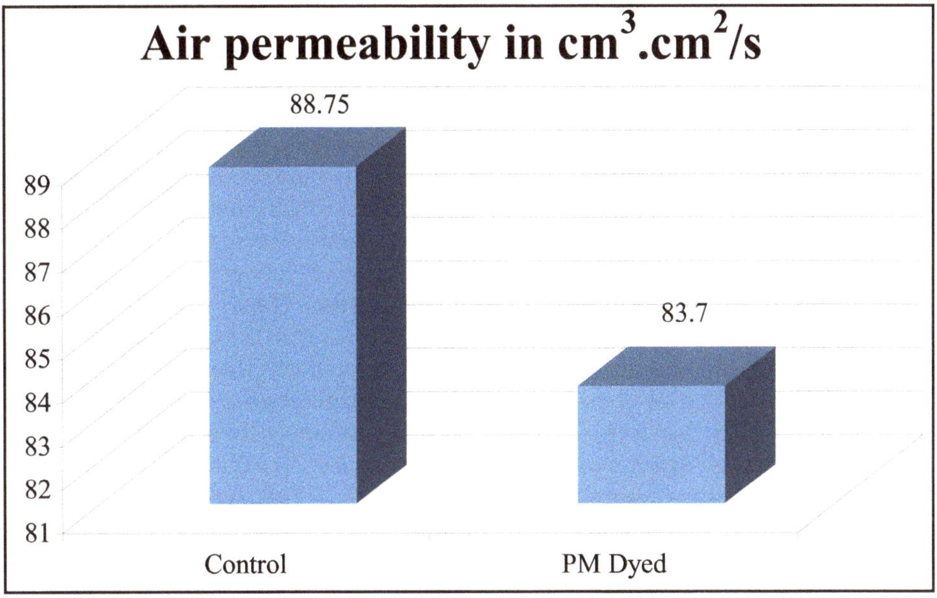

Figure 24.5: Air Permeability Values of Natural Dyed Textile Material.

of the dyeing process is carried out with methanol, which might have degraded the fiber structure to a little extend.

Table 24.9: Evaluation of Bursting Strength for Dyed Fabric

Sl.No.	*Test*	*Control*	*Pre Mordanted Punica granutum*
1.	Bursting Strength (pounds/inch²)	57	50
2.		57	49
3.		56	50
	Avg	**57**	**50**

However, the reduction in bursting strength of the textile material is very meager influence in the aspect of practical clothing application

Conclusion

In this research, natural dyeing of cotton knitted material was tried with a plant material *Punica granutum* L peel (Pomegranate rind). Simple solvent extraction method is used to obtain the colourants from the plant material. The extracted compounds were applied on the textile material in three different ways namely, pre mordanting, post mordanting and simultaneous mordanting. The dyed textile material were analysed for their colour strength values (K/S) using CIE Lab values. From the results it can be noted that, The pre mordanted *Punica granutum* L. peel gave the better and deeper shade (higher K/S) than that of other combinations.

Based on the results, the selected (Pre mordanted *Punica granutum* L peel) dyed fabrics were preceded for the analysis of different properties The finished textiles were also evaluated for their antimicrobial efficacy. The results from agar diffusion test indicates that the extract of *Punica granatum* L rind treated fabrics were having high potential against wide verity of human pathogenic bacteria. 27 -38 mm of inhibition zone was noticed for both water and methanol extract treated textile material. It is concluded from the challenge test results that *Punica granatum* L rind extract treated samples have 99.5 per cent of bacterial reduction against *s.areaus* (Gram positive). But in the case of *E.coli* (Gram negative) the *Punica granatum* rind extract treated material showed 79 per cent of bacterial reduction. The wash durability result indicates that, the effectiveness of the treatment reduces with the increase in washing cycle. Though the tested samples shows poor wash durability, the unwashed samples shows high and clear zone of inhibition against wide spectrum of human pathogenic bacterial stains. With respect to the colour fastness properties, the dyed samples were shown better fastness against washing and perspiration (3-4). But, in the case of light fastness the subjective rating was observed as low (2). The dyed fabric were also analysed for their physical properties like, water vapour permeability, water absorption, air permeability and bursting strength. The results show that, there is no significant deviation in the fabric property after the dyeing process. Hence this study suggest the potential use of natural plant material as a dye for textile dyeing in an eco friendly way without affecting their original properties. The extract can be applied simply by the pad-dry-cure method. If the treated sample is for one time usage like surgical mask, bandage gauze, wound healing bandages, wound management material, dressings etc,. this method very effective and cost effective for commercialization. Since *Punica granatum* L was a natural resources, they were eco friendly and because of their abundant availability, the scope of implementation and commercialization of herbal extract on cotton material is much possible and this antimicrobial treatment definitely has their potential.

References

AATCC Technical Manual, (2007). Standard test methods for textile, 82.

Ashis Kumar Samanta and Adwaita Konar (2011). Dyeing of Textiles with Natural Dyes, Natural Dyes, Dr. Emriye Akcakoca Kumbasar (Ed.), ISBN: 978-953-307-783-3, InTech, DOI: 10.5772/21341. Available from: http: // www.intechopen.com/books/natural-dyes/dyeing-of-textiles-with-natural-dyes.

Bechtold, T., and Turcanu, A. (2003). Natural dyes in modern textile dyehouses – how to combine experiences of two centuries to meet the demands of the future? *J. Clean Prod.*, **11**: 499.

Bechtold, T, Turcanu, A,, Geissler, S., and Gangleberger, E. (2002). Process balance and product quality in the production of natural indigo from *Polygonum tinctorium* Ait. applying low-technology methods, *Bioresour. Technol.*, **81**: 171-177.

Chidambara, M.K., Jayaprakasha, G.K., and Singh, R.P. (2002). Studies on antioxidant activity of pomegranate (*Punica granatum*) peel extract using *in vivo* models. *J. Agric Food Chem.*, **14(50)**: 47-91.

Dutta, B.K., Rahman, I., and Das T.K. (1998).Antifungal activity of Indian plant extracts. *Mycoses*, **41**: 11-12.

Fuentes, V.R., and Exposito, A. (1995). Las encuestas etnobotanicas sobre plantas medicinales en Cuba. *Rev. Jard. Bot. Nacion. Univ. Habana*, **16**: 77-144.

Gracious, R.R., Selvasubramanian, S., and Jayasundar S. (2001). Immunomodulatory activity of *Punica granatum* in rabbits: a preliminary study. *J. Ethnopharmacol.*, **78**: 85-7.

http: / /www.colourphil.co.uk/lab_lch_colour_space.shtml

Hunter Labs (1996). "Hunter Lab Color Scale". Insight on Color 8 9 (August 1-15, 1996). Reston, VA, USA: Hunter Associates Laboratories.

Kim, N.D., Mehta, R., Yu, W., Neman, I., Livney, T., Amichay, A., Poirer, D., Nicholls, P., Kirby, A, Jiang, W., Mansel, R., Ramachandran, C., Rabi, T., Kaplan, B., and Lansky, E. (2002). Chemopreventive and adjuvant tharapeutic potential of pomegranate (*Punica granatum*) for human breast cancer. *Breast Cancer Res.*, **71**: 203-17.

Lamar, A.S., Fonseca, G., Fuentes, J.L., Cozzi, R., Cundari, E., Fiore, M., Ricordy, R., Perticone, P., Degrassi, F., and Salvia, R.D. (2008). Assessment of the genotoxic risk of *Punica granatum* L. (*Punicaceae*) whole fruit extracts. *J. Ethnopharmacol.*, **115**: 416-22.

Melendez, P.A., and Capriles, V.A. (2006). Antibacterial properties of tropical plants from Puerto Rico. *Phytomedicine*, **13**: 272-76.

Murthy, K.N., Reddy, V.K., Veigas, J.M., and Murthy, U.D. (2001). Study on wound healing activity of *Punica granatum* peel. *J. Med. Food*, **7**: 256-59.

Navarro, V., Villarreal, M., Rojas, G., and Lozoya, X. (1998). Antimicrobial Evaluation of some plants used in Mexican traditional medicine for the treatment of infectious diseases.J. *Ethnopharma.*, **53(3)**: 143

Navarro, V., Villarreal, M.L., Rojas, G., and Lozoya, X. (1996). Antimicrobial evaluation of some plants used in Mexican traditional medicine for the treatment of infectious diseases. *J Ethnopharmacol.*, **53**: 143-47.

Paliwal, J. (2001). Effect of mordants on henna, dyed cotton and silk fabrics. *Textile Magazine*, **42 (11)**: 79.

Paul, S., Sharma, A., and Grover, E. (2002). Application of natural dyes on textiles. *Asian Text J.*, **11(11)**: 65.

Prabu, H.G., and Premraj, L. (2001). Study on the use of natural extracts as mordant in basic dyeing, *Man-made Textiles in India*, **44 (12)**: 488.

Prashanth, D., Asha, M.K., and Amit, A. (2001). Antibacterial activity of *Punica granatum*. *Fitoterapia*, **72**: 171.

Prathiba Devi, R., Rathinamoorthy, R., and Jeyakodi Moses, J. (2013). Effect of Jute Proportion on the color strength value of jute/cotton union fabric. *International Journal of Engineering Research*, **2(4)**: 294-299.

Raja, A.S.M., Thilagavathi, G., and Kannaian T. (2010). Synthesis of spray dried polyvinyl pyrrolidone coated silver nanopowder and its application on wool and cotton for microbial resistance. *Indian Journal of Fiber and Textile Research*, **35**: 59-64.

Rathinamoorthy, R., and Thilagavathi, G. (2013). Optimisation of process conditions of cotton fabric treatment with *Terminalia chebula* extract for antibacterial application. *Ind J Fib and Text Res.*, **38**: 293-303.

Rathinamoorthy, R., Udayakumar, S., and Thilagavathi, G. (2011). Antibacterial efficacy analysis of *Punica granatum* L. leaf, rind and *Terminalia chebula* fruit extract treated cotton fabric against five most common human pathogenic bacteria. *Int. J. of Pharm. and Life Sci.*, **2**: 1147- 1153.

Saville, B. P. (1999). Physical Testing of Textiles, Woodhead Publishing, Limited,

Sen, S., and Demirer, G.N. (2003). Anaerobic treatment of real textile wastewater with a fluidized bed reactor. *Water Research*, **37 (8)**: 1868-1878.

Sengupta, S. (2001). Dyeing with natural dyes and mordent. *J Text Assoc.*, **62(4)**: 161.

Shahid Adeel, Shaukat Ali, Ijaz A. Bhatti and Ferenc Zsila, (2009). Dyeing of cotton fabric using pomegranate (*Punica granatum*) aqueous extract. *Asian Journal of Chemistry*, **21(5)**: 3493-3499.

Thilagavathi, G., and Kannaian, T. (2008). Application of prickly chaff (*Achyranth aspera*) Leaves as herbal antimicrobial finish for cotton fabric used in healthcare textile. *Natural Product Radiance*, **4(7)**: 330-34.

Thilagavathi, G., Rajendrakumar, K., and Rajendran, R. (2005). Development of antimicrobial textile finishes from plant species. *Express Textile.*

Weber, E. J., and Adams, R. L. (1995). Chemical and sediment mediated reduction of the azo dye Disperse Blue 79. *Environmental Science and Technology*, **29**: 1163-1170.

Zolinger, H. (1987). Colour chemistry- Synthesis, Properties of organic dyes and pigments, VCH publishers, New York, 92.

Utilisation and Management of Medicinal Plants Vol. 3 (2015) *Pages* **347–365**
Editors: **V.K. Gupta and A. Kaul**
Published by: **DAYA PUBLISHING HOUSE, NEW DELHI**

25

Development of Mosquito Repellent Fabric Using *Leuca indica* Leaves Extracts

R. Surjit[1], K.J. Vishnu Vardhini[2] and R. Rathinamoorthy[1*]

ABSTRACT

Mosquito repellent textiles is one of the revolutionary advancements in the textile field as it provides the much needed features of driving away mosquitoes, especially in the tropical areas avoiding serious mosquito borne diseases. Various natural mosquito repellents like neem oil, clove oil's application on textile materials and their effect on the mosquito repellency is discussed through literatures available. In this research, the aqueous extract of Leuca indica is used as mosquito repellent finish by using microencapsulation technique. Cotton fabric is treated with extracts of Leuca indica with the wall to core ratio of 70:30.

The Microcapsule finished textile were characterised by scanning electron microscope and FTIR analysis. The wash fastness of the finished textile is relatively good because of the microencapsulation. The controlled release of Leuca indica extract from the capsule aids the better performance for the finished textile. The materials were evaluated for the mosquito repellency testing of fabric which was done according to USDA standards. The results indicate that, the finished textiles show a mosquito repellency of more than 75 per cent even after 15 washes. The Bite counts per 3 minutes exposure shows considerable decrease and number of mosquitoes knocked down also increases due to the controlled release of microencapsulated Leuca indica on the fabric. The HPLC analysis of the selected herb revealed that, the presence of ascorbic acid and acetic acid might be the potential reason for the effect and the same was confirmed by FTIR analysis.

Keywords: *Leuca Indica*, Mosquito repellency, Controlled release, HPLC, FTIR.

1 Department of Fashion Technology,
2 Department of Textile Technology,
PSG College of Technology, Coimbatore – 4, Tamil Nadu, India
* *Corresponding author*: E-mail: r.rathinamoorthy@gmail.com

Introduction

Clothing for mankind has served many purposes. Starting as a basic necessity, it has gained versatility in becoming more functional and aesthetic in nature. Textile industry also has diversified into various aspects of technical textiles in this era of modernisation. Protective textile is one area of technical textiles which has started gaining more momentum due to latest technological advancements and closer association of textile products in medical field. Protective textiles refer to those textile products which have a functionality of giving protection from something in some or the other sense (Fradin, 1998). Mosquito repellent textile is one such textile product. Mosquitoes are considered to be one of the deadliest pests known to man. It affects people in the tropical countries to a very great extent. Mosquitoes are responsible for more human suffering than any other organism (Fradin, 1998). Around the globe, mosquitoes are considered to transmit disease to more than 700, 000, 000 people annually and will be responsible for the deaths of 1 of every 17 people currently alive (www.ontariofishingcommunity.com).

An individual can avoid being bitten by mosquitoes with the help of using a repellent, mosquito catcher or by using attractants to lure mosquitoes elsewhere. Malaria results from infection with a protozoan carried by mosquitoes and, according to reports from the World Health Organization, causes as many as 3 000 000 deaths annually. Mosquitoes transmit the arbo viruses responsible for yellow fever, dengue hemorrhagic fever, epidemic poly arthritis, and several forms of encephalitis. Only female mosquitoes bite. Male mosquitoes feed primarily on flower nectar, whereas female mosquitoes require a blood meal to produce eggs. They usually feed every 3 to 4 days; in a single feeding, a female mosquito typically consumes more than its own weight in blood. Some mosquito species are zoophilic (preferring to feed on animals) and others are anthropophilic (showing a preference for human blood). In some mosquito species, seasonal switching of hosts provides a mechanism for transmitting diseases from animal to human

Need for Mosquito Repellent Textile

Because of global warming the distribution of mosquitoes has expanded from tropical regions to northern latitudes, and that leads to a spread in sources of viral infection from mosquitoes. West Nile fever occurs routinely in Africa as its virus was first recognized in humans at the West Nile District of Uganda in 1937. Chickengunia is one more mosquito transmitted disease which was on the rise and lot of people in India got affected by it. Similarly typhoid, malaria, etc., are all spread by mosquitoes. Hence it is essential at this juncture to develop mosquito repellent textiles especially using natural products to prevent any environmental hazards in the long run because of use of synthetic repellents.

Numerous mosquito repellents are available in the market and the most notable one is *N,N*-Diethyl-*meta*-toluamide (DEET). It is estimated that 110 million people use DEET worldwide each year. It is applied as a skin lotion and its effectiveness last for four to five hours (Lupi *et al.*, 2013). DEET cannot be applied on fabrics, as it does not withstand even mild laundering. Mosquito bites are prevented normally by applying mosquito repellents directly on the skin as a lotion. The effectiveness of the repellent remains for few hours and drives away mosquitoes. However those lotions can be harmful to human beings because of direct contact with skin. Hence imparting the repellent agent on the fabric would reduce any harmful effect caused by the lotion. So people are now looking towards mosquito repellent textiles to prevent from mosquito bites. Since the mosquito repellent finish is given to the fabric, it would remain for few weeks and hence they get prolonged repellence effect.

Microencapsulation

Microencapsulation is defined as the process of enclosing micron sized particles of solids or droplets of liquids or gasses in an inert shell, which in turn isolates and protects them from the external environment.

Microencapsulation of insect repellent agents can be employed to provide a longer controlled release effect. Microencapsulation technology is an effective technique used to control the release properties of active ingredients that prolong the functionality of cosmetic textiles. The principle is achieved by simply imparting the cosmetic and pharmaceutical ingredients into the fabric of the clothing so that with the natural movements of the body, the skin is slowly freshened and revitalised. Microencapsulation technology is rapidly developing in the field of chemical finishing because of its versatility and flexibility. One major advantage of using microencapsulation technology is its ability to protect the active ingredients from hazardous environments.

The microcapsules are produced by depositing a thin polymer coating on small solid particles or liquid droplets, or on dispersions of solids in liquids. The core contents are released under controlled conditions to suit a specific purpose (Nelson, 2001). The core content may be released by friction, pressure, change of temperature, diffusion through the polymer wall, dissolution of the polymer wall coating, biodegradation etc (Anon, 2005). In the 21st century, more commercial applications of microencapsulation in the textile industry can be found, particularly in Western Europe, Japan and North America. The technique is being used to develop textiles with new properties and added value, including climate-controlled materials, fragrance release fabrics, cosmetic, therapeutic and medical textiles. Textiles and garment manufacturers are very optimistic that this novel technology will open up exciting market opportunities and a world of possibilities for consumers (Anon, 2005). Micro encapsulated particles are anchored onto the fiber. As the wearer moves, the capsules are activated producing a slow release of the active ingredient. Active ingredients run the gamut including moisturizers, aloe, vitamin E, therapeutic smells and insect repellent.

The release mechanisms of the core contents vary depending on the selection of wall materials and more importantly, its specific end uses. The core content may be released by friction, pressure, change of temperature, diffusion through the polymer wall, dissolution of the polymer wall coating, biodegradation etc.The earliest conception of microencapsulation, which carries or holds a core material trapped within a shell material, possibly dates back to the 1930s by using the spray-drying technique (Nelson, 2001). Many different manufacturing approaches have been adopted for microencapsulation. The most commonly used microencapsulation processes, include (a) Complex Coacervation, (b) Polymer-Polymer Incompatibility, (c) Interfacial Polymerisation and *In situ* Polymerisation, (d) Spray Drying, (e) Centrifugal Extrusion, (f) Air Suspension Coating, (g) Pan Coating and (h) Emulsion Hardening Process (Holme, 2004).

Complex Coacervation method takes advantage of the abilities of cationic and anionic water-soluble polymers to interact with water, forming a liquid, polymer-rich phase called complex coacervation. When the complex coacervate forms, it will be in equilibrium with a dilute solution called the supernatant. The supernatant acts as the continuous phase, whereas the complex coacervate acts as the dispersed phase. As the water-insoluble core materials are dispersed in the system, each droplet or particle of dispersed core material is spontaneously coated with a thin film of coacervate. The liquid film is then solidified to make the capsules harvestable. This method has been applied to encapsulate many water-immiscible liquids and is used in a variety of products.

In an earlier work, the cotton fabric was treated with N, N Diethyl benzamide and its effectiveness was checked by cage test. Cage tests are the quick and cost-effective way to determine the mosquito-repelling qualities of treated textiles. The result they obtained is N-N Diethyl Benz amide treated fabric gave 100 per cent effectiveness on cotton fabric. The treated fabric at 7 per cent and 14 per cent N-N Diethyl Benzamide showed better mosquito repellency. The findings of the study reveal that the 14 per cent concentration of Mosquito repellent treated fabrics found to be very effective and can be used for hygienic and health care application (Krishnaveni, 2009).

In another study, they have applied chrysanthemum oil to the fabric using microencapsulation and found that the fabric treated shows satisfactory performance with nylon and cotton and is less complicated than the padding technique. The major disadvantage associated with this finish is slight yellowing problem. Both cotton and nylon fabrics treated possess good wash fastness (finish remained after 5-10 washes) (Shirish *et al.*, 2007).

In an another work it was discussed that the fibres can be modified such that they get repellent character from within thereby, giving a scope for better growth of this technology. The ecological and economical aspect should also be taken care of and the technology should be imparted through some natural sources other than chemicals which help in maintaining the ecological balance and also in making the goods available to the commons. Neem leaves can be of great importance in this regard (Malik *et al.*,).

Natural mosquito repellents like extracts of neem oil and clove oil were used and it was found that neem oil provided an excellent mosquito repellency of 90 per cent similar to that of NN Diethyl benzamide (Surjit *et al.*, 2011). Hence application of natural extracts on the fabric for mosquito repellency in turn reduces environmental concerns and so can be used for health care and hygiene application.

Testing Method

USDA Laboratory Method

For the treated fabrics, mosquito repellency is calculated by USDA laboratory method which simulates the natural conditions of exposure. An artificially created tent is prepared in which laboratory reared strains of female species of mosquitoes are released in the range of 10 to 30. An hourly observations are made on their behavior, mobility etc. During a 3 hour exposure, hourly data are recorded in terms of total knock down (KD), inability to fly on account of their loosing appendages is recorded.

At frequent intervals, their ability to bite volunteers is also recorded as in Figure 25.1. In the end, the test data are computed in terms of percentage repellency and in terms of percentage protection. Experiments are usually conducted two times and average data is used in calculating mosquito repellency.

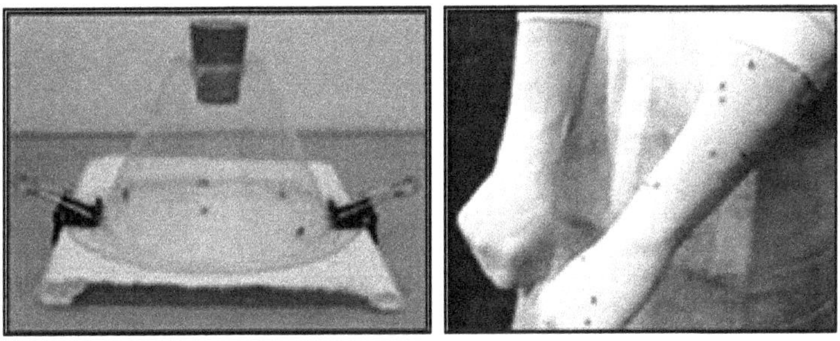

Figure 25.1: USDA Laboratory Method.

Modified WHO/CTD/WHO PES/IC/96.1

Test fabrics with incorporated Mosquito Repellent formulations, prepared by padding method are analyzed using live Culex species of mosquitoes by modified Excito repellency chamber method which takes into account migration and behaviour pattern of mosquitoes. An improved Excito repellency test chamber is used for conducting mosquito Repellency evaluation. A collapsible test chamber is simulated in laboratory like a tunnel to analyze the behavior of Mosquitoes towards the treated fabric. WHO method is useful for finding mosquito repellency and not mosquito mortality effect. 100 per cent Repellency is most desirable.

Modified WHO Cone Bioassay (WHO 1998)

Five susceptible, non-blood fed, 2–5-day old Anopheles7 of mosquitoes will be exposed to netting materials (25 cm x 25 cm) for 3 minutes, under standard WHO cones, after which they are held for 24 hours with access to sugar solution. KD is measured after 60 minutes post-exposure and mortality after 24 hours. At least 50 mosquitoes on each net/fabric will be tested. Mosquitoes exposed to untreated nets/ fabrics shall be used as controls. Bioassays will be carried out at 25 + 2 °C and 75 + 10 per cent RH. Washed Fabric which cause >80 per cent mortality and/or >95 per cent KD meet the criteria.

In this study, the on 100 per cent cotton fabric, microencapsulated *Leuca indica* extract solution were applied as repellent finish. The efficacy of the finished textile against mosquito bites were subjectively calculated as the efficiency of repellent by USDA laboratory method and also by subjective evaluation to check the mosquito repellency percentage.

Materials and Methods

Materials

The 100 per cent cotton plain woven fabric with ends per inch of 82 and picks per inch of 74 with the warp and weft count of 40s were sourced from the commercial outlets of Coimbatore, Tamil nadu, India. Green leaves of *Leuca indica* is sourced from a village near Coimbatore. Other chemicals used in this study were sourced from local supplier in Coimbatore. All the chemicals sourced were of analytical grade and used without any further purification.

Methods

Desizing

Desizing was carried out to remove the size (added) impurities from the woven cotton fabric. In this case grey cotton fabric is weighed accurately and run through the bath of boiling water for 10 minutes. The fabric is taken out, squeezed and immersed into the acid bath for 1 hour at 50 degree. After desizing, the fabric is washed with hot water and then with cold water to remove the impurities and the residual chemicals.

Scouring

Scouring is carried out to remove the natural as well as added impurities in the cotton fabric and it increases the absorbency of the cotton fabric. The desized cotton fabric is weighed and immersed in the solution containing caustic soda, soda ash and wetting agent at 40 degree. The temperature of the solution is increased uniformly to boil and the fabric is worked in the solution at boil for 2 hours. Then the fabric is washed with hot water and then with cold water to remove the impurities and chemicals.

Bleaching

Bleaching is carried out using hydrogen peroxide to remove the natural yellowish grey coloring matter to obtain white fabric and facilitate the production of pale and bright shades on cotton fabric in subsequent process. The desized and scoured fabric is treated with the hydrogen peroxide solution at 40 degree. Fabric bleaching is started at this temperature and it is uniformly increased to boiling temperature within 20 minutes and the fabric is subjected for bleaching process for about 2 hours. Then the fabric is washed with hot water followed by normal water to remove impurities and chemicals.

Leuca indica Extract Preparation

In this case *Leuca indica* extract is prepared by using water. The *Leuca indica* leaves are allowed for shadow drying for about 10 days and they are made as fine powder using a mixer grinder. For preparing 30 per cent concentration 100 ml of water is taken and 30 g of *Leuca indica* leaves powder is added to it then it is allowed for about 12hrs and then it is extracted using filter paper. This extract was directly used for the subsequent processes.

Microencapsulation and Application of *Leuca indica* Extract

Microencapsulation solution of ratio 70:30 is prepared using coacervation method. By taking 500 ml of boiling water at 60 degree and is added with 70 gm of wall material (gum acacia). It is stirred for 15 min and then 50 ml of boiling water is added and 45 degree is maintained. Then 30 ml of core material is added and it is stirred for 20 min. After that 10 ml of 20 per cent sodium sulphate is added to it with continuous stirring for 10 min and finally 10 ml of 17 per cent glycerol (plasticiser) is added to it, it is stirred for 10 min and the encapsulated solution is prepared.

The processed cotton fabrics were made as 25cm X 25cm square. The samples are dipped in the above prepared extract solution for about 12hrs by simple exhaust method. The samples which are dipped in the microencapsulated *Leuca indica* directly allowed to dry after the 12 hrs and used for evaluation.

Characterisation of Finished Textile for Mosquito Repellency

High Performance Liquid Chromatography (HPLC) Analysis of *Leuca indica* Extract

HPLC fingerprints were prepared using waters HPLC equipped with UV-VIS detector. Solvents were pre-filtered and analysis was performed in Symmetry C18 column (4.6 x 250mm). The water extracts of *Leuca indica* were injected in HPLC system. Injection volume was 20 µl. The flow rate was 0.7ml/min and the active spots were identified.

USDA Laboratory Method

Insect Used: Female Mosquito - Anopheles species; No. of Insects Released in Laboratory Tent: 12 per 10-20 grams of Fabric/half a metre; No. of hrs. of Incubation :3 hrs (intermittent)

FTIR Analysis

FT-IR spectra for extract treated and untreated fabric were measured with a SHIMADZU spectrophotometer to identify the presence of active substance in the fabric. The spectra were obtained in the range of 400 to 4000 cm^{-1}.

Results and Discussion

The *Leuca indica* extracts are tested with HPLC to find out the active component it the extracts. The SEM analysis was used to check the capsule formation on the surface of the fabric. Mosquito repellent testing was carried out for the samples and FTIR analysis was done to find the reasons for mosquito repellent activity of *Leuca indica*.

SEM Analysis

The microencapsulated *Leuca indica* Extract solution at 70:30 wall to core ratio concentration were finished on cotton textile. The finished materials were evaluated for the presence of microcapsule on the fabric surface by morphological study.

In Figures 25.2a 2000X magnification, it is found that the micro capsules were deposited on the fabric. In Figure 25.2b at 5000X it can be clearly seen that, the capsules were of good spherical shape and they were not agglomerated. Since the

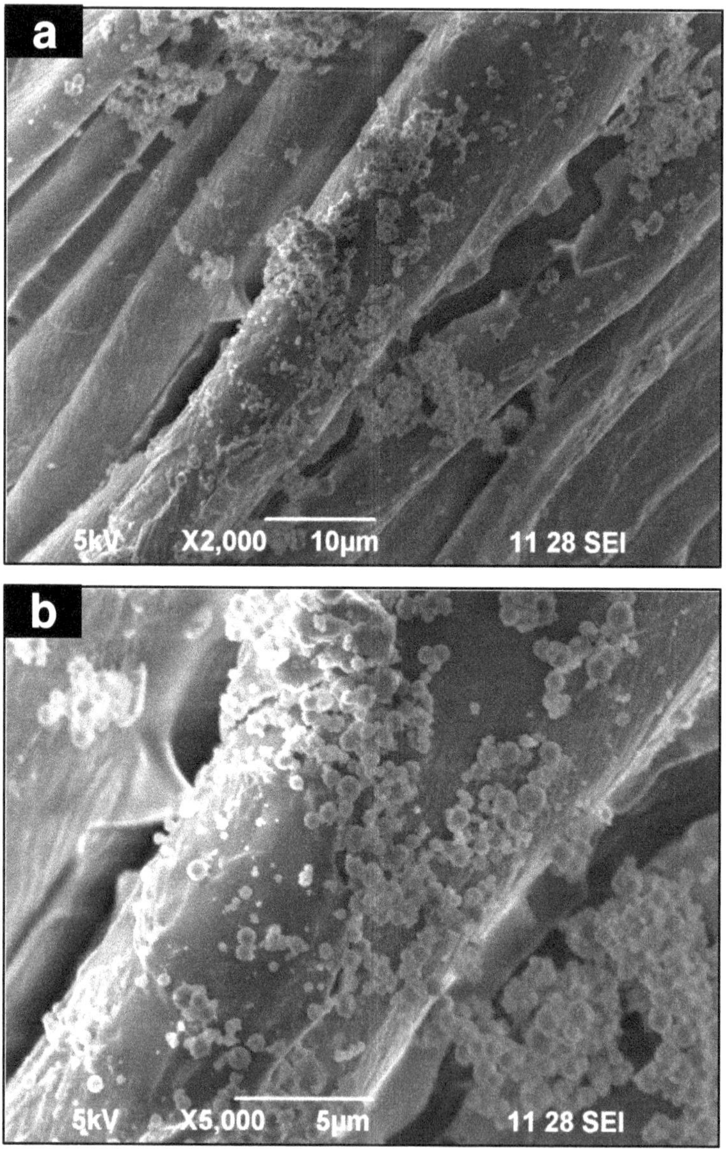

Figure 25.2: (a) The Microencapsulated *Leuca indica* Exract Finished Textile at 2000X Magnification and (b) 5000X Magnification.

micro capsules formed were applied by exhaust method, we don't have any control over the uniformity of the distribution of microcapsule.

Results of Mosquito Repellency by USDA Laboratory Method

The developed samples along with washed samples (after 5, 10 and 15 washes) were allowed for 3 hrs exposure of mosquito. The experiment was carried out in a tent made by the provided fabric. Every 1 hour, the observations were made about the movements, biting tendency and survival of mosquitoes. Following observations have been made as mean average of the experiment conducted triplicate and the results were discussed.

Total Number of Mosquitoes Unable to Fly

The Figure 25.3 shows the number of mosquitoes which were unable to fly during the developed fabric exposure. The microencapsulated *Leuca indica* Extract finished textile found more effective against the mosquitoes. It is found that, direct exposure of the developed sample made 5 mosquitoes were unable to fly, For the sample after 5 and 10 wash 6 mosquitoes were unable to fly. Out of ten mosquitoes 7 mosquitoes are unable to fly after 15 wash. It shows that controlled release is aiding in increasing the count of mosquitoes that are unable to fly.

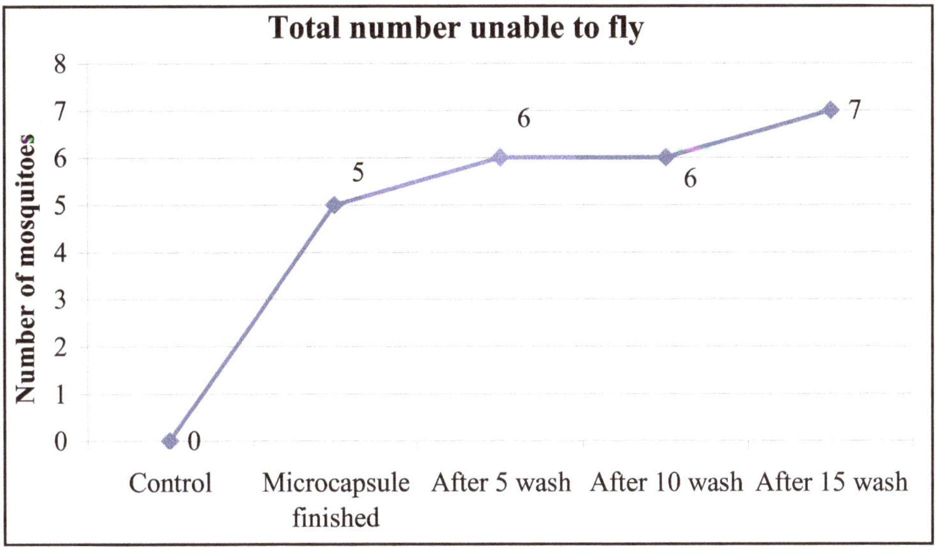

Figure 25.3: Total No. of Mosquitoes which were Unable to Fly.

Total Number of Bite Counts Per 3 Mins Exposure

The results of the total number of bite counts per 3 minutes exposure is measured in presents of finished textile sample. From the results, it is found that the bite counts remained between 2 to 3 for all the 4 samples whereas for the control sample it was 10. The Figure 25.4 shows that the presence of mosquito repellent compounds on the surface of fabric helps in repelling mosquitoes from biting.

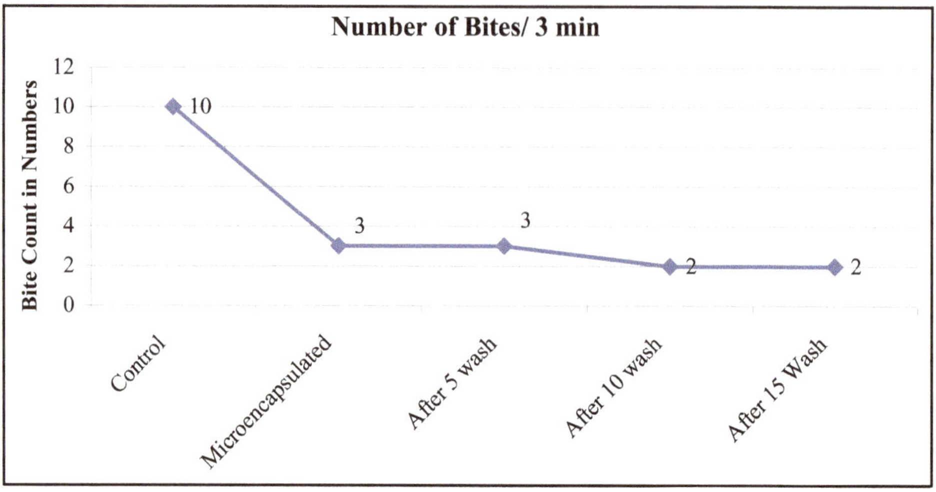

Figure 25.4: Total No. of Bite Counts per 3 Minutes Exposure.

Percentage of Mosquito Repellency

The core material has a slow release nature due to microencapsulation. So, as the number of washes increases, the core material gets slowly released thereby increasing the percentage repellence. When the wash is zero, the percentage repellence is 50 per cent and as the wash is increased to 15, the percentage repellence gets increased to 75 per cent. So there is a slow and controlled release of core material and that leads to increased repellency, this can be clearly seen in the results depicted in Figure 25.5. The presence of ascorbic acid in the *Leuca indica* extracts gives the much need mosquito repellency characteristics to the fabric. The repellency percentage is

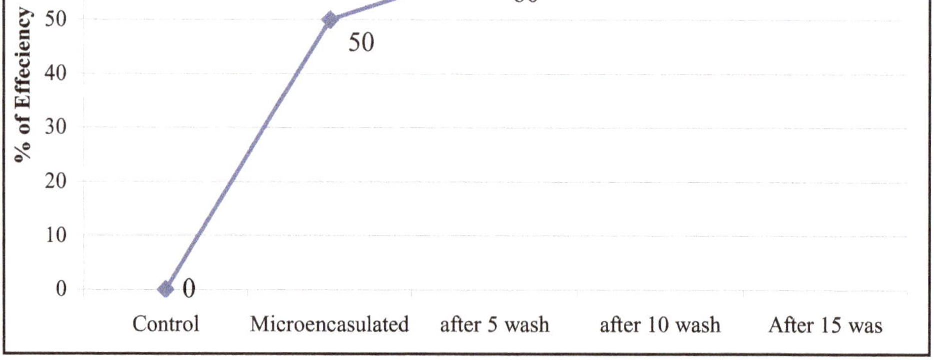

Figure 25.5: Percentage Mosquito Repellency.

good when compared with control sample thereby showing the use of the natural extracted solution of *Leuca indica* as a mosquito repellent.

HPLC Results

The HPLC analysis was performed to identify the active substances in the *Leuca indica* extract and the results were in Table 25.1 and Figure 25.6. This will help to get clear understanding for the mosquito repellent mechanism. The HPLC results reveals that, the major peak area of 59.313 per cent at the retention time of 5.088. This retention time obtained were in line with the standard peak of ascorbic acids [11]. The second major peak was observed in the retention time of 5.997 with area of 28.219 per cent. This retention time of 5.997 is closely associated with the acetic acid content of the *Leuca indica* plant leaves (http://pirika.com/NewHP/PirikaE/CarboxylicAcid.html). The other nominal peaks were noticed at the range of 8.195, 8.492, 8.995 and 9.466 with the areas respectively of 3.078 per cent, 2.834 per cent, 3.224 per cent and 3.322 per cent. These retention times were the representatives of phyto chemical compounds such as maslinic acid (Guo-Liang Li *et al.*, 2011), chlorogenic acid (Anda Natanela Croci *et al.*, 2009), nicotin (https://sites.google.com/site/masonaco/Home/applications/services-offered-2/hplc-nicotine), oleanolic acid (Marina Zacchigna *et al.*, 2009) respectively.

Table 25.1: Different Retention Time Peaks and their Areas for *Leuca indica* Extract

Peak#	Ret. Time	Area	Height	Area %	Height %
1	5.088	21828014	616867	59.313	62.919
2	5.997	10385183	110140	28.219	11.234
3	8.195	1132856	56894	3.078	5.803
4	8.492	1046313	58267	2.843	5.943
5	8.995	1186429	51014	3.224	5.203
6	9.466	1222727	87236	3.322	8.898
Total		36801521	980417	100.000	100.000

From the results, it can seen that, the major compound present in the plant extract is ascorbic acid. Which is one of the most commonly widespread carboxylic acid found in the plant based material with anti oxidant properties (Herbert Baxter *et al.*, 1999). Ascorbic acid is one form of vitamin C found in the plant source. The structure is given in Figure 25.7.

The second major retention peak is the representative of acetic acid (Figure 25.8). Acetic acid is the most important and universal precursor of fatty acids, lipids and many other organic products (Ullrich, 1970). Acetyl coenzyme A, together with malonyl co enzyme A the active forms involved in the bio synthesis of fatty acids from acetate. This acetic acid is also occurs in trace amount in both free and combined in essential oils of many plans. This substance inhibits the tricarboxylic acid cycle at very low concentrations and hence, is toxic to all living organism (Ullrich, 1970).

The other small peaks noticed at the different retention times are because of the components like, maslinic acid, chlorogenic acid, nicotin and oleanolic acid. All these are the phytochemical components of *Leuca indica* Leaf extract. The first component, Maslinic acid is a member of the group of triterpenes known as oleananes.

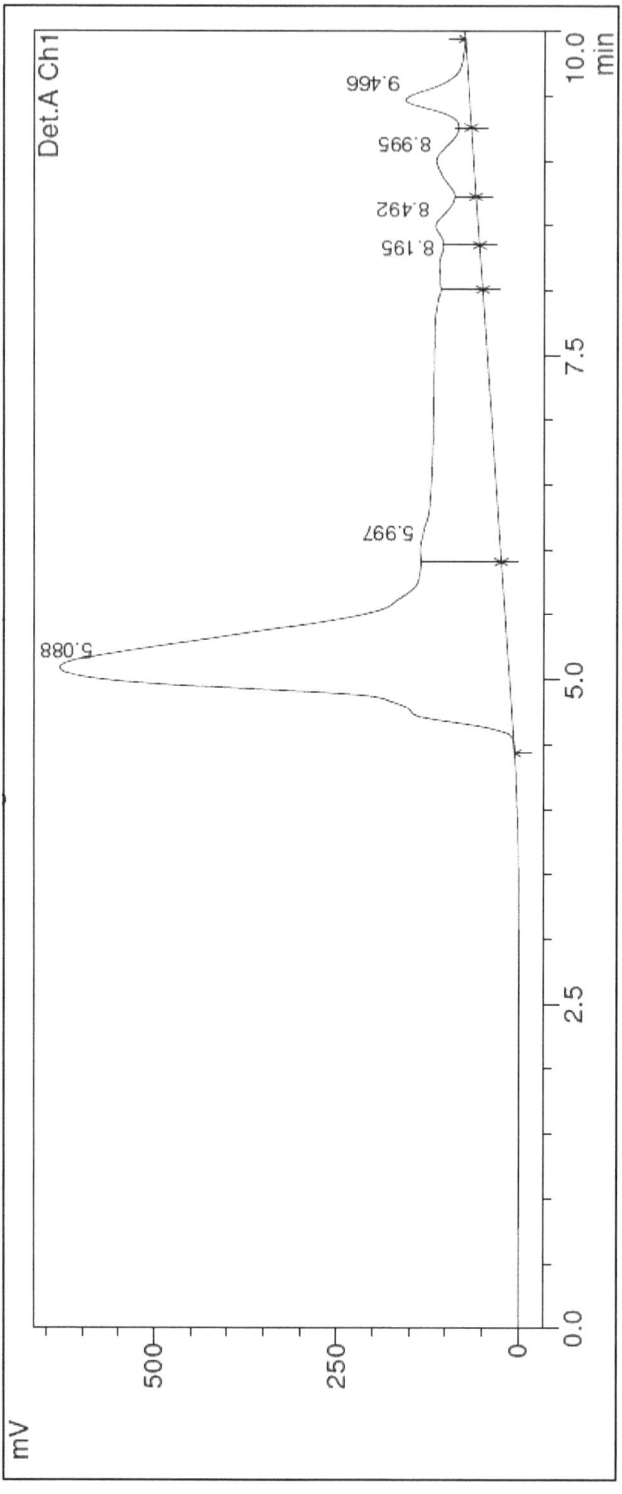

Figure 25.6: HPLC Spectrum for the *Leuca indica* Extract.

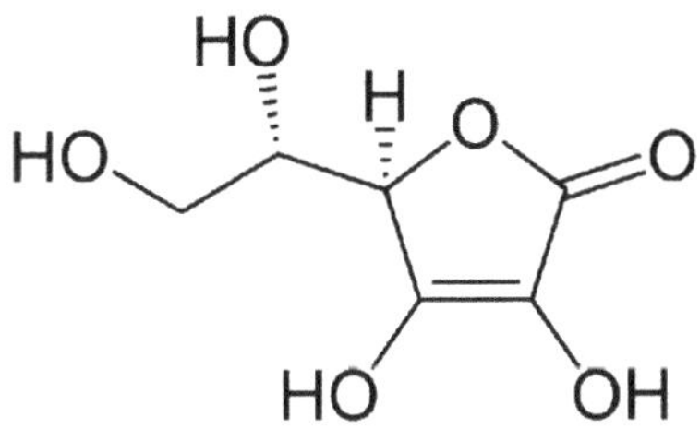

Figure 25.7: Structure of the Ascorbic Acid.

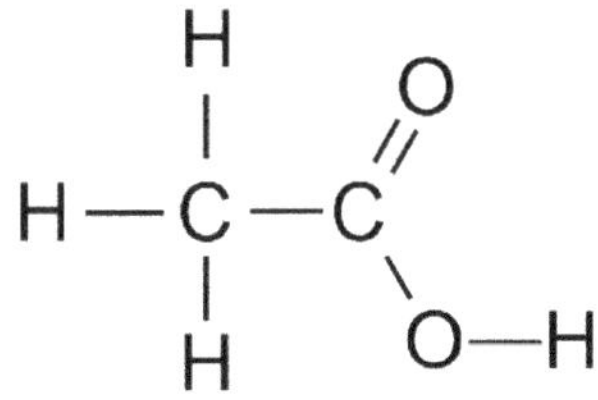

Figure 25.8: Structure of Acetic Acid.

The plant based bio active components has the following structure as shown in Figure 25.9 (http://en.wikipedia.org/wiki/Maslinic_acid).

The another phyto component is Chlorogenic acid (CGA), a natural chemical compound which is the ester of caffeic acid, Chlorogenic acid can be found in amboo *Phyllostachys edulis* as well as in many other plants (Clifford, 2003). It is one of the major phenolic compound which having antihypertensive effect (Zhao *et al.*, 2011). The structure is given in Figure 25.10.

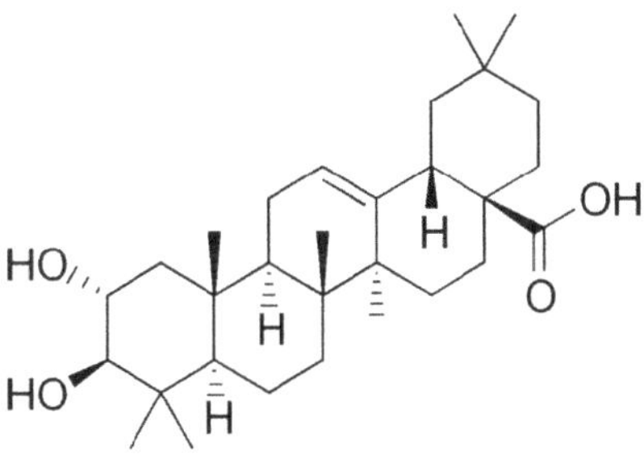

Figure 25.9: Structure of Maslinic Acid.

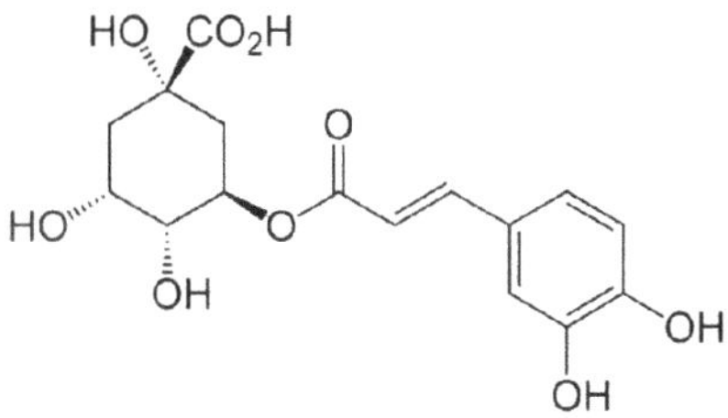

Figure 25.10: Structure of Chlorogenic Acid.

Nicotine is the most familiar pyridine alkaloid. The nicotine occurs in a number of plants, the major source is Nicotiana, so that it and related structures are usually referred to as tobacco alkaloids (Von Euler, 1965). The structure is given in Figure 25.11.

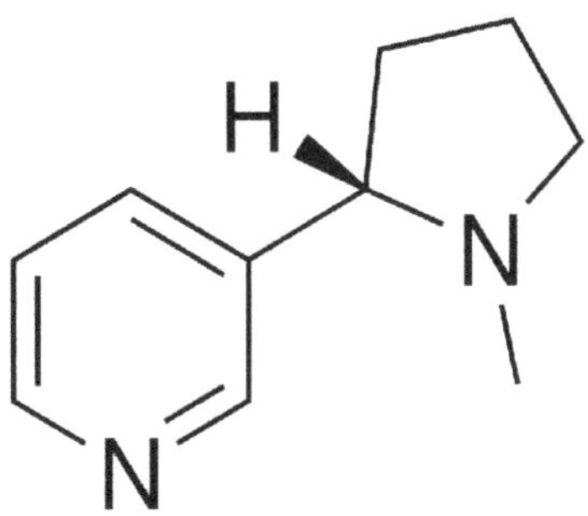

Figure 25.11: Structure of Nicotine.

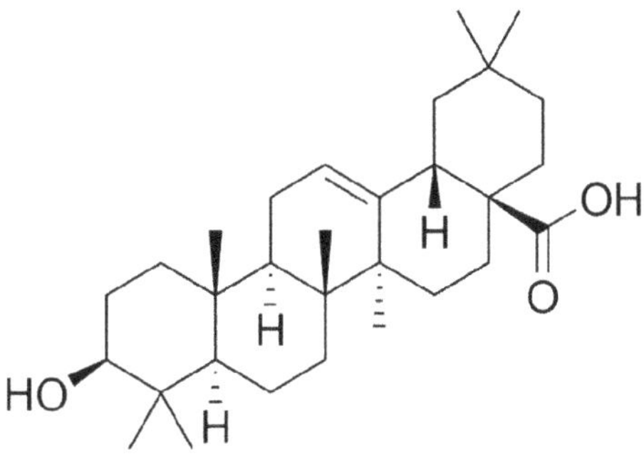

Figure 25.12: Structure of Oleanolic Acid.

Oleanolic acid or oleanic acid is a naturally occurring triterpenoid, widely distributed in food and medicinal plants (Liu, 1995), Oleanolic acid or oleanic acid is a naturally occurring triterpenoid (Mengoni *et al.*, 2002), widely distributed in food and medicinal plants (Yu *et al.*, 2013). Figure 25.12 gives the structure of Oleanolic acid.

Hence, the major content of the *Leuca indiaca* leaves were of highly effective phytochmical components, the mosquito repellent effect may be by the components. The findings of Pichersky *et al.* (2002) and Marta Ferreira Maia *et al.* (2011) supports our results. They have stated that, Most plants contain compounds that they use in preventing attack from phytophagous (plant eating) insects. These chemicals fall into several categories, including repellents, feeding deterrents, toxins, and growth regulators. Most can be grouped into five major chemical categories: (1) nitrogen compounds (primarily alkaloids), (2) terpenoids, (3) phenolics, (4) proteinase inhibitors and (5) growth regulators. Although the primary functions of these compound is defense against phytophagous insects, many are also effective against mosquitoes and other biting Diptera, especially those volatile components released as a consequence of herbivory. (Pichersky *et al.*, 2002; Marta Ferreira Maia *et al.*, 2011)

Kishore *et al.* (2011) reviewed the efficacy of phytochemicals against mosquito larvae according to their chemical nature and described the mosquito larvicidal potentiality of several plant derived secondary materials, such as, alkanes, alkenes, alkynes and simple aromatics, lactones, essential oils and fatty acids, terpenes, alkaloids, steroids, isoflavonoids, pterocarpans and lignans. They also documented the isolation of several bioactive toxic principles from various plants and reported their toxicity against different mosquito species. Rattan (2010) reviewed the mechanism of action of plant secondary metabolites on insect body and documented several physiological disruptions, such as inhibition of acetylecholinestrase (by essential oils), sodium and potassium ion exchange disruption (by pyrethrin) and inhibition of cellular respiration (by rotenone). Such disruption also includes the blockage of calcium channels (by ryanodine), of nerve cell membrane action (by sabadilla), of octopamine receptors (thymol), hormonal balance disruption, mitotic poisioning (by azadirachtin) morphogenesis and alteration in the behaviour and memory of cholinergic system (by essential oil), *etc.*

However, the basic mechanism remains under the activities are explained by Ansari and Razdam (1995), as, the mosquitoes do not understand anything disgusting in the palant extracts to repel them. They were merely not attracted as they could normally be under normal circumstances. Perhaps the active ingredients (alkaloids, flavonoids, saponins, phenolics and tannins), present in the phytochemical extracts from the plant extract s might have exerted some inhibitory effect on lactic acid receptor cells by masking or changing the lactic acids that normally attract them thereby confusing or distracting the mosquitoes. Thus, the blood-feeding contact or response is prevented. Consequently, with the application of the phytochemical extract on the skin, the mosquito could not bite because the active ingredients does not allow it to smell the attractant (lactic acids) and could not therefore identify the human as its source of meal. This suggests that the active ingredients confused the olfactory receptors and the mosquito simply could not smell the host (Effiom *et al.*, 2012)

FTIR Analysis

The Figures 25.13a shows the cotton fabric without the herbal extract finishing and Figure 25.13b shows that FTIR spectrum of the *Leuca indica* extract finished textile. The studies carried out after 10 washes because, the washing process will release the *Leuca indica* extract from the microcapsule. The Spectrum indicates that, there is a considerable amount of increments in the per cent of reflectance of the finished textile than the normal cotton textile. This is perhaps because of the deposits of the extract on the surface of the fabric.

From the spectrum it is identified that the *Leuca indica* treated cotton fabric contain more carboxyl group [C(=O)-OH] due to the presence of active substance like ascorbic acid, maslinic acid, chlorogenic acid, acetic acid and oleanolic acid in their extract. The absorption in the region 3200 cm^{-1} to 3600 cm^{-1} and 1200 $cm^{-1}$to 1700 cm^{-1} of treated sample confirms the presence of -OH group stretching. The presence of –(C=O)- group stretching also confirmed by the absorption, in the region of 1600 cm^{-1} to 1900 cm^{-1} in the treated sample (William Kemp, 1987).

To specific, the peaks at the 2360 cm^{-1} and 2347 cm^{-1} represents the carboxylic acid with broad C-H overlaps. The peaks in the range of 3500-3600 cm^{-1} (3629 cm^{-1} and 3587 cm^{-1}) represents free O-H stretching bonds (Sharp peaks) the alcohols and phenolic groups. The peak at 1772 cm^{-1} represents the stretching of C=O bonds (H bonded), which means the presence of the acid derivatives in the finished fabric. The bending vibration in the region of 1300- 1470 cm^{-1} represents the CH_2 and CH_3 deformation of alkanes. These results confirm the presence phytochemical compounds on the surface of the finished textile material. Thus the presence aids the wearer to protect it from the mosquitoes. (https://www2.chemistry.msu.edu).

Conclusion

In this research, a plant based Mosquito repellent finish was applied using microencapsulation techniques. The water based extracts of *Leuca indica* were obtained from normal solvent extraction method and used for microencapsulation. Four different samples were prepared and subjected to mosquito repellency testing according to USDA standards and it was found that the sample which was given 15 washes exhibited increased mosquito repellent activity. The HPLC studies were performed for herbal extract and the results identified that, the phytochemical compounds like phenolic contents, carboxylic acids and alkanes along with short chain fatty acids were the main active components in *Leuca indica* extracts. The previous research data from various researchers confirms the findings, that the presence of these above stated active compounds are the major reason for the higher mosquito repellency. To confirm the presence of these substances on the finished textile, the FTIR analysis was performed. This indicates that the possible deposition of the components from the *Leuca indica* extracts. Further the results of mosquito repellency test of the sample after 15 washes was found to be 75 per cent which shows that microencapsulation aids in controlled release of chemicals and leads to increased effect of mosquito repellency on the fabric for longer periods of time. The SEM analysis confirms the deposition of developed micro capsules on the surface of the fabric. This evidence that, the herbal extract finish has the potential scope in consumer market.

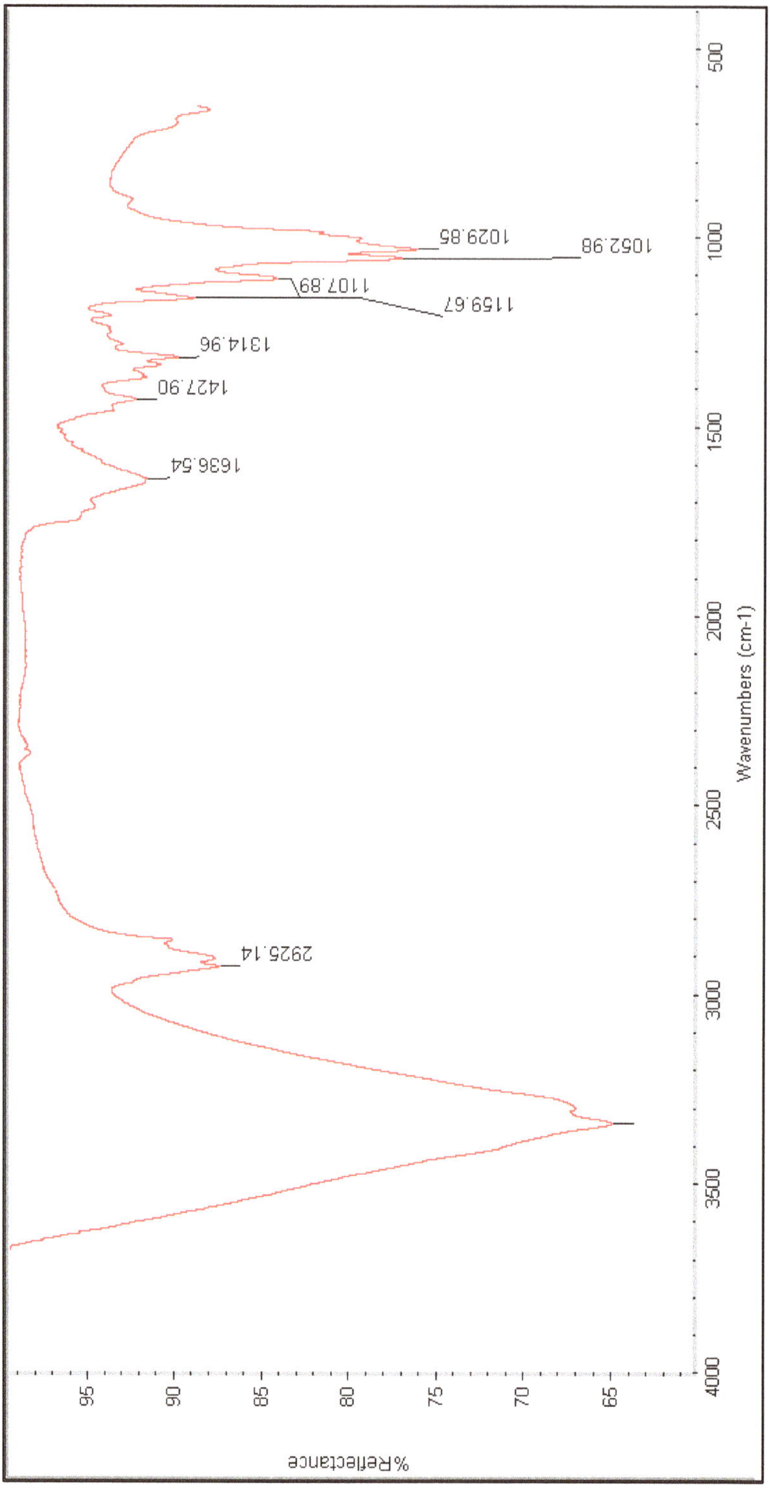

Figure 25.13a: FTIR Spectrum of a Cotton Fabric Treated with Microencapsulated *Leuca indica* Extract Treated Fabric after 10 Washes.

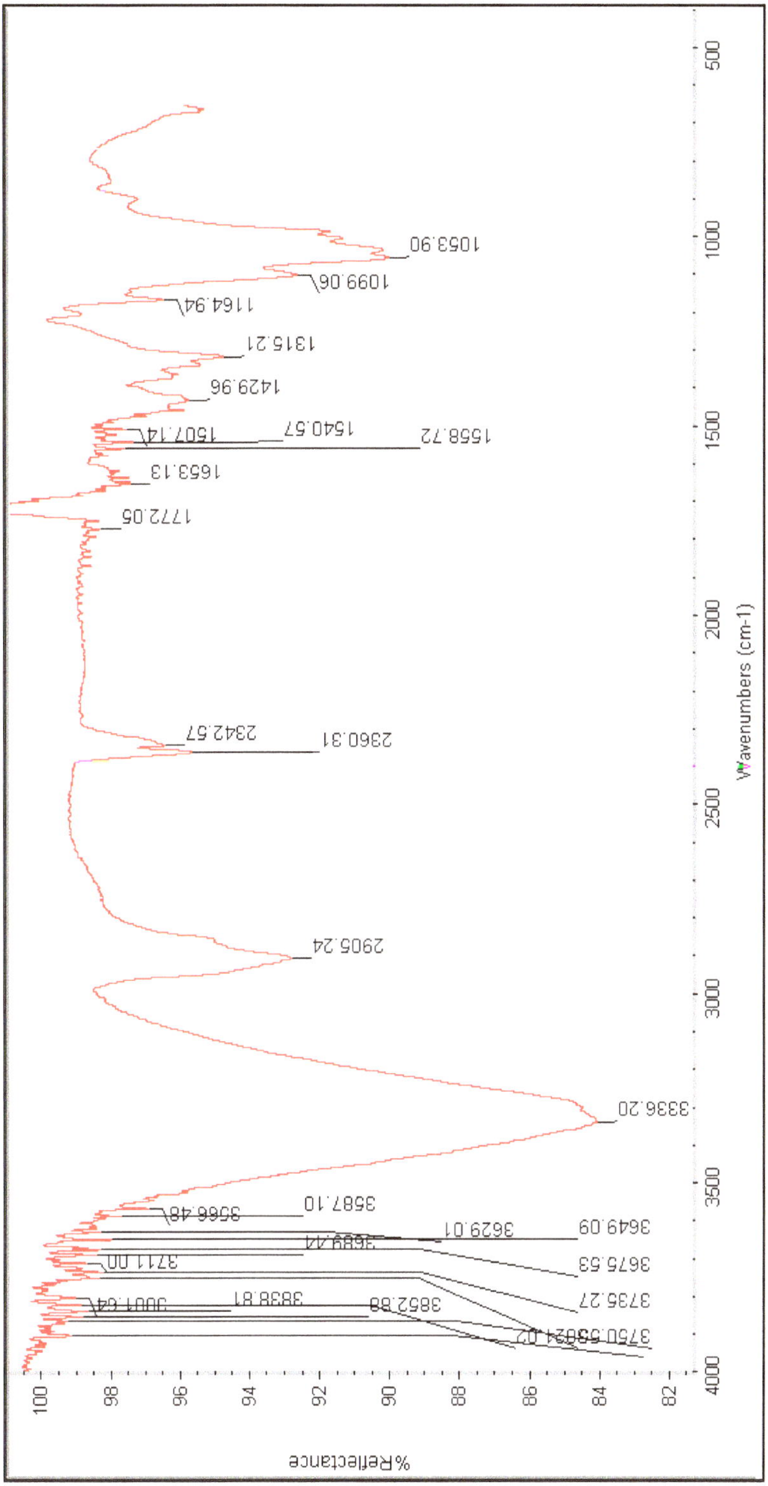

Figure 25.13b: FTIR Spectrum of Cotton Fabric Treated with Microencapsulated *Leuca indica* Extract Treated Fabric after 10 Washes.

References

All about mosquitoes and DEET- A general discussion', Ontario fishing community home, www.ontariofishingcommunity.com, Accessed on 16.04.2014.

Anda Natanela Croci, Bogdan Cioroiu, Doina Lazar, Andreia Corciova, Bianca Ivanescu, and Mihai Ioan Lazar, (2009). HPLC Evaluation of phenolic and polyphenolic acids from propolis.*FARMACIA*,.LVII: 1.

Anon. (2005), Microencapsulation: For enhanced textile performance. *Performance Apparel Market*, **12**: 21-39.

Ansari, M. A., and Razdam, R. K. (1995). Relative efficacy of various oils in repelling mosquitoes. *Indian Journal of Malariology*, **32**: 104.

Clifford, M. N. (2003). The analysis and characterization of chlorogenic acids and other cinnamates. In C. Santos-Buelga and G. Williamson (Eds.).*Methods in Polyphenol Analysis*. Cambridge: Royal Society of Chemistry. 314–337.

Effiom, O. E., Avoaja D. A., and Ohaeri, C. C. (2012). Mosquito repellent activity of phytochemical extracts from peels of citrus fruit species. *Global Journal of Science Frontier Research Interdiciplinary*, **12 (1)**: 5-8.

Fradin, S.M. (1998). Mosquitoes and mosquito repellents: A clinician's guide. *Annals of internal medicine*, **128(11)**: 931-940.

Guo-Liang, Li., Jin-Mao, You., Cui-Hua, Song., Lian, Xia., Jie, Zheng., and You-Rui, Suo. (2011). Development of a new HPLC method with precolumn fluorescent derivatization for rapid, selective and sensitive detection of triterpenic acids in fruits. *J. Agric. Food Chem.*, **59**: 2972–2979

Herbert Baxter, J.B., Harborne, and Gerald P. Moss, (Eds) (1999). Phytochemical Dictionary: A Handbook of Bioactive Compounds from Plants, Tylor and Francis Ltd, London, 29-30.

Holme, I. (2004), Microencapsulation: The changing face of finishing. *Textiles-Magazine*, **31(4)**: 7-10.

http: //en.wikipedia.org/wiki/Maslinic_acid

http: //pirika.com/NewHP/PirikaE/CarboxylicAcid.html

https: //sites.google.com/site/masonaco/Home/applications/services-offered-2/hplc-nicotine

https: //www2.chemistry.msu.edu/faculty/reusch/virttxtjml/Spectrpy/InfraRed/infrared.htm#ir3

Kishore, N., Mishra, B.B., Tiwari, V.K., and Tripathi, V. (2011). A review on natural products with mosquitosidal potentials. In: Tiwari VK, editor. *Opportunity, challenge and scope of natural products in medicinal chemistry*. Kerala: Research Signpost; 335-65.

Krishnaveni, V. (2009). Mosquito Repellent Finish on Textiles, www.fibre2fashion.com, Accessed on 14.04.14.

Liu J. (1995). Pharmacology of oleanolic acid and ursolic acid. *Journal of Ethnopharmacology*, **49 (2)**: 57–68.

Lupi, E., Hatz, C., and Schlagenhauf, P. (2013). The efficacy of repellents against Aedes, *Anopheles, Culex* and *Ixodes* spp. – A literature review. *Travel medicine and infectious disease*, 1-38.

Malik, T., *et al.*, Mosquito repellent textiles: An overview, www.fibre2fashion.com, Accessed on 14.04.14.

Marina Zacchigna, Francesca Cateni, Mariangela Faudale, Silvio Sosa, and Roberto Della Loggia, (2009). Rapid HPLC analysis for quantitative determination of the two isomeric triterpenic acids, oleanolic acid and ursolic acid in *Plantago major*. *Sci Pharm.*, **77**: 79–86.

Marta Ferreira Maia, and Sarah J Moore, (2011). Plant-based insect repellents: a review of their efficacy, development and testing, *Malaria Journal*, **10(1)**: S11.

Mengoni, F., Lichtner, M., Battinelli, L., Marzi, M., Mastroianni, C.M., Vullo, V., and Mazzanti, G. (2002). *In vitro* anti-HIV activity of oleanolic acid on infected human mononuclear cells. *Planta medica*, **68 (2)**: 111–4.

Nelson, G. (2001), Microencapsulation in textile finishing. *Review of Progress in Coloration*, **31**: 57.

Pichersky, E., and Gershenzon, J. (2002). The formation and function of plant volatiles: perfumes for pollinator attraction and defense. *Curr Opinion Plant Biology*, **5**: 237-243.

Rattan, R. S. (2010). Mechanism of action of insecticidal secondary metabolites of plant origin. *Crop Protec.*, **29**: 913-20.

Shirish, G.V.N.K., *et al.* (2007). Recent developments of mosquito repellent textiles, *ATA journal for asia on textile and apparel*.

Surjit, R., *et al.* (2011). Application of mosquito repellent finish using natural oils, *Proceedings of the 7th International conference on Advances in textiles, machinery, nonwovens and technical textiles*, KCT, Coimbatore, India.

Ullrich, S.L. (1970). Organic acids. *Biochemistry of Fruits and Their Products*, Volume 1. (Hulme A.C ed) London: Academic Press, 89-118.

Von Euler, V.S. (Ed), (1965). Tobacco alkaloids and related compounds, Oxford: Pergamon press.

William Kemp. (1987). *Organic spectroscopy*. Macmillan Education Ltd.

Yu Fei, Wang Qi, Zhang Zhen, Peng Yi-yun, Qiu Yun-yan, Shi Yong-Ying, Zheng Yong-Xiang, Xiao Su-Long, Wang Han, Huang Xiaoxi, Zhu Linyi, Chen Kunbo, Zhao Chuanke, Zhang Chuanling, Yu Maorong, Sun Dian, Zhang Lihe, Zhou Demin,. (2013). Development of Oleanane-Type Triterpenes as a New Class of HCV Entry Inhibitors. *Journal of Medicinal Chemistry*, **56 (11)**: 130510090711006.

Zhao, Y., Wang, J., Ballevre, O., Luo, H., and Zhang, W. (2011). Antihypertensive effects and mechanisms of chlorogenic acids. *Hypertens Res.*, **35 (4)**: 370–4.

Utilisation and Management of Medicinal Plants Vol. 3 (2015) *Pages* **367–403**
Editors: **V.K. Gupta and A. Kaul**
Published by: **DAYA PUBLISHING HOUSE, NEW DELHI**

26

Umbeliferae: A Review on Plant Tissue Culture

R.K. Sharma[1]

ABSTRACT

With increasing realization of the health hazards and toxicity associated with the indiscriminate use of synthetic drugs and antibiotics, interest in the use of plants and plant based drugs has revived throughout the world. However, a large number of medicinal plants remain to be investigated for possible pharmacological value. Most of the pharmaceutical industry is highly dependent on wild populations for supply of raw material for extraction of medicinal important compounds. Due to a lack of proper cultivation practices, destruction of plant habitats, and the illegal and indiscriminate collection of plants from these habitats, many medicinal plants are severely threatened. Advance biotechnological methods of culturing plant cells and tissue should provide new means of conservation and rapidly propagating valuable rare and endangered medicinal plants. The members of Umbelliferae are valuable source of different kinds of useful products like spices, condiments, food plants, medicinal herbs and pharmacological products. Micropropagation is an alternative mean of propagation that can be employed in conservation of the flora in relatively shorter time. Tissue culture is useful for multiplying and conserving the species, which are difficult to regenerate by conservation methods and save them from extinction as well for commercial exploitation. This paper describes the work carried out on in vitro propagation of some medicinally important umbellifers.

***Keywords*:** Umbelliferae, Plant tissue culture, Micropropagation.

1 Govt. College for Women, Gandhi Nagar, Jammu – 180 004, J&K, India.

* *Corresponding author*: E-mail: raj66a@gmail.com

Introduction

The family umbelliferae consists of about 350 genera and 3225 species including many medicinal and edible plants (Wake *et al.*, 1995). It is distributed throughout the world mainly in the temperate regions (Kim *et al.*, 1997). In India, the family is represented by approximately 68 genera and 240 species (Mukerjee and Constance, 1993). Todate over 200 publications on tissue culture of some 49 species of this family are found scattered in nearly 50 periodicals. A summary of the literature of *in vitro* studies for each species is presented in Table 26.1. Most of these studies have been designed to establish whole plant regeneration system for the clonal multiplication of plants (Vandemoortele *et al.*, 1996), development of somaclonal variants (Browers and Ortan, 1986; Merrick and Collin. 1982; Metcalf and Coffin, 1978), production of disease resistant plants (Health-Pagliuso, *et al.*, 1988, 1989), conservation of threatened plants (Wakhlu *et al.*, 1990), development of transgenic plants (Catlin *et al.*, 1988; Krens *et al.*, 1997; Pawlicki *et al.*, 1992; Thomas *et al.*, 1989) and production of secondary metabolites (Chang *et al.*, 1992; Metcalf and Coffin, 1978; Szoke *et al.*, 1991). This review is intended to supplement the basic information on each species in Table 26.1.

1. *Actinotus helianthi* Labill

There is only on report on tissue culture of *Actinotus helianthi* (Cathy and Joanne, 1993). Axillary shoot formation was obtained on MS medium supplemented with 5μM BA and 0.2-1μM IBA. Maximum shoot proliferation was observed after 6 weeks of culture. Shoots were rooted on MS medium enriched with IBA. Regenerated plants were transferred to soil with 80 per cent survival rate.

2. *Ammi majus* L.

Callus was initiated from young ovaries of *A. majus* on White's medium fortified with CH, CM, YE and IAA. The polyembryonal mass produced from these calli has been shown to differentiate seedlings on a medium containing 200 mgl^{-1} YE (Sehgal, 1972a). Plant regeneration from hypocotyl calli has been obtained by somatic embryogenesis on MS medium supplemented with 2 mgl^{-1} IAA (Grewal *et al.*, 1976). El-Bahr *et al.* (1988) have differentiated shoots from callus cultures after their transfer from MS medium containing 2 mgl^{-1} IAA and 0.2 mgl^{-1} Kn to a medium that lacked growth regulators. Shoot formation has been reported from cotyledon callus cultures on MS medium enriched with 2 mgl^{-1} Kn, 0.5 mgl^{-1} IAA, 40 mgl^{-1} Ad_2SO_4 and 1000 mgl^{-1} CH (Purohit *et al.*, 1995 a). Plants have been successfully established in the field. The regenerating callus cultures were found to produce an enhanced level of xanthotoxin (Purohit *et al.*, 1995 b). Direct organogenesis was achieved in cultures on MS medium supplemented with 0.5 mgl^{-1} IAA, 2 mgl^{-1} Kn and 1000 mgl^{-1} CH. Mature plants bearing fruits were produced in the culture vials on MS medium supplemented with 0.5 mgl^{-1} IAA, 5 mgl^{-1} Kn, 40 mgl^{-1} adenine and 500 mgl^{-1} CH (Pande *et al.*, 2002). The regenerated shoots were rooted on MS medium supplemented with 2 mgl^{-1} IBA and 100 mgl^{-1} glutmine.

3. *Ammi visnaga* L.

Embryogenesis was induced in *A. visnaga* from hypocotyl derived callus cultures upon their transfer from MS medium containing 2,4-D to a medium lacking growth

Table 26.1: Tissue Culture Studies in Apiaceae

Species	*Explant*	*Medium*	*Growth Regulators (mgl⁻¹)*	*Results*	*Reference*
Actinotus helianthi	Shoot tips	MSWM 0.2-1µM IBA	5µM BA+ Roots	Axillary shoots	Cathy and Joanne (1993)
			IBA YE (2000)	Callus, embryos and plantlets	Sehgal (1972a)
Ammi majus L.	Ovary				
	Hypocotyl	MS	IAA (5)	Callus	Grewal *et al.* (1976)
		MS	IAA (2)	Embryos and plantlets	
	Hypocotyl, root	MS	IAA (2) + Kn (0.2)	Callus	El-Bahr *et al.* (1988)
		MS	None	Shoots	
	Cotyledon	MS	IAA (2) + Kn (5) + CH (1000)	Callus	Purohit *et al.* (1995a and b)
	Nodal segments	MSMS	IAA (2.5) + Kn (2) + Ad_2SO_4 (40) + CH (1000)	Shoots	Pande *et al.* (2002)
			IAA(0.5)+Kn(2.0)+CH(1000)	Shoots	
			IAA(0.5)+Kn(5)+Ad(40)+ CH(500)	Flowers/fruits	
A.visnaga L.	Hypocotyl	MS	2, 4-D(I)	Callus	El-Fiky *et al.* (1989)
		MS	None	Embryos	
Anethum graveolens L.	Ovary	WM	IAA (1) + CH (100-1000) or YE (100-1000)	Callus and embryos	Johri and Sehgal (1965, 1966)
	Embryo	WM	NAA (2)	Callus	Stweard *et al.* (1970)
		WM	CH + CM	Embroyos and plantlets	
	Hypocotyl	MSN-1	2,4-D(0.02) + Kn (0.2)	Callus	Ratnamba and Chopra (1974)
		MSN-1	None	Embroyos and plantlets	
	Cotyledon	MSN-1	2,4-D (2.2)	Callus	
		MSN-1	2,4-D (0.2) + Kn (0.2)	Embryos	

Contd...

Table 26.1–*Contd...*

Species	*Explant*	*Medium*	*Growth Regulators* (mgl^{-1})	*Results*	*Reference*
	Inflor	WM	2,4-D(1) + CM (10 per cent)	Callus	Sehgal (1978)
		WM	YE (1000)	Embryos and plantlets	
		WM	None	Shoots and roots	
Anegelica acutiloba	Pedicel	–	–	Callus, embryos plantlets	Nakagawa *et al.* (1982)
	Cotyledon, hypocoty1	MS	2,4-D(0.5-2) + Kn (0.5-1)	Callus	Ohga *et al.* (1989)
		MS	2,4-D + Kn	Embryos	
		MS	None	Plantlets	
A.koreana Max.	Inflor.	MS	2,4-D(1-2)	Callus	Jang (1994)
		MS	None	Embryos and plantlets	
A.sinensis	Cotyledon, hypocotyls, leaf, petiole, root	B_5	2,4-D(0.7) + NAA (0.5) + BA (0.5)	Callus	Zhang Shi-yu and Chang Kuo-Khang (1982)
		B_5	NAA(0.01) + BA (0.5)	Shoots	
		MS	NAA (0.05) + Kn (0.5)	Embryos	
Anthriscus sylvestris	Stem nodes	MS	2,4-D(1)+ Zeatin(0.5)	Callus	Guang-min *et al.*(1991)
		MS	Zeatin(0.1)	Embryos	
Apium graveolens L.	Hypocotyl	MSW	2,4-D (0.05)	Callus and embryos	Reinert *et al.* (1966)
	Petiole	MS	2,4-D (0.5) + Kn (0.6)	Callus	Williams and Collin (1976a), Zee and Wu (1979)
		MS	Kn (0.6)	Embryos	
		MS	None	Plantlets	
	Stem	MS	2,4-D (0.5) + Kn (0.6)	Callus	Al-Abta and Collin (1979)
		MS	2,4-D (0.5) + Kn (0.6)	Embryos	

Contd...

Table 26.1–*Contd...*

Species	*Explant*	*Medium*	*Growth Regulators* (mgl^{-1})	*Results*	*Reference*
	Stem nodes	MSMS	2,4-D(1)+ Zeatin(0.5)	Callus	Guang-min *et al.*(1991)
			Zeatin(0.1)	Embryos	
	Hypocotyl	MSW	2,4-D (0.05)	Callus and embryos	Reinert *et al.* (1966)
	Petiole	MS	2,4-D (0.5) + Kn (0.6)	Callus	Williams and Collin (1976a), Zee and Wu (1979)
		MS	Kn (0.6)	Embryos	
		MS	None	Plantlets	
	Stem	MS	2,4-D (0.5) + Kn (0.6)	Callus	Al-Abta and Collin (1979)
		MS	2,4-D (0.5) + Kn (0.6)	Embryos	
	Petiole	MS	None	Plantlets	Zww and Wu(1980)
		MS	Kn(0.6)	Embryos	
		MS	None	Plantlets	
	Leaf petiole	MS	2,4-D (0.5) + Kn (0.5)+ CH (500)	Callus	Nadel *et al.* (1989)
		MS	2,4-D (0.5) + BA (0.2)+ CH (500)	Suspension	
		MS	Kn (0.5) + CH (100)	Embryos and plantlets	
	Leaf	MS	2,4-D(2)	callus	Kim and janick(1989b)
	Leaf	MS	None	Embryos and plantlets	Pontfarcy and Bigot (1991)
		B5	2,4-D(2.26μM)+BAP(1.1μM)	Callus	
		MS	2,4-D(2.26μM)	Embryos plantlets	
		MS	None		
	Leaf	MS	2,4-D (0.5) + Kn (0.5) + CH (500)	Callus	Saranga and Janick (1991)
		MS	None	Embryos and plantlets	
		MS	2,4-D (2.26μM)	Embryos	
		MS	None	Plantlets	

Contd...

Table 26.1–*Contd...*

Species	*Explant*	*Medium*	*Growth Regulators* (mgl^{-1})	*Results*	*Reference*
	Stem	MS	2,4-D (0.5) + Kn (0.6)	Callus	Donovan *et al.* (1994)
		MS	None	Shoot and plantlets	
	Hypocotyl	MS	2,4-D (0.5) + Kn (0.5)	Callus	Chen *et al.* (1991)
	Cotyledons	MS	None	Suspension	
		MS	None	Embryos	
	Seed, leaf, cotyledons	MS	2,4-D (1) + Kn (0.5)	Callus	De Xiang *et al.* (1995)
		MS	None	Suspension, embryos	
		MS	None	Embryos	
	Leaf	MS	2,4-D(1)	Suspension	Choi and Soh (1997)
		MS	None	Embryos	
Bunium persicum Boiss	Leaf	MS	Kn ($5x10^{-6}$M)+2,4-D($1x10^{-6}$M)	Callus	Grewal (1996a)
		MS	Kn ($5x10^{-6}$M)	Embryos and plantlets	
	Mericarp	½ MS	Kn (1×10^{-8}-$1x10^{-6}$M)	Tubers	Grewal (1996b)
		MS	2,4-D(1)	Callus	Wakhlu *et al.* (1990)
		MS	None	Embryos and plantlets	
Bupleurum falcatum L.	Floral bud, leaf, stem	LS-2	2,4-D (0.02-2)	Callus	Uomori *et al.* (1974)
		LS-2	2,4-D (0.02-2) + Kn (0.02-2)	embroyos	
	Leaf	LS-2	2,4-D (0.02-2)	Callus and roots	Wang and Huang (1982)
		LS-2	NAA (0.02) + Kn (1)	Embryos	
		LS-2	None	Plantlets	
	Floral bud, leaf, stem	LS-2	2,4-D (0.2-2)	Callus	Hiraoka *et al.* (1983)

Contd...

Table 26.1–*Contd...*

Species	*Explant*	*Medium*	*Growth Regulators* (mgl^{-1})	*Results*	*Reference*
		LS-2	2,4-D (0.2)	Embryos	
		LS-2	None	Plantlets	
B. scorzonerifolium, Willd	Stem node	–		Protoplast, embryos and plantlets	Xia *et al.* (1992)
Carum carvi L.	Petiole	WM	NAA (2) + CH (200)	Callus	Ammirato (1974)
		WM	ABA (0.03)	Embryos	
	Hypocoty1	MSN-2	2,4-D (0.3)	Callus	Bajaj *et al.* (1988)
		MSN-2	None	Embryos	
		MS	IBA (0.5)	Plantlets	
	Hypocoty1	MS	NAA (0.25) + BAP (0.5)	Callus and shoot buds	Krens *et al.* (1997)
	Cotyledons	MS	NAA (0.25) + BAP (0.5)	Shoots	
Conium maculatum L.	Embryo	MS	NAA (2) + CH	Callus	Steward *et al.* (1970)
		MS	None	Embryos	
Coriandrum sativum L.	Root	WM	2,4-D(4) + IAA (4) + CH (150-1000)	Callus	Sardesai and Tipnis (1969)
	Embryo	WM	NAA (2)	Callus	Steward *et al.* (1970)
		WM	None	Embryos	
	Shoot tips	SH	NAA (0.15) + GA_3 (0.5)	flowering	Stephen and Jayabalan (1998)
		WM	None	Plantlets	
	Ovary	WM	Kn (0.5-1) + IAA (1)	Callus	Sehgal (1972b)
		WM	YE (500)	embryos and plantlets	
	Petiole	MS	2, 4-D (0.5)	Embryos	Zee (1981)
		MS	None	Plantlets	

Contd...

Table 26.1–*Contd...*

Species	*Explant*	*Medium*	*Growth Regulators (mgl⁻¹)*	*Results*	*Reference*
	Shoot tips	MS	NAA (3) + Kn (0.5)	Callus	Kumar *et al.* (1982c)
		MS	NAA (0.5) + Kn (2)	Shoot buds and plantlets	
		MS	None	Embryos	
	Anther	MS	Kn (3-5) + 2, 4-D or NAA or IAA (0.5-2)	Multinucleatecellular mass	Egorova and Reznikova (1982)
	Hypocotyl	MS	2,4-D or NAA + BA or Kn	Callus	Mujib *et al.* (1990)
		MS	2, 4-D (1)	Embryos	
		MS	Kn (1)	Plantlets	
	Hypocotyl	MS	2,4-D (1)	Callus and suspension	Kim *et al.* (1996)
	Zygotic embryo	MS	None	Embryos and plantlets	
	Stem	MS	NAA (2) + Kn (0.5)	Callus	Chen *et al.* (1995)
		MS	None	Embryos	
Cuminum cyminum L.	Shoot tip	WT	NAA (3) + Kn (0.05)	Callus	Kumar *et al.* (1982b)
		WT	NAA (0.5) + Kn (1)	Shoots	
		WT	NAA (3) + Kn (0.05) + IBA	roots	
		WT	Kn (1) + CH (500) or CM (15 per cent)	Embryos	Kumar *et al.* (1982a)
		WT	NAA (0.5) + Kn (1)	Plantlets	
	Hypocotyl, leaf	B_5	NAA (4) + Kn (2)	Callus	Jha *et al.* (1983)
		B_5	NAA (0.5) + BA (0.1)	Shoots	
		MS	NAA (0.5) + IBA (0.5)	Roots	
Daucus carota	Root	DM	2, 4-D (0.1-2) + Kn (0.2)	Callus, embryos and plantlets	Halperin and Wetherell (1964)
	Petiole	MS	2, 4-D (0.1)	Callus and embryos	Halperin (1966)

Contd...

Table 26.1–*Contd...*

Species	*Explant*	*Medium*	*Growth Regulators* (mgl^{-1})	*Results*	*Reference*
		MS	Zeatin (0.2)	Plantlets	
	Petiole, root and protoplast	B_5	2, 4-D (0.1)	Callus	Grambow *et al.* (1972)
		B_5	2, 4-D (0.2)	embryos	
		B_5	None	Plantlets	
	Hypocotyl	LS-1	2,4-D (0.1)	Cell suspension	Fujimura and Komamine (1980)
		LS-1	Zeatin (0.2)	Embryos	
	Petiole	LS-2	Zeatin (10^{-7} M)	Embryos	Fujimura and Komamine (1975)
	Hypocotyl	LS-2	2,4-D ($5x10^{-8}$ M) + Zeatin (10^{-6} M)	Embryos	Nomura and Komamine (1985)
Didiscus coreulea DC	–	MS	2,4-D (1)	Shoot buds	Ball (1969)
Foeniculum vulgare Mill	Stem	NB	2, 4-D (5) + Kn (0.2) + CM (15 per cent)	Callus	Mahashwari and Gupta (1965)
		NB	None	Embryos and plantlets	
	Ovary	MS	2, 4-D (2) + Kn (5)	Callus	Nagari and Wakhlu (1989)
		MS	2, 4-D (2) + Kn (0.5)/2, 4-D (2)	Embryos	
		MS	2, 4-D (2) + Kn (0.5)	plantlets	
		WM	CH + (500) + YE(500)	Callus, embryos and plantlets	Johri and Sehgal (1966)
	Petiole and stem	MS	2,4-D(1) + Kn (0.3)	Callus	Hunault (1981, 1984)
		MS	None	Embryos and plantlets	
	Apical buds	SD	BA (0.1) + IAA (0.1)	Axillary shoots	Du Manoir *et al.* (1985)
		SD+MS	IAA (0.1)	Roots	

Contd...

Table 26.1–*Contd...*

Species	*Explant*	*Medium*	*Growth Regulators (mgl^{-1})*	*Results*	*Reference*
	Petiole and protoplast	MS	2,4-D(0.2) + Kn (0.2)	Callus	Miura and Tabata (1986)
		MS	None	Embryos and plantlets	
	Cotyledon and hypocotyl	MS	2,4-D (1-2) + Kn (0.5-1)	Callus	Ohga *et al.* (1989)
		MS	2,4-D	Embryos	
		MS	None	plantlets	
	Buds from seedlings	MS	IAA (0.01) + BAP (1)	Shoots	Garcia-Rodriguez *et al.* (1978)
		MS	NAA (0.01) + BAP (0.1)	Shoots	Paupardin *et al.* (1980)
		MS/2	NAA (0.1) + BAP (0.1)	Shoots	Badoc (1982)
	Fruits	WM	CH (0.1 per cent) + YE (0.1 per cent)	Polyembryony	Sehgal (1964)
	Stem	MS	NAA (2) + Kn (0.5)	Buds	Becker (1970)
	Stem	H	NAA (0.5) + Kn (0.5)	Roots	Paupardin *et al.* (1980)
	Stem	MS	IBA (1) or NAA (1)	Roots	Hunault (1981; 1984)
	Hypocotyl, petiole and cotyledons				
		MS	2,4-D(1) + Kn (0.3)	Embryoids	
	Petiole	SD	2,4-D(1) + BA (0.3)	Embryoids	Simard (1985)
	Petiole	MS	2,4-D(0.2) + Kn (0.2)	Embryoids	Miura *et al.* (1987a,b)

Contd...

Table 26.1–*Contd...*

Species	*Explant*	*Medium*	*Growth Regulators* (mgl^{-1})	*Results*	*Reference*
Glehnia littoralis Fr. Schm.	Zygotic embryos	½ MS	2, 4-D (1µm)	Callus	Hirai *et al.* (1997)
		MS	None	embryos	
	Shoot tips	MS	BAP (10µm) + NAA (3µm)	Shoots	Hiraoka and Oyanagi (1992)
Oenathe aquatic Poir	Seedling	DM	None	Embryos and plantlets	Waris (1959)
Petroselium hortense Hoffm.	Petiole	MS	2,4-D + CM	Callus	Vasil and Hildebrandt (1966)
		MS	Kn (0.04)	Roots	
		MS	Ad_2SO_4 (15-20)	Embryos and plantlets	
Petroselium crispum (Mill.) Nymanm	Shots buds	MS	BAP (5µM) + NAA (0. 5µM)	Axillary shoots	Vandemoortele *et al.* (1996)
		MS	NAA (2.5µm)	Roots	
Pimpinella anisum L.	Shoot tips	MS	2,4-D (1-3) + Kn (0.5)	Callus	Chand *et al.* (1997)
		MS	2, 4-D (1)	Embryos	
		MS	None	Plantlets	
		MS	BAP (0.5)/BAP (1) + IAA (0.5)	Shoots	
		MS	NAA (0.5-1)	Roots	
	Hypocotyl	MS	Kn (0.5) + IAA (1)	Callus	Becker (1970)
		B_5	NAA (5µM)	Callus	Huber *et al.* (1978)
		B_5	None	embryos	
Sium suave Walt.	Embryo	WM	NAA (0.2) + CH (200) + CM (10 per cent)	Callus	Ammirato and Steward (1971)
		WM	CH (200) + CM (10 per cent)	Embryos and plantlets	

Contd...

Table 26.1–*Contd...*

Species	*Explant*	*Medium*	*Growth Regulators* (mgl^{-1})	*Results*	*Reference*
Trachyspermum ammi (L.) Sprague	Ovary	WM	Kn (0.1) + IAA (1)	Callus	Johri and Sehgal (1966)
	Hypocotyls	MS	2,4-D(2) + Kn (0.5)	Callus	Jasrai *et al.* (1992)
		MS	Kn (0.2)	Embryos	
		MS	None	Plantlets	
		MS	NAA (1)	Callus	Sehgal and Abbass (1994)
		MS	None/NAA (0.2)	Embryos	
		MS	ABA (0.5)	Embryos	
		MS	None	Plantlets	

regulators (El-Fiky *et al.*, 1989). Cell suspension cultures have been reported to transform digitoxigenin to periplognin. 1-beta-hydroxy digitoxigenin and uzarigenin (El-Olemy *et al.*, 1994).

4. *Anethum graveolens* L.

A number of reports are available on plant regeneration of *A. graveolens* via somatic embryogenesis (Johri and Sehgal, 1965, 1966; Ratnamba and Chopra, 1974 Schafer *et al.*, 1986; Sehgai, 1968; Steward *et al.*, 1970) and *via* organogenesis (Sehgal, 1978). Embryogenesis from young inflorescence calli has been obtained on WM medium containing 1 mgl^{-1} 2, 4-D (Sehgal, 1978). Regenerated plants were able to produce flowers in the culture vials (Ratnamba and Chopra, 1974). Transformed root cultures induced from aseptically grown seedlings were reported to form full plants which were successfully established in the field (Santos *et al.*, 2002).

5. *Angelica sinensis* (Oliv.) Diels.

Somatic embryogenesis and plantlet formation has been reported in *A. sinensis* (Zhang Shi-yu and Cheng Kuo-Chang, 1982,1986). Callus initiation from hypocotyl, petiole, cotyledon, leaf and root explants was reported on MS medium containing 0.7 mgl^{-1} 2,4-D, 0.5 mgl^{-1} NAA and 0.5 mgl^{-1} BAP. Embryogenic callus was induced on MS medium supplemented with 0.05 mgl^{-1} NAA and 0.5 mgl^{-1} Kn. Embryo and plantlet formation has been obtained on MS medium containing 0.02 mgl^{-1} NAA and 0.1 mgl^{-1} Kn. Shoot bud differentiation and rooting of shoots were reported using a medium that contained 0.01 mgl^{-1} NAA and 0.05 mgl^{-1} BAP. Protoplast isolation and callogenesis from cultured protoplasts has been reported (Zhang Shunpei *et al.*, 1985). Calli derived from root explants were noticed to exhibit chromosomal instability (Zhang Shi-yu and Cheng Kuo-Chang, 1986).

6. *Angelica acutiloba* L.

Somatic embryo and plantlet regeneration has been reported in *A. acutifolia* (Nakagawa *et al.*, 1982; Ohga *et al.*, 1989). Callus formation from hypocotyl and cotyledon explants has been demonstrated on MS medium containing 0.5-2 mgl^{-1} 2, 4-D and 0.5-1.0 mgl^{-1} Kn. Embryogenesis has been reported on a medium supplemented with 2, 4-D and Kn and plant regeneration on MS medium that lacked growth regulators (*Ohga et al.*, 1989).

7. *Angelica koreana* Max.

Somatic embryogenesis and plant regeneration of *A. koreana* has been obtained from callus cultures derived from immature inflorescence explants (Jang, 1994).The embryogenic callus could produce plantlets after one year of subculture on MS medium supplemented with 1-2 mgl^{-1} 2, 4-D.

8. *Anthriscus sylvestris* (L.) Hoffm.

Plant regeneration has been reported in *A. sylvestris* via somatic embryogenesis (Guang-min *et al.*, 1991). Embryogenesis was induced in callus obtained from protoplasts on modified MS medium supplemented with 1 mgl^{-1} 2,4-D and 0.5 mgl^{-1} zeatin. The protoplast colonies were proliferated on MS medium containing

0.5 mgl^{-1} 2,4-D. Mature embryos were formed on MS medium lacking growth regulators or on MS medium containing 0.1 mgl^{-1} zeatin. The regenerated plants were established in the field.

9. *Apium graveolens* L.

The first successful induction of somatic embryogenesis in *A. graveolens* cultures was reported by Reinert *et al.* (1966) but the somatic embryos produced were not able to develop further. Protocol for plant regeneration *via* somatic embryogenesis was developed by Chen (1976) and Williams and Collin (1976a) and *via* organogenesis was developed by Donovan *et al.* (1994). Al-Abta and Collin (1979) and *Zee* and Wu (1980) carried out histological studies and traced the origin of celery somatic embryos to globular structures embedded in the callus. Somatic embryos of celery were produced from a single cell on the surface of callus clumps (Kim and Janick, 1989b, 1991). Celery embryos have been used for encapsulation and production of synthetic seeds (Kitto and Janick, 1985; Kim and Janick, 1989a, 1990; Onishi *et al.*, 1992, 1994; Saranga *et al.*, 1992). Williams and Collin (1976b) have reported that 25 per cent of the regenerated plants were aneuploids which were indistinguishable horticulturally from the normal diploids. Cytological variability has been shown to occur between cultures arising from different plants of the same variety (Browers and Ortan, 1986; Murata and Ortan, 1983; Ortan, 1985). Selection of somaclonal variants *via* somatic embryogenesis has resulted in the formation of celery lines which exhibited resistance to herbicides (Metcalf and Collin, 1978; Merrick and Collins, 1982) and to fungal diseases caused by *Fusarium oxysporum f* sp. *apii* (Health – Pagliuso *et al.*, 1988, 1989) and *Septoria apiicola* (Evenor *et al.*, 1994). Transgenic plants have been regenerated through *Agrobacterium tumefaciens* mediated transformation (Catlin *et al.*, 1988).

10. *Arracacia xanthorrhiza* Bancroft

There is only one report available on complete plant regeneration in *A. anthorrhiza* by organogenesis and axillary shoot proliferation method (Sliva *et al.*, 2010). Axillary shoots were formed on MS medium containing 0.1 mgl^{-1} NAA and 1 mgl^{-1} BAP. Callus was formed on MS medium supplemented with 0.1 mgl^{-1} 2,4-D and 0.2 mgl^{-1} Kn. The regenerated shoots obtained via organogenic and axillary shoot proliferation methods were rooted on MS medium supplemented with 1 mgl^{-1} 2,4-D. The rooted shoots were transferred to soil.

11. *Bunium persicum* Bioss

Somatic embryo and plantlet formation has been reported in B. *persicum* (Wakhlu *et al.*, 1990; Grewal, 1996a). Callus formation,somatic embryogenesis and tuber formation has been reported by Sharma (20011a and b; 2012). Embryogenic calli were produced from mericarp callus cultures on MS medium fortified with 1 mgl^{-1} 2, 4-D. Mature embryo formation and conversion of embryos into plantlets has been obtained on MS medium that lacked growth regulators (Wakhlu *et al.*, 1990). The regenerated plantlets have been reported to be diploids with 2n=14 chromosomes (Nagari *et al.*, 1997).Callus cultures were initiated from hypocotyls parts of embryos on MS medium fortified with 1 mgl^{-1} NAA and 2 mgl^{-1} Kn (Valizadeh *et al.*, 2006,2008). Embryogenic calli were produced on MS medium supplemented with 0.1 mgl^{-1} NAA and 0.5

mgl^{-1} Kn. Shoots were rooted on MS medium supplemented with 2 mgl^{-1} NAA and 2 mgl^{-1} Kn. The regenerated plants established in the field.

12. *Bupleurum distichophyllum* Wight

Multiple shoots were obtained from nodal and shoot tip explants on MS medium supplemented with 1 mgl^{-1} BAP (Karuppusamy and Pullaiah, 2007). Explants produced shoots on MS medium containing 1 mgl^{-1} BAP and 1 mgl^{-1} NAA. The shoots produced roots on MS medium supplemented with 2 mgl^{-1} IBA. The regenerated plantlets were successfully tranfered to soil with a 71 per cent survival rate.

13. *Bupleurum falcatum* L.

Callus has been produced from floral bud, leaf and stem explants of *B. falcatum* on LS medium containing 2, 4-D (Uomori *et al.*, 1974). Callus when grown on a growth regulator free medium or on a medium containing 0.02 $mg1^{-1}$ NAA and 1.0 $mg1^{-1}$ Kn is reported to produce somatic embryos (Wang and Huang, 1982). Embryos were germinated on LS basal medium (Hiraoka *et al.*, 1983). Regenerated plants have been grown to maturity in the field. The roots of regenerated plants contained higher saponin contents as compared to the roots of plants propagated from seed (Hiraoka *et al.*, 1986). The level of saikosaponin by cell suspensions established from leaf callus corresponded to the level of saikosaponin produced by the roots (Katakura *et al.*, 1991). Plants were produced from protoplasts isolated from callus suspension cultures (Bang et *et al.*, 1999).

14. *Bupleurum fruticosum* L.

Micropagation of *B. fruticoum* was achieved by axillary shoot proliferation (Fratenale *et al.*, 2002). Optimal shoot proliferation was obtained on MS medium supplemented with 1 mgl^{-1} IAA and 1.5 mgl^{-1} BA. The shoots formed roots on MS hormone free medium or on MS medium supplemented with 2 mgl^{-1} NAA. The rooted plants were successfully established in the soil.

15. *Bupleurum kaoi* Liu.

Callus was formed from leaf explants of B. kaoi on half strength MS medium supplemented with 4 $mg1^{-1}$ 2,4-D. The growth of the callus was optimal on half strength MS medium supplemented with 0.1-0.2 $mg1^{-1}$ 2,4-D (Chen *et al.*, 2005). Shoot formation from callus cultures was obtained on MS medium supplemented with 0.25 $mg1^{-1}$ IAA. Multiple shoots were formed on nodal explants on MS medium supplemented with 1.0 $mg1^{-1}$ BA and 0.1 $mg1^{-1}$ NAA (Chen *et al.*, 2006). Optimal rooting of shoots occurred on 0.5 $mg1^{-1}$ IBA and 0.2 $mg1^{-1}$ NAA. The hardened plants were successfully transferred to soil.

16. *Bupleurum scorzonerifolium* L.

Protoplasts of B. *scorzonerifolium* were prepared from stem node derived embryogenic calli by using an enzyme mixture containing snailase as an essential component. These protoplasts have been reported to develop into plantlets *via* somatic embryogenesis (Xia *et al.*, 1992).

17. *Carum carvi* L.

Callus was formed from petiole explants of *C. carvi* (Ammirato, 1973, 1974, 1977; Furmanowa *et al.*, 1984). Ammirato (1973, 1983a) reported somatic embryogenesis from suspension cultures on White's basal medium. Inclusion of 0.02 mgl^{-1} ABA and maintenance of dark conditions has been noticed to improve embryogenesis (Ammirato, 1974, 1977). Complete plants were regenerated from hypocotyl derived calli on MS medium containing 0.5 mgl^{-1} IBA and 10 mgl^{-1} Ad_2SO_4 (Furmanowa *et al.*, 1984). Regenerated plants were reported to be diploids and fertile (Bajaj *et al.*, 1988). Transgenic plants were developed through *Agrobacterium turnefaciens* mediated transformation (Krens *et al.*, 1997).

18. *Centella asiatica* L. Urban

Plants of *C. asiatica* were regenerated by axillary shoot proliferation method (Tiwari *et al.*, 2000; Shivkumar *et al.*, 2006). Multiple shoots were formed from stem node explants on MS medium supplemented with 1 mgl^{-1} BAP and 0.5 mgl^{-1} NAA (Hossain *et al.*, 2000). Multiple shoots were formed from shoot tip explants on MS medium supplemented with 17.76μM BAP and 1.44μM GA_3. Maximum roots formation on shoots was obtained on half strength MS medium supplemented with 10.74μM NAA (Nath and Buragohain,2003). Somatic embryos were differentiated from callus cultures on MS medium fortified with 0.45μM 2,4-D and 1.16μM Kn (Martin, 2004 b). Axillary shoots were produced from leaf explants on MS medium fortified with 4 mgl^{-1} BA and 0.1 mgl^{-1} NAA were rooted on MS medium containing 2 mgl^{-1} IBA. Petiole explants cultured on MS medium supplemented with 0.5 mgl^{-1} NAA and 2.0 mgl^{-1} Kn gave best callus initiation response (Shashikala *et al.*, 2009). Stem and stolen explants resulted in optimal callus formation on MS medium supplemented with 0.25-0.5 mgl^{-1} Kn and 2 mgl^{-1} 2,4-D (Rao *et al.*, 1999). Multiple shoots obtained from leaf and nodal explants on MS medium supplemented with 3mg dm^{-3} BAP and 0.05 mg dm^{-3} NAA were rooted on half strength MS medium fortified with 0.5 mg dm^{-3} IBA. The regenerated plans were acclimatized in vermin compost and established in the field (Mohapatra *et al.*, 2008). Somatic embryogenesis was obtained from leaf explants on MS medium supplemented with 9.29M Kn and2.26M 2,4-D (Parmageetham *et al.*, 2004). Multiple shoot proliferation was reported on MS medium supplemented with 2 mgl^{-1} BAP (Karthikeyen *et al.*, 2009) and on MS medium containing 1.5 mgl^{-1} Kn (Naidu *et al.*, 2010). Complete plants were regenerated from leaf callus via organogenesis on MS medium containing 6μM BAP and 4μM NAA (Deshpande *et al.*, 2010). Regeneration of plants from protoplasts was reported (Aziz *et al.*, 2006). IBA and Kn were observed to be effective for inducing callus and shoot formation (Singh *et al.*, 2010). Commercial sugar was used instead of sucrose as a carbon source *in vitro* for raising shoots were rooted *in vivo* resulted for making the making the whole regeneration system cost effective (Raghu *et al.*, 2007). Callus was reported to produce various bioactive compounds (Jacinda *et al.*, 2008; Bonfill *et al.*, 2011) like triterpinoid saponin (Mangas *et al.*, 2008), centelloside (Yoo *et al.*, 2011) and asiatcoside (Loc and Ann, 2010; Nath and Buragohain, 2005).

19. *Conium maculatum* L.

Plant regeneration in C. *maculatum* by somatic embryogensis from zygotic embryo derived callus culture has been reported on MS medium containing CM and 2 mgl^{-1} NAA (Steward *et al.*, 1970).

20. *Coriandrum sativum* L.

Coriander cultures were demonstrated to regenerate plants by somatic embryogenesis (Chen *et al.*, 1995; Kim *et al.*, 1996; Mujib *et al.*, 1990; Sehgal, 1972b and Zee, 1981), by organogenesis (Kumar *et al.*, 1982c) and by axiliary shoot proliferation (Kataeva *et al.*, 1993). Shoot buds and somatic embryos were produced from shoot tip derived callus on MS medium containing 0.5 mgl^{-1} NAA and 2 mgl^{-1} Kn (Kumar *et al.*, 1982c). Development of mature embryos and subsequent plantlet regeneration has been obtained on MS medium devoid of growth regulators or on MS medium enriched with low levels of auxins (Zee, 1981; Chen *et al.*, 1995). Artificial seeds were produced by manipulating callus and somatic embryo formation as well as embryo encapsulation. The survival rate of plantlets produced from artificial seeds has been reported to be 83 per cent (Chen *et al.*, 1992).

21. *Cnidium officinale* Makino

Multiple shoots were obtained from shoot tip explants of *C. officinale* on MS medium supplemented with 10^{-6} M BAP and 10^{-7} M NAA. Plants were regenerated through somatic embryogenesis from callus produced from immature flowers on MS medium supplemented with 1 mgl^{-1} 2,4-D and 1 mgl^{-1} BAP (Lee *et al.*, 2009). Addition of cytokinin and GA_3 improved the conversion rate of somatic embryos to plantlets.

22. *Crithmum maritimum* L.

There is a single report on plant regeneration of *C. maritimum* by adventitious shoot proliferation method (Grigoriadou and Maloupa, 2008). Maximum shoot proliferation was obtained on MS medium containing 2.5 μM BA. The shoots were rooted on MS medium supplemented with BA 2.5μM BA and 1.0-2.5 μM NAA. The regenerated shoots were established in the field.

23. *Cuminum cyminum* L.

Callus was formed from hypocotyls explants of *C. cyminum* on B_5 medium supplemented with 0.5 mgl^{-1} 2,4-D. The regenerated plantlets were observed to produce flowers in the culture vials (Jha *et al.*, 1983). Direct shoot regeneration from mature embryos was reported to be rapid and genotype- independent in cumin (Ebrahimie *et al.*, 2003, 2006, 2007). Optimal shooting was achieved on MS medium supplemented with 0.8 mgl^{-1} NAA and 0.3 mgl^{-1} IAA (Ebrahimie *et al.*, 2007). Optimal callogenesis and rooting of the embryos was reported on B_5 medium containing 2.0 mgl^{-1} and 4.0 mgl^{-1} Kn (Valizadeh *et al.*, 2007). Maximum shoot formation was obtained on B_5 medium containing 0.1 mgl^{-1} NAA and 4 mgl^{-1} Kn. Plants were regenerated from callus *via* somatic embryogenesis (Azza and Noga, 2001; 2002). Genetic transformation of *in vitro* raised embryos has been achieved by bombardment by microprojectile on the embryo axis (Singh *et al.*, 2010).

24. *Changiums myrnioides* L.

There is only one report available on plant regeneration from callus cultures of *C. myrnioides* (Xu *et al.*, 2005). Callus was initiated from stem explants on MS medium supplemented with 0.5 mgl^{-1} NAA and 0.5 mgl^{-1} Kn. Adventitious shoots produced on MS medium supplemented with 1-2 mgl^{-1} BA and 1-2 mgl^{-1} GA_3 were rooted upon their subculture on MS medium supplemented with 0.5-1.0 mgl^{-1} IBA.

25. *Dorema ammoniacum* D

Callus was produced from root, hypocotyls and cotyledon explants of seedlings of *D. ammoniacum* on MS medium supplemented with 2.0 mgl^{-1} BAP and 1 mgl^{-1} NAA (Irvani *et al.*, 2010). Optimal shooting response from hypocotyls derived callus was obtained on MS medium supplemented with 2.5 mgl^{-1} IBA. The regenerated plants were hardened and established in the field.

26. *Daucus carota* L.

Since Reinert (1958) and Steward *et al.* (1958) independently reported *in vitro* indirect somatic embryogenesis in carrot, a number of workers reported methods to increase carrot embryo formation and to obtain high quality embryos with a high germination frequency potential (De Vries *et al.*, 1988; Fujimura and Komamine, 1979; Giuliano *et al.*, 1983; Kamada *et al.*, 1989; Kamada and Harada, 1981; Smith *et al.*, 1987; Wake *et al.*, 1991, 1992ab). The embryogenesis occurred readily in suspension cultures when embryogenic cells growing in a medium containing 2,4-D were transferred to an auxin-free medium (Halperin, 1966). The effects of auxins, cytokinins, gibberellic acid and abscisic acid on embryo formation have been investigated (Fujimura and Komamine, 1975; Kamada and Harada, 1979a). Reduced nitrogen sources like NH_4C1_2, amino acids (glutamine, alanine, glutamic acid) and urea is reported to be required as a supplement to nitrate for efficient embryogenesis (Wetherell and Dougall, 1976; Kamada and Harada, 1979b). Serine and proline had stimulatory effect on embryo production (Nuti-Ronchi *et al*,, 1984). Methods reported for mass production of somatic embryos include synchronization and sieving (Giuliano *et al.*, 1983), abscisic acid treatment (Ammirato, 1983b) and direct embryogenesis by physiological stress without hormone treatment (Karnada *et al.*, 1989). Carrot cells cultured *in vitro* as well as protoplasts isolated enzymatically from these cells form embryos and grow into mature plants (Grambow *et al.*, 1972). Nomura *et al.* (1983) have reported culture conditions and morphological observations on the cultured protoplasts obtained from carrot somatic embryos. Methods for using these embryos as a propagation system similar to seeds have recently been developed (Kitto and Janick, 1985; Onishi *et al.*, 1994; Redenbaugh *et al.*, 1991; Wake *et al.*, 1992b). Kamada *et al.* (1989) reported stress- induced somatic embryogenesis in carrot and applied this method for synthetic seed production. Regeneration of transgenic plants through *Agrobacterium tumefaciens* mediated transformation has been reported (Thomas *et al.*, 1989; Pawlicki *et al.*, 1992).

27. *Didiscus coerulea* DC.

Callus cultures of *D. coerulea* were obtained using MS medium supplemented with 1 mgl^{-1} 2, 4-D (Ball, 1969). Shoot buds were differentiated from callus tissues on a medium of similar composition.

28. *Eryngium foetidum* L.

Plant regeneration from callus cultures of *E. foetidum* has been reported (Arockiasamy and Ignacimuthu, 1998; Ignacimuthu *et al.*, 1999; Martin, 2004). LS medium containing $1mgl^{-1}$ 2,4-D and $1mgl^{-1}$ BAP was optimal for initiating callus cultures from leaf explants. Somatic embryos were induced on the callus on MS medium supplemented with $0.1mgl^{-1}$ 2,4-D, $2mgl^{-1}$ BAP and $1mgl^{-1}$ GA_3. The embryos were converted into plantlets on MS medium fortified with $1mgl^{-1}$ GA_3 and $0.1mgl^{-1}$ BAP. Shoots were differentiated from callus cultures on MS medium containing $0.1mgl^{-1}$ IAA and $1mgl^{-1}$BAP. The regenerated shoots were rooted on $0.1mgl^{-1}$ IAA and $1mgl^{-1}$ GA_3.

29. *Ferula asafetida* L.

Plant regeneration in *F. assafoetida* has been achieved by indirect and direct embryogenesis (Hassani *et al.*, 2008; Zare *et al.*, 2010). Callus was induced on hypocotyl explants from seedlings on MS medium supplemented with $1mgl^{-1}$ Kn and $0.5mgl^{-1}$ NAA. Embryogenic calli were produced on hormone free MS medium. Maximum embryo formation response was on a medium supplemented with 1.5 mgl^{-1} Kn and 1 mgl^{-1} NAA. About 40-50 per cent somatic embryos were converted into plantlets. Direct somatic embryogenesis was induced on intact seedlings on basal MS medium. Callus raised from hypocotyl explants showed good shoot regeneration response on 0.2 mgl^{-1} NAA and 1 mgl^{-1} BA. The shoots were rooted on half strength MS medium containing 2.5 mgl^{-1} IBA. The regenerated plants were successfully transferred to pots.

30. *Ferula gummosa* Boiss

Bernard *et al.* (2007) reported callus initiation from zygotic embryonic axis on MS medium supplemented with NAA. Embryogenesis was induced on MS medium containing 1 mgl^{-1} NAA. The proliferation of the embryos was achieved with suspension cultures on MS medium fortified with 1 mgl^{-1} NAA. Mature embryos were formed on hormone free medium.

31. *Ferula sinkiangensis* Boiss

Callus cultures have been initiated from seedlings of *F. sinkiangensis* (Zhu *et al.*, 2009). Optimal callus induction was obtained from seedling hypocotyl on MS medium containing 1 mgl^{-1} 2,4-D and 1.5 mgl^{-1} Kn. Callus induced from radical explants on MS medium supplemented with 0.5 mgl^{-1} NAA and 0.5 mgl^{-1} BA. The growth of callus was optimal on MS medium supplemented with 1.5 mgl^{-1} NAA and 2.5 mgl^{-1} BA.

32. *Foeniculum vulgare* Mill.

Hypocotyl, ovary, petiole and stem explants of fennel have been demonstrated to initiate callus on MS medium supplemented with different growth regulators (Becker, 1970; Garcia-Rodriguez *et al.*, 1978; Hunault, 1981; Nagari and Wakhlu, 1989; Ohga *et al.*, 1989; Paupardin *et al.*, 1980). Shoot buds were differentiated on MS medium supplemented with 0.01 mgl^{-1} IAA and l mgl^{-1} BAP (Garcia-Rodriguez *et al.*, 1978) and with 0.1 mgl^{-1} NAA or 0.1 mgl^{-1} BAP (Paupardin *et al.*, 1980). Hunault

(1981) obtained somatic embryos and plantlets from liquid cell cultures on MS medium lacking growth regulators. Embryo differentiation from protoplasts has been obtained on MS basal medium (Miura and Tabata, 1986). The regenerated plants were reported to exhibit genetic uniformity (Du Manoir *et al.*, 1985) as well as variability (Hunault, 1987). The frequency of embryogenic callus cultures improved in the prescence of GA_3 (Hunault and Maatar, 1995). Calls characteristics has been reported to be important for its morphogenetic response (Anzidei *et al.*, 1996). The polyamine strength was increased proportionally with the auxin concentration in the medium (Bennici *et al.*, 2004).

33. *Glehnia littorails* Fr. Schm.

Callus has been noticed to regenerate *G. littoralis* plants by somatic embryogenesis (Hirai *et al.*, 1997) and by axillary shoot proliferation (Hiraoka and Oyanagi, 1988). Somatic embryos were produced directly from cells of zygotic embryos on a medium containing 1μM 2, 4-D and plantlets were produced on MS medium that lacked growth regulators (Hirai *et al.*, 1997). Shoot multiplication has been achieved on LS basal medium supplemented with 3 per cent sucrose, 1 per cent agar, 1μM NAA and 10μM BAP (Hiraoka and Oyanagi, 1992). The regenerated shoots were rooted on MS medium containing 3-10μM NAA and 1μM IBA.

34. *Heracleum candicans* Wall.

Callus cultures were initiated from petiole explants of *H. candicans* on MS medium supplemented with 1 mgl^{-1} 2,4-D and 0.5 mgl^{-1} BAP and proliferated by subculturing on MS medium containing double strength of macro-salts, 1 mgl^{-1} 2,4-D and 0.25 mgl^{-1} Kn (Sharma, 2008,2009; Wakhlu and Sharma, 1998a). Embryos were formed on medium lacking cytokinins. Somatic embryos were converted into plantlets on MS medium and the plants were transferred to soil. Axillary shoots from nodal explants were formed on MS medium supplemented wit 0.5 mgl^{-1} BAP and 0.1 mgl^{-1} NAA. The shoots were rooted on MS medium fortified with 1 mgl^{-1} IBA (Wakhlu and Sharma,1998 b). Plant regeneration has also been reported from callus cultures initiated on MS medium fortified with 1 mgl^{-1} 2,4-D and 0.5 mgl^{-1} BAP via organogenesis (Sharma and Wakhlu, 2003). The shoots obtained from petiole derived callus cultures on MS medium supplemented with 1 mgl^{-1} BAP and 0.2 mgl^{-1} NAA were rooted on MS medium supplied with 1 mgl^{-1} IBA and plants were established in the soil. The regenerated plants were cytologically identical to their parent plants with 2n=22 chromosomes.

35. *Hydrocotyle conferata* Wight

There is single report published on axillary shoot proliferation in *H. conferata* (Karuppusamy *et al.*, 2007). Axillary shoots were formed on MS medium supplemented with 6.66μM BA and 5.37μM NAA. The best rooting of shoots was achieved on the MS medium supplemented with 0.54μM NAA.

36. *Hydrocotyle bonariensis* Comm. Ex Lam.

Callus cultures of *H. bonariensis* were established with 2 mgl^{-1} 2,4-D and 1 mgl^{-1} Kn (Masoumian *et al.*, 2011). Callus cultures were observed to produce flavonoids.

37. *Levisticum officinale*

Plants of *L. officinale* were regenerated via somatic embryogenesis (Jing-fen *et al.*, 1989). Protoplasts isolated from spongy callus formed colonies in the liquid N_6 medium supplemented with 2,4-D and kintin. Protoplasts derived colonies upon transfer to agar solidified medium, produced white spongy callus which turned embryogenic on N_6 medium supplemented with BAP and IBA. Plants were obtained on MS medium with NAA and IBA via somatic embryogenesis. Variation of nucleus morphology and chromosome number was observed in callus cultures. The plantlets regenerated from protoplast callus cultures were cytologically stable.

38. *Oenanthe aquatica* Poir.

There is only one report available on *in vitro* propagation of *O. aquatica* (Waris, 1959). The seedlings were noticed to produce nodules on their lateral roots on a standard nutrient solution containing glycine (1000 mgl^{-1}). The nodules ultimately developed into embryos and plantlets. The regenerated plants were originally neomorphs and became normal when transferred to a medium containing low levels of glycine (Waris, 1959).

39. *Ostricum koreanum* L.

Somatic embryogenesis has been reported from leaf derived callus cultures of *O. koreanum* on MS medium fortified with 5.37µM NAA and 0.44µM BAP (Cho *et al.*, 2003). Compared to medium with 5.8 pH, the production of somatic embryos was enhanced to six times on medium with 4.3 pH. The percentage of embryos produced on medium with 4.3 pH was 67 per cent.

40. *Pastinaca sativa* L.

Callus cultures of *P. sativa* obtained on LS medium supplemented with IBA and BAP. Optimal biosynthetic ability of calli was observed on a medium containing 3mgl^{-1} NAA and 1mgl^{-1} BAP (Gomolka and Ekiert,2000).

41. *Petroselinum crispum* Mill. L.

Micropopagation of *P. crispum* has been reported by Vandemoortele *et al.*(1996). Shoots were produced on MS medium containing 5µM Kn and 2.5µM NAA. Regenerated shoots were rooted on MS medium containing 2.5µM NAA.The regenerants werefertile and exhibited no morphological differences from the parent plants. Biosynthetic volatile compounds were separted from callus cultures formed on 2,4-D supplemented MS medium (Gbolade and Lockwood,1990).

42. *Petroselinum hortense* Hoffm.

Callus from petiole explants has been formed on MS medium supplemented with CM and 2, 4-D (Vasil and Hildebrandt, 1966). Mature embryo formation and plantlet regeneration from callus cultures was obtained on MS basal medium fortified with Ad_2SO_4 (15-20 mgl^{-1}).

43. *Peucedanum palustre* L.

There is a single report on somatic embryogenesis and plant regeneration in *P. palustre* (Vuorela *et al.*, 1993). Callus formation was reported from the root cultures on B_5 basal medium supplemented with low IAA levels. Embryogenesis was induced on a medium lacking auxins. The embryos were germinated on B_5 hormone –free medium.

44. *Pimpinella anisum* L.

Callus induction in *P.anisum* has been reported from hypocotyl explants cultured on MS medium supplemented with 0.5 mgl^{-1} Kn and 1 mgl^{-1} IAA (Becker 1970). Somatic embryogenesis from callus cultures has been reported by several researchers (Huber *et al.*, 1978; Kudieka and Theimor, 1983; Ernst *et al.*, 1984 and Chand *et al.*, 1997). Somatic embryos have been differentiated on MS medium containing 1 mgl^{-1} 2. 4-D (Chand *et al.*, 1997) and on MS medium devoid of growth regulators (Huber *et al.*, 1978).

45. *Pimpinella tirupatiensis* Bal and Subr.

Somatic embryogenesis has been reported in P. tirupatiensis (Prakash *et al.*, 2001) by initiating callus cultures from aseptic seedlings cultured on MS medium supplemented with 1 mgl^{-1} TDZ and 5.0 mgl^{-1} NAA. Embryogenesis was induced after subculturing the callus on MS medium containing 1 mgl^{-1} TDZ. Embryo germination occurred on MS medium supplemented with 1 mgl^{-1} TDZ and 1 mgl^{-1} GA_3.

46. *Sium suave* Walt.

Cell suspension cultures of *S. suave* were initiated from zygotic embryos on MS medium supplemented with 2 mgl^{-1}NAA, 10 per cent CM and 200 mgl^{-1} CH (Ammirato and Steward, 1971). Somatic embryos were formed when callus was transferred to MS medium lacking NAA. All plants derived from somatic embryos were observed to be normal.

47. *Thapsia garganica* L.

Plant regeneration in *T. garganica* has been reported from leaflet and petiole explants by indirect and direct organogenesis (Mukunga *et al.*, 2003, 2005, 2006). Medium free of growth regulators was observed to promote rooting and about 25 per cent plantlets were successfully acclimatized. The micropropagated plants were susceptible to fungal rot once removed from the culture vials. Somatic embryogenesis up to the globular stage was achieved in a suspension culture on MS medium containing 1 mgl^{-1} 2,4-D (Jager *et al.*, 1993).

48. *Trachyspermum ammi* L. (Sprague)

Plants of *T. ammi* have been regenerated via somatic embryogenesis (Jasrai. *et al.*, 1992; Sehgal and Abbas, 1994). Callus initiation was reported from ovary explants on MS medium containing 1mgl^{-1} IAA and 0.1 mgl^{-1} Kn (Johri and Sehgal, 1966), from hypocotyl explants on medium supplemented with 2 mgl^{-1} 2, 4-D and 0.5 mgl^{-1} Kn (Jasrai *et al.*, 1992) and from hypocotyl explants on medium containing 1 mgl^{-1} N.AA (Sehgal and Abbas, 1994). Formation of somatic embryos have been reported on MS medium enriched with 0.2 mgl^{-1} Kn (Jasrai *et al.*, 1992) and on MS medium that lacked growth regulators (Sehgai and Abbas, 1994). Plantlets have been successfully transplanted into the field.

49. *Vanasushava pedata* (Wight) Mukh. And Cons.

Regeneration of plants has been achieved in *V. pedata* through axillary shoot proliferation from nodal segments (Karuppusamy *et al.*, 2006). MS medium fortified with 5.0 mgl^{-1} BAP, 0.1 mgl^{-1} IAA and 3 per cent sucrose produced the highest number of shoots. Maximum rooting of shoots (100 per cent) was achieved on half strength MS medium containing 2 mgl^{-1} NAA. The regenerated plants were established in garden soil and forest humus and were identical to the donor plants.

Conclusions

The research work carried out by various groups has been summarized in this review. *In vitro* conservation of traditional medicinal plant germplasm is important to support chemical analysis and pharmacological and genetic transformation studies. With resurgence of public interest in plant based medicine and the rapid expansion of pharmaceutical industries, these *in vitro* techniques and innovative approach will be useful. Using *in vitro* propagation techniques it is now possible to produce a large number of pathogen free uniform clones of elite, rare and threatened medicinal plants.

References

Al-Abta, S., and Collin, H.A. (1979). Endogenous auxin and cytokinin changes during embryoid development in celery tissue cultures. *New Phytol.*, **82** : 29-35.

Ammirato, P.V. (1973). Some effects of abscisic acid on the development of embryos from caraway cells in suspension culture. *Amer. J. Bot.*, **60**: 22-23.

Ammirato, P. V. (1974). The effects of abscisic acid on the development of somatic embryos from cells of caraway (*Carum carvi* L.). *Bot Gaz.*, **135**: 328-337.

Ammirato, P.V. (1977). Hormonal control of somatic embryo development from cultured cells of caraway. Interactions of abscisic acid, zeatin and gibberellic acid. *Plant Physiol.*, **59**: 579-586.

Ammirato, P. V. (1983a). Embryogenesis. In: Evans D.A., Sharp W.R., Yamada, Y. (eds.) Handbook of Plant Cell Culture, vol I. Macmillian New York. pp: 82-123.

Ammirato, P.V. (1983b). The regulation of somatic embryo development in plant cell culture: suspension culture techniques and hormone requirements. *Bio. Technology*, **1**: 68-74.

Ammirato, P.V., and Steward. F. C. (1971). Some effects of the environment on the development of embryos from cultured free cells. *Bot. Gaz.*, **132**: 149-158.

Anzidei, M., Vivona, L., Schiff, S., and Bonnici, A. (1996). *In vitro* culture of *Foeniculum vulgare* : callus characteristics in relation to morphogenesis. *Plant Cell Tiss. Org. Cult.*, **45**: 263-268.

Arockiasamy, S., and Ignacimuthu, S. (1998). Plant regeneration from mature leaf and roots of *Eryngium foetidum* L., a food flavouring agent. *Curr. Sci.*, **75**: 664-665.

Aziz, Z.A., Davey, M.R., Lowe, K.C., and Power, J.B. (2006). Isolation and culture of protoplasts from the medicinal plant *Centella asiatica*. *Rev. Bras. Pl. Med. Botucatu.*, **8**: 105-109.

Azza, T., and Noga, G. (2001). Adventitious shoot proliferation from hypocotyls and internodal stem explants of cumin. *Plant Cell Tiss. Org. Cult.*, **66**: 141-147.

Azza, T., and Noga, G. (2002). Cumin regeneration from seedling derived embryogenic callus in response to the amended kinetin. *Plant Cell. Tiss Org Cult.*, **69**: 35-40.

Badoc A. (1982). Contribution a letude des phenomènes d' organogenèse et de callogense de tissue de Fenouil vulgaire (*Foeniculum vulgare* subsp capillaceum var. vulgare (Mill.) Thellung) cultives *in vitro*; analyse des constituents de l'huile essentielle des explants. D.E.A. Lille Univ., France. pp: 74.

Bajaj, Y.P.S., Furmanowa, M., and Olszowska, O. (1988). Biotechnology of the micropropagation of medicinal and aromatic plants. In: Biotechnology in Agriculture and Forestry, vol. 4 Medicinal and Aromatic Plants. Bajaj, Y.P.S.(ed.). Springer-Verlag, Berlin Heidelberg. pp: 60-103.

Ball, E. (1969). Histology of mixed callus cultures. *Bulletin of the Torrey Botanical Club*, **96** : 52-59.

Becker, H. (1970). Untersuchungen zur Frage der Bildung fluchtiger Stoff wechselprodukte in calluskulturen. *Biochem. Physiol. Pflanz.*, 1**61** : 425-441.

Bennici, A. Anzidei, M., and Vendramin, G.G.(2004). Genetic stability and uniformity of *Foeniculum vulgare* Mill. regenerated plants through organogenesis and somatic embryogenesis. *Plant Science*, **166**: 221-227.

Bernard, F., Bazarnov, H.S., Khatab, L.J., Darbi. A.S., and Sheidai, M. (2007). *Ferula gummosa* Boiss. Embryonic culture and karyological changes. *Pak. J. of Biol. Sci.*, **10**: 1977-1983.

Bonfill, M., Mangas, S., Mayono, E., Cusido, R.M., and Palazon, J. (2001). Production of centellosides and phytosterols in cell suspension cultures of *Centella asiatica*. *Plant Tiss Org. Cult.*, **104**: 61-67.

Browers, M. A., and Ortan, T. J. (1986). Celery (*Apium graveolens* L.), In Bajaj: Y. P. S. (Ed.) Biotechnology in agriculture and forestry, vol. 2: Crops I. Springer Berlin Heidelberg, New York Tokyo. pp 405-420.

Cathy, O., and Joanne, T. (1993). Propagation and cultivation of Flannel flowers. SGAP 17th Biennial Seminar, Robert Menzies College, Sydney, 27th Sept to 1st Oct., 1993.

Catlin, U., Ochoa, O., Mc Cormick, S., and Quiros C. F. (1988). Celery transformation by *Agrobacterium tumefaciens* : cytological and genetic analysis of transgenic plants. *Plant Cell Rep.*, **7**: 100-103.

Chand, S.; Sahrawat, A. K., and Prakash, D. V. S. S. R. (1997). *In vitro* culture of *Pimpinella anisum* L. (anise). *J. Plant-Biochem. Biotechnol.*, **6** : 91-95.

Chen, D.F., Cheng, B.S., Wang, C.X., Meng, X.R., and Li, X.Q. (1991). Study on the yield and quality of celeriac (*Apium graveolens*) somatic embryos. *Chinese Bulletin Botany*, **8**: 26-31.

Chen, R. R., Zhang, J. T., Li, B. P., Guo, S. S., Hao, J. P., and Zhou, X. M. (1992). Studies on the production of artificial seeds of coriander. *Chinese J. Biotechnology*, **7**: 127-134.

Chen, R.R., Zhang, J. T., Ping-Li, B., Guo, S. S., Hao, J. P., and Zhou, X. M. (1995). Somatic embryogenesis and artificial seed in coriander (*Coriandrum sativurn*). In : Bajaj, Y. P. S. (Ed.) Biotechnology in agriculture and forestry. vol 31. Somatic embryogenesis and synthetic seed II. Springer-verlag Berlin Heidelberg. pp: 334-342.

Chen, C. H. (1976). Vegetative propagation of the celery plant by tissue culture. *Proc. S. Dakota Acacl. Sci.*, **55**: 44-48.

Cho, D. Y., Lee, E.K, Lee S., Chung, W., and Soh, W.Y. (2003). Enhanced somatic embryogenesis and plant regeneration in leaf explants cultures of *Ostricum koreanum* on medium of varying pH. *Plant Cell Tiss. Org. Cult.*, **75**: 215-222.

Choi, Y. E., and Soh, W. 1. (1997). Origin and early developmental patterns of somatic embryo from freely suspended single cells in cell culture of celery (*Apium graveolens* L.). *Phytomorphology*, **47**: 107-117.

De Vries, S. C., Booij H., Meyerink, P., Huiman, G., Wilde, H. D., Thomas, T. L., and van Kammen, Q. A. (1988). Acquisition of embryogenic potential in carrot cell-suspension cultures. *Planta*, **176**: 196-204.

De Xiang, Ni, Zhilong-Dang, Yi Qun-Cen, Rumin, Jiang, GuangXian, Yang., and Jing, Yu (1995). Optimization of embryoids for artificial seed of *Apium graveolens*. *Acta-Agriculturae-Shanghai*, **11**: 9-15.

Deshpande, H.A. Chalse, M.N., and Bhalsingh, S.R. (2010). *Centella asiatica* linn. Plant regeneration through leaf derived callus. *J. Herbal Medicine and Toxicology*, **4**: 119-122.

Donovan. A., Collin, H. A., Isaac, S., and Mortimer, A. M. (1994). Analysis of potential sources of variation in tissue culture derived celery plants. *Ann. of Applied Biology*, **124**: 383-398.

Drew, R.L,K. (1979). Effect of activated charcoal on embryogenesis and regeneration of plantlets from suspension cultures of carrot (*Daucus carota* L.). *Ann. Bot.*, **44**: 387-389.

Du Manoir, J., Desmarest, P., and Saussay, R. (1985). *In vitro* propagation of fennel (*Foeniculum vulgare* Miller). *Scientia Hortic.*, **27**: 15-19.

Ebrahimie, E., Habashi, A.A., Ghareyazie B., Ghannadha M.R., and Mohammadie M. (2003). Rapid and efficient method for regeneration of plantlets from embryo explants of cumin (*Cuminum cyminum*). *Plant Cell Tiss Org Cult.*, **75**: 19-25.

Ebrahimie, E., Habashi, A.A., Dehcheshmeh, M.M., Ghannadha M.R., Ghareyazie B., and Yazdi-amadi, B. (2006). Direct shoot regeneration from mature embryos as a rapid and genotype-independent pathway in tissue culture of heterogenous diverse sets of Cumin (*Cuminum cyminum* L.) genotypes. *In vitro Cellular Developmental Biology*, **42**: 455-460.

Ebrahimie, E., Hosseinzadeh, A., Nagavi, M.R., Ghannadha M.R., and Dehcheshmeh, M.M. (2007). Combined direct regeneration protocol in tissue culture of different Cumin genotypes based on pre-existing meristem. *Pak J. bio. Sci.*, **10**: 1360-1370.

Egorova, N. A., and Reznikova, S. A. (1982). Investigation of isolated anther culture in connection with the induction of androgenesis in coriander. *Soviet Plant Physiol.*, **29**: 125-131.

E1-Bahr, M. K., Reda, F. A., Moursy, H.A., and Reda, A. A. (1988). Effect of pro-sowing seed treatments on the growth of callus tissues of *Ammi majus* L. *Egyptian J. Botany*, **31**: 227-238.

El-Fiky, F. K., Remmel, R. P., and Staba, E.J. (1989). *Ammi visnaga* : somatic embryo induction and furanochromone production in embryos, seedlings and plants. *Planta Medica*, **5** : 446-451.

El-Olemy. M. M., Elbag, H., El-Domiaty, M., Abdel-Sattar, E., Al-Azizi, M. M., and Al-Said, M. S.(1994). Biotransformation of digitoxigenin by cell suspension cultures of *Pergularia tomentosa* and *Ammi visaga*. *Saudi-Pharm-Journal*, **2** : 76-83.

Ernst, D., Oestrehelt, D., and Schafer, W. (1984). Endogenous cytokinins during embryogenesis in an anise cell culture (*Pimpinella anisum* L.). *Planta*, **161** : 240-245.

Evenor, D., Pressman, E., Ben-Yephet, Y., and Rappaport, L. (1994). Somaclonal variation in celery and selection by coculturing towards resistance to *Septoria apiicola*. *Plant Cell Tiss. Org. Cult.*, **39**: 203-210.

Fratenale, B., Giamperi, L., Ricci, D., and Rocchi, M.L.(2002). Micropropagation of *Bupleurum fruticosum*: the effect of triacontanol. *Plant Cell Tiss. Org. Cult.*, **69**: 135-140.

Fujimura, T., and Komamine, A. (1975). Effects of various growth regulators on the embryogenesis in a carrot cell suspension culture. *Plant Sci. Lett.*, **5** : 359-364.

Fujimura, T., and Komamine, A. (1979). Involvement of endogenous auxin in somatic embryogenesis in a carrot cell suspension culture. *Z. Pflanzenphysiol.*, **95**: 13-19.

Fujimura, T., and Komamine, A. (1980). The serial observation of embryogenesis in a carrot cell suspension cultures. *New Phytol.*, **86** : 213-218.

Furmanowa, M., Oledzka, H., and Sowinska, D. (1984). Regeneration of plants by embryogenesis with callus cultures of *Carum carvi* L. *J. Plant Physiol.*, **115** : 209-210.

Garcia-Rodriguez, M. J., Paupardin, C., and Saussay, R. (1978), Sur Ia formation d' un tissu secreteur et la synthese d' anethole par des tissus de Fenouil (*Foenicutum vulgare* Mill. Var. dulce) cultives *in vitro*. *CR Acad Sd. Paris*, **287**: 693-696.

Gbolade, A.A., and Lockwood, G.B. (1990). Metabolic studies of volatile constituents in tissue cultures of *Petroselinum crispum* (Mill) nayman. *J. of Plant Physiol.*, **136**: 198-202.

Giuliano, G., Rosellini, D., and Terzi, M. (1983). A new method for purification of the different stages of carrot embryoids. *Plant Cell Rep.*, **2**: 216-218.

Gomolka, E., and Ekiert, H. (2000). Furanocoumarins in *Pastinaca sativa* L. *in vitro* cultures. *Pharmazie*, **8**: 618-620.

Grambow, H.J., Kao, K.N., Miller, R.A., and Gamborg, O.L. (1972). Cell division and plant development from protoplasts of carrot cell suspension cell cultures. *Planta*, **103**: 348-355.

Grewal. S., Sachdeva, U., and Atal, C. K. (1976). Regeneration of plants by embryogenesis from hypocotyl cultures of *Ammi majus* L. *Ind. J. Exp. Biol.*, **14** : 716-417.

Grewal S. (1996 a). Induction of somatic embryogenesis and organogenesis in *Bunium persicum* Boiss callus culture. *Ind. J. Expt. Biol.*, **34** : 356-358.

Grewal, S. (1996 b). Microtubers from somatic embryos of *Bunium persicum*. *Ind. J. Expt. Biol.*, 34 : 813-815.

Grigoriadou, K., and Maloupa, E. (2008). Micropropagation and salt tolerance of *in vitro* grown *Crithmum maritimum* L. *Plant Cell Tiss. Org. Cult.*, **94**: 209-217.

Guang-min, X., Zhong-yi, Jia-ju, Yu., Guang-qig G., and Hui-min, C.(1991). Synchronus embryogenesis and plant regeneration from protoplast culture of *Anthriscus sylvestris*. *Acta Biologiae Expt. Sinica*, **4**: 28-31.

Halperin, W (1966). Alternative morphogenetic events in cell suspension. *Am. J. Bot.*, **53** : 443-453.

Halperin, W., and Wetherell, D. F. (1964). Adventive embryony in tissue cultures of the wild carrot, *Daucus carota*. *Aim J. Bot.*, **51** : 274-283.

Hassani, B., Sabora, A., Radjabian, T., and Fallah Husseini, H. (2008). Somatic embryogenesis of *Ferula asafetida* L. *JUST.*, **33**: 15-23.

Health-Pagliuso, S., Pullman, J., and Rappaport, L. (1988). Somaclonal variation in celery: screening for resistance to *Fusarium oxysporuin* f. sp. apii. Theo. *Appi. Genet.*, **75** : 446-451.

Health-Pagliuso, S., Pullman, J., and Rappaport, L. (1989). UC-T3 : somaclone : celery germplasm resistant to *Fusarium oxysporum* f. sp. apii, race 2. *Hortic Sci.*, **24**: 711-712.

Heller, R. (1953). Recherches sur la nutrition minerale des tissus vegetaux cultives *in vitro*. *Ann. Sci. Nat. Bot. Biol. Vègèt.*, **14**: 1-223.

Hirai,G.; Kasai, N., and Harada, T. (1997). Somatic embryogenesis in mature zygotic. embryo culture of *Glehnia littoralis*. *Plant Cell Tiss. Org. Cult.*, **48**: 175-180.

Hiraoka, N.; Kodama, T. and Tomita, Y. (1983). *In vitro* propagation of the *Bupleurum falcatum*. Shoyakugaku Zasshi. 37 : 62-67.

Hiraoka, N., Kodama T., Oyanagi, M., Nakano S., Tomita Y., Yamada, N., Iida, O., and Stake, M. (1986). Characteristics of *Bupleurum falcatum* plants propagated through somatic embryogenesis of callus cultures. *Plant Cell Rep.*, **5** : 319-321.

Hiraoka, N., and Oyanagi, M. (1988). *In vitro* propagation of *Glehnia littoralis* from shoot tips. *Plant Cell Reps.*, **7** : 39-42.

Hiraoka, N., and Oyanagi, M. (1992). Micropropagation of *Glehnia (Glehnia littoralis* Fr. Schmidt ex Mig.). In : Bajaj, Y. P. S. (ed). Biotechnology in Agriculture and

Forestry, vol. 19. High-tech and Micropropagation III. Springer-Verlag Berlin Heidelberg. pp. 443-452.

Hossain, N.S., Rehman, S., Joydhar, A., Islam, S., and Hossain, M. (2000). *In vitro* propagation of Thankuni (*Centella asiatica L.*). *Tissue Cult.*, 1**0**: 17-23.

Huber, J., Constabel, F., and Gamborg O. L. (1978). A cell counting procedure applied to embryogenesis in cell suspensions cultures of anise (*Pimpinel1a anisum L.*). *Plant Sci. Lett.*, 1**2**: 299-315.

Hunault G. (1981). La culture *in vitro* des tissues do fenouil (*Foeniculum vulgare* Miller). Premieres observations sur le comportment des explantats primitifs et des cals. *CR Acad. Sci. Paris*, **293**: 553-558.

Hunault., G. (1984). *In vitro* culture of fennel tissues (*Foeniculum vulgare* Miller) from cell suspension to mature plant. *Scientia Hortic.*, **22**: 55-65.

Hunault G. (1987). Etat cytogenetique des plantes de fenouil (*Foeniculum vulgare* Miller) obtenues par embryogenese somatique a partir cle suspensions cellulaires polyploids. *Rev Cytol. Biol. Veget. But.*, 1**0**: 155-172.

Hunault G., and Maatar, A. (1995). Enhancement of somatic embryogenesis frequency by gibberellic acid in fennel. *Plant Tiss org Cult.*, **41**: 171-176.

Ignacimuthu, S., Arockiasamy, S., Antonysamy, M., and Rvichandran, P. (1999). Plant regeneration through somatic embryogenesis from mature leaf explants of *Eryngium foetidum*, a condiment. *Plant Cell Tiss Org Cult.*, **56**: 131-137.

Irvani, N., Solouki, M., Omidi, M., Zare, A. R., and Shahnazi, S. (2010) Callus induction and plant regeneration in *Dorema ammoniacum* A- an endangered medicinal plant. *Plant Cell Tiss Og cult.*, 1**00**: 293-299.

Jacinda, T.J., Meyer, R., and Dubery, L.A. (2008). Characterization of two phenotypes of *Centella asiatica* in Southern Africa through the composition of four triterpeniods in callus, cell suspension and leaves. *Plant Cell Tiss Org cult.*, **94**: 91-99.

Jang, G. W. (1994). Somatic embryogenesis and plant regeneration from immature inflorescence segments of *Angelica koreana* Max. cultured *in vitro*. *Special-publication Taichung-District Agricult Improv Station*, **35** : 569-573.

Jasrai, Y, T., Barot, S. M., and Mehta, A. R. (1992). Plant regeneration through somatic embryogenesis in hypocotyl explants of *Trachyspermum ammi* (L.) Sprague. *Plant Cell Tiss. Org. Cult.*, **29**: 57-60.

Jha, T. B., Roy, S. C., and Mitra, G. C. (1983). *In vitro* culture of *Cuminum cyminum* regeneration of flowering shoots from calli of hypocotyl and leaf explants. *Plant Cell Tiss. Org. Cult.*, **2** : 11-14.

Jing-fen, J., Ji-hong. S., Yu-mei, W., and Shi-yu, Z. (1989). Somatic embryogenesis and cytological variation in protoplast cultures of *Levisticum officinale* Koch. *Acta Botanica Sinica*, **31**: 361-366.

John, B.M., and Sehgal, C. B. (1965). *In vitro* production of neomorphs in *Anethum graveolens* L. *Nature*, **205** : 1337.

John, B.M., and Sehgal, C. B. (1966). Growth responses of ovaries of *Anethum, Foeniculum* and *Trachyspermum*. *Phytomorphology*, **16**: 364-37

Jones L (1974). Factors influencing embryogenesis in carrot cultures *Daucus carota*. L.). *Ann. Bot.*, **38**: 1077-1088.

Kamada, H., and Harada, H. (1981). Changes in the endogenous levels and effects of abscisic acid during somatic embryogenesis of *Daucus carota* L. *Plant Cell Physiol.*, **22** : 1423-1429.

Kamada, H., and Harada, H. (1979a). Studies on the organogenesis in carrot tissue cultures. I. Effects of growth regulators on somatic embryogenesis and root formation. *Z. Pflanzenphysiol.*, **91**: 255 - 266.

Kamada, H., and Harada, H. (1979b). Studies on the organogenesis in carrot tissue cultures, II. Effects of amino acids and inorganic nitrogenous compounds on somatic. embryogenesis. *Z. Pflanzenphysml.*, **91**: 453-463.

Kamada, H., Kobayashi, K., Kiyosue, T., and Harada, H. (1989). Stress induced somatic embryogenesis in carrot and its application to synthetic seed production. *In vitro. Cell and Dev. Biol.*, **25**: 1163-1166.

Karthikeyen, K., Chandran, C., and Kulothungan S. (2009). Rapid clonal multiplication through *in vitro* axillary shoot proliferation of *Centella asiatica*. *Indian J. Biotechnology*, **8**: 232-235.

Karuppusamy, S., and Pullaiah, T. (2007). *In vitro* shoot multiplication of *Bupleurum distichophyllum* Weight – a native medicinal plant of Southern India. *Plant Tiss Cult and Biotechnology*, **17**: 115-124.

Karuppusamy, S., Aruna, V., Kiranmai, C., and Pullaiah, T. (2007). *In vitro* propagation of an endemic umbellifer- *Hydrocotyle conferta*. *India J. Biotechnology*, **6**: 541-544.

Karuppusamy, S., Kiranmai, C., Aruna, V., and Pullaiah, T. (2006). Micropropagation of *Vanasushava pedata*- an endangered medicinal plant of south India. *India J. Biotechnology*, **16**: 85-94.

Kataeva, N. V.; Popowich, E. A., and Popavich, E. A. (1993). Maturation and rejuvenation of *Coriandrum sativum* shoot clones during micropropagation. *Plant Cell Tiss. Org. Cult.*, **93** : 141-148.

Katakura, M., Kimura, T., and Endo, I. (1991). Production of saikosaponins by tissue culture of *Bupleuruin falcatum* L. *Bioprocess Eng.*, **7** : 97-100.

Kim, Y. H., and Janick. J. (1989a). ABA and polyox-encapsulation or high humidity increase survival of desiccated somatic embryos of celery. *Hort. Science*, **24**: 674-675.

Kim. Y, H., and Janick, J, (1989b). Origin of somatic embryos in celery tissue culture. *Hort. Science*, **24** : 671-673.

Kim, Y. H., and Janick, J. (1990). Synthetic seed technology : improving desiccation tolerance of somatic embryos of celery. *Acta Hortic.*, **280**: 23-28.

Kim, Y. H., and Janick, J. (1991). Abscisic acid and proline improve desticcation tolerance and increases fatty acid contents of celery somatic embryos. *Plant Cell Tiss. Org. Cult.*, **24**: 83-89.

Kim, S. W., Park, M. K., and Liu, J. R. (1996). High frequency plant regeneration via somatic embryogenesis in cell suspension cultures of coriander (*Coriandrum sativum* L.). *Plant Cell Reps.*, 15: 751-753.

Kim. K.; Whang, S. S., and Sun, B. Y. (1997). Micromorphology of the leaf epidermis of Korean umbelliferous plants 1. Preliminary study. *Phytorphology*, **47**: 87-95.

Kitto, S. L., and Janick, J. (1985). Hardening treatment increased survival of synthetically coated asexual embryos of carrot. *J. Am. Sco. Hortic. Sci.*, 1**10** : 283-286.

Krens, F. A., Keizer. L. C. P., and Capel, L. E. M. (1997). Transgenic caraway, *Carum carvi* L. : model species for metabolic engineering. *Plant Cell Reps.*, 17: 39-43.

Kudielka R. A., and Theimer, R. R. (1983). Somatic embryogenesis in anise (*Pimpinelia anisum* L.) cell cultures in microgravity environment. In : Proc. Worksh. Space biology, Cologne, Germany (ESA SP-206). pp: 63-68.

Kumar, K. B., Pillai, S. K., and Pillai, A. (1982a). Morphogenetic studies on *Cuminum cyminum* L. II Somatic Embryogenesis. *Beitr. Biol. Pflanzen.*, **58**: 47-52.

Kumar, K. B., Pillai, S. K., and Pillai, A. (1982b). Morphogenetic studies an *Cuminum cyminum*. L. 1. Callus development and differentiation. *Beitr. Biol. Pflanzen.*, **57** : 85-91.

Kumar, K. B., Pareek, N., Pillai, S. K., and Pillai, A. (1982c). Callus Development, shoot Formation and somatic embryogenesis in *Coriandrum sativum* L. *in vitro*. *Beitr. Biol. Pflanzen.*, **57**: 369-876.

Lee, C.Y., Kim, Y.K., Kim Y.S., Suh, S.Y., Lee, S.Y., and Park, S.U. (2009). Somatic embryogenesis and plant regeneration in *Cnidium officinale* Makino. *J. Medicinal Plant Research*, **3**: 96-100.

Loc, N.Y., and Ann, N.T.T. (2010) Asiaticoside production from *Centella asiatica* L. under cell culture. *Biotechnology and Bioprocess Engineering*, **15**: 1065-1070.

Maheshwari, S. C., and Gupta, G. R. P. (1965). Production of adventitious embryoids *in vitro* from stem callus of *Foeniculum vulgare*. *Planta*, **67**: 384-386.

Mc William, A. A.; Smith, S.M., and Street, H. E. (1974). The origin and development of embryoids in suspension cultures carrot *Daucus carota*). *Ann. Bot.*, **38**: 243-250.

Merrick, M. M. A., and Collin, H. A. (1982). Selection for asulam resistance in tissue cultures of celery, *Apium graveolens* L, var. dulce cv. New Dwarf White. *Plant Sci. lett.*, **20**: 291-296.

Metcalf, E. C., and Collin, H. E. (1978). The effect of simazine on the growth and respiration of a cell suspension culture of celery. *New Phytol.*, **81**: 243-248.

Mitra, J., Mapes, M,O., and Steward, F.C. (1960). Growth and organised development of cultured cells. IV. The behaviour of the nucleus. *Amer. J. Bot.*, **47**: 357-368.

Miura, Y., and Tabata, M. (1986). Direct somatic embryogenesis from protoplasts of *Foeniculum vulgare. Plant Cell Reps.*, **5**: 310-313.

Miura, Y., Fukui, H., and Tabata, M. (1987a). Clonal propagation of chemically uniform fennel plants through somatic embryoids. *Planta Medica*, **1**: 92-94.

Miura, Y., Ogawa, K., Fukui, H., and Tabata, H. (1987b). Changes in the essential oil components during the development of fennel plants from somatic embryoids. *Planta Medica*, **1**: 95-96.

Mohapatra, H., Barik, P.D., and Rath, S.P. (2008). *In vitro* regeneration of medicinal plant *Centella asiatica. Biologia Plantarum*, **52**: 339-342.

Mujib, A., Bandyopadhyay, S., Banerjee, S., and Ghosh, P.D. (1990). Somatic embryogenesis from hypocotyl callus of coriander (*Coriandrum sativum* L.). *Horti. J.*, **3**: 59-62.

Murata, M., and Orton, T. J. (1983). Chromosome structural changes in cultured celery cells. *In vitro*, **19**: 83-89.

Nadel, B. L., Altman, A., anti Ziv, M. (1989). Regulation of somatic embryogenesis in celery cell suspension. 1. Promoting effects of mannitol on somatic embryo development. *Plant Cell Tiss. Org. Cult.*, **18**: 181-189.

Nadel, B. L., Altman, A., and Ziv, M. (1992). Somatic embryogenesis and synthetic seed in *Apium graveolens* (celery). In: Bajaj, Y.P. S. (ed.) Biotechnology in Agriculture and Forestry, vol 31 Somatic embryogenesis and Synthetic seed II. Springer-Verlag Berlin, Heidelberg. pp: 306-320.

Nagar, S., and Wakhlu, A K (1989). Karyotypic studies of plants regenerated from cultured tissues of fennel (*Foeniculum vulgare* Miller). In : Manna, G. K and Sinha, U. (eds.), *Perspectives in cytology and Genetics*. **6**: 155-158.

Nagari, S., Wakhlu, A. K., and Sharma, R. K. (1997). Cytological analysis of embryogenic callus and regenerated plants of *Bunium persicum* Bioss. *Chromosome Sci.*, **1**: 69-71.

Naidu, T.B., Rao, S.N., and Mani, N.S. (2010). Conservation of endangered medicinal plant *Centella asiatica* through plant tissue culture. *Drug Invention Today*, **2**: 17-31.

Nakagawa, K., Miura, Y., Fukui, H., and Tabata, M. (1982). Clonal propagation of medicinal plants through the induction of somatic embryogenesis from cultured cells. In : Fujiwara, A. (ed.). *Plant tissue culture, Maruzen, Tokyo*, PP : 701-702.

Nath, S., and Buragohain, A.K. (2003). *In vitro* method for propagation of *Centella asiatica* L. by shoot tip culture. *J Plant Biochemistry and Biotechnology,* **12**: 167-169.

Nath, S., and Buragohain, A.K. (2005). Establishment of callus and cell suspension cultures of *Centella asiatica. Biologia Plantarum*, **49**: 411-413.

Nickle, T.C., and Yeung, E.C. (1993). Failure to establish a functional shoot meristem may be a cause of conversion failure in somatic embryos of *Daucus carota* (Apiaceae). *Am. J. Bot.*, **80**: 1284-1291.

Nomura, K., Fukuei, K., and Nitta, T. (1983). Culture of protoplasts isolated from somatic embryos of carrot. *Plant Sci. lett.*, **29**: 1-7.

Nomura, K., and Komainine, A. (1985). Identification and isolation of single cells that produce somatic embryos at a high frequency in a carrot suspension culture. *Plant Physiol.*, **79** : 988-991.

Nuti-Ronchi, V., Caligo, M. A., Nozzolini, M., and Luccarini, G. (1984). Stimulation of carrot embryogenosis by proline and serine. *Plant Cell Rep.*, **3**: 210-214.

Ohga, Y., Ono, M., and Furuno, K. (1989). Somatic embryogenesis and plant regeneration from hypocotyls and cotyledons of *Angelica acutiloba* and *Foeniculum vulgare*. *Report-Kyushu-Branch-Crop Sci Soc Japan*, **56**: 89-91.

Onishi, N., Mashiko, T., Okamoto, A., Hayashi, M., Kano, A., and Goto, E. (1992). Cultural system producing encapsulatable units of synthetic seeds in celery. *Acta Horticulture*, **319**: 1 13-124.

Onishi, N., Sakamoto, Y., and Hirosawa, T. (1994). Synthetic seeds and application of mass production of somatic embryos. *Plant Cell Tiss, Org. Cult.*, **39** : 137-145.

Orton, T. J. (1984) *Celery*. In : Sharp W.R., Evan, D. A.; Ammirato, P.V.; Yamada. Y. (eds) Handbook of plant cell culture. Vol 2, Macmillian. New York. pp. 243-267.

Orton, T. J. (1985). Genetic instability during embryogenic cloning of celery. *Plant Cell Tiss. Org. Cult.*, **4** : 159-169.

Pande, D., Purohit, M., and Srivastava, P.S. (2002) Variation in xanthotoxin content in Ammi majus L. cultures during *in vitro* flowering and fruiting. *Plant Science*, **162**: 583-587.

Pant, B., and Manandhar, S. (2007). *In vitro* propagation of carrot (*Daucus carota*) L. *Scientific World*, **5**: 51-53.

Pant, B., Kohda, H., and Namera, A. (1996). Clonal propagation of Cnidium officinale by shoot-tip culture. *Planta Medica*, **62**: 281-283.

Paramageetham, C., Babu, G.P., and Rao, S.V.J. (2004). Somatic embryogenesis in *Centella asiatica* L., important medicinal and neutracetuical plant of India. *Plant Cell Tiss Org Cult.*, **79**: 19-24.

Paupardin. C., Garcia-Rodriguez, M. J., and Bricout, J. (1980). Multiplication vègètative de quelques plantes aromatiques : problems poses par la production d'essence. *CR Acad, Agr. France*, **66** : 658-666.

Pawlicki, N., Sangwan, R. S. and Sangwan-Norreel, B. S. (1992). Factors influencing the *Agrobucterium tumefaciens* - mediated transformation of carrot (*Daucus carota* L.). *Plant Cell Tiss. Org. Cult.*, **31** : 129-139.

Prakash, E., Khan, S.V., Meru, E., and Rao, K.R. (2001). Somatic embryogenesis in *Pimpinella tirupatiensis* Bal and Subr., an endangered medicinal plant of Triumala hills. *Curr. Sci.*, **81**: 1239-1242.

Purohit, M., Pande, D., Datta, A., and Srivastava, P. S. (1995a). *In vitro* flowering and high xanthotoxin in *Ammi majus* L. *J. Plant Biochem and Biotechno.*, **4**: 73-76.

Purohit, M., Pande, D., Datta, A., and Srivastava, P. (1995b). Enhanced xanthotoxin content in regenerated cultures of *Ammi majus* and micropropagation. *Planta Med.*, **61**: 481-482.

Purohit, S., and Kothari, S. L. (2007). Direct somatic embryogenesis from cotyledons and cotyledonary node explants in Bishop's weed *Trachyspermum ammi* L. *In vitro Cell and Dev. Biol.*, **43**: 154-158.

Raghu, A. V., Martin, G., Priya, V. Geetha, S.P., and Balachandran, I (2007). Low cost alternative for the micropropagation of *Centella asiatica*. *J. Plant Sci.*, **6**: 592-599.

Rao, P.K. Srinath, S., and Sadanandam, M. (1999). Tissue culture studies of *Centella asiatica*. *Indian J Pharma. Sci.*, **61**: 392-394.

Ratnamba, S. P., and Chopra, R. N. (1974). *In vitro* induction of embryoids from hypocotyls and cotyledons of *Anethum graveolens* seedlings. *Z. Pflanzenphysiol.*, **73**: 452-455.

Redenbaugh, K., Fujii, J., Slade, D., Viss, P., and Kossler, M. (1991). Artificial seeds-encapsulated somatic embryos. In : Bajaj, Y.P.S (Ed.) Biotechnology in agriculture and forestry vol. 17. High-tech and micropropagation I. Springer, Berlin Heidelberg New York. pp: 395-416.

Reinert J. (1958). Untersuchungen uber die Morphogenese an Gewbekulturen, Ber, Dtsch. *Bot. Ges.*, **71**: 15.

Reinert. J., Backs, P., and Krosing, M. (1966). Faktoren dcii Embryogenese in Gewebekulturen aus Kulturformen von Umbelhferen. *Planta*, **68** : 375-378.

Santos, P.A. G., Figueriredo, A. C., Lourenco, P.M.L., Barroso, J.G., Pedro, L. G., Oliveira, M.M., Schripsema, J., Beans, S.G., and Scheffer, J.J.C. (2002). Hairy root cultures of *Anethum graveolens* (dill): Establishment, growth, time course study of their essential oils and its comparison with parent plant oil. *Biotechlogy lett.*, **24**: 1031-1036.

Saranga, Y., and Janick, J. (1991). Celery somatic embryo production and regeneration : improved protocols. *Hort. Science*, **26** : 1335.

Saranga. Y., Kim, K. H., and Janick, J. (1992). Changes in amino acid composition associated with desiccation tolerance of celery somatic embryos. *J. Am. Soc. Hort. Sci.*, **117**: 337-341.

Sardesai, P. L., and Tipnis, H. P. (1969). Production of flavouring principles by tissue culture of *Coriandrum sativum*. *Curr. Sci.*, **38** : 545.

Schafer F., Li, T., and Neumann, K. H. (1986). Petiole explants and intact seedlings as sources of somatic embryos of some Apiaceae *(Daucus carota, Anthum graveolens, Pimpineila anisum, Coriandrum sativum, Petroselinurn crispnum Carum carvi).* In : VIth Int, Cong. Of Plant Tissue and Cell Culture. Minneapolis, MN, Abstr. pp : 346.

Sehgal, C. B. (1964). Artificial induction of polyembrony in *Foeniculum* vulgare Mill. *Cur. Sci.*, **83:** 373-374.

Sehgal, C. B. (1968). Response of young infloresences of *Anethum graveolens* L. to growth substances. *Curr. Sci.*, **37** : 425-426.

Sehgal, C. B. (1972a). *In vitro* induction of polyembryony in *Ammi majus* L. *Cnrr Sci.*, **41**: 263-264.

Sehgal, C. B. (1972b). Experimental induction of zygotie multiple embryos in *Coriandrum sativum* L. *Ind. J. Exp. Biol.*, **10** : 457-459.

Sehgal. C. B. (1978). Differentiation of shoot-bud and embryoids from inflorescence of *Anethum graveolens* in cultures. *Phytomorphology*, **28** : 291-297.

Sehgal, C. B., and Abbas, N. S. (1994). Somatic embryogenesis and plant regeneration from hypocotyl tissue of *Trachyspermum ammi* (L.) Sprague. *Phytomorphology*, **44** : 265-271.

Sharma, R.K. (2008). Effect of carbohydrates and vitamins on axillary shoots proliferation in *Heracleum candicans* Wall. *Ind. J. Bot. Res.*, **4(2)**: 177-182.

Sharma, R.K. (2009). Caulogenesis in *Heracleum candicans* Wall. *Internat. J. Plant Sciences*, **4**: 354-356.

Sharma, R.K. (2011a). Somatic embryogenesis in Kalazira (*Bunium persicum* Bioss). *Asian J. Bio. Sci.*, **6**: 59-61.

Sharma, R.K. (2011b). *In vitro* tuberization and plant regeneration in *Bunium persicum* Bioss. *Internat. J. Plant Sci.*, **6**: 193-195.

Sharma, R.K. (2012). Induction and maintenance of callus cultures in *Bunium persicum* Boiss. *Internat. J. Plant Sci.*, **7**: 90-92, 2012.

Sharma, R.K., and Wakhlu, A.K. (2003). Regeneration of *Heracleum candicans* Wall plants from callus cultures through organogenesis. *J Plant Biochem and Biotech*, **12**: 71-72.

Shashikala, C.M., Shashidhara, S., and Rajshekharan, P.E. (2009). *In vitro* callus culture of *Centella asiatica L.* and its HPLC analysis. *Biosci and Biotech Research Asia,* **6**: 863-868.

Simard M. H. (1985). Methodologies de selection de plants de Fenouil amer (*Foeniculum vulgare*) tolerants au *Phomopisis foeniculi* et a ses principes toxiques. Rapport de stage, Ecole Nationale Superieure d' Horticulture, versailles, France.

Singh, S.K., Rathore, Z., and Saxena, O.P. (20010). Multiple shoot regeneration from callus cultures of *Centella asiatica* under the influence of various conc. of PGRS. *J. of Pure and Applied Sciences*, **18**: 18-20.

Shivkumar, G., Alagumanian, S., and Rao, M.V. (2006). High frequency *in vitro* multiplication of *Centella asiatica*: an important industrial herb. *Engineering in Life Sci.*, **6**: 597-601.

Sliva, S. Viehmannova, I., and Vitamvas, J, (2010). Micropropagation and morphogenesis of arracacha (*Arracacia xanthorrhiza* Bancroft). *Agricultura Tropica et Subtropica*, **43**: 206-211.

Smith, D. L., and Krikorian, A. D, (1988). Production of somatic embryos from carrot tissue in hormone free medium. *Plant Sci.*, **58**: 103-110.

Smith, D. L., and Krikorian, A. D. (1989). Release of somatic embryogenic potential from excised zygotic embryos of carrot and maintenance of proembryogenic cultures in hormone-free medium. *Am. Bot.*, **76**: 1832-1843.

Smith, D. L., and Krikorian, A. D. (1990). Low external pH replaces 2,4-D in maintaining and multiplying 2,4-D initiated embryogenic cells of carrot. *Physiol. Plant*, **80** : 329-336.

Smith, R. H., Kerns, H. A., Anthony, J. L., and Wild, J. R. (1987). Methotrexate and aminopterin effects on growth and regeneration in *Daucus carota*. *Plant Cell Rep.*, **6**: 60-62.

Stephen, R., and Jayabalan, N. (1998). *In vitro* flowering and seed setting formation of coriander (*Coriandrum sativum* L.). *Curr. Sci.*, **74**: 195-197.

Steward, F. C., Mapes, M. O., and Mears, K. (1958). Growth and organized development of cultured cells. II. Organization in cultures grown from freely suspended cells. *Am. J. Hot.*, **45**: 705-708.

Steward, F. C., Amrnirato, P. V., and Mapes, M. O. (1970). Growth and development of totipotent cells: some problems, procedures and perspectives. *Ann. Hot.*, **34** : 761-787.

Szoke, E., Mensi, F. S., Oroszlan, P., and Petri, G. (1991). Furanocoumarin and Chromone production and the effects of growth regulators in tissue cultures of *Ammi majus* and *Ammi visnaga*. *Planta-Medica*, **57**: 132-133.

Takeda, T., Mizukami, M., and Matsuoka, H. (2008). Characterization of two step direct somatic embryogenesis in carrot. *Biochemical Engineering Journal*, **38**: 206-211.

Thomas, J. C.; Guiltinan, M. J.; Bustos, S.; Thomas, T. and Nessler, C. (1989). Carrot (*Daucus carota*) hypocotyl transformation using *Agrobacterium tumefaciens*. Plant Cell Rep. **8**: 354-357.

Tiwari, K.N., Sharma, C.N., Tiwari, V., and Singh, D.B. (2000). Micropropagation of *Centella asiatica* L., a valuable medicinal herb. *Plant Cell Tiss Org Cult.*, **63**: 179-185.

Uomori. A., Seo, S., and Tomita, Y. (1974). Studies on the constituents in tissue culture of *Bupleurum falcatum* L. *Shoyakugaku Zasshi*, **29** : 152-180.

Valizadeh, M., Kazemitabar, S.K., and Nematzadeh, G.A. (2007). Effect of plant growth regulators on callus induction and regeneration of *Cuminum cuminum*. *Indian J. Crop Sci.*, **2**: 221-214.

Valizadeh, M., Safernejad, A., Nematzadeh, G.A., and Kazemitabar, S.K. (2008). Regeneration of plantlets from embryo explants of *Bunium persicum* Boiss. *J. Sci and Techno of Agri. and Natural Resources*, **11**: 122-129.

Vandemoortel, J. L., Billard, J. P., Boucaud, J., and Gaspar, Th. (1996). Micropropagation of parsley through axilary shoot proliferation. *Plant Cell Ties. Org. Cult.*, **44**: 25-80.

Vasil, I. K., and Hildebrandt, A. C. (1966). Variations of morphogenetic behaviour in plant tissue cultures. II. *Petroselinum hortense. Amer. S. J. Bot.*, **53**: 869-874.

Vuorela, P. Oksman-Caldentey, K.M., Lipponen, J., and Hiltunen, R. (1993). Spontaneous somatic embryogenesis and plant regeneration from root cultures of *Peucedanum palustre. Plant Cell Reports*, **12**: 453-456.

Wake, H., Akasaka,A;Umetsu, H., Ozeki, Y., Shimomura, K., and Matsunaga, T. (1992 a). Promotion of plantlet formation from somatic embryos of carrot treated with a high molecular weight extract from a marine cyanobacterium. *Plant Cell Reps.* **11**: 62-65.

Wake, H., Akasaka, A., Umetsu, A., Ozeki, Y., Shimomura, K., and Matsunaga, T. (1992 b). Enhanced germination of artificial seeds by marine cyanobacterial extract. *Appl Microbiol Biotechnol.*, **36**: 684-688.

Wake, H., Umetsu, K., Ozeki, Y., Shimomura, K., and Matsunaga, T. (1991). Extracts of marine cyanobacteria stimulated somatic embryogenesis of *Daucus carota* L. *Plant Cell Rep.*, **9**: 655-658.

Wake, H., Umetsu, H., and Matsunaga, T. (1995). Somatic embryogenesis and Artificial seed in carrot (*Daucus carota* L.). In: Bajaj, Y. P. S. (Ed,) Biotechnology in Agriculture and Forestry, Vol. 31. Somatic embryogenesis and synthetic seed II. Springer-Verlag Berlin Heidelberg. pp: 170-182.

Wakhlu, A. K., and Sharma R.K. (1998a). Somatic embryogenesis and plant regeneration in *Heracleum candicans* Wall. *Plant Cell Reports*, **17**: 866-869.

Wakhlu, A. K., and Sharma R.K. (1998b). Micropropagation of *Heracleum candicans* Wall: A rare medicinal berb. *In Vitro Cell Dev Biol-Plant*, **35**: 79-81.

Wakhlu, A. K., Nagari, S., and Barna, K. S. (1990). Somatic embryogenesis a and plant regeneration from callus cultures of *Bunium persicum* Boiss. *Plant Cell Rep.*, **9**: 137-138.

Wang, P.J., and Huang, C.I. (1982). Production of saikosaponins by callus and redifferenliated organs of *Bupleurum falcatum* L. In: Fujiwara, A. (ed.). Plant Tissue Culture, Maruoen, Tokyo. pp : 71-72.

Waris, H. (1 959). Nomorphosis in seed plants induced by amino acids. I. *Oenanthe aquatica*. Physiol. *Plant*, **12**: 753-766.

Watin de Pontfarcy, C., and Bigot, C. (1991). Somatic embryogenesis and regeneration of plants from protoplasts derived from cell suspensions of celery (*Apium graueoiens*). *Can. J. Bot.*, **69** : 1583-1592.

Wetherell, D. F., and Dougall, D. K. (1976). Sources of nitrogen supporting growth and embryogenesis in cultured wild carrot tissue. *Physiol. Plant.*, **37**: 97-103.

Williams, L., and Collin, H.A. (1976a). Embryogenesis and plantlet formation in tissue cultures of celery. *Ann. Bot.*, **40**: 325-332.

Williams, L., and Collin. H.A. (1976b). Growth and cytology of celery plants derived from tissue cultures. *Ann. Bot.*, **40**: 333-338.

Wochok, Z, S., and Wetherell, D. F. (1971). Suppression of organised growth in cultured wild carrot tissue by 2-chloroethyl-phosphoric acid. *Plant Cell Physiol. Tokyo*, **12**: 771 -774.

Xia, G.M. Li, Z., Guo, G.Q., and Chen, H.M. (1992). Direct somatic embryogenesis and plant regeneration from protoplasts of *Bupleurum scorzonerifolium* Wild. *Plant Cell Rep.*, **11**: 155-158.

Xu, X., Yan, C., and Yu, R. (2005). Plant regeneration from callus cultures of *Changiums myrnioides*. *Zhong Yao Chi.*, **28**: 971-973.

Yoo, N.H., Kim, O.T., Kim, J.B., Kim, Y.C., Bang, K.H., hyun, D.Y., Cha, S.W., Kim, M.Y., and Hong, B. (2011). Enhancement of centelliocide production from cultured plants of *Centella asiatica* by combination of thidiazuron and methyl jasmonate. *Biotechnology Report*, **5**: 283-287.

Zare, A.R., Solouki, M., Omidi, M. Irvani, N., Nezad, M., and Rezazadeh, S.H. (2010). Callus induction and plant regeneration in *Ferula assafoetida* L.(Asafetida), an endangered medicinal plant. *Trakia J of Sci.*, **8**: 11-18.

Zee, S. Y. (1981). Studies on adventive embryo formation in the petiole explants of coriander (*Coriandrum sativum*). *Protoplasma*, **107**: 2 1-26.

Zee, S.Y., and Wu. S.C. (1979). Embryogenesis in the petiole explants of Chinese celery. *Z. Pflanzenphysiol.*, **93**: 325-335.

Zee, S.Y.,and Wu. S.C. (1980). Somatic embryogenesis in leaf explants of Chinese celery. *Aust. J. Bot.*, **28**: 429-436.

Zhang, Shi-yu, and Cheng, Kuo-chang (1982). Callus induction and plantlet regeneration from several organs of *Angelica sinensis* (Oliv.). *Diel. Acta Bot Sin.*, **24**: 512-518.

Zhang, Shi-yu, and Cheng, Kuo-chang (1986). Induction of embryogenic callus and histocytological study on embryoid development of *Angelica sinensis* (Oily.) Dials. *Acta Bot. Sin.*, **28**: 241-244.

Zhang Shunpei., Jigfen Ja., Haori Li., and Zang Guo-Chang (1985). Protoplast isolation, culture and callus formation of Chinese *Angelica* Kexue. *Tongbao.*, **31**: 635-638.

Zhu. J., Li, X.J., Kaisa, S., and Jia, X.G. (2009). Study on callus induction of *Ferula sinkiangensis*. *Zhong Yao Cai.*, **32**: 1655-1658.

Utilisation and Management of Medicinal Plants Vol. 3 (2015) *Pages* **405–412**
Editors: **V.K. Gupta and A. Kaul**
Published by: **DAYA PUBLISHING HOUSE, NEW DELHI**

27

Recent Trends in *Gymnema sylvestre* Tissue and Cell Suspension Culture

C. Subathra Devi[1*] and V. Mohana Srinivasan[1]

ABSTRACT

Use of Gymnema sylvestre, commonly known as periploca of the woods, has increased recently due to the pharmaceutical potential of gymnemic acids, found in its leaves. Gymnemic acid has been reported to effect a natural treatment for diabetes melliteus. G. sylvestre natural strands are fast disappearing and are threatened with extinction due to its indiscriminate collection, over exploitation of natural resources for commercial purposes and to meet the requirements of the pharmaceutical industry. Somatic embryogenesis and whole plant regeneration were achieved in callus culture derived from hypocotyl, cotyledon and leaf explants excised from seedlings of Gymnema sylvestre. A cell culture system for Gymnema sylvestre was developed for large scale production of gymnemic acid. Combination of external phytohormone, shaking speeds, pH of the medium, played important roles in growth and gymnemic acid production in suspension cell cultures. Physical and chemical factors greatly favor the production of gymnemic acid in callus and suspension cultures of G. sylvestre. The recent development of elicitation has opened a new avenue for the production of secondary compounds. The influence of various elicitors on secondary metabolite production in cell suspension cultures of Gymnema sylvestre was investigated. Fungal and bacterial elicitors showed significant stimulation in the gymnemic acid production. Advance biotechnological methods culturing plant cells and tissues should provide new means of conserving and rapid propagation of valuable, rare, and endangered medicinal plants.

Keywords: *Gymnema sylvestre*, Gymnemic acid, *In vitro* propagation, Tissue culture, Cell suspension culture.

1 School of Biosciences and Technology, Industrial Biotechnology Division, VIT University, Vellore – 632 014, Tamil Nadu, India.

* *Corresponding author*: E-mail: csubathradevi@vit.ac.in; subaresearch@rediffmail.com

Introduction

The world health organization predicts that diabetes will be a major health issue in the 21st century, with approximately 80 per cent of all new cases in developing countries. At present Asia has the world's largest diabetes populations, India is estimated to have 35 million people with diabetes. Emerging diabetes population and increased demands for herbal medicines, this had renewed interest of the multinational pharmaceutical industry in bioprospecting.

Use of *Gymnema sylvestre,* commonly known as periploca of the woods, has increased recently due to the pharmaceutical potential of gymnemic acids, found in its leaves. Gymnemic acid has been reported to effect a natural treatment for diabetes melliteus. At present 90 per cent collection of medicinal plants is from the wild. This will not augur the sustainable use of medicinal plants and also it will not fulfill the needs of the majority of the population (Subathra Devi *et al.,* 2006).

G. sylvestre natural strands are fast disappearing and are threatened with extinction due to its indiscriminate collection, over exploitation of natural resources for commercial purposes and to meet the requirements of the pharmaceutical industry (Chowdhury, 1988). The attempt for its production and conventional propagation is hampered due to its poor seed viability, low rate of germination and poor rooting ability of vegetative cuttings. Alternative propagation methods would be beneficial in accelerating large scale multiplication, improvement and conservation of the plant. Due to lack of proper cultivation practices, destruction of plant habitats and the illegal, indiscriminate collection of plants from these habitats, many medicinal plants are severely threatened. Advance biotechnological methods culturing plant cells and tissues should provide new means of conserving and rapid propagation of valuable, rare, and endangered medicinal plants (Nalawade *et al.,* 2003).

To overcome those problems, we are in need of alternate technologies. Plant cell culture represent a potential renewable source of valuable compounds, flavours, colorants, which cannot be produced by microbial cells or chemical synthesis (Subathra Devi *et al.,* 2006). This review describes the medicinal properties, *in vitro* propagation, callus and suspension culture of *Gymnema sylvestre.*

Gymnema sylvestre

Gymnema sylvestre is a potent antidiabetic plant and used in folk, ayurvedic and homeopathic systems of medicine (Kapoor, 1990; Ravi and Wahi, 1995; Mitra *et al.,* 1995). It is also used in the treatment of asthma, eye complaints, inflammations, family planning and snake bite (Anonymous, 1956; Uniyal, 1993; Selvanayagam *et al.,* 1995). In addition, it possesses antimicrobial, antihypercholesterolemic (Bishayee and Chatterjee, 1994), hepatoprotective (Rana and Avadhoot, 1992) and sweet suppressing (Kurihara, 1989) activities. It also acts as feeding deterrents to caterpillar, *Prodenia eridania* (Granich *et al.,* 1974), prevent dental caries caused by *Streptococcus mutans* (Hiji Yasutake, 1990) and used in skin cosmetic (Maeda *et al.,* 1996). The various reports on its multiple uses attracted attention for utilization of the plant for its active principle, gymnemic acid.

It actually contains lot of molecules brought together under the name of gymnemic acid, which has properties of regenerating pancreas cells secreting insulin (the hormone responsible for the body's assimilation of sugar) (Shanmugasundram *et al.*, 1990) and improving glucose assimilation by increasing enzyme activity at this level. Furthermore *Gymnema* slows down the stimulating action of adrenalin on the liver, thus reducing glucose production and therefore controls the blood sugar levels. Finally, eliminating the perception of sweet flavours, *Gymnema* is very useful for those who nibble bars of chocolates (Shanmugasundram *et al.*, 1990). It also acts against non-insulin dependent Diabetes mellitus (Baskaran *et al.*, 1990). Since the gymnemic acid fraction is allegedly beneficial for treating diabetes and obesity, preparations containing *Gymnema* plants of Indian origin are now commercially available in Japan, Germany and the USA as health foods (Wen-Caiye *et al.*, 2000).

The plant is also reported to be astringent, acrid, thermogenic, antiinflammatory, anodyne, digestive, liver tonic, emetic, diuretic, stimulant, antihelmenthics, laxative, cardiotonic, expectorant, antipyretic and uterine tonic. It is useful in dyspepsia, constipation, jaundice, haemorrhoids, renal and vesical calculi, cardiopathy, asthma, bronchitis, amenoerhoea, conjuctivitis and leucoderma (Nadkarni, 1993).

G. sylvestre leaf contains more than 20 saponin glycosides. The major saponin fraction comprises of gymnemic acid (the anti-sweet principle) which is a complex mixture of at least nine closely related acidic glycosides (Yoshikawa *et al.*, 1991), also 23-hydroxylongispinogenin, gymneastrogenin (pentahydroxyoleane-12-ene) and a few dammarane derivatives as aglycones, that makes the purification and characterization of gymnemic acid a difficult task. Neverthless, a quantitative analysis of gymnemic acid by HPLC analysis of gymnemagenin, the common aglycone obtained on hydrolysis has been reported by Toshihiro *et al.*(1994). Growing commercial interest in the production of GA for the development of pharmacological agents posed the interest to investigate the *in vitro* production of active principles in *G. sylvestre* cultures (Veerashree *et al.*, 2011)

In vitro Propagation of *Gymnema sylvestre*

Over exploitation of natural stands has caused depletion of these plants in nature. Propagation of *Gymnema sylvestre* is by seeds. However poor germination restricts its multiplication (Reddy *et al.*, 1998). An alternate method, the *in vitro* propagation via organogenesis has been reported (Reddy *et al.*, 1998; Komalavalli and Rao, 2000). Somatic embryogenesis and whole plant regeneration were achieved in callus culture derived from hypocotyl, cotyledon and leaf explants excised from seedlings of *Gymnema sylvestre*. The first report in regeneration of *G. sylvestre* through somatic embryogenesis was given by Ashokkumar *et al.* (2002). Limited tissue culture work has been done on *Gymnema* species (Komalavalli and Rao, 1997; Reddy *et al.*, 1998). Komalavalli and Rao (2000) reported an efficient and rapid propagation of *G. sylvestre* using various explants from *in vitro* grown seedling explants, but the multiplication rate achieved was very low. The nature of explant, seedling age, medium type, plant growth regulators, complex extracts, and antioxidants markedly influenced the *in vitro* propagation of *G. sylvestre* (Komalavalli and Rao, 2000). Reddy *et al.* (1998) developed a rapid system for regenerating shoots from mature nodal explants of *G.*

sylvestre. Maximum numbers of shoots were observed on the medium containing BAP (5 mg/l) and NAA (0.2 mg/l). Regenerated shoots were rooted on half strength MS without supplementation of any growth regulators. Relatively very few studies have been published on the mass *in vitro* clonal propagation of some latex producing milk weeds like *Tylophora, Hemidesmus, Asclepias* (Reddy *et al.*, 1998). The need of MS salts for shoot sprouting and proliferation shows the high salt requirement for the growth of *Gymnema sylvestre*. Influence of plant growth regulators, IAA, BA, 2,4-D and kinetin on shoot sprouting was investigated. Synergistic effect of Vitamin B_2 was also reported (Subathra Devi and Mohanasrinivasan, 2008). To overcome phenolic exudation, the effects of antioxidants, the effect of activated charcoal, citric acid and ascorbic acid were investigated. Incorporation of citric acid (100 mg L^{-1}) to the medium prevented phenolic exudation and increased the production of healthy normal shoots and shoot bud differentiation in *Gymnema sylvestre*. There was a considerable improvement in rooting as about 53 per cent shoots could be induced to root on 1/2 MS medium within 45 days with a fairly good length and number of roots per shoot. MS medium containing 1 mg L^{-1} BA+0.5 mg L^{-1} IAA + 100 mg L^{-1} Vitamin B_2+100 mg L^{-1} citric acid is best for shoot proliferation and 1/2 strength MS medium with IBA 3 mg L^{-1} is best for root induction (Subathra Devi and Mohanasrinivasan, 2008). Micropropagation by axillary bud proliferation has proved to be the most reliable method for large scale production of many forest and medicinal plants.

Gymnema sylvestre Callus and Suspension Culture

A cell culture system for *in vitro* growth of *Gymnema sylvestre* is one of the possible technologies for large scale production of gymnemic acid. The first report in suspension culture of *G. sylvestre* was given by Subathra Devi *et al.* (2006). Leaf explants grown in Muragish and Skoog salts supplemented with IAA 1.5 mg L^{-1} and BA 0.5 mg L^{-1} gave maximum percentage callus formation compared to other medium treatments evaluated. Combination of external phytohormone, shaking speeds, pH of the medium, played important roles in growth and gymnemic acid production in suspension cell cultures (Subathra Devi *et al.*, 2006). The presence of active principles gymnemic acid and gymnemagenin was successfully investigated in the cultured undifferentiated cells of *G. sylvestre* (Gopi and Vatsala, *2006*). Optimization studies for the enhanced production of gymnemic acid has also been reported. Praveen *et al.* (2011) investigated the effect of inoculum densities (2.5–20.0g/L), the strength of the Murashige and Skoog (MS) medium (0.25–2.0), carbon source (sucrose, glucose, fructose, maltose), and the concentration of the sucrose (1–8 per cent w/v) to determine their effects on biomass accumulation and production of gymnemic acid. 10g/L of inoculum density, full-strength MS medium supplemented with 2,4-dichlorophenoxy acetic acid (2.0mg/L) and Kinetin (0.1mg/L), and 3 per cent w/v sucrose was found best for the accumulation of biomass and gymnemic acid content (9.95mg/g dry weight) (Praveen *et al.*, 2011). Physical and chemical factors greatly favors the production of gymnemic acid in callus and suspension cultures of *G. sylvestre*. Factors such as light, temperature, sucrose, and photoperiod were studied to observe their effect on GA production. 5 per cent sucrose,12 h photoperiod and blue light favored the gymnemic acid production. (Bakrudheen Ali Ahmed *et al.*, 2012). Abiotic stressers like blue light and fluorescence light have been shown to significantly influence the higher rate of gymnemic acid production in callus cultures (Ali Ahmed *et al.*, 2009). The effect of plant growth regulators in combination with salt stress was reported to influence the gymnemic acid accumulation in callus cultures (Kumar *et al.*, 2010).

Elicitors and Gymnemic Acid Production

The recent development of elicitation has opened a new avenue for the production of secondary compounds. Secondary metabolite synthesis and accumulation in cell cultures can be triggered by the application of elicitors to the culture medium. Elicitors are microbe derived molecules which can enhance secondary metabolite production in cultured cells (Dicosmo and Misawa, 1985). Elicitors of biological origin are called biotic elicitors, which include polysaccharides, proteins, glycoproteins or cell wall fragments derived from bacteria, fungi and even plants. Elicitors of non-biological origin are called abiotic elicitors, which include metal ions, chemically defined compounds and UV light. For the laboratory scale studies of the abiotic elicitor it is cost effective but when used in pilot scale production of the bioactive compound, it becomes highly economic due to the cost of the metal as well as the purification steps involved in the downstream processing. The improper purification of these metals may also lead to lethal effects for humans. These problems can be overcome by using biotic elicitor such as microorganism for giving stress to the callus. (Subathra Devi *et al.*, 2012). Only very few studies has been reported on the elicitor-enhanced production of secondary metabolites in *G. sylvestre*.

The influence of various elicitors on secondary metabolite production in cell suspension cultures of *Gymnema sylvestre* was investigated. Cell suspension cultures of *Gymnema sylvestre* treated with four different elicitors, methyl jasmonate (MJ), yeast extract, chitin and pectin were studied for the production of gymnemic acid. It was evident from the study reported by Veerashree *et al.* (2011) that yeast extract and MJ markedly promoted the gymnemic acid production to 5.25-folds and 2.8-folds respectively. Pectin at 0.6 per cent 20 days of culture period has enhanced gymnemic acid production by 2.65-folds, while chitin at 100 mg l^{-1} increased the gymnemic acid production by 2.62-folds after 15 days of treatment (Veerashree *et al.*, 2011). Fungal and bacterial elicitors showed significant stimulation in the gymnemic acid production. *Aspergillus niger* cell extract was used as an elicitor to stimulate the production of secondary metabolite. Comparatively 9 fold increase of gymnemic acid yield was obtained in elicited cultures (Subathra Devi *et al.*, 2011). The HR protein from *Xanthomonas* sp. was used as an elicitor for the production of gymnemic acid. When compared to non-elicited cultures, two fold increase of gymnemic acids yield in elicited cultures was observed. The hypersensitivity reaction which is caused in the plant cells in presence of microorganism leads to the production of the secondary metabolite as a defense mechanism is the main principle behind this stress. Addition of HR elicitor to cell suspension cultures of *G. sylvestre* enhanced the production of gymnemic acid (Subathra Devi *et al.*, 2012).

Conclusion

As we know India has lot of *Gymnema* plants, so lots of works should be done to explore the gymnemic acid activity, because any metabolite with antidiabetic activity have been a great demand. Production of plant derived metabolites in bioreactor will serve to extend and enhance the continued medicinal compounds in future. Also plant tissue culture is widely used to conserve rare or endangered medicinal plants.

References

Ali Ahmed, A.B., Rao, A.S., Rao, M.V., and Taha, R.M. (2012). Different wavelengths light to induce physiological changes callus for the biosynthesis of gymnemic acid in *Gymnema sylvestre*. *Agro food industry hitech*., **23 (3)**: 31-34.

Ali Ahmed, A.B, Rao, A.S., and Rao, M.V. (2009). *In vitro* production of gymnemic acid from *Gymnema sylvestre* (Retz) R. Br. Ex Roemer and Schultes through callus culture under abiotic stress conditions. *Methods Mol Biol.*, **547(1)**: 93–105.

Anonymous, (1956). The Wealth of India Raw Materials, Publication and Information Directorate, CSIR, New Delhi, IV: 276-277.

Ashok Kumar, H.G., Murthy, H.N., and Paek, K.Y. (2002). Somatic embryogenesis and plant regeneration in *Gymnema sylvestre. Plant Cell Tissue and Organ Culture*, **71**: 85-88.

Baskaran, K., Ahamath, B.K., Shanmugasundaram, K.R., and Shanmugasundaram, E.R.B. (1990). Antidiabetic effect of a leaf extract from *Gymnema sylvestre* in non-insulin dependent *Diabetes mellitus* patients. *J. Ethanopharmacol.*, **30**: 295-305.

Bishayee, A., and Chatterjee, M. (1994). Hypolipidemic and antihypercholesterolemic effects of oral *Gymnema sylvestre* R.Br. leaf extract in albino rats fed on a high fat diet. *Phytothera. Res.*, **8**: 118-120.

Chowdhury, B.P. (1988). Assessment and conservation of medicinal plants of Bubhaneswar and its neighbourhood. *In* Indigenous Medicinal Plants, Today and Tomorrow's Printers and Publishers, New Delhi, India, pp.211-219.

Dicosmo, F., and Misawa, M. (1985). Eliciting secondary metabolism in plant cell culture. *Trends Biotech.*, **3(12)**: 318-322.

Gopi, C., and Vatsala, T.M. (2006). *In vitro* studies on effects of plant growth regulators on callus and suspension biomass yield from *Gymnema sylvestre* R.Br. *African Journal of Biotechnology*, **5(12)**: 1215-1219.

Granich, S.M., Halperin, B.P., and Eisner, T. (1974). Gymnemic acids: secondary plant substances of dual defensive action? *J. Insect. Physiol.*, **20**: 435-439.

Kapoor, L.D. (1990). CRC Handbook of Ayurvedic Medicinal Plants, CRC Press, Boca Raton, Fl., pp.200-201.

Komalavalli, N., and Rao, M.V. (1997). *In vitro* micropropagation of *Gymnema elegans* W and A. a rare medicinal plant. *Indian J. Exp. Biol.*, **35**: 1088-1092.

Komalavalli, N., and Rao, M.V. (2000). *In vitro* micropropagation of *Gymnema sylvestre* – A multipurpose medicinal plant. *Plant Cell Tissue and Organ Culture*, **61**: 97-105.

Kumar, U., Singh, I., and Priyanka, V.Y. (2010). *In vitro* salt stress induced production of gymnemic acid in callus cultures of *Gymnema sylvestre* R. Br. *Afr J Biotechnol.*, **9(31)**: 4904–4909.

Kurihara, Y. (1989). Antisweet activity of gymnemic acid A_1 and its derivatives. *Life Sci.*, **8**: 537-543.

Mitra, S.K., Gopu Madhavan, S., Muralidhar, T.S., Anturlikar, S.D., and Sujatha, M.B. (1995). Effect of D-400 herbomineral preparation on lipid profile glycated haemoglobin and glucose tolerance in streptozotocin induced diabetes and rats. *Indian J. Exp. Biol.*, **33**: 798-800.

Nadkarni, K.M. (1993). Indian Materia Medica, Popular Prakashan. Bombay, **1**: 596-599.

Nalawade, S.M., Sagare, A.P., Lee, C.Y., and Kao, C.L. (2003). Studies on tissue culture of Chinese medicinal plant resources in Taiwan and their sustainable utilization. *Botanical Bull. Academia Sinica.*, **45**: 79-98.

Praveen Nagella, Ill-Min chung, Hosakatte Niranjana Murthy. (2011). *In vitro* production of gymnemic acid from cell suspension cultures of *Gymnema sylvestre* R. Br. *Engineering in life sciences,* **11 (5)**: 537–540.

Rana, A.C., and Avadhoot, Y. (1992). Experimental evaluation of hepatoprotective activity of *Gymnema sylvestre* and *Curcuma zedoaria. Filoterap.*, **63**: 60-62.

Ravi, J., and Wahi, A.K. (1995). Antidiabetic studies of *Daemia extensa*, International Seminar on recent trends in pharmaceutical sciences, Ootacamund, Abstract, No.A44.

Reddy, P.S., Gopal, G.R., and Sita, G.L. (1998). *In vitro* multiplication of *Gymnema sylvestre* R. Br., An important medicinal plant. *Curr. Sci.*, **75**: 843-845.

Selvanayagam, Z.E., Ganavendhan, S.G., Chandrasekaran, P., Balakrishav, K., and Rao, R.B. (1995). Plants with antisnake venom activity – a review on pharmacological and clinical studies. *Filoterap.*, **65**: 99-111.

Shanmugasundaram, E.R.B., Gopinath, R.L., Shanmugasundaram, K.R., and Rajendran, V.M. (1990). Possible regeneration of the islets of langerhans in streptozotocin – diabetic rats given *Gymnema sylvestre* leaf extracts. *j. Ethanopharmacol.*, **30**: 265-279.

Subathra Devi,C., Murugesh, S., and Mohanasrinivasan, V. (2006). Gymnemic acid production in suspension cell cultures of *Gymnema sylvestre. J. Appl. Sci.*, **6(10)**: 2263-2268.

Subathra Devi. C., and Mohanasrinivasan, V. (2008). *In vitro* propagation of *Gymnema sylvestre. Asian Journal of Plant Sciences*, **7 (7)**: 660-665

Subathra Devi, C., and Mohana Srinivasan, V. (2011). *In vitro* studies on stimulation of gymnemic acid production using fungal elicitor in suspension and bioreactor based cell cultures of *gymnema sylvestre* R.Br. *Recent Research in Science and Technology,* **3(4)**: 101-104.

Subathra Devi. C, Ipsita Nandi, Mohana srinivasan, V., and Sriram Kalyan, P.(2012). Enhanced production of gymnemic acid using HR bioelicitor extracted from *Xanthomonas* spp. *Research journal of Biotechnology,* **7(4)**: 162-167.

Uniyal, M.R. (1993). Some popular and traditional ayurvedic herbs useful in family planning. *Sachitra Ayurved.*, **45**: 665-668.

V. Veerashree, C. M., Anuradha, and Vadlapudi Kumar (2012). Elicitor-enhanced production of gymnemic acid in cell suspension cultures of *Gymnema sylvestre* R. Br. *Plant Cell. Tissue and Organ Culture (PCTOC),* **108(1)**: 27-35.

Wen-Caiye, Qing-WenZhang, Xiu Liu, Chun-Taoche and Shou-Xun Zhao. (2000). Oleanane saponins from *Gymnema sylvestre. Phytochem.*, **53**: 893-899.

Yoshikawa, K., Arihara, S., and Matsuura, K. (1991). New type of antisweet principles occurring in *Gymnema sylvestre. Tetrahedron. Lett.*, **32(6)**: 789.

Utilisation and Management of Medicinal Plants Vol. 3 (2015) *Pages* **413–421**
Editors: **V.K. Gupta and A. Kaul**
Published by: **DAYA PUBLISHING HOUSE, NEW DELHI**

28

Efficient and Effective Micropropagation of *Aloe barbadensis*

Ujjwala Supe[1*]

ABSTRACT

An efficient and effective protocol is described for rapid and large scale in vitro propagation of the valuable medicinal herb, Aloe barbadensis by bud proliferation on young leaves and shoot tip, isolated from field-grown mature plants. This was achieved on MS solid and liquid medium with 4 mg/l BAP and 1 mg/l IAA within three weeks of inoculation. Each leaf gave rise to a large number of shoot buds (36 shoots) from all over the surface, while Shoot tip gave rise to a clump of shoots (20 shoots). The solid medium was more effective for bud proliferation from the leaf while the liquid medium proved more suitable for Shoot tip. Elongation of shoots and subsequent root induction were achieved on the same proliferation medium only. On an average, within a period of three subcultures, different explants like leaf bit and shoot tip explants generated 678, 438 shoots, respectively thereby favoring the economics of the cost of the materials and time factors. The regenerated plants resembled the mother plants in general habit without any morphological variation. HPLC analysis of the regenerated shoots revealed a phytochemical profile similar to that of the market sample and mother plants. A reproducible, very simple - one step procedure for in vitro propagation of Aloe barbadensis has been established. This protocol can be used to generate foundation stocks of elite planting material for large scale cultivation.

Keywords: *Aloe barbadensis*, Leaf culture, Micropropagation, Medicinal plant.

1 Plant tissue culture Laboratory, Department of Biotechnology and Microbiology, St. Thomas College, Bhilai, Dist. Durg – 490 006 Chattisgargh, India.

* *Corresponding author*: E-mail: ujsupe@ gmail.com

Introduction

Historical evidences point to Africa as origin center of *Aloe barbadensis*, a species cultured by Egyptian and Mediterranean people thousands of years B.C. Phytochemical studies have shown the presence of several bioactive compounds originated from *Aloe barbadensis* primary and secondary metabolism pathways, which have been widely used in formulations (gels and juices) in commercial scale all over the world. More than 160 chemicals or secondary metabolites found in *Aloe barbadensis* leaves. As example of those compounds is worth mentioning enzymes (lipases, bradykinases and proteases), monopolysaccharides (glucomannans), amino acids, vitamins (A, B_{12}, C and D), anthraquinones (aloin and emodine), saponins, salicilic acid, lignin and steroids (lupeol, campesterol, and sitosterol). A number of biological activities have been claimed to those substances, such as antiseptic (saponins and anthraquinones), anti-tumoral (mucopolysaccharides), anti-inflammatory (steroids and salicilic acid), antioxidant (vitamins), and immune-regulator (glucomannans) effects (Joshi, 1998).

Due to the wide spectrum of application in human health, the products containing *Aloe barbadensis* compounds have showed a strong demand in both national and international markets. *Aloe barbadensis* is usually propagated through lateral shoots, which the donor-plant produces mostly over the growth season. The number of lateral shoots/donor plant is low and variable over time, becoming difficult to plan in a rational basis a production system in commercial scale for obtainment of plant propagation material.

Aloe barbadensis is used internally as laxatives, antihelminthic, hemorrhoid remedy and uterine stimulant as menstrual regulator. *Aloe barbadensis* gel is an active ingredient in hundreds of skin lotions, sun blocks and cosmetics (Gridley, 1986). The gel used in cosmetics has been claimed that it has similar antiaging effects to Vitamin A derivatives (Danhof, 1993).

Aloe drug has a wide range of therapeutic action such as alternative, stomachic, emmerenjogic, astrigent, antihelminthic, stimulant hair growth restoration of sensation and no scaring. It has also show remarkable results on sun damaged skin and UV damaged skin. (Rowe, 1941; Afzal, 1991; Bunyapraphatsara, 1996; Mantle *et al.*, 2001) and anti-inflammatory activity (Davis, 1994).

The present communication reports an effective, efficient, rapid, cost-effective protocol for large-scale *in vitro* multiplication of *Aloe barbadensis.*

Materials and Methods

Juvenile shoots were obtained from three-month-old mature plants of *Aloe barbadensis* growing in the Botanical Garden of the St. Thomas College, Bhilai. Young leaves and shoot tip were used as explants.

The explants were thoroughly washed under running tap water (30 min) and treated with 0.2 per cent (v/v) Tween 20 for 15 min followed by repeated rinsing with distilled water. Further sterilization was done under aseptic conditions inside a laminar Airflow Hood (Lab Services, India).

Explants were surface sterilized with 50 per cent (v/v) ethanol (1 min) followed by a 3 min treatment with 0.1 per cent (w/v) HgCl2. Finally, the explants were washed thoroughly (4 - 5 times) with sterilized distilled water. Throughout the experiments, MS (Murashige and Skoog, 1962) medium with 3 per cent (w/v) sucrose and gelled with 0.8 per cent (w/v) agar (Hi-media, India) was used. The pH of all media was adjusted to 5.8 before autoclaving at 121°C (15 min). The cultures were incubated in a culture room at 25 ± 1°C under 16 h photoperiod (50 µEm-2s-1) provided by cool white fluorescent tubes (Phillips, India).

For initiation, various explants as described above were inoculated on both solid and liquid MS medium supplemented with different concentrations of BAP (1-6 mg/l) alone and with IAA (1 mg/l). The regenerated shoots were subcultured every three weeks in the same medium.

Experiments were also carried out to check the effect of different nodes (2 – 7 nodes) from *in vitro* developed shoots on MS media separately with different concentrations of BAP (1- 6 mg/l) alone and in combination with IAA (1 mg/l).

The experiments were performed in replicates of ten for each type of explants and all experiments were repeated three times. The growth responses of the explants were studied at weekly intervals in terms of the initiation and distribution sites of shoots and root regeneration.

Phytochemical evaluation was carried out by HPLC for two-month-old micropropagated plants, market samples and field-grown plants. All the samples were air-dried and crushed to a fine powder form. Optimal extraction was achieved in soxlet apparatus in 150 ml methanol for 5 h. The extract was cooled, dried transferred acetonitrile solution (20 ml).

Separation and determination of anthraquinone was performed with HPLC column (250 mm × 4.6 mm) that contained ODS (18) packing (Sigma-Aldrich Hypersil ODS 5mm). The solvent flow-rate was 1 ml/min and separated components were monitored by UV (220 nm). For acclimation, the regenerated plantlets were transferred to small earthen pots containing sand and soil in the ratio of 1: 1. Initially, high humidity was maintained with water spray at regular intervals (Jasrai *et al.*, 1999) and then transferred to the field for further growth.

Results and Discussion

Earlier reports available on *Aloe barbadensis* demonstrated plant regeneration through shoot tips and young leaves on medium with high concentrations of cytokinin and auxins (Supe, 2007; Kumar, 2011; Diwedi, 2014; Barna, 2004). However, we report here a one-step medium with low concentrations of cytokinin and auxin that were found suitable in all the types of explants for a rapid and large scale multiplication at effective level. The shoot tip (Figure 28.1a) implanted on MS medium supplemented with only BAP (4 mg/l) showed multiple shoot (3-5) within two weeks of incubation. Several workers have reported multiple shoot induction with cytokinins in the growth medium (Clog *et al.*, 1990; Kalimuthu, 2010). Addition of IAA (1 mg/l) with BA (4 mg/l) enhanced the number of shoots (17 - 18 shoots) from the shoot tip (Figure 28.1b)

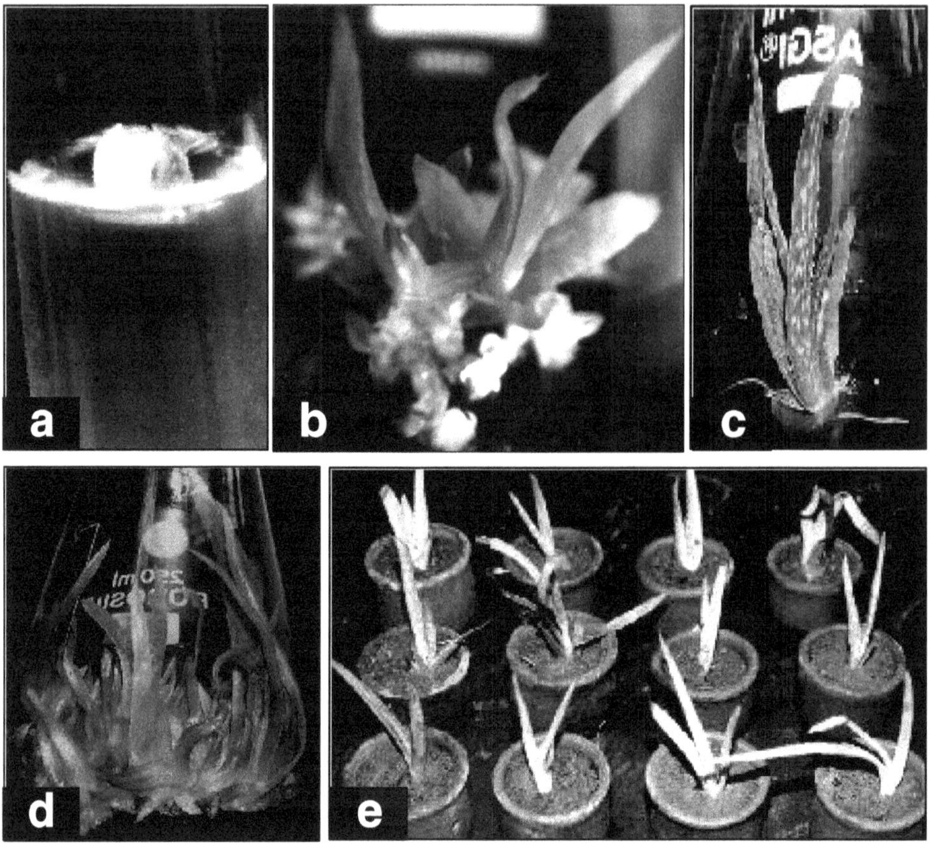

Figure 28.1: Micropropagation of *Aloe barbadensis.*

a: Inoculation of shoot tip in MS medium with Plant hormones; b: Induction of shoots from shoot tip in MS medium + BAP (4 mg/l) + IAA (1 mg/l); c.: Induction of shoots from leaf bit in MS medium + BAP (4 mg/l) + IAA (1 mg/l); d: Induction of roots in regenerated shoot; e: Well acclimatized plantlets ready to transfer in field.

in both liquid and solid MS medium. Choudhari (2001) has also reported shoot regeneration potential of IAA in *Aloe barbadensis.*

Proliferation of shoot buds and elongation of shoots was comparatively higher in the liquid medium than solidified medium for shoot tip. In the liquid medium regeneration, response was uniform; a higher biomass with twenty shoots from the shoot tip explants was recorded compared to that in the solid medium with twelve shoots (Table 28.1). Further, increase in shoot length was faster in the liquid medium, 5- 6 cm within 15 - 18 days than that observed in the solid medium (20 - 25 days). This might be due to better uptake of nutrients as large surface areas of explants were in contact with the liquid medium, thereby increasing the growth and multiplication. Furthermore, the liquid medium helps in maintaining O_2: CO_2 balance (Biondi and Thorpe, 1981). The slow growth and fewer shoots on agar solidified medium might be partly due to: (a) the specific habitat needs of this medicinal plant species, (b) a lower diffusion rate of molecules passing through the medium to the regenerant, (c) growth inhibition by an undefined agar-borne inhibitor and (d) reduced availability of water to tissues growing on agar solidified medium (Kohlenbach and Wernicke, 1978; Hashemabadi, 2008). Earlier, no report has been published on suitability of the liquid medium for this plant species.

Table 28.1: Effect of different Combination of BAP and IAA on Oraganogenesis from Shoot Tip Explants

Concentration (mg/l)		*Number of Shoot/ Ex. (S.E.)*	*Response*	*Shoot Length (cm)*	*Survival*
BAP	*IAA*				
			Solid medium		
0.0	1.0	2.4 + 0.3	60	3.1 + 1.5	70
2.0	1.0	8.2 + 0.2	75	4.5 + 0.5	80
4.0	**1.0**	**12.8 + 0.4**	**100**	5.5 + 0.2	**100**
6.0	1.0	9.8 + 0.5	80	3.8 + 0.2	85
			Liquid medium		
0.0	1.0	3.2+ 0.6	65	2.1 + 0.01	73
2.0	1.0	9.5 + 0.04	80	3.4 + 0.5	87
4.0	**1.0**	**20.5 + 0.2**	100	**5.9 + 1.2**	100
6.0	1.0	11.7 + 0.07	84	2.8 + 0.7	93

Performance of the shoot tip explants with reference to their position on *in vitro* developed shoots was evaluated in optimal liquid medium. It was observed that upper nodes (2 - 3) were slower in growth, while lower nodes (4 - 7) were found to be more promising and resulted in multiplication of greater number of shoots (8 shoots). Leaf bit explants were also used; however, proliferation response was found to be poor with only a single shoot. The shoot tip explants, when cultured on the optimal solid medium, yielded a large number of shoot (20) regenerants within a period of two and a half weeks (Figure 28.1c). This is in agreement with the results reported by Zeng *et al.* (2000) with (12) shoots on a higher concentration of BA. In contrast, Hirimburegama (1995) observed fewer shoots from axillary bud explants on a higher concentration of BAP and IAA, while Zhao (1990) recorded 15 shoots from stem disc explants after a six-week incubation period but without any growth hormone supplement. A large number of shoot buds were also observed on *in vitro* leaf explants on liquid and agar based MS media with BAP (4 mg/l) and IAA (1 mg/l) without the intervention of callus. Shoot, tip proliferation was observed initially from the base, which subsequently extended all over the surface. A maximum of 36 shoots was observed within three weeks of incubation in the first sub-cycle (Table 28.2).

Research is said to be more successful if it is cost effective. The number of shoots per subculture and media quantity per subculture was standardized from the commercial point of view. Within a period of three subcultures, the number of shoots at each subculture generated from young leaves and shoot tip were: 678, 438 shoots, respectively. The system demonstrated a continuous supply of shoots up to ten cycles without any decline in their number in subsequent subcultures. Subcultures were performed frequently (3 weeks), as delayed subcultures (more than four weeks) in the liquid medium were found to cause vitrification of shoots similar to that observed in the tissue culture-raised carnation plantlets (Ziv *et al.*, 1983). Furthermore, 20 ml of basal media for leaf and 40 ml liquid media for shoot tip explants with ten explants were found to be optimal.

Table 28.2: Effect of different Combination of BAP and IAA on Organogenesis from Leaf Bit Explants

Concentration (mg/l)		*Number of Shoot/ Ex. (S.E.)*	*Response*	*Shoot Length (cm)*	*Survival*
BAP	*IAA*				
0.0	1.0	3.4 + 0.1	53	1.1 + 0.3	64
2.0	1.0	11.3 + 0.09	69	4.1 + 0.4	80
4.0	1.0	36.9 + 0.2	100	5.7 + 0.8	100
6.0	1.0	13.8 + 0.3	80	3.1+ 1.3	92

Well-grown shoots (3 - 4 cm) were isolated and transferred to the basal and the optimal medium for root induction (Figure 28.1d). Root induction was found to be better in the optimal medium as compared to the basal medium. The approximately 5 – 6 cm long shoots with 3 - 4 cm roots were transferred to trays containing sand, soil and in the ratio of 1 : 1 and kept under shade for hardening.

All plants regenerated from different explants were hardened directly in the greenhouse. Initially high humidity was maintained. The plantlets so hardened for two weeks in green house (Figure 28.1e) were subsequently transferred to open beds with 100 and 98 per cent survival rate for shoo tip and leaf based explants, respectively. No morphological variation of any nature was observed among the *in vitro* raised plants when compared with the mother stock.

For HPLC, different solvent strengths of mobile phase were used; a mixture of acetonirile and water in the proportion of 33 : 67 was found to be suitable for the separation of anthraquinone in *Aloe barbadensis* plant extract. A matching profile of representative chromatograms of *in vitro* generated plants, market sample and field-grown plants of *Aloe barbadensis* was observed (Figure 28.2). We report high level of shoot bud regeneration from various explants with continuous proliferation and elongation of shoot buds and root induction on MS medium supplemented with 4 mg/l BAP and 1 mg/l IAA. Thus, a commercially viable protocol has been established for mass micropropagation of *Aloe barbadensis*. The procedure described here will go a long way to meeting on one hand the ever-increasing demands of the pharmaceutical industries and on the other save this species from extinction.

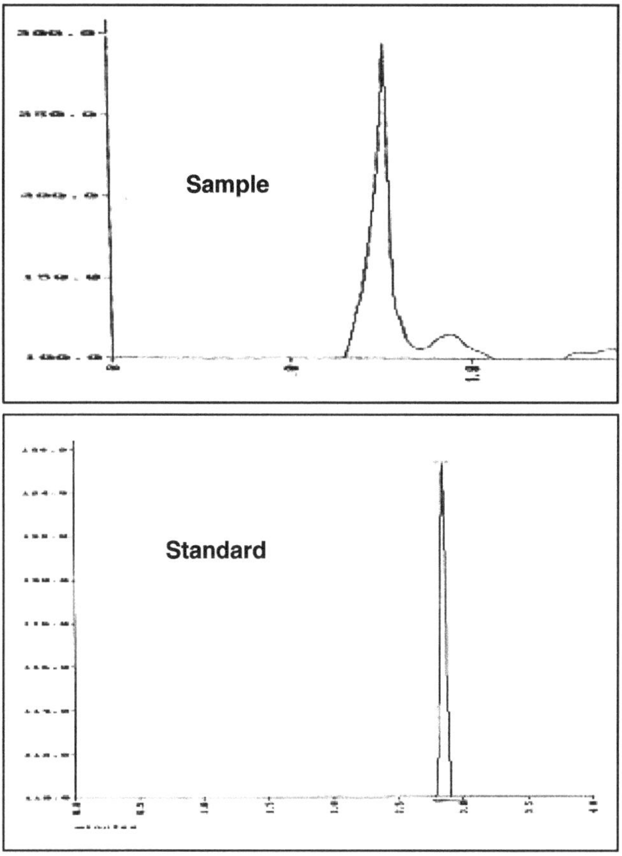

Figure 28.2: Separation of Anthraquinone: A present in the methanol extracts of *Aloe barbadensis* by HPLC using mobile phase acetonitrile-Water (33: 67) of *in vitro* generated plants, Standard.

References

Afzal, M., Ali, M. and Hassan, R.H.A. (1991). Identification of some prostanoids on Aloe extracts. *Plant. Med.*, **72**: 201-205

Barna, K. S. and Agarwal, D. (2004). Tissue culture propagation of elite plant of *Aloe vera* Linn. *J. Plant Biochemistry and Biotechnology*, **13**: 77-79.

Biondi, S. and Thorpe, T. A. (1981). Requirement for a tissue culture facility. *In*: Plant Tissue Culture: Methods and Applications in Agriculture, TA Thorpe Academic Press (Eds), New York, USA, pp. 1-20.

Bunyapraphjatsara, N. S., Jirakukhaiiwing, S.C. and Thirawarapan (1996). The efficiency of *Aloe vera* cream in the treatment of first, Second and third degree burns. *Phytomedicine*, **2(3)**: 247-251.

Choudhary, S., and Mukundan, U. (2001). *Aloe vera* L. - Micropropagation and characterization of its gel. *Phytomorphology*, **51(2)**: 155-157.

Clog, E., Bass, P., and Walter, B.(1990). Plant regeneration by organogenesis in *Vitis* root stock species. *Plant Cell Reporter.*, **8**: 726-728.

Danhof, I. (1993). Potential reversal of Chronological and photoaging of skin by topical application of natural substances. *Phytotherapy Research*, **7**: S53-S56.

Davis, S. D. (1994). Anti-inflammatory and wound healing activity of growth substances in Aloe vera. J. *Am. Pod. Med. Asso.*, **84**: 74-81.

Dwivedi, N.K., Indiradevi, A., Asha1, K.I., Nair, R. A., and Suma, A. (2014). A protocol for micropropagation of *Aloe vera* L. (Indian Aloe) – a miracle plant. *Research in Biotechnology*, **5(1)**: 01-05.

Gridley, D., and Reynold, T. (1986). *Aloe vera* phenomenon A review of the properties and modern use of leaf parenchyma gel. *J. of Ethanoparmacology*, **16**: 117-151.

Hashemabadi, D., and Kaviani, B. (2008) Rapid micro-propagation of *Aloe vera* L. via shoot multiplication. *Afr J Biotech.*, **7**: 1899–1902

Hirimburegama, K., and Gamage, N. (1955). *In vitro* multiplication of *Aloe vera* meristem tips for mass propagation. *Horti. Science*, **27(3/4)**: 15-18.

Jasrai, Y.T., Kannan, V.R., and George, M. M. (1999). *Ex vitro* survival of *in vitro* derived banana plants without greenhouse facilities. *Plant Tissue Cult.*, **9**: 127-132.

Joshi, S.P. (1998). Chemical constituents and biological activity of *Aloe barbadensis* –A review. J. of *Medicinal and Aromatic plant sciences*, **20**: 768-773.

Kalimuthu, K., Vijayakumar, S., Senthilkumar, R.R., and Sureshkumar, M. (2010). Micropropagation of Aloe vera Linn-A medicinal plant. *Biotech and Biochem.*, **6**: 405-410.

Kohlenbach, H. W., and Wernicke (1978). Investigation as the inhibitory effect of agar and the function of active carbon in anther culture. *Z. Pflanenphysiol.*, **86**: 463-472.

Kumar, M., Singh, S., and Singh, S. (2011). *In vitro* Morphogenesis of a medicinal plant –*Aloe vera* L. *Asian Journal of Plant Science and Research*, **1(1)**: 31-40.

Mantle, D., Gok, M.A., and Lennard, T.W (2001). Adverse and beneficial effects of plant extracts on skin and skin disorders. *Adverse Drug Reat Toxicolo. Rev.*, **20 (2)**: 89-103.

Murashige, T., and Skoog, F. (1962). A revised medium for rapid growth and bioassays with tobacco tissue cultures. *Physiol plant*, **15**: 473- 497.

Rowe, T.D. (1941). Effect of fresh *Aloe vera* gel in treatment of third degree roentgen reaction on white rats, preliminary report. *J. Am. Pharm.*, **29**: 266-269.

Supe, U.J. (2007). *In vitro* regeneration of *Aloe barbadensis*. *Biotechnology*, **6(4)**: 601-603.

Zeng, S., and Peng, X. (2000). Tissue culture and rapid propagation of *Aloe arborescens*. *Zhong Yao Cai.*, **23(2)** : 63-65.

Zhao, M. (1990). Rapid propagation of *Aloe variaeta* by tissue culture. *J. of shenyang Agricultural University*, **21(2)**: 339-340.

Ziv, M, Meir, G., and Halvey, A. H. (1983). Factors influencing the production of hardened glaucous carnation plantlets *in vitro*. *Pl. Cell Tiss. Org. Cult.*, **2**: 55-65.

Utilisation and Management of Medicinal Plants Vol. 3 (2015) *Pages* **423–433**
Editors: **V.K. Gupta and A. Kaul**
Published by: **DAYA PUBLISHING HOUSE, NEW DELHI**

29

Hydrogen Peroxide: A Natural Plant Produce against Environmental Stresses

Nooris Naqvi[1*], Mohammad Mazid[1], Khalil Khan[2] and Gulnaz Sabri[3]

ABSTRACT

Abiotic stresses are major constraint to agricultural production worldwide. The plants have an inbuilt mechanism to respond to fluctuations in circadian and seasonal environmental conditions. Abiotic stresses disrupt the cellular redox homeostasis which leads to the oxidative stress or generation of ROS. H_2O_2 is the two electron reduction product of O_2. It is potentially reactive oxygen, but not a free radical, generated as a result of oxidative stress via superoxide (O_2•-), presumably in a non-controlled manner during electron transport processes such as photosynthesis and mitochondrial respiration. H_2O_2 generation via electron transport is increased in response to environmental stresses such as excess excitation energy, drought and cold and also induced in plants following exposure to a wide variety of abiotic and biotic stimuli. Although, in the last two decades, the potential of H_2O_2 as a stress induced signal to abiotic stress in plants has positioned for much attention. H_2O_2 is produced in response to various stimuli and mediates cross-talk between signaling pathways and is an attractive signaling molecule contributing to the phenomenon of cross-tolerance in which exposure of

1 Environmental Science Divison, Bareilly College, Rohilkhand University, Bareilly – 248 392, Uttar Pradesh, India.

2 Chandra Sekhar Azad University of Agriculture and Technology, Kanpur – 208 020, Uttar Pradesh, India.

3 Department of Microbiology, Hospital laboratory, Gitam University, Visakhapatnam, India

* *Corresponding author*: E-mail: biologydone@gmail.com

plants to one stress offers protection towards another. As it is well cleared in present review, recent studies on H_2O_2 functionality to induce tolerance in plants to abiotic stresses has been unravelled and very impressive.

Keywords: Antioxidant, Hydrogen peroxide, Nitric oxide, Signaling molecules, UV-radiation

Introduction

H_2O_2 has been implicated as a key factor mediating programmed cell death. Plants exposed to abiotic stresses can produce a systemic signal, a component of which may be H_2O_2 which sets up an acclamatory response in unstressed regions of plants. Also H2O2 is an ancient signaling molecule that not only played a key role in inducing evolution of oxygenic photosynthesis but also modulates many physiological events, such as stomatal movement, hypersensitive responses, and programmed cell death. Increasing evidence now indicates that H_2O_2 acts as a local and systemic signal that directly regulates expression of numerous genes. During normal metabolism in a plant cell, H_2O_2 is generated in chloroplast, mitochondria and peroxisome and is kept in homeostasis by complicated and effective scavenging systems that have developed over the course of evolution. H_2O_2 has a long lifespan, is able to cross biological membranes, and rapidly diffuses from cell to cell or can be transported long distances from its sites of origin in plants. Also, its production quickly responds to various environment stimuli. Thus, H_2O_2 has all of the characteristic features of an intercellular signaling molecule and, for this reason, has received increasing attention in recent years (Mazid *et al.*, 2011a; b). Several recent reviews have described the biological activities of ROS, placing special emphasis on the signaling role of H_2O_2. In particular, H_2O_2 that is produced by cytosolic membrane-bound NADPH oxidases has been implicated as a signal in a wide range of biotic and abiotic stress responses. These responses include defense reactions against pathogens and herbivores, the closure of stomata and the regulation of cell expansion and plant development.

Moreover, understanding the role of H_2O_2 in plant growth requires models that accommodate the large number of ways in which it can be formed and degraded at any given time, and that ROS produced by one source may be the drivers or substrates for a second. Whatever, the source of ROS, it is now apparent that H_2O_2 acts as a signal to induce a range of molecular, biochemical responses with in cells and plants. In addition, exposure to low levels of one stress can induce tolerance towards subsequent higher levels of exposure to the same stress, a phenomenon termed acclimation tolerance. The cellular responses of H_2O_2 differ according to its site of synthesis or perception for example, whether the H_2O_2 is synthesized in plastids or at the plasma membrane.

Environmental Stress

The generation of ROS in different location of the plants cell causes injury and cell death (Mano, 2002). On the other hand, ROS play a vital role in intracellular redox signaling, activating antioxidant resistance mechanisms. It is now well established that virtually all stresses induce or involve oxidative stress to some degree and the ability of plants to control oxidant levels is highly correlated with stress tolerance. For this purpose, they are equipped with complex processes such as perception, transduction and transmission of stress stimuli. Since, all parts of plant body use oxygen to produce energy. Oxidative stress is mediated by excess ROS which include OH^-, O_2^- and H_2O_2 can lead to destruction of cellular components including lipid, protein and DNA. H_2O_2 can induce the expression of genes potentially involved in its synthesis such as NADPH oxidase and also of those encoding proteins involved in its degradation, implying a complex mechanism for cellular regulation of oxidative status. H_2O_2 induced the expression of genes encoding ascorbate peroxidise in germinating rice embryos and in *Arabidopsis* leaves, and wounding induced the expression of gene encoding a catalase via H_2O_2 in embryos and leaves of maize (Abogadallah, 2010). Induction of peroxisome biogenesis genes by various abiotic stresses like temperature, UV and mechanical injury (also generates H_2O_2) and exogenous H_2O_2, places H_2O_2 as a key signal molecule mediating various cellular responses to biotic and abiotic stresses. Further studies is needed to understand the actual function of GPX in chilling stress during H_2O_2 pre-treatment as observed during the chilling resistance (Desikan *et al.*, 1998). However, the fact that oxidative stress is a common facet of many cellular stress responses means that elucidating those intracellular signaling, processes mediating H_2O_2 signaling.

Signaling Molecules and Stresses Response

H_2O_2 is an important signal molecule that induces stomatal closure. To cope with increased levels of ROS, plants have evolved at least four kinds of ROS scavenging enzymes including CAT, SOD, APX and GPX (Gechev and Hille, 2005). It has been shown that H_2O_2 is scavenged by GPX in the guard cells. Recently, Arabidopsis mutants lacking either or both a cytosolic and chloroplatic APX, which are responsible for H_2O_2 removal, are found to be more tolerant to salinity stress. So far, the genes that regulate the expression of H_2O_2 scavenger's enzymes have not been identified and the mechanism of regulation of H_2O_2 levels especially in the stomata is largely unknown. Since H_2O_2 induces leaf stomatal closure. Results of Bienert *et al.* (2006) suggested that accumulation of H_2O_2 in guard cells probably underlies the increased stomatal closure in *dst* mutants and also propose that accumulation of H_2O_2 in the *dst* mutant may be the result of down-regulation of the genes related to H_2O_2 homeostasis.

The ABA is a vital signal that modulates a variety of growth and developmental processes and responses to environmental stress such as drought, salt and cold. When plants are drought-stressed, ABA acts as a key regulator of stomatal apertures to restrict transcription and reduces water loss. H_2O_2 is an essential signal in ABA induced stomatal closure. Exogenous ABA stimulated apoplast CuAo activity, increased H_2O_2 production and $[Ca^{2+}]$ cyt levels in *vicia faba* guard cells and induced stomatal closure. ABA-induced H_2O_2 production and the H_2O_2 activated Ca^{2+} channels in the plasma membrane is important mechanism for ABA induced stomatal closing (Halliwell *et al.*, 2000). Molecular genetic evidence shows that in Arabidopsis, the plasma membrane-associated NADPH oxidase catalytic subunits AtrobohD and

AtrbohF are implicated in the ROS-dependent activation of Ca^{2+} channels and cytosolic Ca^{2+} increase. The results of An *et al.* (2008) suggests that CuAO in *vicia faba* guard cells is an essential enzymatic sources for H_2O_2 production in ABA-induced stomatal closure via the degradation of putrescine. The products from putrescine oxidation by CuAO include pyrroline that CuAo and PAO participate in plant development and defence responses via their reaction products.

Salinity Stress and H_2O_2

In sodic soils, Na^+ binds to negatively charged clay particles, causing swelling and dispersal, thus making the soil less unfit for crop growth. High salt concentration causes osmotic and ionic stress in plants. It limits growth and development of plants by affecting several key metabolic processes. Further, salinity alters the activities of many enzymes involve in nitrate and sulphate assimilation pathways in plants which lowers their energy status and increase the demand for nitrogen and sulphur. Exogenous H_2O_2 possibly functioned directly or may have induced intracellular H_2O_2 generation to act as a signaling molecule under abotic stresses. Pretreatment of wheat seeds with H_2O_2 has been shown to improve the subsequent salt tolerance of the seedlings (Wahid *et al.*, 2007). The control of Na^+ involvement across the PM and tonoplast to maintain a low Na^+ concentration in adaptation to salt stress (Rausch *et al.*, 1996). Further results indicated that H_2O_2 content increased greatly under salt stress. The results suggested that H_2O_2 inducing an increased K to Na ratio. Therefore, it is clear from summing up this new assay that NO may be regulate the H_2O_2 generation (Wang *et al.*, 2007).

Metal Pollution and H_2O_2

Because of their high reactivity they can directly influence growth, senescence and energy synthesizing processes. Tolerance to heavy metals in plants may be defined as the ability to survive in a soil that is toxic to other plants, and is manifested by an interaction between a genotype and its environment. The effects of their toxic influence on plants is largely a strong and fast inhibition of growth processes of the above and underground parts, as well as the active decrease of the photosynthetic apparatus, often correlated with progressing senescence processes (Mazid *et al.*, 2010). High H_2O_2 and O_2^- had been reported earlier in the case of various other plants under Cr, Zn, Pb etc. (Panda *et al.*, 2003). The increase in SOD activity indicated higher H_2O_2 level seen by the increase in total H_2O_2 content in leaves, which tallies with those observed in the case of *Brassica juncea* and *Vigna radiata* under Zn and Al treatment. However, in barley plants only Mn increased the H_2O_2 content after 5 days but not Cu. This difference may indicate that H_2O_2 accumulation developed differently during a larger stress action. After a long time of Cd action, decrease in SOD activity observed. More recently, Lin *et al.* (2005) have shown that Cu can act through changes in H_2O_2-dependent peroxidase activity followed by cell wall stiffening due to the formation of cross-linking among the cell wall polymers. Also, Cd enhances H_2O_2 accumulation. NADPH oxidase is involved in plant growth and plant response to Cu. In fact, accumulation of H_2O_2 has been observed in Cd-exposed roots and in Cd exposed tobacco suspension cultures. Taking all these observation together, a hypothetical framework may be suggested that Cd induces a transient loss in antioxidative capacity perhaps accompanied by a stimulation of oxidant producing enzymes, which results in intrinsic H_2O_2 accumulation. H_2O_2, then, would act as a signaling molecule triggering secondary defenses. These, in turn, would cause an ultimately cell wall rigidification and lignifications, thereby, decreasing cellular viability and finally resulting in cell death (Yeh *et al.*, 2003).

Drought and H_2O_2

Lipid per-oxidation and membrane deterioration in plants leads to an imbalance between antioxidant defenses and the amount of ROS resulting in oxidative stress. Apart from morphological structures contributing to drought stress tolerance, plants have evolved a variety of physiological and biochemical processes, which act as components of drought tolerance. The accumulation of H_2O_2 observed at the onset of drought in the mesophyll cell walls of *C. albidus* and *C. clusii* may be associated with its function in cellular signaling at the first stage of drought or with drought-induced changes in cell wall structure or both. Accumulation of H_2O_2 in mesophyll cell walls occurred as the first symptom of drought in both species, indicating that H_2O_2 may play a role in inter or intracellular signaling or both. Moreover, diurnal variations in H_2O_2 suggest a particular role for H_2O_2 in the response of plants to a combination of stresses. After a general introduction to the concept of drought and oxidative stress and its relationship, Jubany *et al.* (2010) describe the role of H_2O_2 in drought stress responses, emphasizing the importance of studies in H_2O_2 subcellular localization, needed for a better understanding of its role in plant responses to stress. The role of H_2O_2 in intracellular signaling in drought-stressed plants and in the putative regulation of antioxidant synthesis needs to be investigated also. Furthermore, during drought stress, ABA-mediated stomatal closure is a mechanism that plants use to adapt to water deficiency. H_2O_2 that widely generated under abiotic stress have been proposed to function as second messengers in ABA signaling. Early work has shown that H_2O_2 can induce stomata closure. In guard cells, ABA-stimulated ROS accumulation activates plasma membrane.

Heat Stress and H_2O_2

High temperature induces oxidative stress, lipid peroxidation, membrane injury, protein degradation, enzyme inactivation, pigment bleaching and DNA strands disruption in plants. Similarly low temperature (cold stress) caused many changes in biochemical and physiological processes and ROS-homoeostasis in plants. The initial perturbations in plant metabolism are followed by alteration in plant development and morphology with a final result achievement of maximum freezing tolerance. Studies of Apostolova *et al.* (2008) reported that levels of endogenous H_2O_2 strongly increased in the spring wheat cultivar in response to cold hardening, and to a lesser extent in the winter wheat. Results of Wang *et al.* (2010) showed that pre-treatment with H_2O_2 at appropriate concentration may improve the tolerance of warm-season Zoysia grasses to chilling stress and that manilagrass had better tolerance to chilling, as evaluated by lower MDA and EL, and better turfgrass quality, regardless of the pre-treatment applied. Pre-treatment with H_2O_2 has been shown to induce chilling tolerance in normally chill-sensitive maize seedlings (Neto *et al.*, 2005). Similarly regenerated potato nodal explants treated with H_2O_2 became significantly more thermo tolerant compared with untreated control (Lopez- Delgodo*et al.*,1998). However, if the phenolic substrates are replaced by NADPH or related reduced compounds, a chain reaction starts that provides the basis for the H_2O_2 producing NADPH-oxidase activity of peroxidases. Because H_2O_2 is a mild oxidant that can oxidize thiol residues, it has been speculated that H_2O_2 is sensed via modification of thiol groups in certain proteins. So, therefore, H_2O_2 activates several MAPK cascades have important functions in plant stress responses and development and are key players in ROS signaling and in innate immunity. In Arabidopsis, the transmission of ROS signaling by MAPKs involves the coordinated activation of MPK6 and MPK3. Moreover, in Arabidopsis, H_2O_2 activates the MAPKs, MAK3 and MAK6 via MAPKKKANP1. Over-expression of ANP1 in transgenic plants resulted in increased tolerance to heat shock, freezing and oxidative stress. It was also exhibited that pre-treatment of H_2O_2 induces not only ROS-scavenging enzyme activities but also expression of transcripts for oxidative stress related gene encoding sucrose phosphate synthase, D-pyrroline-5-carboxylate synthase and small heat shock protein 26 (HSP26). Microarray analysis of H_2O_2-induced gene expression in Arabidopsis indicates potential H_2O_2-responsive cis-elements in genes regulated by H_2O_2. One of these elements, the as-1 promoter element, also has high homology with redox-sensitive mammalian AP-1 *cis*–element. However, recent analysis of transgenic plants indicates that ROS other than H_2O_2 activates this as-1 element. Further analysis will reveal whether similarity exists among plant, animal and fungal regulatory cis-elements of ROS-responsive genes. In contrast to $O_2^{\cdot-}$, H_2O_2 can diffuse in to cells and activate many of the plant defences, including PCD.The suppression of ROS detoxifying mechanisms is crucial for the onset of PCD. Furthermore, recent work shows that H_2O_2 function as a second messenger mediating the systematic expression of various defense related genes in tomato plants. A transient increase in H2O2 was suggested to signal activation of protective mechanism for acclimation to chilling. Exogenous application of H2O2 can induce tolerance to chilling, high temperature and various other abiotic stresses, all of which cause elevated endogenous H2O2 production.

U-V Stress and H_2O_2

Most types of protein oxidations are especially irreversible, whereas, a few involving sulphur containing amino acids are reversible. Protein oxidation is wide spread and often used as a diagnostic marker for oxidative stress. Concerns about potential impacts of stratospheric ozone depletion contributed to spark-interest in studies of plant responses to enhanced UV-B levels during the last two decades. The direct effects of UV radiation are mostly damaging, because UV photons have enough energy to create lesions in important UV-absorbing biomolecules such as nucleic acids and proteins. In addition, many of the effects of UV-B involve the differential regulation of gene expression. Responses to UV-B are mediated by both non-specific signaling pathways, involving DNA damage, ROS (*e.g.*, H_2O_2) and various other defensive signaling molecules, and UV-B specific pathways that mediate photo-morphogenic responses to low levels of UV-B (Jenkins, 2009). In the leaves of tropical tress, the ambient UV-B radiation might contribute to the reversible decline in potential PS-ÉÉ efficiency observed upon exposure to full and direct sunlight. Progress in the understanding of the mechanism that mediates these effects of UV-B has been slow for various reasons. Moreover, UV-B may activate a variety of molecular sensitizers and give rise to enhanced H_2O_2 levels, which may lead to convergence with the wound induced cascade downstream of NADPH oxidase. Arabidopsis leaves pre-treated with H_2O_2 have been shown to develop increased tolerance to excess light.

Stress Induction of Signaling Molecules

Previous studies have shown that H_2O_2 function as stress signal in plants, mediating a range of resistance mechanisms in plants under stress conditions. The plasma membrane is an important protective barrier against extracellular H_2O_2. In higher plant plasma membranes are known for their high sterol content, which suggests that native plant membranes are relatively impermeable to non-electrolytes. The work of Phillips *et al.* (2005) had suggested that algal plasma-membrane may contain pores for H_2O_2 signaling and defence-related small neutral solute like H_2O_2. Gradient of H_2O_2 have been reported across the membranes from mammalian cell lines, bacteria and yeast, suggesting that native biological membranes generally prevent the free diffusion of H_2O_2. The physiological role of H_2O_2 permeability across the tonoplast remains somewhat unclear. Although H_2O_2 diffusion across membranes is often considered as a passive property of lipid bilayers native membranes represent significant barriers for H_2O_2. H_2O_2 was detected by DAB (diamoinobenzidine) staining. The calculative identity differences in the conduction between the substrates and reveal channel residues critically involved in H_2O_2 conductions. The results of Dynowski *et al.* (2008) strongly suggest that plasma membrane aquaporin pores determine the efficiency of H_2O_2 signaling between cells. Cellular H_2O_2 redox chemistry is mainly restricted to compartments surrounded by membranes such as peroxisome and mitochondria and may also be found in other endomembrane compartment. Among the ROS, H_2O_2 seems best suited to play the role of signaling molecule due to its higher stability and longer half-life. For example, in winter wheat (*Triticum aestivum*) leaf, chilling at 4 °C caused the H_2O_2 level to increase to threefold that of control within 1 min. Recently, PCD triggered in barley aleurone by the phytohormone gibberellins also found to be mediated by H_2O_2, implying a role for H_2O_2 in developmental PCD in addition to that induced by a stress factor. Specific H_2O_2 generation in plants involves plasma membrane NADPH oxidases that produce the O^{2-} by enabling the transfer of electrons from the cytosolic NADPH across the membrane to external molecular oxygen. The spontaneous dismutation of the short-lived O^{2-} will produce external H_2O_2 followed by H_2O_2 influx and the subsequent activation of Ca^{2+} channels. Such NADPH-dependent processes have been identified in the plant defence against oxide burst in the regulation of ABA signaling and stomatal movements (Hancock *et al.*, 2001). Under normal conditions, H_2O_2 can be generated in normal metabolism via the Mehler reaction in chloroplasts electron transport system in mitochondria and photorespiration in peroxisomes (Neill *et al.*, 2002). Furthermore, studies of Wi *et al.* (2010) suggest that H_2O_2 is important signaling molecule involves in response to and abiotic stresses and developmental and physiological processes and stress tolerance of H_2O_2 treated transgenic plants resulted from reduced ethylene biosynthesis, which decreased ROS accumulation via increased gene expression and activity of ROS-detoxifying enzymes. Different stimuli activate specific H_2O_2-generating pathways and produce the signal leading to specific physiological processes (Neil *et al.*, 2002).

Although it was hypothesized that H_2O_2 produced intracellularly diffuses to other cells for use by POX and other defensive enzymes, it non appearance more probable that intracellularly produced H_2O_2 is consumed quickly and locally, and that extracellular metabolism uses H_2O_2 produced extracellularly. H_2O_2 localization with in tissues, sometimes to portions of cell wall in root hairs, or in epidermal cells in association with wounding or stomatal movements, indicates as does compartmentation at the tissue level with in the leaves, in vascular tissues, and in areas of regulation (Naves *et al.*, 1998). During oxidative stress H_2O_2 was shown to induce stomatal closure (McAnish *et al.*, 1996) showed that elicitors could induce H_2O_2 production and Lee *et al.* (1999), demonstrated that both these responses were linked: elicitors caused H_2O_2 production which, in turn, caused stomatal closure. For H_2O_2 to really to be a specific signaling molecule, a mechanism must exist to perceive the elevation of H_2O_2 in cells. H_2O_2 can interact with cysteine residues with in proteins and this redox modulation of protein could potentially alter protein conformation, affecting protein activity and therefore initiating subsequently cellular responses. In the last five years, this response network has been repeatedly extended and summarized, not only for stomatal responses but for responses to other abiotic stresses, and the signaling cascade have been shown to have many similarities. Furthermore, recent work has shown that H_2O_2 induces the expression of genes encoding proteins required for peroxisome are important sources of ROS as well as antioxidant and NO, and are thus important regulators of the cellular redox state. Calmodulin regulates NAD kinase activity, which generates NADPH for NADPH oxidase activity. Thus, cross-talk between H_2O_2 and calcium could regulate specifity and/or cross tolerance towards various abiotic stresses. To date, though, no calcium-dependent protein kinases have been shown to be regulated by H_2O_2, although H_2O_2-regulated genes encoding protein kinases and phosphatases have been discovered. More lines of evidence concerning the relationship between H_2O_2 and Ca^{2+} signals is provided by the study of H_2O_2 homeostasis in Arabidopsis.In last, a protein phosphorylation cascade that has been shown to be activated by H_2O_2 in a MAPK cascade. This kinase then phosphorylated a MAPKK which in turn activates a MAPK by dual phosphorylation on both threonine and tyrosine residues in a conserved T-X-Y motif. Different previous observations indicate that H_2O_2 activating of a MAPK cascade is a central response mediating tolerance of various stresses. (1) That H_2O_2 generation occurs in response to diverse abiotic stresses. (2) That exposure to one stress offers cross-tolerance towards another. (3) That there exist commonalities in defence responses to various stresses. (4) That activation of a H_2O_2-regualed MAPK pathway mediates multiple stress tolerance.

Future Challenges

The intracellular signaling cascades that transduce H_2O_2 perception in to cellular responses has so far been characterized only superficially. Finally, they raise the question of how H_2O_2 is detected by cells. Such perception could conceivably involve direct interaction of H_2O_2 and other signaling molecule with cellular proteins, such as transcription factors, ion channels or enzymes. H_2O_2 is a strong toxic oxidant causing cell damage or even cell death. At the same time it serves conversely as a signaling molecule to activate a defense system for restoring the redox homeostasis in plant cells.The data available to date show that H_2O_2 is a key factor involved in responses to a number of abiotic stimuli. They indicate that endogenous H_2O_2 is a key factor in the tolerance of cells to the oxidative stress induced by a range of abiotic conditions, and that this probably involves the enhanced expression of genes encoding antioxidant enzymes. Of course, authors expect that numerous unanswered questions and important areas for further research will be open in future. The mechanisms by which H_2O_2 is generated are still largely unresolved and elucidation of how it is made by different plant cells in different situations is clearly a research priority.

References

Abogadallah, G.M. (2010). Antioxidative defense under salt stress. *Plant Signaling and Behaviour,* **5 (4)**: 369-374

An, Z., Jing, W., Liu, Y., and Zhang, W. (2008). Hydrogen peroxide generated by copper amine oxidase is involved in abscisic acid-induced stomatal closure in Viciafaba. *Journal of Experimental Botany,* **59 (4)**: 815–825

Apostolova, P., Yordanova, R., and Popova, L. (2008). Response of antioxidative defence system to low temperature stress in two wheat cultivars. *General and Applied Plant Physiology,* **34**: 281-294.

Bienert, G.P., Schjoerring, J.K., and Jahn. T.P. (2006). Membrane transport of hydrogen peroxide. *Biochimica Biophysica Acta,* **1758**: 994–1003

Cominelli, E., Galbiati, M., Vavasseur, A., Conti, L., Sala, T., Vuylsteke, M., Leonhardt, N., Dellaporta, S.L., and Tonelli, C. (2005). A guard-cellspecific MYB transcription factor regulates stomatal movements and plant drought tolerance. *Current Biology,* **15**: 1196–1200

Desikan, R., Reynolds, A., Hancock, J.T., and Neill, S.J. (1998). Harpin and hydrogen peroxide both initiate programmed cell death but have differential effects on gene expression in *Arabidopsis* suspension cultures. *Biochemical Journal,* **330**: 115-120

Dynowski, M., Schaff, G., Loque, D., Moran, O., and Ludewig, U. (2008). Plant plasma membrane water channels conduct the signalling molecule H_2O_2. *Biochemical Journal,* **414**: 53–61

Gechev, T.S., and Hille, J. (2005). Hydrogen peroxide as a signal controlling plant programmed cell death. *Journal of Cell Biology,* **168**: 17–20

Halliwell, B., Clement, M., and Long, L. (2000). Hydrogen peroxide in the human body. *FEBS Letters,* **486**: 10-13

Hancock, J.T., Desikan, R., and Neill, S.J. (2001). Role of reactive oxygen species in cell signalling pathways. *Biochemical Society Transactions,* **29**: 345–350

Jenkins, (2009). Signal transduction in responses to UV-B radiation. *Annual Reviews of Plant Biology*, **60**: 407-431

Jubany, M.T., Munne, B.S., and Alegre, L. (2010). Redox regulation of water stress responses in field-grown plants Role of hydrogen peroxide and ascorbate. *Plant Physiology and Biochemistry*, **48 (5)**: 351-358

Lee, S., Choi, H., Suh, S., Doo, I.S., Oh, K.Y., Choi, E.J., Taylor, A.T.S., Low, P.S., and Lee, Y. (1999). Oligogalacturonic acid and chitosan reduce stomatal aperature by inducing the evolution of reactive oxygen species from guard cells of tomato and *Commelina communis*. *Plant Physiology*, **121**: 147-152

Li, F., Liu, P., Wang, T., Bian, P., Wu, Y., Wu, L., and Yu, Z. (2010). The induction of by stander mutagenic effects *in vivo* by alpha-particle irradiation in whole *Arabidopsis thaliana* plants. *Radiation Research*, **174 (2)**: 228-237

Lopez-delgado, H., Dat, J.F., Foyer, C.H., and Scott, I.M. (1998). Induction of thermo tolerance in potato micro plants by acetylsalicylic acid and H_2O_2. *Journal of Experimental Botany*, **49**: 713-720

Mano, J. (2002). Early events in environmental stresses in plants. Induction mechanisms of oxidative stress (Ed.) Oxidative stress in plants. Taylor and Francis, London, pp. 217-246

Mazid, M., Ali, B., Hayat, S., and Ahmad, A. (2010). The effect of 4-chloroindole-3-acetic acid on some growth parameters in mung bean under cadmium stress. *Turkey journal of Biology*, **34**: 9-13

Mazid, M., Khan, T.A., and Mohammad, F. (2011a). Potential of NO and H_2O_2 as signaling molecules in tolerance to abiotic stress in plants. *Journal of Industrial Research and Technology*, **1(1)**: 56-68

Mazid, M., Khan, T.A., and Mohammad, F. (2011b). Role of nitric oxide in regulation of H_2O_2 mediating tolerance of plants to abiotic stress: A synergistic signaling approach. *Journal of Stress Physiology and Biochemistry*, **7(2)**: 34-74.

McAinsh, M.R., Clayton, H., Mansfield, T.A., and Hetherington, A.M. (1996). Changes in stomatal behavior and guard cell cytosolic free calcium in response to oxidative stress. *Plant Physiology*, **111**: 1031–1042

Muller, H. E. (1985). Detection of hydrogen peroxide produced by microorganism on ABTS-peroxidase medium. *ZentralblBakteriolMikrobiolHyg*, **259**: 151–158.

Neill, S.J., Desikan, R., Clarke, A., and Hancock, J.T. (2002). Nitric oxide is a novel component of abscisic acid signalling in stomatal guard cells. *Plant Physiology*, **128**: 13–16.

Neto, A.D.A, Prisco, J.T., Eneas-Filho, J., Jand-Venes, R., and Gomes-Filho, E. (2005). Hydrogen peroxide pre-treatment induces salt stress acclimation in maize plants. *Journal of Plant Physiology*, **162**: 1114-1122.

Neves, C., Sa, M.C., and Amancio, S. (1998). Histochemical detection of H_2O_2 by tissue printing as a precocious marker of rhizogenesis in grapevine. *Plant Physiologyand Biochemistry*, **36**: 817-824.

Panda, S.K, Chaudhury, I., and Khan, M.H. (2003). Heavy metal induced lipid peroxidation and affects antioxidant in wheat leaves. *Biologia Plantarum,* **46**: 289-294.

Phillips, J.C., Braun, R., Wang, W., Gumbart, J., Tajkhorshid, E., Villa, E., Chipot, C., Skeel, R.D., Kale, L., and Schulten, K. (2005). Scalable molecular dynamics with NAMD. *Journal of Computational Chemistry,* **26**: 1781–1802.

Rausch, T., Kirsch, M., Low, R., Lehr, A., Vierck, R., and Zhigang, A. (1996). Salt stress response of higher plants: the role of proton pumps and Na^+/H^+ antiporters. *Journal of Plant Physiology,* **148**: 425-433.

Wahid, A., Perveen, M., Geelani, S., and Basra, S.M.A. (2007). Pretreatment of seed with H_2O_2 improves salt tolerance of wheat seedlings by alleviation of oxidative damage and expression of stress proteins. *Journal of Plant Physiology,* **164 (30)**: 283-294.

Wang, L., Yang, L., Yang, F., Li, X., Song, Y., Wang, X., and Hu, X. (2010). Involvements of H_2O_2 and metallothionein in NO-mediated tomato tolerance to copper toxicity. *Journal of Plant Physiology,* **167 (15)**: 1298-1306.

Wang, T., Zhang, X., and Li, C. (2007). Growth abscisic acid content and carbon isotope composition in wheat cultivars grown under different soil moisture. *Biologia Plantarum,* **51**: 181-184.

Wi, S.J., Jang, S.J., and Park, K.Y. (2010). Inhibition of biphasic ethylene production enhances tolerance to abiotic stress by reducing the accumulation of reactive oxygen species in *Nicotiana tabacum*. *Molecules and Cells,* **30 (1)**: 37-49

Wolfe, S.A., Nekludova, L., and Pabo, C.O. (2000). DNA recognition by Cys (2)His(2) zinc finger proteins. *Annual Reviews of Biophysics and Biomolecular Structure,* **29**: 183–212

YehCh-M, Hung, W-Ch., and Huang, H-J. (2003). Copper treatment activates mitogen-activated protein kinase signalling in rice. *Physiologia Plantarum,* **119**: 392–399.

Utilisation and Management of Medicinal Plants Vol. 3 (2015) *Pages* **435–446**
Editors: **V.K. Gupta and A. Kaul**
Published by: **DAYA PUBLISHING HOUSE, NEW DELHI**

30

Physiological and Yield Variations in the Response of Sarsoon Cultivars to Saline Circumstances

Mohammad Mazid[1*] and Khalil Khan[2]

ABSTRACT

An experiment on mustard (Brassica juncea L. Czern and Coss.) cultivars Varuna, Kalamoti, Babafarm and Anmol was conducted in 'rabi' season of 2008-2009 in the net house of the Department of Botany, Aligarh Muslim University, U.P. The seeds were surface sterilized with 0.0l per cent $HgCl_2$ and washed several times with deionized water and sown in pots filled with 5 kg homogenous mixture of soil and farmyard manure in the ratio of 4:1. Plants grown under non-saline and saline conditions (100 mM NaCl). The plants were sampled at 40 and 80 days after sowing (DAS), for analysis parameters viz., plant dry mass, leaf area (LA), root and shoot length, leaf chlorophyll content, percent pheophytin and total carbohydrate content. NaCl treatment decreased shoot length; leaf area and plant dry matter. For root length and shoot length, maximum reductions were reported in cultivar, Anmol while minimum in cultivar, Varuna. Similarly, cultivar Anmol also reported maximum reduction in plant dry matter and leaf area and least by cultivar, Varuna.

Keywords: Brassica, Cultivars, Growth, Yield, Physiological, Sress, toxicity.

1 Department of Agricultural Sciences, MAJU, Rampur – 244 901, Uttar Pradesh, India.

2 Chandra Sekhar Azad University of Agriculture and Technology, Kanpur – 208 020, Uttar Pradesh, India.

* *Corresponding author*: E-mail: mazidmohd699@gmail.com

Introduction

One of the major environmental factors limiting the productivity of crops is salt stress which negatively affects the metabolism of plants and causes important modification in different biochemical and molecular processes. The negative effect of salinity on plant growth is a result of low osmotic potential of soil solution, changes in nutrient uptake and specific sodium (Na^+) and chloride ion (Cl^-) effects. These changes affect plant growth and development at different levels of plant organization (Munns, 2002). In saline soils, water availability and nutrient uptake by plant roots is limited because of high osmotic potential and toxicity of Na^+ and Cl^- ions (Al-Karaki, 1996; Mazid *et al.*, 2014). Thus excessive uptake of Na^+ and Cl^- may lead to ionic disturbance of whole plants and this increase in a Na^+ and Cl^- concentration is detrimental to plants. The negative effect of the various environmental stresses is at least partially due to the generation of ROS and the inhibition of the system which defends against them (Asada, 1997; Mazid *et al.*, 2012). ROS include super oxide (O_2), hydroxyl radicals (OH^-) and hydrogen peroxide (H_2O_2) can cause oxidative damage to many cellular components including membrane lipids, proteins, nucleic acids and chlorophyll (Halliwell *et al.*, 1989). A significant inhibition of photosynthesis by high salinity seems to be associated with the photosystem II complex. Salinity changes the photosynthetic carbon metabolism, leaf chlorophyll content as well as photosynthetic efficiency (Cengiz and David, 2002). The chlorophyll content of salt stressed plants depends on the biological processes and developmental stages of the plant and also on the type and concentration of the salts (Khan *et al.*, 2011a; Mazid *et al.*, 2013 a, b). The growth of root and shoot reduces abruptly in salt sensitive plants (Munns, 2002). Salinity also reduces the leaf area and chlorophyll content as the salinity level increases with no significant effect on leaf water content. Shoot and root length and fresh and dry weight of the seedling reduces in salinity stress which may be due to the metabolic disorders induced by stress and generation of ROS (Marcum and Murdoch, 1994). Osmotic regulators in plants include potassium (K), soluble sugar, and proline, are molecules are important physiological indicators for evaluating osmotic adjustment ability and drought resistance (Khan *et al.*, 2011b). Carbohydrates are accumulated in plant tissues under saline stress and these substances are suspected of contributing to osmotic adjustment. This may act as an osmotic adjustment or osmotic protectant factor (Sanchez *et al.*, 1998; Mazid *et al.*, 2011 a, b). The present work was planned with the objective of determining changes in various morphological and physiological characters of different mustard (*Brassica juncea* L.) cultivars exposed to stress caused by salinity.

Material and Methods

An experiment on four mustard cultivars namely, Varuna, Kalamoti, Babafarm and Anmol was conducted in 'rabi' season of 2008-2009 in the net house of the Department of Botany, Aligarh Muslim University. The details of the experimentation are given below.

Experimental Design and Materials Used

The experiment was carried out in 5 cm diameter thermocol cups containing soil composed of peat and compost (4:1) mixed with sand (4:1) in the rabi season under natural day-night condition with average day and night temperature of 21±3 and 12±2°C, respectively. Relative humidity was 56 ± 6 per cent and photosynthetically active radiation was 876±25 µmol m^{-2} s^{-1}. The experiment was conducted with 24 cups established in three rows with each row having 8 cups; 2 cups from each row represent for each cultivar. From these 2 cups of each row represent the control and the other one represent the NaC1 treatment. The seeds obtained from local market of Aligarh and were surface sterilized with 0.0l per cent $HgCl_2$ and washed several times with deionized water. Six healthy seeds of each cultivar sown in every cup. After germination three plants per cup were maintained. The NaCl treatment was given after germination. The treatments were given at interval of two days After 30 DAG (days of germination). Parameters studied are plant dry mass, leaf area (LA), root and shoot length, leaf chlorophyll content, percent pheophytin and total carbohydrate content. The plants were uprooted and washed under running tap water and dried in hot air oven at 80°C till constant weight was obtained. The samples were weighed to obtain dry mass with the help of electronic balance. LA was measured by using leaf area meter. Plants were uprooted, washed and dried on blotting paper. Length of root and shoot were measured using meter scale.

Total Chlorophyll content was estimated and calculated by the method of Arnon (1949). 1g finely cut leaves from each replicate was homogenized in a sufficient amount of 80 per cent acetone (Appendix) using a mortar and pestle. The extract was filtered through Whatman No. 42 filter paper and the filtrate was collected into a 100 ml volumetric flask. Finally, the volume was made up to 100 ml using 80 per cent acetone. 5ml extract from the 100 ml volumetric flask was transferred into a 50 ml volumetric flask and the volume was made up to the mark using 80 per cent acetone. From 50 ml volumetric flask, 5 ml chlorophyll extract was transferred into a cuvette and the optical density (OD) was read at 645 and 663 nano metre (nm) on a spectrophotometer (Spectronic 20 D, Milton Roy, USA). The total chlorophyll content was calculated using the following formula:

$$\text{Chlorophyll content} = \frac{20.2 \times (\text{OD}645) + 8.02(\text{OD}663) \times \text{V} \times \text{W}}{1000} \text{ mg g}^{-1}$$

(leaf fresh mass)

where,

V: Volume of the extract in ml

W: Weight of the fresh leaves used for the extraction of the pigment in g.

The method of Basalah *et al.* (1991) was followed for the determination of pheophytin as percent degradation of Chl to pheophytin. Fresh leaf tissues (1 g) were ground with sufficient amount of 80 per cent acetone and centrifuge at 10,000 rpm for 5 mm and the increase in the absorbance was recorded at 553 and 665 nm.

Determination of total carbohydrate by Anthrone Method. Carbohydrates are first hydrolysed into simple sugars using dilute hydrochloric acid. In hot acidic medium glucose is dehydrated to hydroxymethyl furfural. This compound forms with anthrone a green coloured product with an absorption maximum at 630 nm. Amount of carbohydrate present in 100 mg of the sample:

$$= \frac{\text{mg of glucose}}{\text{Volume of test sample}} \times 100$$

Results

The data on the effect of NaCI on growth parameters measured in terms of plant root length and shoot length, leaf area and plant dry mass (PDM). NaC1 treatment decreased the growth characteristics. All the cultivars showed decrease in plant root and shoot length, dry mass and leaf area. Among all the cultivars 'Anmol' showed the maximum reduction of 63.49 per cent and 55.6 per cent in root and shoot length respectively. The minimum reduction of 23.26 per cent and 20.96 per cent was reported in 'Varuna' cultivar. 'Kalamoti' and 'Babafarm' showed a reduction of 32.02 per cent and 33.06 per cent and 3 5.57 per cent and 40.49 per cent in root length and shoot length respectively compared to the control. Plant dry mass and leaf area were also significantly reduced at 200 mM NaC1 concentration. The maximum reduction of 50.78 per cent and 50.70 per cent in PDM and leaf area was observed respectively in 'Anmol' cultivar. The least reduction of 18.04 per cent and 19.35 per cent was reported in 'Varuna' cultivar. 'Kalamoti' and Babafarm cultivars showed reduction of 27.46 per cent and 29.2 1 per cent and 40.57 per cent and 36.36 per cent in PDM and leaf area (LA) respectively. Chlorophyll content was reported to decrease with salinity. The maximum reduction of 55.46 per cent was observed in 'Anmol' cultivar compared to the control. Varuna showed the minimum reduction of 27.61 per cent and Kalamoti and Babafarm revealed a reduction of 34.75 per cent and 44.09 per cent respectively compared to the control. Unlike chlorophyll content, pheophytin content was found to increase with increase in salinity. The maximum increase was reported in Anmol cultivar while minimum increase was found in Varuna. Kalamoti and Babafarm results were in between the two cultivars. Anmol, Varuna, Kalamoti and Babafarm showed an increase of 5832.43 per cent, 4073.08 per cent, 4730.5 1 per cent and 5302.99 per cent respectively compared to the control. Like pheophytin content, carbohydrate content was also found to increase with an increase in NaC1 concentration. However, maximum increase was observed in Varuna and least in Anmol cultivar. Kalamoti and Babafarm revealed intermediate increases. The increase in carbohydrate content was 9.13 per cent, 19.27 per cent, 32.45 per cent and 44.42 per cent in Anmol, Babafarm, Kalamoti and Varuna cultivars respectively.

Discussion

The sustainability of crop productivity is primarily a function of various environmental stress factors, such as high temperature, cold, drought, salinity or biotic factors and is also associated with the fertility status of soil. Soil fertility is adversely affected by salinity, which has emerged as one of the most serious factors limiting plant growth and productivity and soil health. It is expected to result in the loss of up to 50 per cent fertile land by the middle of the 21st century. Salinity is a worldwide problem causing serious constraints in arid and semi-arid regions of the world for growth and yield of crops (Hussain *et al.*, 2009). The salinity induced reduction in photosynthesis, growth and development of plants is associated with ionic/osmotic effects, nutritional imbalance or oxidative stress. Plants are exposed to various types of environmental stresses. Among these stresses, osmotic stress, in particular, due to drought and salinity is the most serious problem that limits plant growth and crop productivity in agriculture. A soil is said to be saline when it has enough of soluble salt to affect plant growth. Saline reduces the uptake of water by plants, restrict root growth, cause marginal burning of the foliage, inhibit flowering, limit seed germination and reduce fruit and vegetable yields.

The stress imposed by salt excess is an important restriction for the productive use of lands as it reduces plant growth and productivity at soil conductivity over 50 mM. High salt content, especially chloride and sodium sulphates, affects plant growth by modifying their morphological, anatomical and physiological traits. Such growth impairment is due to osmotic effects and ionic imbalances affecting plant metabolism. However, plants differ in their tolerance to salinity. Differences in salt tolerance exist not only among different genera and species, but also within the different organs of the same species.

Many studies showed that the performances of the chlorophyll pigment contents in the leaves of tolerant varieties were maintained better than in the sensitive varieties. There is increasing evidence that enhanced salinity changes photosynthetic parameters, osmotic and leaf-water potential, transpiration rate, leaf temperature and relative leaf water content (Khan *et al.*, 2014). Salinity alters a wide array of metabolic processes in growing plants and induces changes in contents and activities of many enzymes. It also affects photosynthetic components such as enzymes, chlorophylls and carotenoids. Changes in these parameters depend on the severity and duration of stress. Salinity increases carbohydrate contents especially in the salt tolerance plants. Salt stress causes a significant effect on the growth characteristics such as plant dry mass, leaf area, shoot length and root/shoot ratio. It is known to induce ionic stress, which causes premature abscission and senescence of adult leaves, thus reducing the available photosynthetic area. The reduction in fresh and dry weight of leaves may be due to suppressing cell enlargement and division and also to the inhibition of enzyme activities by salts, especially Na^+. Further, it was observed that the reduction in leaf-area can be caused either by reduced cell division or expansion, since both of them are equally sensitive to environmental stresses (Jimenz *et al.*, 1997)). Leaf fresh weight, leaf dry weight and leaf area were reported to decrease in wheat by about 32 per cent, 41 per cent and 33 per cent respectively in comparison with control plants.

However, it was observed that salinity causes reduction in leaf area that leads to a drastic reduction in net CO_2 assimilation and also an increase in thickness of the leaf. Decreasing photosynthetic activity under saline conditions was found to decrease the plant growth, leaf area and chlorophyll content in *Phaseolus vulgaris* L. Under salt stress and showed that plant height, number of leaves per plant, LA, fresh and dry weight of leaves and dry weight of stern had slightly increased than the control plants whereas decreased than other treatments under salinity 1000 ppm (El-Quesni, 2010). Salinity had a highly significant effect on growth attributes of plants. A significant reduction in the root length and shoot length under saline conditions in Chaksu plant. A significant reduction in root and shoot length has been observed in sugar beet and Amaranth but root length is effected more than shoot length. The reduction in root and shoot development is thought to the due to toxic effects of NaCI used as well as unbalanced nutrient uptake by the seedlings (Amirjani, 2011). Salinity alters a wide array of metabolic processes in growing plants and induces changes in contents and activities of enzymes. Photosynthetic pigments are also found to be affected adversely by salinity. Since, salinity has been regarded as one of the limiting factors in crop production, it must have negative effects on photosynthesis and the same was reported in. A reduction synthetic assimilation associated to a reduction of chlorophyll content (Figure 30.2).

The decrease in the chlorophyll content in salt sensitive plants on exposure to salinity may be either due to degradation of the chlorophyll pigments by pigment degrading enzymes like chlorophyllase, or by the interference of Na^+ and Cl^- ions with the biosynthetic precursors and pathways of these pigments. Decrease in chlorophyll content in plants grown from NaC1 treated seeds may be the consequence of the action of chiorophyllase, the enzyme that degrades chlorophyll which is activated by various abiotic stresses. A reduction in the chlorophyll content of strawberry plantlets treated with NaC1. This reduction in the chlorophyll content is believed to be due to interference of Na^+ and Cl^- ions with the enzymes associated with chlorophyll biosynthetic pathway or due to the disturbance in the integration of chlorophyll molecules into stable complexes (Amjad *et al.*, 2008). The application of nitrates has increased the Chi a and b and total chlorophyll content under salinity. During the course of salt stress, accumulation of compatible solutes such as amino acids, polyamines and carbohydrates is claimed to be an effective stress tolerance mechanism. Carbohydrate changes are of particular importance because of their direct relationship with physiological processes as photosynthesis, translocation and respiration.

Salinity affects positively or negatively the carbohydrate content of plants. Shonjani (2002) observed increase in glucose, fructose and maltose content in maize. Under stress a considerable decrease in the contents of total sugar, reducing sugars, non-reducing sugars was observed in two finger millet varieties. On the other hand, starch content showed reverse trend *i.e.* increased when salinity increases in both varieties. The effect of salt stress on soluble carbohydrates and proline content of sorghum (Bezerra, 1992; Figure 30.1). Soluble carbohydrates increased in all genotypes of Sorghum due to the addition of 100 mol m^{-3} NaC1. Further studies showed that sorghum grown in highly saline soil, showed 28 per cent increase in soluble sugar content in leaves (Chavan and Karadje, 1986). However, sugar content of tomato leaf sap was affected differently by sodium chloride stress. Sucrose increased and fructose decreased while glucose did not change as an effect of sodium chloride stress. The effect of salt stress on yield, yield components and carbohydrate content in four Hulliess Barley (*Hordeum vulgare* L.) cultivars and observed a decrease in starch content but increase in sucrose content. Carbohydrates such as sugars and starch accumulate under salt stress, playing a leading role in osmoprotection, osmotic adjustment, carbon storage and radical scavenging. A decrease in starch content and an increase in both reducing and non-reducing sugars have been reported in leaves of *Bruguiera pravflora* (Panda *et al.*, 2002; Mazid *et al.*, 2011 c).

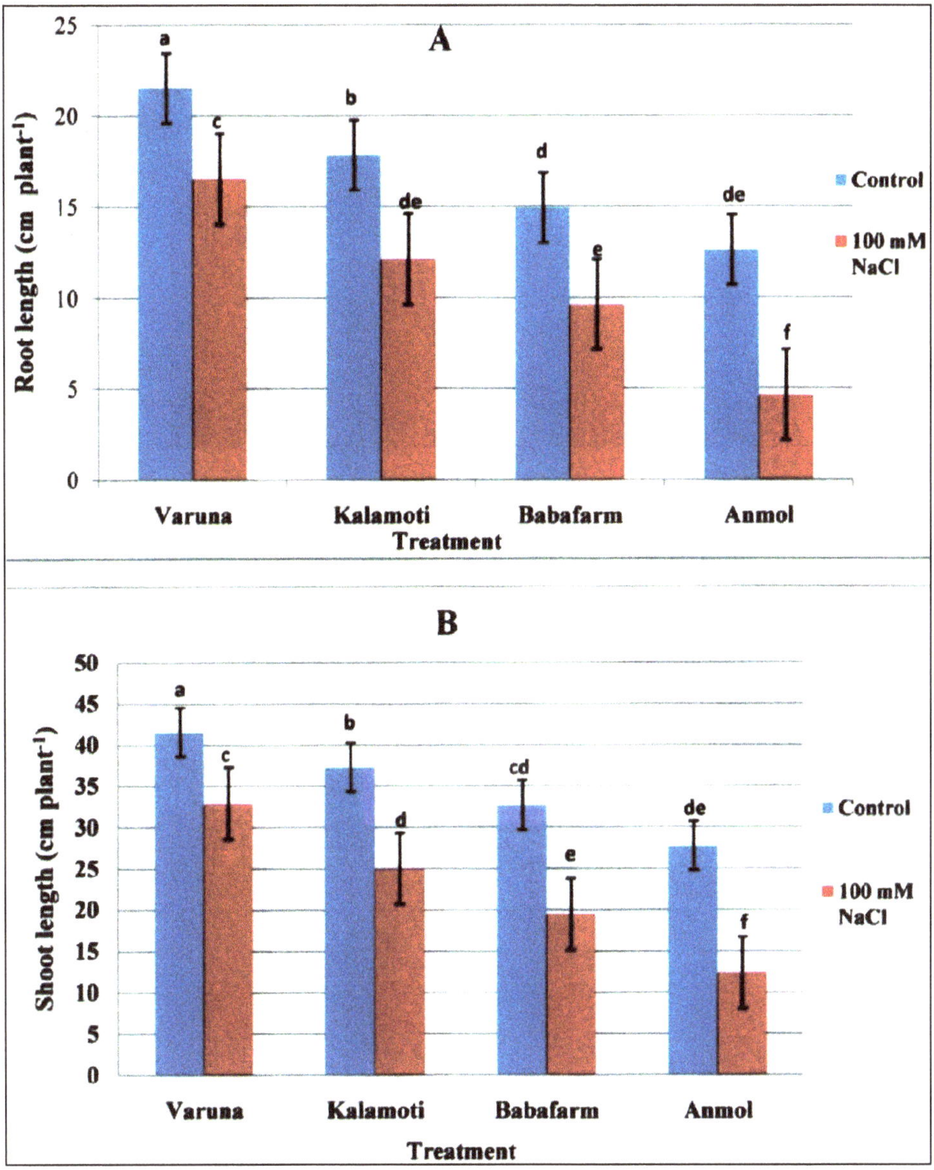

Figure 30.1: Root Length (A) and Shoot Length (B) of Mustard (*Brassica juncea* L.) Cultivars Grown with 100 mM NaCl at 30 Days after Sowing. Treatment mean ± standard error (SE) (n=3).

Santos (2004) observed that at low and moderate salinity levels, sugars and consequently the total carbohydrates are decreased. The accumulation of carbohydrates varied among the three wheat organs. In roots, the total carbohydrates remained mostly unchanged even at the highest salinity level. The similar effect was found in shoots while in spikes the soluble fixation increased significantly, the insoluble slightly reduced and the total carbohydrates decreased significantly

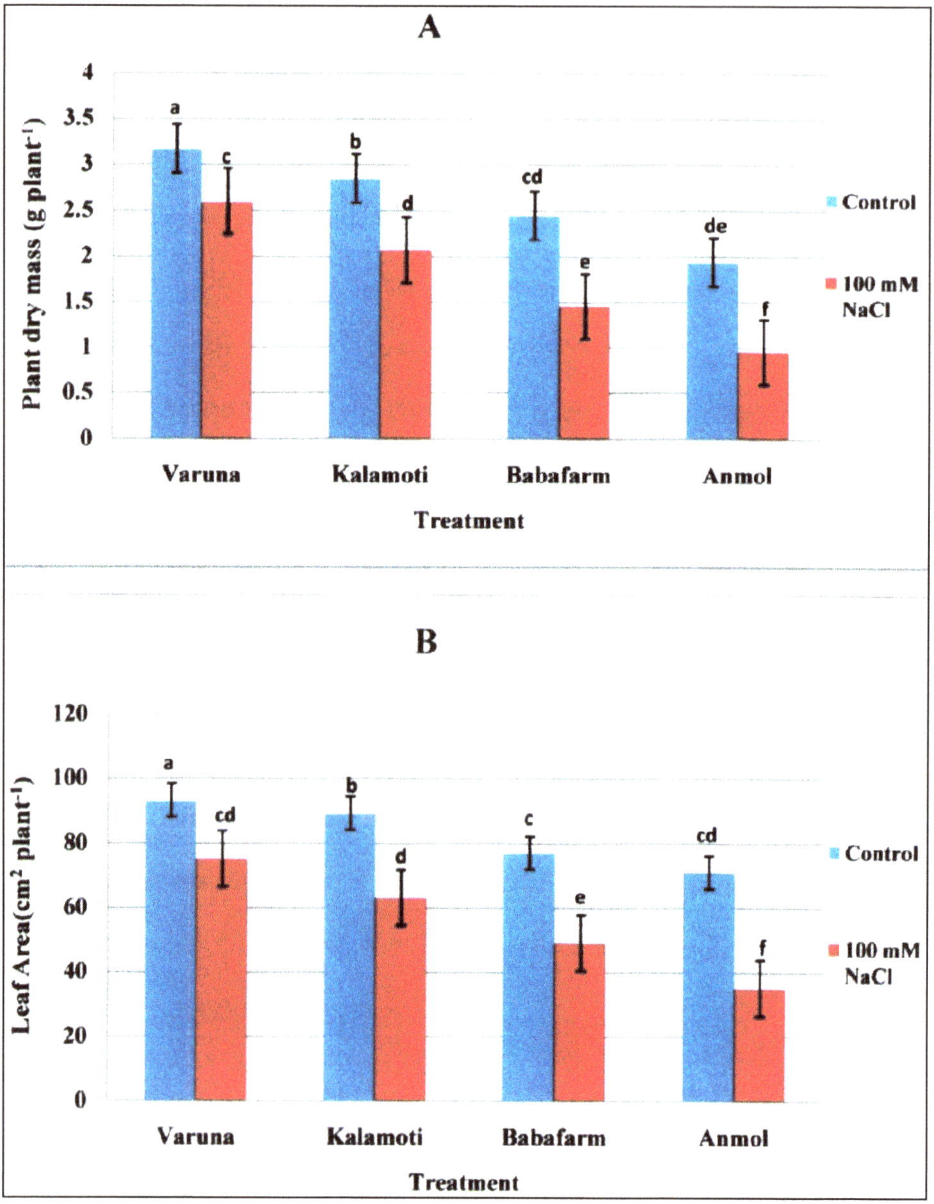

Figure 30.2: Plant Dry Mass (A) and Leaf Area (B) of Mustard (*Brassica juncea* L.) Cultivars Grown with 100 mM NaCl at 30 Days after Sowing. Treatment mean ± standard error (SE) (n=3).

(Tammam *et al.*, 2008). Plants respond by means of osmotic adjustment, normally by increasing the concentration of Na^+ and $C1^-$ in their tissue, although such accumulation of inorganic ions may produce important toxic effects and cell damage and inactivate both photosynthetic and respiratory electron transport. Leaves are the main organs for photosynthesis and transpiration and play an important role in the growth and development of higher plants especially under salt stress conditions (Munns, 2002). The decline in leaf growth under salt stress is due to an inhibition of H^+ pumping activity and increase in apoplastic pH^+ of leaves (Pittan *et al.*, 2009), or may be attributed to the changes in plant water relations due to stomatal closure, which also induced reduction in enzymatic activity, cell wall expansion as well as leaf expansion. Furthermore, high salinity is known to induce ionic stress (excessive accumulation of toxic ions including Na^+ and $C1^-$) which causes senescence and abscission of old leaves, thus reducing the available photosynthetic area (Munns, 2002; Mazid, 2014). The reduction in leaf area can be caused either by reduced cell division or expansion, since both of them are equally sensitive to environmental stress.

The reduction in the root and shoot dry weight leads to the overall reduction in plant growth under salinity stress. Salinity stress increases the content of carbohydrate in all the cultivars with maximum increase in tolerant cultivar. Similar results have been reported in many plant species exposed to salinity (Flowers, 2004; Mazid *et al.*, 2011d). According to Stoop and Pharr (1994) the increase in carbohydrate content induced by salinity appeared to be due to decreased demand of carbon. It is believed that under salinity stress accumulation of sugars along with other compatible solutes contribute to an osmotic adjustment and allows the plants to maximize sufficient storage reserves to support basal metabolism under stressed environment. The expression of Rubisco could be repressed by excess amount of sugars in cytoplasm (Sairam *et al.*, 1996; Figure 30.1). Therefore reduction of photosynthesis and metabolic alteration by sugar accumulation could contribute to salt sensitivity, which limits growth of the salt-sensitive cultivar under salt stress conditions. Starch accumulation perhaps resulted from the increased activity of alkaline invertase activity which hydrolyzes sucrose. Starch may be synthesized from such sugars. High starch accumulation was reported in mature leaves of the salt-tolerant cultivars of tomato (Balibrea *et al.*, 2000; Mazid *et al.*, 2011e). With increasing NaC1 concentration both root and shoot length. In this study, higher decrease occurred in sensitive than tolerant cultivar. Therefore, it may be summarized that salinity adversely affects plants growth however; greater reduction occurs in sensitive cultivars compared to tolerant cultivars. Further, accumulation of carbohydrate under salinity stress is an osmotic adjustment that 'helps plants to survive under stress and greater accumulation in Varuna and least in Anmol helps in categorizing this as a tolerant cultivar.

References

Al-Karaki, G.N. (1986). Response of three tomato cultivars to increasing salt stress. *Multan Journal of Research Studies*, **1**: 23-38.

Amirjani, M.R. (2010). Effect of salinity stress on growth, mineral composition, proline content, antioxidant enzymes of Soybean. *American Journal of Plant Physiology*, **5**: 350-360.

Amjad, M.H., and Malik, Z.A. (2008). Over expression of tobacco osmotin gene leads to salt stress tolerance in strawberry (*Fragaria x ananassa* Duch.) plants. *Indian Journal of Biotechnology*, **7**: 465-471.

Arnon, D.I. (1949). Copper enzymes in isolated chloroplasts. Polyphenol oxidase in *Beta vulgaris. Plant Physiology*, **24**: 1-15.

Asada (1997). The role of ascorbate peroxidase and mono dehydroascorbate reductase in H_2O_2 scavenging in plants. *In*: Ed. By scandalios, J.G. oxidative stress and the molecular biology of antioxidant defences, cold spring harbor laboratory press, New York, pp. 7 15-735.

Balibrea, M.E., Amico, D.J., Bolarin, M.C., and Perez, A.F. (2000). Carbon partitioning and sucrose metabolism in tomato plants growing under salinity. *Physiology of Plant*, **110**: 503-511.

Basalah, M.O. (1991). Action of salinity on seed germination and seedling growth of *Solanum melongena* L. *Engineerong and Technology*, **71**: 569-571.

Bezerra, N. E. (1992). *Salt tolerance in tomatoes.* Bangor, University College of North Wales. p. 161. Tese de Doutorado.

Cengiz, K., and David, H. (2002). Calcium nitrate as a remedy for salt-stressed cucumber plants. *Journal of Nutrition*, **25**: 861-871.

Chevan, P.D., and Karadge, B.A. (1986). Growth, mineral nutrition, organic constituents and rate of photosynthesis in *Sesbania grandflora* L. grown under saline conditions. *Plant and Soil*, **93**: 395-404.

El-Quesni, F.E.M., Sahar, M., Zaghioul and Hanan, Siam, S. (2010). Effect of microbien and compost on growth and chemical composition of *Schefflera arboricola* L. under salt stress. *Journal of American Science*, **6**: 234-237.

Flowers, T. J. (2004). Improving crop salt tolerance. *Journal of Experimental Botany*, **55**: 307-319.

Halliwell, B., and Gutteridge, J.M.C. (1989). Free Radicals in Biology and Medicine. Ed. 2 Clarendon Press, Oxford U.K

Hussain, K., Majeed, A., Nawaz, K., Bhatti, K.H., and Nisar, F.K. (2009). Effect of different levels of salinity on growth and ion contents of black seeds (*Nigella sativa* L.). *Current Research Journal of Biological Sciences*, **1**: 135-138.

Jimenez, M.S., Gonzalez-Rodriguez, A.M., Morales, D., Cid, M.C., Socorro, A.R., and Caballero, M. (1997). Evaluation of chlorophyll fluorescence as a tool for salt stress detection in roses. *Photosynthetica*, **33**: 291 – 301.

Khan, T. A., Mazid, M., and Mohammad, F. (2011a). Sulphur management: An agronomic and transgenic approach. *Journal of Industrial Research and Technology*, **1 (2)**: 147-161.

Khan, T. A., and Mazid, M. (2011b). Nutritional significance of sulphur in pulse cropping system. *Biology and Medicine*, **3 (2)**: 114-133.

Khan, T.A., Mazid, M., and Quddusi, S. (2014). Role of organic and inorganic chemicals in plant-stress mitigation. *In*. Approaches to Plant Stress and their Managemen, Ed. By Gaur, P.K. and Sharma, P. DOI-10.1007/978-81-322-1620-9-3. Springer, India, 2014. pp, 39-52.

Marcum, K.B., and Murdoch, C.L. 1994. Salinity tolerance mechanisms of six C4 turfgrasses. *J. Amer. Soc. Hort. Sci.*, 1**19**: 779-784.

Mazid, M., Khan, T. A., and Mohammad, F. (2011a). Role of secondary metabolites in defense mechanisms of plants. *Biology and Medicine,* **3(2)**: 232-249.

Mazid, M., Khan, T. A., Zeba, H. K., Quddusi, S., and Mohammad, F. (2011b). Occurrence, biosynthesis and potentialities of ascorbic acid in plants. *International Journal of Plant, Animal and Environmental Sciences,* **1(2)**: 167-184.

Mazid, M., Zeba, H. K., Quddusi, S., Khan, T. A., and Mohammad, F. (2011c). Significance of sulphur nutrition against metal induced oxidative stress in plants. *Journal of Stress Physiology and Biochemistry*, **7(3)**: 165-184.

Mazid, M., Khan, T. A., and Mohammad, F. (2011d). Response of crop plants under sulphur stress tolerance: A holistic approach. *Journal of Stress Physiology and Biochemistry,* **7(3)**: 23-57.

Mazid, M., Khan, T. A., and Mohammad, F. (2011e). Effect of abiotic stress on synthesis of secondary plant products: a critical review. *Agricultural Reviews,* **32(3)**: 172-182.

Mazid, M., Khan, T.A., and Mohammad, F. (2012). Role of nitrate reductase in nitrogen fixation under photosynthetic regulation. *World Journal of Pharmaceutical Research,* **1(3)**: 386-414.

Mazid, M., Jamil, B., Khan, T.A., and Mohammad, F. (2013a). Glucosinolate engineering: Revisiting an old problem with solution of enhanced anti-herbivore defense. Paper presented in Internatioal conference on Bio engineering, held at Rajalakshmi Engineering College, Rajalakshmi Nagar, Thandalam, Chennai-602105, 2-4 January, India, (ICBE 2013); ISBN-978-1-62590-606-9.

Mazid, M., Khan, T.A., Mohammad, F., and Nooris, N. (2013b). Physico-biochemical effects of pre-sowing seed treatment with gibberellic acid on *Helianthus annus* L.- Paper presented at National conference of Plant Physiology (NCPP-2013) "Current Trends in Plant Biology Research": 895-896. Junagarh, Gujarat

Mazid, M. (2014). *Response of cicer arietinum L. to plant growth regulators and mineral nutrients.* Ph.D. thesis, Aligarh Muslim University, Aligarh, India.

Mazid, M., Khan, T.A., and Qudassi, S. (2014). Effects of climate change on crops: Food security and Biotechnology: *In*: Crop Improvement in the Era of Climate Change, Ed. By Rajib Roychowdhury, I.K. International Publication Pvt. Ltd, India, pp. 117-199.

Munns, R. (2002). Comparative physiology of salt and water stress. *Plant Cell Environment*, **25**: 239 – 250.

Pittan, B., Schubert, S., and Muhling, K.H. (2009). Decline in leaf growth under salt stress is due to an inhibition of H pumping activity and increase in apoplastic pH of maize (*Zea mays* L.) leaves. *Journal of Plant Nutrition and Soil Science,* **172**: 535-543.

Roben, M., and Dorries, W. (1932). Einige Versuche fiber den EinfluB von schwefliger Saure auf *Elodea canadensis.- Ber. Deutsch. Botaniaca Genetics*, **50**: 53-58.

Sairam, R.K., Srivastava, G.C., Agarwal, S., and Meena, R.C. (2005). Differences in antioxidant activity in response to salinity stress in tolerant and susceptible wheat genotypes. *Biologia Plantarum*, **49**: 85-9 1.

Sanchez, F.J., Manzanares, M., Andres, E.F., Tenorio, J.L., Ayerbe, L., and Andres, E.F. (1998). Turgor maintenance, osmotic adjustment and soluble sugar and proline accumulation in 49 pea cultivars in response to water stress. *Field Crops Research*, **59**: 225-235.

Santos, C.V. (2004). Regulation of chlorophyll biosynthesis and degradation by salt stress in sunflower leaves. *Scientia Horticulture*, **103**: 93-99.

Shonjani, S. (2002). Salt sensitivity of rice, maize, sugar beet and cotton during germination and early vegetative growth. Inaugural dissertation.

Stoop, J.M.H., and Pharr, D.M. (1994). Mannitol metabolism in celery stressed by excess macronutrients. *Plant Physiology*, **106**: 503 – 511.

Tammam, A.A., Abou Ahmad, M.F., and Hemeda, M.M. (2008). Study of salt tolerance in wheat (*Triticum aestivum* L.) cultivar Banysoif. *Australian Journal Crop Science*, **1**: 115-125.

Utilisation and Management of Medicinal Plants Vol. 3 (2015) *Pages* **447–458**
Editors: **V.K. Gupta and A. Kaul**
Published by: **DAYA PUBLISHING HOUSE, NEW DELHI**

31

Effect of Inorganic Fertilizers on Growth, Physiology, Yield and Quality Parameters of Garlic (*Allium sativum* L.)

Mohammad Mazid[1*] and Saima Quddusi[2]

ABSTRACT

A pot experiment was conducted on garlic in 'rabi' (winter) season of 2009-2010 in the net house of the Department of Botany, Aligarh Muslim University, Aligarh, U.P. India with an aim to study the effect of potassium (K) application on performance of the garlic. Forty earthen pots of equal size are filled, each with five kg mixture of soil and organic manure. Spices possess a variety of phytochemicals that gaining wide attention for their well-recognized potential health benefits against advanced physiological threats. Among the spices, garlic holds a unique position for its well recognized therapeutic potential. The remedic role of garlic recognized particularly with respect to aged-garlic extract (AGE) has been the subjects of much interest as well as emergent focus for a large number of research papers so. The use of traditional remedies may encounter more frequently due to a broad myriad of scientific evidence in their remedic favour. Five levels of K, viz. 0, 15, 25, 35 and 45 mg K per kg soil designated as $K_{0,}$ K_{15} $K_{25,}$ $K_{35,}$ and K_{45} were applied. The plants were sampled at 30 and 60 days after sowing (DAS), for analysis of shoot length per plant, leaf number per plant, leaf area per plant, fresh weight plant per, dry weight per plant, carbonic anhydrase activity (CA) in leaves, nitrogen (N) P (phosphorus) and K content in leaves, bulb diameter, weight per bulb, cloves per bulb, 100-cloves weight, bulb yield and harvest index (HI) at harvest. All parameters were found to be significantly enhanced by the K effect, with maximum stimulation being noted following a K_{35} treatment.

Keywords: Disease, Garlic, *Allium sativum* L., Nitrogen, Phosphorus, Productivity, Nitrate reductase, Potassium, Yield.

1 Department of Agricultural Sciences, MAJU, Rampur – 244 901, Uttar Pradesh, India.
2. Amity Institute of Biotechnology, Amity University, Lucknow, Uttar Pradesh, India.
* *Corresponding author*: E-mail: mazidmohd699@gmail.com

Introduction

Garlic (*Allium sativum* L.) is a commercially grown important spice crop of India and is known by other common names like ajo, cropleek, cultivated garlic, lasan, poor man's treacle and rosina. It is a member of the family Liliaceae and widely cultivated in the Latin countries bordering on the Mediterranean. It has been used as a medicine and a food since the time of the Egyptian Pharaoks and the earliest Chinese dynasties. To be effective, garlic preparations must smell of allicin. This oil is rich in sulphur (S), but contains no oxygen. India ranks second in the production with a productivity of 4.5 tonnes per ha (Pandey, 2000), and an important foreign exchange earner for India. Present production of garlic is not adequate to meet the increasing demands of garlic within the country and also for export. Garlic has a very long folk history of use in a wide range of ailments, particularly ailments such as ringworm, candida and vaginitis where its fungicidal, antiseptic, tonic and parasitic have proved of benefit.

It is said to have anticancer activity and has also been shown that garlic aids detoxification of chronic lead poisoning. Daily use of garlic in the diet has been shown to have beneficial effect on the body, especially the proper blood vascular system and the heart. It reduces glucose metabolism in diabetics, slows the development of arteriosclerosis and lowers the risk of repeated heart attacks in myocardial infarct patients. Its juice is an excellent antiseptic for treating wounds and fresh bulb is much more effective medicinally than stored bulbs. The bulb is said to be anti-asthmatic, anti-cholesterolemic, antiseptic, antispasmodic, diaphoretic, expectorant, febrifuge, repellent, stimulant, stings, stomachic, and vasodilator also. It is used in flavoring a wide range of foods (Mazid *et al.*, 2011 a). It is wonderfully nutritious and health giving addition to the diet, but it has a very strong flavor so used in very small quantities as a flavoring in salads and cooked foods. Among the several factors for low productivity and poor economic yield of garlic in India, nutrients such as nitrogen, phosphorus and potassium are considered to be the most limiting factors (Verma *et al.*, 1996).

One of the earliest experiments on nutrition designed to elucidate the source of increase in the weight of a growing plant was suggested by Nicolas de Cusa in 1450. However, the seventeenth century is thought to be the beginning of extensive researches on plant nutrition. Plant nutrition is probably one of the earliest of the many aspects of plant physiology and biochemistry. The modem concept of plant nutrition dates back in the beginning of the 19th century and the credit of which goes to de Saussure (1804), mentioned that mineral elements, including N supplied by the soil were essential for growth and development of plants (Bould, 1963). The beginning of the 20th century marked the intensification of research on soil solution and ion exchange in soils (Khan *et al.*, 2014). Moreover, nutrients are necessary for maintenance of the physical organization and activities of living cells by virtue of their function in the generation of energy building and repair of protoplasm of metabolic processes (Nason and McElery, 1963). Three sources of K, *viz.*, muriate of potash, nitrate of potash and sulphate of potash are used to apply as to crops (Jones, 1987). When K is added to the soil, a part of it goes from the soluble to the solid phase. Part of the retained ions became exchangeable and others remained "fixed". It has been observed that part of the fixed K passes to the vegetables as residual K contributing to the future plant nutrition. K is an important macronutrient in plants although it does not form structural constituent in the cell (Mazid, 2014). However, it is essential for respiration and photosynthesis. K is an activator of many enzymes needed to form starch and proteins. Information regarding the effect of K on garlic is lacking, this work was carried out to produce informations for enhanced garlic productivity under influence of K fertilisation.

Materials and Methods

A pot experiment was conducted on garlic in 'rabi' season of 2009-2010 in the net house of the Department of Botany, Aligarh Muslim University, Aligarh, U. P. India with an objective to study the influence of K application on growth and yield performance of the garlic. Forty earthen pots of equal size (25 cm height and 25 cm diameter) are filled in each with five kg mixture of soil with organics (4:1). The soil of pot had a texture; sandy loam, pH 7.8, conductivity – 0.57 dSm^{-1}' and available N – 0.14 g N, P – 0.016 g P and K– 0.23 g K kg^{-1} soil. Five levels of K as 0, 15, 25, 35 and 45 mg K kg^{-1} soil designated as $K_{0,}$ $K_{15,}$ $K_{25,}$ K_{35} and K_{45} were applied along with a recommended form of N 125 kg N and 26.3 kg P/ha. N, P and K were applied as urea, sodium dihydrogen ortho-phosphate (Na_2HPO_4) and muriate of potash respectively. N was applied as two equal splits, *i.e.* of at the time of sowing and remaining half at 30 DAS.

Three garlic cloves of equal size were sown in each pot at suitable distance on 29th October 2009 after the application of K-fertilizers to the pots. There were 6 pots for each treatment. Of 6 pots, three were used to the growth characteristics and the rest three for yield characteristics. Thus, each treatment possesses three replicates. The pots were irrigated as and when required. The performance of growth was assessed at two stages of growth *i.e.*, 30 and 60 DAS. Yield characteristics were studied on full maturity. The growth and yield characteristics were studied are shoot length per plant, leaf number per plan^t, leaf area per plant, fresh weight per plant, dry weight per plant, CA in leaves, NPK content in leaves, bulb diameter, weight per bulb, cloves per bulb, 100-cloves weight, bulb yield and HI. The shoot length was pointed out to assess meristematic activity, leaf number as a measure of differentiation and fresh and dry weights for total productivity as rates of increase of weight, volume and dry matter accumulation. CA activity was determined in fresh leaves collected randomly from each replicate. The enzyme CA catalyzes the reversible hydration of CO_2 to give the bicarbonate ion (HCO_3^-)

$$H_2O + CO_2 \xrightleftharpoons{CA} H^+ + HCO_3^-$$

The activity of the enzyme was estimated by adopting the method of Dwivedi and Randhawa (1974). The collected leaves were cut into the small pieces ($1cm^2$) at a temperature below 20 celsius (C). After mixing them, 200 mg leaf pieces were weighed and cut further into smaller pieces keeping them in 100 ml 0.2 M aqueous cysteine hydrochloride solution in a Petridish at 0 to 4°C for 20 minutes (min). The solution adhering on their surface was thin removed with the help of a blotting paper. This was followed by transfer immediately into a test tube having 4 ml phosphate buffer of pH 6.8. To this, 4 ml 0.2 M sodium bicarbonate in 0.2 M sodium hydroxide (NaOH) solution and 0.2 ml 0.002 per cent bromothymol blue indicator were added. After shaking, the tubes were kept at 0-4°C for 20 min. CO_2 liberated during catalytic action of the enzyme on sodium bicarbonate was estimated by titrating the reaction mixture against 0.05 N hydrochloric acid (HCl) using methyl red as an indicator. A control reaction mixture was also titrated against 0.05 N HCl. The difference of the sample reading and blank reading was noted for further calculations of enzyme activity. The activity of the enzyme was calculated by the following formula:

$$\frac{0.5 \times V \times N}{W \times T} \text{ m mol } (CO_2) \text{ mg}^{-1} \text{ (leaf fresh mass) min}^{-1}.$$

where,

V: Difference in volume (ml) of HCl used in blank and test sample titration

N: Normality of HCl

W: Fresh weight of tissue in mg

T: Duration of the catalytic action of the enzyme (min). Finally, the activity of the enzyme was expressed in terms of mol CO_2 kg^{-1} (leaf fresh mass) s^{-1}.

NPK Content

N, P and K were estimated in dried powder of leaves obtained from each replicate. The sampled plant leaves were dried in an oven at 80°C for 24 h. The dried leaves from each sample were finally powdered and then passed through a 72-mesh screen. For the estimation of these nutrients the leaf powder was first digested according to the standard technique described below.

Digestion

100 mg oven dried powder of leaf material was transferred into a digestion tube to which 2 ml sulphuric acid (H_2SO_4) was added. The tube was then kept on a digestion assembly at 80C for about 2 h to allow the complete reduction of NO_3^- present in the plant material by the organic matter itself. Initially, dense white fumes were given off and then the content of the tube turned black. After cooling the tube for about 15 min, 0.5 ml 30 per cent hydrogen peroxide (H_2O_2) was added drop by drop and the tube was heated again till the colour of the solution changed from black to light yellow. The digestion tube cooled for 10 min. and an additional amount (2-3 drops) of 30 per cent H_2O_2 was added followed by gentle heating for about 15 min to get a clear and colourless solution. At this stage, care was taken in the addition of H_2O_2 because its excess might oxidize ammonia in the absence of organic matter. The H_2O_2 digested material was diluted with DDW and transferred with three washings into a 100 ml volumetric flask and finally the volume was made up to the mark with DDW. The details of methods employed for the estimation of N, P and K are given below.

Nitrogen

N was estimated according to the method of Lindner (1944). A 10 ml H_2SO_4-H_2O_2 digested material was taken into a 50 ml volumetric flask and the excess of the acid was neutralized by the addition of 2 ml 2.5 N NaOH. 1 ml 100 per cent sodium silicate was added to prevent turbidity and finally, the volume was made up with DDW. Into a 10 ml graduated test tube, 5ml this solution was taken and 0.5 ml Nessler's reagent was added. The content of the test tube were allowed to stand for 5 min for maximum colour developed. The solution was transferred into a calorimetric tube and OD was read at 525 nm, using a blank on the spectrophotometer. N content was determined with the help of the standard curve and was expressed in terms of percentage on dry weight basis.

Standard Curve

50 mg ammonium sulphate was dissolved 100 ml DDW and the final volume was made 1l with DDW. From this stock solution, 0.1, 0.2, 0.3, 0.4, 0.5, 0.6, 0.7, 0.8, 0.9 and 1.0 ml aliquots were pipetted into ten different test tubes. The solution in each test tube was diluted to 5 ml with DDW. In each test tube, 0.5 ml Nessler's reagent was added. After 5 min, the solution was transferred into a calorimetric tube and OD was read on the spectrophotometer at 525 nm. A blank was run with each determination. A curve was plotted for various concentrations of ammonium sulphate solution versus OD.

Phosphorus

Leaf P content in the H_2SO_4-H_2O_2 digested material was estimated by the method of Fiske and Subbarow (1925). 5 ml H_2SO_4-H_2O_2 digested material was taken into a 10 ml graduated test tube and 1 ml molybdic acid was carefully added followed by the addition of 0.4 ml 1-amino-2-naphthol-4-sulphonic acid. The colour of the solution turned blue. The final volume in the tubes was made up to 10 ml with DDW. After mixing throughly, the contents of the tube were allowed to stand for 5 min. They were then transferred into a calorimetric tube and OD was read at 620 nm on the spectrophotometer. A blank was run simultaneously for each determination. P content was computed with the help of the standard curve and was expressed in terms of percentage on dry weight basis.

Standard Curve

0.351 g potassium dihydrogen orthophosphate was dissolved in 100 ml DDW followed by the addition of 10 ml 10 N H_2SO_4. The final volume was made up to 1 l with DDW. From this stock solution, 0.1, 0.2, 0.3, 0.4, 0.5, 0.6, 0.7, 0.8, 0.9 and 1.0 ml aliquots were taken in separate test tubes. 1 ml molybdic acid and 0.4 ml 1-amino-2-naphthol-4-sulphonic acid were added in each test tube. The final volume in each test tube was made up to 10 ml with DDW. After 5 min, OD of the developed colour was read at 620 nm on the spectrophotometer. A blank was run with each set of determination. A curve was plotted for various concentrations of potassium dihydrogen orthophosphate solution versus OD.

Potassium

It was estimated flame photometrically. 10 ml H_2SO_4-H_2O_2 digested material was taken into a vial and was run into a flame-photometer (Model,C 150, AIMIL, India) using the filter for K. A blank was run side by side. K content was computed with the help of a standard curve and was expressed in terms of percentage on dry weight basis.

Standard Curve

1.91 g potassium chloride (KCl) was dissolved in 50 ml DDW followed by dilution to 100 ml. 1 ml of this solution was diluted to 1 l. The resultant stock solution would contain 10 ppm K, from which 1, 2, 3, 4, 5, 6, 7, 8, 9 and 10 ml aliquots were transferred into 10 different vials. The solution in each vial was diluted to 10 ml. The diluted solution of each vial was run separately. A blank was also run with each set of determination. Standard curve was prepared for different dilutions of KCl solution versus readings on the scale of the galvanometer. Yield characteristics, *viz.*, bulb diameter, bulb weight, clove number of a bulb 100-clove weight were studied to assess their attribution towards bulb yield. HI was calculated to assess the partitioning of photosynthesis to the bulb.

Determination of various Characteristics

One plant from each pot was taken and length of the shoot measured in cm. Leaf number was counted in the sampled plant. Area of a randomly selected leaf from each sampled plant was recorded graphically. It was determined by keeping individual leaf on graph paper and marking an outline. Total number of squares (cm^2) was counted. Fresh weight of the sampled plant was determined with the help of an electrical balance. After drying it at 80°C for 24 hour in an oven, dry weight of the sampled plant was determined with the help of an electrical balance.

Results

Effect of graded levels of K was significant on all parameters studied at various stages of growth, except leaf number per plant and leaf P content at 40 DAS. Increasing levels of K upto K_{35} increased shoot length per plant linearly. Treatment K_{35} increased this parameter by 11.15 per cent at 40 and 46 per cent at 60 DAS over the control. Potassium did not affect Leaf number per plant at 30 DAS. At 60 DAS, K_{35} gave maximum value, however its effect was at par with that of K_4, K_2 and K_1. Treatment K_{35} gave 33.9 per cent higher leaf number than the control (K_0). Increasing levels of K upto K_{35} increased leaf area per plant linearly at both stages. However, at 30 DAS, its effect was at par with that of K_2. K_3 increased leaf area by 27.73 and 100.7 over the K_0 and 30 and 60 DAS respectively. Increasing levels of K up to K_{35} improved fresh matter production per plant linearly. K_{35} gave 47.38 and 185.76 per cent more fresh matter at 30 and 60 DAS respectively than K_0. At both stages, application of K_{35} gave maximum value of dry weight per plant; however its effect was at par with that of K_{25}. K at K_{35} gave 46.60 and 145.5 per cent more fresh matter than K_0 at 30 and 60 DAS respectively.

A progressive increase in CA was recorded with increasing levels of K at both stages. The values recorded for all the treatments critically differed with one another at both stages. Treatment K_{45} increased CA by 259.6 and 368.9 per cent at 30 and 60 DAS respectively over K_0. Treatment K_{35} gave maximum N content. However, its effect was at par with that of K_{45}. Treatment K_{35} gave 23.3 and 20.3 per cent higher value at 30 and 60 DAS respectively than the control. K did not affect P content at 30 DAS. At 60 DAS, K_{35} gave maximum P content, however its effect was at par with that of K_{45}.

Table 31.1: Effect of Graded Levels of Potassium on Growth Characteristics, Carbonic Anhydrase (CA) and Leaf NPK Content of *Allium sativum* L. at Two Stages of Growth

Treatment (mg K kg^{-1} soil)	*DAS*	*Shoot Length per Plant (cm)*	*Leaf Number per Plant*	*Area per Leaf (cm^2)*	*Fresh Weight per Plant (g)*	*Dry Weight per Plant (g)*	*CA (n mol CO_2 per mg 20 per min.)*	*N (per cent)*	*P (per cent)*	*K (per cent)*
0	40	8.34	6.0	5.16	4.29	0.65	1.45	3.46	0.24	2.74
	70	37.13	6.1	5.56	4.78	0.76	1.47	3.13	0.22	2.27
15	40	8.83	7.0	5.72	4.38	0.60	2.93	3.68	0.25	3.05
	70	45.86	7.0	5.77	8.63	1.16	4.45	3.27	0.23	2.49
25	40	9.07	7.2	6.88	4.62	0.74	3.34	3.95	0.28	3.34
	70	52.70	7.0	10.04	12.13	1.62	5.53	3.67	0.24	2.78
35	40	9.27	7.1	6.58	6.32	0.87	3.65	4.65	0.26	3.90
	70	54.26	8.2	11.33	13.63	1.76	5.64	3.65	0.28	3.56
45	40	8.55	7.3	5.19	4.66	0.64	5.13	4.65	0.27	3.65
	70	48.30	7.2	8.93	10.55	1.35	6.64	4.23	0.26	3.65
C.D. at 5 per cent	40	0.608	NS	0.73	0.976	0.23	0.142	4.67	0.25	3.64
	70	1.45	1.70	0.42	1.25	0.79	0.267	0.53	0.27	3.26

N.S.: Non-significant.

Table 31.2: Effect of Graded Levels of Potassium on Yield Characteristics of *Allium sativum* L. at Harvest

Treatment (mg K kg^{-1} soil)	*Bulb Diameter (cm)*	*Weight $Bulb^{-1}$ (cm)*	*Clove Number $Bulb^{-1}$*	*100 Clove Weight (g)*	*Bulb Yield Pot^{-1} (g)*	*Harvest*
0	5.7	6.8	12.3	35.34	22.78	90.34
15	7.3	10.8	26.6	39.53	36.27	92.87
25	8.6	14.0	30.6	43.63	56.20	96.37
35	12.0	14.2	38.6	69.37	68.37	94.87
45	9.7	13.7	32.0	54.76	78.35	95.38
C.D. at 5 per cent	1.42	0.36	3.55	1.67	1.86	1.68

Treatment K_{35} gave 18.0 per cent higher P content than K_0. Increasing levels of K upto K_{35} increased K content linearly. However at 30 DAS, its effect was at par with that of K_{45}. Treatment K_{35} gave 37.8 and 36.3 per cent higher value at 30 and 60 DAS than K_0. Increasing levels of K upto K_{35} increased bulb diameter linearly, with K_{30} giving 109 per cent higher value than K_0.

Like bulb diameter, increasing levels of K upto K_{35} increased Weight per bulb linearly. K_{35} gave 118.4 per cent higher value than the control. Like above yield parameters a progressive increase in clove number per bulb with increase in K level upto K_{35} was noted. K_{35} enhanced clove number per bulb by 216.6 per cent than K_0. 100 clove weight also increased with increasing levels of K upto K_{35}. K_{35} gave 105.5 per cent higher value than K_0. Like above yield parameters, bulb yield per pot increased with increasing levels of K upto K_{35} which gave 206.2 per cent higher value than K_0. Increasing levels of K upto K_{35} increased HI linearly. However its effect was at par with that of K_{25} and K_{45}. K_{35} gave 7.29 per cent higher value than K_0.

Discussion

According to the WHO, non-communicable chronic disorders, such as heart diseases, stroke, cancer, chronic respiratory diseases, diabetes, are by far the principle leading cause of mortality in the world reporting 60 per cent of all death. In 2004, the 192 member states of the WHO endorsed the global strategy on diet, physical activity and health addressing two main risk factors. Diet is a central pivotal link with human health and food rich in phytochemicals often associated with proper functionality of immune system (Mazid *et al.*, 2011 c, d and e). Food levels are widely used to convey the health benefits of conventional foods or a component of food to the consumer. The increasing amount of health information available regarding the health benefits of food has resulted in consumer interest in the health issues and has become a leading factor in purchasing decisions. The effect of four levels of N (0, 50, and 150 kg/ha), three levels of P, *viz.* 40 and 80 kg P_2O_5/ha (0, 17.5 and 34.9 kg P)/ha and two levels of K, *i.e.* 0 and 50 kg K_2O (0 and 41.5 kg K)/ha on growth, yield and quality of garlic (Verma *et al.*, 1996; Mazid *et al.*, 2011 b)). They concluded also that garlic responded to the application of N, P and K up to 100 K N, 40 kg P_2O_5 (17.5 kg P) and 50kg K_20/ ha (41.5 kg K) with regard to yield of bulbs. P and S influence differentiation in plants (Marschner, 2002; Mazid *et al.*, 2012). The improvement in growth, physiological and biochemical parameters resulted from the spray of these nutrients together with enhancement in differentiation may lead to the improvement in yield parameters. The effects of planting time and different levels of K on the growth and yield of The experiment consisted of two planting dates and four levels of K, *viz.*, 0, 50, 100 and 200 kg 1 (Talukder *et al.*, 2002; Mazid *et al.*, 2013a). It was revealed that early planting favoured better growth and yield of and highest level of K yielded the best. The effect of four levels of S (0, 20, 40 and 60 kg S/ha) and K at 0, 80 and 120 kg K_20 (0, 33.2, 66.4 and 99.6 kg K)/ha on the growth characters, yield attributes and quality of garlic in sandy loam soil. The application of 60 kg S/ha significantly increased the diameter of stem and bulb, number of cloves per bulb, yield per plant, and percent dry weight of bulb over the control (Khan *et al.*, 2011 a and b). The application of 120 kg K 20/ha significantly increased the diameter of stem and bulb only. Yield per plant, percent dry weight of bulb and bulb yield also increased significantly with 120 K_{20}/ha (99.6 kg K/ha) (Nagaich *et al.*, 2003).

An improvement in shoot length, leaf number and leaf area in comparison with K_0 was observed with the application of K (Nagaich *et al.*, 2003; Tiwari and Agarwal, 2003). The reason for the favourable effect of K application on these parameters is not far to seek. Although K is not constituent of metabolites but it plays important roles in metabolism of plants. For example, K acts as an activator of many enzymes, including fructokinase, pyruvic acid kinase and transacetylase (Mazid *et al.*, 2013 b). It is required in adenine synthesis, glycolysis, opening and closing of stomata, osmoregulaton, oxidative phosphorylaton, photophosphorylation and translocation (Marschner, 2002). Thus, it is involved directly or indirectly in the production and enlargement of new cells, tissues and organs. These, in turn, are responsible for the observed increase in shoot length, leaf number and leaf area of treated plants. As would be expected, improvement in shoot length, leaf number and leaf area contributed in enhancing the ability of treated plants for fresh and dry matter production. Improvement in yield parameters resulted from the application of K over K_0 was also noted (Talukder *et al.*, 2002; Nagaich *et al.*, 2003; Tiwari and Agarwal, 2003; Mazid *et al.*, 2014).

The cumulative effect of increased CA and adequate leaf NPK manifested in the positive response of growth parameters. Thus, enhanced dry matter production by the treated plants would have positive effect on the yield parameters. Thus, it may be concluded that the ameliorative effect of applied K on growth parameters CA, NPK content augmented the growth and yield attributes leading to the observed increase in yield.

References

Bould, C. (1963). Mineral nutrition of plants in soil. *In*: Plant Physiology – A Treatise, Vol. Ill. Ed. By Steward, F.C, Academic Press Inc., New York, pp.16-91.

Dwivedi, R.S., and Randhawa, N.S. (1974). Evaluation of rapid test for the hidden hunger of zinc in plants. *Plant Soil.*, **40**: 445-451.

Fiske, C.H., and Subbarow, Y. (1925). The colorimetric determination of phosphorus. *Journal of Bioogical Chemistry*, **66**: 375-400.

Khan, T. A., Mazid, M., and Mohammad, F. (2011a). Sulphur management: An agronomic and transgenic approach. *Journal of Industrial Research and Technology*, **1 (2)**: 147-161.

Khan, T. A., and Mazid, M. (2011b). Nutritional significance of sulphur in pulse cropping system. *Biology and Medicine*, **3 (2)**: 114-133.

Khan, T.A., Mazid, M. and Quddusi, S. (2014). Role of organic and inorganic chemicals in plant-stress mitigation. *In.* Approaches to Plant Stress and their Management, Ed. By Gaur, P.K. and Sharma, P. DOI-10.1007/978-81-322-1620-9-3. Springer, India, 2014. pp, 39-52.

Lindner, R.C. (1944). Rapid analytical methods for some of the more common inorganic constituents of plant tissues. *Plant Physiology*, **19**: 76-89.

Marschner, H. (2002). Mineral Nutrition of Higher Plants. Second edition, Academic Press, London, pp. 265-277.

Mazid, M., Khan, T. A., and Mohammad, F. (2011a). Role of secondary metabolites in defense mechanisms of plants. *Biology and Medicine,* **3(2)**: 232-249.

Mazid, M., Khan, T. A., Zeba, H. K., Quddusi, S., and Mohammad, F. (2011b). Occurrence, biosynthesis and potentialities of ascorbic acid in plants. *International Journal of Plant. Animal and Environmental Sciences,* **1(2)**: 167-184.

Mazid, M., Zeba, H. K., Quddusi, S., Khan, T. A., and Mohammad, F. (2011c). Significance of Sulphur nutrition against metal induced oxidative stress in plants. *Journal of Stress Physiology and Biochemistry,* **7(3)**: 165-184.

Mazid, M., Khan, T. A., and Mohammad, F. (2011d). Response of crop plants under sulphur stress tolerance: A holistic approach. *Journal of Stress Physiology and Biochemistry,* **7(3)**: 23-57.

Mazid, M., Khan, T. A., and Mohammad, F. (2011e). Effect of abiotic stress on synthesis of secondary plant products: a critical review. *Agricultural Reviews,* **32(3)**: 172-182.

Mazid, M., Khan, T.A., Mohammad, F. (2012). Role of nitrate reductase in nitrogen fixation under photosynthetic regulation.*World Journal of Pharmaceutical Research,* **1(3)**: 386-414.

Mazid, M., Jamil, B., Khan, T.A., and Mohammad, F. (2013a). Glucosinolate engineering: Revisiting an old problem with solution of enhanced anti-herbivore defense. Paper presented in Internatioal conference on Bio engineering, held at Rajalakshmi Engineering College, Rajalakshmi Nagar, Thandalam, Chennai-602105, 2-4 January, India, (ICBE 2013); ISBN-978-1-62590-606-9.

Mazid, M., Khan, T.A., Mohammad, F., and Nooris, N. (2013b). Physico-biochemical effects of pre-sowing seed treatment with gibberellic acid on *Helianthus annus* L.- Paper presented at National conference of Plant Physiology (NCPP-2013) "Current Trends in Plant Biology Research": 895-896. Junagarh, Gujarat

Mazid, M. (2014). *Response of cicer arietinum L. to plant growth regulators and mineral nutrients.* Ph.D. thesis, Aligarh Muslim University, Aligarh, India.

Mazid, M., Khan, T.A., and Qudassi, S. (2014). Effects of climate change on crops: Food security and Biotechnology: *In*: Crop Improvement in the Era of Climate Change, Ed. By Rajib Roychowdhury, I.K. International Publication Pvt. Ltd, India, pp. 117-199.

Nagaich, K.N., 'Frivedi, S.K., Lekhi, R., and Lekhi, R. (2003). Effect of sulphur and potash on growth, yield and quality of garlic (*Allium sativum* L.). *Scientific Horticulture,* **8**: 143-147.

Nason, A., and McElroy, W.D. (1963). Modes of action of the essential mineral elements. *In*: Plant Physiology – A Treatise. Vol. III, Ed. By Steward, F.C Academic Press Inc., New York, pp. 451-536

Pandey, B.P. (2000). *Economic Botany*. 6th Ed. S. Chand and Company Ltd., New Delhi.

Quddasi, S., Khan, S., Mazid, M., Khan, T.A., Shakeel.A., and Arshad, M. (2014). Quality status of nutrients, vitamins and hormones under pathological threats in tomato: Basic scenario. *Medicinal Chemistry Analalysis*, **4(1)**: 12-21.

Talukder, A.F., Rahim, M.A., and Alam, M.S. (2002). Effect of planting time and different levels of potassium on the growth and yield of garlic. *Bangladesh Journal of Training and Development*, **15(2)**: 175-180.

Tiwari, R.S., and Agarwal, A. (2003). Influence of NPK levels on yield of garlic (*Allium sativum* L.). *Journal of Potassium Research*, **19**: 119-123.

Verma, D. P. Sharma, B.R., Chadha, A.P.S., Bajpai, H.K., and Bhadauria,U.P.S.(1996). Response of garlic (*Allium sativum* L.) to nitrogen, phosphorus and potassium levels. *Advances in Plant Sciences*, **9(2)**: 37-41.

Utilisation and Management of Medicinal Plants Vol. 3 (2015) *Pages* **459–533**
Editors: **V.K. Gupta and A. Kaul**
Published by: **DAYA PUBLISHING HOUSE, NEW DELHI**

32

Orchid: An Ornamental Treasure

Ipshita Chattopadhyaya[1], Sumeet Gupta[1*], Manisha Bhatia[1], Nazia Mustaq[1], Saikat Ghosh[1] and Vipin Saini[1]

ABSTRACT

Orchids are one of the nature's beauties but with that, they hold one of the highest potential in medicine. They contain 1000 genera and about 25000-35000 species growing from arid regions to wet coastal regions. They contain an array of chemical components from complex steroidal moieties to alkaloids such as gastrdin. Orchids contain compounds which show potential antimicrobial, neuroprotective, hepatoprotective, antiplatelet aggregation properties and have been used since ancient times and till today they are used in allopathic medicines and will continue to be a medicinal treasure in future.

Keywords: Orchid, Medicine, Allopathic, Ethnic uses.

Introduction

Herbal plants are staging a dramatic comeback with herbal 'renaissance' worldwide. In the vast array of medicinal plants available, orchids (flowering plants) have their own value by virtue of their great contribution in medicine. The term orchid was coined by Theophrastus as anatomy of the plants resemble with testicles. Greek word orchid literally means testicles (Singh *et al.*, 2012). This may account for use of orchids as aphrodisiacs in ancient civilizations. Orchidaceae is one of the most highly evolved family among monocotyledons with nearly about 1,000 genera and 25,000-35,000 species and exhibit an incredible range of diversity in size, shape and colour of their flowers (Singh *et al.*,2012). According to the botanists Orchidaceae

1 M.M. College of Pharmacy, Maharishi Markandeshwar University, Mullana-Ambala – 133 207, Haryana, India.

* *Corresponding author*: E-mail: sumeetgupta25@gmail.com

contains the largest number of species than any other family of flowering plants (Pillon *et al.*, 2006; Chase *et al.*,2003; Chase,2005). Mostly known for their beautiful flowers orchids are resource of great economic importance in the global horticultural industry, with great ornamental value. One of the most famous ornamental orchids is the black orchid, *Paphiopedilum wardii*, whose flowers are light green leaves mottled dark green, with red purple spotting on the underside of the leaf. Orchids grow most abundantly in tropical and subtropical forests, where they are largely epiphytic. This diversity increases towards the tropic; where the epiphytic species predominate that almost constitute 73 per cent of the family (Dressler, 1981). India has a very large variety of orchids and hilly regions have one or the other orchid flowering almost throughout the year (Anitha *et al.*, 2013). Presently there are about 1,150 species distributed in various states. However, Arunachal Pradesh alone accounts for 601 species; almost 52 per cent of the total species known in India. Some species in the genera like *Arundina, Cymbidium, Coelogyne, Dendrobium, Paphiopedilum, Renanthera,* and *Vanda* are almost extinct. The provisional list of 150 endangered plants of India includes many orchids like *Acanthephippium sylhetense, Anoectochilus sikkimensis, Aphyllorchis montana, Arachnanthe clarkei, Arundina graminifolio, Cymbidium macrorhizon, Dendrobium densiforum, Didiciea cunninghamii, Eria crassicaulis, Galeola lindleyana, Gastrodia exilis, Paphiopedilum fairanum, P. druryi, Pleione humilis, Renanthera imschootiana, Vanda coerulea, V. pumila* and *V. roxburghi* (http://www.orchidsasia.com/asco.htm).

Botany and Taxonomy of Orchids

Nomenclature

The International Code of Botanical Nomenclature (I.C.B.N.) and the International Code of Nomenclature of Cultivated Plants (I.C.N.C.P.) govern the naming of orchids. The classification of orchids is shown in Figure 32.1. The I.C.B.N. has standardized classification of plants (Table 32.1) and imposed the following endings:

Table 32.1: I.C.B.N Nomenclature System for Orchids

Family	Aceae	Example: Orchidaceae
Tribe	Eae	Example: Vandeae
Subtribe	Inae	Example: Sarcanthinae

Taxonomy

Taxonomy is showing in Figure 32.1.

Botany

The primary characteristics that distinguish the orchids as a group are found in the flower. At the bottom of an unspecialized non-orchid flower is the stem that supports it, called the pedicel. Directly above, and at the base of the flower itself, is a whorl of green, leaf like organs called sepals. Above and inside the sepals is a second whorl of coloured petals. Together the sepals and petals are called the perianth, which constitute the nonreproductive parts of the flower. The perianth protects the

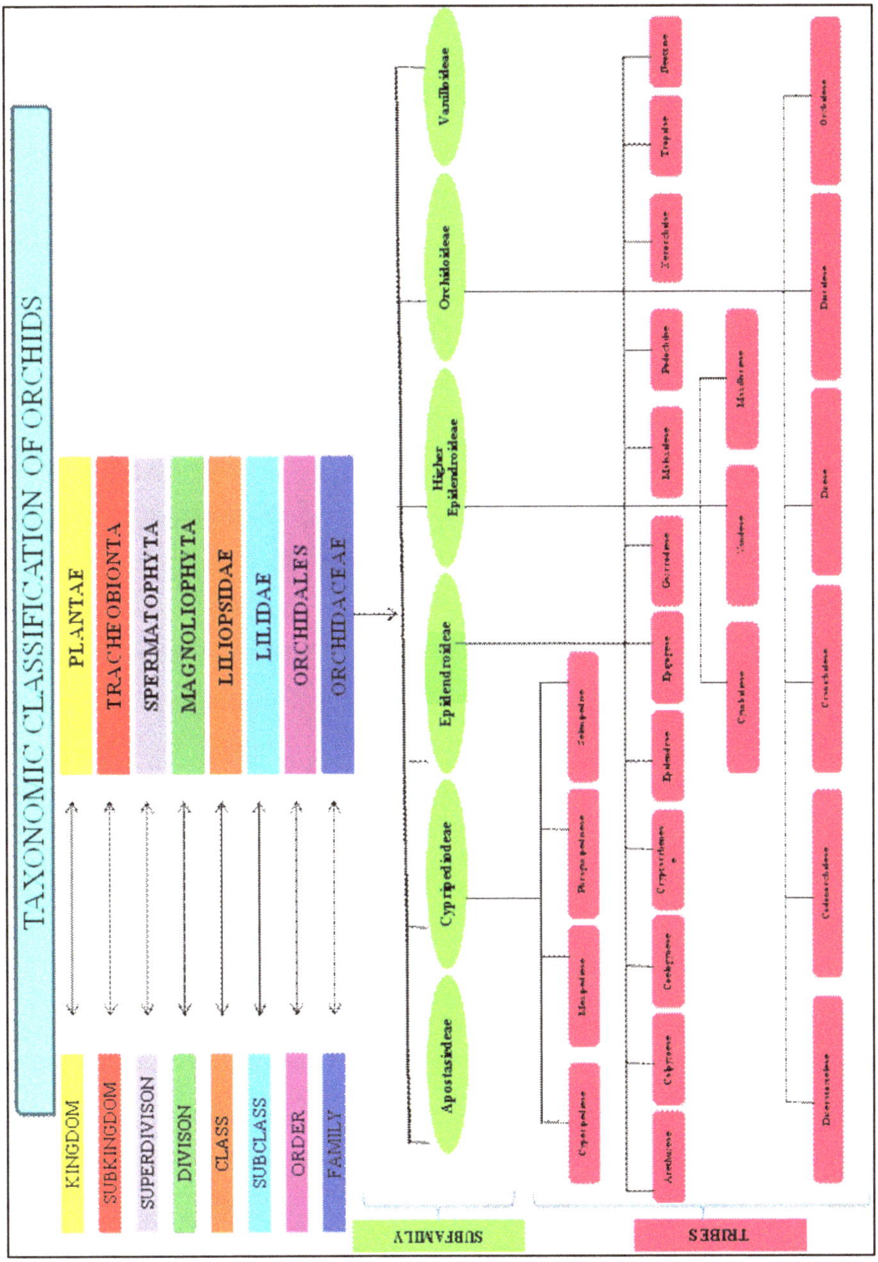

Figure 32.1: Classification of Orchids.

flower or attracts pollinators or both. Inside (also arranged in whorls) are the sexual portions of the flower. First are the pollen-producing stamens in up to several whorls; each stamen consists of an anther on a long slender filament. In the centre of the flower is the female pistil, which consists of an enlarged inferior ovary topped by a stalk like style with a stigma at its apex. The sepals and petals are usually similar, often highly coloured, and in sets of three. One petal is developed as a landing platform for the pollinator and is called the lip (or labellum) (Figure 32.2).

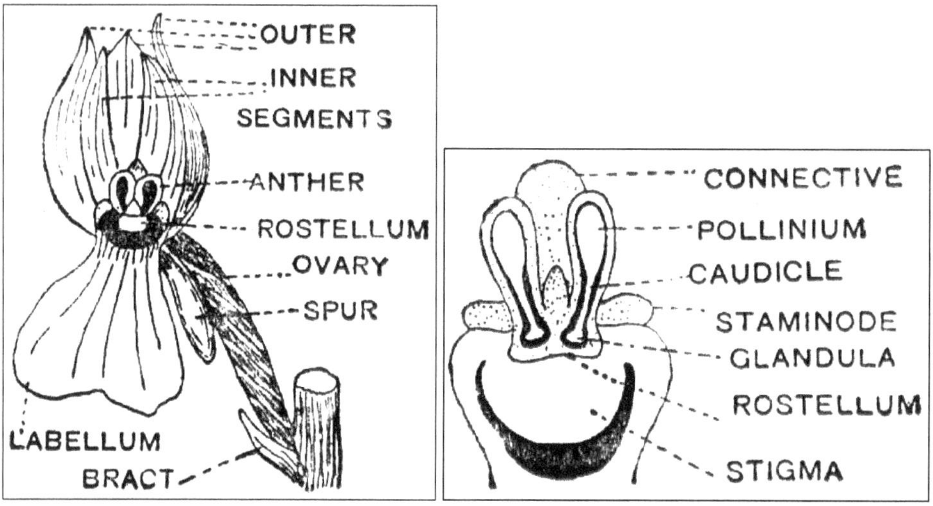

Figure 32.2: Orchid Anatomy.

History

As early as 1892, E de Wildeman had already begun investigation of orchid alkaloids in domesticated European orchid species as well as *Dendrobium nobile* and *Phalaenopsis lueddemanniana*. From then until 1896, E de Droog analyzed 104 species in 78 genera. In late 1890s, W Boorsma studied orchid alkaloids at the Bogor Botanical Gardens and detected some in *Paphiopedilum javanicum* and *Liparis parviflora*, among other species.

Not only for their beauty, but they are also known for their medicinal value, the reason why ethnic communities used it regularly in their traditional system of medicine (Jalal *et al.*, 2009). The use of orchids in herbal medicine has a very long history. Even Vedas have also depicted their importance since ancient times. A total of 365 plants, including several orchids are listed in the earliest known Chinese Materia Medica (Shen Nung Pen-tsao Ching). Confucius, the Chinese philosopher, grew orchids in the fifth century BC and wrote a poem about them. Dioscorides, a Greek botanist, physician and pharmacologist mentioned orchids in his work "De Materia Medica" "Of Medical Maters" published around 60 AD. This work remained a reference manual till the Middle Age (1,400 - 1,500 AD) (http://www.orchidsusa.com/6fertilizing.htm). In the first century AD, Dioscorides, who was a Greek working as a Roman military physician, wrote his De Materia Medica, including two terrestrial orchids (Berliocchi *et al.*, 2004). He adopted and promoted the 'Doctrine of Signatures' whereby plants were used for medicinal purposes according to their resemblance to parts of the human anatomy, for example by shape or colour. Naturally this led to orchid tubers being used to heal diseases of the testicles, and to stimulate lust. Moreover if given to men as whole fat new tubers this was supposed to produce male progeny, and if the shrivelled old tubers were given to women, this should produce female children (De materia medica 1543). In the history of ancient Alternative systems of medicine Ayurveda and Traditional Chinese Medicine (TCM) are on the forefront (Stuart, 1984). The Chinese were the first to write books about orchids. In 1233, Chao Shih-Keng wrote Chin Chan Lan P'u, and described 20 species and how to grow them. In 1247, Wang Kuei-hsueh wrote his Treatise on Chinese orchids, and described 37 species.

Ethnic Uses

India

Traditional Culture

Orchids are one of the important part of culture and beliefs of the local people of Northeastern region like Tai people, Deoris of Assam and Arunachal Pradesh who used *Rhynchostylis retusa, Papilionanthe teres, Vanda roxburghii, Aerides* etc. The flower *Vanda roxburghii* and *Coelogynae nitida* are used as hair accessories while *Papilionanthe teres* flowers is offered to Lord Buddha and spirits by the Khamtis and other Tai ethnics of Assam and Arunachal Pradesh.

Traditional Food

Local people of Northeastern region use some orchids as their delicacies *e.g.*, leaves of *Cymbidium* species are taken as food while shoots are grounded to sauce and pseudobulbs of these orchids are eaten like potatoes with curry or boiled pseudobulb, roots and rhizomes of many wild orchid species *Cephalanthera ensifolia, Habenaria acuminate, Habenaria susannae, Orchis latifolia, Microstylis wallachii, Pholidata articulate* and *Satyrium* species are used as food

Traditional Medicine

Northeastern natives also use many orchids as medicines and cues as orchids are rich phytochemically like *Dendrobium fimbriatum* has been used for liver upsets and nervous deability. *Papilionanthe teres* leaf paste is applied on forehead to keep it cool during high fever. *Eria muscicola* is used to cure diseases of chest, heart, lungs, eyes, ears and nervous system. *Eulophia camprestris* rhizome is used in folk medicine as a tonic, cough problems, paralysis and also used as a aphrodisiac. *Satyrium nepalense* is used as a tonic to cure dysentery and medicine. *Liparis odorata* is used in burns, cancerous ulcers and gangrene. *Orchis latifolia* tubers and roots are used nerve tonic and aphrodisiac. *Vanda coerulea* is used for eye diseases.

Worldwide

Bletilla striata, also known as *Limodorum striatum, Epidendrum tuberosum, Bletia hyacinthina* and Jimensia. In Chinese known as Dai Chi(Baiji). It is used to treat tuberculosis, hemoptysis, gastric and duodenal ulcers, bleeding and cracked skin on the feet and hands, euphoria, purification of blood, strengthening and consolidation of lungs, as well as the treatment of pus, boils, abscesses, malignant swellings, ulcers, and breast cancer, treatment of the flatulence, dispepsia, dysentery, fever, malignant ulcers, gastrointestinal disorders, hemorrhoids, anthrax, malaria, eye diseases, tinea, ringworm, tumors, and necrosis, silicosis, traumatic injuries, coughs, chest pain, tuberculosis, vomiting of blood, gastrorrhagia, enterorrhagia, internal bleeding, inflammation, and chopped skin. Powdered roots mixed with oil have been used as an emollient for burns and skin diseases. Whole plant preparations are tonic and treatment against leucorrhea, hemoptysis, and purulent coughs. Leaves collected in the autumn are reported to cure lung disease. *Dendrobium Sw* (Shih-hu) Among 1400 *Dendrobium* species in China, frequently used Shih-hu includes such orchids as Chin chai Shih-hu (Golden Hairpin *Dendrobium*), Er Huan Shih-hu (Earring *Dendro bium*), Ma poen Shih-hu (Horse Whip *Dendrobium*), Huan tsao Shih-hu (Yellow Herb *Dendrobium*), and Yukue Shih-hu (Melon Flickingeria). The Er Huan Shih-hu (Earring *Dendrobium*), among them showed the potential to cure stomachic, night sweats,strengthen the kidneys, impotence, anti-pyretic, tonic, and peptic having tonic, stomachic, pectoral, and antiphlogistic. *Dendrobium nobile* (Lindl) is the most commonly used species in the preparation of Shih-hu and used as a tonic and a strengthening medicine to impart longevity and serve as an aphrodisiac, stems are used to alleviate thirst, calm restlessness, accelerate convalescence, and reduce dryness of the mouth. Additional properties of Shih-hu include stomachic, pectoral, antiphlogistic, analgesic, and antipyretic medication and was also used to treat rheumatism, excessive perspiration, weakness brought about by thirst, excessive perspiration, impotence, entropion, leucorrhea, and menstrual pain. *Gastrodia elate* Blume (Chih Chien, Tien Ma) is native to Madagascar, tropical Asia, Oceania, Japan, China, and parts of the Soviet Union consisting of 20 species, five of which are found in China. Known to control of the internal movement of wind according to the ancient Chinese medicine theory the rhizomes, stalks, and dried tubers (known as Tianma) are used to treat headaches, dizziness, blackouts, numbness of the limbs, hemiplegia, epilepsy, limb cramps, spasms, migraine, expulsion of poisonous effluvia, rheumatism, vertigo, neuralgia, facial paralysis, dysphasia, infantile convulsions, lumbago, fever, and numerous afflictions. The herb was also used to give strength and virility, improve circulation and enhance memory while the stalks were considered to be aphrodisiacs, the tubers were used to treat nervous disorders, prevent the common cold and serve as a tonic. as a fortifier, to treat weak nervous systems and headaches. Dried plants were used in Japan to treat vertigo, headaches, and nervous diseases, especially in children. Native to West Indies and tropical Central and South American are present 40 species of epiphytic herbs *Dichaea*in genus *e.g., Dichaea muricata* (Sw) Lindley. A wash prepared from this orchid is valued by the Kofans for treating eye infections, like conjunctivitis. *Elleanthus* Presl has approx 50 species native to tropical America and the West Indies, *e.g., Elleanthus* sp. whose large purple flowers are crushed and added to chicha for flavoring. *Epidendrum* (Linnaeus) has nearly 500 species native to the area from the South Eastern United States to the American tropics which are mostly epiphytic. Among them more than 50 species are cultivated as horticultural plants, *e.g., Epidendrum* sp. Mucilage from the pseudo bulbs is collected for treating sores on the lips. *Eriopsis* (Lindley) Half of the species of this genus are tropical

Central and South American epiphytes, *e.g.*, *Eriopsis sceptrum* Reichenbach fil. Et Warscewicz, Bonpl 2 (1854) 98. The basal stems of this clumped epiphyte are boiled in water to extract the copious mucilage which is applied to sores of the gums and mucous membranes of the mouth for relief from discomfort. Masdevallia (Ruiz et pavon) comprises of at least300 species of this genus, all epiphytic distributed from Mexico to South America, *e.g.*, *Masdevallia* sp. A tea of the entire plant is recommended by the Kamsa medicine men to facilitate urination in pregnant women and to reduce inflammation of the bladder. *Oncidium* (Swartz) about 50 species of this genus are either erect or hanging epiphytes or terrestial plants native to tropical and subtropical parts of the New World, *e.g.*, *Oncidium pusillum* L Reichenbachfil, The Kofans treat lacerations with a wash prepared by boiling the plant in water. *Phragmipedium* (Rolfe) is a tropical South American genus of a dozen terrestial species, *e.g.*, *Phragmipedium ecuadorense* Garay, Boiled in water this orchid makes a tea used to allay "stomach trouble". In many African regions they are also used as food source (Hamisy, 2007).

Phytochemistry of Orchids

An array of pharmacologically active phytochemicals have been isolated from orchids such as alkaloids, bibenzyl derivatives, flavonoids, phenanthrenes etc. Recent works have reported isolation of anthocyanins, stilebnoids and triterpenoids from orchids. Orchinol, hircinol, cypripedin, jibantine, nidemin and loroglossin are some important phytochemicals reported from orchids (Fossen *et al.*, 2003). Pharmacologically most studied chemical component in orchids are mainly alkaloids, bibenzyl derivatives, flavonoids, phenanthrenes and terpenoids which are present in leaves, roots, flowers and whole plant (Guttirrez *et al.*, 2010). Heptacosane ($C_{27}H_{56}$) and octacosanol ($C_{27}H_{58}O$) has been isolated from *Vanda roxburghii* (Anitha and Tamilarasi, 2013) which has been found to show marked anti-inflammatory activity in carrageenan induced oedema in rats and mice (Griesbach *et al.*, 2002). Calanthoside isolated from *Calanthe discolor* and *Calanthe liukiuensis* (http://www.orchidsasia.com/asco.htm) showed an activating effect on skin blood flow. *Dendrobium densiflorum* (Chen *et al.*,1994) which contains scopoletin and scoparone exhibited *in vitro* ant-platelet aggregation activity. Rhizomes of *Gastrodia elata* contained gastrol which isreported to have relaxant effects on smooth muscle preparations of guinea pig ileum. Compounds such as nidemin and 9,19-Cyclolanosta-24,24- dimethyl-25-en-3_-yl-trans-p-hydroxycinnamate has also been isolated from *Scaphyglottis livida* and *Nidema boothi* (An *et al.*, 2003) and agrostophyllinol and agrostophyllinone has been isolated from *Agrostoohyllum brevipes* (Khasin *et al.*,1999). Antioxidatve properties has been shown by the alkyl ferulates isolated from methanolic extract of *Dendrobium monoliforme* (Fossen *et al.*, 2003) while some other antioxidants such as cis and trans-melilotoside, dihydromelilotoside has been isolated from *Dendrobium* sp. (http://www.orchidsusa.com/6fertilizing.htm). Compounds with anti tumor activities have also been isolated from *Bulbophyllum kwangtungense* (Stuart, 1984). Cymbidine A from *Cymbidium goeringii* (Berliochhi *et al.*, 2004) with hypotensive and diuretic activities from and cyclobalanone and 5_-lanosta- 24,24-dimethyl-9(11),25-dien-3_-ol from *Scaphyglottis livida* which produces dose dependent antinociceptive and anti inflammatory effect has also been isolated (http://www.orchidsusa.com/6fertilizing.htm).Kinsenoside has been isolated from *Anoectochilus formosanus* with significant antihepatotoxic activity (http://www.orchidlady.com). Three phenanthrenes were isolated from *Dendrobium plicatile* (http://www.orchidlady.com) whose structures are 2,5-dihydroxy-4,9,10-trimethoxyphenanthrene; 2,5-dihydroxy-4-methoxyphenanthrene and 2,5,9-trihydroxy-4-methoxy-9,10-dihydrophenanthrene while seven stilbenoids were isolated from the same plant (http://www.orchidsusa.com/6fertilizing.htm) which are 3- methoxy-3',5-dihydroxybibenzyl ; 3,3',4'-trimethoxy-5- hydroxybibenzyl ; 2-methoxy-4,7-dihydroxy-9,10-dihydrophenanthrene; 2,4-dimethoxy-3,7dihydroxyphenanthrene; 3,4-dimethoxy-2,7-dihydroxy-9 ; 10-dihydrophenanthrene ; ephemeranthoquionone, and 2,2'-dimethoxy -4,4',7,7' - tetra hydrox y-9,9 ',10,1 0' - tetrahydro-1,1'-biphenanthrene. (N Saito *et al*) An acylated anthocyanin from the red-purple flower of *Dendrobium* 'Pramot'(*phalaenopsis* type cv) was isolated in 1994, whose structure was determined to becyanidin-3-*O*-(6-malonyl)-*D*-glucopyranoside)7,3'-di-*O*-[6-glucopyranosyl) oxybenzoyl]-*D*-glucopyranoside) (Guttierez, 2010).

The *Dendrobium loddigesii* methanol extract analysis contained compounds moscatilin, moscatin, and moscatilin diacetate that can inhibit the aggregation of rabbit platelets induced by arachidonic acid and collagen (Fossen *et al.*, 2004). It was also found that *Dendrobium loddigesii* rolfe contained shihunidine, shihunine and dendrophenol (4,4'-dihydroxy-3,3',5-trimethoxybibenzyl). Chemically shihunidine was isolated product derived from shihunine and both were shown to be inhibitors of Na+, K+-ATPase in the rat kidney (Anuradha *et al.*,1998). Two bibenzyl derivatives cumulatin and tristi, were isolated from *Dendrobium cumulatum* and *Bulbophyllum triste* respectively which were chemically 3,3'-dihydroxy-4,4',5,5'-tetramethoxy-bibenzyl and 3,4',5-trihydroxy-3'-methoxybibenzyl, respectively (Chen *et al.*,1994). They also isolated 2,7-dihydroxy-3,4,6-trimethoxy-phenanthrene,2,7-dihydroxy-3,4,6-trimethoxy-9,10-dihydrophenanthrene, nudol, moscatin, batatasin-III and 2,5, 9-trihydroxy-4-methoxy-9,10-dihydrophenan-threne from same plant (Ho *et al.*, 2003). From *Dendrobium crepidatum* and *Dendrobium moscatum*, Majumder isolated bibenzyl derivatives crepidatin and moscatilin (4,4'-dihydroxy-3,3',5-trimethoxybi-benzyl) (Uniyal, 1977). Denfigenin, a diosgenin derivative with the structure (25*R*)-22-*O*-spirost-5-ene- 3,16,17-triol was isolated from the whole plant *Dendrobium* fimbriatum, along with diosgenin anddefuscin (*n*-triacontyl-*p*-coumarate) (Uniyal, 1975) and amoenumin with the structure 9,10-dihydro-5*H*-phenanthropyran was isolated from the orchid *Dendrobium amoenum*. From the tuber of *Bletill* five antibacterial compounds: 3,3'-dihydroxy-2',6'-*bis*(phydroxybenzyl)-5-methoxybibenzyl;2,6-*bis*(*p*-hydroxybenzyl)-3'5-dimethoxy-3-hydroxybibenzyl; 3,3'-dihydroxy-5-methoxy-2,5'6'-*tris*(*p*hydroxybenzyl) bibenzyl; 4,7-dihydroxy-2-methoxy-9,10-dihydrophenanthrene; and 4,7-dihydroxy-1-*p*-hydroxybenzyl-2-methoxy-9,10dihydrophenanthrene were isolated. Later on 4,4'-dimethoxy-9,9',10,10'-tetrahydro-[1,1'-biphenanthrene]-2,2',7,7'-tetrol; 4,4' -dimethoxy-9,10-dihydro-[1,1'-biphenanthrene]-2,2',7,7'-tetrol; and 4,4'-dimethoxy-[1,1'-biphenanthrene]-2,2',7,7'-tetrol were also isolated with a series of *bis*(dihydrophenanthrene) ethers, benzylphenanthrenes, dihydrophenanthropyrans, phenanthrene with spirolactone ring, and stilbenoids (methylated stilbenoids) were also isolated from from *Bletilla striata* (Parkinson,1640). Saito *et al.*, also isolated and elucidated four acylated anthocyanins from the purple-red flowers of *Bletilla striata* [35] which were found to be cyaniding 3-*O*-[6-*O*-(malonyl)-β-*D*-glucopyranoside]-7-

Table 32.2: Important Phytochemicals Isolated from Orchids

Sl.No.	*Name*	*Category*	*Source*
1.	AERIDIN	PHENANTHROPYRAN	*Aerides crispum*

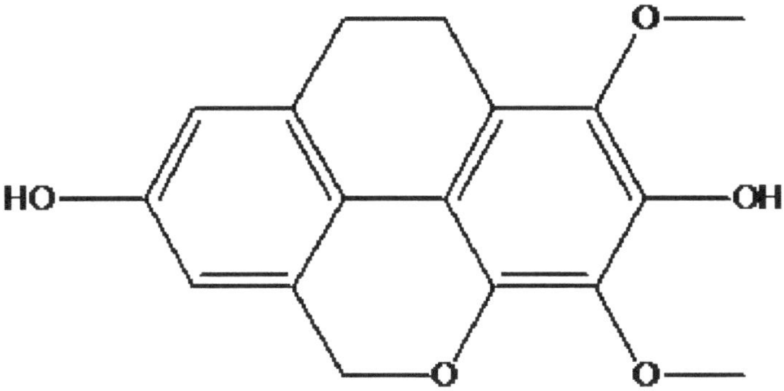

2.	AGROSTOPHYLLINOL	TRITERPENOID	*Agrostophyllum brevipes*

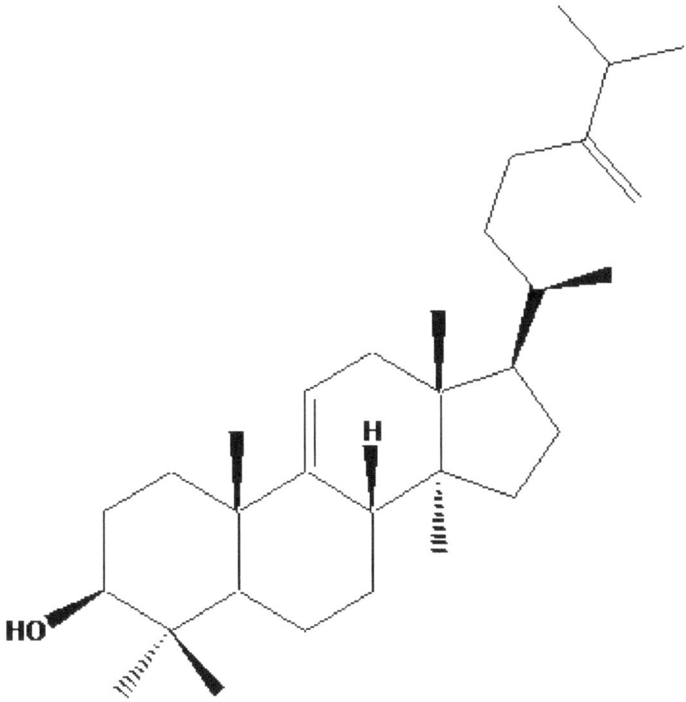

Contd...

Table 32.2–*Contd...*

Sl.No.	*Name*	*Category*	*Source*
3.	ISOAGROSTOPHYLLOL	TRITERPENOID	*Agrostophyllum callosum*
	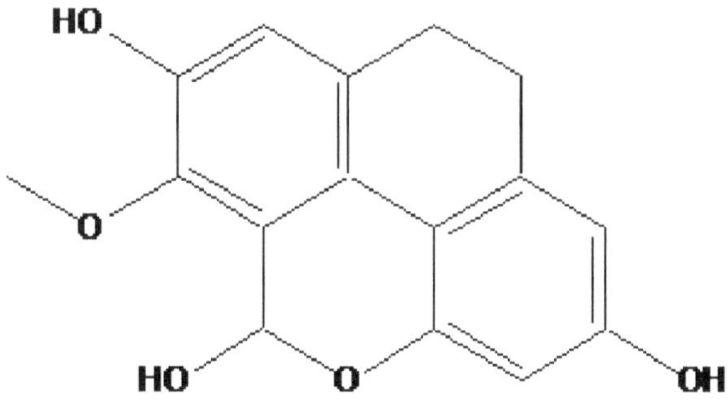		
4.	ORCHINOL	STILLBENOIDS	*Agrostophyllum callosum*
	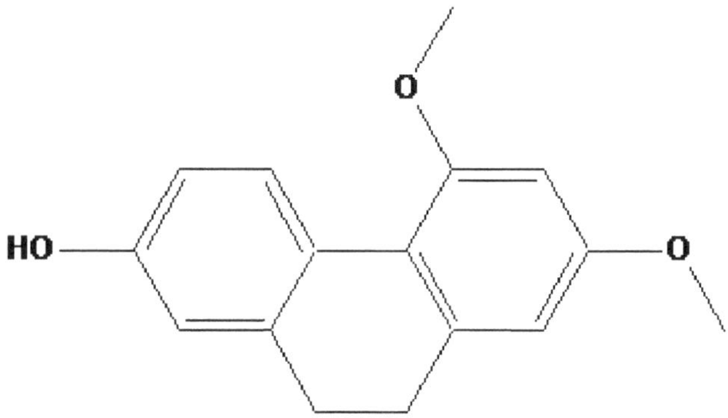		
5.	ARUNIDAN	STILLBENOIDS	*Arundina graminifolia*
	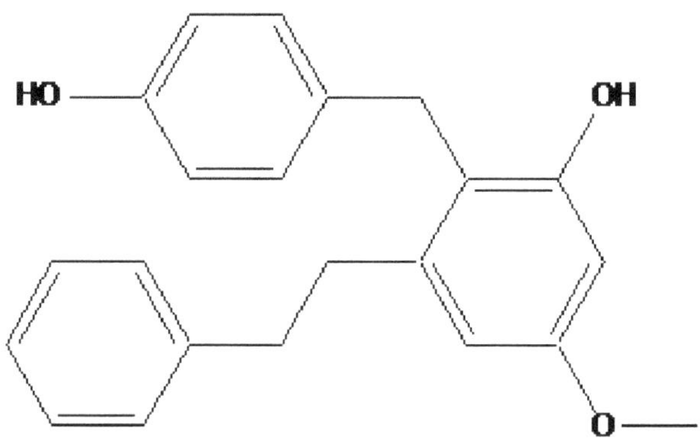		

Contd...

Table 32.2–*Contd...*

Sl.No.	*Name*	*Category*	*Source*
6.	CYPRIPEDIN	1-4-PHENANTHROQUINONE	*Cypripedium calceolus pubscens*

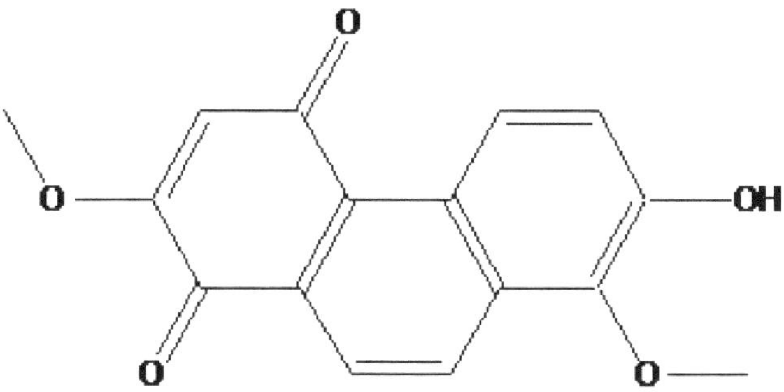

7.	LOROGLOSSIN	GLUCOSIDE	*Orchis latifolia*

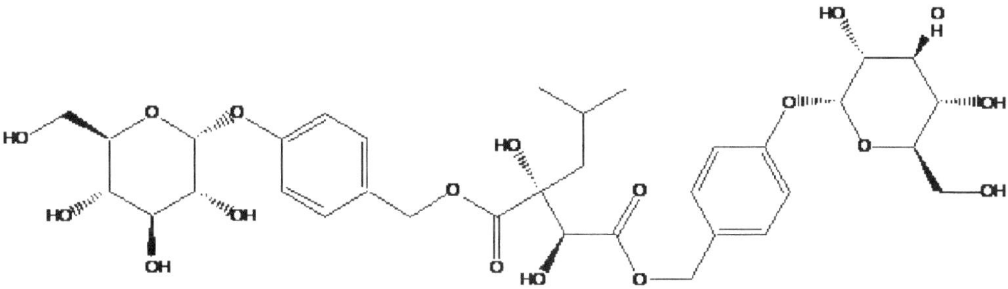

8.	GIGANTOL	BIBENZYL	*Dendrobium nobile*

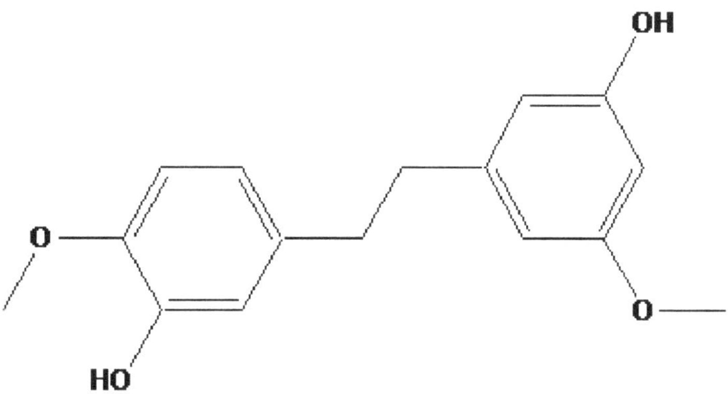

Contd...

Table 32.2–*Contd...*

Sl.No.	*Name*	*Category*	*Source*
9.	NUDOL	PHENANTHRENE	*Eulophia nuda*

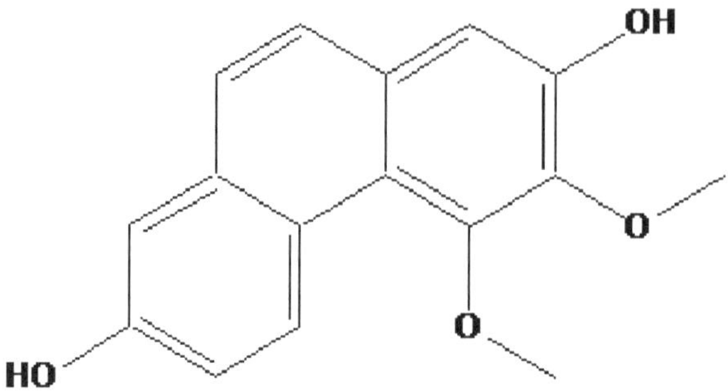

10.	MELIANIN	GLYCOSIDE	*Vanda roxburghii*

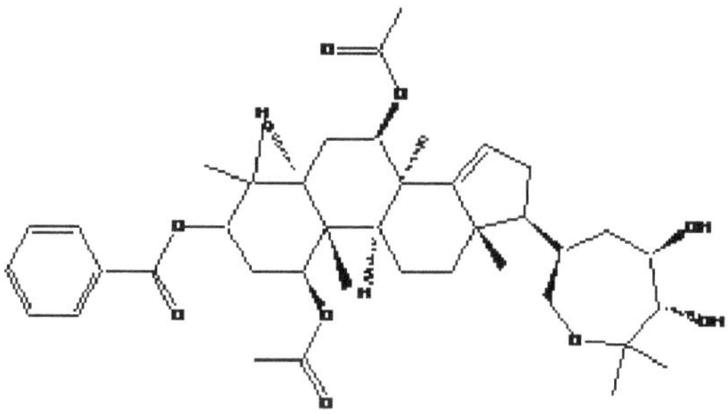

11.	NIDEMIN	TRITERPENOID	*Nidema boothi* and *Scaphyglottis*

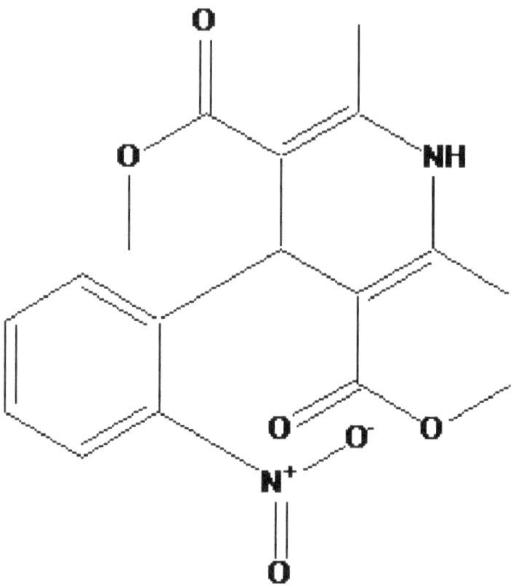

Contd...

Table 32.2–*Contd...*

Sl.No.	Name	Category	Source
12.	KINSENOSIDE	GLYCOSIDE	*Anoectochilus formosanus*

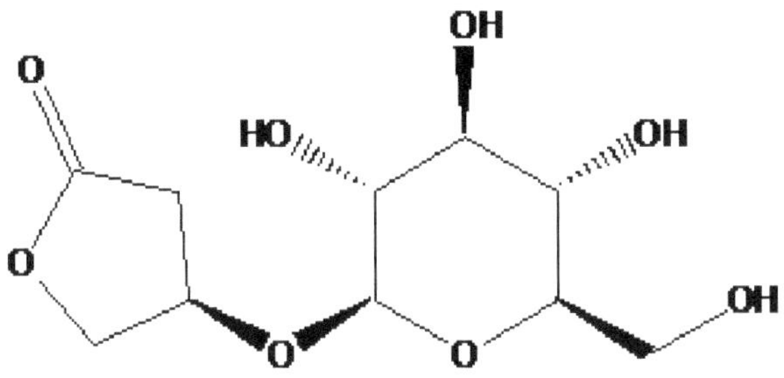

Sl.No.	Name	Category	Source
13.	ROTUNDATIN	Phenanthrene	*Dendrobium moscatum*

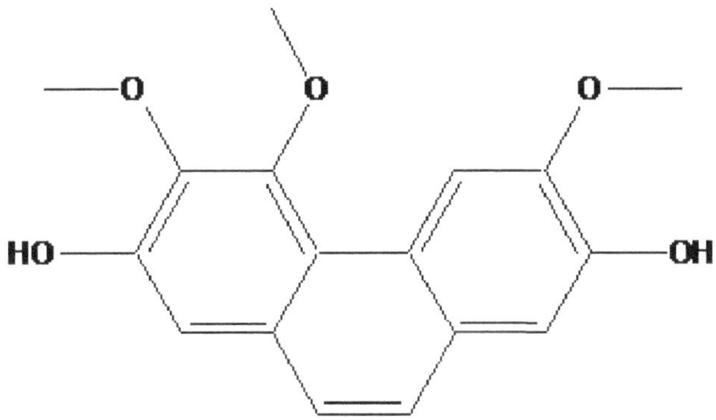

Sl.No.	Name	Category	Source
14.	MOSCATIN	PHENANTHRENE	*Dendrobium moscatum*

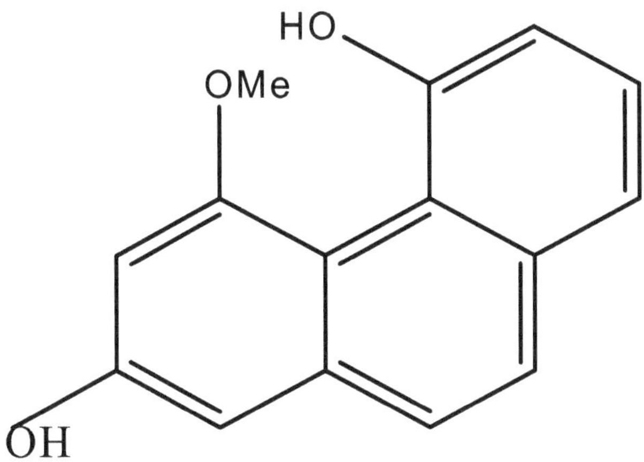

Contd...

Table 32.2–*Contd...*

Sl.No.	*Name*	*Category*	*Source*
15.	GYMOPUSIN	PHENANTHRENE	*Bulbophyllum gymopus*

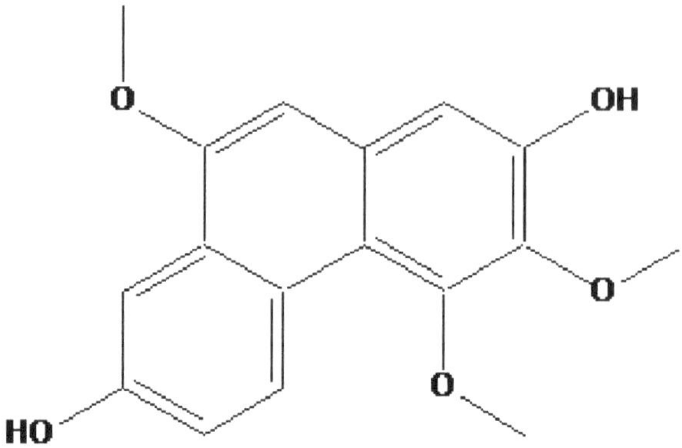

Sl.No.	*Name*	*Category*	*Source*
16.	SHIHUNIDINE	PYRROLE ALKALOID	*Dendrobium loddigessi*

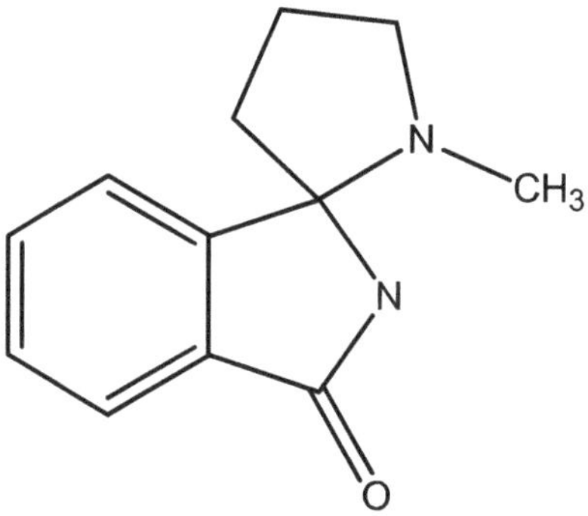

SHIHUNIDINE

Sl.No.	*Name*	*Category*	*Source*
17.	SHIHUNINE	Phathalidepyrrolidine alkaloid	*Dendrobium loddigessi*

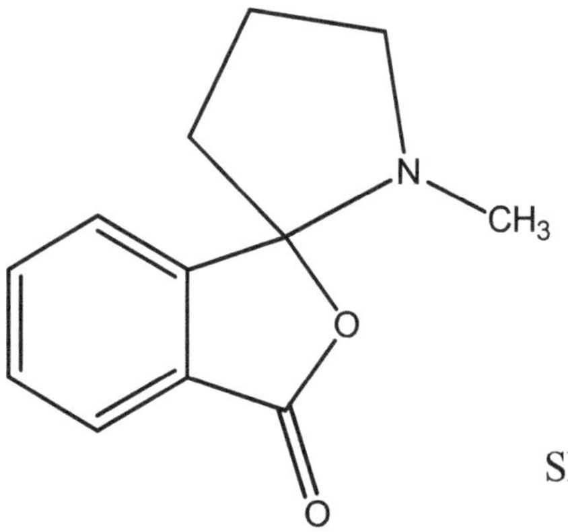

SHIHUNINE

Contd...

Table 32.2–*Contd...*

Sl.No.	Name	Category	Source
18.	N^6-(4HYDROXYBENZYL) ADENINE RIBOSIDE	Adenosine	*Gastrodia elata*

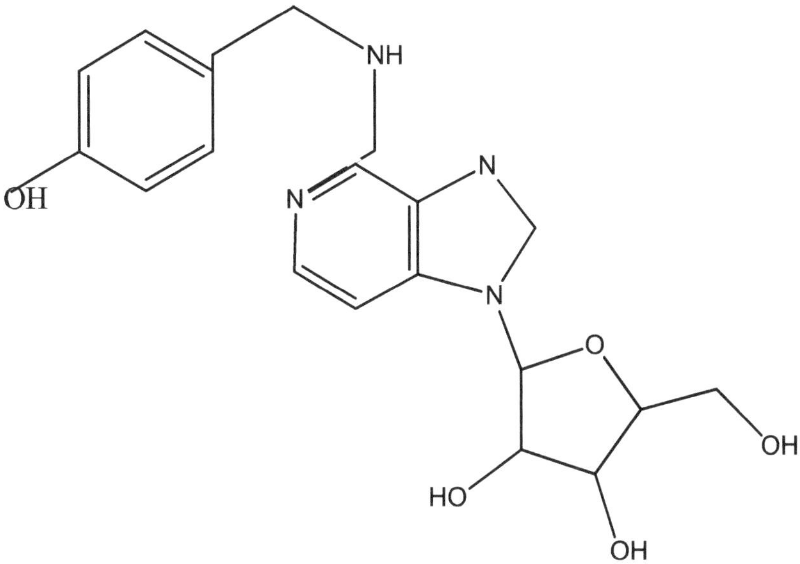

Sl.No.	Name	Category	Source
19.	CREMASTRINE	Pyrrolizidine alkaloid	*Cremastra appendiculata*

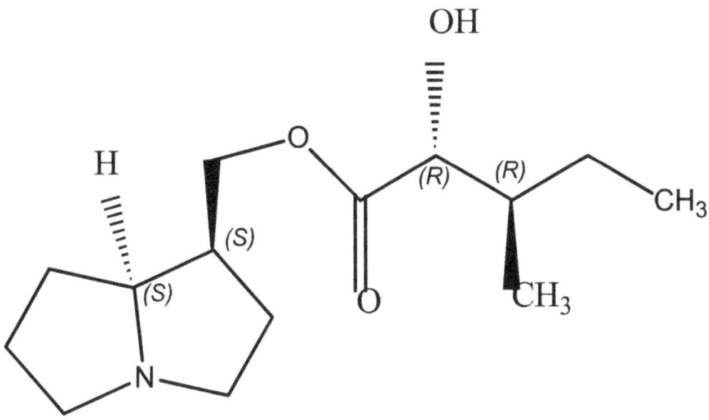

Contd...

Table 32.2–*Contd...*

Sl.No.	*Name*	*Category*	*Source*
20.	CHYSIN A	Pyrrolizidine alkaloid	*Chysis bractescens*
		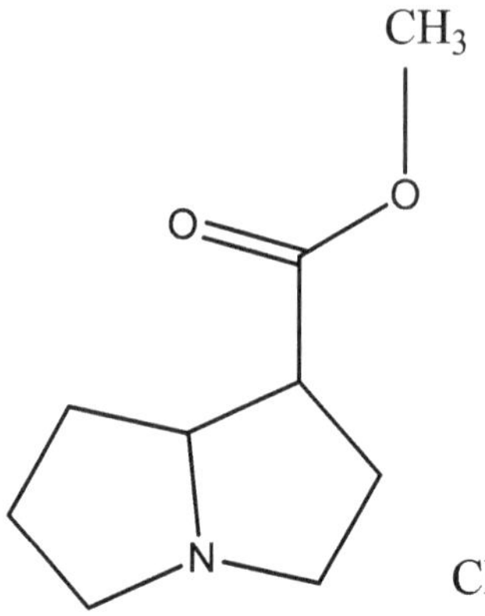 CHYSIN A	
21.	Chysin B	Pyrrolizidine alkaloid	*Chysis bractescens*
		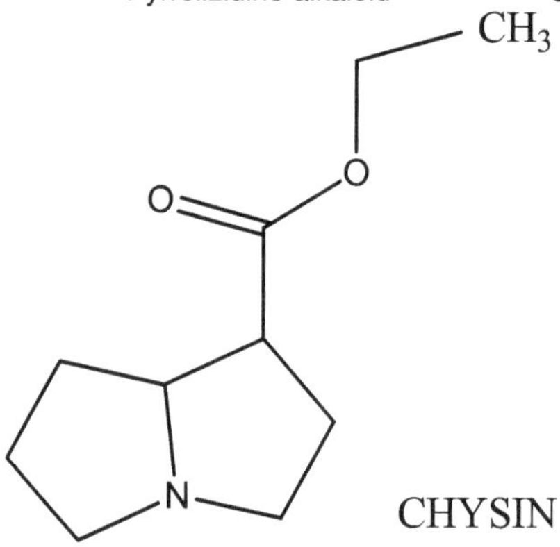 CHYSIN B	
22.	ERIANIN	PHENOL	*Dendrobium chyrosotoxum*
		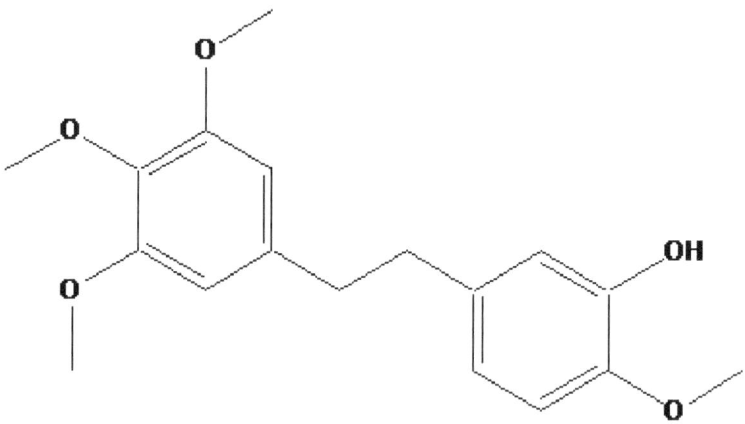	

Contd...

Table 32.2–*Contd...*

Sl.No.	*Name*	*Category*	*Source*
23.	ISOAMOENYLIN	dihydrostillbene	*Dendrobium amoenum*
	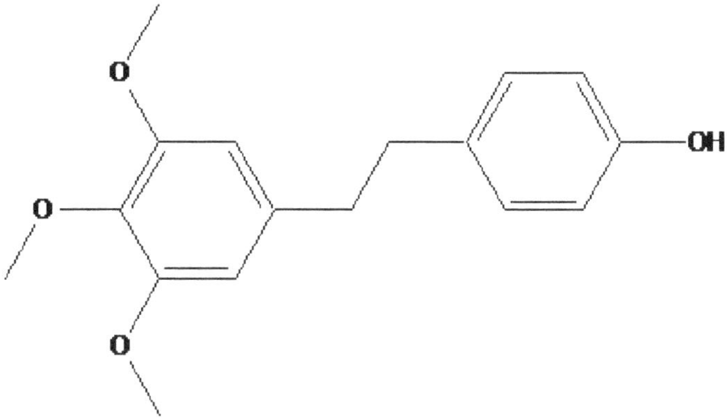		
24.	MOSCATILIN	bibenzyl	*Dendrobium loddigesii*
	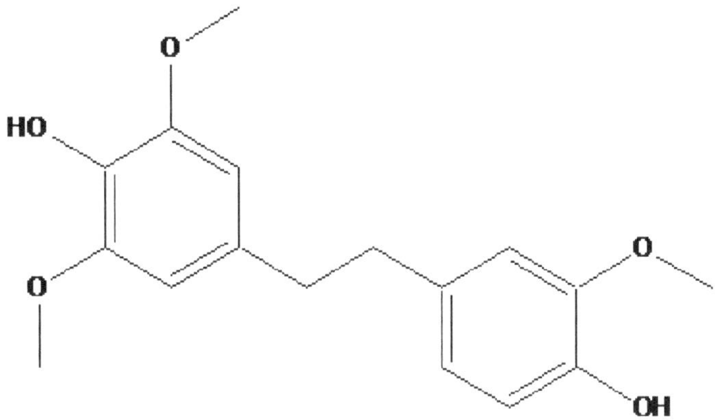		
25.	MOSCATILIN DIHYDROACETATE	BIBENZYL	*Dendrobium loddigesii*
	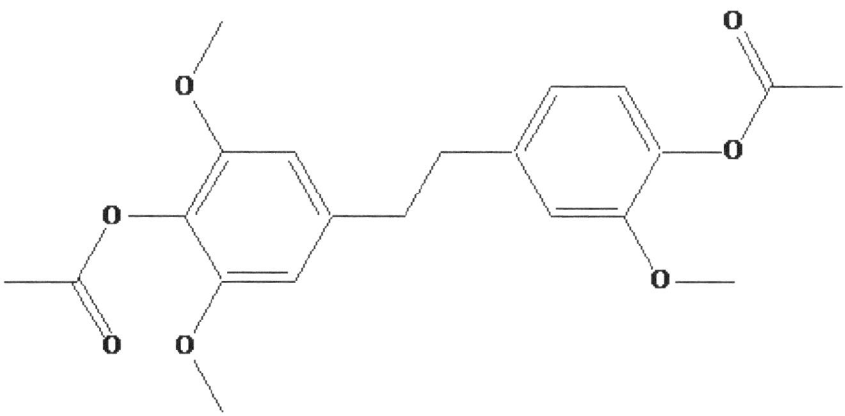		

Contd...

Table 32.2–*Contd...*

Sl.No.	*Name*	*Category*	*Source*
26.	NOBILIN D	BIBENZYL	*Dendrobium nobile*
	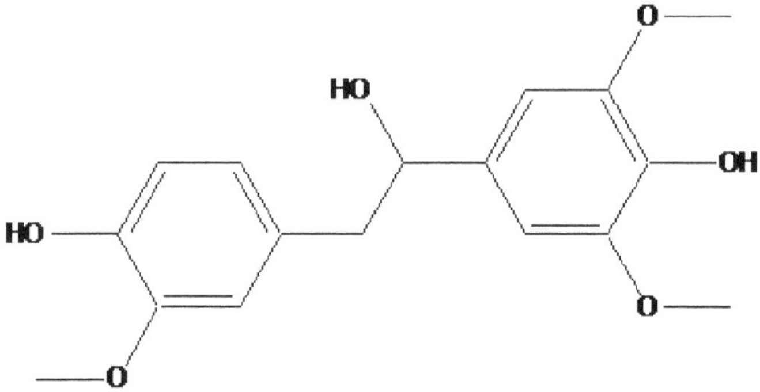		
27.	NOBILIN E	BIBENZYL	*Dendrobium nobile*
	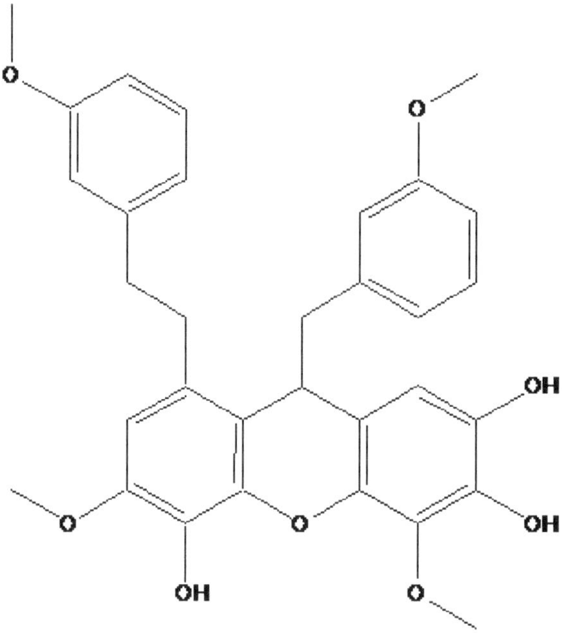		
28.	BATATASIN III	BIBENZYL	*Epiendrum Riggidum*
	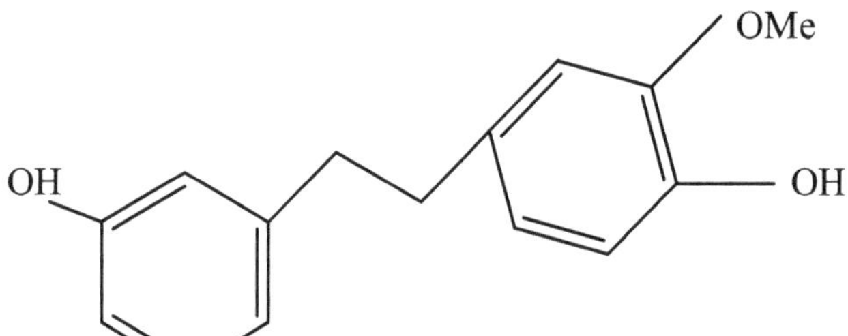 BATATASIN III		

Contd...

Table 32.2–*Contd...*

Sl.No.	*Name*	*Category*	*Source*
29.	PHOLIDOTOL A	dihydrostillbene	*Pholidota chinesis*

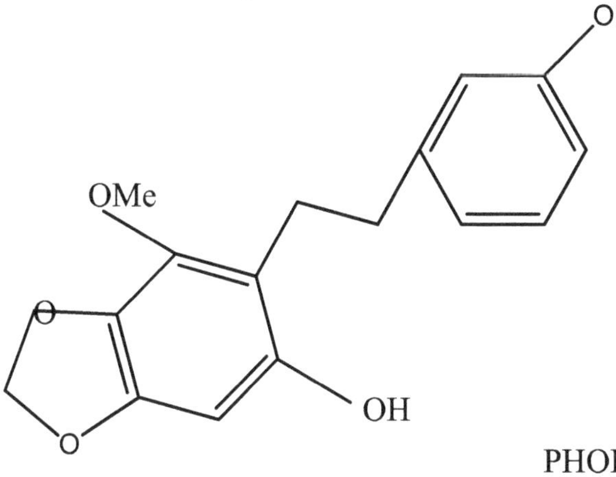

PHOLIDOTOL A

Sl.No.	*Name*	*Category*	*Source*
30.	PHOLIDOTOL B	Dihydrostillbene	*Pholidota chinesis*

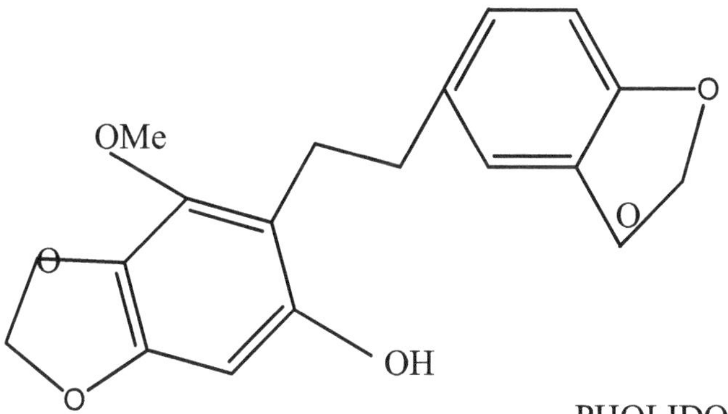

PHOLIDOTOL B

Sl.No.	*Name*	*Category*	*Source*
31.	CUMULATIN	BIBENZYL	*Bulbophyllum kwangtungense*

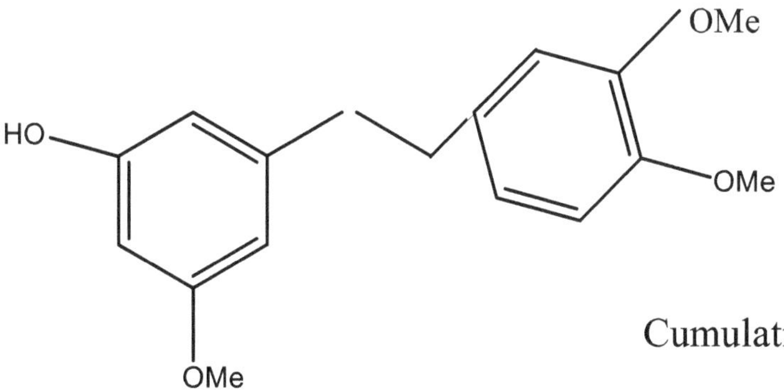

Cumulatin

Contd...

Table 32.2–*Contd...*

Sl.No.	*Name*	*Category*	*Source*
32.	DENSIFLOROL A	Dibenz[b,f]oxepin	*Bulbophyllum kwangtungense*
	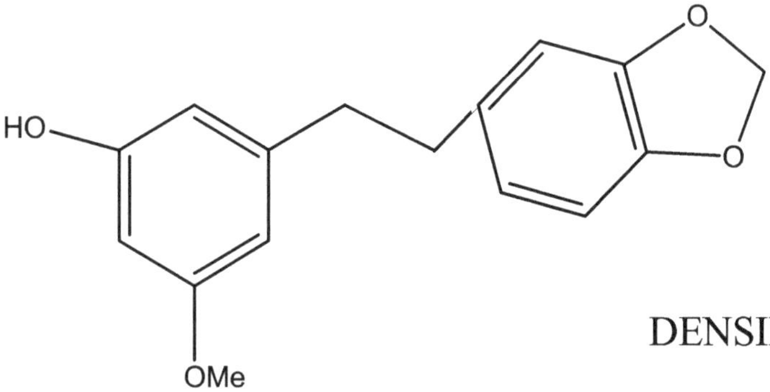 DENSIFLOROL A		
33.	3-METHYL GIGANTOL	Bibenzyl	*Ephemerantha lonchophyllla*
	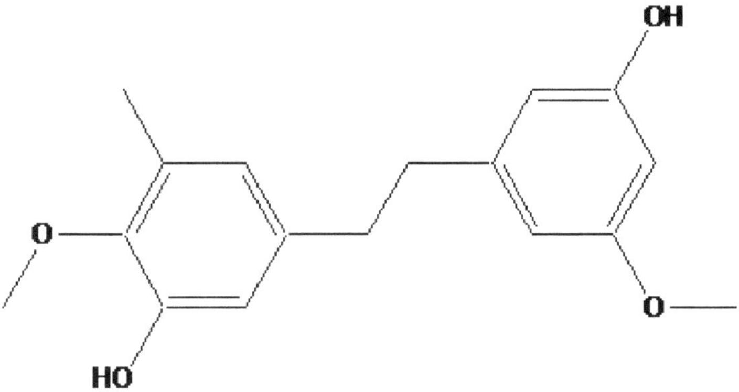		
34.	ALOIFOL	BIBENZYL	*Nidema boothi*
	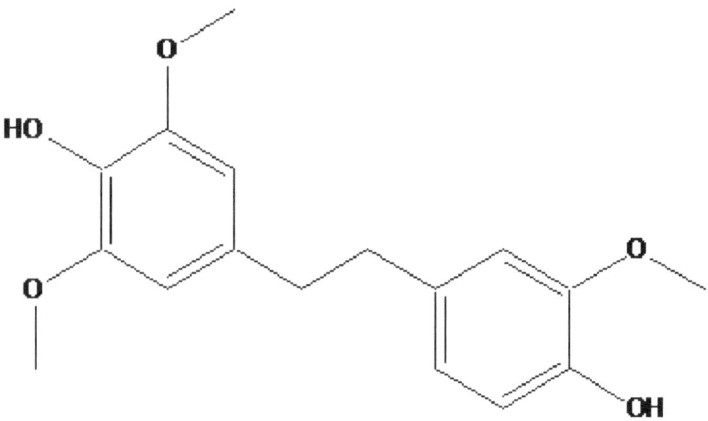		

Contd...

Table 32.2–*Contd...*

Sl.No.	*Name*	*Category*	*Source*
35.	GYMCONOPIN	DIHYDROSTILLBENE	*Gymnadenia conopsea*

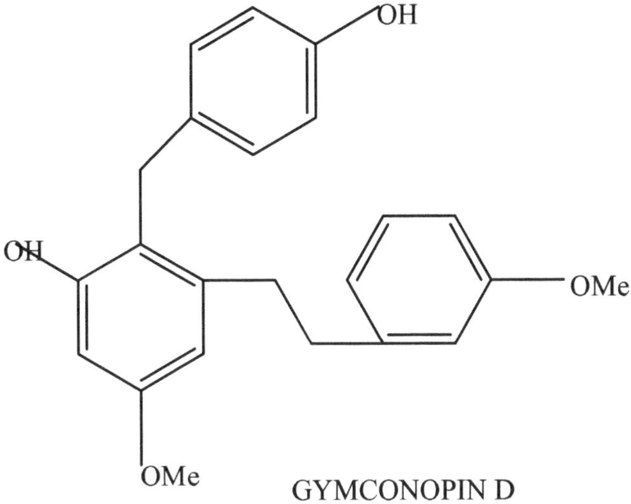

GYMCONOPIN D

36.	Dihydroxy-2,6-bis(4-hydroxybenzyl)-5-methoxybibenzyl	bibenzyl	*Gymnadenia conopsea*

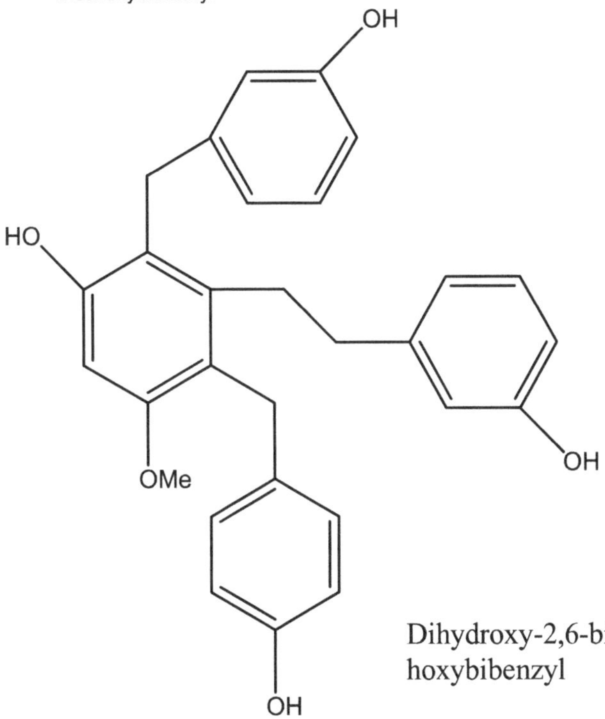

Dihydroxy-2,6-bis(4-hydroxybenzyl)-5-met hoxybibenzyl

Contd...

Table 32.2–*Contd...*

Sl.No.	*Name*	*Category*	*Source*
37.	Quercetin -7-o β-d-[6"-o-(trans-feruloyl)]-glucopyranoside	Benzopyranone	*Anoectochilus roxburghii*

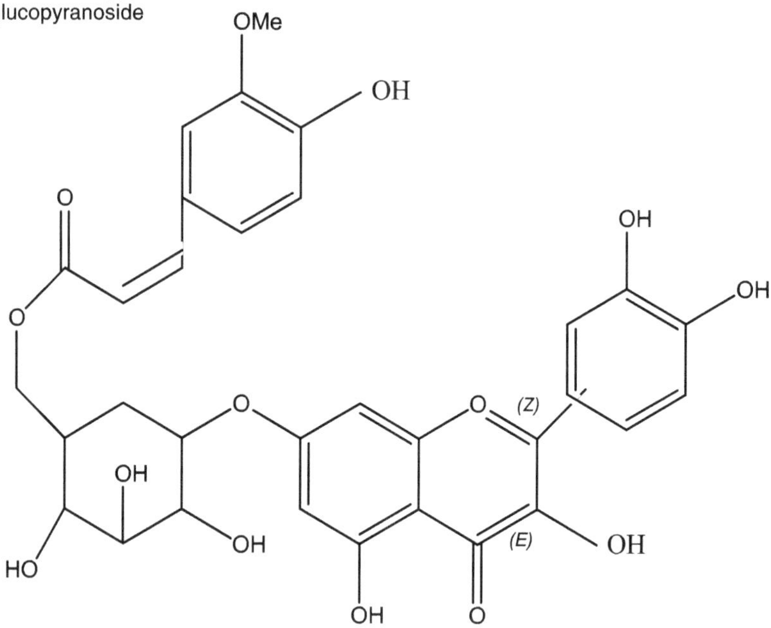

Sl.No.	*Name*	*Category*	*Source*
38.	8-C-P-hydroxy-benzylquercetin	BENZOPYRANONE	*Anoectochilus roxburghii*

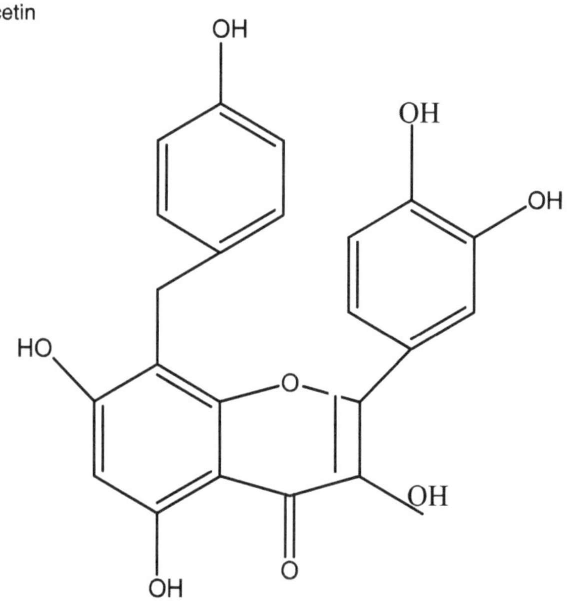

Contd...

Table 32.2–*Contd...*

Sl.No.	*Name*	*Category*	*Source*
39.	Isorhamnetin-7-o-β-D-glucopyranoside	Bibenzyl	*Anoectochilus roxburghii*

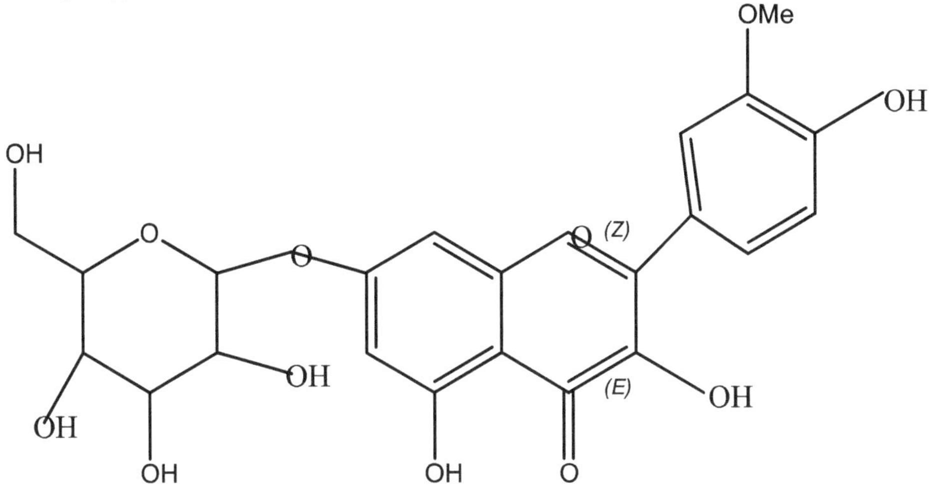

Sl.No.	*Name*	*Category*	*Source*
40.	Isorhamnetin-3-o-β-D-glucopyranoside	Bibenzyl	*Anoectochilus roxburghii*

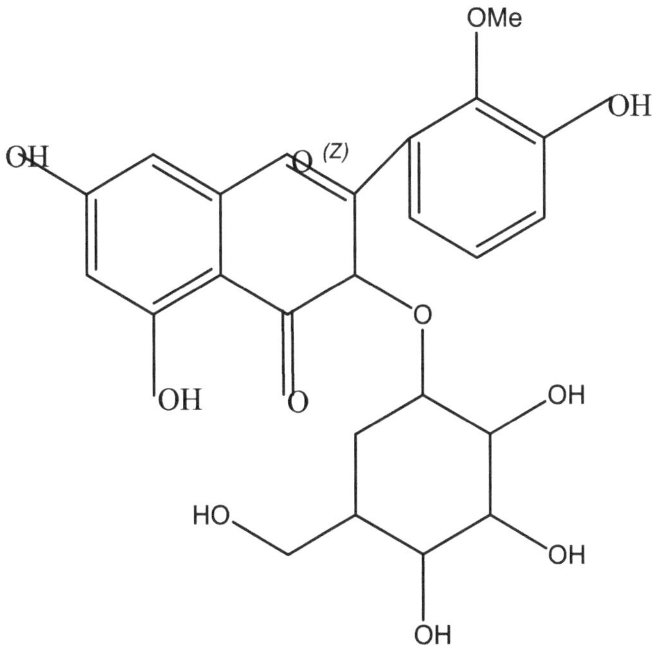

Contd...

Table 32.2–*Contd...*

Sl.No.	*Name*	*Category*	*Source*
41.	Kaempeferol -3-0- β-D-glucopyranoside	Bibenzyl	*Anoectochilus roxburghii*

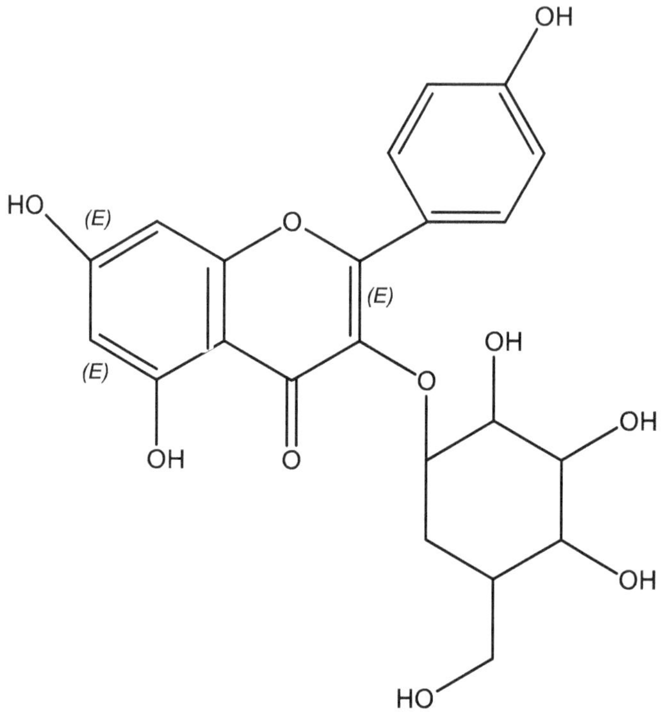

Sl.No.	*Name*	*Category*	*Source*
42.	Kaempeferol -7-0- β-D-glucopyranoside	Bibenzyl	*Anoectochilus roxburghii*

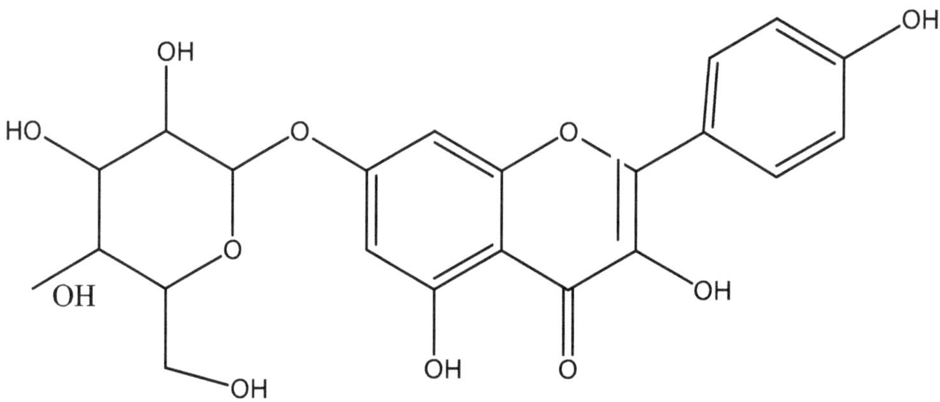

Contd...

Table 32.2–*Contd...*

Sl.No.	Name	Category		Source
43.	5-Hydroxy-3',4',7-trimethoxyflavanol-3 –β-D-[rutinoside]	Bibenzyl	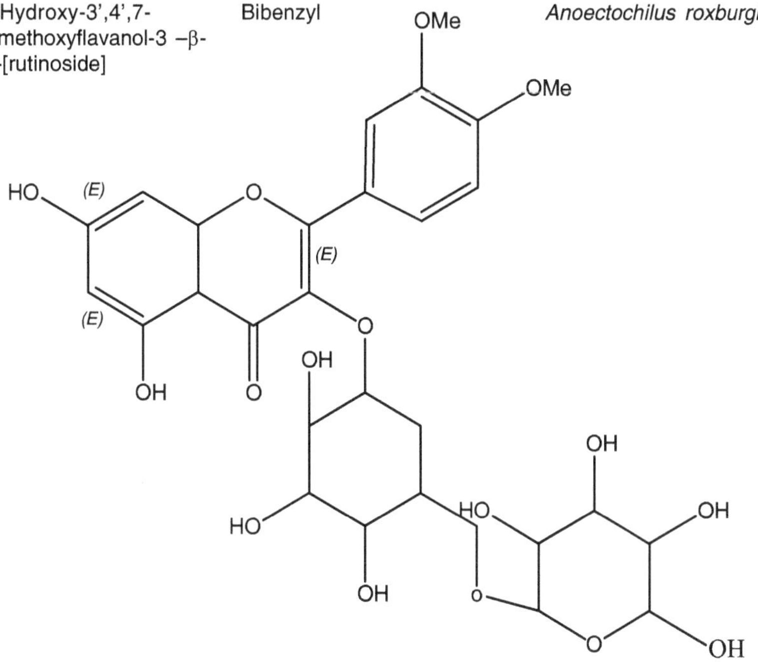	*Anoectochilus roxburghii*
44.	ISORHAMNETIN -3-β-D-[RUTINOSIDE]	BIBENZYL	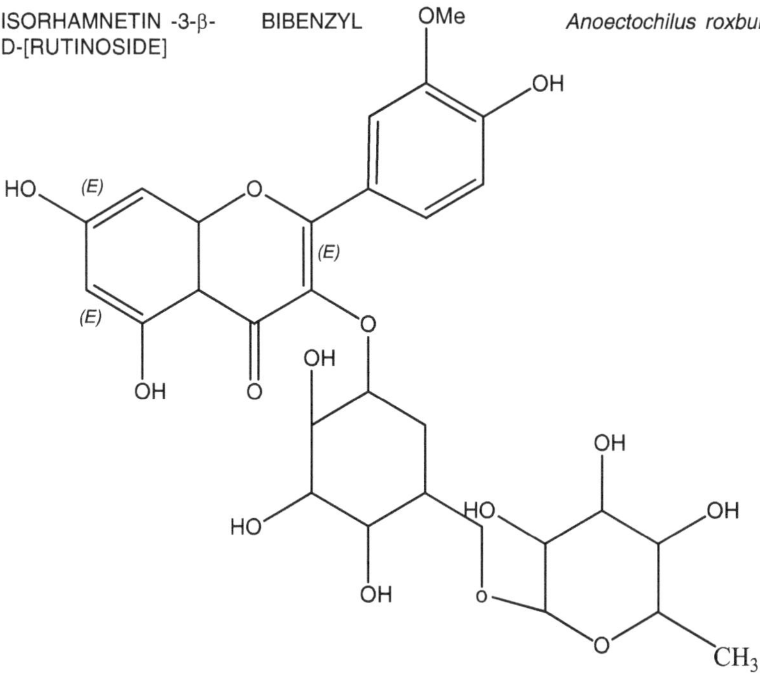	*Anoectochilus roxburghii*

Contd...

Table 32.2–*Contd...*

Sl.No.	*Name*	*Category*	*Source*
45.	Quercetin	Benzopyranone	*Dendrobium tosaense*

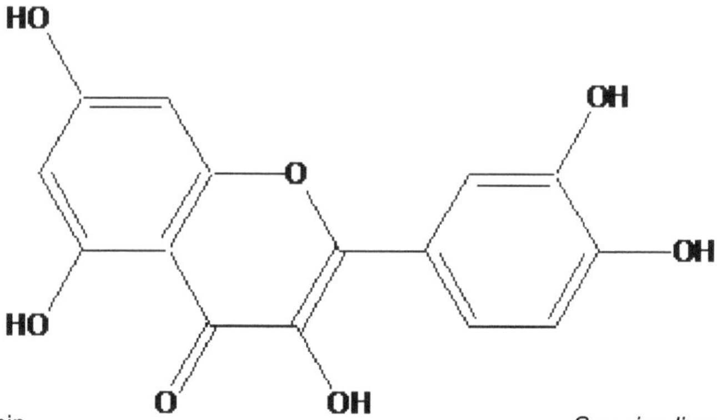

46.	Chrysin		*Cyperipedium acranthos*

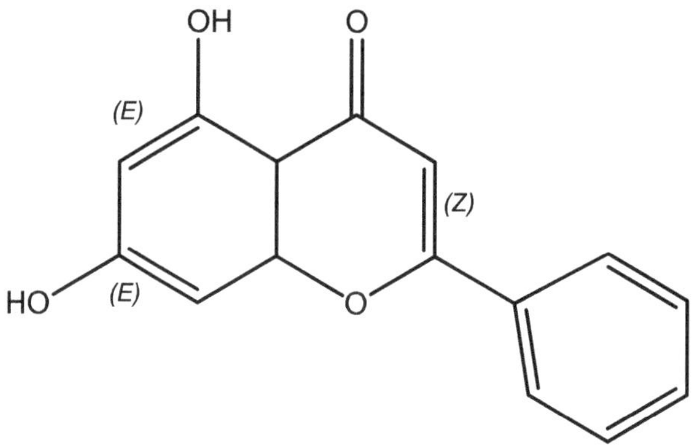

47.	Homoeriodictyol	Bibenzyl	*Dendrobium densiflorum*

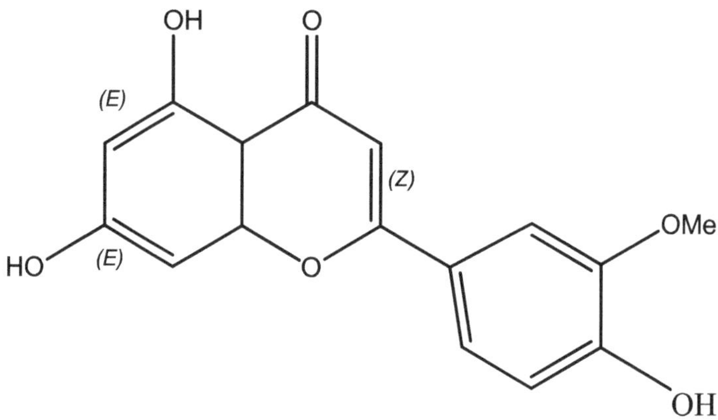

Contd...

Table 32.2–*Contd...*

Sl.No.	*Name*	*Category*	*Source*
48.	(2S)-5',2',6'-trihydroxy-6-lavandulyyl-4"-(y,y-dimethylallyl)-2",2"-dimethylpyrano-[5",6": 7,8}-flavone		*Spiranthes australis*
	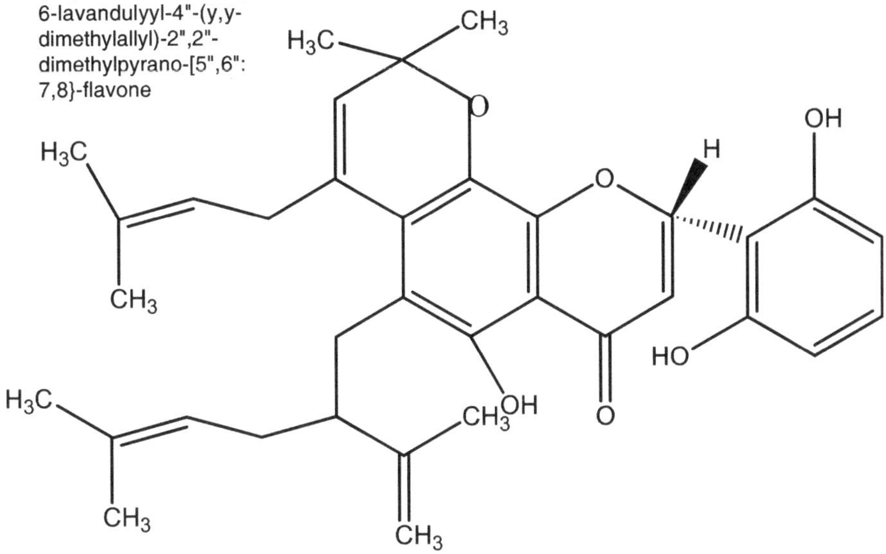		
49.	Homoisoflavanone		*Cremastra appendiculata*
	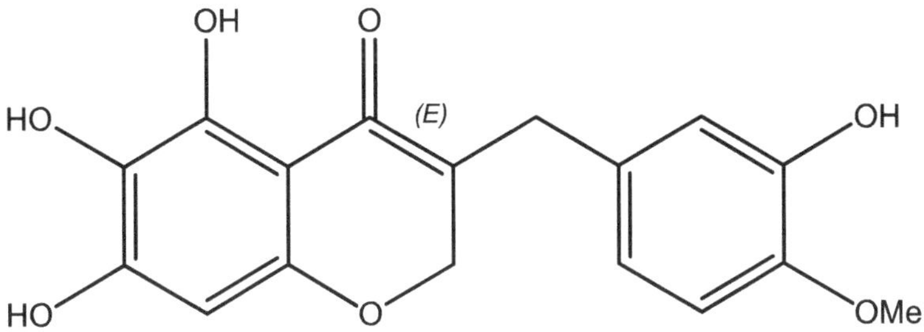		
50.	Coeloginanthridin		*Coelogyne cristata*
	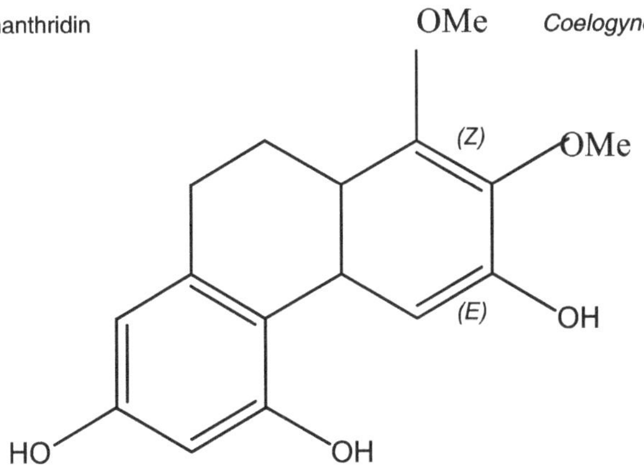		

Contd...

Table 32.2–*Contd...*

Sl.No.	*Name*	*Category*	*Source*
51.	Coeloginanthrinin	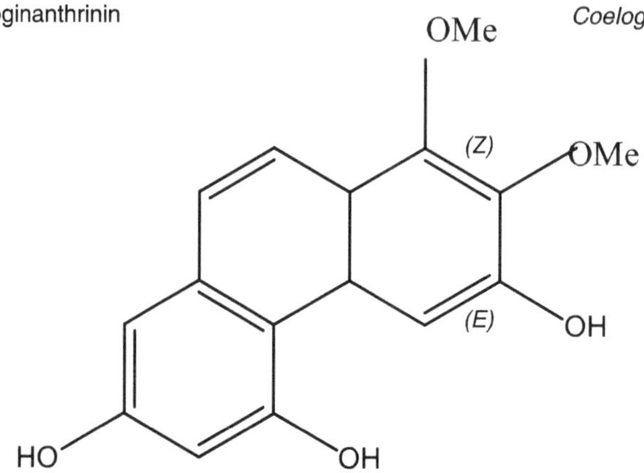	*Coelogyne cristata*
52.	Lusianthrin	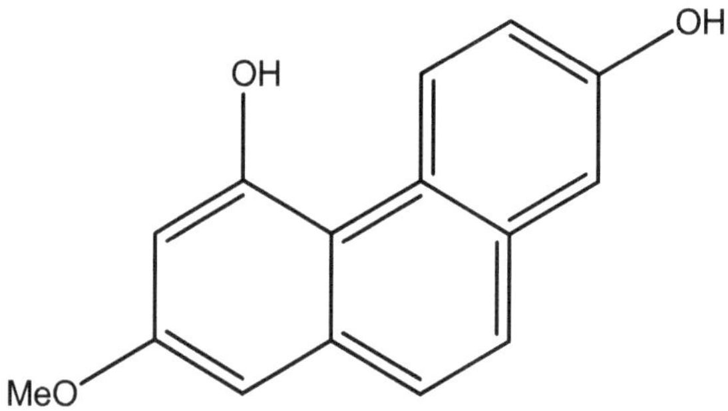	*Cypripedium macranthos*
53.	Epemeranthone	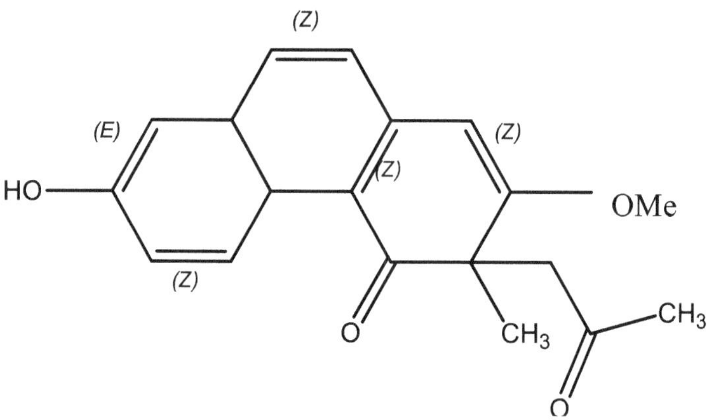	*Epemerantha lonchophylla*

Contd...

Table 32.2–*Contd...*

Sl.No.	*Name*	*Category*	*Source*
54.	Fimbriol A	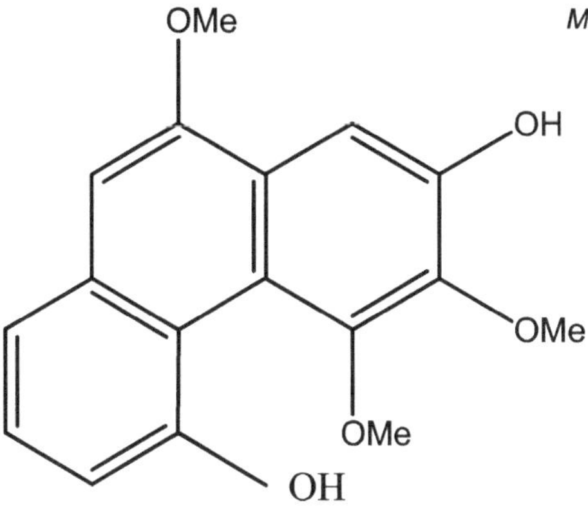	*Maxillaria densa*
55.	Erianthridin	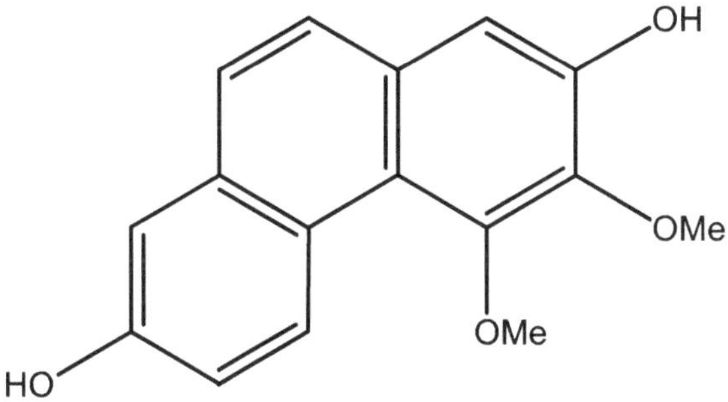	*Maxillaria densa*
56.	4,7 dihydroxy-2-methoxy-9,10-dihydrophenanthrene	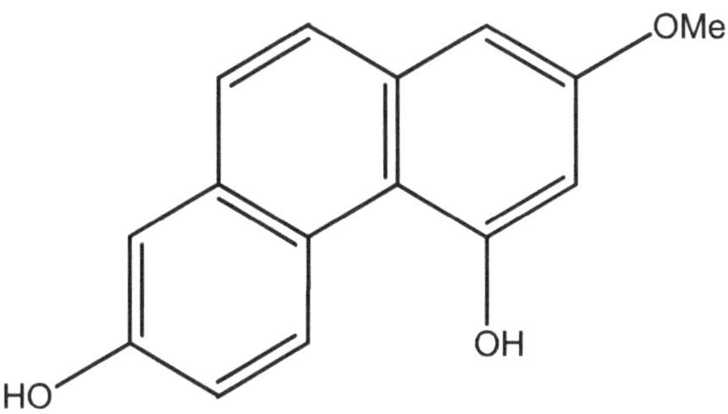	*Dendrobium nobile*

Contd...

Table 32.2–*Contd...*

Sl.No.	*Name*	*Category*	*Source*
57.	Denbinobin		*Dendrobium nobile*
	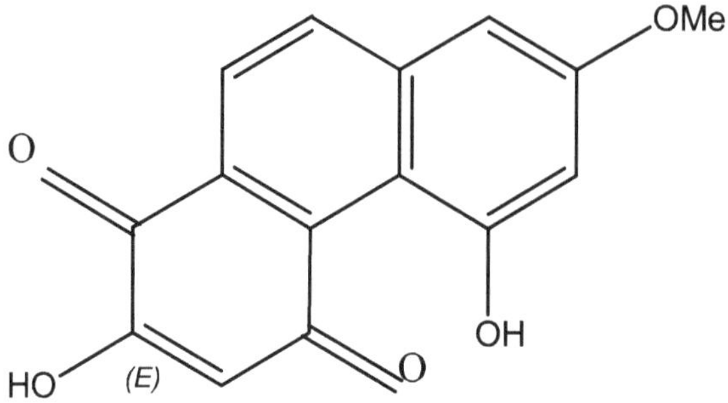		
58.	Monoliformin		*Dendrobium monoliforme*
	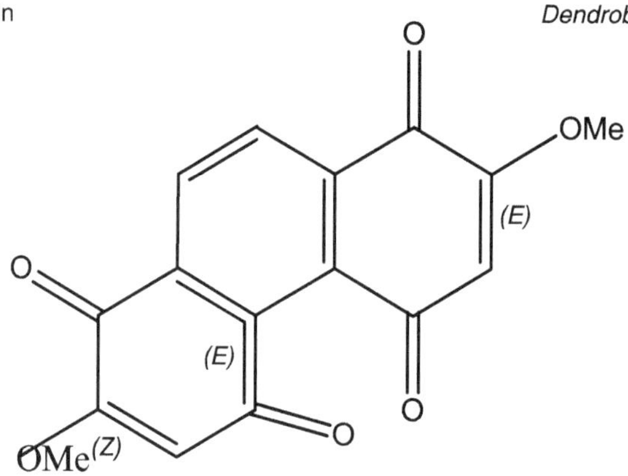		
59.	2,6-dimethoxy-1,4, 5,8-phenanthridiquinone		*Dendrobium monoliforme*
	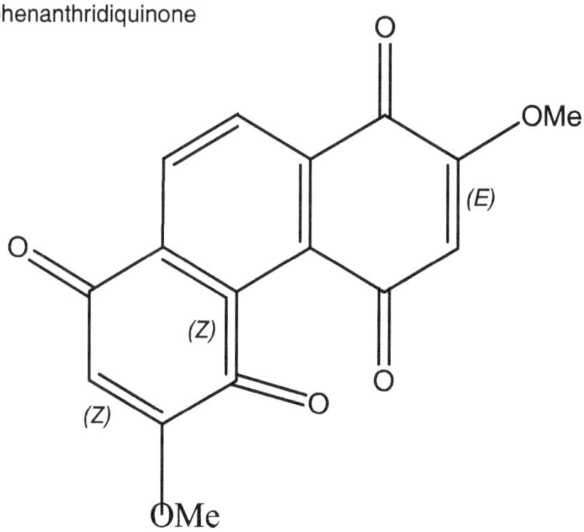		

Contd...

Table 32.2–*Contd...*

Sl.No.	*Name*	*Category*	*Source*
60.	7-hydroxy-5,6dimethoxy-1,4-phenanthrenequinone	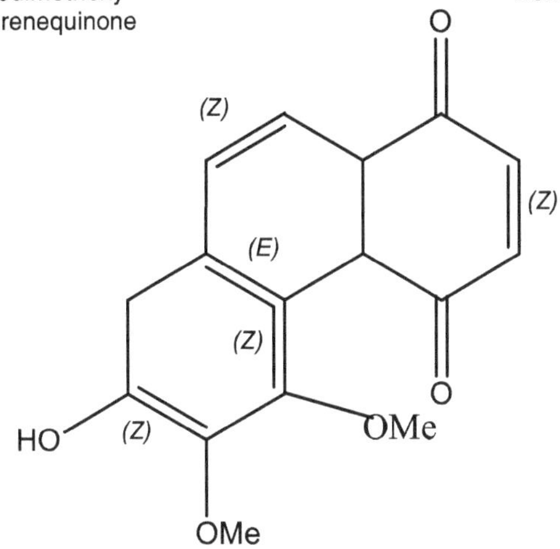	*Dendrobium monoliforme*
61.	2,3-Dimethoxy-9,10-dihydrophenanthrne-4,7 -diol	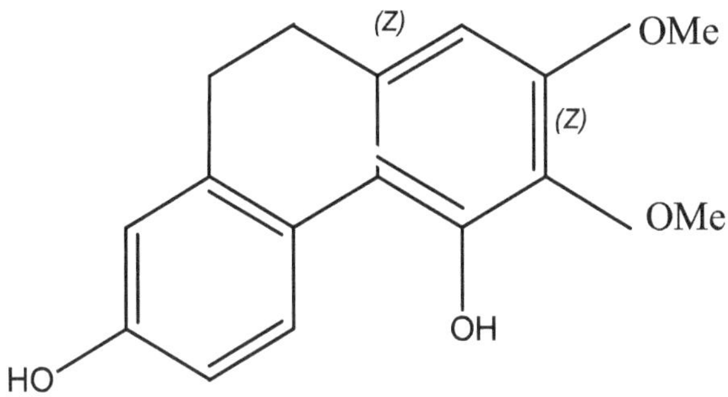	*Epidendrum rigidum*
62.	3,4,9,-trimethoxyphe-nanthrnen-2,5-diol	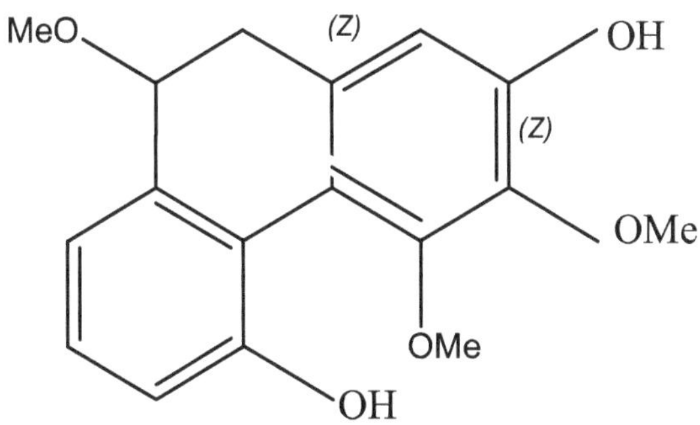	*Epidendrum rigidum*

Contd...

Table 32.2–*Contd...*

Sl.No.	*Name*	*Category*	*Source*
63.	4-Methoxy-9,10-dihydro-phenanthrene-1,2,7-triol	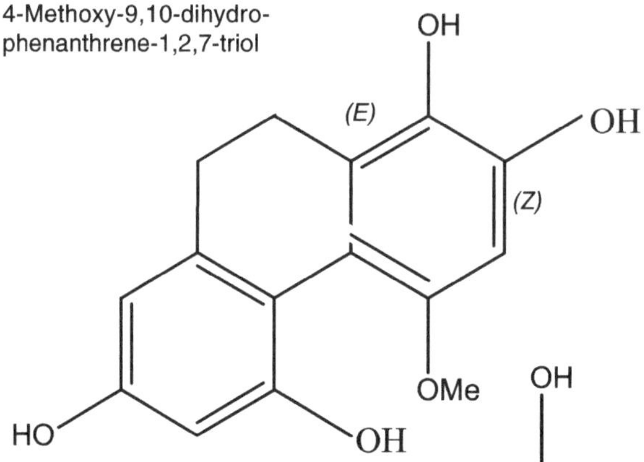	*Bletilla formosana*
64.	1-(4-Hydroxybenzyl)-4,7-dimethoxy-9,10-dihydro-phenanthrene-2-ol	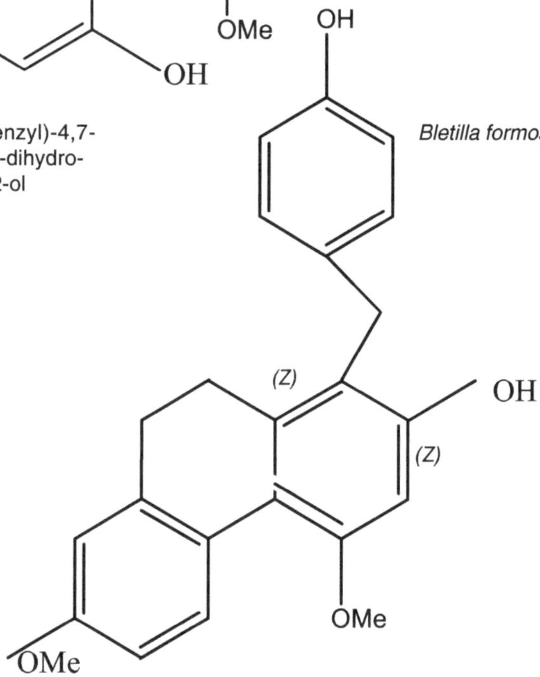	*Bletilla formosana*
65.	1,3,6-tri(4-Hydroxybenzyl)-4-methoxydihydrophen-anthrene-2,7-diol	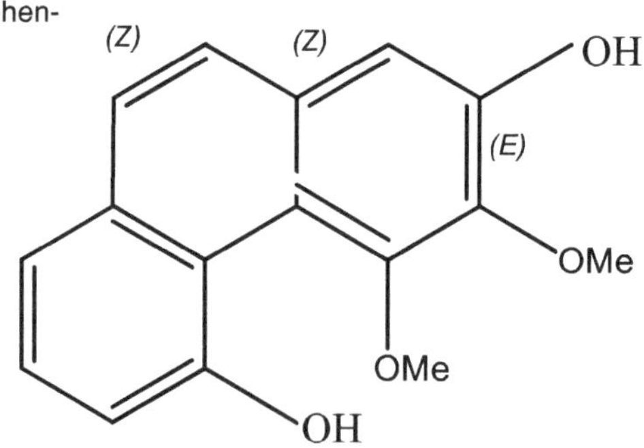	*Bletilla formosana*

Contd...

Table 32.2–*Contd...*

Sl.No.	Name	Category	Source
66.	2,5-dihydroxy-3,4-dimethoxyphenanthrene		*Maxillaria densa*
67.	9,10-dihydro-2,5-dihydroxy-3,4-dimethoxyphenanthrene 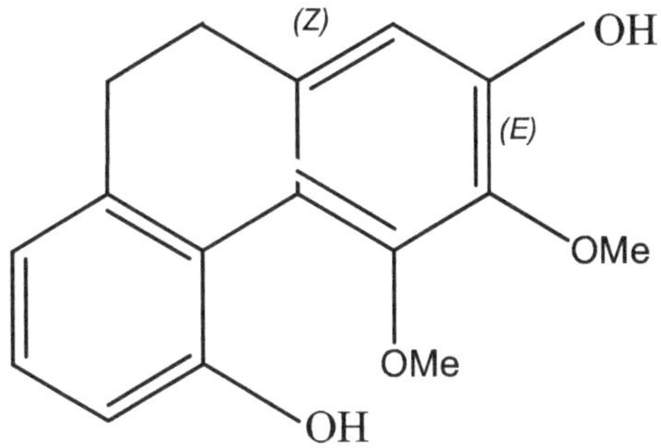		*Maxillaria densa*
68.	Plicatol B 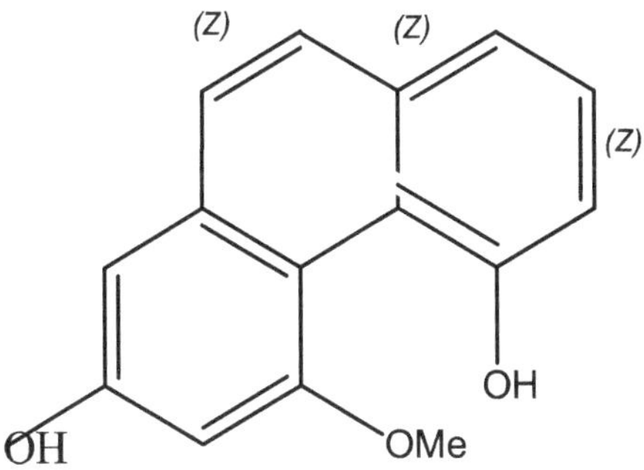		*Bulbophyllum kwangtungnense*
69.	Nudol 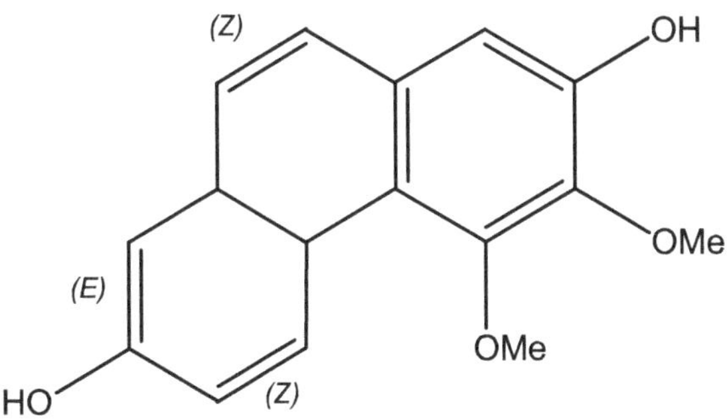		*Maxillaria densa*

Contd...

Table 32.2–*Contd...*

Sl.No.	*Name*	*Category*	*Source*
70.	Dendrochrysanene	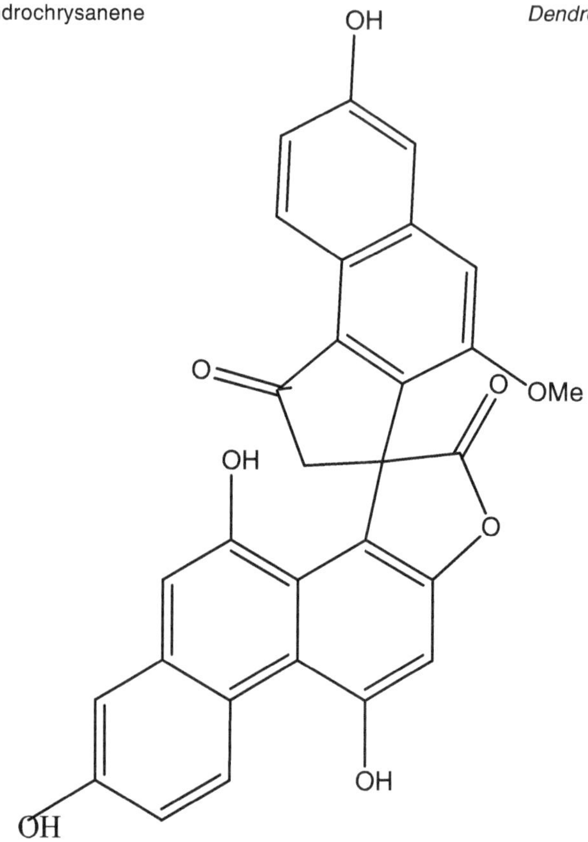	*Dendrobium chrysotoxum*
71.	OchronneA	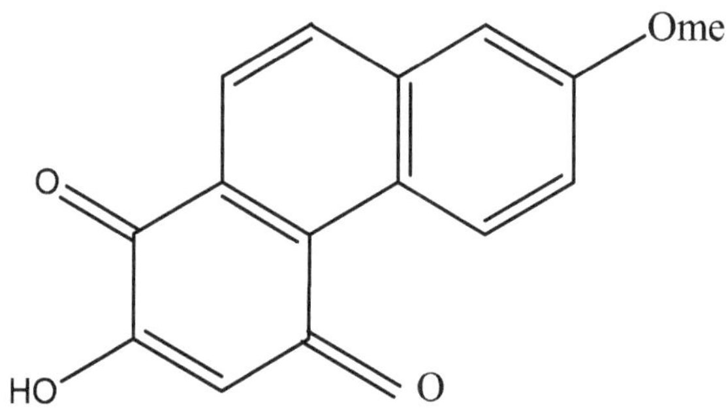	*Coleogyne ochracea*

Contd...

Table 32.2–*Contd...*

Sl.No.	Name	Category	Source
72.	Callosuminin		*Agrostophyllum callosum*
73.	3,3,-DIHYDROXY-2'6'-BIS(P-HYDROXYBENZYL)-5-METHOXYBIBENZYL		*Bletilla striata*

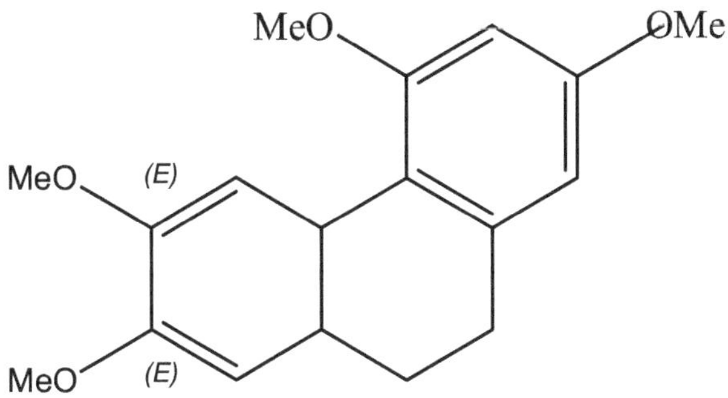

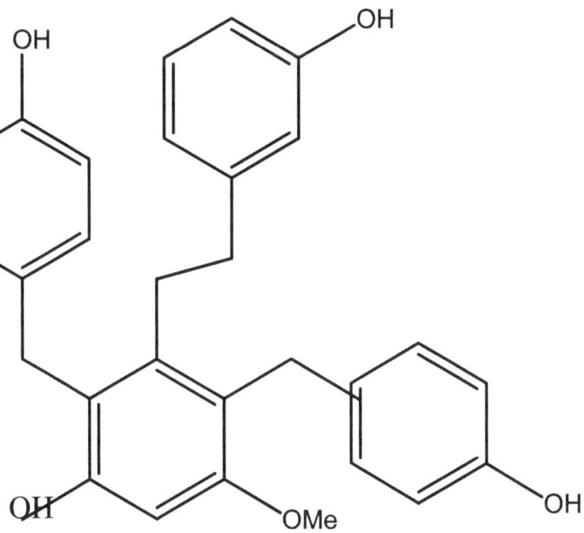

Contd...

Table 32.2–*Contd...*

Sl.No.	*Name*	*Category*	*Source*
74.	3',5-Dihydroxy-2-(p-hydroxybenzyl)-3-methoxybibenzyl	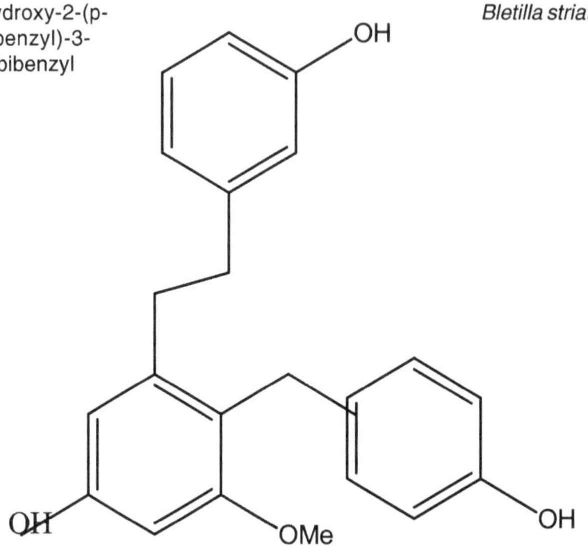	*Bletilla striata*
75.	1-(p-Hydroxybenzyl)-4,8-dimethoxyphenanthrene-2,7-diol	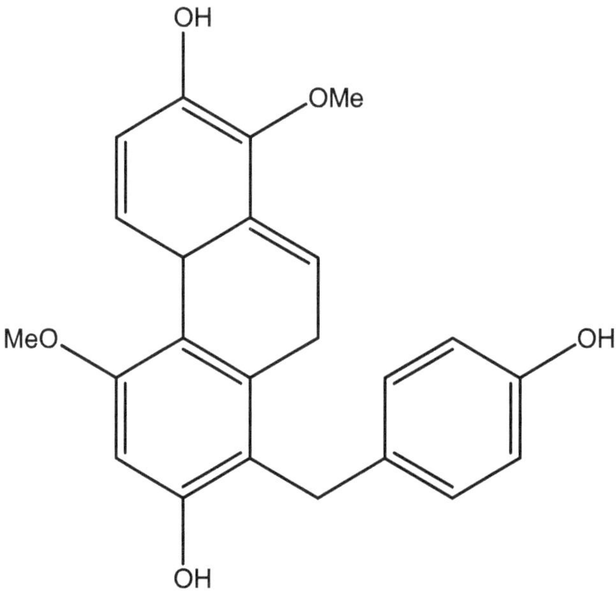	*Bletilla striata*

Contd...

Table 32.2–*Contd...*

Sl.No.	*Name*	*Category*	*Source*
76.	2,7-Dihydroxy-1,3-bis(p-hydroxybenzyl)-4-methoxy-9,10-dihydroxyphenanthrene	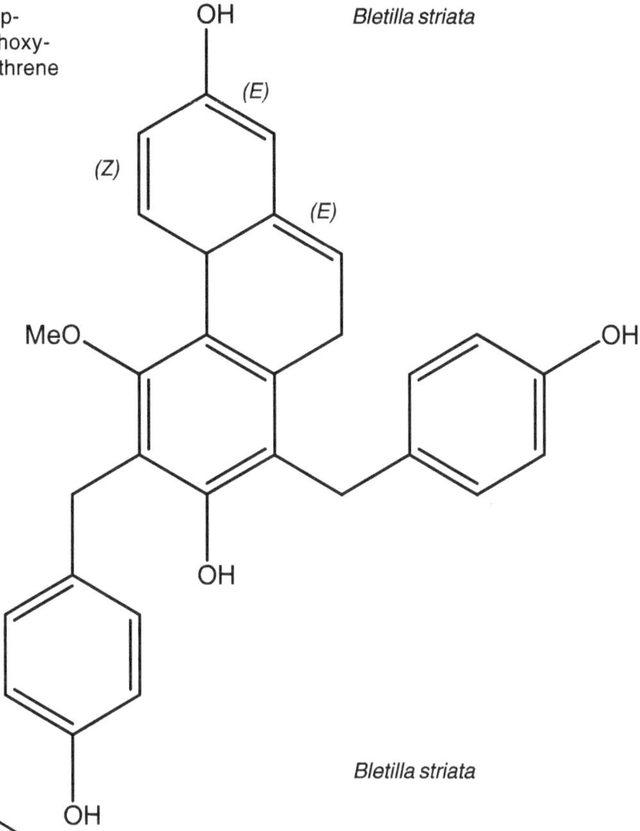	*Bletilla striata*
79.	Blestriarene B	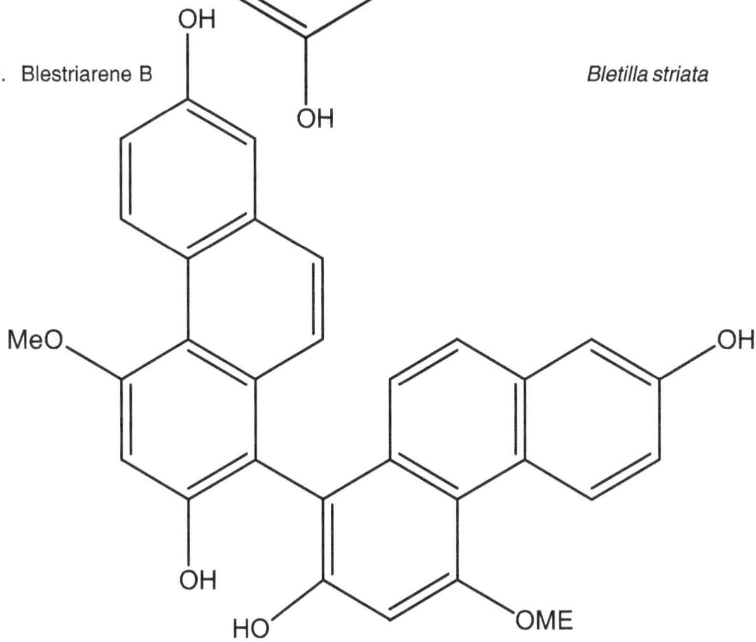	*Bletilla striata*

Contd...

Table 32.2–*Contd...*

Sl.No.	*Name*	*Category*	*Source*
80.	Blestriarene C	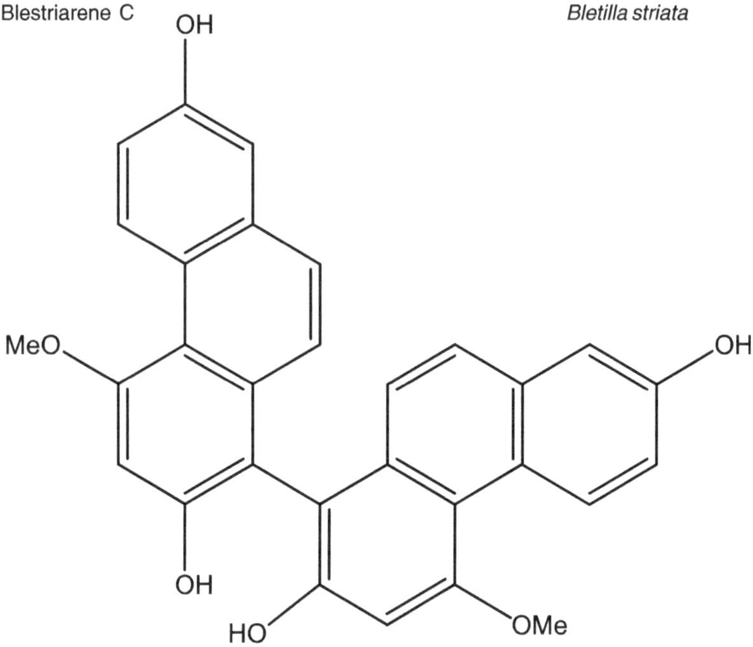	*Bletilla striata*
81.	Blestrianol A	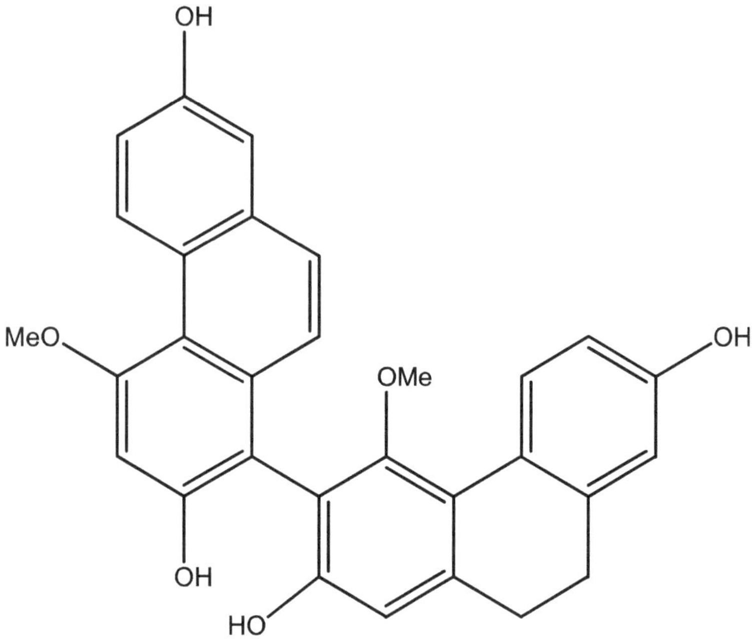	*Bletilla striata*

Contd...

Table 32.2–*Contd...*

Sl.No.	Name	Category	Source
82.	Lusianthridin		*Nidema boothi*

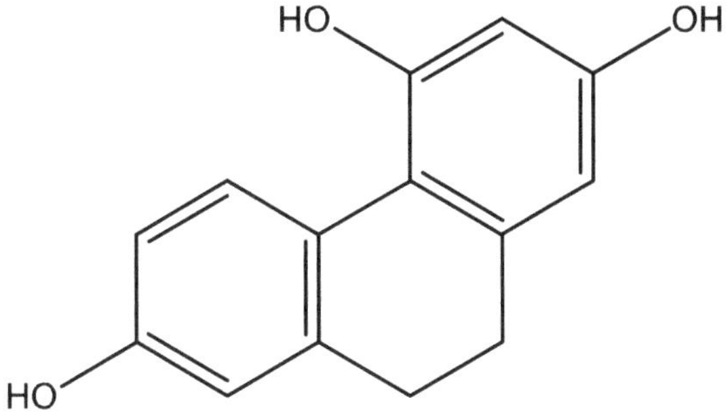

Sl.No.	Name	Category	Source
83.	2,4,7-Trihydroxy-9,10 dihyrophenanthrene		*Pholiodota yunnanensis*

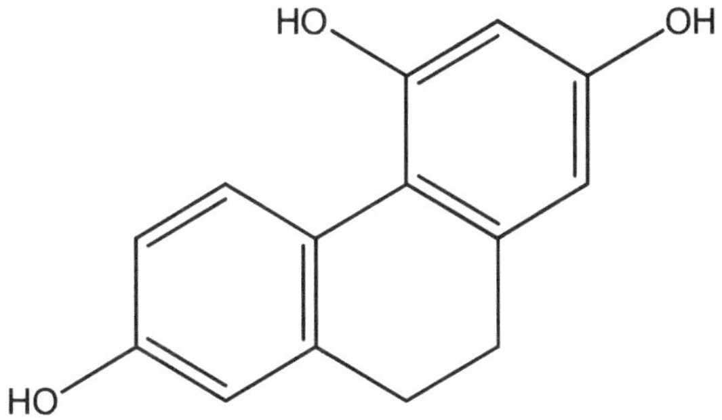

Sl.No.	Name	Category	Source
84.	3,7-dihydroxy-2,4,8-trimethoxyphenanthrene		*Pholiodota yunnanensis*

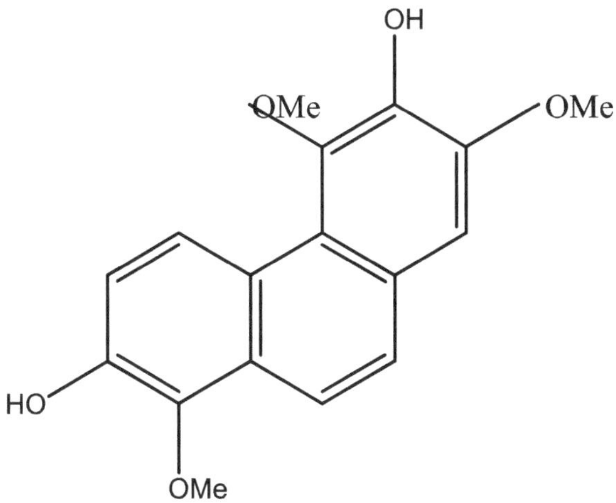

Contd...

Table 32.2–*Contd...*

Sl.No.	*Name*	*Category*	*Source*
85.	Coelonin		*Pholiodota yunnanensis*

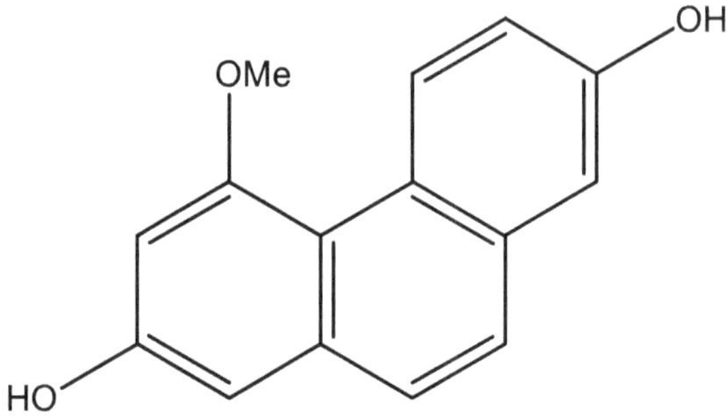

86.	3,7 –dihydroxy-2,4-dimethoxyphenanthrene		*Pholiodata yunnanensis*

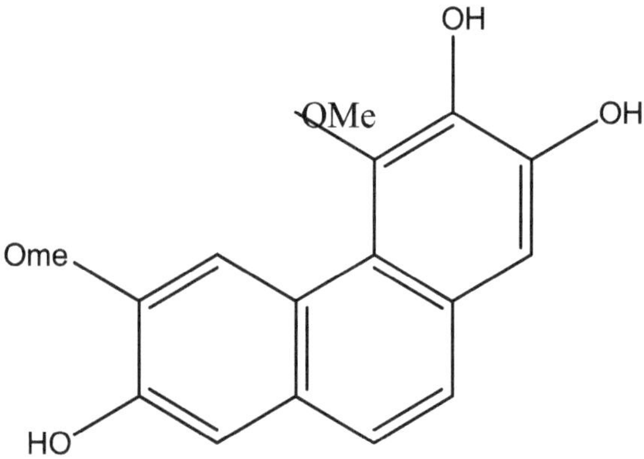

87.	3,7 –dihydroxy-2,4,6-trimethoxyphenanthrene		*Pholiodata yunnanensis*

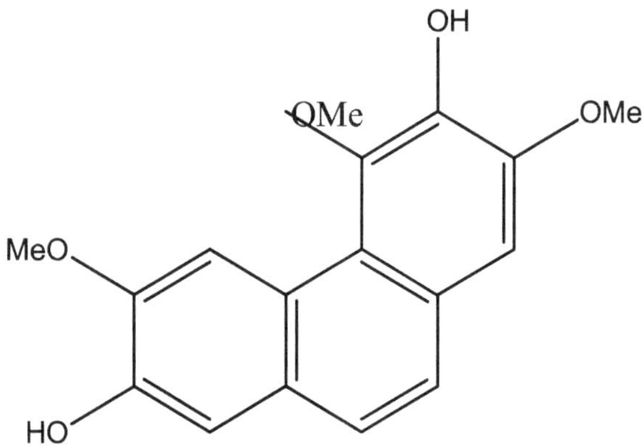

Contd...

Table 32.2–*Contd...*

Sl.No.	*Name*	*Category*	*Source*
88.	2,7,2',7',2'-Pentahydroxy-4,4,', 4",7"-tetramethoxy-1,8,1', 1"-triphenanthrene	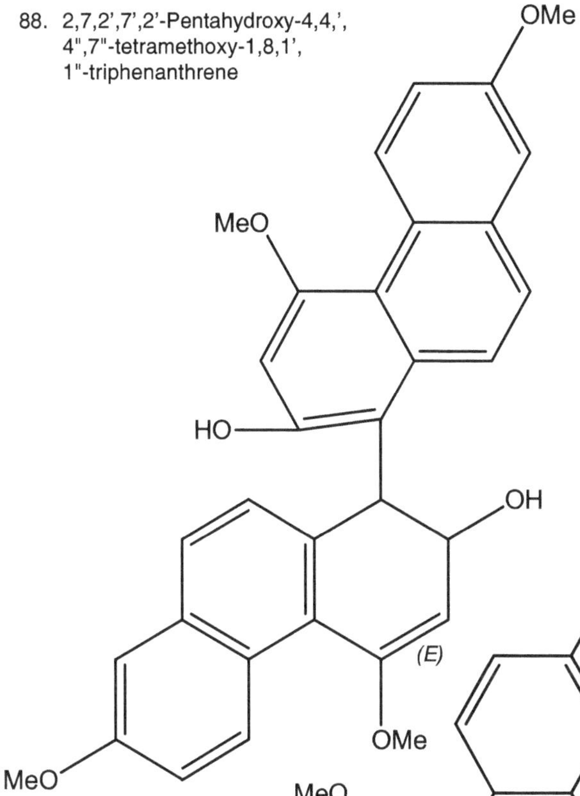	*Cremastra appendiculata*
89.	Cirrhopetalanthin	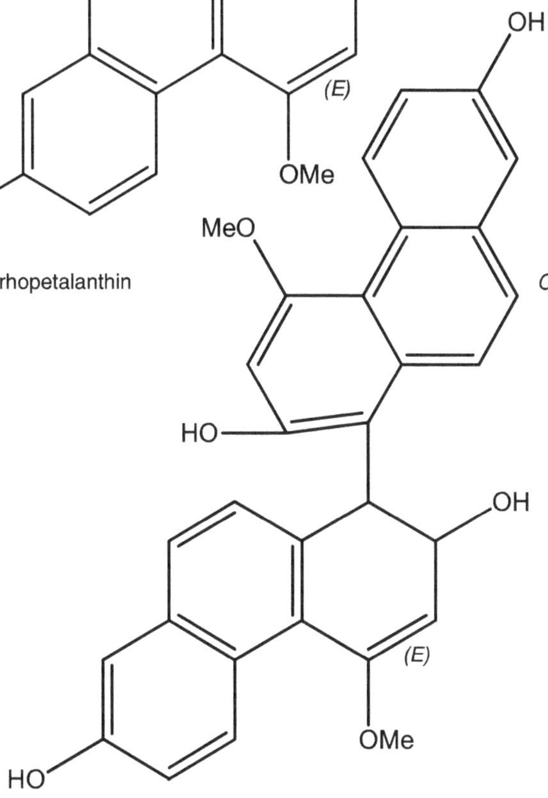	*Cremastra appendiculata*

Contd...

Table 32.2–*Contd...*

Sl.No.	*Name*	*Category*	*Source*
90.	Gymnocopin A	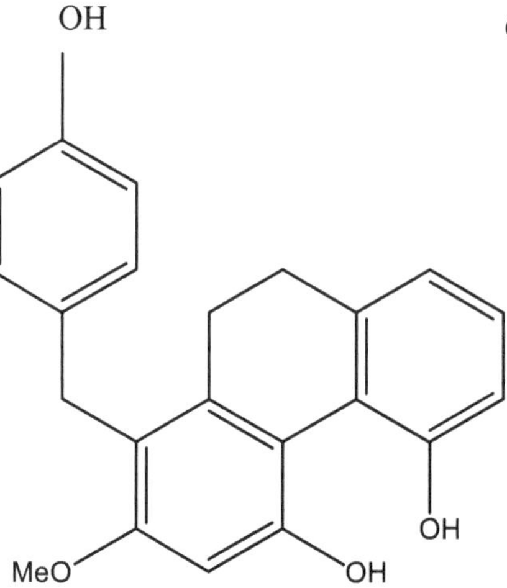	*Cremastra appendiculata*
91.	Gymnocopin B	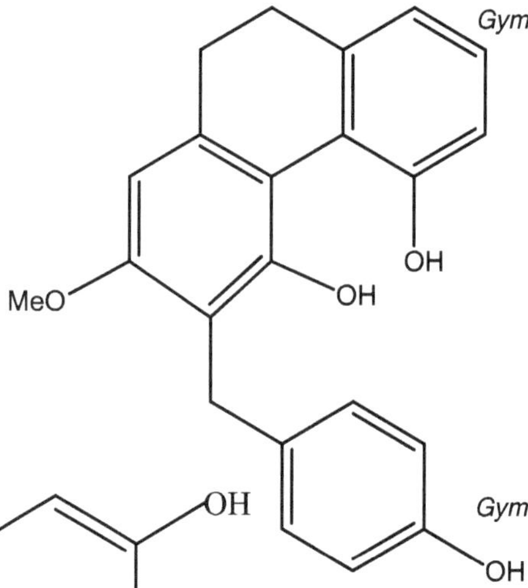	*Gymnadenia conopsea*
92.	Gymnocopin D	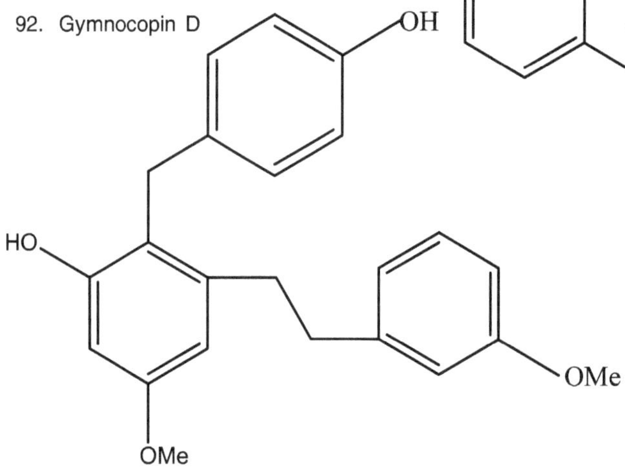	*Gymnadenia conopsea*

Contd...

Table 32.2–*Contd...*

Sl.No.	Name	Category	Source
93.	Dihydroxy-2,6-bis(4-hydroxybenzyl)-5-methoxybibenzyl	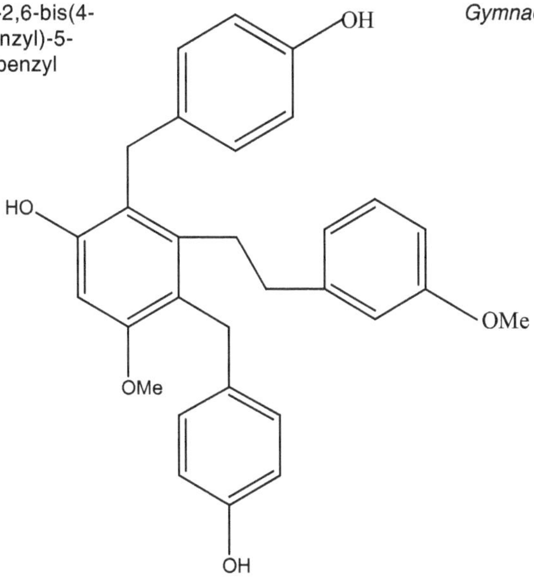	*Gymnadenia conopsea*
94.	Spiranthesol	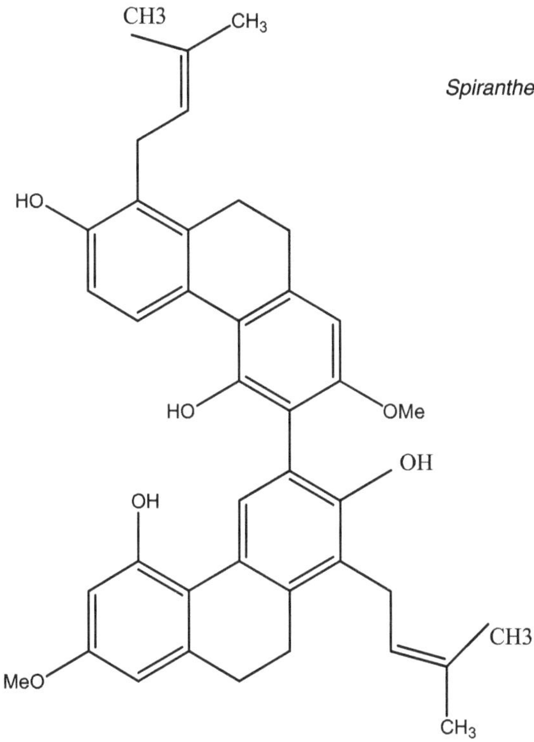	*Spiranthes sinesis*

Contd...

Table 32.2–*Contd...*

Sl.No.	*Name*	*Category*	*Source*
95.	Spiranthoquinone	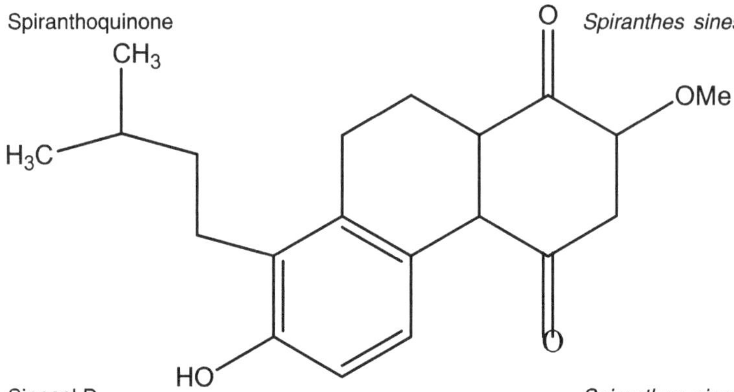	*Spiranthes sinesis*
96.	Sinesol D	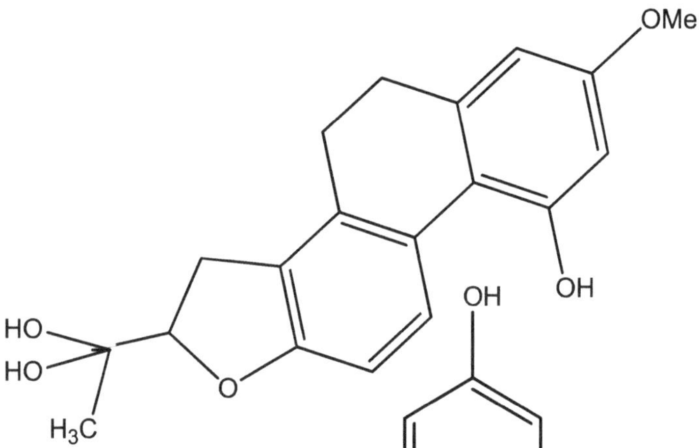	*Spiranthes sinesis*
97.	Sinesol E	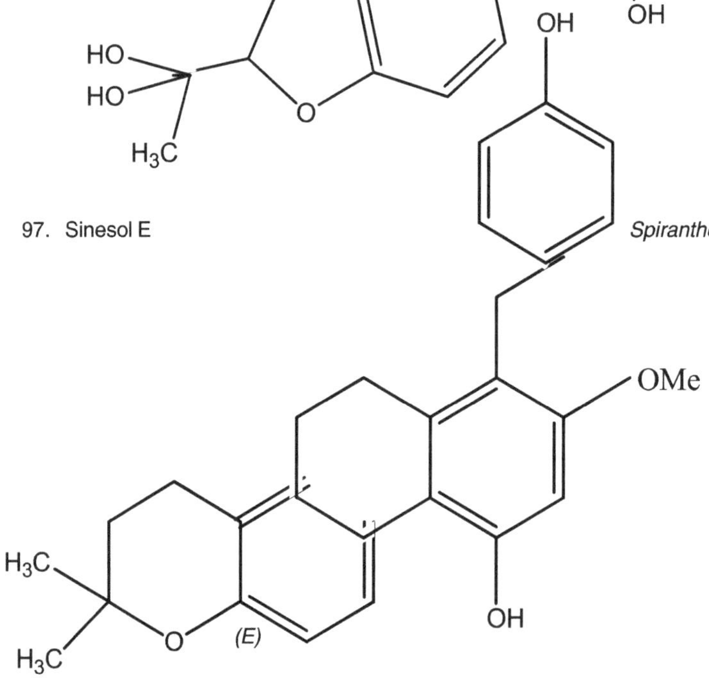	*Spiranthes sinesis*

Contd...

Table 32.2–*Contd...*

Sl.No.	*Name*	*Category*	*Source*
98.	Sinesol F	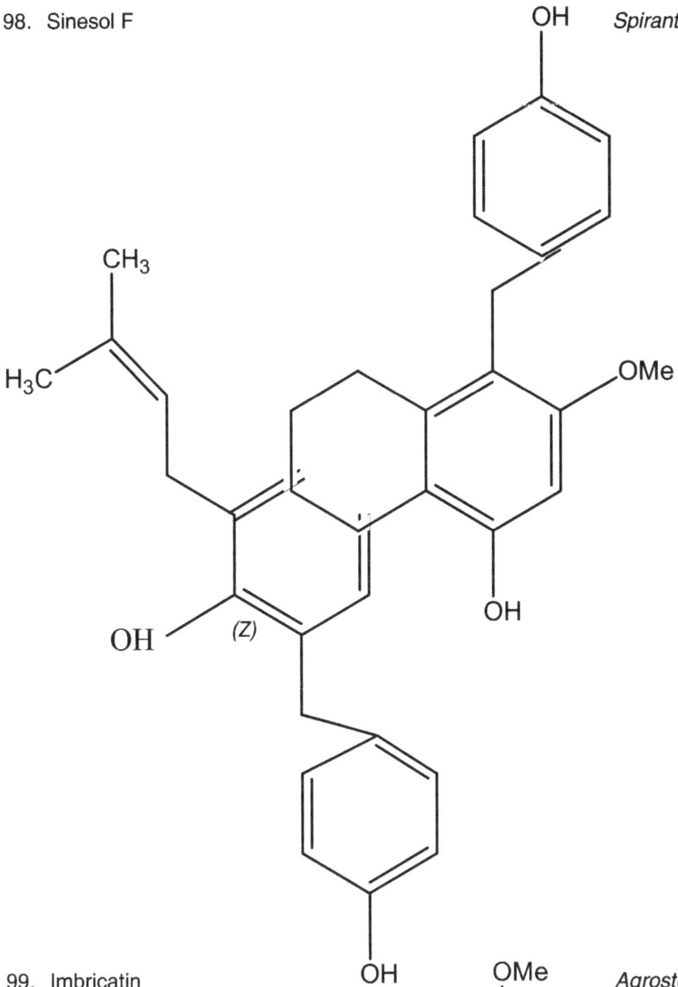	*Spiranthes sinesis*
99.	Imbricatin	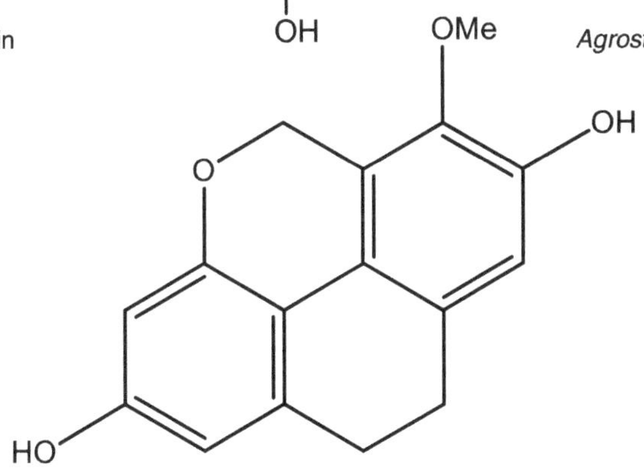	*Agrostophyllum brevipes*

Contd...

Table 32.2–*Contd...*

Sl.No.	*Name*	*Category*	*Source*
100.	Flaccidin	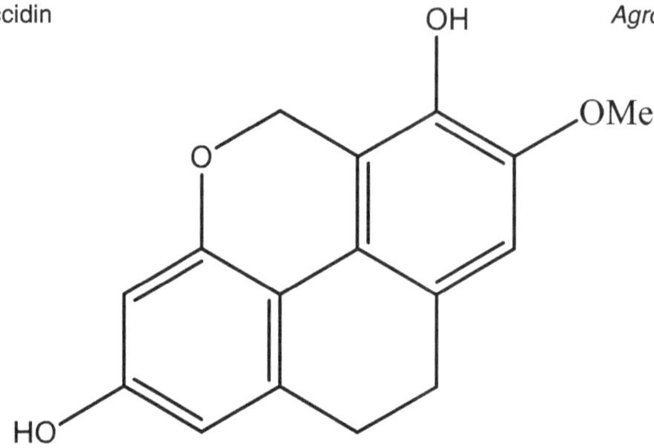	*Agrostophyllum brevipes*
101.	Lonchophylloid A	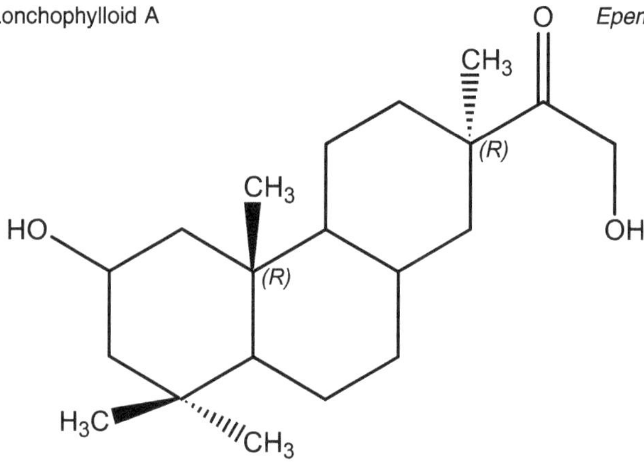	*Epemerantha lonchophylla*
102.	Lonchophylloid B	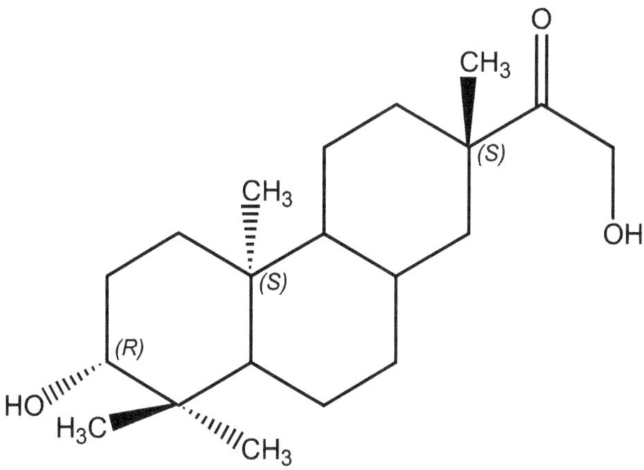	*Epemerantha lonchophylla*

Contd...

Table 32.2–*Contd...*

Sl.No.	Name	Category	Source
103.	Dendroside D		*Dendrobium nobile*

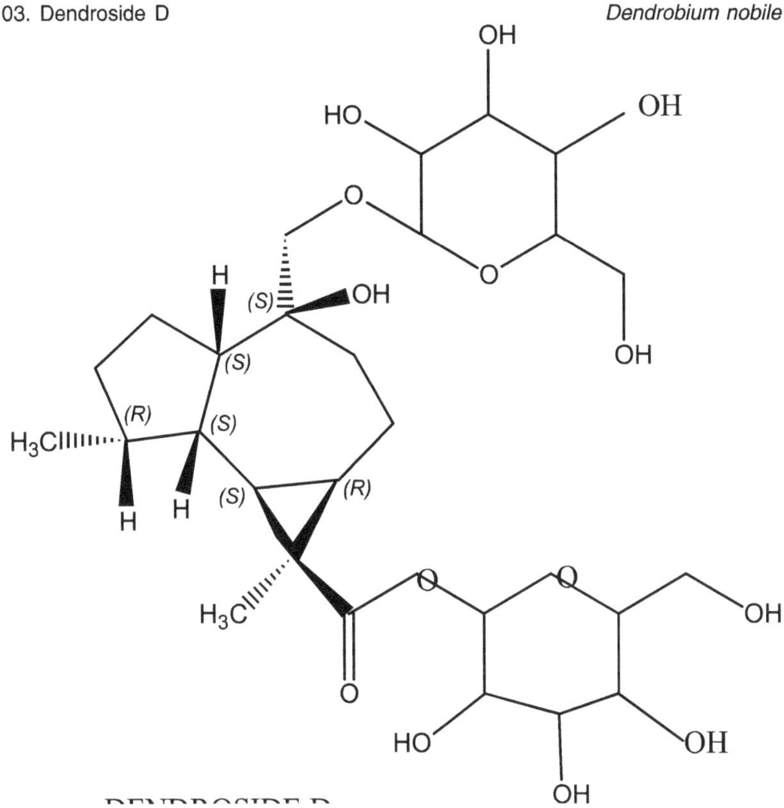

Sl.No.	Name	Category	Source
104.	Gastrol		*Gatrodia elata*

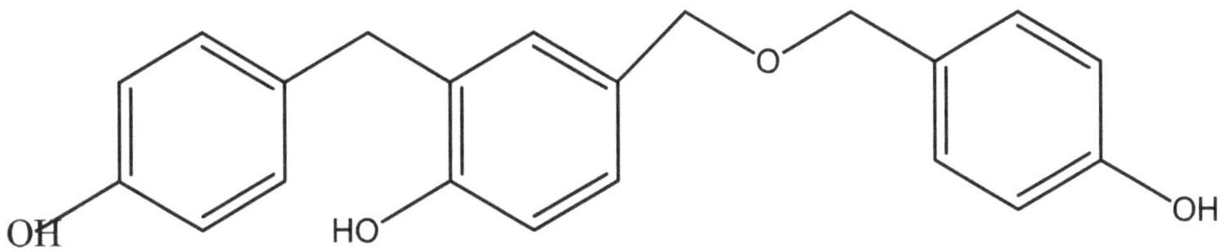

Contd...

Table 32.2–*Contd...*

Sl.No.	Name	Category	Source
105.	Dendroside F	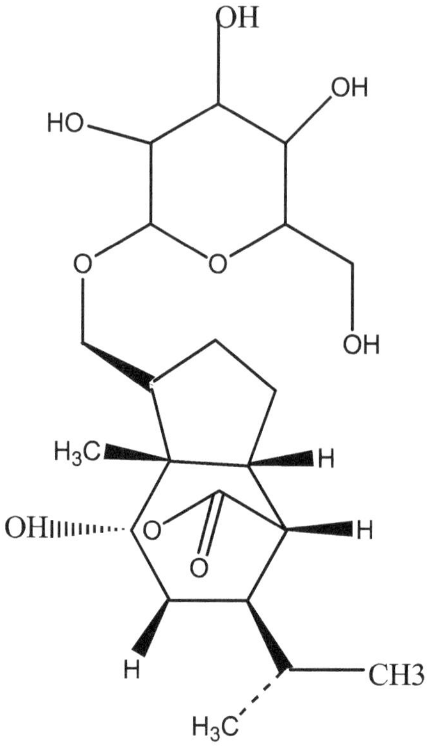	*Dendrobium nobile*
106.	Kinsenoside	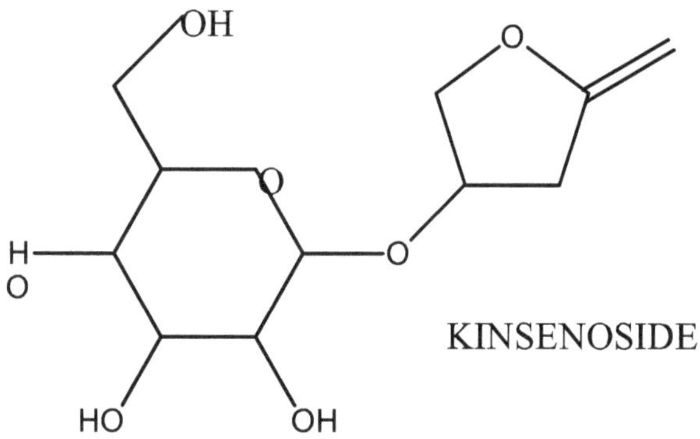	*Anoectochilus formosanus*

Contd...

Table 32.2–*Contd...*

Sl.No.	*Name*	*Category*	*Source*
107.	Cirrhopetalathrin		*Cremastra appendiculata*
	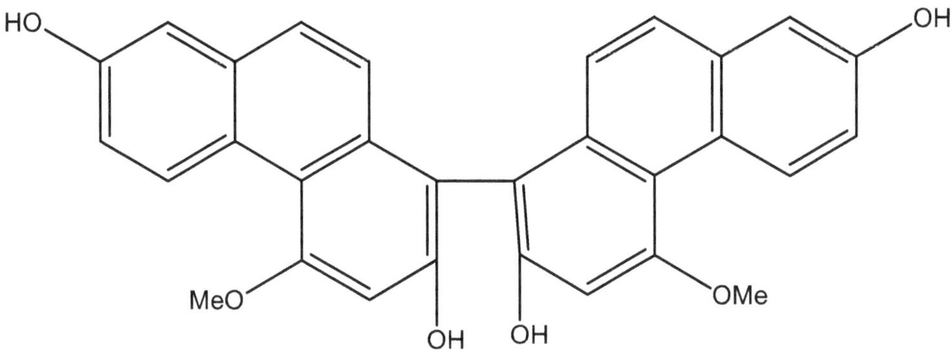		
108.	Cis-Meliotoside		*Dendrobium auranticum*
	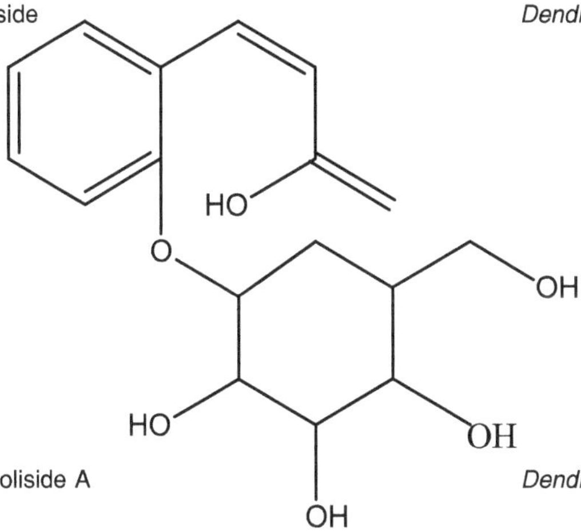		
109.	Dendromonoliside A		*Dendrobium monoliforme*
	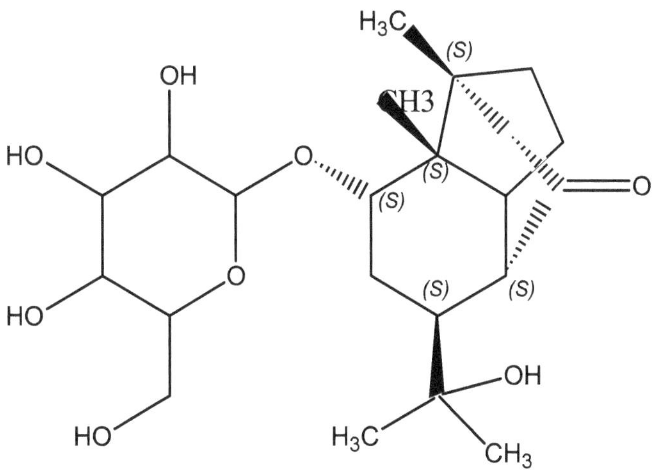		

Contd...

Table 32.2–*Contd...*

Sl.No.	Name	Category	Source
110.	Dendroside G	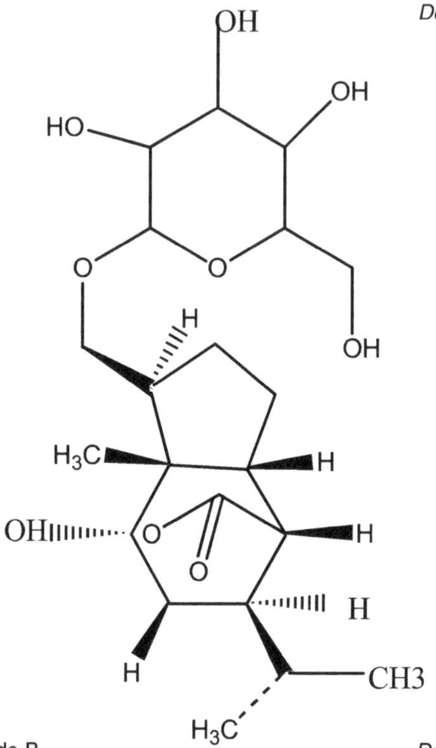	*Dendrobium nobile*
111.	Dendromonoliside B	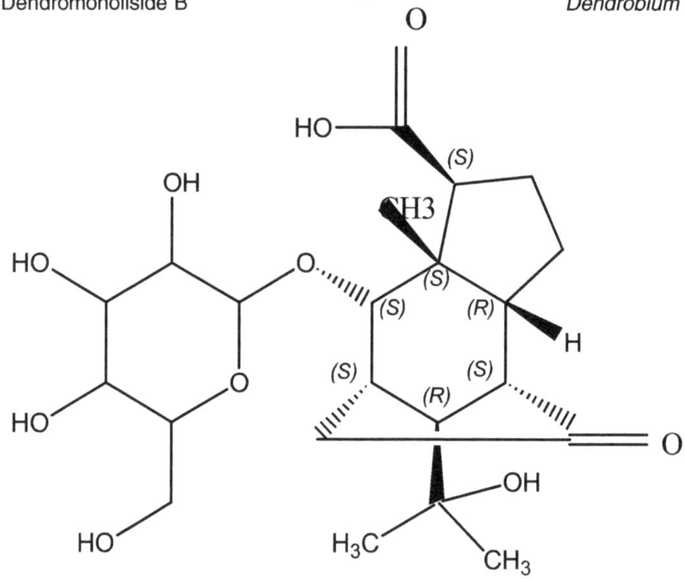	*Dendrobium monoliforme*

Contd...

Table 32.2–*Contd...*

Sl.No.	*Name*	*Category*	*Source*
112.	Dendromonoliside C	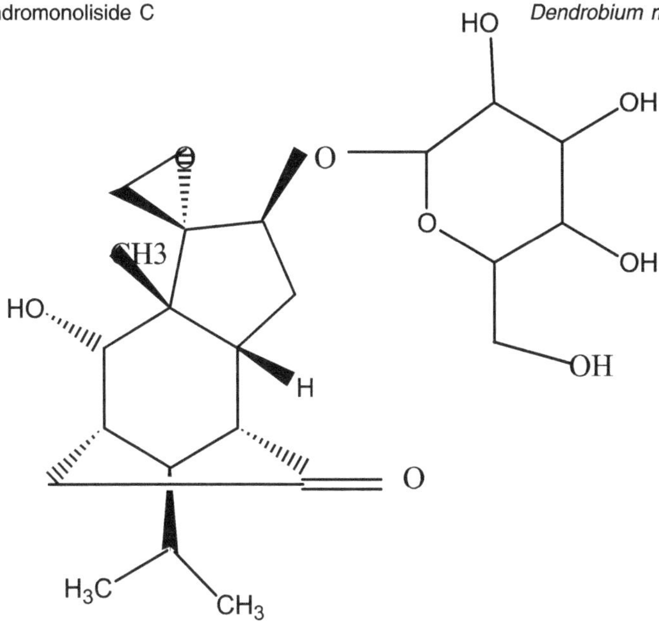	*Dendrobium monoliforme*
113.	Spiranthol C	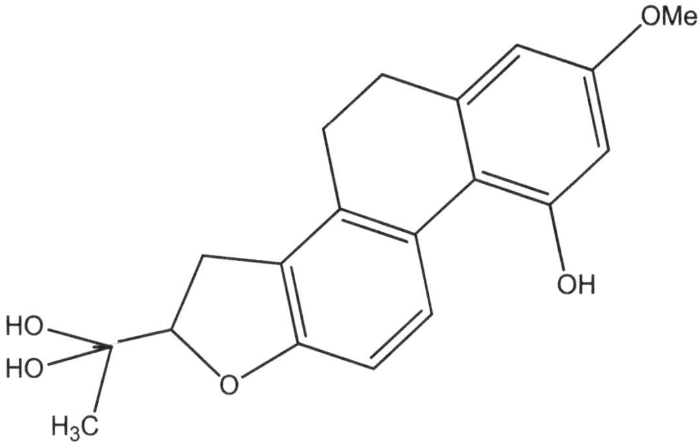	*Spiranthes sinesis*

Contd...

Table 32.2–*Contd...*

Sl.No.	*Name*	*Category*	*Source*
114.	Spirasineol B	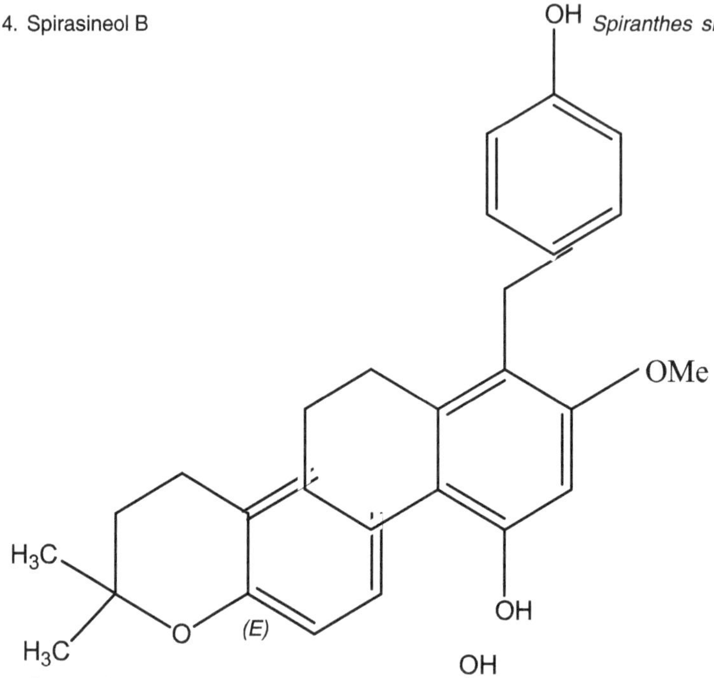	*Spiranthes sinesis*
115.	Sinesol A	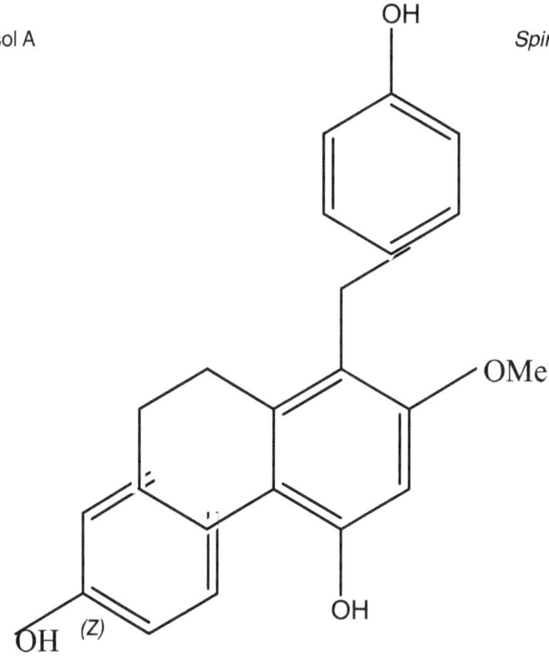	*Spiranthes sinesis*

Contd...

Table 32.2–*Contd...*

Sl.No.	Name	Category	Source
116.	Sinesol B	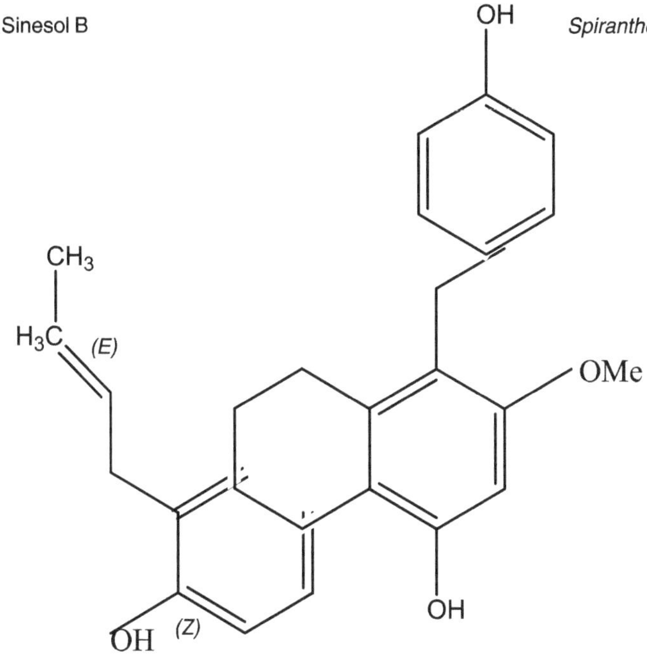	*Spiranthes sinesis*
117.	Sinesol C	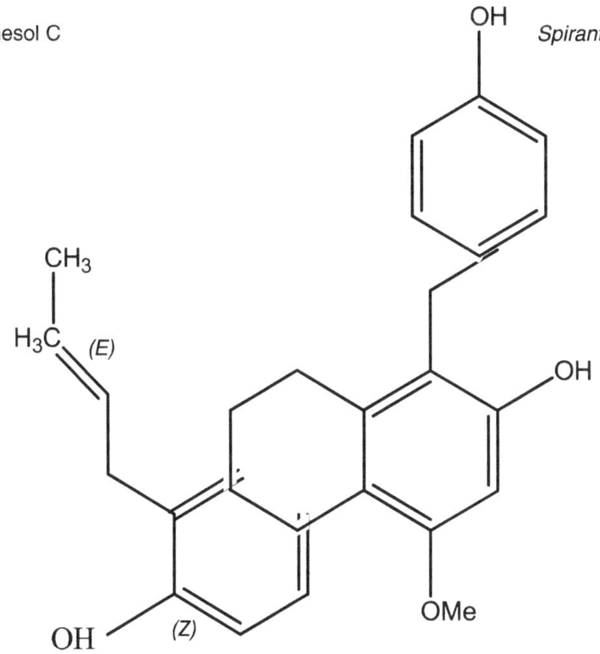	*Spiranthes sinesis*

Contd...

Table 32.2–*Contd...*

Sl.No.	*Name*	*Category*	*Source*
118.	Callosnin	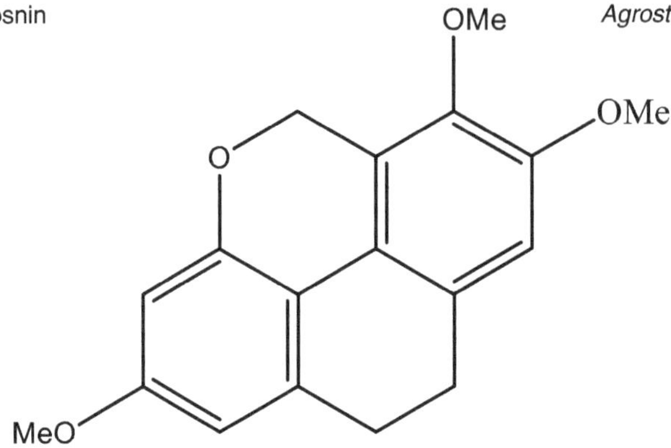	*Agrostophyllum brevipes*
119.	Callosumin	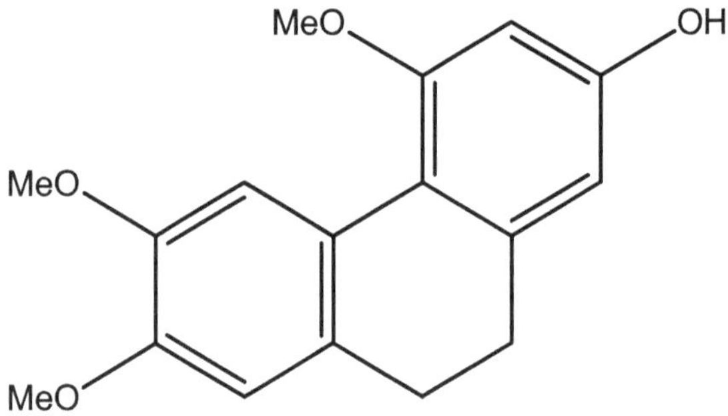	*Agrostophyllum brevipes*
120.	Trans-mellitoside	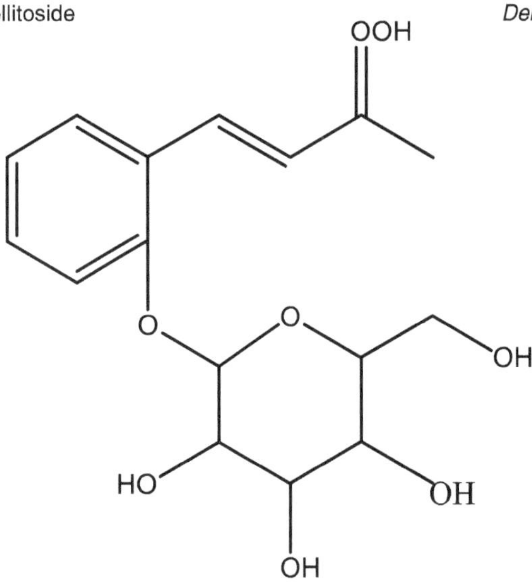	*Dendrobium auranticum*

Contd...

Table 32.2–*Contd...*

Sl.No.	Name	Category	Source
121.	Dendroside A	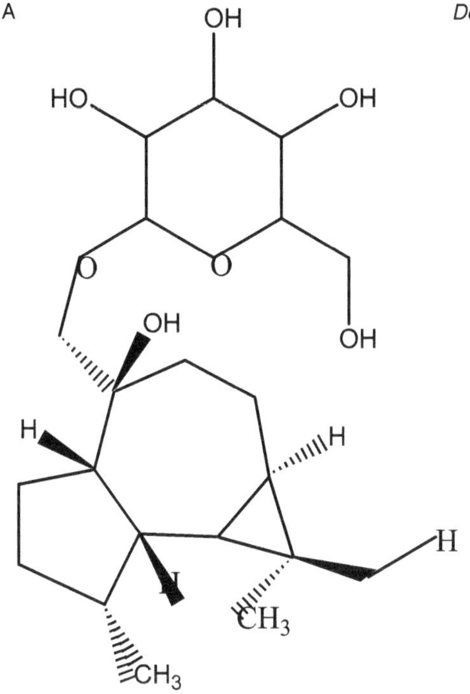	*Dendrobium auranticum*
122.	Bis(4-hydroxybenzyl)sulfide	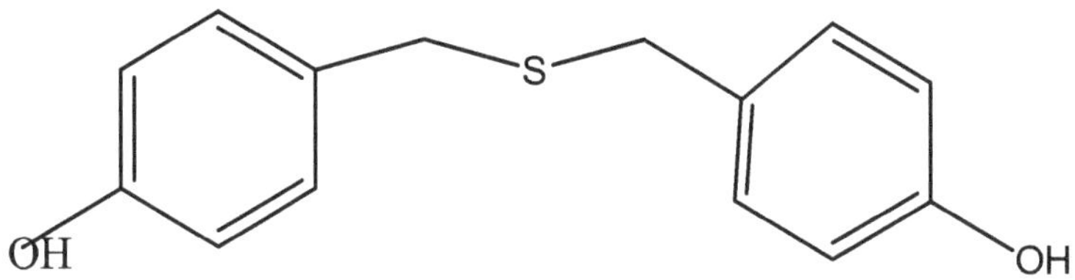	*Gastrodia elata*
123.	Scoparone	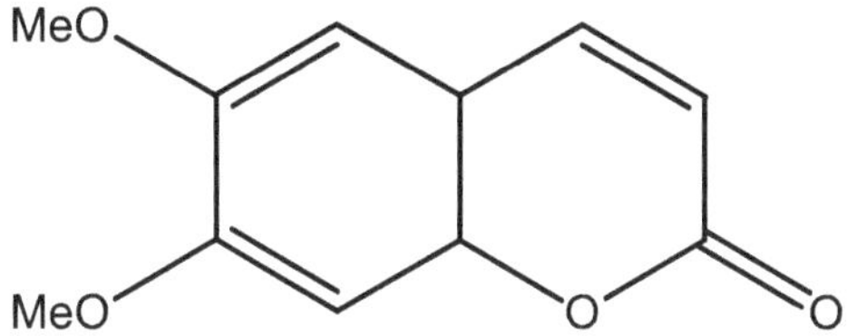	*Dendrobium densiiflorum*

Contd...

Table 32.2–*Contd...*

Sl.No.	*Name*	*Category*	*Source*
124.	Scopoletin	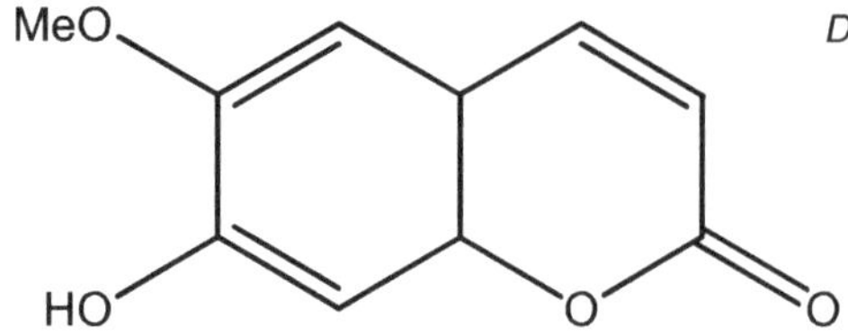	*Dendrobium densiiflorum*
125.	Nobilone	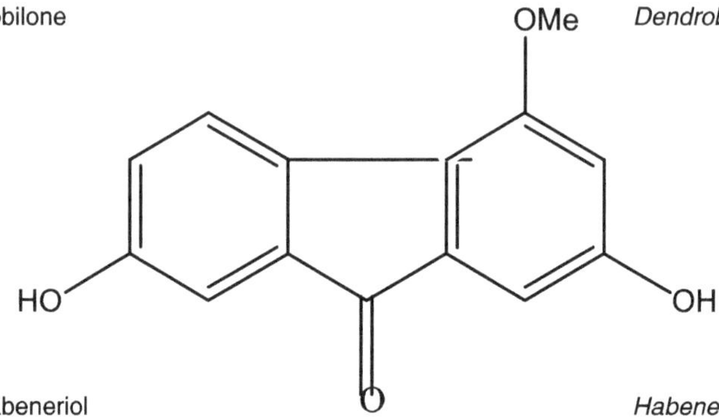	*Dendrobium nobile*
126.	Habeneriol	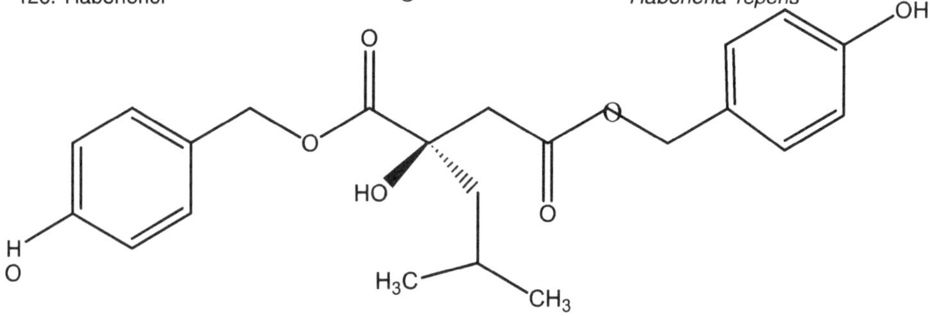	*Habeneria repens*
127.	Cymbidine A	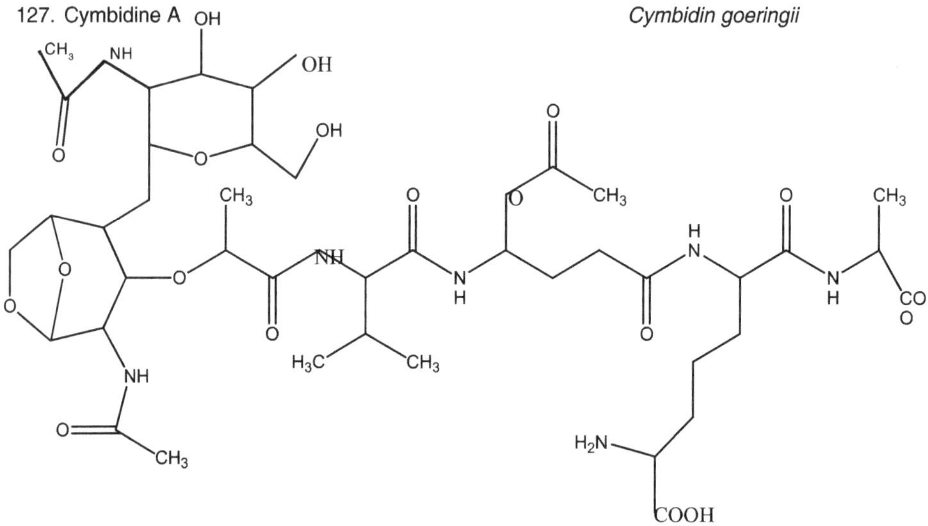	*Cymbidin goeringii*

Contd...

Table 32.2–*Contd...*

Sl.No.	*Name*	*Category*	*Source*
128.	7,8-dihydro-5-hydroxy-12,13-methylenedioxy-11-methoxyl dibenzoxepin 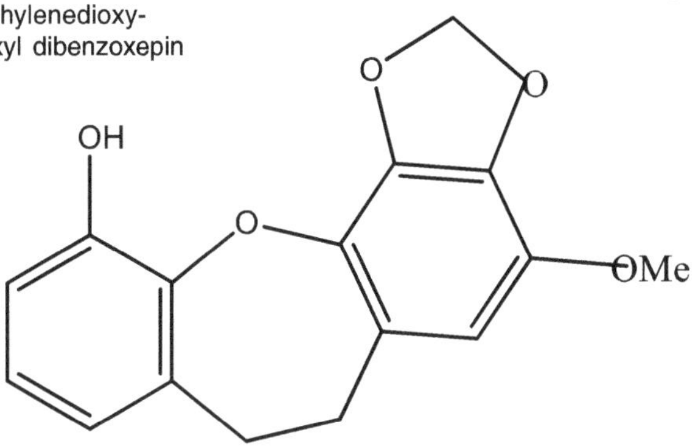		*Bulbophyllum kwangαtungense*
129.	7,8-dihydro-4-hydroxy-12,13-methylenedioxy-11-methoxyl dibenzoxepin 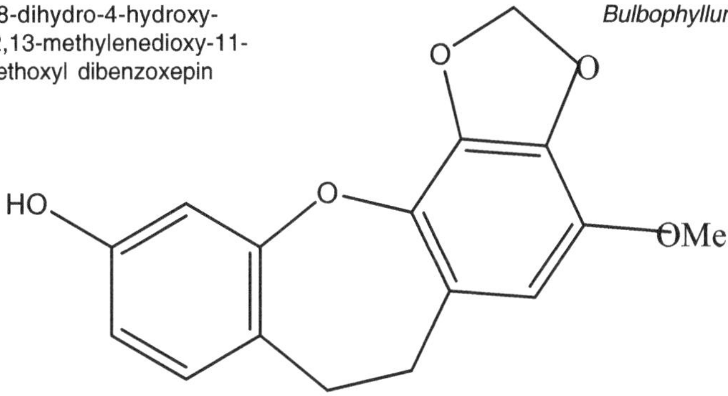		*Bulbophyllum kwangtungense*
130.	7,8-dihydro-3-hydroxy-12,13-methylenedioxy-11-methoxyl dibenzoxepin 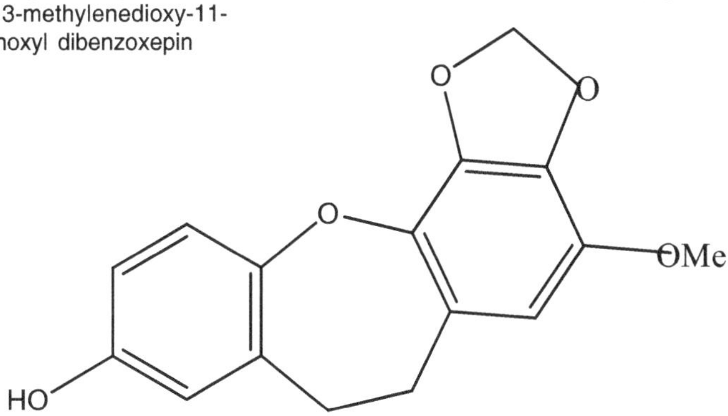		*Bulbophyllum kwangtungense*

Contd...

Table 32.2–*Contd...*

Sl.No.	*Name*	*Category*	*Source*
131.	Cyclobalanone		*Scaphyglottis livida*

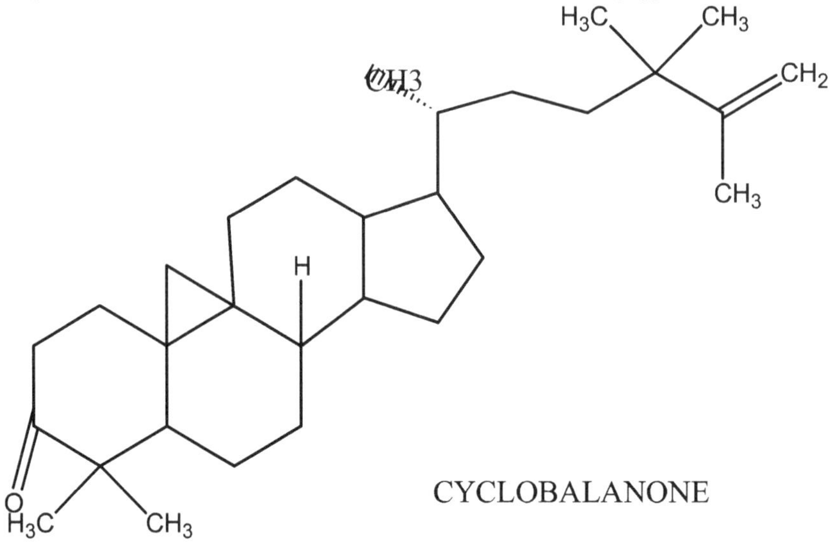

CYCLOBALANONE

Sl.No.	*Name*	*Category*	*Source*
132.	5α-lanosta -24,24-dimethyl-99110,25-diene-3β-ol		*Scaphyglottis livida*

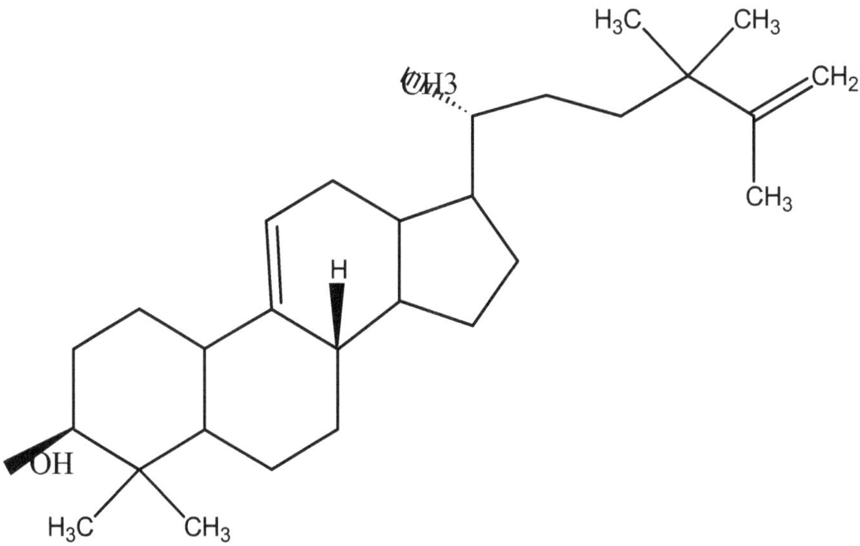

Contd...

Table 32.2–*Contd...*

Sl.No.	*Name*	*Category*	*Source*
133.	Calanthoside	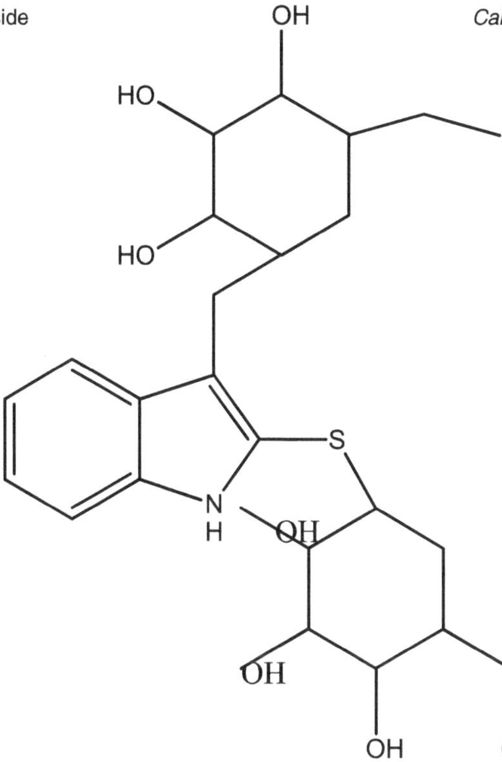	*Calanthe discolor*
134.	Agrostophyllinol	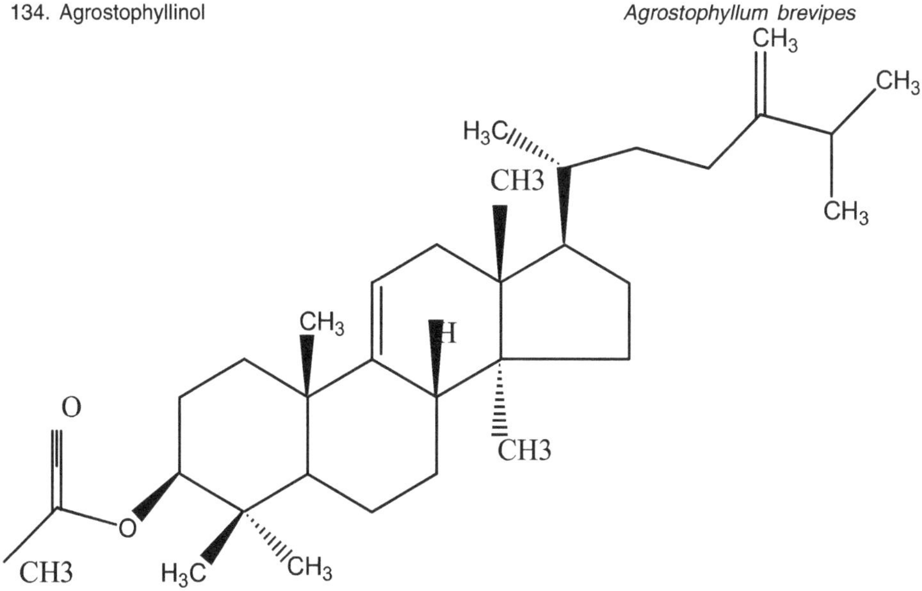	*Agrostophyllum brevipes*

Contd...

Table 32.2–*Contd...*

Sl.No.	*Name*	*Category*	*Source*
135.	Agrostophyllinone		*Agrostophyllum brevipes*

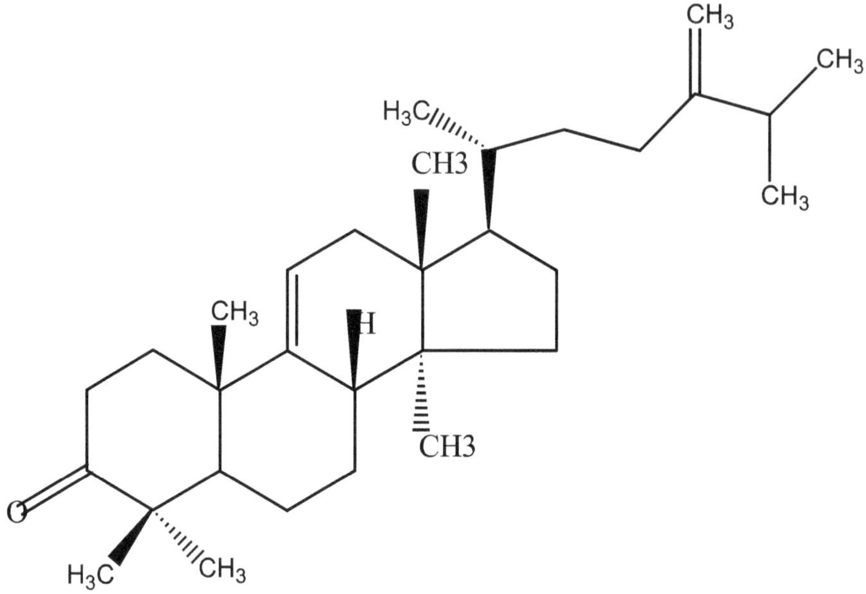

136.	Nidemin		*Scaphyglottis livida* and *Nidema boothi*

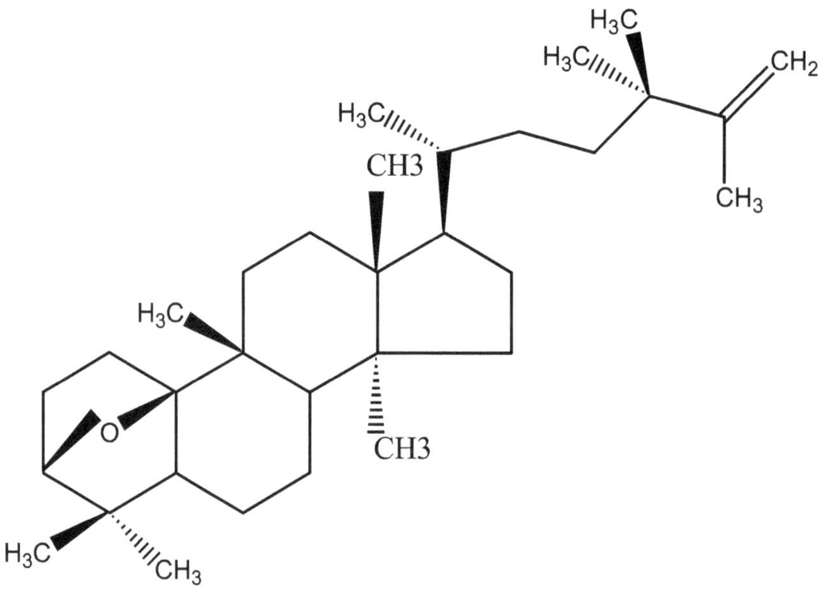

Contd...

Table 32.2–*Contd...*

Sl.No.	Name	Category	Source
137.	9.19.-Cyclolanosta-24,24-dimethyl-25-en-3β-yl-trans-p-hydroxycinnmate		*Scaphyglottis livida* and *Nidema boothi*

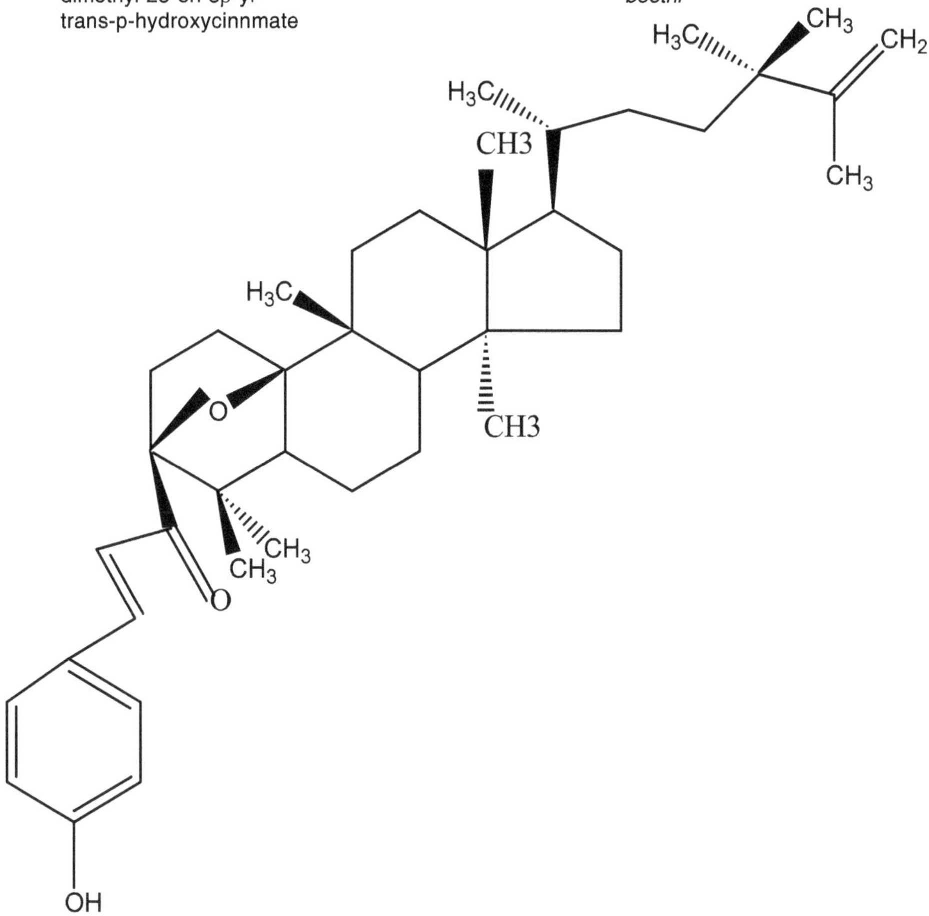

O-[6-*O*-(*trans*-*p*-coumaryl)-β-*D*-glucopyranoside] -3'-*O*-[6-*O*-(*trans*-4-*O*-(6-*O*-(*trans*-4-*O*-(α β-*D*-glucopyranosyl)-*p*-coumaryl)-β-*D*-glucopyranosyl)-*p*coumaryl)-β-*D*-gluco-pyranoside] and demalonyl cyanidin3-*O*-[6-*O*-((malonyl)-β-*D*-glucopyranoside]-7-*O* [6-*O*-(*trans*-*p*-coumaryl)-β-*D*-glucopyranoside]-3'-*O*-[6-*O*-(*trans*-4-*O*-(6-*O*-(*trans*-4-*O*-(α β-*D*-glucopyranosyl)-*p*-coumaryl)-β-*D*-glucopyranosyl)-*p*coumaryl)-β-*D*-glucopyranoside]; cyanidin 3-*O*-[6-*O*-(malonyl)-β-*D*-glucopyranoside]-7-*O*-[6-*O*-(*trans*caffeyl)-β-*D*-glucopyranoside]-3′-*O*-[6-*O*-(*trans*-4-*O*-(6-*O*-(*trans*-4-*O*-(β-*D*-glucopyranosyl)-caffeyl)-β-*D*glucopyranosyl)-caffeyl)-β-*D*-glucopyranoside] anddemalonyl cyanidin 3-*O*-[6-*O*-(malo nyl)-β-*D*glucopyranoside]-7-*O*-[6-*O*-(*trans*-caffeyl)-β-*D*glucopyranoside]-3′-*O*-[6-*O*-(*trans*-4-*O*-(6-*O*-(*trans*-4-*O*-(β α-*D*-glucopyranosyl)-caffeyl)-βα-*D*-glucopyranosyl)-caffeyl)-β-*D*-glucopyranoside] respectively. Several compounds with anti-inflammatory activity were isolated from *Dendrobium monoliforme* (Parkinson, 1640). -hydroxy-5,6-dimethoxy-1,4 phenanthrenequinone isolated from *Dendrobium moniliforme* have shown inhibitory effect on VHR dual specificity protein tyrosine phosphatase activity (http://www.botanical.com/botanical/mgmh/o/orchid13.html). *Gastrodia elata* has also been a good source of compounds with anti-inflammatory, analgesic and anti-angiogenic activity (Hamisy, 2007). Many phenolic compounds such as 4-hydroxybenzaldehyde, 4-hydroxybenzyl alcohol, benzyl alcohol, bis-(4-hydroxyphenyl) methane, 4(4'-hydroxybenzyloxy) benzyl methyl ether, 4-hydroxy-3-methoxybenzyl alcohol, 4-hydroxy- 3-methoxybenzaldehyde and 4-hydroxy-3-methoxybenzoic acid with anti-inflammatory as well as analgesic activity has been isolated from *Gastrodia elata* (Pillon *et al.*, 2006).

Medicinal Profile of Orchids in Different Regions of World

Asia

Orchids basically grow in tropical countries like Asia and show maximum diversity in humid tropical forest of India, Burma, South China, Thailand, Malaysia and Philippines, amongst them china, India and Japan are the leaders in orchid producers in Asia. Modern orchid industry in the world basically thrives on Indian dendrobiums, cymbidiums and Vandas (Indian orchid) (Chen *et al.*, 1994; Ho *et al.*, 2003; www.puretango.com/library/gastrodia.html).

Table 32.3: Orchids in Chinese Medicine

Medicine	*Orchid*	*Use*
Shihu (herba Dendrobii)	Several Dendrobium species*	Nourish stomach (promotes peristalis and gastric secretions)
		Replenish body fluid (treats dry tongue, thirst, red tongue)
		Reduce fever
		Increase white cells
		'Reduce fidgets'
Shihu Yeguang Wan	Dendrobium a minor part	Pseudomyopic
		Tired eyes
		Amblyopia
		'Central retinitis'
Moscatilin	*Dendrobium loddigesii*	Antiplatelet agent (in animals)
		Anti-cancer (stomach and lung cell lines)
Tian Ma	*Gastrodi elata*	Treat allergies
		Headache
		Fatigue

* *Diequiao Shihu—Dendrobium denneanum; Jichai Shihu—Dendrobium Nobile.* May also include *Flickingeria* and *Pholidota.*

Some medicinally useful orchids in India are *Eulophia campestris*, *Orchis latifolia*, *Vanda roxburgii* (Uniyal, 1977). Asthavarga, an important ingredient of Chavyanprasa contains four species of orchids out of eight components which are bulbs of *Malaxis muscifrea* (Jivaka), pseudo-bulbs of *Malaxis acuminate* (*Rishbhaka*) and roots of *Habenaria intermedia* (*Riddhi*), and *Habenaria edgeworthii* (*Vriddhi*) (Uniyal,1975).

Table 32.4: Orchids in Indian Medicine

Medicine	*Orchid*	*Use*
Ulat kambal	*Abroma augulata* Lam	Used as uterine tonic and emmenagogue; regulates menstrual flow
Rati, Kunch	*Abrus minor*	Roots: emetic. Decoction of roots and leaves: for cough, cold and colic. Seeds: purgative, emetic, tonic, used in nervous disorder and cattle poisoning
Brahmi	*Herpestis monniera*	Plant-extract: In experiments with mice, rat and dog, it has been proved to be tranquilizer, musculature relaxant, antispasmodic, anticancer
Hingan	*Balanites aegyptica*	cytotoxic and antitumour
Kesuth	*Eclipta alba*	Herb: in skin diseases Plant (powder) : curative of infective hepatitis, jaundice and viral hepatitis
Chhoti elaichi	*Elettaria cardamomum*	Essential oil from seed: antimicrobial; oil is used in several pharmaceutical preparations.
Gurmar, Merasingi	*Gymnema sylvestre*	Aerial parts (50 per cent EtOH extract) : spasmolytic, hypoglycaemic, *in vitro* antiviral against influenza A2 virus
Kurchi	*Holarrhena pubescence*	Fruit (50 per cent EtOH extract) : anticancer, antiprotozoa, hypoglycaemic, astringent, febrifuge, useful in diarrhoea, intestinal worms, and to regulate menstruation
Keyu	*Costus speciosus*	In carcinogenic tumours

Europe

William Turner in the first English Herbal (1568) reported orchids to be useful in alcoholic gastritis (Turner, 1568). Later on, Williams Langham also familiarized with its anti-pyretic, anti-consumption and anti-diarrhoeal effects (Langham, 1597). The Ottomans extracted from the dried tubers were used by the ottomans to extract 'Sahlep' which was later on pronounced as Salep. In the East, Salep was (and is) mainly made from *Orchis morio*, but in UK it is also made from *Orchis mascula*, the early purple orchid or from *Orchis maculata* or *Orchis latifolia*.

Table 32.5: Use of European Orchids

Author and Year	*Proposed Use of Orchid (Preparation)*	*Indication*
Turner (1568)	Tender tuber and goats milk, Dry, Topical, Tuber	Aphrodisiac, Anti-aphrodisiac, Antiseptic, Gastro-intestinal problems arising from too much wine
Langham (1579)	Tuber	Anti-pyretic, Anti-consumption, Anti-diarrhoea
Parkinson (1640)	Tuber, Tuber ('Salep')	Increase fertility in men

Africa

In African countries such as Tanzania, Zambia, and Malawi orchids are mostly cultivated for their tubers which are used as a source of food, and processed into a meatless sausage locally called chinaka, chikande and kikande which is consumed as relish or just as a snack. *Cyrtorchis arcuata* is also employed to treat diabetes or skin infections and *Eulophia cucullata* to prevent epilepsy. An infusion of the leaves and pseudobulbs of *Bulbophyllum maximum* is used to protect against sorcery, and *Tridactyle tricuspis* to treat madness (Morris, 2003).

America

Most famous and widely used orchid in America is vanilla (Figure 32.3), is the most famous orchid product. The most important vanilla species is *Vanilla planifolia* whose aromatic oil from seed pods is used (Beriolichhi *et al.*, 2004). Vanilla is basically used now for relieving nausea and improving food intake in patients receiving chemotherapy (Menashian *et al.*,1992) and as a diagnostic aroma for Alzheimer's disease etc.

Table 32.6: Use of American Orchid

Author	*Year*	*Proposed Use of vanilla Derivatives*
Aztec herbal	1552	Used to remove fear from human heart, Flavoring and perfume
Menashian *et al.*	1992	Improve food intake,reduce nausea and vomiting in patients given chemotherapy
Fladby *et al.*	2004	Diagnostic of Alzheimer's disease (patients cannot smell vanilla)
Fitzgerald *et al.*	2004	Antimicrobial against *Eschericha coli*, *Lactobacillus planatarum* and *Listeria innocua*

Role of Orchids in Various Diseases

Antimicrobial Activity

A protein named Gastrodianin, isolated from *Gastrodia elata* have shown *in vitro* activity against plant pathogenic fungi (Pillon *et al.*, 2006) and methanolic leaf extract of *Spiranthus mauritianum* has shown inhibitory effect against gram positive bacteria and anti-inflammatory activity (Gutierezz *et al.*, 2010). A broad spectrum antibacterial

Figure 32.3: *Vanilla planifolia.*

activity against gram positive and gram negative bacteria was shown by methylene chloride extract from the leaves and stem bark of *Galeola foliate*, (Fossen *et al.*, 2003). Vanillin has shown to be effective against *Escherichia coli, Lactobacillus plantarum* and *Listeria innocua* (Anuradha *et al.*, 1998). Ethanolic extract of seedlings of *Cypripedium macranthos* var. *rebunense* contained antifungal compounds lusianthrin and chrysin. Antimicrobial activity was shown by methanolic extract from the leaves of*Acanthephippium bicolor* Lindley against *Staphylococcus aureus, Streptococcus foecalis,*

Bacillus cereus, Proteus vulgaris, Proteus mirabilis, Pseudomonas aeruginosa, Klebsiella pneumonia, Shigella dysenteriae, Escherichia coli, Microsporum audouinii, Microsporum fulvum, Candida albicans and *Trichophyton rubrum*.

Anti-inflammatory Activity

The carrageenan induced oedema model in rats and mice was inhibited by *Vanda roxburghii*. The ethanolic extract from the leaves of *Anoectochilus formosanus* have showed delayed onset of anti-inflammatory activity starting from 4 hours post carrageenan administration.

Antitumor Activity

Cremastra appendiculata contains a compound Cirrhopetalanthrin which potentially initiates cytotoxicity against human colon cancer (HCT-8), human hepatoma (Bel7402), human stomach cancer (BGC-823) human lung adenocarcinoma (A549), human breast cancer (MCF-7) and human ovarian cancer (A2780) cell lines is isolated from (Stewart, 1995). A chemical category of stillbenoids isolated from methanolic extract of *Bletilla striata* showed inhibitory effect of tubulin polymerization at IC50 10 ìM. and were chemically reported as 3,3'-dihydroxy-2',6'-bis(p-hydroxybenzyl)-5-methoxybibenzyl and 3',5-dihydroxy-2-(p-hydroxybenzyl)-3-methoxybibenzyl. *Dendrobium* species contains an array of anticancer compounds with anti-tumor activity like Denbinobin and 4,7-Dihydroxy-2-methoxy-9,10-dihydrophenanthrene isolated from *Dendrobium nobile* which showed cytotoxicity against human lung carcinoma, human ovary adenocarcinoma and human promyelocytic leukemia cell lines (Morris, 2003). Another *Dendrobium* species *Dendrobium chrysanthum* contained Dendrochrysanene which was found to suppress the Mrna level of TNF-alpha, IL8, IL10 and iNOS in murine peritoneal macrophages (Satish *et al.*, 2003). *Dendrobium fimbriatum* contained fimbriatone which was found to be potent inhibitor of BGC cell line (Fan *et al.*,1997). *Bulbophyllum kwangtungense* contained a number of potent antitumour compounds which were active against Hela and K562 human tumor cell lines such as7,8-dihydro-4-hydroxy-12,13-methylenedioxy 11methoxyldibenz[bf]oxepin,7,8dihydro-4-hydroxy-12,13 methylenedioxy11methoxyl dibenz [bf]oxepin,7,8-Dihydro 5 hydroxy- 12, 13methylendioxy-11methoxyldibenz [bf]oxepin, cumulating, densiflorol A and plicatol B. Methanolic extract of *Anoectochilus formosanus* have shown to induce apoptosis of MCF-7 cells (Lin *et al.*, 2005). From the tuber of *Cremastra appendiculata* compounds isolated are cirrohopetalanthrin and 2,7,2′,7′,2″-pentahydroxy-4,4′4″7″-tetramethoxy-1,8,1′,1″-triphenanthrene which were found to have moderate cytotoxicity against human colon cancer, human stomach cancer, human hepatoma, human breast cancer, human lung adenocarcinoma and human ovarian cancer cell lines (Stewart *et al.*,1995, Chang *et al.*, 1977). Compounds lonchophylloids A and lonchophylloids B present in the stems of *Ephemerantha lonchophylla* sensitized those cells which have expressed the multidrug resistance phenotype to the toxicity of the anticancer drug doxorubicin (Wu *et al.*, 2006). *Gastrodia elata* methanolic extract was found to be preventive against serum deprived apoptosis through activation of serine/threonine kinase dependent pathway and suppression of JNK activity (Yoshikawa *et al.* 1998), while the ethanolic extract from the rhizomes have shown potent anti tumor activity *in vitro* in a dose dependent manner *Spiranthes australis* (R. Brown) Lindl contains potent anticancer dihyroflavanoid compound namely (2S)-5, 2′,6′-trihydroxy-6-lavandulyl-4″-(_,_-dimethylallyl)-2″,2″-dimethylpyrano-[5″,6″,7,8] which is shown to inhibit human cancer cells growth including A498, A549, BEL-7402,SGC-7901,MCF-7, HT-29 and K562 cell lines (Wu *et al.*,2006). *Bulbophyllum odoratissimum* contained a phenanthrene derivative3,7-dihydroxy-2,4,6 trimethoxyphenanthrene showed cytotoxicity against the human cancer cell lines such as human leukemia cell lines K562 and HL-60, human hepatoma BEL-7402, human lung adenocarcinoma A549 and human stomach cancer cell lines SGC-7901 (Zhang *et al.*, 2006).

Antioxidative Property

Ephemerantha lonchophylla contained a dihydrostilbene compound ephemeranthone, obtained from the leaf ethanolic extract was found to have strong antioxidative property for *in vitro* inhibition of human low density lipoprotein (Lin *et al.*, 1985). Yet another dihydrostilbene isoamoenylin obtained from the roots of *Dendrobium amoenum* var. *denneanum* showed moderate antioxidative property (Watanbe *et al.*, 2007). Stem of *Dendrobium aurantiacum* contained cismelilotoside, dihydromelilotoside and trans-melilotoside which were found to be potent antioxidants (http://www.orchidsusa.com/6fertilizing.htm). Antioxidative compounds like alkyl ferulates and quercetin from *Dendrobium monoliforme* and *Dendrobium tosaense* (Fossen *et al.*, 2003), kinsenone from *Anoectochilus formosanus* were reported to have antioxidative property (Venkateswarulu *et al.*, 2002). Hydroxybenzyl alcohol, hydroxybenzaldehyde, vanillin and vanillyl alcohol obtained from leaf methanolic extract of *Gastrodia elata* were reported to possess antioxidative property (Du *et al.*, 2002). 60 per cent ethanolic extract of air dried plant of *Pholidota yunnanensis* contained phenanthrenes which *were* found to show the DPPH free radical scavenging activity.

Convulsive Diseases

G. elata Methanol extract of rhizomes of the plant is shown to protect the hippocampal neuronal damage induced by transient global ischemia in a gerbil model (Kim *et al.*, 2003). while the aqueous extract of rhizomes of the plant exhibited in preventing PC12 cell apoptosis induced by serum deprivation through suppression of the JNK pathway The methanol extract of the above said plant was significantly shown to inhibit the recovery time and severity induced by pentylenetetrazole (PTZ) treatment 4-hydroxybenzaldehyde, an analogue of *p*-hydroxybenzyl alcohol.

Diuretic

Cymbidine A a monomeric peptidoglycan-related involving four amino acids (D-alanin, meso-diaminopimelic acid, Dgultamic acid, and L-valine) and two amino sugars (Nacetylglucosamine and 1,6-anhydro-N-acetylmuramic acid) isolated from *Cymbidium goeringii*, orchid showed hypotensive and diuretic activities (Watanabe *et al.*, 2007). Gigantol showed potent inhibitory effects on LPS induced nitric oxide (NO) and prostaglandin E production in RAW 264.7 cells. Similar findings with the compound suggested gigantol suppressed expression of inducible nitric oxide synthase (iNOS) and cyclooxygenase-2 (COX-2) at the protein and mRNA levels in RAW 264.7 cells in a concentration-dependent manner. Also that the compound is a potent inhibitor of tumor necrosis factor-alpha (TNF alpha), interleukin-1beta (IL-1beta) and interleukin-6 (IL- 6) release and influenced the mRNA expression levels of these cytokines in a dose-dependent manner (Won *et al.*, 2006).

Antihepatotoxic

Another plant *Anoectochilus formosancus* fresh crude extract showed inhibition of chronic hepatitis induced by CCl_4 in mice. A component namely Kinsenoside isolated from *A. formosanus* was reported to be significantly showing antihepatotoxic activity (Wu *et al.*, 2007). While the aqueous extract of the same plant can reduce liver fibrogensis in rats induced by CCl_4 (Shih *et al.*, 2005). In accordance with the above said antihepatotoxic activity a study conducted by (Shih *et al.* 2005) also showed aqueous extract of *A. formosanus* protected the liver against dimethylnitrosamine (DMN) induced fibrosis, repair of liver injury, improvement of fibrotic changes and promotion of liver regeneration. While other study of the extract of whole plants of *in vitro* cultured *A. formosanus* showed hepatoprotective and antihyperliposis activities. The extract also showed antihyperliposis using aurothioglucose-induced obese mice, as the extract significantly reduced the increase in weights of body and liver, ameliorated triglyceride levels in the liver and serum, and also the deposition of adipose tissue (Du *et al.*, 2003). *Goodyera* species, *G. schlechtendaliana* Reichb. fil., *G. matsumurana* Schltr. and *G. discolor* were used to isolate hepatoprotective aliphatic glycosides 3-(S)-3-β-Dglucopyranosyloxybutanolide and its congener, 3-(S)-3-β-D-glucopyranosyloxy-4-hydroxybutanoic acid.

Neuroprotective

Coeloglossum viride (L.) Hartm. var. *bracteatum* (Willd.) contains an active compound dactylorhin B supposedly possessing neuroprotective effects. Ethyl ether extracts of leaves *Gastrodia elata* Blume containing contains *p*-hydroxybenzyl alcohol (HBA) was used to evaluate brain damage and transcriptional levels of protein disulfide isomerase (PDI) and 1-Cys peroxiredoxin (1-Cys Prx) genes known to play a role in antioxidant systems after transient focal ischemia in rat brain and shown to possess neuroprotective effect on neuronal cell death induced by amyloid beta-peptide in IMR-32 neuroblastoma cells (Kim *et al.*, 2003).

Anti-platelet Aggregation

Methanol extract of stem and the chemical entities moscatilin moscatin and moscatilin diacetate presentin the plant *Dendrobium loddigesii* was found to inhibit aggregation of rabbit platelets induced by arachidonic acid and collagen as well as antiplatelet aggregation activity respectively (Chen *et al.*, 1994). In a study conducted on another species *D. densifiorum* Lindl. ex Wall the phytochemicals moscatilin, homoeriodictyol, scoparone, scopoletin and gigantol were identified from stems and were found to exhibit anti-platelet aggregation activity *in vitro*, scoparone being the most potent (Fan *et al.*, 1997). Further experiment with the ethanolic extract of *Ephemerantha lonchophylla* led to bioactivity-directed separation of 3,7-dihydroxy-2,4-dimethoxyphenanthrene, 3-methylgigantol, and erianthridin from. The compounds exhibited significant anti-aggregation activities against AA-induced aggregation with estimated IC50, values in this regard for were 24, 30 and 9 ìM respectively (Chen *et al.*, 1994). Methanol extract of rhizomes of *Gastrodia elata* contain polysaccharide which potentiated inhibition of platelet aggregation and antithrombosis activity (Ding *et al.*, 2013).

Other Medicinal Properties

Dendrobium species has a lot of species included in it which are medicinally important like methanolic extract of rhizomes of *Dendrobium monoliforme* shows potential antipyretic action in rabbits (Wang *et al.*, 2006) while the stem of *Dendrobium nobile* contains potent phytochemicals such as sesquiterpene glycosides with aloo aroma dendrane, emmotin and picrotoxane type aglycones which possess immunomodulatory activity (Peng *et al.*, 2007). Another bibenzyl, derivative from the above said plant moscatilin, is shown to possess antimutagenic activity on various mutagens such as UV rays, furylfuramide,4-nitroquinoline-1-oxide, benzo[a]pyrene, aflatoxinB(1) (Lans *et al.*,2001). Methanolic extract from *Calanthe* species specially *Calanthe discolour* (LINDL) and *Calanthe liukiuensis* (SCHLTR) are reported to contain S, Obisdesmoside, calanthoside, glucoindican, calalikiuenoside, calaphenanthrenol, tryptanthrin, indirubin, isatin shows hair restoring and skin blood flow promoting activities (Tezuka *et al.*,1990). Ethanolic extract from *Vanda roxburghii* possess wound healing property (Lin *et al.*, 2000). Moscatilin, homoeridictyol, scoparone, scopoletin and gigantol obtained from the stem of *Dendrobium densiflorum* Lindl. ex Wall were found to possess *in vitro* anti platelet aggregation activity. The aqueous extract from the rhizomes of *Anoectochilus formosanus* possess anti hyperglycemic and antioxidant property. *Dendrobium candidum* showed antihyperglycemic effect and stimulates secretion of insulin from beta cells and also inhibits secretion of glucagon from cells. Cymbidine A, a monomeric peptidoglycan isolated from *Cymbidium goeringii* has shown hypotensive and diuretic activity. Gigantol isolated from the same source have shown potent inhibitory effect on LPS-induced nitric oxide and production of prostaglandin E in RAW 264.7 cells. Kinsenoside obtained from *Anoectochilus formosanus* shows antihepatotoxic activity. It was also found that the aqueous extract of *Anoectochilus formosanus* can lead to reduction of liver fibrinogenesis in rats induced by CCl_4.

More recent ethno pharmacological studies show that orchids are used in many parts of the world and in treatment of a number of diseases (Table 32.7).

Some Interesting Facts about Orchids

1. Orchid flowers come in every colour of the rainbow and combination thereof. Contrary to the old movie, there is no black orchid. Floral fragrances are equally diverse, from quite sweet to the smell of rotting liver.
2. *Grammatophyllum speciosum* is the giant of the orchid world. It can reach 25 feet in height with leaves up to 2½ feet long. The flower stalk can reach 8 feet, often bearing more than 100 flowers, each 5-6 inches across.
3. *Vanilla planifolia*, a native of Mexico and Central America, produces a long seedpod that is the chief commercial source of natural vanilla. Vanilla is a

Table 32.7: Ethnopharmacological Uses of Orchids in Different Parts of World

Sl.No.	*Species*	*Medicinal Use*	*References*
1.	*Anoectochilus formosanus*	Chest and abdominal paints, diabetes, fever, nephritis, hypertension, impotence, liver spleen disorders, and pleurodynia, antiinflammatory agent	Satish *et al.* (2003)
2.	*Anoectochilus roxburghii*	Treatment of fever, pleurodynia, snake bite, lung and liver disease, hypertension, and malnourished children	Fan *et al.* (1997).
3.	*Bletilla Formosana*	Is associated with the lung, stomach and liver meridians and has a bitter taste and cool properties	Lin *et al.* (2005)
4.	*Bletilla striata*	Treatment of sores, ulcers and chapped skin, heal wounds, reduce swelling, and promote regeneration of tissue. Have been used to treat pulmonary tuberculosis and as hemostatic agent	Luo *et al.* (2007), Chang (1977)
5.	*Bulbophyllum kwangtungense*	Treat pulmonary b tuberculosis and as hemostatic agent, promote the production of body liquid and reduce fever	Wu *et al.* (2006)
6.	*Bulbophyllum odoratissimum*	Treat tuberculosis, chronic inflammation and fracture	Chen *et al.* (2007)
7.	*Calanthe discolor*	hair restoring	Yoshikawa *et al.* (1998)
8	*Calanthe liukiuensis*	hair restoring	Yoshikawa *et al.* (1998)
9.	*Catasetum barbatum*	Febrifuge, anti inflammatory	Shimizu *et al.* (1988)
10.	*Cephalanceropsis gracilis*	Cancer	Wu *et al.* (2006)
11.	*Coeloglossum viride*	Memory deficits	Zhang *et al.* (2006)
12.	*Cremastra appendiculata*	Is associated with the liver, Spleen and stomach meridians. Its can be used internally, to fight tumors and cancers of the breast, cervix and uterus in women. Externally, it treats boils and skin lesions, and can be applied to affected parts of the body as part of a poultice or paste	Lin and Namba (1985)
13.	*Cymbidium goeringii*	Used as hypotensive and diuretic activities	Watanabe *et al.* (2007)
14.	*Cypripedium macranthos*	Used for skin diseases	Shimura *et al.* (2007)
15.	*Dendrobium amoenum*	Skin diseases	Venkateswarlu *et al.* (2002)
16.	*Dendrobium aurantiacum*	Diabetes	Yang *et al.* (2006).
17.	*Dendrobium candidum*	Diabetes	Wu *et al.* (2004)

Contd...

Table 32.7–*Contd...*

Sl.No.	*Species*	*Medicinal Use*	*References*
18.	*Dendrobium chrysanthum*	Antipyretic, eyes-benefiting,immunoregulatory purposes, kin diseases	Li *et al.* (2001)
19.	*Dendrobium densifiorum*	Promotes the production of body fluid	Fan *et al.* (2001)
20.	*Dendrobium fimbriatum*	Promotes the production of body fluid. Paste applied on fracturated area to set bone	Bi *et al.* (2003)
21.	*Dendrobium nobile*	Antitumor and Antimicrobial activities as well as the inhibition of *in vitro* lipid peroxidation	Uma Devi *et al.* (2008)
21.	*Epipactis helleborine*	It is used as aphrodisiac	Balzarini *et al.* (1992)
22.	*Epidendrum rigidum*	Replenish body fluid	Hernandez-Romero *et al.* (2005)
23.	*Goodyera schlechtendaliana*	Tonic for internal injuries and to improve circulation	Du *et al.* (2002)
24.	*Listera ovata*	Stomach disease. Externally skin tone	Olof (1972)
25.	*Maxillaria densa*	Treatment of painful complaints. Relaxant agent	Déciga-Campos *et al.* (2007a)
26.	*Pholidota chinensis*	Is taken for scrofula, feverish stomachache and toothache, chronic bronchitis, and duodenal ulcer	Wang *et al.* (2006)
27.	*Pholidata articulate*	Bone jointer	Jalal *et al.* (2009)
28.	*Spiranthes australis*	Used for urinary problems and diabetes mellitus. Treatment ofbacterial and inflammatory diseases, cancer, blood, andchest disorders	Lans (2006), Peng *et al.* (2007)
29.	*Spiranthes mauritianum*	Used for snakebites,scorpion stings, for injuries and mange ofdogs and to facilitate hunting success	Lans *et al.* (2001)
30.	*Spiranthes sinensis varamoena*	Aphrodisiac, treatment of hemoptysis, epistaxis, headache, chronic dysentery and meningitis	Tezuka *et al.* (1990), Lin *et al.* (2000)
31.	*Vanda roxburghii*	The paste es applied to the body to bring down fever. The juice is dropped in the ear for the treatment of otitis. The roots are used in dyspepsia, bronchitis, rheumatim and siatica	Chawla *et al.*

genus of climbing, vining orchids, capable of growing 30 to 50 feet in length.

4. The male Crestless Gardener Bowerbird of New Guinea is the only animal, besides man, that uses orchids for their beauty. The orchid blossoms are used to decorate the entrance of the bower, made of sticks and vines, which the male builds for his courtship display. The orchids are placed on a carpet of moss laid down at the bower entrance and the Bowerbird periodically replaces the wilted blossoms with fresh one, thus impressing and winning the female of the species.
5. Orchid flowers often last for several weeks – unless they are pollinated. Once fertilization occurs, the blossoms wilt very quickly.
6. Many Asian recipes include orchid blooms in their list of ingredients.

References

An, S.J., Park, S.K., Hwang, I.K., Choi, S.Y., Kim, S.K., Kwon, O.K., Jung, S.J., Baek, N.I., Lee, H.Y., Won, M.N., and Kang, T.C. (2003). Gastrodin decreases immuno reactions of gama-amino butyric acid shunt enzymes in hippocampus of seizure – sensitive gerbils. *J. Neurosa. Res.*, **71**: 534-43.

Anitha, T.S., and Tamilarasi, V.S.(2013). Introduction to floral diversity of Arunachal Pradesh *In*, floral diversity of Arunachal Pradesh Press Luvjoychoker (http: / / www.scribd.com/doc/186947116/Floral-Diversity-of-Arunachal-Pradesh).

Anuradha, V., and Prakash, N,S. (1998) Aeridin: A phenanthropyran from *Aerides crispum*. *Phytochemistry*, **48(1)**: 185-186.

Berliocchi, L., and Griffiths, M. (2004). The Orchid in Lore and Legend *In*: Portland OR Press Timber, pp128-132.

Berliocchi. L., and Griffiths, M. (2003). The Orchid in Lore and Legend In: Portland OR Vol. 24, *Press Timber, pp 7–21.*

Chase, M.W., Cameron, K.R., Barrett, R.L., and Freudenstein, J.V. (2003). DNA data and Orchidaceae systematics: A new phylogenetic classification.; Orchid conservation. *Natural History Publications*, **4**: 69–89.

Chase, M.W. (2005). Classification of Orchidaceae in the age of DNA data *Curtis's Botanical Magazine*, **22(1)**: 2-7.

Chen, L.G., Wu, F.N., and Teng, C.M. (1994). Antiplatelet aggregation principles of *Dendrobrium loddigesil. Journal of Nat prod.*, **57**: 1271-4.

Ding, K., and Qiu, H. (2013). Sulfated derivative of *Gastrodia elata* polysaccharide, preparation method and antitumor use thereof, Shanghai Institute of Materia Medica, Chinese Academy of Sciences, Patent granted on 5/14/2013.

Dioscorides Pedanius. De Materia Medica *Frankfurt edition 1543 or De Materia Medica libra quinque 572 (In Imperial Library at Vienna)*

Dressler, R. L. (1981). The Orchids *In*: Natural history and classification. Harvard University Press.Cambridge MA.

Du, X.M., Sun, N.Y., Takizawa, N., Guo. Y.T., and Shoyama, Y. (2004). Sedative and anticonvulsant activities of goodyerin, a flavonol glycoside from *Goodyera schlechtendaliana*. *Phytotherapy Re.*, **16**: 261-26.

Fan, Z.N., Xiao, H.S., Fan, X.H., and Wu, W.S. (1997). Study on tissue culture of *Anoectochilus roburghii*. *Journal Fujian Normal University*, **13**: 82-87.

Fitzgerald, D.J., Stratford, M., Gasson, M.J., Ueckert, J., Bos, A., and Narbad, A.(2004). Mode of antimicrobial action of vanillin against *Escherichia coli*, *Lactobacillus plantarum* and *Listeria innocua*. *J Appl Microbiol.*, **97**: 104–13.

Fladby, T., Bryhn, G., Halvorsen, O., Rose, I., Wahlund, M., Wiig, P.,and Wetterberg, L. (2004). Olfactory response in the temporal cortexof the elderly measured with near infrared spectroscopy: a preliminary feasibility study. *J Cereb Blood Flow Metab.*, **24**: 677–80.

Fossen, T., and Ovstedal, D. (2003). Anthocyanins from flowers of orchids *Dracula chimaera* and *Dracula cordobae*. *Phytochem.*, **63(7)**: 783-7.

Griesbach, R.J. (2002). Development of Phalaenopsis Orchids for the Mass-Market. p. In: J. Janick and A. Whipkey (eds.), Trends in new crops and new uses. ASHS Press, Alexandria, VA.pp 458–465

Gutiérrez, R.M.P. (2010). Orchids A review of uses in traditional medicine, its phytochemistry and pharmacology. *Journal of Medicinal Plants Research*, **4(8)**: 592-638.

Hamisy, C.W. (2007). Development of conservation strategies for the wild edible orchid in Tanzania, Progress report for the Rufford Small Grants Foundation. 458–465.

Ho, C.K., and Chen, C.C. (2003). Moscatillin from the orchid *Dendrobrium loddigesii* is a potential anticancer agent. *Cancer Invest.*, **21**: 729–36.

http: //www.botanical.com/botanical/mgmh/o/orchid13.html A modern herbal by Mrs M. Grieve. 1995-2014 Botanical.com.

http: //www.orchidsasia.com/asco.htm, Orchid Asia, Orchidsasia-Vanam Orchids Portal Site,1998-2005, Pavitram Bali, Bali, Indonesia

http: //www.orchidsusa.com/6fertilizing.htm, Venamy Orchids, Copyright 2002 Venamy Orchids. A Comprehensive Guide to orchid.

Introduction To Orchids.http: //www.orchidlady.com, Linda's orchid pages. 1996-2002 Linda Fortner.

Jalal, J.S., Tewari, L.M., and.Pangtey, Y.P.S. (2009). *Pholidata articulata* Lindl., an orchid used in bone jointing in Kumaun region, Western Himalaya. *Ethnobotanical Leaflets*, **13**: 1047-50.

Khasin, S.M., and Mohana, P.Y. (1999). Medicinal Importance of Orchids. *The Botanica*, **49**: 86-91.

Kim, H.J., Lee, S.R., and Moon, K.D.(2003). Ether fraction of methanol extracts of *Gastrodia elata*, medicinal herb protects against neuronal cell damage after transient glaobal ischemia in gerbils. *Phytother Res.*, **17(8)**: 909-12.

Langham, W. (1957). The Garden of Health. London.

Lans, C., Harper, T., Georges, K., and Bridgewater, E. (2001). Medicinal and ethnoveterinary remedies of hunters in Trinidad. *BMC Complementary and Alternative Medicine*, **1**: 10-15.

Lin, C.C., and Namba, T. (1985). Historical and herbological studies on the traditional Japanese and Chinese crude drugs. *Journal of the Shan-Ci-Gu Yakushigaku Zasshi*, **20**: 88-98.

Lin, Y.L., Huang, R., Don, M., and Kuo, Y. (2000). Dihydrophenanthrenes from *Spiranthes sinensis*. *J. Nat. Prod.*, **63**: 1608-1610.

Lin, Y.L., Chen, W.P., and Macabalang, A.D. (2005). Dihydrophenanthrenes from *Bletilla formosana*. *Chem Pharm Bull.*, **53**: 1111-1113.

Menashian, L., Flann, M., Douglas-Paxton, D., and Raymond, J.(1992). Improved food intake and reduced nausea and vomiting in patients given a restricted diet while receiving cisplatin chemotherapy. *J Am Diet Assoc.*, **92**: 58–61.

Morris, B. (2003). Children of the Wind—Orchids as Medicines in Malawi. The Orchid Review, **111**: 271–7.

Olof, T.C. (1972). Survival and flowering of some perennial herbs II. The behavior of some orchids on permanent plots. *Oikos*, **23**: 23-28.

Orchids Asia-Vanam. Orchids Portal site (1998-2005)

Parkinson, J. (1640). Theatrum Botanical The Theatre of Plants. An Universall and Complete Herball. London.

Peng, J., Xu, Q., Xu, Y., Qi, Y., Han, X., and Xu, L. (2007). A new anticancer dihydroflavanoid from the root of *Spiranthes australis* (R. Brown) Lindl. *Nat. Prod. Res.*, **21**: 641-645.

Pillon, Y., and Chase, M.W. (2006). Taxonomic exaggeration and its effects on orchid conservation. *Conserv. Biol.*, **21(1)**: 263-265.

Satish, M.N., Abhay, P.S., Chen-Yue, L., Chao-Lin, K., and Hsin-Sheng, T. (2003). Studies on tissue culture of Chinese medicinal plant resources in Taiwan and their sustainable utilization. *Bot. Bull. Acad.*, **44**: 79- 98.

Shih, C.C., Wu, Y.W., Hsieh, C.C., and Lin, W.C. (2004). Effect of Anoectochilus formosanus on fibrosis and regeneration of the liver in rats. *Clinical and Experimental Pharmacology and Physiology*, **31**: 620-625.

Singh, S., Singh, A.K., Kumar, S., Kumar, M.P. Andey, K.P., and Singh, K.M.C. (2012). Medicinal properties and uses of orchids: a concise review. *Applied botany*,.**52** : 11627-11634

Stuart, R.G.A. (1984). Chinese Materia Medica Taipei. Southern Materials Centre, A translation of an ancient Chinese herbal.

Tezuka, Y., Li, J., Hirano, H., Ueda, H., Nagashima, K., and Kikuchi, T. (1990). Studies on the constituents of Orchidaceous plants IX. Constituents of *Spiranthes sinensis* (PERS.) AMES var. *amoena* (M. Bieberson) HARA. Structures of spiranthesol, spiranthoquinone, spiranthol C and spirasineol B, new isopentenyldi hydrophenantrenes. *Chem. Pharm. Bull.,* **38**: 629-635.

Turner, W. (1568). The first and seconde partes of the Herbal of William Turner, doctor in Phisick, lately oversene, corrected and enlarged with the Third Parte. *Cologne.*

Uniyal, M.R. (1975). Astavarga Sandigdha Vanaushadhi. Dhanwantri Partrika. Sri Jwala Ayurevd Bhawan Aligarh pp. 123-124.

Uniyal, M.R. (1977). Medicinal Plants and Minerals of Uttarakhand Himalaya. Baidyanath, Ayurved Shodh Sanstha, Patna.

Venkateswarlu, S., Raju, M.R., and Subbaraju, G.V. (2002). Synthesis and biological activity of isoamoenylin, a metabolite of *Dendrobium amoenum. Biosci. Biotechnol. Biochem.,* **66**: 2236-2238.

Wang, J., Matsuzaki, K., and Kitanaka, S. (2006). Stilbene derivatives from *Pholidota chinensis* and their anti-inflammatory activity. *Chem. Pharm. Bulletin,* **54**: 1216-1218.

Watanabe, K., Tanaka, R., Sakurai, H., Iguchi, K., Yamada, Y., Hsu, C.S., Sakuma, C., Kikuchi, H., Shibayama, H., and Kawai, T. (1995). Structure of cymbidine A, a monomeric peptidoglycan-related compound with hypotensive and diuretic activities, isolated from a higher plant, *Cymbidium goeringii* (Orchidaceae). *Chemical Pharmaceutic Bulletin,* **55**: 780-783.

Won, J.H., Kim, J.Y., Yun, K.J., Lee, J.H., Back, N.I., Chung, H.G., Chung, S.A., Jeong, T.S., Choi, M.S., and Lee, K.T.(2006). Gigantol isolated from the whole plants of cymbidium goeringii inhibits the LPS induced I NOS and COX -2 expression viaNF-kappa b inactivation in RAW 264.7 macrophage cells. *Plant Med.,* **72(13)**: 1181-1187.

Wu, B., He, S., and Pan, Y.J. (2006). New dihydrodibenzoxepins from Bulbophyllum kwangtungense. *Planta Medica,* **72**: 1244-1247.

Wu, H., Hsu,Y., and Jao, C. (2007). Indol alkaloids from *Cephalanceropsis gracili. J. Nat. Prod.,* **69**: 1467-1470.

www.puretango.com/library/gastrodia.html

Yoshikawa, M., Murakami, T., Kishi, A., Sakurama, T., Matsuda, H., Nomura, M., Matsuda, H., and Kubo, M. (1996). Novel indole S,O-bisdesmoside, calanthoside, the precursor glycoside of tryptanthrin, indirubin, and isatin, with increasing skin blood flow promoting effects, from two *Calanthe* species (Orchidaceae) *Chem. Pharm. Bull.,* **46**: 886-888.

Zhang, D., Zhang, Y., Liu, G., and Zhang, J (2006). Dactylorhin B reduces toxic effects of beta-amyloid fragment (25-35) on neuron cells and isolated rat brain mitochondria. *Naunyn Schmiedebergs Archives Pharmacol.,* **374**: 117-125.

Utilisation and Management of Medicinal Plants Vol. 3 (2015) *Pages* **535–545**
Editors: **V.K. Gupta and A. Kaul**
Published by: **DAYA PUBLISHING HOUSE, NEW DELHI**

33

A Review on Medicinal Plant *Adhatoda* spp. and its Modern Viewpoint

Jitendriya Panigrahi[1]* and Illa Patel[2]

ABSTRACT

Adhatoda vasica and A. beddomei belongs to family Acanthaceae are traditional Indian medicinal shrubs widely used in the treatment of many medications in India. It is an important drug popularly known as "Vasaca". A. beddomei is now included in the IUCN Threatened Plants List of India. The leaves and roots contain several alkaloids. These alkaloids have been used in combination with an acid that has been named adhatodic acid. Its local use quinazoline alkaloid, (vasicine and vasicinone, vasicinolone and vasicol), which may have a bronchodilator effect of the bronchi gives relief in pyorrhoea and in bleeding gums. This plant soothes vitiated pitta, cough, bronchitis, asthma, inflammation, haemorrhage, haemorrhoids, bleeding of diarrhoea, and disease of eyes. The fresh flowers are used in ophthalmic, lessen strangury and jaundice. It has many therapeutic applications in Ayurvedic, Unani, homeopathic and Allopathic system of medicine. In this review paper, an idea has been generated to explore various scope of this plant products which will be beneficial to the society.

Keywords: *Adhatoda beddomei, Adhatoda vasica,* Alkaloid, Adusa, Ardusi, Vasicine.

Introduction

Adhatoda beddomei and *Adhatoda vasica* belongs to the Acanthaceae Family and is an important medicinal plant. Acanthaceae is a large panatropical family of about

1 Department of Biotechnology, Shri A.N. Patel P.G. Institute, Anand – 388 001, Gujarat, India.

2 Department of Life Sciences, Hemchandracharya North Gujarat University, Patan – 384 265, Gujarat, India.

* *Corresponding author*: E-mail: jitenp2001@gmail.com

250 genera and 2,500 species. It's a rare evergreen shrub having restricted distribution at elevation of 1000 m in the hills of southern India. In Hindi, it is known as Adusa, Arusa. Mature plants are the important source of drug vasaca and it is considered therapeutically important to the allied species *Adhatoda vasica*. It is a large shrub or small tree grows up to 3-4 meters in height. This plant grows wild in abundance all over Sri Lanka, Nepal, India, and the Pothohar region of Pakistan, particularly in the Pharwala area. It commonly grows in waste places. This shrub grows on the plains of India and in the lower Himalayans, up to a range of 1000 meters above sea level. It will grow well in low moisture areas and dry soil.

Now a days natural forest are the only source of *A. beddomei*. Natural regeneration and conventional propagation through the vegetative cuttings are slow and insufficient for conversion. Non availability of plants in other parts of India and rapid depletion due to over-exploitation of the resource in natural habitats have forced the local industry to depend on the more common *A. vasica* for raw material supply. *A. beddomei* is now included in the IUCN threatened plants list of India.

Due to its medicinal importance and potential, there is a need of mass propagates of this plant species. Plant tissue culture offers an effective method for rapid multiplication to overcome these natural impediments in propagation of this species (Fay, 1994).

Figure 33.1: (A) *Adhatoda beddomie* and (B) *Adhatoda vasica* (Courtesy: Anand Agriculture University, Anand, Gujarat, India).

Adhatoda vasica is a well-known plant drug in Ayurvedic and Unani medicine. Adhatoda leaves have been used extensively in Ayurvedic medicine primarily for respiratory disorders. The medicinal properties of *Adhatoda vasica*, called Vasa or Vasaka in Sanskrit have been known in India and several other countries for thousands of years.

Classification

Kingdom: Plantae

Division: Angiosperms

Class: Eudicots

Order: Lamiales

Family: Acanthaceae

Genus: *Adhatoda*

Species: *beddomei* and *vasica*

Adhatoda beddomei is a rare evergreen shrub with restricted distribution at an elevation of 1000 m in the hills of South Travancore in southern India. Mature plants are an important source of the drug vasica in the indigenous of medicine (Aiyer and Kolammal, 1963), where it is considered therapeutically superior to the allied species, *A. vasica*.

At present, natural forests are the only source of *A. beddomei*. The plants rarely set seeds; natural regeneration and conventional propagation through vegetative cuttings are slow and insufficient for conservation. Non availability of the plants in other parts of India and rapid depletion due to over exploitation of the resource in natural habitats have forced the local industry to depend on more common *A. vasica* for a raw material supply.

Sanskrit synonyms: Vasa, Vasaka, Vrisha, Vrishaka, Simhasya, Vishnu, Vajidanta, Adaroosha.

Plants Name in Different Languages

English : Malabar nut

Hindi : Adusa, Arusa

Malayalam : Adalotakam, Chittadalatokam

Distribution: In southern parts of India, especially Kerala

Medicinal Properties

Plant pacifies vitiated pitta, kapha, cough, bronchitis, asthma, inflammation, hemorrhage, hemorrhoids, diseases of eyes, and bleeding diarrhea.

Useful part : Whole plant

This is a shrub with lance-shaped leaves 10 to 15 centimeters in length by four wide. They are oppositely arranged, smooth-edged, and borne on short petioles. When dry they look dull brownish-green color. They are bitter in taste. They have a smell similar to strong tea. The wood of the stem is soft, and makes a great charcoal for gunpowder. The flower has large, attractive, white petals, streaked with purple on the lower lip. The fruit has a small capsule with 4 seeds. The leaves, roots, flowers and stem bark of this plant are used in medicinal applications.

Chemical Components

Herbs are staging a comeback and herbal resurgence is happening all over the Globe. The herbal products today symbolize safety in contrast to the synthetics that are regarded as unsafe to human and environment. Although herbs had been priced for their medicinal, flavouring and aromatic qualities for centuries, the synthetic products of the modern age surpassed their importance, for a while. However, the blind dependence on Synthetics is over and people are returning to the naturals with hope of safety and security.

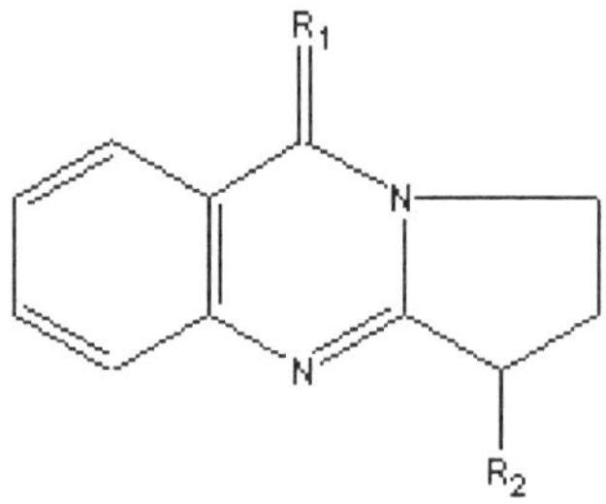

vasicine(peganine) R_1=H_2,R_2=OH

vasicinone R_1=O,R_2=OH

desoxyvasicinone R_1=O,R_2=H

Global estimates indicate that 80 per cent of about 4 billion population cannot afford the products of the Western Pharmaceutical Industry and have to rely upon the use of traditional medicines which are mainly derived from plant material. This fact is well documented in the inventory of medicinal plants, listing over 20,000 species. In spite of the overwhelming influences and our dependence on modern medicine and tremendous advances in synthetic drugs, a large segment of the world still likes drugs from plants. In many of the developing countries the use of plant drugs is increasing because modern life saving drugs are beyond the reach of three quarters of the third world's population although many such countries spend 40-50 per cent of their total wealth on drugs and health care. The oldest root of medicinal plant biotechnology has grown from cell technology, specifically plant Tissue culture. In fifty years since its inception as a tool in Physiology (Chatterji, 1963) tissue culture has found most visible application in plant regeneration from *in vitro* cultured cells. One commercial by product of this technology is micropropagation (Dalal, 1995).

The leaves contained an essential oil and alkaloids quinazoline, vasicine, vasicinone and deoxyvasicine (Shinawie, 2002). Their roots contained vasicinolone, vasicol, peganine and 2′-hydroxy-4-glucosyl-oxychalcone. The flowers contained D-glucoside, kaempferol, glucosides, and the bioflavonoid namely quercetine. The alkaloids, vasicine and vasicinone present in the leaves, possess respiratory stimulant activity; whereas, vasicine, at low concentrations, induced bronchodilation and relaxation of the tracheal muscle. The plant has antispasmodic, antiseptic, antihelminthic, insecticidal and parasiticidal properties (Gulfraz *et al.*, 2004).

Few of the main chemical constituents of this plant are vasicine (derived from leaves), 2'-hydroxy-4- glucosyloxychalcone, vasicol (from leaves), vasicinone (from leaves, stem and roots), vasicinol (contained in stem and roots), and deoxyvasicinone (from leaves) etc. It was first isolated from by Sen and Ghose in 1924. Different morphotypes of this species of the plant are found in nature. These species are classified in different groups and can be characterized with the help of fingerprint profile and vasicine content (S.K. Chauhan and G.P. Kimothi 1999). *Adhatoda vasica* (AV) is found in tropical regions of Southeast Asia, including India.

Pharmacological properties of this plant are known since very old times. It is well established now that, vasicine is the major, as well as, the most important active principle of this medicinal plant. It is reported to be responsible for most of its activities including: antioxidant, anti-inflammatory and bronchodilatory activity. It is an optically active molecule in its natural condition but, gets racemized when extracted. Although, its importance has been recognized since years and it is being used extensively in many herbal formulations for respiratory diseases and disease related to female reproductive system, limited data is available for its molecular mechanism of action.

Realizing its medicinal importance, Shivanna (2009) has worked upon its pollination biology. There are studies describing various methods for extraction of vasicine from the plant (Chattopadhyay, 2003). HPLC determination of vasicine with photo diode (PD) detector is not well established (Srivastava, 2001). C18 column (pH 3.9) with acetonitrile - phosphate – acetic acid (15:85:1, v/v/v) buffer can be used to separate vasicinone and vasicine. Soni *et al.* (2008) compared different methods of extraction of vasicine including the classical method in which, the crushed leaves are subjected to wet heat followed by squeezing the extract as well as modernization of the same by exposing the leaves to steam at 15 lb. The total alkaloid content varied from 0.3 mg/ml to 5.93 mg/ml and that of vasicine from 0.2 mg/ml to 5.64 mg/ml for different methods. International Application Published under the Patent Cooperation Treaty (PCT) has published method for the production of vasicine by Chattopadhyay *et al.* (2003).

Although *Adhatoda vasica* is the main source of obtaining vasicine, other plants were also worked upon to isolate this molecule. In 1975, vasicine was isolated from a different plant *Sida cordifolia* by Ghoshal *et al.* (1975). Susag *et al.* (2003) isolated vasicine from two species of Afrogalega: *Galega battiscombei* (Bak.f) Gillett and *Galega lindblomi* (Harms) Gillett. Bagchi *et al.* (2003) reported its isolation from *A. beddomei* and the content of vasicine was compared to *Adhatoda zeylanica/vasica*. They reported that *Adhatoda zeylanica/vasica* showed wide seasonal variation throughout the year. The variation of vasicine was found from 1.22 per cent to 2.57 per cent twice a year. Peganine/vasicine hydrochloride was isolated from *Peganum harmala* seeds in dihydrated form by Khaliq *et al.* (2009).

Tissue Culture Studies

This plant has medicinal uses, mainly antispasmodic, fever reducer, anti-inflammatory, anti-bleeding, branchodilator, antidibetic, antihelminthic, disinfectant, anti-jaundice, antiseptic, oxitocic and expectorant and has many other medicinal applications (Patel and Bhatt,1984; Chakraborty and Brantner, 2001; Wakhloo *et al.*, 1980). There is a considerable demand of this plant in India and this demand is met from the natural habitat. This plant show low seed germination and conventional propagation through cutting is slow (The wealth of India,1985; Mathew and Patel, 1998). This leads to rapid depletion of plant material due to over exploitation of this important plant. Pant tissue and cell culture system are being exploited for the accumulation of the variety of natural products (Kishor, 1987; Komaraiah *et al.*, 2007). The tissue culture systems for a number of medicinal plants have been established, and this enables the analysis of callus and suspension for the presence of the various secondary metabolites (Rao and Ravishankar, 2002).

Generally, the plant is propagated by seed as well as stem cutting but very low seed germination and labour intensive vegetative propagation by stem cutting are constrains for large scale production of *A. vasica*. Attempts have been made in the past for regeneration of *A. vasica* through tissue culture (Jaiswal *et al.*, 1989; Azad *et al.*, 2003; Abhyankar and Reddy, 2007; Khalekuzzaman *et al.*, 2008; Kumar *et al.*, 2007). The alkaloid content of the plant has been reported to vary with the genotype and environment, so clonal propagation has been recommended (Duster, 1985).

The young juvenile nodal explants of *Adhatoda vasica* cultured on Murashige and Skoog's (MS) medium supplemented with CW (15 per cent v/v) + BAP (5 mg/l) started proliferating multiple shoot buds (14±2) in 3 to 4 weeks and the shoots became 3 cm long in 6 to 8 weeks. The regenerated plants produced well developed roots on transfer to MS + IBA 1 mg/l in 10 to 12 days. The rooted plantlets were acclimatized before transfer to soil and 80 per cent of the transferred plants survived (Raageeva and Shahnawaz, 2011). Callus culture of *Adhatoda vasica* Nees were established from leaf explant. Different growth regulators greatly influenced the growth of callus cultures. Leaf derived callus grown on Murashige and Skooge (MS) medium fortified with different concentration (1, 1.5, 2,2.5 ppm) of Indole Acetic Acid (IAA), 6-Benzyladenine (BA), Kinetin (Kn), 2, 4- Dichlorophenoxy acetic Acid (2, 4-D) (Sunita Maurya and Dhananjay Singh, 2010). Micropropagation study of *Adhatoda vasica* Nees from shoot tip has already done by Sangeeta and Alak (2005).

Pharmacological Study

Antitussive and Bronchodilatory Effect

One of the derivatives of pure vasicine is Bisolvon/bromhexine (N-cyclo-N-methyl-(2-amino-3, 5-dibromo-benzyl) amine, hydrochloride), which has been reported to possess mucus liquefying/expectorant activity by Amin *et al.* (1959) and Sharafkhaneh *et al.* (2007). Antitussive effect of *Adhatoda vasica* extract was also investigated in mechanical or chemical induced coughing in guinea-pigs (Dhuley, 1999). Oral administration of AV extract (ED50: 75.6-200 mg/kg) inhibited chemically induced coughing, in dose dependent manner (up to 75 per cent) and was comparable to codeine. An i.v administration of the extract (5-20mg/kg) shown to have almost complete inhibition (at 20mg/kg) of the coughing induced by mechanical methods. Electrically induced coughing was also found to be inhibited by i.v. administration of AV extract (ED50: 15.5 mg/kg) but was not comparable to codeine. Another derivative of vasicine: 2,4-diethoxy- 6,7,8,9,10,12-hexahydroazepino[2,1-b]quinazolin-12-one, exhibited marked bronchodilator activity as evaluated on contracted trachea or constricted tracheo-bronchial tree. Cumulative dose response study with acetylcholine and histamine indicated its non-specific direct effect on smooth muscle (Zabeer, 2006).

Hepatoprotective Activity

AV was reported to be hepatoprotective and, it is believed to be based on its antioxidant property. Pandit *et al.* (2004) investigated the antioxidant property of *Adhatoda* on carbon tetrachloride induced hepatotoxicity in rats. They had shown that AV (100 mg/kg and 200 mg/kg) increased the activity of protective enzymes in the liver. Doses from 50 to100 mg/kg, per os, of AV leaf have also been shown to have significant hepatoprotective effect on liver damage induced by d-galactosamine in rats (Bhattacharyya, 2005).

Antioxidant and Anti-inflammatory Activity

Methanolic extract of *Adhatoda* has been reported to have antiinflammatory activity. At a dose of 50 ìg/pellet alkaloid fraction, Chakraborty *et al.* (2001) had shown its potent activity in modified hen's egg chorioallantoic membrane test. Srinivasrao *et al.* (2006) have worked upon the antioxidant and anti-inflammatory activity of vasicine against ovalbumin and aluminum hydroxide induced lung damage in rats. They had shown that, vasicine treatment has effect on superoxide dismutase, catalase, glutathione peroxidase and "reduced glutathione".

Abortifacient Activity

Vasicine, at a dose of 1 mg itself has been shown to have no effect on contractions (Gupta *et al*, 1978). It has stimulatory effect on F2α and PGE1 evoked contractions in isolated rat uterus. Vasicine increased the contractions when it was given before PGE1. Further Gautam *et al.* (1982) studied the effect of vasicine and (+) sotalol and deoxysotalol on oxytocin induced contractions. They found that, (+) sotalol (10 ìg/ml), deoxysotalol (10 ìg/ml), and vasicine (1 ìg/ml) produced marked potentiation in the contractile responses of oxytocin, while (+) sotalol and deoxysotalol did not potentiate the responses of oxytocin on mammary strip (selective for uterus tissue), vasicine HCL potentiate these responses. No explanation was available for this "non-selectivity" of vasicine for this action at that time. Further aqueous or 90 per cent ethanolic extracts of leaf of *Adhatoda vasica* has been reported to be 100 per cent abortive on rats after 10th day of insemination, at a doses equivalent to 175 mg/kg of the starting dry material (Nath *et al.*, 1992).

Anti-diabetic Activity

Bromhexine, as it found to have effect on mucus glycoproteins, was tried on diabetic patients and it was reported by Clamp *et al.* (1979) that, it can restore the balance in glucose level in the urine of diabetic patients but, has no effect on normal patient. They suggested that, this change may be due to reduction in the amount of glycoprotein and related material in the body, or from a change in the catabolism of these materials.

A different study by Gao *et al.* (2008) highlights the role of vasicine in sucrose metabolism.

Conclusion and Future Prospects

The medicinal plant *Adhatoda sps* and its constituent: Vasicine and its derivatives are comprehensively being used for bronchodilatory/mucolytic measures ever since history till date. The antioxidant properties of this herb and its components including vasicine are reported to be responsible for anticancer, anti-aging, antidiabetic, hepatoprotective, genoprotective, cardiotonic properties. Literature review suggests us that, vasicine might have multiple objectives which are accountable for its valuable effect for the treatment of respiratory ailments for example: mucolytic property, anti-microbial activity, antioxidant activity etc. which, can improve the pathological conditions synergistically. It is also likely that, vasicine is supportive in regulating the inflammatory reaction as well and, other than that, its smooth muscle stimulating activities might also be useful to remove the phlegm from the respiratory tract. It has become necessary now to explore molecular mechanism of this potent herb and its active principles. This plant might also have a specific application in case of induction of childbirth in diabetic women due to its anti-diabetic potential for optimized doses, and might also help to control post-partum haemorrhage. Being antibacterial it can further help to prevent infection during the above mentioned cases. In short, *Adhatoda vasica* and its phytoconstituents including: vasicine and its derivatives have great potentials to develop safe medications for respiratory and reproductive medicine. Our lab is currently in process of finding out the molecular fidelity study of these plants. We are hoping to identify few molecular aspects so that, further strategy can be developed regarding these plants for utilization and clarification of this molecule for pharmaceutics.

Acknowledgements

Thanks to Medicinal and Plant Aromatic and Research Center, Anand Agriculture University, Anand, Gujarat, to provide access to images of the plant, for a part of the literature.

References

Abhyankar, G., and Reddy, V.D. (2007). Rapid micropropagation via axillary bud proliferation of *Adhatoda vasica* Nees from nodal segments. *Indian J. Exp. Biol.*, **45**: 268-271.

Aiyer, N.K., and Kolammal, M. (1963). Pharmacognosy of Ayurvedic Drugs. Dept. of Pharmacognosy, university of Kerala, Trivandrum.

Amin, A.H., and Mehta, D.R. (1959). A bronchodilator alkaloid (vasicinone) from *Adhatoda vasica* Nees. *Nature*, **184**: 1317.

Azad, M.A.K., Amin, M.N., and Begum, F. (2003). Rapid clonal propagation of a medicinal plant – *Adhatoda vasica* Nees using tissue culture technique. *Online J. Biol. Sci.*, **3(2)**: 172-182.

Bagchi, G.D., Dwivedi, P.D., Haider, J. *et al.* (2003). Seasonal variation in vasicine content in *Adhatoda* Species grown under North Indian plain conditions. *J. Med. Arom. Plant Sci.*, **25**: 37-40.

Bhattacharyya, D., Pandit, S., Janab, U., *et al.* (2005). Hepatoprotective activity of *Adhatoda vasica* aqueous leaf extract on D-galactosamine induced liver damage in rats. *Fitoterapia*, **76**: 223-225.

Chakraborty, A., and Brantner, A.H. (2001). Study of alkaloids from *Adhatoda vasica* Nees on their anti-inflammatory activity. *Phytother. Res.*, **15**: 532-534.

Chatterji, N., Rastogi, R. P., and Dhar, M.L. (1963). Chemical examination of *Bacopa monnieri* wattst. Part 1.

Chattopadhyay, S.K., Bagchi, G.D., Dwivedi, P.D., *et al.* (2003). An improved process for the production of vasicine. WO/2003/080618.

Chauhan, S.K., and Kimothi, G.P. (1999). Development of HPLC method for vasicine and vasicinone in *Adhatoda vasica* Nees. *Indian J. Nat. Prod.*, **15**: 21-24.

Clamp, J.R., Hartog, M., and Shelley, J.H. (1979). Carbohydrate-containing materials in urine from normal and diabetic subjects. *Clin. Sci.* (*Lond*). **56**: 193-196.

Dalal. K. C. (1995). All about Gugglu, Amruth.

Dhuley, J.N. (1999). Antitussive effect of *Adhatoda vasica* extract on mechanical or chemical stimulation-induced coughing in animals. *J. Ethnopharmacol.*, **67**: 361-365.

Duster, J. (1985). Medicinal Plants of India and Pakistan. D B Taraporevala Sons and Com. Pvt. Ltd, Bombay.

Fay, M. F. (1994). In What situations is *in vitro* culture appropriate to plant conversion? *Biodiversity conservation*, **31**: 76-183

Gao, H., Huang, Y.N., Gao, B. *et al.* (2008). Inhibitory effect on α-glucosidase by *Adhatoda vasica* Nees. *Food Chem.*, **108**: 965–972.

Gautam, C.S., and Sharma, P.L. (1982). Potentiation of oxytocin evoked responses by (+) sotalol and deoxysotalol and vasicine HCL on isolated rat and rabbit uterus. *Indian J. Med. Res.*, **76**: 107-114.

Ghosal, S., Ballav, R., Chauhan, P.S., *et al.* (1975). Alkaloids of *Sida cordifolia*. *Phytochemistry*, **14**: 830-832.

Gulfraz, M., Arshad, M., Nayyer, N., Kanwal, N., and Nisar, U. (2004). *Ethnobotanical Leaflets.*

Gupta, O.P., Ghatak, H.J.R., and Atal, C.K. (1978). Vasicine alkaloid of *Adhatoda vasica* a promising uterotonic abortifacient. *Indian J. Exp. Biol.*, **16**: 1075-1077.

Jaiswal, V.S., Narayan, P., and Lal, M. (1989). Micro propagation of *Adhatoda vasica* Nees. through nodal segment culture. In: Kukreja *et al.* (eds.) Tissue culture and Biotechnology of Medicinal and Aromatic plants. CIMAP, Lucknow, India, pp. 7-11.

Khalekuzzaman, M., Rahman, M.S., Rashid, M.H., and Hossain, M.S. (2008). High frequency *in vitro* propagation of *Adhatoda vasica* Nees through shoot tip and nodal explants culture. *J. Biosci.*, **16**: 35-39.

Khaliq, T., Misra, P., Gupta, S., *et al.* (2009). Peganine hydrochloride dihydrate an orally active antileishmanial agent. *Bioorg. Med. Chem. Lett.*, **19**: 2585-2586.

Kishor, K., and Mehta, P.B (1987). Ergot. *Current Science*, **56**: 781-783.

Komaraiah, P., Ramakrishna, S.V., Reddanna, P., Kishor, K., and Mehta, P.B. (2003). *J. Biotechnol*, **101**: 181-187.

Kumar, M., Samarth, R., Kumar, M., Selvan, R.S., Saharan, B., and Kumar, A. (2007). Protective effect of *Adhatoda vasica* Nees. against radiation – Induced damage at cellular, biochemical and chromosomal levels in Swiss mice. *Evid Based Complement Alternat. Med.*, **4(3)**: 343-350.

Mathew, A.S., and Patel, K.N. (1998). *Ind. J. Nat. Prod.*, **14**: 11–16.

Nath, D., Sethi, N., Singh, R.K., *et al.* (1992). Commonly used Indian abortifacient plants with special reference to their teratologic effects in rats. *J. Ethnopharmacol.*, **36**: 147-154.

Pandit, S., Sur, T.K., Jana, U., *et al*, (2004). Prevention of carbon tetrachloride induced hepatotoxicity in rats by *Adhatoda vasica*. *Indian J. Pharmacol.*, **5**: 312-320.

Patel, P.K., and Bhatt, P.H. (1984), *Ind. J. Med. Sci.*, **38**: 70-72.

Raageeva Bimal and Shahnawaz, M.D. (2012). Plant regeneration from nodal explants of *Adhatoda vasica* Nees. *Journal of Medicinal Plants Research*, **6(7)**: 1229-1233.

Raja, S.S., Unnikrishnan, K.P., George, S., *et al.* (2008). Variation in vasicine content and pharmacognostic characters of morphotypes of *Adhatoda zelylanica*. *Medic. J. Plant Sci.*, **3**: 61-68.

Rao, S.R., and Ravishankar, G.A. (2002), *Biotechnol. Adv.*, **20**: 101-153.

Sangeeta Nath and Alak K Buragohain (2005). Micropropagation of *Adhatoda vasica* Nees- A woody medicinal Plant by shoot tip culture. *Indian Journal of Biotechnology*, **4**: 396-399.

Sen, J.N., and Ghose, T.P. (1924). Alkaloid from leaves of *Adhatoda vasica*. *J. Indian Chem. Soc.*, **1**: 315.

Sharafkhaneh, A., Velamuri, S., Badmaev, V., *et al.* (2007). Review: The role of natural agents in treatment of airway inflammation. *Ther. Adv. Respir. Dis.*, **1**: 105-120.

Shivanna, K.R. (2009). Pollination biology, breeding system and reproductive success of *Adhatoda vasica*, an important medicinal plant. *Curr. Sci.*, **96**: 408-412.

Soni, S., Anandjiwala, S., Patel, G., *et al.* (2008). Validation of different methods of preparation of *Adhatoda vasica* leaf juice by quantification of total alkaloids and vasicine. *Indian J. Pharm. Sci.*, **70**: 36-42.

Srinivasarao, D., Jayarraj, I.A., Jayraaj, R., *et al*, (2006). A study on antioxidant and anti-inflammatory activity of vasicine against lung damage in rats. *Indian J. Allergy Asthma Immunol.*, **20**: 1-7.

Srivastava, S., Verna, R.K., Gupta, M.M., *et al.* (2001). HPLC determination of vasicine in *Adhatoda vasica* with PDA detector. *J. Liq. Chrom. and Rel. Technol.*, **24**: 153-159.

Sunita Maurya and Dhananjay Singh (2010). *In vitro* Callus culture of *Adhatoda vasica*: A medicinal Plant. *Annals of Biological Research*, **1 (4)**: 57-60

Susag, L., Mathenge, S., and Benn, M. (2003). The alkaloids of two species of *Afrogalega.* Biochem. *Syst. Ecol.*, **31**: 645–647.

The wealth of India (1985): *Raw Materials*, CSIR New Delhi, I, pp.76.

Wakhloo, R.L., Kaul, G., Gupta, O.P., and Atal, C.K. (1980). *Indian Journal of pharmacology*, **13**: 129.

Zabeer, A., Bhagat, A., Gupta, O.P., *et al.* (2006). Synthesis and bronchodilator activity of new quinazolin derivative. *Eur. J. Med. Chem.*, **41**: 429-434.

Utilisation and Management of Medicinal Plants Vol. 3 (2015) *Pages* **547–563**
Editors: **V.K. Gupta and A. Kaul**
Published by: **DAYA PUBLISHING HOUSE, NEW DELHI**

34

Aloe barbadensis Mill. Protects against Acetaminophen Coupled by Diethyl Ether Induced Hepatotoxicity

B.K. Chandan[1*], N. Sharma[1], Jyotsna Suri[3], D.K. Gupta[1], A. Kaul[1], Shanu Magotra[1], A.K. Saxena[1] and Sangeeta Shukla[2]

ABSTRACT

The use of Aloe barbadensis Mill. (Family- Liliaceae) in variety of diseases including liver disorders is well documented in traditional Indian system of medicine. Present study was carried out to confirm the hepatoprotective efficacy of aqueous extract (AB-4) of Aloe barbadensis against acetaminophen (APAP) hepatotoxicity coupled by diethyl ether. Prophylactic and curative treatments with AB-4 against APAP induced hepatotoxicity were used to determine the prophylactic and therapeutic efficacy of AB-4. Alanine aminotransferase (ALT), aspartate aminotransferase (AST), lactate dehydrogenase (LDH), alkaline phosphatase (ALP) in serum; lipid peroxidation (LP), glutathione (GSH), glucose-6-phosphatase (G-6-P), triglycerides (TG) and protein in whole liver homogenate and amidopyrine N-demethylase (N-de) and aniline hydroxylase (AH) in microsomes were estimated for the evaluation of hepatoprotective activity. AB-4 exhibited significant hepatoprotective activity against APAP induced hepatotoxicity as evident by restoration of ALT, AST, LDH, ALP, TG and LP levels. Hepatoprotective potential was further confirmed by recoupment in the levels of protein, GSH, G-6-P and microsomal AH and N-de towards near normal. Histopathology of the liver tissue further supports the biochemical findings. Present study exhibited significant efficacy of the aqueous extract of Aloe barbadensis against acetaminophen coupled by diethyl ether

1 Indian Institute of Integrative Medicine (CSIR), Canal Road, Jammu – 180 016, J&K, India.

2 School of Studies in Zoology, Jiwaji University, Gwalior, M.P., India.

3 Department of Pathology, Government Medical College, Jammu – 180 001, J&K, India.

* *Corresponding author*: E-mail: bkrishan@iiim.ac.in

induced hepatotoxicity. It appears that AB-4 is orally absorbed, well tolerated and safe at therapeutic doses, thus may be of significant therapeutic value in the management of hepatic diseases.

Keywords: Hepatoprotective activity, Acetaminophen (APAP), Diethyl ether, Hepatotoxicity, *Aloe barbadensis.*

Introduction

Plants based traditional systems of medicines are playing important role in providing health care to large section of population, especially in developing countries. The interest in herbal products is increasing globally including the developed nations. *Aloe barbadensis* Mill. Syn. *A. vera* Tourn. ex Linn. (Liliaceae) commonly known as Ghee kanwar (Hindi), found in semi wild state and can be cultivated in all parts of India (Mhaskar *et al.*, 2000). In "Indian System of Medicine" fresh juice of leaves of *Aloe barbadensis* is used in diseases of eye and enlargement of spleen and liver (The wealth of India, 2000). It is also useful for wounds, burns, dermatitis, cutaneous leishmaniasis and other skin disorders (Mhaskar *et al.*, 2000). It is also reported for antiviral (Saoo *et al.*, 1996) and anti-inflammatory (Beatriz *et al.*, 1996) and anti-diabetic (Annamalai *et al.*, 2012) activities. The methanol extract of *Aloe vera* gel is reported for antibacterial and antioxidant activities (Saritha *et al.*, 2010). Chemically, the presence of barbaloin, chrysophanol glycoside and the aglycone, aloe-emodin, glucose, galactose, mannose and galacturonic acid in addition to aldopentose and proteins with eighteen amino acids have been reported. It also contains aloesone and aloesin (The wealth of India, 2000). It is well documented that *Aloe vera* is widely used in the cosmetics and alternative medicine industries, being marketed as having rejuvenating, healing or soothing properties. There is, however, little scientific evidence of the effectiveness or safety of *Aloe vera* extracts for either cosmetic or medicinal purposes, and whatever positive evidence is available is frequently contradicted by other studies (Vogler, 1999; Ernst, 2000). Based on its diversified pharmacological properties, its use in traditional Indian System of Medicine for liver diseases and our previous study against carbon tetrachloride induced hepatotoxicity (Chandan *et al.*, 2007), present study was carried out to confirm the hepatoprotective efficacy of aqueous extract (AB-4) of *Aloe barbadensis* against acetaminophen induced hepatotoxicity coupled by diethyl ether exposure in mice.

Materials and Methods

Plant Material, Extraction and Chromatography

The plant was collected from the botanical garden of Indian Institute of Integrative Medicine, Jammu and was authenticated at source by taxonomist of the institute. A voucher specimen has been deposited at the herbarium of the Institute vide RRL collection No. 17875, Acc. No. 19650. The authenticated and freshly collected aerial parts of the plant were chopped and dried in shade. The coarse plant material (185 g) was extracted successively with petroleum ether (1000 ml x 4), chloroform (900 ml x 4) and methanol (900 ml x 5) at room temperature to obtain petroleum ether

(AB-1 2g; yield 1.08 per cent, w/w), chloroform (AB-2, 1.9 g; yield 1.03 per cent, w/w) and methanol extracts (AB-3, 10g; Yield 5.40 per cent, w/w). The plant marc was finally extracted with distilled water (1000 ml x 3) and lyophilised in freeze drier to furnish aqueous extract (AB-4, 31 g; yield 16.75 per cent, w/w). Quantification of the barbaloin in the extracts was carried by HPLC (Shimadzu make HPLC system with LC-10A TVP pumps, DGU-14A degassing unit, SIL-10A PDA detector and CTO-10AS column oven. Data was acquired and analyzed by using Shimadzu Class- VP software version 6.10) using mobile phase: methanol: water (45 : 55); column: RP-8 (5µm), flow rate: 1 ml/min, detection: 254 nm., injection volume: 50 µl and 100 µl, Barbaloin was used as standard for quantification of barbaloin in the extracts. Methanolic and aqueous extracts were found to contain 0.0127 per cent and 2.09 per cent (w/w) barbaloin respectively. Barbaloin in petroleum ether (AB-1) and chloroform (AB-2) extracts were below detection limits.

Chemicals

Acetaminophen (APAP) from Sigma Chemical Co., USA. All other chemicals used were of analytical grade and obtain locally.

Experimental Animals

Swiss Albino mice (20-25 g) were used in the present study. Animals were housed under standard conditions (23.0±2°C, 60-70 per cent humidity, and 12h photo period) and allowed free access to food (Ashirwad India Ltd., Chandigarh, India) and water. The number of animals and protocol used in the study were approved by Institutional Animal Ethics Committee IIIM, Jammu.

Induction of Liver Injury

Liver injury was induced by injecting paracetamol (200 mg/kg, i.p.) in mice six hours after giving exposure to diethyl ether (5 ml/group) in a closed chamber till the animals fall asleep.

Grouping and Treatment

Six groups with six mice/group were used in the study. First three groups were fed AB-4 125, 250 and 500mg/kg, p.o. respectively + APAP (200mg/kg, i.p.). Fourth group was fed with reference drug silymarin (50 mg/kg, p.o.) + APAP. Group five was given proportionate volume of vehicle + same dose of APAP while the last 6th group was given proportionate volume of vehicle only.

Prophylactic Study

AB-4, reference drug (silymarin) and vehicle were given 74 h, 50 h, 26 h, 2 h before and 6 h after ether exposure. APAP was administered 6 hour after ether exposure. Liver and blood samples were collected 18 hour after APAP administration (Singh *et al.*, 2006).

Curative Study

Test, reference drug and vehicle were given 6 h, 24h, and 48 h after exposure with ether. APAP was administered 6h after exposure to ether. Liver and blood samples were collected 2 hour after last treatment with the test material (Singh *et al.*, 2006).

Blood Biochemistry

Blood samples were collected in glass tubes from orbital sinus to obtain haemalysis free clear serum for the analysis of ALT and AST (Reitman and Frankel, 1957), ALP (Walter and Schutt, 1974), LDH (Wroblewski, 1964) by standard methods.

Preparation of Liver Homogenate and Microsomes

All the animals were sacrificed and livers were quickly excised washed and perfused with chilled normal saline, minced and homogenized in ice bath in chilled 10mM Tris-HCl buffer (pH 7.4) to obtain 10 per cent liver homogenate for the estimation of protein (Lowry *et al.*, 1951), GSH (Ellman, 1959), LP (Buege and Aust, 1978), gluose-6-phosphatase (Baginski *et al.*, 1974) and TG (Neri and Frings, 1973) by using standard methods. Liver microsomes were prepared (Schenkman and Cinti, 1978) and stored in liquid nitrogen for the estimation of Aniline hydroxylase and Amidopyrine-N-demethylase (Anand *et al.*,1997).

Histopathology

From all the animals used for curative study a small portion of liver tissue was fixed in 10 per cent formaline saline, processed and embedded in paraffin wax to obtain 4 to 5 micron thick sections using a microtome. These sections were subjected to histopathological examination after processing in alcohol - xylene series and haematoxylin and eosin staining (Krajian, 1963).

Hepatoprotective Activity Per cent

The hepatoprotective activity (H) percent protection, was calculated as follows:

$$H = \left(1 - \frac{T - V}{C - V}\right)100$$

where,

'T' is mean value of drug and Toxin. 'C' is mean value of Toxin alone and 'V' is the mean value of vehicle treated group.

Overall hepatoprotective activity = Mean ± SE of H of all the parameters

Statistical Analysis

The data obtained is expressed as Mean ± SE and analyzed by One Way analysis of variance (ANOVA) followed by Tukey's multiple comparison test using computerized program. p - value< 0.05 was taken as the criterion of significance.

Results

Prophylactic Study

Normal levels of serum ALT (111.91±8.41 μ mole pyruvate /min/L), AST (109.32±5.40 μ mole pyruvate /min/L), LDH (161.40±5.15 μ mole pyruvate /min/L), ALP (21.06±1.56 μ mole PNP/min/L), and hepatic TG (11.22±0.56 per cent), LP (21.90±1.25 n mole MDA/g liver), GSH (7.01±0.04 μ mole GSH /g liver), G-6-P (4.65±0.17 μ mole pi/min/g liver), N-de (189.00±3.03 nmole HCHO/min/g liver),

AH (8.46±0.29 n mole PAP/g liver) and protein (173.78±6.17 mg/g liver) were estimated in vehicle treated mice. Results of the study showed significant ($p<0.001$) increase in the levels of ALT (1590.15±102.13), AST (1086.47±54.76), LDH (4314.95±181.58), ALP (56.40±3.33), hepatic TG (37.96±1.97), LP (97.24±3.46) and decrease in the levels of GSH (3.92±0.24), G-6-P (3.01±0.07), N-de (74.67±4.67), AH (4.67±0.37), and hepatic protein (140.22 ± 5.66) with acetaminophen (200mg/kg, i. p., 6h after exposure with ether) administration. Prophylactic treatments of AB4 125, 250 and 500mg/kg, p.o., reversed these parameters significantly in a dose dependant manner (Figures 34.1 and 34.2).

Curative Study

Normal levels of serum ALT (98.32 ± 5.20 μ mole pyruvate /min/L), AST (117.49±8.26 μ mole pyruvate /min/L), LDH (171.87±5.72 μ mole pyruvate /min/L), ALP (21.42±2.04 μ mole PNP/min/L), and hepatic TG (10.10±0.55mg per cent), LP (23.75±1.24 n mole MDA/g liver), GSH (7.44±0.32 μ mole GSH /g liver), G-6-P (4.80±0.08 μ mole pi/min/g liver), N-de (188.00±5.60 nmole HCHO/min/g liver), AH (8.44±0.30 n mole PAP/g liver) and protein (173.54±4.96 mg/g liver) were estimated in vehicle treated mice. A significant ($p<0.001$) increase in the levels of ALT (1102.10±46.98), AST (839.79±59.17), LDH (3043.94±120.69), ALP (40.54±3.61), hepatic TG (34.55±2.77), LP (78.77±7.13) and decrease in the levels of GSH (3.95±0.16), G-6-P (3.34±0.12), N-de (87.83±6.14), AH (4.88±0.19), and hepatic protein (153.42±5.50) was observed after acetaminophen (200mg/kg, i. p., 6h after exposure with ether) administration. Curative treatments of AB4 125, 250 and 500mg/kg, p.o., reversed these parameters significantly in a dose dependant manner (Figures 34.1 and 34.2).

Overall hepatoprotective activity exhibited by AB-4 at doses 125, 250 and 500 mg/kg, p.o. was 28.75±1.58, 53.03±2.01 and 77.78±2.05 percent respectively with prophylactic treatment and 27.15±0.95, 46.44±1.69 and 74.15±1.38 respectively with curative treatment. Hepatoprotective activity exhibited by AB-4 (500 mg/kg) was significantly ($p<0.05$) better than silymarin (50 mg/kg, p.o.) with both treatments.

Histopathology

Microscopic examination of the vehicle treated animals showed normal architecture of hepatic parenchyma. Treatment with APAP alone showed extensive centrilobular necrosis of hepatic parenchyma, along with evidence of hepatic steatosis and scattered lymphomononuclear (LMN) cell infiltrate. AB-4 showed dose dependent hepatoprotective effect when given after APAP administration in the form of centrilobular necrosis involving smaller foci of hepatic parenchyma and evidence of regeneration in the form of binuclear hepatocytes adjacent to area of necrosis. The hepatoprotective effect of AB-4 500mg/kg was comparable to that of silymarin (Figures 34.3A-E).

Discussion

Acetaminophen is a hepatotoxicant, which is primarily metabolized by sulfation and glucuronidation to unreactive metabolites and then activated by the cytochrome P_{450} system to produce liver injury. Acetaminophen is bioactivated to a toxic

Figure 34.1A-D: Effect of AB-4 (Prophylactic and Curative Treatments) against APAP Induced Liver Injury in Mice. Blood Biochemistry:

Values represent the mean ± SE of six animals in each group; P Value Vs Vehicle b*< 0.05, b**< 0.001, b***< 0.001; P Value Vs CCl_4- c*< 0.05, c**< 0.01, c***< 0.001, NS not significant.

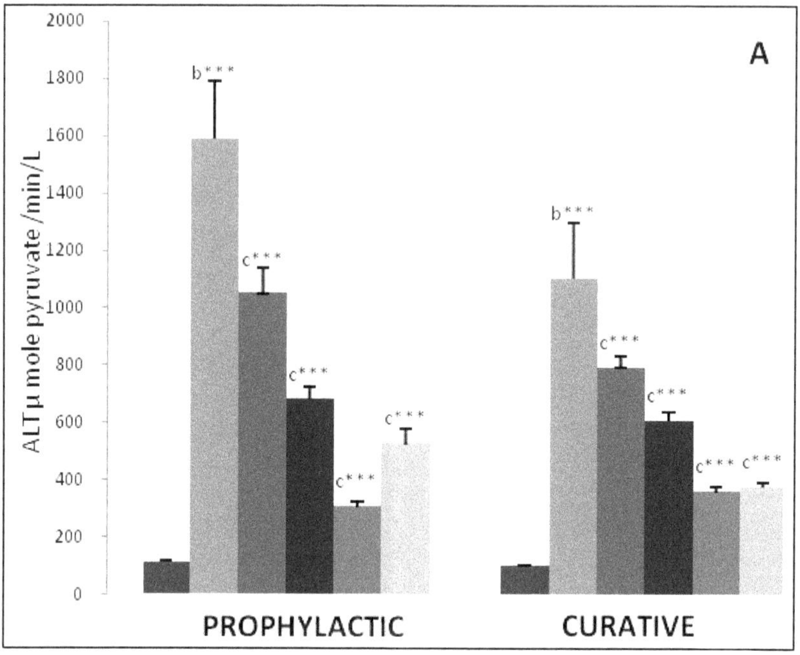

A: Representation of alanine aminotransferase (ALT)

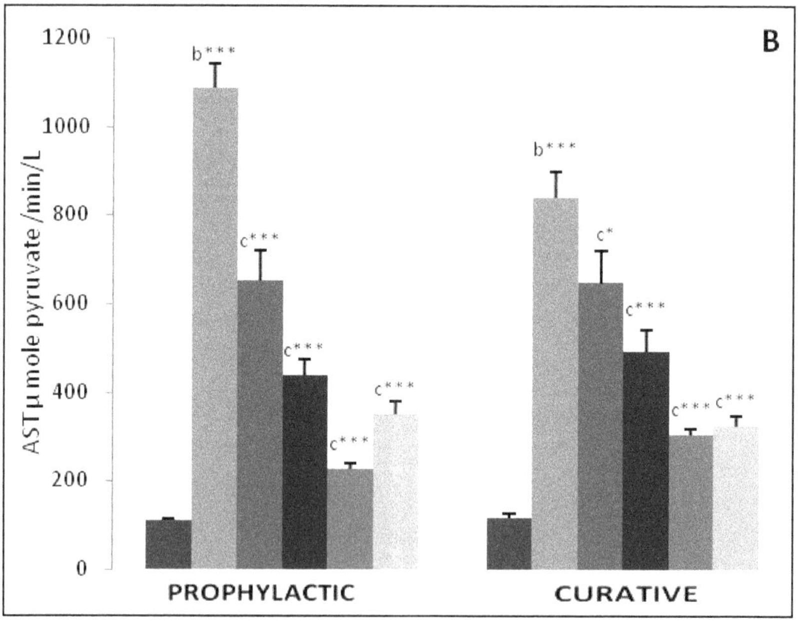

B: Representation of aspartate aminotransferase (AST)

Contd...

Figure 34.1–*Contd...*

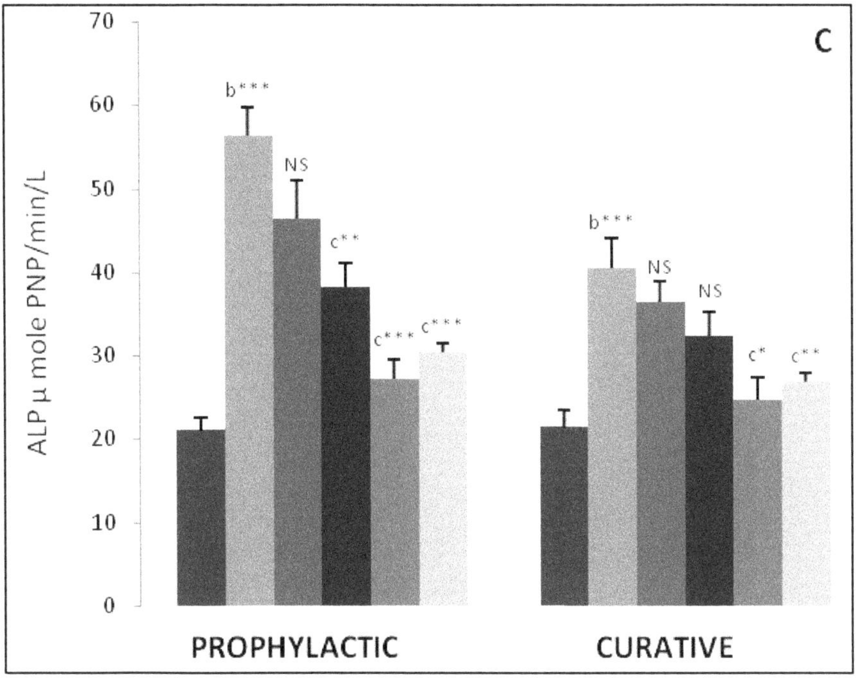

C: Representation of alkaline phosphatase (ALP)

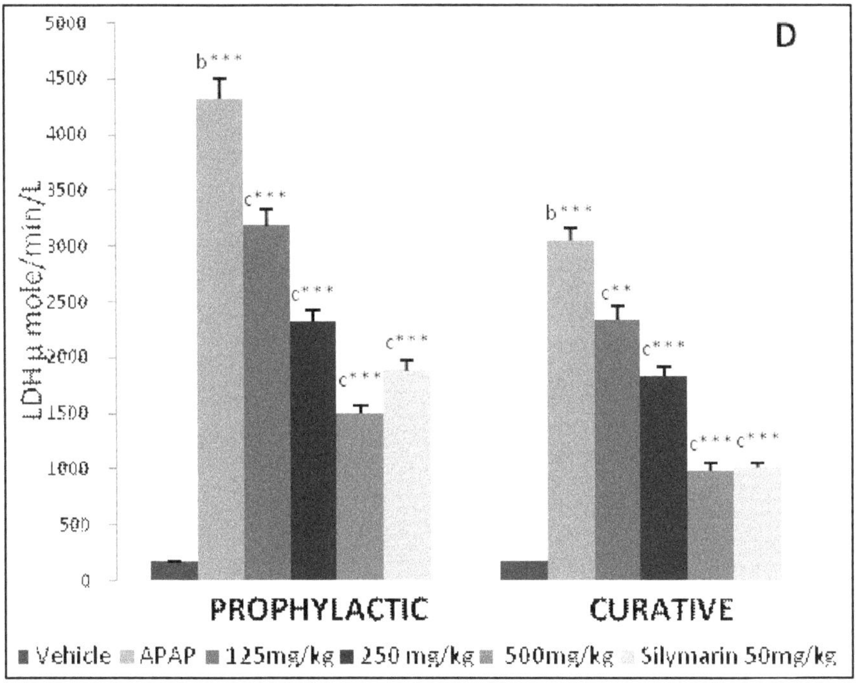

D: Representation of Lactate dehydrogenase (LDH)

Figure 34.2A-G: Effect of AB-4 (Prophylactic and Curative treatments) against APAP Induced Liver Injury in Mice. Liver Biochemistry.

Values represent the mean ± SE of six animals in each group; P Value Vs Vehicle b*< 0.05, b**< 0.001, b***< 0.001; P Value Vs CCl_4- c*< 0.05, c**< 0.01, c***< 0.001, NS not significant.

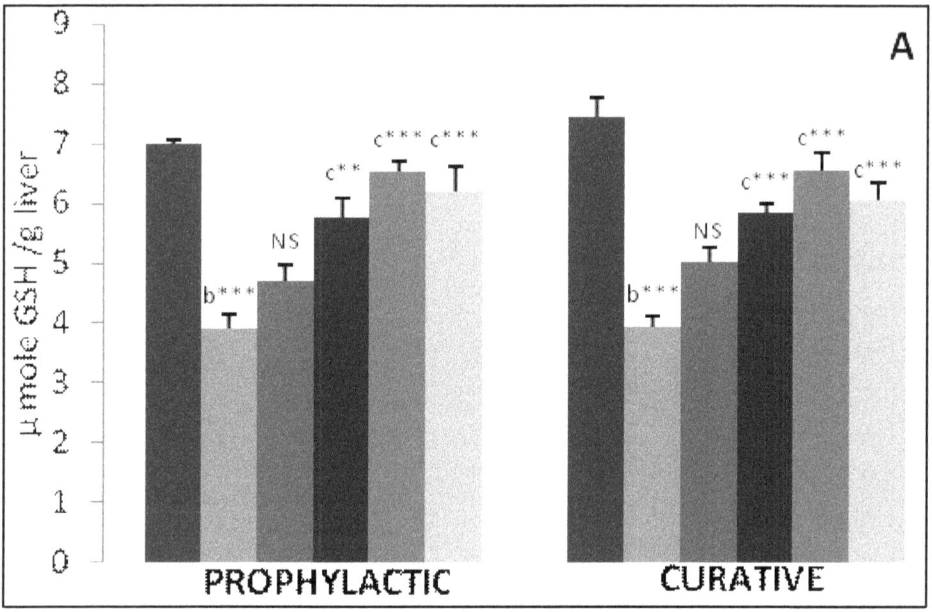

A: Representation of glutathione (GSH).

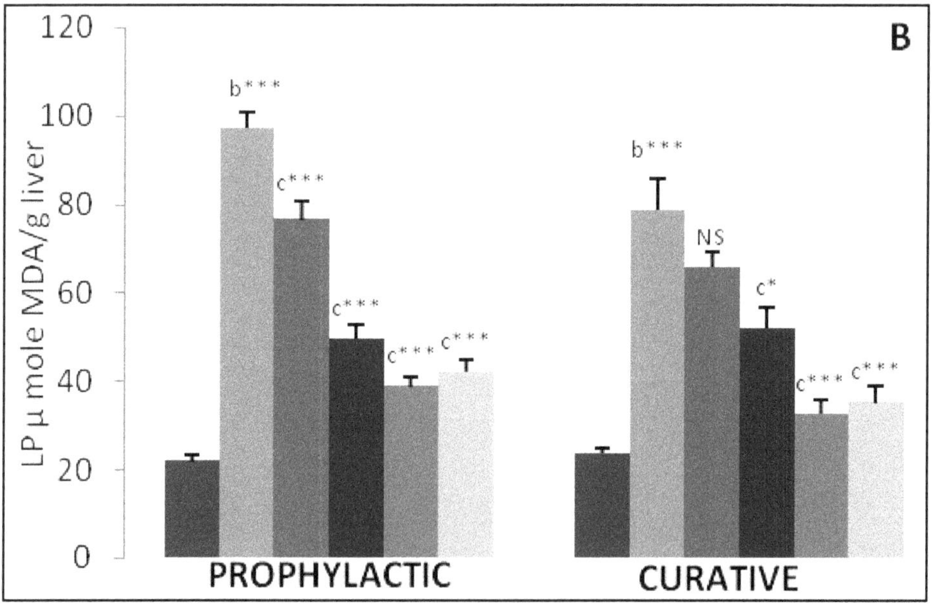

B: Representation of lipid peroxidation (LP).

Contd...

Figure 34.2–*Contd...*

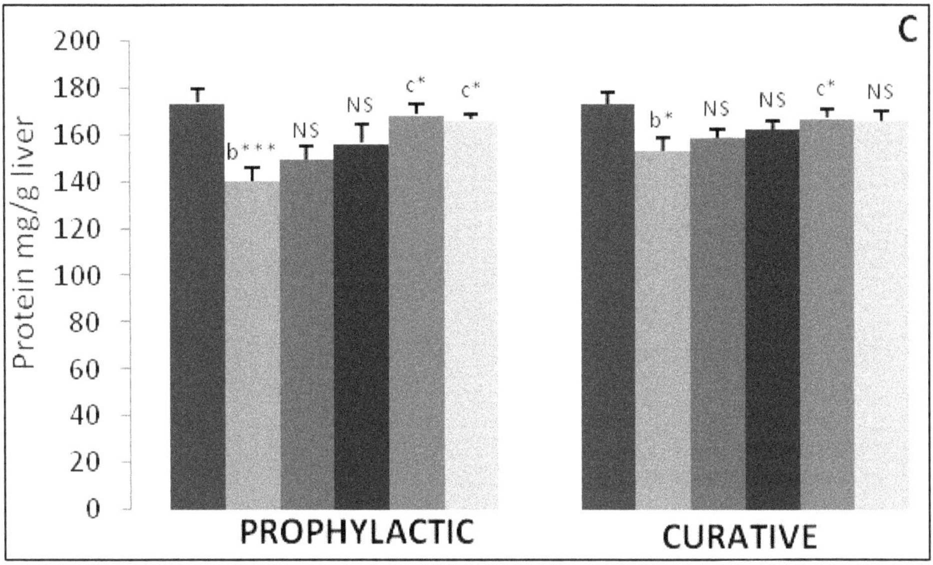

C: Representation of Total Protein (Protein).

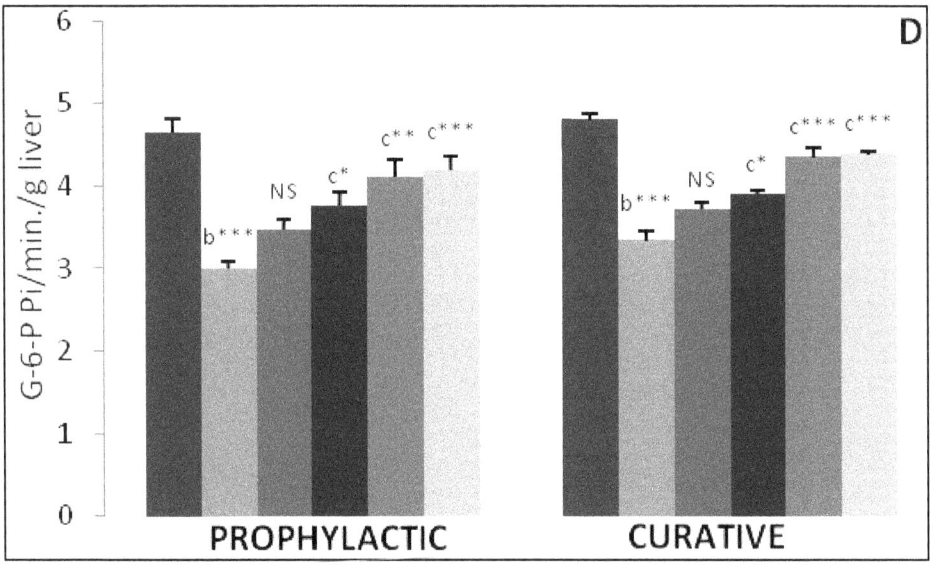

D: Representation of Glucose-6-phosphatase (G-6-P).

Contd...

electrophile, N-acetyl-p-benzoquinoneimine (NAPQI) which binds covalently to tissue macromolecules and probably also oxidizes lipids or the critical sulphydryl groups (protein thiols) and alters homeostasis of calcium. NAPQI produces liver injury unless it is conjugated readily with GSH to form mercapturic acid (Shenoy *et al.,* 2002).

Figure 34.2–*Contd...*

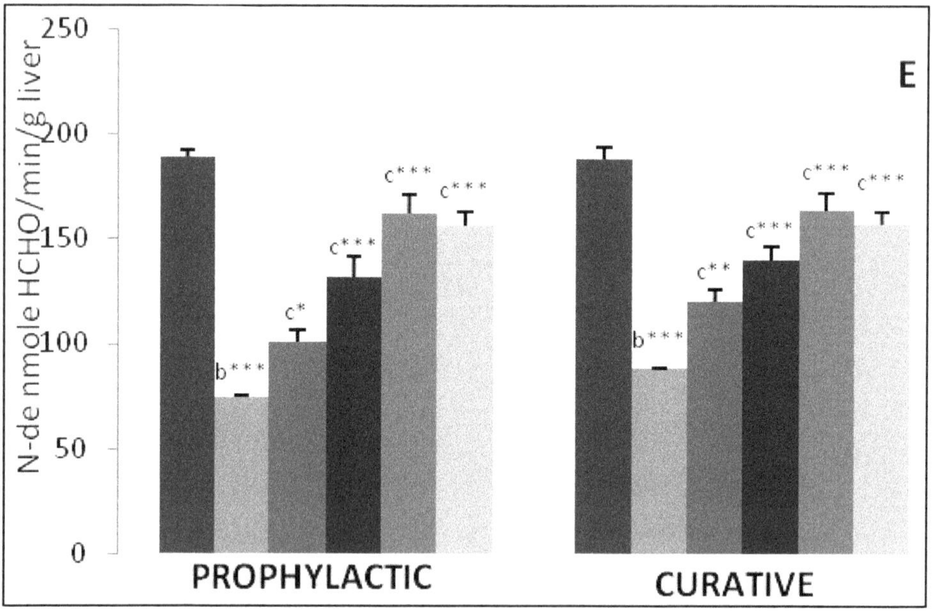

E: Representation of Amidopyrine-N-demethylase (N-de).

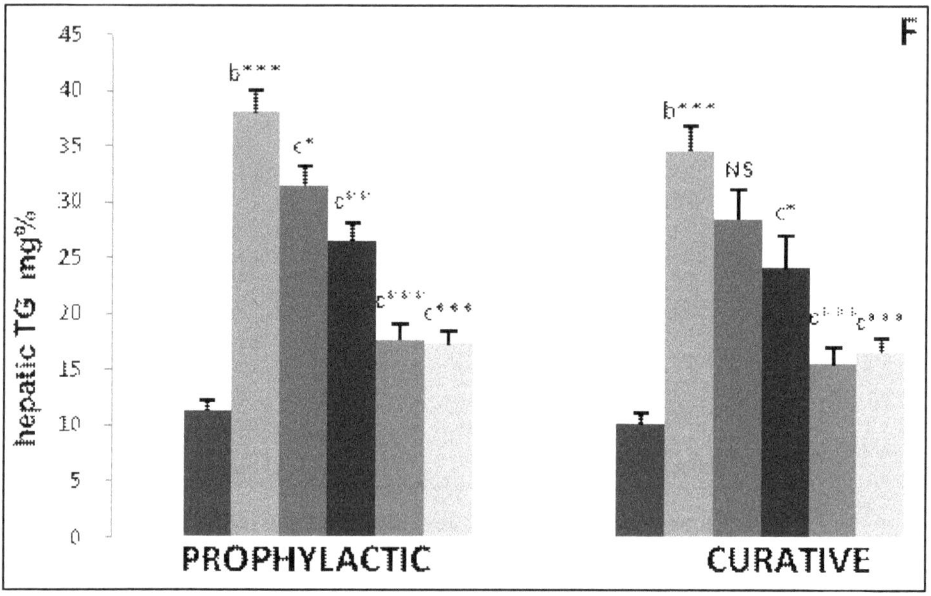

F: Representation of Triglycerides (TG).

Figure 34.2—*Contd...*

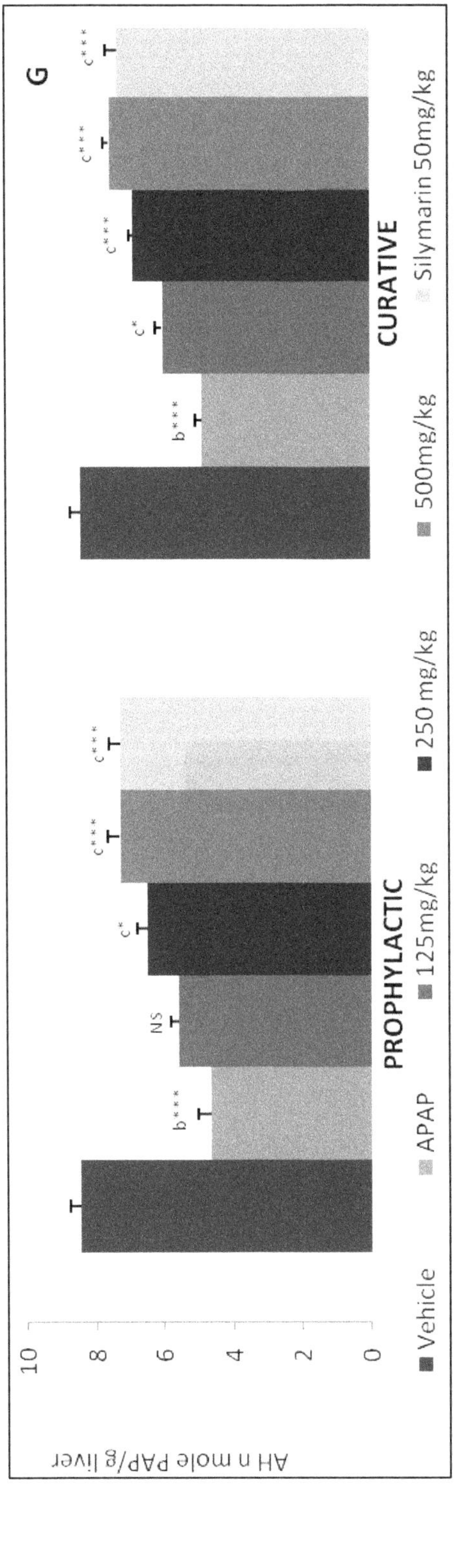

G: Representation of aniline hydroxylase (AH).

Figure 34.3A-E: Hematoxylin and Eosin Stain Sections of Liver (10x).

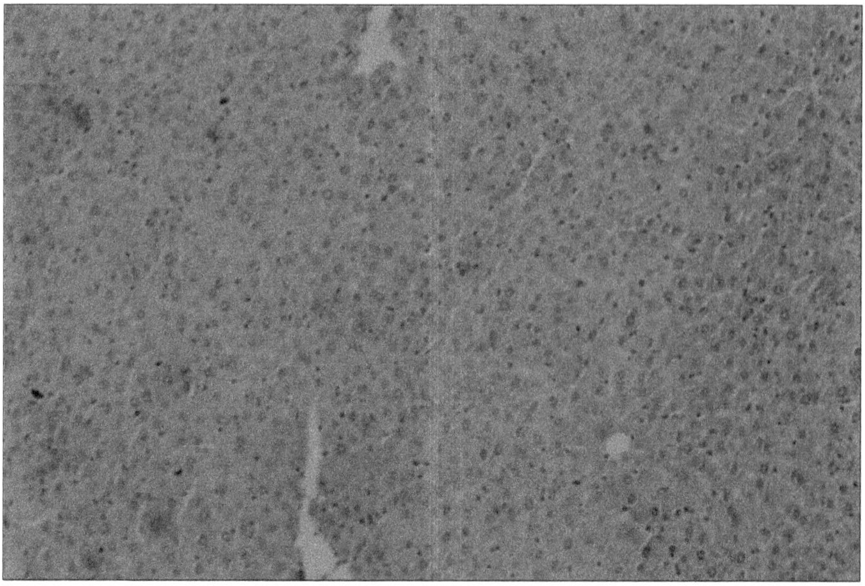

A. Photomicrograph of vehicle treated mouse liver section showing normal hepatic architecture.

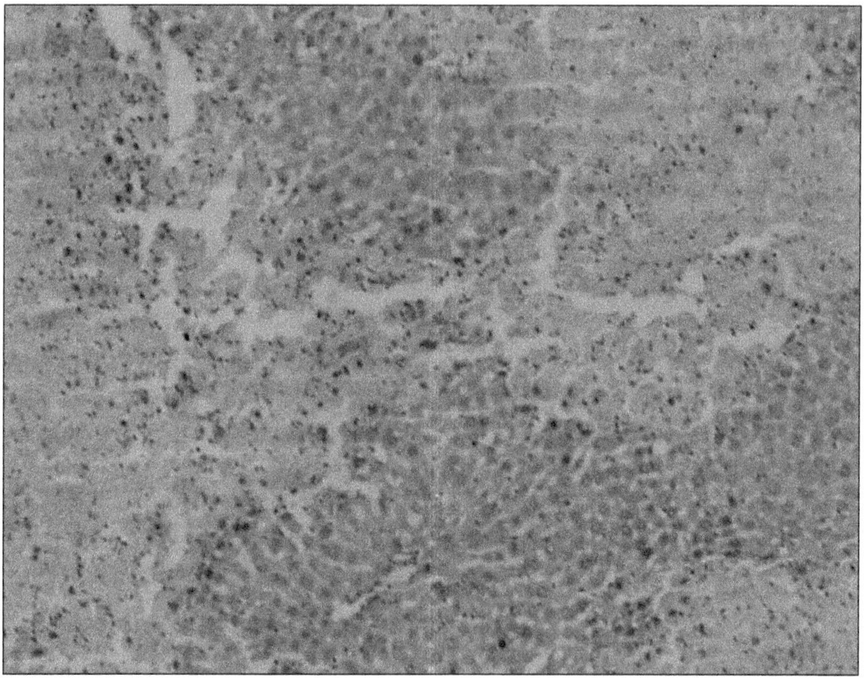

B. Photomicrograph of APAP treated mouse liver section showing extensive area of centrilobular necrosis and evidence of hepatic steatosis with scattered LMN cell infiltrates.

Contd...

Figure 34.3–*Contd...*

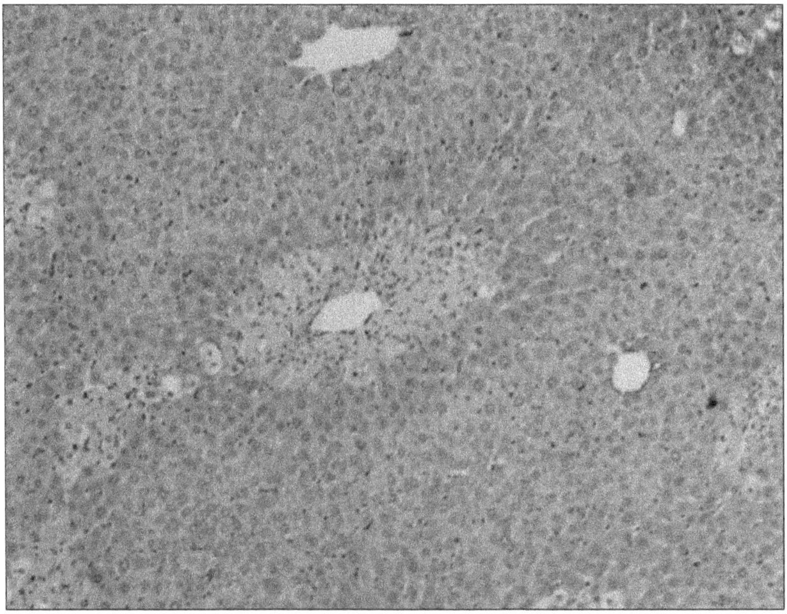

C. Photomicrograph of APAP + $AB_{4,}$ 500 mg/kg, p. o treated mouse liver section showing hepatoprotective effect in the form of smaller foci of centrilobular necrosis and evidence of regeneration of hepatocytes.

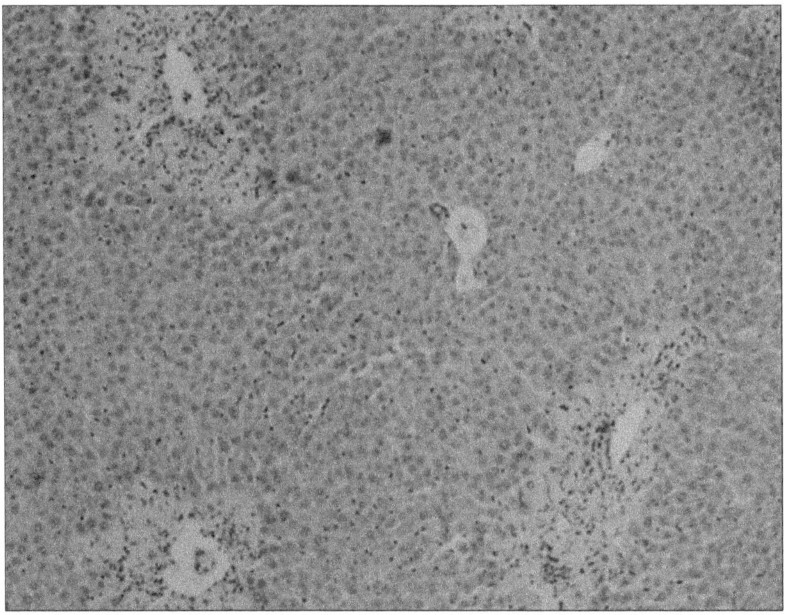

D. Photomicrograph of APAP + $AB_{4,}$ 250 mg/kg, p. o treated mouse liver section showing areas of centrilobular necrosis (less as compared to APAP alone), scattered LMN cell infiltrate in the hepatic parenchyma.

Contd...

Figure 34.3–*Contd...*

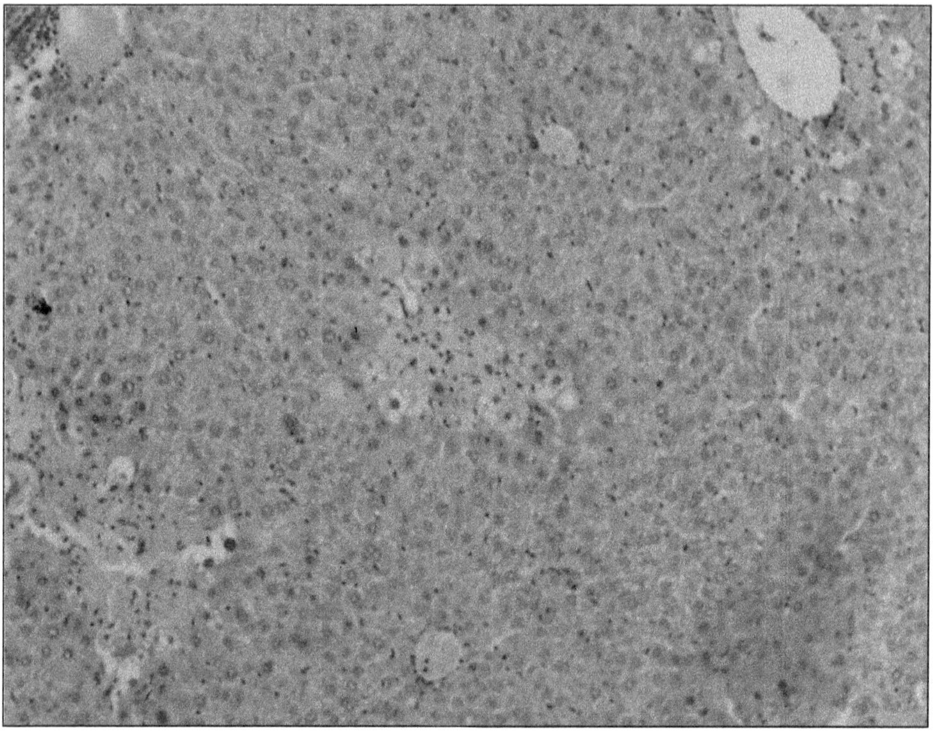

E. Photomicrograph of APAP + Silymarin 50 mg/kg, p. o treated rat liver section showing protection in the form of smaller foci of centrilobular necrosis, LMN cell infiltrate in the hepatic parenchyma and evidence of regeneration of hepatocytes.

Pretreatment with ether anesthesia greatly potentiates acetaminophen hepatotoxicity (Wells *et al.*,1986). Ether could initially inhibit both the non-toxifying glucurnonidation pathway and the toxifying or bioactivating cytochrome P_{450} pathway (Johannessen *et al.*, 1981). The toxifying P_{450} pathway might recover first. In that case, acetaminophen hepatotoxicity would be potentiated by anesthesia only if acetaminophen was given during the time window when P450 activity had recovered while glucuronidation remained inhibited.

The increased levels of serum enzymes (ALT, AST, LDH and ALP) may be interpreted as a result of liver cell destruction or changes in membrane permeability. Our experiments with APAP elevated all these enzymes significantly ($p<0.001$). These enzymes are characteristic of liver cell damage therefore their release into serum after APAP administration confirms liver damage (Wilhelm, 2009). Treatment with AB-4 counteracts the toxicity of APAP dose dependently and the activity of transaminases, serum ALP and LDH were restored towards normal depicting a marked protective effect. Total protein levels are known to be depressed by APAP induced hepatotoxicity due to defective protein biosynthesis in liver (Choi *et al.*, 2011). The hepatotoxins induced fatty liver is mainly due to the accumulation of triglycerides which is caused by decrease in the synthesis of apoprotein, a component of very low density lipoprotein (Singh *et al., 1986*). Both, prophylactic and curative treatments with AB-4 significantly reversed acetaminophen induced increase in triglycerides indicating protective and curative effects against fatty liver.

An accelerated lipid peroxidation and drastic fall in hepatic GSH content were observed with APAP administration as reported by others (Brien *et al.*, 2000). The damage of the cellular membranes due to lipid peroxidation also leads to decrease in the activity of membrane bound enzymes such as glucose-6-phosphatase (Chandan *et al.*, 2007). Treatments (prophylactic and curative) with AB-4 significantly protected liver by recouping the increased lipid peroxidation and maintaining hepatic GSH and glucose-6-phosphatase levels.

Aniline hydroxylase and amidopyrine–N-demethylase activities are generally determined to assess the metabolizing potential of drug metabolizing enzymes (DMEs) and in turn functional status of endoplasmic reticulum. A single dose of the hepatotoxin studied, induced the inhibition of these enzyme activities (Anand *et al.*,1997). Our results indicate that prophylactic and curative treatments with AB4 counteract APAP induced reduction of aniline hydroxylase and amidopyrine –N-demethylase activities dose dependently, thereby maintain subcellular architecture and functional integrity of cells.

Microscopic examination of histopathological section confirms dose dependent hepatoprotective efficacy of AB-4 against APAP. This was evident as smaller area of necrosis, and regeneration in the form of binucleation of hepatocytes at margin of area of necrosis.

Hepatoprotective action of aqueous extract (AB-4) of *Aloe barbadensis* may be due to its protective effect against oxidant free radical generation exhibited by protection against increased lipid peroxidation and restoration of glutathione status. Rest of the biochemical and histopathological parameters studied indicate the structural and functional integrity of the cells and provide further support to the suggestive mechanism of action. The exact hepatoprotective mechanism can be further established by different mechanistic studies. Since AB-4 exhibited no mortality up to 2g/kg, p.o. in mice (Chandan *et al.*, 2007) and with significant efficacy in the present study, may be of significant therapeutic value in the management of hepatic diseases.

References

Anand, K.K., Singh, B., Saxena, A.K., Chandan, B.K., Gupta, V.N., and Bhardwaj, V. (1997). 3,4,5-trihydroxy benzoic acid (gallic acid), the hepatoprotective principle in the fruits of *Terminalia Belerica*-bioassay guided activity. *Pharmacological Research*, **36:** 315-321.

Annamalai, A., Christina, V.L., and Lakshmi, P.T.V. (2012). A comparative study on the anti-diabetic potential of *Aloe vera* gel and fenugreek seeds on experimentally induced diabetic rats. *Pharmacognosy Communications*, **2**: 57-61.

Baginski, E.S., Foa, P.P., and Zak, B. (1974). Glucose-6-phosphate. In: Hans Ulrich Bergmeyer (Eds.) *Methods of Enzymatic Analysis*, Vol. 2. New York Academy Press Inc, pp. 876-880.

Beatriz, V., Gillermo, A., David, S., and Bruno, E. (1996). Anti-inflammatory activity of extracts from *Aloe vera* gel. *Journal of Ethanopharmacology,* **55**: 69-75.

Brien, P.J., Slaughter, M.R., Swami, A., Birmingham, J.M., Greenhill, R.W., Elcock, F., and Bugelski, P.J. (2000). Repeated acetaminophen dosing in rats: adaptation of hepatic antioxidants system. *Human and Experimental Toxicology,***19**: 277-283.

Buege, J.A., and Aust, S.D. (1978). Microsomal lipid peroxidation. In: Fleischer S, Packer L, (Eds.) *Methods in Enzymology,* Vol. 52. New York Academy Press Inc., pp. 302-310.

Chandan, B.K., Saxena, A.K., Shukla, S., Sharma, N., Gupta, D.K., Suri, K.A., Suri, J., Bhadauria, M., and Singh, B. (2007). Hepatoprotective potential of *Aloe barbadensis* Mill. against carbon tetrachloride induced hepatotoxicity. *Journal of Ethnopharmacology,* **111**: 560-566.

Choi, J.H., Kang, J.W., Kim, D.W., Sung, Y.K., and Lee, S.M. (2011). Protective effects of Mg-CUD against D-galactosamine-induced hepatotoxicity in rats. *European Journal of Pharmacology,* **657**: 138-143.

Ellman, G.L. (1959). Tissue sulfhydral group ABB. *Archives of Biochemistry and Biophysics,* **82**: 70-77.

Ernst, E. (2000). Adverse effects of herbal drugs in dermatology. *British Journal of dermatology,* **143**: 923-929.

Johannessen, W., Gadeholt, G., and Aarbakke, J. (1981). Effect of diethyl ether anaesthesia on the pharmacokinetics of antipyrine and paracetamol in the rat. *J Pharm Pharmacol.,* **33:** 365-368.

Krajian, A.A. (1963). Tissue cutting and staining. In: Sam Frankel and Stanley Reitman (eds.) *Gradwohl's Clinical Laboratory Methods and Diagnosis,* 6th edn., The CV. Mosby Co. Saint Louis (USA), pp1639.

Lowry, D.H., Rosebrough, N.J., Farr, A.L., and Randall, R.J. (1951). Protein measurement with the pholin phenol reagent. *Journal of Biological Chemistry,* **193:** 265-275.

Mhaskar, K.S., Blatter, E., and Caius, J.F. (2000). Kiritikar and Basu's Illustrated Indian Medicinal Plants, their usages in Ayurveda and Unani Medicines. Vol. 10. Sri Satguru publications New Delhi pp. 3457-3459.

Neri, B.P., and Frings, C.S. (1973.) Improved method for determination of triglycerides in serum. *Clinical Chemistry,* **19**: 1201.

Reitman, S., and Frankel, S. (1957). A colorimetric method for the determination of serum glutamic oxaloacetic and glutamic pyruvic transaminases. *American Journal Clinical Pathology,* **28**: 56-63.

Saoo, K., Miki, H., Ohmori, M., and Winters, W.D. (1996). Antiviral activity of Aloe extracts against cytomegalovirus. *Phytotherapy Research,* **10**: 348-358.

Saritha, V., Anilakumar, K.R., and Farhath, K. (2010). Antioxidant and antibacterial activity of *Aloe vera* gel extracts. *International journal of Pharmaceutical and Biological Archives,* **4**: 376-384.

Schenkman, J.B., and Cinti, D.L. (1978). Preparation of microsomes with calcium. In: Fleischer, S., Packer, L. (eds.) *Methods in Enzymalogy*, Vol. 52. New York, San Francisco, London Academic press,, pp. 83-89.

Shenoy, K.A., Somayaji, S.N., and Bairy, K.L. (2002). Evaluation of hepatoprotective activity of *Ginkgo biloba* in rats. *Indian J Physiol. Pharmacology,* **46**: 167-174.

Singh, B., Chandan, B.K., Prabhakar, A., Taneja, S.C., Singh, J., and Qazi, G.N. (2005). Chemistry and hepatoprotective activity of an active fraction from *Barleria prionitis* Linn. in experimental animals. *Phytotherapy Research,* **19**: 391-404.

Singh, B., Saxena, A.K., Chandan, B.K., Sharma, N., Bhardwaz, V., Satti, N.K., Gupta, V.N., Gupta, B.D., Suri, K.A., and Suri, O.P. (2006). Isolation, structure elucidation and *in vivo* hepatoprotective potential of trans- tetracos-15-enoic acid from *Indigofera tinctoria* Linn. *Phytotherapy Research,* **20**: 831-838.

The Wealth of India (2000). A Dictionary of Indian Raw Materials and Industrial Products. National Institutes of Science Communication New Delhi. Vol. 1, A-Ci. Council of Scientific and Industrial Research; pp. 47-49.

Vogler, B.K. (1999). *Aloe vera* : a systematic review of its clinical effectiveness. *British Journal of General Practice,* **49**: 823-828.

Walter, K., and Schutt, C. (1974). Acid and Alkaline Phosphatases in serum. In: Ulrich Bergmeyer H. (eds.) *Methods of Enzymatic Analysis*, vol. 2. New York Academic Pres Inc., pp. 856-864.

Wells, P.G., Ramji, P., and Ku, M.S.W. (1986). Delayed enhancement of acetaminophen hepatotoxicity by general anesthesia using diethyl ether or halothane. *Fundam Appl Toxicol.,* **6**: 299-306.

Wilhelm, E.A., Jesse, C.R., Leite, M.R., and Nogueira, C.W. (2009). Studies on preventive effects of diphenyl diselenide on acetaminophen-induced hepatotoxicity in rats. *Pathophysiology,* **16:** 31-37.

Wroblewski, L.D. (1964). Colorimetric method for LDH. In: Wootton I.D.P (ed.). *Micro analysis in Medical Biochemistry*, vol. 104, London: J. and A. Churchill Ltd., pp. 115-118.

Previous Volumes Contents

— Volume 1 —

2010, xii+391p., col. plts., figs., tabls., ind., 25 cm **Rs. 1800**

ISBN 978-81-7035-625-7

— Volume 2 —

2014, xiii+507p., col. plts., figs., tabls., ind., 25 cm **Rs. 2995**

ISBN 978-81-7035-875-6

Index

A

B

M

N

O

P

Y

Z

www.ingramcontent.com/pod-product-compliance
Ingram Content Group UK Ltd.
Pitfield, Milton Keynes, MK11 3LW, UK
UKHW022218051026
15624UKWH00001B/214

9 789351 308867